Praise for Steven Pinker's *The Better Angels of Our Nature*

A *New York Times* Notable Book
A *Library Journal* Best Book of the Year
One of Amazon's 100 Best Books of the Year
A NetGalley Best of 2011

"For anyone interested in human nature, the material is engrossing, and when the going gets heavy, Pinker knows how to lighten it with ironic comments and a touch of humor. . . . A supremely important book. To have command of so much research, spread across so many different fields, is a masterly achievement."

—*The New York Times Book Review*

"An extraordinary range of research . . . a masterly effort."

—*The Wall Street Journal*

"It is quite a story, and Pinker tells it ably. There are stimulating thoughts on nearly every page." —*New York*

"*Better Angels* is a monumental achievement. His book should make it much harder for pessimists to cling to their gloomy vision of the future. Whether war is an ancient adaptation or a pernicious cultural infection, we are learning how to overcome it." —Slate.com

"Classic Pinker, jammed with facts, figures, and points of speculative departure; a big, complex book, well worth the effort for the good news that it delivers." —*Kirkus Reviews*

"This long, well-researched, comprehensive tour de force provides a helpful look at the human condition." —*Booklist*

"A hugely important work and major contribution to historiography."

—Niall Ferguson, professor of history, Harvard University, and author of *Civilization: The West and the Rest*

PENGUIN BOOKS

THE BETTER ANGELS OF OUR NATURE

Steven Pinker is the Johnstone Professor of Psychology at Harvard University. His research on language and cognition has won prizes from the National Academy of Sciences, the Royal Institution of Great Britain, the American Psychological Association, and the Cognitive Neuroscience Society. He has also received several teaching awards, eight honorary doctorates, and many prizes for his nine books, including *The Language Instinct*, *How the Mind Works*, *The Blank Slate*, and *The Better Angels of Our Nature*. His most recent book, *The Sense of Style*, was a *New York Times* bestseller. He has been named Humanist of the Year and has been listed among *Foreign Policy* magazine's "The World's Top 100 Public Intellectuals" and *Time*'s "The 100 Most Influential People in the World Today." He is currently Chair of the Usage Panel of *The American Heritage Dictionary* and writes frequently for *The New York Times*, *Time*, *The New Republic*, and other publications.

ALSO BY STEVEN PINKER

Language Learnability and Language Development

Learnability and Cognition

The Language Instinct

How the Mind Works

Words and Rules

The Blank Slate

The Stuff of Thought

EDITED BY STEVEN PINKER

Visual Cognition

Connections and Symbols (with Jacques Mehler)

Lexical and Conceptual Semantics (with Beth Levin)

The Best American Science and Nature Writing 2004

THE BETTER ANGELS OF OUR NATURE

WHY VIOLENCE HAS DECLINED

STEVEN PINKER

PENGUIN BOOKS

PENGUIN BOOKS
Published by the Penguin Group
Penguin Group (USA) Inc., 375 Hudson Street, New York, New York 10014, U.S.A. • Penguin Group (Canada), 90 Eglinton Avenue East, Suite 700, Toronto, Ontario, Canada M4P 2Y3 (a division of Pearson Penguin Canada Inc.) • Penguin Books Ltd, 80 Strand, London WC2R 0RL, England • Penguin Ireland, 25 St. Stephen's Green, Dublin 2, Ireland (a division of Penguin Books Ltd) • Penguin Books Australia Ltd, 250 Camberwell Road, Camberwell, Victoria 3124, Australia (a division of Pearson Australia Group Pty Ltd) • Penguin Books India Pvt Ltd, 11 Community Centre, Panchsheel Park, New Delhi—110 017, India • Penguin Group (NZ), 67 Apollo Drive, Rosedale, Auckland 0632, New Zealand (a division of Pearson New Zealand Ltd) • Penguin Books (South Africa) (Pty) Ltd, 24 Sturdee Avenue, Rosebank, Johannesburg 2196, South Africa

Penguin Books Ltd, Registered Offices: 80 Strand, London WC2R 0RL, England

First published in the United States of America by Viking Penguin,
a member of Penguin Group (USA) Inc. 2011
Published in Penguin Books 2012

20 19 18 17 16

Excerpts from "MLF Lullaby," "Who's Next?," and "In Old Mexico" by Tom Lehrer. Reprinted by permission of Tom Lehrer.

Excerpt from "It Depends on What You Pay" by Tom Jones. Reprinted by permission of Tom Jones.

Excerpt from "Feel Like I'm Fixin' to Die Rag," words and music by Joe McDonald. © 1965, renewed 1993 by Alkatraz Corner Music Co.

THE LIBRARY OF CONGRESS HAS CATALOGED THE HARDCOVER EDITION AS FOLLOWS:
Pinker, Steven, 1954–
The better angels of our nature: why violence has declined / Steven Pinker.
p. cm.
Includes bibliographical references and index.
ISBN 978-0-670-02295-3 (hc.)
ISBN 978-0-14-312201-2 (pbk.)
1. Violence—Psychological aspects. 2. Violence—Social aspects. 3. Nonviolence—Psychological aspects. I. Title.
HM1116.P57 2011
303.609—dc22
2011015201

Printed in the United States of America
Designed by Carla Bolte
Charts rendered by Ilavenil Subbiah

TO

Eva, Carl, and Eric

Jack and David

Yael and Danielle

and the world they will inherit

What a chimera then is man! What a novelty, what a monster, what a chaos, what a contradiction, what a prodigy! Judge of all things, feeble earthworm, repository of truth, sewer of uncertainty and error, the glory and the scum of the universe.

—*Blaise Pascal*

CONTENTS

LIST OF FIGURES

Figure

PREFACE

This book is about what may be the most important thing that has ever happened in human history. Believe it or not—and I know that most people do not—violence has declined over long stretches of time, and today we may be living in the most peaceable era in our species' existence. The decline, to be sure, has not been smooth; it has not brought violence down to zero; and it is not guaranteed to continue. But it is an unmistakable development, visible on scales from millennia to years, from the waging of wars to the spanking of children.

No aspect of life is untouched by the retreat from violence. Daily existence is very different if you always have to worry about being abducted, raped, or killed, and it's hard to develop sophisticated arts, learning, or commerce if the institutions that support them are looted and burned as quickly as they are built.

The historical trajectory of violence affects not only how life is lived but how it is understood. What could be more fundamental to our sense of meaning and purpose than a conception of whether the strivings of the human race over long stretches of time have left us better or worse off? How, in particular, are we to make sense of *modernity*—of the erosion of family, tribe, tradition, and religion by the forces of individualism, cosmopolitanism, reason, and science? So much depends on how we understand the legacy of this transition: whether we see our world as a nightmare of crime, terrorism, genocide, and war, or as a period that, by the standards of history, is blessed by unprecedented levels of peaceful coexistence.

The question of whether the arithmetic sign of trends in violence is positive or negative also bears on our conception of human nature. Though theories of human nature rooted in biology are often associated with fatalism about violence, and the theory that the mind is a blank slate is associated with progress, in my view it is the other way around. How are we to understand the natural state of life when our species first emerged and the processes of history began? The belief that violence has increased suggests that the world we made has contaminated us, perhaps irretrievably. The belief that it has decreased suggests that we started off nasty and that the artifices of civilization have moved us in a noble direction, one in which we can hope to continue.

This is a big book, but it has to be. First I have to convince you that violence really has gone down over the course of history, knowing that the very idea invites skepticism, incredulity, and sometimes anger. Our cognitive faculties predispose us to believe that we live in violent times, especially when they are stoked by media that follow the watchword "If it bleeds, it leads." The human mind tends to estimate the probability of an event from the ease with which it can recall examples, and scenes of carnage are more likely to be beamed into our homes and burned into our memories than footage of people dying of old age.[1] No matter how small the percentage of violent deaths may be, in absolute numbers there will always be enough of them to fill the evening news, so people's impressions of violence will be disconnected from the actual proportions.

Also distorting our sense of danger is our moral psychology. No one has ever recruited activists to a cause by announcing that things are getting better, and bearers of good news are often advised to keep their mouths shut lest they lull people into complacency. Also, a large swath of our intellectual culture is loath to admit that there could be anything good about civilization, modernity, and Western society. But perhaps the main cause of the illusion of ever-present violence springs from one of the forces that drove violence down in the first place. The decline of violent behavior has been paralleled by a decline in attitudes that tolerate or glorify violence, and often the attitudes are in the lead. By the standards of the mass atrocities of human history, the lethal injection of a murderer in Texas, or an occasional hate crime in which a member of an ethnic minority is intimidated by hooligans, is pretty mild stuff. But from a contemporary vantage point, we see them as signs of how low our behavior can sink, not of how high our standards have risen.

In the teeth of these preconceptions, I will have to persuade you with numbers, which I will glean from datasets and depict in graphs. In each case I'll explain where the numbers came from and do my best to interpret the ways they fall into place. The problem I have set out to understand is the reduction in violence at many scales—in the family, in the neighborhood, between tribes and other armed factions, and among major nations and states. If the history of violence at each level of granularity had an idiosyncratic trajectory, each would belong in a separate book. But to my repeated astonishment, the global trends in almost all of them, viewed from the vantage point of the present, point downward. That calls for documenting the various trends between a single pair of covers, and seeking commonalities in when, how, and why they have occurred.

Too many kinds of violence, I hope to convince you, have moved in the same direction for it all to be a coincidence, and that calls for an explanation. It is natural to recount the history of violence as a moral saga—a heroic struggle of justice against evil—but that is not my starting point. My approach is scientific in the broad sense of seeking explanations for why things happen. We may discover that a particular advance in peacefulness was brought about by moral entrepreneurs and their movements. But we may also discover that the explanation is more prosaic, like a change in technology, governance, commerce, or knowledge. Nor can we understand the decline of violence as an unstoppable force for progress that is carrying us toward an omega point of perfect peace. It is a collection of statistical trends in the behavior of groups of humans in various epochs, and as such it calls for an explanation in terms of psychology and history: how human minds deal with changing circumstances.

A large part of the book will explore the psychology of violence and nonviolence. The theory of mind that I will invoke is the synthesis of cognitive science, affective and cognitive neuroscience, social and evolutionary psychology, and other sciences of human nature that I explored in *How the Mind Works, The Blank Slate*, and *The Stuff of Thought*. According to this understanding, the mind is a complex system of cognitive and emotional faculties implemented in the brain which owe their basic design to the processes of evolution. Some of these faculties incline us toward various kinds of violence. Others—"the better angels of our nature," in Abraham Lincoln's words—incline us toward cooperation and peace. The way to explain the decline of violence is to identify the changes in our cultural and material milieu that have given our peaceable motives the upper hand.

Finally, I need to show how our history has engaged our psychology. Everything in human affairs is connected to everything else, and that is especially true of violence. Across time and space, the more peaceable societies also tend to be richer, healthier, better educated, better governed, more respectful of their women, and more likely to engage in trade. It's not easy to tell which of these happy traits got the virtuous circle started and which went along for the ride, and it's tempting to resign oneself to unsatisfying circularities, such as that violence declined because the culture got less violent. Social scientists distinguish "endogenous" variables—those that are inside the system, where they may be affected by the very phenomenon they are trying to explain—from the "exogenous" ones—those that are set in motion by forces from the outside. Exogenous forces can originate in the practical realm, such as changes in technology, demographics, and the mechanisms of commerce and governance. But they can also originate in the intellectual realm, as new ideas are conceived and disseminated and take on a life of their own. The most satisfying explanation of a historical change is one that identifies an exogenous trigger. To the best that the data allow it, I will try to identify exogenous forces that have engaged our mental faculties in different ways at different times and that thereby can be said to have caused the declines in violence.

The discussions that try to do justice to these questions add up to a big book—big enough that it won't spoil the story if I preview its major conclusions. *The Better Angels of Our Nature* is a tale of six trends, five inner demons, four better angels, and five historical forces.

Six Trends (chapters 2 through 7). To give some coherence to the many developments that make up our species' retreat from violence, I group them into six major trends.

The first, which took place on the scale of millennia, was the transition from the anarchy of the hunting, gathering, and horticultural societies in which our species spent most of its evolutionary history to the first agricultural civilizations with cities and governments, beginning around five thousand years ago. With that change came a reduction in the chronic raiding and feuding that characterized life in a state of nature and a more or less fivefold decrease in rates of violent death. I call this imposition of peace the Pacification Process.

The second transition spanned more than half a millennium and is best documented in Europe. Between the late Middle Ages and the 20th century, European countries saw a tenfold-to-fiftyfold decline in their rates of homicide. In his classic book *The Civilizing Process*, the sociologist Norbert Elias attributed this surprising decline to the consolidation of a patchwork of feudal territories into large kingdoms with centralized authority and an infrastructure of commerce. With a nod to Elias, I call this trend the Civilizing Process.

The third transition unfolded on the scale of centuries and took off around the time of the Age of Reason and the European Enlightenment in the 17th and 18th centuries (though it had antecedents in classical Greece and the Renaissance, and parallels elsewhere in the world). It saw the first organized movements to abolish socially sanctioned forms of violence like despotism, slavery, dueling, judicial torture, superstitious killing, sadistic punishment, and cruelty to animals, together with the first stirrings of systematic pacifism. Historians sometimes call this transition the Humanitarian Revolution.

The fourth major transition took place after the end of World War II. The two-thirds of a century since then have been witness to a historically unprecedented development: the great powers, and developed states in general, have stopped waging war on one another. Historians have called this blessed state of affairs the Long Peace.[2]

The fifth trend is also about armed combat but is more tenuous. Though it may be hard for news readers to believe, since the end of the Cold War in 1989, organized conflicts of all kinds—civil wars, genocides, repression by autocratic governments, and terrorist attacks—have declined throughout the world. In recognition of the tentative nature of this happy development, I will call it the New Peace.

Finally, the postwar era, symbolically inaugurated by the Universal Declaration of Human Rights in 1948, has seen a growing revulsion against aggression on smaller scales, including violence against ethnic minorities, women, children, homosexuals, and animals. These spin-offs from the concept of human rights—civil rights, women's rights, children's rights, gay rights, and animal rights—were asserted in a cascade of movements from the late 1950s to the present day which I will call the Rights Revolutions.

Five Inner Demons (chapter 8). Many people implicitly believe in the Hydraulic Theory of Violence: that humans harbor an inner drive toward aggression (a death instinct or thirst for blood), which builds up inside us and must periodically be discharged. Nothing could be further from a contemporary scientific understanding of the psychology of violence. Aggression is not a single motive, let alone a mounting urge. It is the output of several psychological systems that differ in their environmental triggers, their internal logic, their neurobiological basis, and their social distribution. Chapter 8 is devoted to explaining five of them. *Predatory* or *instrumental violence* is simply violence deployed as a practical means to an end. *Dominance* is the urge for authority, prestige, glory, and power, whether it takes the form of macho posturing among individuals or contests for supremacy among racial, ethnic, religious, or national groups. *Revenge* fuels the moralistic urge toward retribution, punishment, and justice. *Sadism* is pleasure taken in another's suffering. And *ideology* is a shared belief system, usually involving a vision of utopia, that justifies unlimited violence in pursuit of unlimited good.

Four Better Angels (chapter 9). Humans are not innately good (just as they are not innately evil), but they come equipped with motives that can orient them away from violence and toward cooperation and altruism. *Empathy* (particularly in the sense of sympathetic concern) prompts us to feel the pain of others and to align their interests with our own. *Self-control* allows us to anticipate the consequences of acting on our impulses and to inhibit them accordingly. The *moral sense* sanctifies a set of norms and taboos that govern the interactions among people in a culture, sometimes in ways that decrease violence, though often (when the norms are tribal, authoritarian, or puritanical) in ways that increase it. And the faculty of *reason* allows us to extricate ourselves from our parochial vantage points, to reflect on the ways in which we live our lives, to deduce ways in which we could be better off, and to guide the application of the other better angels of our nature. In one section I will also examine the possibility that in recent history *Homo sapiens* has literally evolved to become less violent in the biologist's technical sense of a change in our genome. But the focus of the book is on transformations that are strictly environmental: changes in historical circumstances that engage a fixed human nature in different ways.

Five Historical Forces (chapter 10). In the final chapter I try to bring the psychology and history back together by identifying exogenous forces that favor our peaceable motives and that have driven the multiple declines in violence.

The *Leviathan*, a state and judiciary with a monopoly on the legitimate use of force, can defuse the temptation of exploitative attack, inhibit the impulse for revenge, and circumvent the self-serving biases that make all parties believe they are on the side of the angels. *Commerce* is a positive-sum game in which everybody can win; as technological progress allows the exchange of goods and ideas over longer distances and among larger groups of trading partners, other people become more valuable alive than dead, and they are less likely to become targets of demonization and dehumanization. *Feminization* is the process in which cultures have increasingly respected the interests and values of women. Since violence is largely a male pastime, cultures that empower women tend to move away from the glorification of violence and are less likely to breed dangerous subcultures of rootless young men. The forces of *cosmopolitanism* such as literacy, mobility, and mass media can prompt people to take the perspective of people unlike themselves and to expand their circle of sympathy to embrace them. Finally, an intensifying application of knowledge and rationality to human affairs—the *escalator of reason*—can force people to recognize the futility of cycles of violence, to ramp down the privileging of their own interests over others', and to reframe violence as a problem to be solved rather than a contest to be won.

As one becomes aware of the decline of violence, the world begins to look different. The past seems less innocent; the present less sinister. One starts to appreciate the small gifts of coexistence that would have seemed utopian to our ancestors: the interracial family playing in the park, the comedian who lands a zinger on the commander in chief, the countries that quietly back away from a crisis instead of escalating to war. The shift is not toward complacency: we enjoy the peace we find today because people in past generations were appalled by the violence in their time and worked to reduce it, and so we should work to reduce the violence that remains in our time. Indeed, it is a recognition of the decline of violence that best affirms that such efforts are worthwhile. Man's inhumanity to man has long been a subject for moralization. With the knowledge that something has driven it down, we can also treat it as a matter of cause and effect. Instead of asking, "Why is there war?" we might ask, "Why is there peace?" We can obsess not just over what we have been doing wrong but also over what we have been doing right. Because we *have* been doing something right, and it would be good to know what, exactly, it is.

Many people have asked me how I became involved in the analysis of violence. It should not be a mystery: violence is a natural concern for anyone who studies human nature. I first learned of the decline of violence from Martin Daly and Margo Wilson's classic book in evolutionary psychology, *Homicide*, in which they examined the high rates of violent death in nonstate societies and the decline in homicide from the Middle Ages to the present. In several of my previous books I cited those downward trends, together with humane developments such as the abolition of slavery, despotism, and cruel punishments in the history of the West, in support of the idea that moral progress is compatible with a biological approach to the human mind and an acknowledgment of the dark side of human nature.[3] I reiterated these observations in response to the annual question on the online forum www.edge.org, which in 2007 was "What Are You Optimistic About?" My squib provoked a flurry of correspondence from scholars in historical criminology and international studies who told me that the evidence for a historical reduction in violence is more extensive than I had realized.[4] It was their data that convinced me that there was an underappreciated story waiting to be told.

My first and deepest thanks go to these scholars: Azar Gat, Joshua Goldstein, Manuel Eisner, Andrew Mack, John Mueller, and John Carter Wood. As I worked on the book, I also benefited from correspondence with Peter Brecke, Tara Cooper, Jack Levy, James Payne, and Randolph Roth. These generous researchers shared ideas, writings, and data and kindly guided me through fields of research that are far from my own specialization.

David Buss, Martin Daly, Rebecca Newberger Goldstein, David Haig, James Payne, Roslyn Pinker, Jennifer Sheehy-Skeffington, and Polly Wiessner read most or all of the first draft and offered immeasurably helpful advice and criticism. Also invaluable were comments on particular chapters offered by Peter Brecke, Daniel Chirot, Alan Fiske, Jonathan Gottschall, A. C. Grayling, Niall Ferguson, Graeme Garrard, Joshua Goldstein, Capt. Jack Hoban, Stephen Leblanc, Jack Levy, Andrew Mack, John Mueller, Charles Seife, Jim Sidanius, Michael Spagat, Richard Wrangham, and John Carter Wood.

Many other people responded to my inquiries with prompt explanations or offered suggestions that were incorporated into the book: John Archer, Scott Atran, Daniel Batson, Donald Brown, Lars-Erik Cederman, Christopher Chabris, Gregory Cochran, Leda Cosmides, Tove Dahl, Lloyd deMause, Jane Esberg, Alan Fiske, Dan Gardner, Pinchas Goldschmidt, Cmdr. Keith Gordon, Reid Hastie, Brian Hayes, Judith Rich Harris, Harold Herzog, Fabio Idrobo, Tom Jones, Maria Konnikova, Robert Kurzban, Gary Lafree, Tom Lehrer, Michael Macy, Steven Malby, Megan Marshall, Michael McCullough, Nathan Myhrvold, Mark Newman, Barbara Oakley, Robert Pinker, Susan Pinker, Ziad Obermeyer, David Pizarro, Tage Rai, David Ropeik, Bruce Russett, Scott Sagan, Ned Sahin, Aubrey Sheiham, Francis X. Shen, Lt. Col. Joseph Shusko, Richard Shweder, Thomas Sowell, Håvard Strand, Ilavenil Subbiah, Rebecca Sutherland, Philip Tetlock, Andreas Forø Tollefsen, James Tucker, Staffan Ulfstrand, Jeffrey Watumull, Robert Whiston, Matthew White, Maj. Michael Wiesenfeld, and David Wolpe.

Many colleagues and students at Harvard have been generous with their expertise, including Mahzarin Banaji, Robert Darnton, Alan Dershowitz, James Engell, Nancy Etcoff, Drew Faust, Benjamin Friedman, Daniel Gilbert,

Edward Glaeser, Omar Sultan Haque, Marc Hauser, James Lee, Bay McCulloch, Richard McNally, Michael Mitzenmacher, Orlando Patterson, Leah Price, David Rand, Robert Sampson, Steve Shavell, Lawrence Summers, Kyle Thomas, Justin Vincent, Felix Warneken, and Daniel Wegner.

Special thanks go to the researchers who have worked with me on the data reported in these pages. Brian Atwood carried out countless statistical analyses and database searches with precision, thoroughness, and insight. William Kowalsky discovered many pertinent findings from the world of public opinion polling. Jean-Baptiste Michel helped develop the Bookworm program, the Google Ngram Viewer, and the Google Books corpus and devised an ingenious model for the distribution of the magnitude of wars. Bennett Haselton carried out an informative study of people's perceptions of the history of violence. Esther Snyder assisted with graphing and bibliographic searches. Ilavenil Subbiah designed the elegant graphs and maps, and over the years has provided me with invaluable insight about the culture and history of Asia.

John Brockman, my literary agent, posed the question that led to the writing of this book and offered many helpful comments on the first draft. Wendy Wolf, my editor at Penguin, offered a detailed analysis of the first draft that did much to shape the final version. I'm enormously grateful to John and Wendy, together with Will Goodlad at Penguin UK, for their support of the book at every stage.

Heartfelt thanks go to my family for their love and encouragement: Harry, Roslyn, Susan, Martin, Robert, and Kris. My greatest appreciation goes to Rebecca Newberger Goldstein, who not only improved the book's substance and style but encouraged me with her belief in the value of the project, and who has done more than anyone to shape my worldview. This book is dedicated to my niece, nephews, and stepdaughters: may they enjoy a world in which the decline of violence continues.

1

A FOREIGN COUNTRY

The past is a foreign country: they do things differently there.

—L. P. Hartley

If the past is a foreign country, it is a shockingly violent one. It is easy to forget how dangerous life used to be, how deeply brutality was once woven into the fabric of daily existence. Cultural memory pacifies the past, leaving us with pale souvenirs whose bloody origins have been bleached away. A woman donning a cross seldom reflects that this instrument of torture was a common punishment in the ancient world; nor does a person who speaks of a *whipping boy* ponder the old practice of flogging an innocent child in place of a misbehaving prince. We are surrounded by signs of the depravity of our ancestors' way of life, but we are barely aware of them. Just as travel broadens the mind, a literal-minded tour of our cultural heritage can awaken us to how differently they did things in the past.

In a century that began with 9/11, Iraq, and Darfur, the claim that we are living in an unusually peaceful time may strike you as somewhere between hallucinatory and obscene. I know from conversations and survey data that most people refuse to believe it.[1] In succeeding chapters I will make the case with dates and data. But first I want to soften you up by reminding you of incriminating facts about the past that you have known all along. This is not just an exercise in persuasion. Scientists often probe their conclusions with a sanity check, a sampling of real-world phenomena to reassure themselves they haven't overlooked some flaw in their methods and wandered into a preposterous conclusion. The vignettes in this chapter are a sanity check on the data to come.

What follows is a tour of the foreign country called the past, from 8000 BCE to the 1970s. It is not a grand tour of the wars and atrocities that we already commemorate for their violence, but rather a series of glimpses behind deceptively familiar landmarks to remind us of the viciousness they conceal. The past, of course, is not a single country, but encompasses a vast diversity of cultures and customs. What they have in common is the shock of the old: a backdrop of violence that was endured, and often embraced, in ways that startle the sensibilities of a 21st-century Westerner.

HUMAN PREHISTORY

In 1991 two hikers stumbled upon a corpse poking out of a melting glacier in the Tyrolean Alps. Thinking that it was the victim of a skiing accident, rescue workers jackhammered the body out of the ice, damaging his thigh and his backpack in the process. Only when an archaeologist spotted a Neolithic copper ax did people realize that the man was five thousand years old.[2]

Ötzi the Iceman, as he is now called, became a celebrity. He appeared on the cover of *Time* magazine and has been the subject of many books, documentaries, and articles. Not since Mel Brooks's 2000 Year Old Man ("I have more than 42,000 children and not one comes to visit me") has a kilogenarian had so much to tell us about the past. Ötzi lived during the crucial transition in human prehistory when agriculture was replacing hunting and gathering, and tools were first made of metal rather than stone. Together with his ax and backpack, he carried a quiver of fletched arrows, a wood-handled dagger, and an ember wrapped in bark, part of an elaborate fire-starting kit. He wore a bearskin cap with a leather chinstrap, leggings sewn from animal hide, and waterproof snowshoes made from leather and twine and insulated with grass. He had tattoos on his arthritic joints, possibly a sign of acupuncture, and carried mushrooms with medicinal properties.

Ten years after the Iceman was discovered, a team of radiologists made a startling discovery: Ötzi had an arrowhead embedded in his shoulder. He had not fallen in a crevasse and frozen to death, as scientists had originally surmised; he had been murdered. As his body was examined by the the CSI Neolithic team, the outlines of the crime came into view. Ötzi had unhealed cuts on his hands and wounds on his head and chest. DNA analyses found traces of blood from two other people on one of his arrowheads, blood from a third on his dagger, and blood from a fourth on his cape. According to one reconstruction, Ötzi belonged to a raiding party that clashed with a neighboring tribe. He killed a man with an arrow, retrieved it, killed another man, retrieved the arrow again, and carried a wounded comrade on his back before fending off an attack and being felled by an arrow himself.

Ötzi is not the only millennia-old man who became a scientific celebrity at the end of the 20th century. In 1996 spectators at a hydroplane race in Kennewick, Washington, noticed some bones poking out of a bank of the Columbia River. Archaeologists soon recovered the skeleton of a man who had lived 9,400 years ago.[3] Kennewick Man quickly became the object of highly publicized legal and scientific battles. Several Native American tribes fought for custody of the skeleton and the right to bury it according to their traditions, but a federal court rejected their claims, noting that no human culture has ever been in continuous existence for nine millennia. When the scientific studies resumed, anthropologists were intrigued to learn that Kennewick Man was anatomically very different from today's Native Americans. One report argued that he had European features; another that he matched the Ainu, the aboriginal inhabitants of Japan. Either possibility would imply that the Americas had been peopled by several independent migrations, contradicting DNA evidence suggesting that Native Americans are descendants of a single group of migrants from Siberia.

For plenty of reasons, then, Kennewick Man has become an object of fascination among the scientifically curious. And here is one more. Lodged in Kennewick Man's pelvis is a stone projectile. Though the bone had partially healed, indicating that he didn't die from the wound, the forensic evidence is unmistakable: Kennewick Man had been shot.

These are just two examples of famous prehistoric remains that have yielded grisly news about how their owners met their ends. Many visitors to the British Museum have been captivated by Lindow Man, an almost perfectly preserved two-thousand-year-old body discovered in an English peat bog in 1984.[4] We don't know how many of his children visited him, but we do know how he died. His skull had been fractured with a blunt object; his neck had been broken by a twisted cord; and for good measure his throat had been cut. Lindow Man may have been a Druid who was ritually sacrificed in three ways to satisfy three gods. Many other bog men and women from northern Europe show signs of having been strangled, bludgeoned, stabbed, or tortured.

In a single month while researching this book, I came across two new stories about remarkably preserved human remains. One is a two-thousand-year-old skull dug out of a muddy pit in northern England. The archaeologist who was cleaning the skull felt something move, looked through the opening at the base, and saw a yellow substance inside, which turned out to be a preserved brain. Once again, the unusual state of preservation was not the only noteworthy feature about the find. The skull had been deliberately severed from the body, suggesting to the archaeologist that it was a victim of human sacrifice.[5] The other discovery was of a 4,600-year-old grave in Germany that held the remains of a man, a woman, and two boys. DNA analyses showed that they were members of a single nuclear family, the oldest known to science. The foursome had been buried at the same time—signs, the archaeologists said, that they had been killed in a raid.[6]

What is it about the ancients that they couldn't leave us an interesting corpse without resorting to foul play? Some cases may have an innocent explanation based in taphonomy, the processes by which bodies are preserved over long spans of time. Perhaps at the turn of the first millennium the only bodies that got dumped into bogs, there to be pickled for posterity, were those that had been ritually sacrificed. But with most of the bodies, we have no reason to think that they were preserved only because they had been murdered. Later we will look at the results of forensic investigations that can distinguish how an ancient body met its end from how it came down to us. For now, prehistoric remains convey the distinct impression that The Past is a place where a person had a high chance of coming to bodily harm.

HOMERIC GREECE

Our understanding of prehistoric violence depends on the happenstance of which bodies were accidentally embalmed or fossilized, and so it must be radically incomplete. But once written language began to spread, ancient people left us with better information about how they conducted their affairs.

Homer's *Iliad* and *Odyssey* are considered the first great works of Western literature, and occupy the top slots in many guides to cultural literacy. Though these narratives are set at the time of the Trojan War around 1200 BCE, they were written down much later, between 800 and 650 BCE, and are thought to reflect life among the tribes and chiefdoms of the eastern Mediterranean in that era.[7]

Today one often reads that total war, which targets an entire society rather than just its armed forces, is a modern invention. Total war has been blamed on the emergence of nation-states, on universalist ideologies, and on technologies that allow killing at a distance. But if Homer's depictions are accurate (and they do jibe with archaeology, ethnography, and history), then the wars in archaic Greece were as total as anything in the modern age. Agamemnon explains to King Menelaus his plans for war:

> Menelaus, my soft-hearted brother, why are you so concerned for these men? Did the Trojans treat you as handsomely when they stayed in your palace? No: we are not going to leave a single one of them alive, down to the babies in their mothers' wombs—not even they must live. The whole people must be wiped out of existence, and none be left to think of them and shed a tear.[8]

In his book *The Rape of Troy,* the literary scholar Jonathan Gottschall discusses how archaic Greek wars were carried out:

> Fast ships with shallow drafts are rowed onto beaches and seaside communities are sacked before neighbors can lend defensive support. The men are usually killed, livestock and other portable wealth are plundered, and women are carried off to live among the victors and perform sexual and menial labors. Homeric men live with the possibility of sudden, violent death, and the women live in fear for their men and children, and of sails on the horizon that may harbinger new lives of rape and slavery.[9]

We also commonly read that 20th-century wars were unprecedentedly destructive because they were fought with machine guns, artillery, bombers, and other long-distance weaponry, freeing soldiers from natural inhibitions against face-to-face combat and allowing them to kill large numbers of faceless enemies without mercy. According to this reasoning, handheld weapons are not nearly as lethal as our high-tech methods of battle. But Homer vividly described the large-scale damage that warriors of his day could inflict. Gottschall offers a sample of his imagery:

> Breached with surprising ease by the cold bronze, the body's contents pour forth in viscous torrents: portions of brains emerge at the ends of quivering spears, young men hold back their viscera with desperate hands, eyes are knocked or cut from skulls and glimmer sightlessly in the dust. Sharp points forge new entrances and exits in young bodies: in the center of foreheads, in temples, between the eyes, at the base of the neck, clean through the mouth or cheek and out the other side, through flanks, crotches, buttocks, hands, navels, backs, stomachs, nipples, chests, noses, ears, and chins. . . . Spears, pikes, arrows, swords, daggers, and rocks lust for the savor of flesh and blood. Blood sprays forth and mists the air. Bone fragments fly. Marrow boils from fresh stumps. . . .
>
> In the aftermath of battle, blood flows from a thousand mortal or maiming wounds, turns dust to mud, and fattens the grasses of the plain. Men plowed into the soil by heavy chariots, sharp-hoofed stallions, and the sandals of men are past recognition. Armor and weaponry litter the field. Bodies are everywhere, decomposing, deliquescing, feasting dogs, worms, flies, and birds.[10]

The 21st century has certainly seen the rape of women in wartime, but it has long been treated as an atrocious war crime, which most armies try to prevent and the rest deny and conceal. But for the heroes of the *Iliad*, female flesh was a legitimate spoil of war: women were to be enjoyed, monopolized, and disposed of at their pleasure. Menelaus launches the Trojan War when his wife, Helen, is abducted. Agamemnon brings disaster to the Greeks by refusing to return a sex slave to her father, and when he relents, he appropriates one belonging to Achilles, later compensating him with twenty-eight replacements. Achilles, for his part, offers this pithy description of his career: "I have spent many sleepless nights and bloody days in battle, fighting men for their women."[11] When Odysseus returns to his wife after twenty years away, he murders the men who courted her while everyone thought he was dead, and when he discovers that the men had consorted with the concubines of his household, he has his son execute the concubines too.

These tales of massacre and rape are disturbing even by the standards of modern war documentaries. Homer and his characters, to be sure, deplored the waste of war, but they accepted it as an inescapable fact of life, like the weather—something that everyone talked about but no one could do anything about. As

Odysseus put it, "[We are men] to whom Zeus has given the fate of winding down our lives in painful wars, from youth until we perish, each of us." The men's ingenuity, applied so resourcefully to weapons and strategy, turned up empty-handed when it came to the earthly causes of war. Rather than framing the scourge of warfare as a human problem for humans to solve, they concocted a fantasy of hotheaded gods and attributed their own tragedies to the gods' jealousies and follies.

THE HEBREW BIBLE

Like the works of Homer, the Hebrew Bible (Old Testament) was set in the late 2nd millennium BCE but written more than five hundred years later.[12] But unlike the works of Homer, the Bible is revered today by billions of people who call it the source of their moral values. The world's bestselling publication, the Good Book has been translated into three thousand languages and has been placed in the nightstands of hotels all over the world. Orthodox Jews kiss it with their prayer shawls; witnesses in American courts bind their oaths by placing a hand on it. Even the president touches it when taking the oath of office. Yet for all this reverence, the Bible is one long celebration of violence.

In the beginning God created the heaven and the earth. And the Lord God formed man of the dust of the ground, and breathed into his nostrils the breath of life; and man became a living soul. And the Lord God took one of Adam's ribs, and made he a woman. And Adam called his wife's name Eve; because she was the mother of all living. And Adam knew Eve his wife; and she conceived, and bare Cain. And she again bare his brother Abel. And Cain talked with Abel his brother: and it came to pass, when they were in the field, that Cain rose up against Abel his brother, and slew him. With a world population of exactly four, that works out to a homicide rate of 25 percent, which is about a thousand times higher than the equivalent rates in Western countries today.

No sooner do men and women begin to multiply than God decides they are sinful and that the suitable punishment is genocide. (In Bill Cosby's comedy sketch, a neighbor begs Noah for a hint as to why he is building an ark. Noah replies, "How long can you tread water?") When the flood recedes, God instructs Noah in its moral lesson, namely the code of vendetta: "Whoso sheddeth man's blood, by man shall his blood be shed."

The next major figure in the Bible is Abraham, the spiritual ancestor of Jews, Christians, and Muslims. Abraham has a nephew, Lot, who settles in Sodom. Because the residents engage in anal sex and comparable sins, God immolates every man, woman, and child in a divine napalm attack. Lot's wife, for the crime of turning around to look at the inferno, is put to death as well.

Abraham undergoes a test of his moral values when God orders him to take his son Isaac to a mountaintop, tie him up, cut his throat, and burn his body as a gift to the Lord. Isaac is spared only because at the last moment an angel stays his father's hand. For millennia readers have puzzled over why God insisted on this horrifying trial. One interpretation is that God intervened not because Abraham had passed the test but because he had failed it, but that is anachronistic: obedience to divine authority, not reverence for human life, was the cardinal virtue.

Isaac's son Jacob has a daughter, Dinah. Dinah is kidnapped and raped—apparently a customary form of courtship at the time, since the rapist's family then offers to purchase her from her own family as a wife for the rapist. Dinah's brothers explain that an important moral principle stands in the way of this transaction: the rapist is uncircumcised. So they make a counteroffer: if all the men in the rapist's hometown cut off their foreskins, Dinah will be theirs. While the men are incapacitated with bleeding penises, the brothers invade the city, plunder and destroy it, massacre the men, and carry off the women and children. When Jacob worries that neighboring tribes may attack them in revenge, his sons explain that it was worth the risk: "Should our sister be treated like a whore?"[13] Soon afterward they reiterate their commitment to family values by selling their brother Joseph into slavery.

Jacob's descendants, the Israelites, find their way to Egypt and become too numerous for the Pharaoh's liking, so he enslaves them and orders that all the boys be killed at birth. Moses escapes the mass infanticide and grows up to challenge the Pharaoh to let his people go. God, who is omnipotent, could have softened Pharaoh's heart, but he hardens it instead, which gives him a reason to afflict every Egyptian with painful boils and other miseries before killing every one of *their* firstborn sons. (The word *Passover* alludes to the executioner angel's passing over the households with Israelite firstborns.) God follows this massacre with another one when he drowns the Egyptian army as they pursue the Israelites across the Red Sea.

The Israelites assemble at Mount Sinai and hear the Ten Commandments, the great moral code that outlaws engraved images and the coveting of livestock but gives a pass to slavery, rape, torture, mutilation, and genocide of neighboring tribes. The Israelites become impatient while waiting for Moses to return with an expanded set of laws, which will prescribe the death penalty for blasphemy, homosexuality, adultery, talking back to parents, and working on the Sabbath. To pass the time, they worship a statue of a calf, for which the punishment turns out to be, you guessed it, death. Following orders from God, Moses and his brother Aaron kill three thousand of their companions.

God then spends seven chapters of Leviticus instructing the Israelites on how to slaughter the steady stream of animals he demands of them. Aaron and his two sons prepare the tabernacle for the first service, but the sons slip up and use the wrong incense. So God burns them to death.

As the Israelites proceed toward the promised land, they meet up with the Midianites. Following orders from God, they slay the males, burn their city, plunder the livestock, and take the women and children captive. When they return to Moses, he is enraged because they spared the women, some of whom had led the Israelites to worship rival gods. So he tells his soldiers to complete the genocide and to reward themselves with nubile sex slaves they may rape at their pleasure: "Now therefore kill every male among the little ones, and kill every woman that hath known man by lying with him. But all the women children, that have not known a man by lying with him, keep alive for yourselves."[14]

In Deuteronomy 20 and 21, God gives the Israelites a blanket policy for dealing with cities that don't accept them as overlords: smite the males with the edge of the sword and abduct the cattle, women, and children. Of course, a man with a beautiful new captive faces a problem: since he has just murdered her parents and brothers, she may not be in the mood for love. God anticipates this nuisance and offers the following solution: the captor should shave her head, pare her nails, and imprison her in his house for a month while she cries her eyes out. Then he may go in and rape her.

With a designated list of other enemies (Hittites, Amorites, Canaanites, Perizzites, Hivites, and Jebusites), the genocide has to be total: "Thou shalt save alive nothing that breatheth: But thou shalt utterly destroy them . . . as the Lord thy God has commanded thee."[15]

Joshua puts this directive into practice when he invades Canaan and sacks the city of Jericho. After the walls came tumbling down, his soldiers "utterly destroyed all that was in the city, both man and woman, young and old, and ox, and sheep, and ass, with the edge of the sword."[16] More earth is scorched as Joshua "smote all the country of the hills, and of the south, and of the vale, and of the springs, and all their kings: he left none remaining, but utterly destroyed all that breathed, as the Lord God of Israel commanded."[17]

The next stage in Israelite history is the era of the judges, or tribal chiefs. The most famous of them, Samson, establishes his reputation by killing thirty men during his wedding feast because he needs their clothing to pay off a bet. Then, to avenge the killing of his wife and her father, he slaughters a thousand Philistines and sets fire to their crops; after escaping capture, he kills another thousand with the jawbone of an ass. When he is finally captured and his eyes are burned out, God gives him the strength for a 9/11-like suicide attack in which he implodes a large building, crushing the three thousand men and women who are worshipping inside it.

Israel's first king, Saul, establishes a small empire, which gives him the opportunity to settle an old score. Centuries earlier, during the Israelites' exodus from Egypt, the Amalekites had harassed them, and God commanded the Israelites to "wipe out the name of Amalek." So when the judge Samuel anoints Saul as king, he reminds Saul of the divine decree: "Now go and smite Amalek, and utterly destroy all that they have, and spare them not; but slay both man and woman, infant and suckling, ox and sheep, camel and ass."[18]

Saul carries out the order, but Samuel is furious to learn that he has spared their king, Agag. So Samuel "hewed Agag in pieces before the Lord."

Saul is eventually overthrown by his son-in-law David, who absorbs the southern tribes of Judah, conquers Jerusalem, and makes it the capital of a kingdom that will last four centuries. David would come to be celebrated in story, song, and sculpture, and his six-pointed star would symbolize his people for three thousand years. Christians too would revere him as the forerunner of Jesus.

But in Hebrew scripture David is not just the "sweet singer of Israel," the chiseled poet who plays a harp and composes the Psalms. After he makes his name by killing Goliath, David recruits a gang of guerrillas, extorts wealth from his fellow citizens at swordpoint, and fights as a mercenary for the Philistines. These achievements make Saul jealous: the women in his court are singing, "Saul has killed by the thousands, but David by the tens of thousands." So Saul plots to have him assassinated.[19] David narrowly escapes before staging a successful coup.

When David becomes king, he keeps up his hard-earned reputation for killing by the tens of thousands. After his general Joab "wasted the country of the children of Ammon," David "brought out the people that were in it, and cut them with saws, and with harrows of iron, and with axes."[20] Finally he manages to do *something* that God considers immoral: he orders a census. To punish David for this lapse, God kills seventy thousand of his citizens.

Within the royal family, sex and violence go hand in hand. While taking a walk on the palace roof one day, David peeping-toms a naked woman, Bathsheba, and likes what he sees, so he sends her husband to be killed in battle and adds her to his seraglio. Later one of David's children rapes another one and is killed in revenge by a third. The avenger, Absalom, rounds up an army and tries to usurp David's throne by having sex with ten of his concubines. (As usual, we are not told how the concubines felt about all this.) While fleeing David's army, Absalom's hair gets caught in a tree, and David's general thrusts three spears into his heart. This does not put the family squabbles to an end. Bathsheba tricks a senile David into anointing their son Solomon as his successor. When the legitimate heir, David's older son Adonijah, protests, Solomon has him killed.

King Solomon is credited with fewer homicides than his predecessors and is remembered instead for building the Temple in Jerusalem and for writing the books of Proverbs, Ecclesiastes, and the Song of Songs (though with a harem of seven hundred princesses and three hundred concubines, he clearly didn't spend all his time writing). Most of all he is remembered for his eponymous virtue, "the wisdom of Solomon." Two prostitutes sharing a room give birth a few days apart. One of the babies dies, and each woman claims that the surviving boy is hers. The wise king adjudicates the dispute by pulling out a sword and threatening to butcher the baby and hand each woman a piece of the bloody corpse. One woman withdraws her claim, and Solomon awards the baby to her. "When all Israel heard of the verdict that the king had rendered, they stood in awe of the king, because they saw that he had divine wisdom in carrying out justice."[21]

The distancing effect of a good story can make us forget the brutality of the world in which it was set. Just imagine a judge in family court today adjudicating a maternity dispute by pulling out a chain saw and threatening to butcher the baby before the disputants' eyes. Solomon was confident that the more humane woman (we are never told that she was the mother) would reveal herself, and that the other woman was so spiteful that she would allow a baby to be slaughtered in front of her—and he was right! And he must have been prepared, in the event he was wrong, to carry out the butchery or else forfeit all credibility. The women, for their part, must have believed that their wise king was capable of carrying out this grisly murder.

The Bible depicts a world that, seen through modern eyes, is staggering in its savagery. People enslave, rape, and murder members of their immediate families. Warlords slaughter civilians indiscriminately, including the children. Women are bought, sold, and plundered like sex toys. And Yahweh tortures and massacres people by the hundreds of thousands for trivial disobedience or for no reason at all. These atrocities are neither isolated nor obscure. They implicate all the major characters of the Old Testament, the ones that Sunday-school children draw with crayons. And they fall into a continuous plotline that stretches for millennia, from Adam and Eve through Noah, the patriarchs, Moses, Joshua, the judges, Saul, David, Solomon, and beyond. According to the biblical scholar Raymund Schwager, the Hebrew Bible "contains over six hundred passages that explicitly talk about nations, kings, or individuals attacking, destroying, and killing others. . . . Aside from the approximately one thousand verses in which Yahweh himself appears as the direct executioner of violent punishments, and the many texts in which the Lord delivers the criminal to the punisher's sword, in over one hundred other passages Yahweh expressly gives the command to kill people."[22] Matthew White, a self-described atrocitologist who keeps a database with the estimated death tolls of history's major wars, massacres, and genocides, counts about 1.2 million deaths from mass killing that are specifically enumerated in the Bible. (He excludes the half million casualties in the war between Judah and Israel described in 2 Chronicles 13 because he considers the body count historically implausible.) The victims of the Noachian flood would add another 20 million or so to the total.[23]

The good news, of course, is that most of it never happened. Not only is there no evidence that Yahweh inundated the planet and incinerated its cities, but the patriarchs, exodus, conquest, and Jewish empire are almost certainly fictions. Historians have found no mention in Egyptian writings of the departure of a million slaves (which could hardly have escaped the Egyptians' notice); nor have archaeologists found evidence in the ruins of Jericho or neighboring cities of a sacking around 1200 BCE. And if there was a Davidic empire stretching from the Euphrates to the Red Sea around the turn of the 1st millennium BCE, no one else at the time seemed to have noticed it.[24]

Modern biblical scholars have established that the Bible is a wiki. It was compiled over half a millennium from writers with different styles, dialects, character names, and conceptions of God, and it was subjected to haphazard editing that left it with many contradictions, duplications, and non sequiturs.

The oldest parts of the Hebrew Bible probably originated in the 10th century BCE. They included origin myths for the local tribes and ruins, and legal codes adapted from neighboring civilizations in the Near East. The texts probably served as a code of frontier justice for the Iron Age tribes that herded livestock and farmed hillsides in the southeastern periphery of Canaan. The tribes began to encroach on the valleys and cities, engaged in some marauding every now and again, and may even have destroyed a city or two. Eventually their myths were adopted by the entire population of Canaan, unifying them with a shared genealogy, a glorious history, a set of taboos to keep them from defecting to foreigners, and an invisible enforcer to keep them from each other's throats. A first draft was rounded out with a continuous historical narrative around the late 7th to mid-6th century BCE, when the Babylonians conquered the Kingdom of Judah and forced its inhabitants into exile. The final edit was completed after their return to Judah in the 5th century BCE.

Though the historical accounts in the Old Testament are fictitious (or at best artistic reconstructions, like Shakespeare's historical dramas), they offer a window into the lives and values of Near Eastern civilizations in the mid-1st millennium BCE. Whether or not the Israelites actually engaged in genocide, they certainly thought it was a good idea. The possibility that a woman had a legitimate interest in not being raped or acquired as sexual property did not seem to register in anyone's mind. The writers of the Bible saw nothing wrong with slavery or with cruel punishments like blinding, stoning, and hacking someone to pieces. Human life held no value in comparison with unthinking obedience to custom and authority.

If you think that by reviewing the literal content of the Hebrew Bible I am trying to impugn the billions of people who revere it today, then you are missing the point. The overwhelming majority of observant Jews and Christians are, needless to say, thoroughly decent people who do not sanction genocide, rape, slavery, or stoning people for frivolous infractions. Their reverence for the Bible is purely talismanic. In recent millennia and centuries the Bible has been spin-doctored, allegorized, superseded by less violent texts (the Talmud among Jews and the New Testament among Christians), or discreetly ignored. And *that* is the point. Sensibilities toward violence have changed so much that religious people today compartmentalize their attitude to the Bible. They pay it lip service as a symbol of morality, while getting their actual morality from more modern principles.

THE ROMAN EMPIRE AND EARLY CHRISTENDOM

Christians downplay the wrathful deity of the Old Testament in favor of a newer conception of God, exemplified in the New Testament (the Christian Bible) by his son Jesus, the Prince of Peace. Certainly loving one's enemies and turning the other cheek constitute an advance over utterly destroying all that breatheth. Jesus, to be sure, was not above using violent imagery to secure the loyalty of his flock. In Matthew 10:34–37 he says:

> Think not that I am come to send peace on earth: I came not to send peace, but a sword. For I am come to set a man at variance against his father, and the daughter against her mother, and the daughter in law against her mother in law. And a man's foes shall be they of his own household. He that loveth father or mother more than me is not worthy of me: and he that loveth son or daughter more than me is not worthy of me.

It's not clear what he planned to do with that sword, but there's no evidence that he smote anyone with the edge of it.

Of course, there's no direct evidence for anything that Jesus said or did.[25] The words attributed to Jesus were written decades after his death, and the Christian Bible, like the Hebrew one, is riddled with contradictions, uncorroborated histories, and obvious fabrications. But just as the Hebrew Bible offers a glimpse into the values of the middle of the 1st millennium BCE, the Christian Bible tells us much about the first two centuries CE. Indeed, in that era the story of Jesus was by no means unique. A number of pagan myths told of a savior who was sired by a god, born of a virgin at the winter solstice, surrounded by twelve zodiacal disciples, sacrificed as a scapegoat at the spring equinox, sent into the underworld, resurrected amid much rejoicing, and symbolically eaten by his followers to gain salvation and immortality.[26]

The backdrop of the story of Jesus is the Roman Empire, the latest in a succession of conquerors of Judah. Though the first centuries of Christianity took place during the Pax Romana (the Roman Peace), the alleged peacefulness has to be understood in relative terms. It was a time of ruthless imperial expansion, including the conquest of Britain and the deportation of the Jewish population of Judah following the destruction of the Second Temple in Jerusalem.

The preeminent symbol of the empire was the Colosseum, visited today by millions of tourists and emblazoned on pizza boxes all over the world. In this stadium, Super Bowl–sized audiences consumed spectacles of mass cruelty. Naked women were tied to stakes and raped or torn apart by animals. Armies of captives massacred each other in mock battles. Slaves carried out literal enactments of mythological tales of mutilation and death—for example, a man playing Prometheus would be chained to a rock, and a trained eagle would pull out his liver. Gladiators fought each other to the death; our thumbs-up and thumbs-down gestures may have come from the signals flashed by the crowd to a victorious gladiator telling him whether to administer the coup de grâce to his opponent. About half a million people died these agonizing deaths to provide Roman citizens with their bread and circuses. The grandeur that was Rome casts our violent entertainment in a different light (to say nothing of our "extreme sports" and "sudden-death overtime").[27]

The most famous means of Roman death, of course, was crucifixion, the source of the word *excruciating*. Anyone who has ever looked up at the front of a church must have given at least a moment's thought to the unspeakable agony of being nailed to a cross. Those with a strong stomach can supplement their imagination by reading a forensic investigation of the death of Jesus Christ, based on archaeological and historical sources, which was published in 1986 in the *Journal of the American Medical Association*.[28]

A Roman execution began with a scourging of the naked prisoner. Using a short whip made of braided leather embedded with sharpened stones, Roman soldiers would flog the man's back, buttocks, and legs. According to the *JAMA* authors, "The lacerations would tear into the underlying skeletal muscles and produce quivering ribbons of bleeding flesh." The prisoner's arms would then be tied around a hundred-pound crossbar, and he would be forced to carry it to a site where a post was embedded in the ground. The man would be thrown onto his shredded back and nailed through the wrists to the crossbar. (Contrary to the familiar depictions, the flesh of the palms cannot support the weight of a man.) The victim was hoisted onto the post and his feet were nailed to it, usually without a supporting block. The man's rib cage was distended by the weight of his body pulling on his arms, making it difficult to exhale unless he pulled his arms or pushed his legs against the nails. Death from asphyxiation and loss of blood would come after an ordeal ranging from three or four hours to three or four days. The executioners could prolong the torture by resting the man's weight on a seat, or hasten death by breaking his legs with a club.

Though I like to think that nothing human is foreign to me, I find it impossible to put myself in the minds of the ancients who devised this orgy of sadism. Even if I had custody of Hitler and could mete out the desert of my choice, it would not occur to me to inflict a torture like that on him. I could not avoid wincing in sympathy, would not want to become the kind of person who could indulge in such cruelty, and could see no point in adding to the world's reservoir of suffering without a commensurate benefit. (Even the practical goal of deterring future despots, I would reason, is better served by maximizing the expectation that they will be brought to justice than by maximizing the gruesomeness of the penalty.) Yet in the foreign country we call the past, crucifixion was a common punishment. It was invented by the Persians, carried back to Europe by Alexander the Great, and widely used in Mediterranean empires. Jesus, who was convicted of minor rabble-rousing, was crucified along with two common thieves. The outrage that the story was meant to arouse was not that petty crimes were punishable by crucifixion but that Jesus was treated like a petty criminal.

The crucifixion of Jesus, of course, was never treated lightly. The cross became the symbol of a movement that spread through the ancient world, was adopted by the Roman Empire, and two millennia later remains the world's most recognizable symbol. The dreadful death it calls to mind must have made it an especially potent meme. But let's step outside our familiarity with Christianity and ponder the mindset that tried to make sense of the crucifixion. By today's sensibilities, it's more than a little macabre that a great moral movement would adopt as its symbol a graphic representation of a revolting means of torture and execution. (Imagine that the logo of a Holocaust museum was a shower nozzle, or that survivors of the Rwandan genocide formed a religion around the symbol of a machete.) More to the point, what was the lesson that the first Christians drew from the crucifixion? Today such a barbarity might galvanize people into opposing brutal regimes, or demanding that such torture never again be inflicted on a living creature. But those weren't the lessons the early Christians drew at all. No, the execution of Jesus is The Good News, a necessary step in the most wonderful episode in history. In allowing the crucifixion to take place, God did the world an incalculable favor. Though infinitely powerful, compassionate, and wise, he could think of no other way to reprieve humanity from punishment for its sins (in particular, for the sin of being descended from a couple who had disobeyed him) than to allow an innocent man (his son no less) to be impaled through the limbs and slowly suffocate in agony. By acknowledging that this sadistic murder was a gift of divine mercy, people could earn eternal life. And if they failed to see the logic in all this, their flesh would be seared by fire for all eternity.

According to this way of thinking, death by torture is not an unthinkable horror; it has a bright side. It is a route to salvation, a part of the divine plan. Like Jesus, the early Christian saints found a place next to God by being tortured to death in ingenious ways. For more than a millennium, Christian martyrologies described these torments with pornographic relish.[29]

Here are just a few saints whose names, if not their causes of death, are widely known. Saint Peter, an apostle of Jesus and the first Pope, was crucified upside down. Saint Andrew, the patron saint of Scotland, met his end on an X-shaped cross, the source of the diagonal stripes on the Union Jack. Saint Lawrence was roasted alive on a gridiron, a detail unknown to most Canadians who recognize his name from the river, the gulf, and one of Montreal's two major boulevards. The other one commemorates Saint Catherine, who was broken on the wheel, a punishment in which the executioner tied the victim to a wagon wheel, smashed his or her limbs with a sledgehammer, braided the shattered but living body through the spokes, and hoisted it onto a pole for birds to peck while the victim slowly died of hemorrhage and shock. (Catherine's wheel, studded with spikes, adorns the shield of the eponymous college at Oxford.) Saint Barbara, namesake of the beautiful California city, was hung upside down by her ankles while soldiers ripped her body with iron claws, amputated her breasts, burned the wounds with hot irons, and beat her head with spiked clubs. And then there's Saint George, the patron saint of England, Palestine, the republic of Georgia, the Crusades, and the Boy Scouts. Because God kept resuscitating him, George got to be tortured to death many times. He was seated astride a sharp blade with weights on his legs, roasted on a fire, pierced through the feet, crushed by a spiked wheel, had sixty nails hammered into his head, had the fat rendered out of his back with candles, and then was sawn in half.

The voyeurism in the martyrologies was employed not to evoke outrage against torture but to inspire reverence for the bravery of the martyrs. As in the story of Jesus, torture was an excellent thing. The saints welcomed their torments, because suffering in this life would be rewarded with bliss in the next one. The Christian poet Prudentius wrote of one of the martyrs, "The mother was present, gazing on all the preparations for her dear one's death and showed no signs of grief, rejoicing rather each time the pan hissing hot above the olive wood roasted and scorched her child."[30] Saint Lawrence would become the patron saint of comedians because while he was lying on the gridiron he said to his tormenters, "This side's done, turn me over and have a bite." The torturers were straight men, bit players; when they were put in a bad light it was because they were torturing *our* heroes, not because they used torture in the first place.

The early Christians also extolled torture as just deserts for the sinful. Most people have heard of the seven deadly sins, standardized by Pope Gregory I in 590 CE. Fewer people know about the punishment in hell that was reserved for those who commit them:

Pride: Broken on the wheel
Envy: Put in freezing water
Gluttony: Force-fed rats, toads, and snakes
Lust: Smothered in fire and brimstone
Anger: Dismembered alive
Greed: Put in cauldrons of boiling oil
Sloth: Thrown in snake pits[31]

The duration of these sentences, of course, was infinite.

By sanctifying cruelty, early Christianity set a precedent for more than a millennium of systematic torture in Christian Europe. If you understand the expressions *to burn at the stake, to hold his feet to the fire, to break a butterfly on the wheel, to be racked with pain, to be drawn and quartered, to disembowel, to flay, to press, the thumbscrew, the garrote, a slow burn,* and *the iron maiden* (a hollow hinged statue lined with nails, later taken as the name of a heavy-metal rock band), you are familiar with a fraction of the ways that heretics were brutalized during the Middle Ages and early modern period.

During the Spanish Inquisition, church officials concluded that the conversions of thousands of former Jews didn't take. To compel the conversos to confess their hidden apostasy, the inquisitors tied their arms behind their backs, hoisted them by their wrists, and dropped them in a series of violent jerks, rupturing their tendons and pulling their arms out of their sockets.[32] Many others were burned alive, a fate that also befell Michael Servetus for questioning the trinity, Giordano Bruno for believing (among other things) that the earth went around the sun, and William Tyndale for translating the Bible into English. Galileo, perhaps the most famous victim of the Inquisition, got off easy: he was only *shown* the instruments of torture (in particular, the rack) and was given the opportunity to recant for "having held and believed that the sun is the center of the world and immovable, and that the earth is not the center and moves." Today the rack shows up in cartoons featuring elasticized limbs and bad puns (Stretching exercises; Is this a wind-up? No pain no gain). But at the time it was no laughing matter. The Scottish travel writer William Lithgow, a contemporary of Galileo's, described what it was like to be racked by the Inquisition:

> As the levers bent forward, the main force of my knees against the two planks burst asunder the sinews of my hams, and the lids of my knees were crushed. My eyes began to startle, my mouth to foam and froth, and my teeth to chatter like the doubling of a drummer's sticks. My lips were shivering, my groans were vehement, and blood sprang from my arms, broken sinews, hands, and knees. Being loosed from these pinnacles of pain, I was hand-fast set on the floor, with this incessant imploration: "Confess! Confess!"[33]

Though many Protestants were victims of these tortures, when they got the upper hand they enthusiastically inflicted them on others, including a hundred thousand women they burned at the stake for witchcraft between the 15th and 18th centuries.[34] As so often happens in the history of atrocity, later centuries would treat these horrors in lighthearted ways. In popular culture today witches are not the victims of torture and execution but mischievous cartoon characters or sassy enchantresses, like Broom-Hilda, Witch Hazel, Glinda, Samantha, and the Halliwell sisters in *Charmed.*

Institutionalized torture in Christendom was not just an unthinking habit; it had a moral rationale. If you really believe that failing to accept Jesus as one's savior is a ticket to fiery damnation, then torturing a person until he acknowledges this truth is doing him the biggest favor of his life: better a few hours now than an eternity later. And silencing a person before he can corrupt others, or making an example of him to deter the rest, is a responsible public health measure. Saint Augustine brought the point home with a pair of analogies: a good father prevents his son from picking up a venomous snake, and a good gardener cuts off a rotten branch to save the rest of the tree.[35] The method of choice had been specified by Jesus himself: "If a man abide not in me, he is cast forth as a branch, and is withered; and men gather them, and cast them into the fire, and they are burned."[36]

Once again, the point of this discussion is not to accuse Christians of endorsing torture and persecution. *Of course* most devout Christians today are thoroughly tolerant and humane people. Even those who thunder from televised pulpits do not call for burning heretics alive or hoisting Jews on the strappado. The question is why they don't, given that their beliefs imply that it would serve the greater good. The answer is that people in the West today compartmentalize their religious ideology. When they affirm their faith in houses of worship, they profess beliefs that have barely changed in two thousand years. But when it comes to their actions, they respect modern norms of nonviolence and toleration, a benevolent hypocrisy for which we should all be grateful.

MEDIEVAL KNIGHTS

If the word *saintly* deserves a second look, so does the word *chivalrous*. The legends of knights and ladies in King Arthur's time have provided Western culture with some of its most romantic images. Lancelot and Guinevere are the archetypes of romantic love, Sir Galahad the embodiment of gallantry. Camelot, the name of King Arthur's court, was used as the title of a Broadway musical, and when word got out after John F. Kennedy's assassination that he had enjoyed the sound track, it became a nostalgic term for his administration. Kennedy's favorite lines reportedly were "Don't let it be forgot that once there was a spot / For one brief shining moment that was known as Camelot."

As a matter of fact, the knightly way of life *was* forgot, which is a good thing for the image of the knightly way of life. The actual content of the tales of medieval chivalry, which were set in the 6th century and written between the 11th and the 13th, was not the stuff of a typical Broadway musical. The medievalist Richard Kaeuper tallied the number of acts of extreme violence in the most famous of these romances, the 13th-century *Lancelot*, and on average found one every four pages.

> Limiting ourselves to quantifiable instances, at least eight skulls are split (some to the eyes, some to the teeth, some to the chin), eight unhorsed men

> are deliberately crushed by the huge hooves of the victor's war-horse (so that they faint in agony, repeatedly), five decapitations take place, two entire shoulders are hewn away, three hands are cut off, three arms are severed at various lengths, one knight is thrown into a blazing fire and two knights are catapulted to sudden death. One woman is painfully bound in iron bands by a knight; one is kept for years in a tub of boiling water by God, one is narrowly missed by a hurled lance. Women are frequently abducted and we hear at one point of forty rapes. . . .
>
> Beyond these readily enumerable acts there are reports of three private wars (with, in one case, 100 casualties on one side, and with 500 deaths with poison in another). . . . In one [tournament], to provide the flavor, Lancelot kills the first man he encounters with his lance and then, sword drawn, "struck to the right and the left, killing horses and knights all at the same time, cutting feet and hands, heads and arms, shoulders and thighs, striking down those above him whenever he met them, and leaving a sorrowful wake behind him, so that the whole earth was bathed in blood wherever he passed."[37]

How did the knights ever earn their reputation for being gentlemen? According to *Lancelot*, "Lancelot had the custom of never killing a knight who begged for mercy, unless he had sworn beforehand to do so, or unless he could not avoid it."[38]

As for their vaunted treatment of the ladies, one knight woos a princess by pledging to rape the most beautiful woman he can find on her behalf; his rival promises to send her the heads of the knights he defeats in tournaments. Knights do protect ladies, but only to keep them from being abducted by other knights. According to *Lancelot*, "The customs of the Kingdom of Logres are such that if a lady or a maiden travels by herself, she fears no one. But if she travels in the company of a knight and another knight can win her in battle, the winner can take a lady or maiden in any way he desires without incurring shame or blame."[39] Presumably that is not what most people today mean by the word *chivalry*.

EARLY MODERN EUROPE

In chapter 3 we will see that medieval Europe settles down a bit when the knightly warlords are brought under the control of monarchs in centralized kingdoms. But the kings and queens were hardly paragons of nobility themselves.

Commonwealth schoolchildren are often taught one of the key events in British history with the help of a mnemonic:

> King Henry the Eighth, to six wives he was wedded:
> One died, one survived, two divorced, two beheaded.

Beheaded! In 1536 Henry had his wife Anne Boleyn decapitated on trumped-up charges of adultery and treason because she gave him a son that did not survive, and he had become attracted to one of her ladies-in-waiting. Two wives later he suspected Catherine Howard of adultery and sent her to the ax as well. (Tourists visiting the Tower of London can see the chopping block for themselves.) Henry was clearly the jealous type: he also had an old boyfriend of Catherine's drawn and quartered, which is to say hanged by the neck, taken down while still alive, disemboweled, castrated, decapitated, and cut into four.

The throne passed to Henry's son Edward, then to Henry's daughter Mary, and then to another daughter, Elizabeth. "Bloody Mary" did not get her nickname by putting tomato juice in her vodka but by having three hundred religious dissenters burned at the stake. And both sisters kept up the family tradition for how to resolve domestic squabbles: Mary imprisoned Elizabeth and presided over the execution of their cousin, Lady Jane Grey, and Elizabeth executed another cousin, Mary Queen of Scots. Elizabeth also had 123 priests drawn and quartered, and had other enemies tortured with bone-crushing manacles, another attraction on display in the Tower. Today the British royal family is excoriated for shortcomings ranging from rudeness to infidelity. You'd think people would give them credit for not having had a single relative decapitated, nor a single rival drawn and quartered.

Despite signing off on all that torture, Elizabeth I is among England's most revered monarchs. Her reign has been called a golden age in which the arts flourished, especially the theater. It's hardly news that Shakespeare's tragedies depict a lot of violence. But his fictional worlds contained levels of barbarity that can shock even the inured audiences of popular entertainment today. Henry V, one of Shakespeare's heroes, issues the following ultimatum of surrender to a French village during the Hundred Years' War:

> why, in a moment look to see
> The blind and bloody soldier with foul hand
> Defile the locks of your shrill-shrieking daughters;
> Your fathers taken by the silver beards,
> And their most reverend heads dash'd to the walls,
> Your naked infants spitted upon pikes.[40]

In *King Lear*, the Duke of Cornwall gouges out the eyes of the Earl of Gloucester ("Out, vile jelly!"), whereupon his wife, Regan, orders the earl, bleeding from the sockets, out of the house: "Go thrust him out at gates, and let him smell his way to Dover." In *The Merchant of Venice*, Shylock obtains the right to cut a pound of flesh from the chest of the guarantor of a loan. In *Titus Andronicus*, two men kill another man, rape his bride, cut out her tongue, and amputate her hands. Her father kills the rapists, cooks them in a pie, and feeds them to their mother, whom he then kills before killing his own daughter for having gotten raped in the first place; then he is killed, and his killer is killed.

Entertainment written for children was no less grisly. In 1815 Jacob and Wilhelm Grimm published a compendium of old folktales that had gradually been adapted for children. Commonly known as *Grimm's Fairy Tales*, the collection ranks with the Bible and Shakespeare as one of the bestselling and most respected works in the Western canon. Though it isn't obvious from the bowdlerized versions in Walt Disney films, the tales are filled with murder, infanticide, cannibalism, mutilation, and sexual abuse—grim fairy tales indeed. Take just the three famous stepmother stories:

- During a famine, the father and stepmother of Hansel and Gretel abandon them in a forest so that they will starve to death. The children stumble upon an edible house inhabited by a witch, who imprisons Hansel and fattens him up in preparation for eating him. Fortunately Gretel shoves the witch into a fiery oven, and "the godless witch burned to death in a horrible way."[41]
- Cinderella's stepsisters, when trying to squeeze into her slippers, take their mother's advice and cut off a toe or heel to make them fit. Doves notice the blood, and after Cinderella marries the prince, they peck out the stepsisters' eyes, punishing them "for their wickedness and malice with blindness for the rest of their lives."
- Snow White arouses the jealousy of her stepmother, the queen, so the queen orders a hunter to take her into the forest, kill her, and bring back her lungs and liver for the queen to eat. When the queen realizes that Snow White has escaped, she makes three more attempts on her life, two by poison, one by asphyxiation. After the prince has revived her, the queen crashes their wedding, but "iron slippers had already been heated up for her over a fire of coals. . . . She had to put on the red-hot iron shoes and dance in them until she dropped to the ground dead."[42]

As we shall see, purveyors of entertainment for young children today have become so intolerant of violence that even episodes of the early Muppets have been deemed too dangerous for them. And speaking of puppetry, one of the most popular forms of children's entertainment in Europe used to be the Punch and Judy show. Well into the 20th century, this pair of bickering glove puppets acted out slapstick routines in ornate booths in English seaside towns. The literature scholar Harold Schechter summarizes a typical plot:

> It begins when Punch goes to pet his neighbor's dog, which promptly clamps its teeth around the puppet's grotesquely oversized nose. After prying the dog loose, Punch summons the owner, Scaramouche and, after a bit of crude banter, knocks the fellow's head "clean off his shoulders." Punch then calls for his wife, Judy, and requests a kiss. She responds by walloping him in the

face. Seeking another outlet for his affection, Punch asks for his infant child and begins to cradle it. Unfortunately, the baby picks that moment to dirty itself. Always the loving family man, Punch reacts by beating the baby's head against the stage, then hurling its dead body into the audience. When Judy reappears and discovers what's happened, she is understandably upset. Tearing Punch's stick from his hands, she begins to lay into him. He wrestles the cudgel away from her, pummels her to death, and then breaks into a triumphant little song:

Who'd be plagued with a wife
That could set himself free
With a rope or a knife
Or a good stick, like me?[43]

Even Mother Goose nursery rhymes, which mostly date from the 17th and 18th centuries, are jarring by the standards of what we let small children hear today. Cock Robin is murdered in cold blood. A single mother living in substandard housing has numerous illegitimate children and abuses them with whipping and starvation. Two unsupervised children are allowed to go on a dangerous errand; Jack sustains a head injury that could leave him with brain damage, while Jill's condition is unknown. A drifter confesses that he threw an old man down the stairs. Georgie Porgie sexually harasses underage girls, leaving them with symptoms of post-traumatic stress disorder. Humpty Dumpty remains in critical condition after a crippling accident. A negligent mother leaves a baby unattended on a treetop, with disastrous results. A blackbird swoops down on a domestic employee hanging up laundry and maliciously wounds her nose. Three vision-impaired mice are mutilated with a carving knife. And here comes a candle to light you to bed; here comes a chopper to chop off your head! A recent article in the *Archives of Diseases of Childhood* measured the rates of violence in different genres of children's entertainment. The television programs had 4.8 violent scenes per hour; the nursery rhymes had 52.2.[44]

HONOR IN EUROPE AND THE EARLY UNITED STATES

If you have an American ten-dollar bill handy, look at the man portrayed on it and give a moment's thought to his life and death. Alexander Hamilton is one of American history's most luminous figures. As a coauthor of the *Federalist Papers*, he helped to articulate the philosophical basis of democracy. As America's first secretary of the treasury, he devised the institutions that support modern market economies. At other times in his life he led three battalions in the Revolutionary War, helped launch the Constitutional Convention, commanded a national army, established the Bank of New York, served in the New York legislature, and founded the *New York Post*.[45]

Yet in 1804 this brilliant man did something that by today's standards was astonishingly stupid. Hamilton had long exchanged bitchy remarks with his rival Vice President Aaron Burr, and when Hamilton refused to disavow a criticism of Burr that had been attributed to him, Burr challenged him to a duel. Common sense was just one of many forces that could have pulled him away from a date with death.[46] The custom of dueling was already on the wane, and Hamilton's state of residence, New York, had outlawed it. Hamilton had lost a son to a duel, and in a letter explaining his response to Burr's challenge, he enumerated five objections to the practice. But he agreed to the duel anyway, because, he wrote, "what men of the world denominate honor" left him no other choice. The following morning he was rowed across the Hudson to face Burr on the New Jersey Palisades. Burr would not be the last vice president to shoot a man, but he was a better shot than Dick Cheney, and Hamilton died the following day.

Nor was Hamilton the only American statesman to be drawn into a duel. Henry Clay fought in one, and James Monroe thought the better of challenging John Adams only because Adams was president at the time. Among the other faces on American currency, Andrew Jackson, immortalized on the twenty-dollar bill, carried bullets from so many duels that he claimed to "rattle like a bag of marbles" when he walked. Even the Great Emancipator on the five-dollar bill, Abraham Lincoln, accepted a challenge to fight a duel, though he set the conditions to ensure that it would not be consummated.

Formal dueling was not, of course, an American invention. It emerged during the Renaissance as a measure to curtail assassinations, vendettas, and street brawls among aristocrats and their retinues. When one man felt that his honor had been impugned, he could challenge the other to a duel and cap the violence at a single death, with no hard feelings among the defeated man's clan or entourage. But as the essayist Arthur Krystal observes, "The gentry . . . took honor so seriously that just about every offense became an offense against honor. Two Englishmen dueled because their dogs had fought. Two Italian gentlemen fell out over the respective merits of Tasso and Ariosto, an argument that ended when one combatant, mortally wounded, admitted that he had not read the poet he was championing. And Byron's great-uncle William, the fifth Baron Byron, killed a man after disagreeing about whose property furnished more game."[47]

Dueling persisted in the 18th and 19th centuries, despite denunciations by the church and prohibitions by many governments. Samuel Johnson defended the custom, writing, "A man may shoot the man who invades his character, as he may shoot him who attempts to break into his house." Dueling sucked in such luminaries as Voltaire, Napoleon, the Duke of Wellington, Robert Peel, Tolstoy, Pushkin, and the mathematician Évariste Galois, the last two fatally. The buildup, climax, and denouement of a duel were made to order for fiction writers, and the dramatic possibilities were put to use by Sir Walter Scott, Dumas père, de Maupassant, Conrad, Tolstoy, Pushkin, Chekhov, and Thomas Mann.

The career of dueling showcases a puzzling phenomenon we will often encounter: a category of violence can be embedded in a civilization for centuries and then vanish into thin air. When gentlemen agreed to a duel, they were fighting not for money or land or even women but for honor, the strange commodity that exists because everyone believes that everyone else believes that it exists. Honor is a bubble that can be inflated by some parts of human nature, such as the drive for prestige and the entrenchment of norms, and popped by others, such as a sense of humor.[48] The institution of formal dueling petered out in the English-speaking world by the middle of the 19th century, and in the rest of Europe in the following decades. Historians have noted that the institution was buried not so much by legal bans or moral disapproval as by ridicule. When "solemn gentlemen went to the field of honor only to be laughed at by the younger generation, that was more than any custom, no matter how sanctified by tradition, could endure."[49] Today the expression "Take ten paces, turn, and fire" is more likely to call to mind Bugs Bunny and Yosemite Sam than "men of honor."

THE 20th CENTURY

As our tour of the history of forgotten violence comes within sight of the present, the landmarks start to look more familiar. But even the zone of cultural memory from the last century has relics that feel like they belong to a foreign country.

Take the decline of martial culture.[50] The older cities in Europe and the United States are dotted with public works that flaunt the nation's military might. Pedestrians can behold statues of commanders on horseback, beefcake sculptures of well-hung Greek warriors, victory arches crowned by chariots, and iron fencing wrought into the shape of swords and spears. Subway stops are named for triumphant battles: the Paris Métro has an Austerlitz station; the London Underground has a Waterloo station. Photos from a century ago show men in gaudy military dress uniforms parading on national holidays and hobnobbing with aristocrats at fancy dinners. The visual branding of long-established states is heavy on aggressive iconography, such as projectiles, edged weapons, birds of prey, and predatory cats. Even famously pacifistic Massachusetts has a seal that features an amputated arm brandishing a sword and a Native American holding a bow and arrow above the state motto, "With the sword we seek peace, but under liberty." Not to be outdone, neighboring New Hampshire adorns its license plates with the motto "Live Free or Die."

But in the West today public places are no longer named after military victories. Our war memorials depict not proud commanders on horseback but weeping mothers, weary soldiers, or exhaustive lists of names of the dead. Military men are inconspicuous in public life, with drab uniforms and little prestige among the hoi polloi. In London's Trafalgar Square, the plinth across from the big lions and Nelson's column was recently topped with a sculpture that is about as far from military iconography as one can imagine: a nude, pregnant artist who had been born without arms and legs. The World War I battlefield in Ypres, Belgium, inspiration for the poem "In Flanders Fields" and the poppies worn in Commonwealth countries on November 11, has just sprouted a memorial to the thousand soldiers who were shot in that war for desertion—men who at the time were despised as contemptible cowards. And the two most recent American state mottoes are Alaska's "North to the Future" and Hawaii's "The life of the land is perpetuated in righteousness" (though when Wisconsin solicited a replacement for "America's Dairyland," one of the entries was "Eat Cheese or Die").

Conspicuous pacifism is especially striking in Germany, a nation that was once so connected to martial values that the words *Teutonic* and *Prussian* became synonyms for rigid militarism. As recently as 1964 the satirist Tom Lehrer expressed a common fear at the prospect of West Germany participating in a multilateral nuclear coalition. In a sarcastic lullaby, the singer reassures a baby:

> Once all the Germans were warlike and mean,
> But that couldn't happen again.
> We taught them a lesson in 1918
> And they've hardly bothered us since then.

The fear of a revanchist Germany was revived in 1989, when the Berlin Wall came down and the two Germanys made plans to reunite. Yet today German culture remains racked with soul-searching over its role in the world wars and permeated with revulsion against anything that smacks of military force. Violence is taboo even in video games, and when Parker Brothers tried to introduce a German version of Risk, the board game in which players try to dominate a map of the world, the German government tried to censor it. (Eventually the rules were rewritten so that players were "liberating" rather than conquering their opponents' territories.)[51] German pacifism is not just symbolic: in 2003 half a million Germans marched to oppose the American-led invasion of Iraq. The American secretary of defense, Donald Rumsfeld, famously wrote the country off as part of "Old Europe." Given the history of ceaseless war on that continent, the remark may have been the most flagrant display of historical amnesia since the student who complained about the clichés in Shakespeare.

Many of us have lived through another change in Western sensibilities toward military symbolism. When the ultimate military weapons, nuclear bombs, were unveiled in the 1940s and 1950s, people were not repelled, even though the weapons had recently snuffed out a quarter of a million lives and were threatening to annihilate hundreds of millions more. No, the world found them charming! A sexy bathing suit, the bikini, was named after a Micronesian atoll that had been vaporized by nuclear tests, because the designer compared the onlookers' reaction to an atomic blast. Ludicrous "civil defense" measures like backyard fallout shelters and duck-and-cover classroom drills encouraged the delusion that a nuclear attack would be no big deal. To this day triple-triangle fallout shelter signs rust above the basement entrances of many American apartment buildings and schools. Many commercial logos from the 1950s featured mushroom clouds, including Atomic Fireball Jawbreaker candies, the Atomic Market (a mom-and-pop grocery store not far from MIT), and the Atomic Café, which lent its name to a 1982 documentary on the bizarre nonchalance with which the world treated nuclear weapons through the early 1960s, when horror finally began to sink in.

Another major change we have lived through is an intolerance of displays of force in everyday life. In earlier decades a man's willingness to use his fists in response to an insult was the sign of respectability.[52] Today it is the sign of a boor, a symptom of impulse control disorder, a ticket to anger management therapy.

An incident from 1950 illustrates the change. President Harry Truman had seen an unkind review in the *Washington Post* of a performance by his daughter, Margaret, an aspiring singer. Truman wrote to the critic on White House stationery: "Some day I hope to meet you. When that happens you'll need a new nose, a lot of beefsteak for black eyes, and perhaps a supporter below." Though every writer can sympathize with the impulse, today a public threat to commit aggravated assault against a critic would seem buffoonish, indeed sinister, if it came from a person in power. But at the time Truman was widely admired for his paternal chivalry.

And if you recognize the expressions "ninety-seven-pound weakling" and "get sand kicked in your face," you are probably familiar with the iconic ads for the Charles Atlas bodybuilding program, which ran in magazines and comic books starting in the 1940s. In the typical storyline, an ectomorph is assaulted on the beach in front of his girlfriend. He skulks home, kicks a chair, gambles a ten-cent stamp, receives instructions for an exercise program, and returns to the beach to wreak revenge on his assailant, restoring his standing with the beaming young woman (figure 1–1).

When it came to the product, Atlas was ahead of his time: the popularity of bodybuilding soared in the 1980s. But when it came to marketing, he

FIGURE 1–1. Everyday violence in a bodybuilding ad, 1940s

belonged to a different era. Today the ads for gyms and exercise paraphernalia don't feature the use of fisticuffs to restore manly honor. The imagery is narcissistic, almost homoerotic. Bulging pectorals and rippling abdominals are shown in arty close-up for both sexes to admire. The advantage they promise is in beauty, not might.

Even more revolutionary than the scorn for violence between men is the scorn for violence against women. Many baby boomers are nostalgic for *The Honeymooners*, a 1950s sitcom featuring Jackie Gleason as a burly bus driver whose get-rich-quick schemes are ridiculed by his sensible wife, Alice. In one of the show's recurring laugh lines, an enraged Ralph shakes his fist at her and bellows, "One of these days, Alice, one of these days . . . POW, right in the kisser!" (Or sometimes "Bang, zoom, straight to the moon!") Alice always laughs it off, not because she has contempt for a wife-beater but because she knows that Ralph is not man enough to do it. Nowadays our sensitivity to violence against women makes this kind of comedy in a mainstream television program unthinkable. Or consider the *Life* magazine ad from 1952 in figure 1–2.

Today this ad's playful, eroticized treatment of domestic violence would put it beyond the pale of the printable. It was by no means unique. A wife is also spanked in a 1950s ad for Van Heusen shirts, and a 1953 ad for Pitney-Bowes postage meters shows an exasperated boss screaming at a stubborn secretary with the caption "Is it always illegal to kill a woman?"[53]

FIGURE 1–2. Domestic violence in a coffee ad, 1952

And then there's the longest-running musical, *The Fantasticks*, with its Gilbert-and-Sullivan-like ditty "It Depends on What You Pay" (whose lyrics were based on a 1905 translation of Edmond Rostand's play *Les Romanesques*). Two men plot a kidnapping in which the son of one will rescue the daughter of the other:

> You can get the rape emphatic.
> You can get the rape polite.
> You can get the rape with Indians:
> A very charming sight.
> You can get the rape on horseback;
> They'll all say it's new and gay.
> So you see the sort of rape
> Depends on what you pay.

Though the word *rape* referred to abduction rather than sexual assault, between the opening of the play in 1960 and the end of its run in 2002 sensibilities about rape changed. As the librettist Tom Jones (no relation to the Welsh singer) explained to me:

> As time went on, I began to feel anxious about the word. Slowly, ever so slowly, things began to register on me. Headlines in the papers. Accounts of brutal gang rapes. And of "date rapes" too. I began to think: "this isn't funny." True, we weren't talking about "real rape," but there is no doubt that part of the laughter came from the shock value of using the word in this comic manner.

In the early 1970s, the producer of the play refused Jones's request to rewrite the lyrics but allowed him to add an introduction to the song explaining the intended meaning of the word and to reduce the number of repetitions of it. After the play closed in 2002 Jones rewrote the lyrics from scratch for a 2006 restaging, and he has legally ensured that only the new version may be performed in any production of *The Fantasticks* anywhere in the world.[54]

Until recently, children too were legitimate targets of violence. Parents not only spanked their children—a punishment that today has been outlawed in many countries—but commonly used a weapon like a hairbrush or paddle, or exposed the child's buttocks to increase the pain and humiliation. In a sequence that was common in children's stories through the 1950s, a mother warned a naughty child, "Wait till your father gets home," whereupon the stronger parent would remove his belt and use it to flog the child. Other commonly depicted ways of punishing children with physical pain included sending them to bed without dinner and washing their mouths out with soap. Children who were left to the mercy of unrelated adults were treated even more brutally. Within recent memory, many schoolchildren were disciplined in ways that today would be classified as "torture" and that would put their teachers in jail.[55]

People today think of the world as a uniquely dangerous place. It's hard to follow the news without a mounting dread of terrorist attacks, a clash of civilizations, and the use of weapons of mass destruction. But we are apt to forget the dangers that filled the news a few decades ago, and to be blasé about the good fortune that so many of them have fizzled out. In later chapters I will present numbers that show that the 1960s and 1970s were a vastly more brutal and menacing time than the one in which we live. But for now, in keeping with the spirit of this chapter, I will make the case impressionistically.

I graduated from university in 1976. Like most college alumni, I have no memory of the commencement speech that sent me into the world of adulthood. This gives me license to invent one today. Imagine the following forecast from an expert on the state of the world in the mid-1970s.

> Mr. Principal, members of the faculty, family, friends, and Class of 1976. Now is a time of great challenges. But it is also a time of great opportunities. As you embark on your lives as educated men and women, I call on you to give something back to your community, to work for a brighter future, and to try to make the world a better place.
>
> Now that we have that out of the way, I have something more interesting to say to you. I want to share my vision of what the world will be like at the time of your thirty-fifth reunion. The calendar will have rolled over into a new millennium, bringing you a world that is beyond your imagination. I am not referring to the advance of technology, though it will have effects you can barely conceive. I am referring to the advance of peace and human security, which you will find even harder to conceive.
>
> To be sure, the world of 2011 will still be a dangerous place. During the next thirty-five years there will be wars, as there are today, and there will be genocides, as there are today, some of them in places no one would have predicted. Nuclear weapons will still be a threat. Some of the violent regions of the world will continue to be violent. But superimposed on these constants will be unfathomable changes.
>
> First and foremost, the nightmare that has darkened your lives since your early memories of cowering in fallout shelters, a nuclear doomsday in a third world war, will come to an end. In a decade the Soviet Union will declare peace with the West, and the Cold War will be over without a shot being fired. China will also fall off the radar as a military threat; indeed, it will become our major trading partner. During the next thirty-five years no nuclear weapon will be used against an enemy. In fact, there will be no wars between major nations at all. The peace in Western Europe will continue

indefinitely, and within five years the incessant warring in East Asia will give way to a long peace there as well.

There is more good news. East Germany will open its border, and joyful students will sledgehammer the Berlin Wall to smithereens. The Iron Curtain will vanish, and the nations of Central and Eastern Europe will become liberal democracies free of Soviet domination. The Soviet Union will not only abandon totalitarian communism but will voluntarily go out of existence. The republics that Russia has occupied for decades and centuries will become independent states, many of them democratic. In most of the countries this will happen with not a drop of blood being spilled.

Fascism too will vanish from Europe, then from much of the rest of the world. Portugal, Spain, and Greece will become liberal democracies. So will Taiwan, South Korea, and most of South and Central America. The generalissimos, the colonels, the juntas, the banana republics, and the annual military coups will depart the stage in most of the developed world.

The Middle East also has surprises in store. You have just lived through the fifth war between Israel and Arab states in twenty-five years. These wars have killed fifty thousand people and recently threatened to drag the superpowers into a nuclear confrontation. But within three years the president of Egypt will hug the prime minister of Israel in the Knesset, and they will sign a peace treaty that will last into the indefinite future. Jordan too will make a lasting peace with Israel. Syria will engage in sporadic peace talks with Israel, and the two countries will not go to war.

In South Africa, the apartheid regime will be dismantled, and the white minority will cede power to the black majority. This will happen with no civil war, no bloodbath, no violent recriminations against the former oppressors.

Many of these developments will be the results of long and courageous struggles. But some of them will just happen, catching everyone by surprise. Perhaps some of you will try to figure out how it all happened. I congratulate you on your accomplishments and wish you success and satisfaction in the years ahead.

How would the audience have reacted to this outburst of optimism? Those who were listening would have broken out in snickers and shared a suspicion that the speaker was still tripping on the brown acid from Woodstock. Yet in every case the optimist would have been right.

No sightseer can understand a country from a city-a-day tour, and I don't expect this skitter across the centuries to have convinced you that the past was more violent than the present. Now that you're back home, you are surely filled with questions. Don't we still torture people? Wasn't the 20th century the bloodiest in history? Haven't new forms of war replaced the old ones? Aren't we living in the Age of Terror? Didn't they say that war was obsolete in 1910? What about all the chickens in factory farms? And couldn't nuclear terrorists start a major war tomorrow?

These are excellent questions, and I will try to answer them in the rest of the book with the help of historical studies and quantitative datasets. But I hope that these sanity checks have prepared the ground. They remind us that for all the dangers we face today, the dangers of yesterday were even worse. Readers of this book (and as we shall see, people in most of the rest of the world) no longer have to worry about abduction into sexual slavery, divinely commanded genocide, lethal circuses and tournaments, punishment on the cross, rack, wheel, stake, or strappado for holding unpopular beliefs, decapitation for not bearing a son, disembowelment for having dated a royal, pistol duels to defend their honor, beachside fisticuffs to impress their girlfriends, and the prospect of a nuclear world war that would put an end to civilization or to human life itself.

2

THE PACIFICATION PROCESS

Look, life is nasty, brutish, and short, but you knew that when you became a caveman.

—*New Yorker* cartoon[1]

Thomas Hobbes and Charles Darwin were nice men whose names became nasty adjectives. No one wants to live in a world that is Hobbesian or Darwinian (not to mention Malthusian, Machiavellian, or Orwellian). The two men were immortalized in the lexicon for their cynical synopses of life in a state of nature, Darwin for "survival of the fittest" (a phrase he used but did not coin), Hobbes for "the life of man, solitary, poor, nasty, brutish, and short." Yet both men gave us insights about violence that are deeper, subtler, and ultimately more humane than their eponymous adjectives imply. Today any understanding of human violence must begin with their analyses.

This chapter is about the origins of violence, in both the logical and the chronological sense. With the help of Darwin and Hobbes, we will look at the adaptive logic of violence and its predictions for the kinds of violent impulses that might have evolved as a part of human nature. We will then turn to the prehistory of violence, examining when violence appeared in our evolutionary lineage, how common it was in the millennia before history was written down, and what kinds of historical developments first reduced it.

THE LOGIC OF VIOLENCE

Darwin gave us a theory of why living things have the traits they have, not just their bodily traits but the basic mindsets and motives that drive their behavior. A hundred and fifty years after the *Origin of Species* was published, the theory of natural selection has been amply verified in the lab and field, and has been augmented with ideas from new fields of science and mathematics to yield a coherent understanding of the living world. These fields include genetics, which explains the replicators that make natural selection possible, and game theory, which illuminates the fates of goal-seeking agents in a world that contains other goal-seeking agents.[2]

Why should organisms ever evolve to seek to harm other organisms? The answer is not as straightforward as the phrase "survival of the fittest" would suggest. In his book *The Selfish Gene*, which explained the modern synthesis of evolutionary biology with genetics and game theory, Richard Dawkins tried to pull his readers out of their unreflective familiarity with the living world. He asked them to imagine animals as "survival machines" designed by their genes (the only entities that are faithfully propagated over the course of evolution), and then to consider how those survival machines would evolve.

> To a survival machine, another survival machine (which is not its own child or another close relative) is part of its environment, like a rock or a river or a lump of food. It is something that gets in the way, or something that can be exploited. It differs from a rock or a river in one important respect: it is inclined to hit back. This is because it too is a machine that holds its immortal genes in trust for the future, and it too will stop at nothing to preserve them. Natural selection favors genes that control their survival machines in such a way that they make the best use of their environment. This includes making the best use of other survival machines, both of the same and of different species.[3]

Anyone who has ever seen a hawk tear apart a starling, a swarm of biting insects torment a horse, or the AIDS virus slowly kill a man has firsthand acquaintance with the ways that survival machines callously exploit other survival machines. In much of the living world, violence is simply the default, something that needs no further explanation. When the victims are members of other species, we call the aggressors predators or parasites. But the victims can also be members of the same species. Infanticide, siblicide, cannibalism, rape, and lethal combat have been documented in many kinds of animals.[4]

Dawkins's carefully worded passage also explains why nature does not consist of one big bloody melee. For one thing, animals are less inclined to harm their close relatives, because any gene that would nudge an animal to harm a relative would have a good chance of harming a copy of *itself* sitting inside that relative, and natural selection would tend to weed it out. More important, Dawkins points out that another organism differs from a rock or a river because *it is inclined to hit back*. Any organism that has evolved to be violent is a member of a species whose other members, on average, have evolved to be just as violent. If you attack one of your own kind, your adversary may be as strong and pugnacious as you are, and armed with the same weapons and defenses. The likelihood that, in attacking a member of your own species, you will get hurt is a powerful selection pressure that disfavors indiscriminate pouncing or lashing out. It also rules out the hydraulic metaphor and most folk theories of violence, such as a thirst for blood, a death wish, a killer instinct, and other destructive itches, urges, and impulses. When a tendency toward violence evolves, it is always *strategic*. Organisms are selected to deploy violence only in circumstances where the expected benefits outweigh the expected costs. That discernment is especially true of intelligent species, whose large brains make them sensitive to the expected benefits and costs in a particular situation, rather than just to the odds averaged over evolutionary time.

The logic of violence as it applies to members of an intelligent species facing other members of that species brings us to Hobbes. In a remarkable passage in *Leviathan* (1651), he used fewer than a hundred words to lay out an analysis of the incentives for violence that is as good as any today:

> So that in the nature of man, we find three principal causes of quarrel. First, competition; secondly, diffidence; thirdly, glory. The first maketh men invade for gain; the second, for safety; and the third, for reputation. The first use violence, to make themselves masters of other men's persons, wives, children, and cattle; the second, to defend them; the third, for trifles, as a word, a smile, a different opinion, and any other sign of undervalue, either direct in their persons or by reflection in their kindred, their friends, their nation, their profession, or their name.[5]

Hobbes considered competition to be an unavoidable consequence of agents' pursuing their interests. Today we see that it is built into the evolutionary process. Survival machines that can elbow their competitors away from finite resources like food, water, and desirable territory will out-reproduce those competitors, leaving the world with the survival machines that are best suited for such competition.

We also know today why "wives" would be one of the resources over which men should compete. In most animal species, the female makes a greater investment in offspring than the male. This is especially true of mammals, where the mother gestates her offspring inside her body and nurses them after they are born. A male can multiply the number of his offspring by mating with several females—which will leave other males childless—while a female cannot multiply the number of her offspring by mating with several males. This makes female reproductive capacity a scarce resource over which the males of many species, including humans, compete.[6] None of this, by the way, implies that men are robots controlled by their genes, that they may be morally excused for raping or fighting, that women are passive sexual prizes, that people try to have as many babies as possible, or that people are impervious to influences from their culture, to take some of the common misunderstandings of the theory of sexual selection.[7]

The second cause of quarrel is diffidence, a word that in Hobbes's time meant "fear" rather than "shyness." The second cause is a consequence of the first: competition breeds fear. If you have reason to suspect that your neighbor is inclined to eliminate you from the competition by, say, killing you, then you will be inclined to protect yourself by eliminating him first in a preemptive strike. You might have this temptation even if you otherwise wouldn't hurt a fly, as long as you are not willing to lie down and be killed. The tragedy is that your competitor has every reason to crank through the same calculation, even if *he* is the kind of person who wouldn't hurt a fly. In fact, even if he *knew* that you started out with no aggressive designs on him, he might legitimately worry that you are tempted to neutralize him out of fear that he will neutralize you first, which gives you an incentive to neutralize him before that, ad infinitum. The political scientist Thomas Schelling offers the analogy of an armed homeowner who surprises an armed burglar, each being tempted to shoot the other to avoid being shot first. This paradox is sometimes called the Hobbesian trap or, in the arena of international relations, the security dilemma.[8]

How can intelligent agents extricate themselves from a Hobbesian trap? The most obvious way is through a policy of deterrence: Don't strike first; be strong enough to survive a first strike; and retaliate against any aggressor in kind. A credible deterrence policy can remove a competitor's incentive to invade for gain, since the cost imposed on him by retaliation would cancel out the anticipated spoils. And it removes his incentive to invade from fear, because of your commitment not to strike first and, more importantly, because of your reduced incentive to strike first, since deterrence reduces the need for preemption. The key to the deterrence policy, though, is the credibility of the threat that you will retaliate. If your adversary thinks that you're vulnerable to being wiped out in a first strike, he has no reason to fear retaliation. And if he thinks that once attacked you may rationally hold back from retaliation, because at that point it's too late to do any good, he might exploit that rationality and attack you with impunity. Only if you are committed to disprove any suspicion of weakness, to avenge all trespasses and settle all scores, will your policy of deterrence be credible. Thus we have an explanation of the incentive to invade for trifles: a word, a smile, and any other sign of undervalue. Hobbes called it "glory"; more commonly it is called "honor"; the most accurate descriptor is "credibility."

The policy of deterrence is also known as the balance of terror and, during the Cold War, was called mutual assured destruction (MAD). Whatever peace a policy of deterrence may promise is fragile, because deterrence reduces violence only by a threat of violence. Each side must react to any nonviolent sign of disrespect with a violent demonstration of mettle, whereupon one act of violence can lead to another in an endless cycle of retaliation. As we shall see in chapter 8, a major design feature in human nature, self-serving biases, can make each side believe that its own violence was an act of justified retaliation while the other's was an act of unprovoked aggression.

Hobbes's analysis pertains to life in a state of anarchy. The title of his masterwork identified a way to escape it: the Leviathan, a monarchy or other government authority that embodies the will of the people and has a monopoly on the use of force. By inflicting penalties on aggressors, the Leviathan can eliminate their incentive for aggression, in turn defusing general anxieties about preemptive attack and obviating everyone's need to maintain a hair trigger for retaliation to prove their resolve. And because the Leviathan is a disinterested third party, it is not biased by the chauvinism that makes each side think its opponent has a heart of darkness while it is as pure as the driven snow.

The logic of the Leviathan can be summed up in a triangle (figure 2–1). In every act of violence, there are three interested parties: the aggressor, the victim, and a bystander. Each has a motive for violence: the aggressor to prey upon the victim, the victim to retaliate, the bystander to minimize collateral damage from their fight. Violence between the combatants may be called war; violence by the bystander against the combatants may be called law. The Leviathan theory, in a nutshell, is that law is better than war. Hobbes's theory makes a testable prediction about the history of violence. The Leviathan made its first appearance in a late act in the human pageant. Archaeologists tell us that humans lived in a state of anarchy until the emergence of civilization some five thousand years ago, when sedentary farmers first coalesced into cities and states and developed the first governments. If Hobbes's theory is right, this transition should also have ushered in the first major historical decline in violence. Before the advent of civilization, when men lived without "a common power to keep them all in awe," their lives should have been nastier, more brutish, and shorter than when peace was imposed on them by armed authorities, a development I will call the Pacification Process. Hobbes claimed that "savage people in many places in America" lived in a state of violent anarchy, but he gave no specifics as to whom he had in mind.

In this data vacuum, anyone could have a go at speculating about primitive people, and it did not take long for a contrary theory to turn up. Hobbes's opposite number was the Swiss-born philosopher Jean-Jacques Rousseau

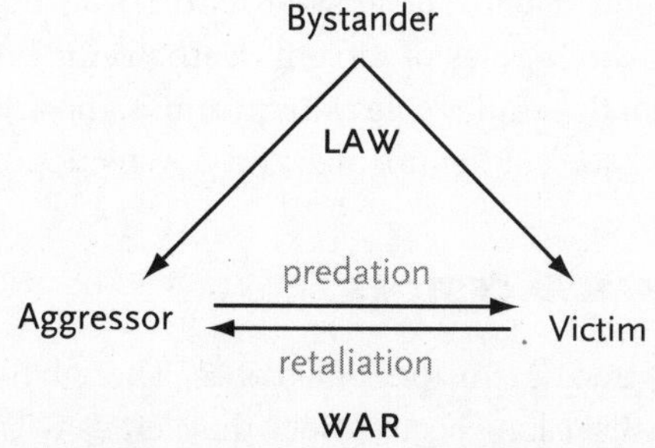

FIGURE 2–1. The violence triangle

(1712–78), who opined that "nothing can be more gentle than [man] in his primitive state. . . . The example of the savages . . . seems to confirm that mankind was formed ever to remain in it, . . . and that all ulterior improvements have been so many steps . . . towards the decrepitness of the species."[9]

Though the philosophies of Hobbes and Rousseau were far more sophisticated than "nasty brutish and short" versus "the noble savage," their competing stereotypes of life in a state of nature fueled a controversy that remains with us today. In *The Blank Slate*, I discussed how the issue has accumulated a heavy burden of emotional, moral, and political baggage. In the second half of the 20th century, Rousseau's romantic theory became the politically correct doctrine of human nature, both in reaction to earlier, racist doctrines about "primitive" people and out of a conviction that it was a more uplifting view of the human condition. Many anthropologists believe that if Hobbes was right, war would be inevitable or even desirable; therefore anyone who favors peace must insist that Hobbes was wrong. These "anthropologists of peace" (who in fact are rather aggressive academics—the ethologist Johan van der Dennen calls them the Peace and Harmony Mafia) have maintained that humans and other animals are strongly inhibited from killing their own kind, that war is a recent invention, and that fighting among native peoples was ritualistic and harmless until they encountered European colonists.[10]

As I mentioned in the preface, I think the idea that biological theories of violence are fatalistic and romantic theories optimistic gets everything backwards, but that isn't the point of this chapter. When it came to violence in pre-state peoples, Hobbes and Rousseau were talking through their hats: neither knew a thing about life before civilization. Today we can do better. This chapter reviews the facts about violence in the earliest stages of the human career. The story begins before we were human, and we will look at aggression in our primate cousins to see what it reveals about the emergence of violence in our evolutionary lineage. When we reach our own species, I will zero in on the contrast between foraging bands and tribes who live in a state of anarchy and peoples who live in settled states with some form of governance. We will also look at how foragers fight and what they fight over. This leads to the pivotal question: Is the warring of anarchic tribes more or less destructive than that of people living in settled states? The answer requires a switch from narratives to numbers: the per capita rates of violent death, to the best we can estimate them, in societies that live under a Leviathan and in those that live in anarchy. Finally we will take a look at the upsides and downsides of civilized life.

VIOLENCE IN HUMAN ANCESTORS

How far back can we trace the history of violence? Though the primate ancestors of the human lineage have long been extinct, they left us with at least one kind of evidence about what they might have been like: their other descendants, chimpanzees. We did not, of course, evolve from chimps, and as we shall see it's an open question whether chimpanzees preserved the traits of our common ancestor or veered off in a uniquely chimp direction. But either way, chimpanzee aggression holds a lesson for us, because it shows how violence can evolve in a primate species with certain traits we share. And it tests the evolutionary prediction that violent tendencies are not hydraulic but strategic, deployed only in circumstances in which the potential gains are high and the risks are low.[11]

Common chimpanzees live in communities of up to 150 individuals who occupy a distinct territory. As chimpanzees forage for the fruit and nuts that are unevenly distributed through the forest, they frequently split and coalesce into smaller groups ranging in size from one to fifteen. If one group encounters another group from a different community at the border between their territories, the interaction is always hostile. When the groups are evenly matched, they dispute the boundary in a noisy battle. The two sides bark, hoot, shake branches, throw objects, and charge at each other for half an hour or more, until one side, usually the smaller one, skulks away.

These battles are examples of the aggressive displays that are common among animals. Once thought to be rituals that settle disputes without bloodshed for the good of the species, they are now understood as displays of strength and resolve that allow the weaker side to concede when the outcome of a fight is a foregone conclusion and going through with it would only risk injury to both. When two animals are evenly matched, the show of force may escalate to serious fighting, and one or both can get injured or killed.[12] Battles between groups of chimpanzees, however, do not escalate into serious fighting, and anthropologists once believed that the species was essentially peaceful.

Jane Goodall, the primatologist who first observed chimpanzees in the wild for extended periods of time, eventually made a shocking discovery.[13] When a group of male chimpanzees encounters a smaller group or a solitary individual from another community, they don't hoot and bristle, but take advantage of their numbers. If the stranger is a sexually receptive adolescent female, they may groom her and try to mate. If she is carrying an infant, they will often attack her and kill and eat the baby. And if they encounter a solitary male, or isolate one from a small group, they will go after him with murderous savagery. Two attackers will hold down the victim, and the others will beat him, bite off his toes and genitals, tear flesh from his body, twist his limbs, drink his blood, or rip out his trachea. In one community, the chimpanzees picked off every male in a neighboring one, an event that if it occurred among humans we would call genocide. Many of the attacks aren't triggered by chance encounters but are the outcome of border patrols in which a group of males quietly seek out and target any solitary male they spot. Killings can also occur within a community. A gang of males may kill a rival, and a strong female, aided by a male or another female, may kill a weaker one's offspring.

When Goodall first wrote about these killings, other scientists wondered whether they might be freak outbursts, symptoms of pathology, or artifacts of the primatologists' provisioning the chimps with food to make them easier to observe. Three decades later little doubt remains that lethal aggression is a part of chimpanzees' normal behavioral repertoire. Primatologists have observed or inferred the killings of almost fifty individuals in attacks between communities, and more than twenty-five in attacks within them. The reports have come from at least nine communities, including ones that have never been provisioned. In some communities, more than a third of the males die from violence.[14]

Does chimpicide have a Darwinian rationale? The primatologist Richard Wrangham, a former student of Goodall's, has tested various hypotheses with the extensive data that have been amassed on the demography and ecology of chimpanzees.[15] He was able to document one large Darwinian advantage and one smaller one. When chimpanzees eliminate rival males and their offspring, they expand their territory, either by moving into it immediately or by winning subsequent battles with the help of their enhanced numerical advantage. This allows them to monopolize access to the territory's food for themselves, their offspring, and the females they mate with, which in turn results in a greater rate of births among the females. The community will also sometimes absorb the females of the vanquished community, bringing the males a second reproductive advantage. It's not that the chimps fight directly over food or females. All they care about is dominating their territory and eliminating rivals if they can do so at minimal risk to themselves. The evolutionary benefits happen indirectly and over the long run.

As for the risks, the chimpanzees minimize them by picking unfair fights, those in which they outnumber their victim by at least three to one. The foraging pattern of chimpanzees often delivers an unlucky victim into their clutches because fruiting trees are distributed patchily in the forest. Hungry chimps may have to forage in small groups or on their own and may sometimes venture into no-chimp's-land in pursuit of their dinner.

What does this have to do with violence in humans? It raises the possibility that the human lineage has been engaged in lethal raiding since the time of its common root with chimpanzees around six million years ago. There is, however, an alternative possibility. The shared ancestor of humans and common chimpanzees (*Pan troglodytes*) bequeathed the world a third species, bonobos or pygmy chimps (*Pan paniscus*), which split from their common cousins around two million years ago. We are as closely related to bonobos as we are to common chimps, and bonobos never engage in lethal raiding. Indeed, the difference between bonobos and common chimpanzees is one of the best-known facts in popular primatology. Bonobos have become famous as the peaceable, matriarchal, concupiscent, herbivorous "hippie chimps." They are the namesake of a vegetarian restaurant in New York, the inspiration for the sexologist Dr. Suzy's "Bonobo Way of Peace Through Pleasure," and if the *New York Times* columnist Maureen Dowd had her way, a role model for men today.[16]

The primatologist Frans de Waal points out that in theory the common ancestor of humans, common chimpanzees, and bonobos could have been similar to bonobos rather than to common chimps.[17] If so, violence between coalitions of males would have shallower roots in human evolutionary history. Common chimpanzees and humans would have developed their lethal raiding independently, and human raiding may have developed historically in particular cultures rather than evolutionarily in the species. If so, humans would have no innate proclivities toward coalitional violence and would not need a Leviathan, or any other institution, to keep them away from it.

The idea that humans evolved from a peaceful, bonobolike ancestor has two problems. One is that it is easy to get carried away with the hippie-chimp story. Bonobos are an endangered species that lives in inaccessible forests in dangerous parts of the Congo, and much of what we know about them comes from observations of small groups of well-fed juveniles or young adults in captivity. Many primatologists suspect that systematic studies of older, hungrier, more populous, and freer groups of bonobos would paint a darker picture.[18] Bonobos in the wild, it turns out, engage in hunting, confront each other belligerently, and injure one another in fights, perhaps sometimes fatally. So while bonobos are unquestionably less aggressive than common chimpanzees—they never raid one another, and communities can mingle peacefully—they are certainly not peaceful across the board.

The second and more important problem is that the common ancestor of the two chimpanzee species and humans is far more likely to have been like a common chimpanzee than like a bonobo.[19] Bonobos are very strange primates, not just in their behavior but in their anatomy. Their small, childlike heads, lighter bodies, reduced sex differences, and other juvenile traits make them different not only from common chimpanzees but from the other great apes (gorillas and orangutans) and different as well from fossil australopithecines, who were ancestral to humans. Their distinctive anatomy, when placed on the great ape family tree, suggests that bonobos were pulled away from the generic ape plan by neoteny, a process that retunes an animal's growth program to preserve certain juvenile features in adulthood (in the case of bonobos, features of the cranium and brain). Neoteny often occurs in species that have undergone domestication, as when dogs diverged from wolves, and it is a pathway by which selection can make animals less aggressive. Wrangham argues that the primary mover in bonobo evolution was selection for reduced aggression in males, perhaps because bonobos forage in large groups without vulnerable loners, so there are no opportunities for coalitional aggression to pay off. These considerations suggest that bonobos are the odd-ape-out, and we are descended from an animal that was closer to common chimpanzees.

Even if common chimps and humans discovered coalitional violence independently, the coincidence would be informative. It would suggest that lethal raiding can be evolutionarily advantageous in an intelligent species that fissions into groups of various sizes, and in which related males form coalitions and can assess each other's relative strength. When we look at violence in humans later in the chapter, we will see that some of the parallels are a bit close for comfort.

It would be nice if the gap between the common ancestor and modern humans could be filled in by the fossil record. But chimpanzees' ancestors have left no fossils, and hominid fossils and artifacts are too scarce to provide direct evidence of aggression, such as preserved weapons or wounds. Some paleoanthropologists test for signs of a violent temperament in fossil species by measuring the size of the canine teeth in males (since daggerlike canines are found in aggressive species) and by looking for differences in the size of the males and the females (since males tend to be larger in polygynous species, the better to fight with other males).[20] Unfortunately the small jaws of hominids, unlike the muzzles of other primates, don't open wide enough for large canines to be practical, regardless of how aggressive or peaceful these creatures were. And unless a species was considerate enough to have left behind a large number of complete skeletons, it's hard to sex them reliably and compare the size of the males and the females. (For these reasons many anthropologists are skeptical of the recent claim that *Ardipithecus ramidus*, a 4.4-million-year-old species that is probably ancestral to *Homo*, was unisex and small-canined and hence monogamous and peaceable.)[21] The more recent and abundant *Homo* fossils show that the males have been larger than the females for at least two million years, by at least as great a ratio as in modern humans. This reinforces the suspicion that violent competition among men has a long history in our evolutionary lineage.[22]

KINDS OF HUMAN SOCIETIES

The species we belong to, "anatomically modern *Homo sapiens*," is said to be 200,000 years old. But "behaviorally modern" humans, with art, ritual, clothing, complex tools, and the ability to live in different ecosystems, probably evolved closer to 75,000 years ago in Africa before setting out to people the rest of the world. When the species emerged, people lived in small, nomadic, egalitarian bands of kinsmen, subsisted by hunting and gathering, and had no written language or government. Today the vast majority of humans are settled in stratified societies numbering in the millions, eat foods cultivated by agriculture, and are governed by states. The transition, sometimes called the Neolithic (new stone age) Revolution, began around 10,000 years ago with the emergence of agriculture in the Fertile Crescent, China, India, West Africa, Mesoamerica, and the Andes.[23]

It's tempting, then, to use the 10,000-year horizon as a boundary between two major eras of human existence: a hunter-gatherer era, in which we did most of our biological evolving and which may still be glimpsed in extant hunter-gatherers, and the era of civilization thereafter. That is the dividing line that figures in theories of the ecological niche to which humans are biologically adapted, which evolutionary psychologists call "the environment of evolutionary adaptedness." But that is not the cut that is most relevant to the Leviathan hypothesis.

For one thing, the 10,000-year milestone applies only to the *first* societies that farmed. Agriculture developed in other parts of the world later and spread outward from those cradles only gradually. Ireland, for example, was not lapped by the wave of farming that emanated from the Near East until around 6,000 years ago.[24] Many parts of the Americas, Australia, Asia, and Africa were populated by hunter-gatherers until a few centuries ago, and of course a few still are.

Also, societies cannot be dichotomized into hunter-gatherer bands and agricultural civilizations.[25] The nonstate peoples we are most familiar with are the hunters and gatherers living in small bands like the !Kung San of the Kalahari Desert and the Inuit of the Arctic. But these people have survived as hunter-gatherers only because they inhabit remote parts of the globe that no one else wants. As such they are not a representative sample of our anarchic ancestors, who may have enjoyed flusher environments. Until recently other foragers parked themselves in valleys and rivers that were teeming with fish and game and that supported a more affluent, complex, and sedentary lifestyle. The Indians of the Pacific Northwest, known for their totem poles and potlatches, are a familiar example. Also beyond the reach of states are hunter-horticulturalists, such as peoples in Amazonia and New Guinea who supplement their hunting and gathering by slashing and burning patches of forest and growing bananas or sweet potatoes in small gardens. Their lives are not as austere as those of pure hunter-gatherers, but they are far closer to them than they are to sedentary, full-time farmers.

When the first farmers settled down to grow grains and legumes and keep domesticated animals, their numbers exploded and they began to divide their labors, so that some of them lived off the food grown by others. But they didn't develop complex states and governments right away. They first coalesced into tribes connected by kinship and culture, and the tribes sometimes merged into chiefdoms, which had a centralized leader and a permanent entourage supporting him. Some of the tribes took up pastoralism, wandering with their livestock and trading animal products with sedentary farmers. The Israelites of the Hebrew Bible were tribal pastoralists who developed into chiefdoms around the time of the judges.

It took around five thousand years after the origin of agriculture for true states to appear on the scene.[26] That happened when the more powerful chiefdoms used their armed retinues to bring other chiefdoms and tribes under their control, further centralizing their power and supporting niches for specialized classes of artisans and soldiers. The emerging states built strongholds, cities, and other defensible settlements, and they developed writing systems that allowed them to keep records, exact taxes and tributes from their subjects, and codify laws to keep them in line. Petty states with designs on their neighbors' assets sometimes forced them to become states in defense, and bigger states often swallowed smaller states.

Anthropologists have proposed many subtypes and intermediate cases among these kinds of societies, and have noted that there is no cultural escalator that inevitably turns simpler societies into more complex ones. Tribes and chiefdoms can maintain their ways indefinitely, such as the Montenegrin tribes in Europe that lasted into the 20th century. And when a state breaks down, it can be taken over by tribes, as in the Greek dark ages (which followed the collapse of the Mycenaean civilization and in which the Homeric epics were set) and the European dark ages (which came after the fall of the Roman Empire). Even today, many parts of failed states, such as Somalia, Sudan, Afghanistan, and the Democratic Republic of the Congo, are essentially chiefdoms; we call the chiefs warlords.[27]

For all these reasons, it makes no sense to test for historical changes in violence by plotting deaths against a time line from the calendar. If we discover that violence has declined in a given people, it is because their mode of social organization has changed, not because the historical clock has struck a certain hour, and that change can happen at different times, if it happens at all. Nor should we expect a smooth reduction in violence along the continuum from simple, nomadic hunter-gatherers to complex, sedentary hunter-gatherers to farming tribes and chiefdoms to petty states to large states. The major transition we should expect is at the appearance of the first form of social organization that shows signs of design for reducing violence within its borders. That would be the centralized state, the Leviathan.

It's not that any early state was (as Hobbes theorized) a commonwealth vested with power by a social contract that had been negotiated by its citizens. Early states were more like protection rackets, in which powerful Mafiosi extorted resources from the locals and offered them safety from hostile neighbors and from each other.[28] Any ensuing reduction in violence benefited the overlords as much as the protectees. Just as a farmer tries to prevent his animals from killing one another, so a ruler will try to keep his subjects from cycles of raiding and feuding that just shuffle resources or settle scores among them but from his point of view are a dead loss.

The topic of violence in nonstate societies has a long and politicized history. For centuries it was conventional wisdom that native peoples were ferocious barbarians. The Declaration of Independence, for instance, complained that the king of England "endeavoured to bring on the inhabitants of our frontiers, the merciless Indian Savages whose known rule of warfare, is an undistinguished destruction of all ages, sexes and conditions."

Today the passage seems archaic, indeed offensive. Dictionaries warn against using the word *savage* (related to *sylvan*, "of the forest") to refer to native peoples, and our awareness of the genocides of Native Americans perpetrated by European colonists makes the signatories seem like a black pot in a glass house casting the first stone. A modern concern with the dignity and rights of all peoples inhibits us from speaking too frankly about rates of violence in preliterate peoples, and the "anthropologists of peace" have worked to give them a Rousseauian image makeover. Margaret Mead, for example, described the Chambri of New Guinea as a sex-reversed culture because the men were adorned with makeup and curls, omitting the fact that they had to earn the right to these supposedly effeminate decorations by killing a member of an enemy tribe.[29] Anthropologists who did not get with the program found themselves barred from the territories in which they had worked, denounced in manifestoes by their professional societies, slapped with libel lawsuits, and even accused of genocide.[30]

To be sure, it is easy to come away from tribal battles with the impression that they are fairly harmless in comparison with modern warfare. Men with a grievance against a neighboring village challenge its men to appear at a given time and place. The two sides face off at a distance at which their missiles can barely reach each other. They talk trash, cursing and insulting and boasting, and fire arrows or chuck spears while dodging those from the other side. When a warrior or two are injured or killed, they call it a day. These noisy spectacles led observers to conclude that warfare among primitive peoples was ritualistic and symbolic, very different from the glorious carnage of more advanced peoples.[31] The historian William Eckhardt, who is often cited for his claim that violence has vastly increased over the course of history, wrote, "Bands of gathering-hunters, numbering about 25 to 50 people each, could hardly have made much of a war. There would not have been enough people to fight, few weapons with which to fight, little to fight about, and no surplus to pay for the fighting."[32]

Only in the past fifteen years have scholars with no political ax to grind, such as Lawrence Keeley, Steven LeBlanc, Azar Gat, and Johan van der Dennen, begun to compile systematic reviews of the frequency and damage of fighting in large samples of nonstate peoples.[33] The actual death counts from primitive warfare show that the apparent harmlessness of a single battle is deceptive. For one thing, a skirmish may escalate into all-out combat that leaves the battlefield strewn with bodies. Also, when bands of a few dozen men confront each other on a regular basis, even one or two deaths per battle can add up to a *rate* of casualties that is high by any standard.

But the main distortion comes from a failure to distinguish the two kinds of violence that turned out to be so important in studies of chimpanzees: battles and raids. It is the sneaky raids, not the noisy battles, that kill in large numbers.[34] A party of men will slink into an enemy village before dawn, fire arrows into the first men who emerge from their huts in the morning to pee, and then shoot the others as they rush out of their huts to see what the commotion is about. They may thrust their spears through walls, shoot arrows through doorways or chimneys, and set the huts on fire. They can kill a lot of drowsy people before the villagers organize themselves in defense, by which time the attackers have melted back into the forest.

Sometimes enough attackers show up to massacre every last member of the village, or to kill all the men and abduct the women. Another stealthy but effective way to decimate an enemy is by ambuscade: a war party can hide in the forest along a hunting route and dispatch enemy men as they walk by. Still another tactic is treachery: the men can pretend to make peace with an enemy, invite them to a feast, and at a prearranged signal stab the unsuspecting guests. As for any solitary man who blunders into their territory, the policy is the same as it is with chimpanzees: shoot on sight.

Men in nonstate societies (and they are almost always men) are deadly serious about war, not just in their tactics but in their armaments, which include chemical, biological, and antipersonnel weapons.[35] Arrowheads may be coated with toxins extracted from venomous animals, or with putrefied tissue that causes the wound to fester. The arrowhead may be designed to break away from its shaft, making it difficult for the victim to pull it out. Warriors often reward themselves with trophies, especially heads, scalps, and genitals. They literally take no prisoners, though occasionally they will drag one back to the village to be tortured to death. William Bradford of the *Mayflower* pilgrims observed of the natives of Massachusetts, "Not being content only to kill and take away life, [they] delight to torment men in the most bloody manner that may be, flaying some alive with the shells of fishes, cutting off members and joints of others by piecemeal and broiling on the coals, eat collops of their flesh in their sight while they live."[36]

Though we bristle when we read of European colonists calling native people savages, and justly fault them for their hypocrisy and racism, it's not as if they were making the atrocities up. Many eyewitnesses have brought back tales of horrific violence in tribal warfare. Helena Valero, a woman who had been abducted by the Yanomamö in the Venezuelan rain forest in the 1930s, recounted one of their raids:

> Meanwhile from all sides the women continued to arrive with their children, whom the other Karawetari had captured. . . . Then the men began to kill the children; little ones, bigger ones, they killed many of them. They tried to run away, but they caught them, and threw them on the ground, and stuck them with bows, which went through their bodies and rooted them to the ground. Taking the smallest by the feet, they beat them against the trees and rocks. . . . All the women wept.[37]

In the early 19th century an English convict named William Buckley escaped from a penal colony in Australia and for three decades lived happily with the Wathaurung aborigines. He provided firsthand accounts of their way of life, including their ways of war:

> On approaching the enemy's quarters, they laid themselves down in ambush until all was quiet, and finding most of them asleep, laying about in groups, our party rushed upon them, killing three on the spot and wounding several others. The enemy fled precipitately, leaving their war implements in the hands of their assailants and their wounded to be beaten to death by boomerangs, three loud shouts closing the victors' triumph. The bodies of the dead they mutilated in a shocking manner, cutting the arms and legs off, with flints, and shells, and tomahawks.
>
> When the women saw them returning, they also raised great shouts, dancing about in savage ecstasy. The bodies were thrown upon the ground, and beaten about with sticks—in fact, they all seemed to be perfectly mad with excitement.[38]

It was not just Europeans gone native who recounted such episodes but the natives themselves. Robert Nasruk Cleveland, an Iñupiaq Inuit, provided this reminiscence in 1965:

> The next morning the raiders attacked the camp and killed all the women and children remaining there. . . . After shoving sheefish into the vaginas of all the Indian women they had killed, the Noatakers took Kititiġaaġvaat and her baby, and retreated toward the upper Noatak River. . . . Finally, when they had almost reached home, the Noatakers gang-raped Kititiġaaġvaat and left her with her baby to die. . . .
>
> Some weeks later, the Kobuk caribou hunters returned home to find the rotting remains of their wives and children and vowed revenge. A year or two after that, they headed north to the upper Noatak to seek it. They soon located a large body of Nuataaġmiut and secretly followed them. One morning the men in the Nuataaġmiut camp spotted a large band of caribou and went off in pursuit. While they were gone, the Kobuk raiders killed every woman in the camp. Then they cut off their vulvas, strung them on a line, and headed quickly toward home.[39]

Cannibalism has long been treated as the quintessence of primitive savagery, and in reaction many anthropologists used to dismiss reports of cannibalism as blood libels by neighboring tribes. But forensic archaeology has recently shown that cannibalism was widespread in human prehistory. The evidence includes human bones that bear human teethmarks or that had been cracked and cooked like those of animals and thrown out in the kitchen trash.[40]

Some of the butchered bones date back 800,000 years, to the time when *Homo heidelbergensis,* a common ancestor of modern humans and Neanderthals, first appears on the evolutionary stage. Traces of human blood proteins have also been found in cooking pots and in ancient human excrement. Cannibalism may have been so common in prehistory as to have affected our evolution: our genomes contain genes that appear to be defenses against the prion diseases transmitted by cannibalism.[41] All this is consistent with eyewitness accounts, such as this transcription by a missionary of a Maori warrior taunting the preserved head of an enemy chief:

> You wanted to run away, did you? But my war club overtook you. And after you were cooked, you made food for my mouth. And where is your father? He is cooked. And where is your brother? He is eaten. And where is your wife? There she sits, a wife for me. And where are your children? There they are, with loads on their backs, carrying food, as my slaves.[42]

Many scholars have found the image of harmless foragers to be plausible because they had trouble imagining the means and motives that could drive them to war. Recall, for example, Eckhardt's claim that hunter-gatherers had "little to fight about." But organisms that have evolved by natural selection always have something to fight about (which doesn't, of course, mean that they will always fight). Hobbes noted that humans in particular have three reasons for quarrel: gain, safety, and credible deterrence. People in nonstate societies fight about all three.[43]

Foraging peoples can invade to gain territory, such as hunting grounds, watering holes, the banks or mouths of rivers, and sources of valued minerals like flint, obsidian, salt, or ochre. They may raid livestock or caches of stored food. And very often they fight over women. Men may raid a neighboring village for the express purpose of kidnapping women, whom they gang-rape and distribute as wives. They may raid for some other reason and take the women as a bonus. Or they may raid to claim women who had been promised to them in marriage but were not delivered at the agreed-upon time. And sometimes young men attack for trophies, coups, and other signs of aggressive prowess, especially in societies where they are a prerequisite to attaining adult status.

People in nonstate societies also invade for safety. The security dilemma or Hobbesian trap is very much on their minds, and they may form an alliance with nearby villages if they fear they are too small, or launch a preemptive strike if they fear that an enemy alliance is getting too big. One Yanomamö man in Amazonia told an anthropologist, "We are tired of fighting. We don't want to kill anymore. But the others are treacherous and cannot be trusted."[44]

But in most surveys the most commonly cited motive for warfare is vengeance, which serves as a crude deterrent to potential enemies by raising the anticipated long-term costs of an attack. In the *Iliad,* Achilles describes a feature of human psychology that can be found in cultures throughout the world: revenge "far sweeter than flowing honey wells up like smoke in the breasts of man." Foraging and tribal people avenge theft, adultery, vandalism, poaching, abduction of women, soured deals, alleged sorcery, and previous acts of violence. One cross-cultural survey found that in 95 percent of societies, people explicitly endorse the idea of taking a life for a life.[45] Tribal people not only feel the smoke welling up in their breasts but know that their enemies feel it too. That is why they sometimes massacre every last member of a village they raid: they anticipate that any survivors would seek revenge for their slain kinsmen.

RATES OF VIOLENCE IN STATE AND NONSTATE SOCIETIES

Though descriptions of violence in nonstate societies demolish the stereotype that foraging peoples are inherently peaceful, they don't tell us whether the level of violence is higher or lower than in so-called civilized societies. The annals of modern states have no shortage of gruesome massacres and atrocities, not least against native peoples of every continent, and their wars have death tolls that reach eight digits. Only by looking at numbers can we get a sense as to whether civilization has increased violence or decreased it.

In absolute numbers, of course, civilized societies are matchless in the destruction they have wreaked. But should we look at absolute numbers, or at *relative* numbers, calculated as a proportion of the populations? The choice confronts us with the moral imponderable of whether it is worse for 50 percent of a population of one hundred to be killed or 1 percent of a population of one billion. In one frame of mind, one could say that a person who is tortured or killed suffers to the same degree regardless of how many other people meet such a fate, so it is the sum of these sufferings that should engage our sympathy and our analytic attention. But in another frame of mind, one could reason that part of the bargain of being alive is that one takes a chance at dying a premature or painful death, be it from violence, accident, or disease. So the number of people in a given time and place who enjoy full lives has to be counted as a moral good, against which we calibrate the moral bad of the number who are victims of violence. Another way of expressing this frame of mind is to ask, "If *I* were one of the people who were alive in a particular era, what would be the chances that I would be a victim of violence?" The reasoning in this second frame of mind, whether it appeals to the proportion of a population or the risk to an individual, ends in the conclusion that in comparing the harmfulness of violence across societies, we should focus on the rate, rather than the number, of violent acts.

What happens, then, when we use the emergence of states as the dividing line and put hunter-gatherers, hunter-horticulturalists, and other tribal peoples (from any era) on one side, and settled states (also from any era) on the other? Several scholars have recently scoured the anthropological and historical literature for every good body count from nonstate societies that they could find. Two kinds of estimates are available. One comes from ethnographers who record demographic data, including deaths, in the people they study over long stretches of time.[46] The other comes from forensic archaeologists, who sift through burial sites or museum collections with an eye for signs of foul play.[47]

How can one establish the cause of death when the victim perished hundreds or thousands of years ago? Some prehistoric skeletons are accompanied by the stone-age equivalent of a smoking gun: a spearhead or arrowhead embedded in a bone, like the ones found in Kennewick Man and Ötzi. But circumstantial evidence can be almost as damning. Archaeologists can check prehistoric skeletons for the kinds of damage known to be left by assaults in humans today. The stigmata include bashed-in skulls, cut marks from stone tools on skulls or limbs, and parry fractures on ulnar bones (the injury that a person gets when he defends himself against an assailant by holding up his arm). Injuries sustained by a skeleton when it was inside a living body can be distinguished in several ways from the damage it sustained when it was exposed to the world. Living bones fracture like glass, with sharp, angled edges, whereas dead bones fracture like chalk, at clean right angles. And if a bone has a different pattern of weathering on its fractured surface than on its intact surface, it was probably broken after the surrounding flesh had rotted away. Other incriminating signs from nearby surroundings include fortifications, shields, shock weapons such as tomahawks (which are useless in hunting), and depictions of human combat on the walls of caves (some of them more than six thousand years old). Even with all this evidence, archaeological death counts are usually underestimates, because some causes of death—a poisoned arrow, a septic wound, or a ruptured organ or artery—leave no trace on the victim's bones.

Once researchers have tallied a raw count of violent deaths, they can convert it to a rate in either of two ways. The first is to calculate the percentage of all deaths that are caused by violence. This rate is an answer to the question, "What are the chances that a person died at the hands of another person rather than passing away of natural causes?" The graph in figure 2–2 presents this statistic for three samples of nonstate people—skeletons from prehistoric sites, hunter-gatherers, and hunter-horticulturalists—and for a variety of state societies. Let's walk through it.

The topmost cluster shows the rate of violent death for skeletons dug out of archaeological sites.[48] They are the remains of hunter-gatherers and hunter-horticulturalists from Asia, Africa, Europe, and the Americas and date from 14,000 BCE to 1770 CE, in every case well before the emergence of state societies or the first sustained contact with them. The death rates range from 0 to 60 percent, with an average of 15 percent.

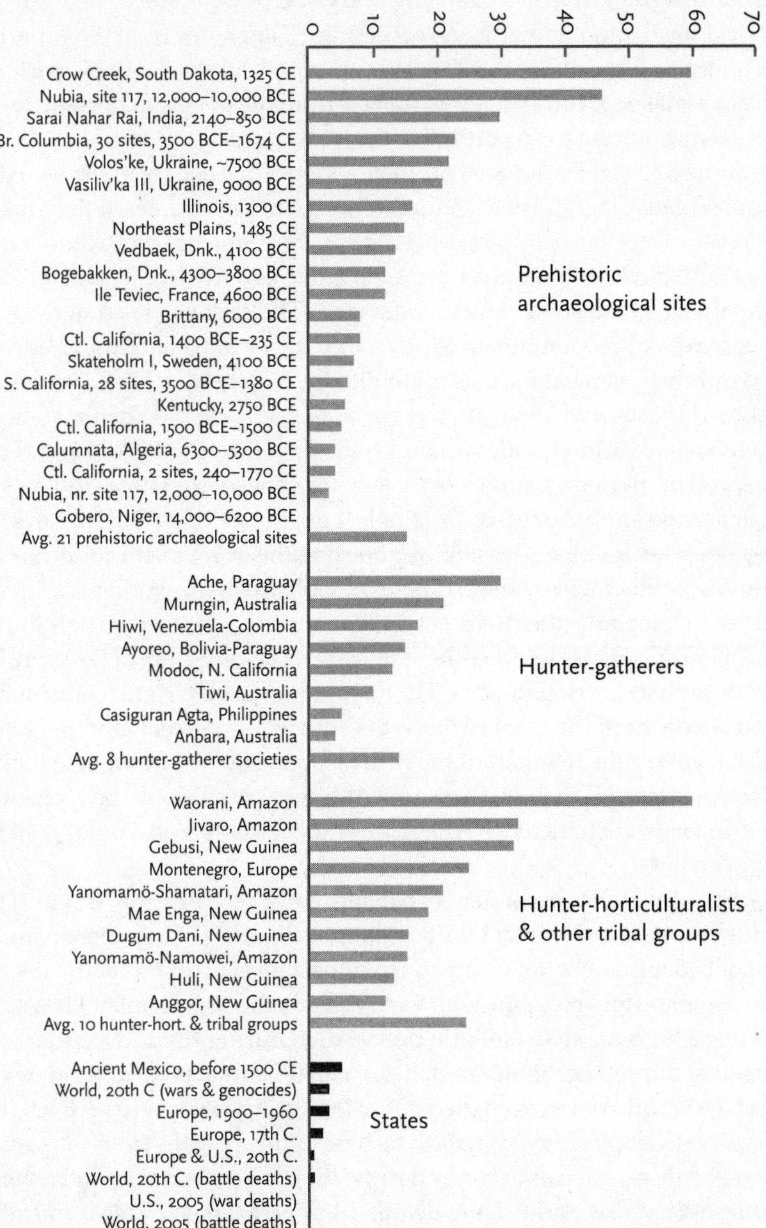

FIGURE 2–2. Percentage of deaths in warfare in nonstate and state societies

Sources: Prehistoric archaeological sites: Bowles, 2009; Keeley, 1996. Hunter-gatherers: Bowles, 2009. Hunter-horticulturalists and other tribal groups: Gat, 2006; Keeley, 1996. Ancient Mexico: Keeley, 1996. World, 20th-century wars & genocides (includes man-made famines): White, 2011. Europe, 1900–60: Keeley, 1996, from Wright, 1942, 1942/1964, 1942/1965; see note 52. Europe, 17th-century: Keeley, 1996. Europe and United States, 20th century: Keeley, 1996, from Harris, 1975. World, 20th-century battle deaths: Lacina & Gleditsch, 2005; Sarkees, 2000; see note 54. United States, 2005 war deaths: see text and note 57. World, 2005 battle deaths: see text and note 58.

Next are figures from eight contemporary or recent societies that make their living primarily from hunting and gathering.[49] They come from the Americas, the Philippines, and Australia. The average of the rates of death by warfare is within a whisker of the average estimated from the bones: 14 percent, with a range from 4 percent to 30 percent.

In the next cluster I've lumped pre-state societies that engage in some mixture of hunting, gathering, and horticulture. All are from New Guinea or the Amazon rain forest, except Europe's last tribal society, the Montenegrins, whose rate of violent death is close to the average for the group as a whole, 24.5 percent.[50]

Finally we get to some figures for states.[51] The earliest are from the cities and empires of pre-Columbian Mexico, in which 5 percent of the dead were killed by other people. That was undoubtedly a dangerous place, but it was a third to a fifth as violent as an average pre-state society. When it comes to modern states, we are faced with hundreds of political units, dozens of centuries, and many subcategories of violence to choose from (wars, homicides, genocides, and so on), so there is no single "correct" estimate. But we can make the comparison as fair as possible by choosing the *most* violent countries and centuries, together with some estimates of violence in the world today. As we shall see in chapter 5, the two most violent centuries in the past half millennium of European history were the 17th, with its bloody Wars of Religion, and the 20th, with its two world wars. The historian Quincy Wright has estimated the rate of death in the wars of the 17th century at 2 percent, and the rate of death in war for the first half of the 20th at 3 percent.[52] If one were to include the last four decades of the 20th century, the percentage would be even lower. One estimate, which includes American war deaths as well, comes in at less than 1 percent.[53]

Recently the study of war has been made more precise by the release of two quantitative datasets, which I will explain in chapter 5. They conservatively list about 40 million battle deaths during the 20th century.[54] ("Battle deaths" refer to soldiers and civilians who were directly killed in combat.) If we consider that a bit more than 6 billion people died during the 20th century, and put aside some demographic subtleties, we may estimate that around 0.7 percent of the world's population died in battles during that century.[55] Even if we tripled or quadrupled the estimate to include indirect deaths from war-caused famine and disease, it would barely narrow the gap between state and nonstate societies. What if we added the deaths from genocides, purges, and other man-made disasters? Matthew White, the atrocitologist we met in chapter 1, estimates that around 180 million deaths can be blamed on all of these human causes put together. That still amounts to only 3 percent of the deaths in the 20th century.[56]

Now let's turn to the present. According to the most recent edition of the *Statistical Abstract of the United States*, 2,448,017 Americans died in 2005. It was one of the country's worst years for war deaths in decades, with the armed forces embroiled in conflicts in Iraq and Afghanistan. Together the two wars killed 945 Americans, amounting to 0.0004 (four-hundredths of a percent) of American deaths that year.[57] Even if we throw in the 18,124 domestic homicides, the total rate of violent death adds up to 0.008, or eight-tenths of a percentage point. In other Western countries, the rates were even lower. And in the world as a whole, the Human Security Report Project counted 17,400 deaths that year that were directly caused by political violence (war, terrorism, genocide, and killings by warlords and militias), for a rate of 0.0003 (three-hundredths of a percent).[58] It's a conservative estimate, comprising only identifiable deaths, but even if we generously multiplied it by twenty to estimate undocumented battle deaths and indirect deaths from famine and disease, it would not reach the 1 percent mark.

The major cleft in the graph, then, separates the anarchical bands and tribes from the governed states. But we have been comparing a motley collection of archaeological digs, ethnographic tallies, and modern estimates, some of them calculated on the proverbial back of an envelope. Is there some way to juxtapose two datasets directly, one from hunter-gatherers, the other from settled civilizations, matching the people, era, and methods as closely as possible? The economists Richard Steckel and John Wallis recently looked at data on nine hundred skeletons of Native Americans, distributed from southern Canada to South America, all of whom died before the arrival of Columbus.[59] They divided the skeletons into hunter-gatherers and city dwellers, the latter from the civilizations in the Andes and Mesoamerica such as the Incas, Aztecs, and Mayans. The proportion of hunter-gatherers that showed signs of violent trauma was 13.4 percent, which is close to the average for the hunter-gatherers in figure 2–2. The proportion of city dwellers that showed signs of violent trauma was 2.7 percent, which is close to the figures for state societies before the present century. So holding many factors constant, we find that living in a civilization reduces one's chances of being a victim of violence fivefold.

Let's turn to the second way of quantifying violence, in which the rate of killing is calculated as a proportion of living people rather than dead ones. This statistic is harder to compute from boneyards but easier to compute from most other sources, because it requires only a body count and a population size, not an inventory of deaths from other sources. The number of deaths per 100,000 people per year is the standard measure of homicide rates, and I will use it as the yardstick of violence throughout the book. To get a feel for what these numbers mean, keep in mind that the safest place in human history, Western Europe at the turn of the 21st century, has a homicide rate in the neighborhood of 1 per 100,000 per year.[60] Even the gentlest society will have the occasional young man who gets carried away in a barroom brawl or an old woman who puts arsenic in her husband's tea, so that is pretty much as low as homicide rates ever go. Among modern Western countries, the United States lies at the dangerous end of the range. In the worst years of the 1970s and 1980s, it had a homicide rate of around 10 per 100,000, and its notoriously violent cities, like Detroit, had a rate of around 45 per 100,000.[61] If you were living in a society with a homicide rate in that range, you would notice the danger in everyday life, and as the rate climbed to 100 per 100,000, the violence would start to affect you personally: assuming you have a hundred relatives, friends, and close acquaintances, then over the course of a decade one of them would probably be killed. If the rate soared to 1,000 per 100,000 (1 percent), you'd lose about one acquaintance a year, and would have a better-than-even lifetime chance of being murdered yourself.

Figure 2–3 shows war death rates for twenty-seven nonstate societies (combining hunter-gatherers and hunter-horticulturalists) and nine that are ruled by states. The average annual rate of death in warfare for the nonstate societies is 524 per 100,000, about half of 1 percent. Among states, the Aztec empire of central Mexico, which was often at war, had a rate about half that.[62] Below that bar we find the rates for four state societies during the centuries in which they waged their most destructive wars. Nineteenth-century France fought the Revolutionary, Napoleonic, and Franco-Prussian Wars and lost an average of 70 people per 100,000 per year. The 20th century was blackened by two world wars that inflicted most of their military damage on Germany, Japan, and Russia/USSR, which also had a civil war and other military adventures. Their annual rates of death work out to 144, 27, and 135 per 100,000, respectively.[63] During the 20th century the United States acquired a reputation as a warmonger, fighting in two world wars and in the Philippines, Korea, Vietnam, and Iraq. But the annual cost in American lives was even smaller than those of the other great powers of the century, about 3.7 per 100,000.[64] Even if we add up all the deaths from organized violence for the entire world for the entire century—wars, genocides, purges, and man-made famines—we get an annual rate of around 60 per 100,000.[65] For the year 2005, the bars representing the United States and the entire world are paint-thin and invisible in the graph.[66]

So by this measure too, states are far less violent than traditional bands and tribes. Modern Western countries, even in their most war-torn centuries, suffered no more than around a quarter of the average death rate of nonstate societies, and less than a tenth of that for the most violent one.

Though war is common among foraging groups, it is certainly not universal. Nor should we expect it to be if the violent inclinations in human nature are a strategic response to the circumstances rather than a hydraulic response to an inner urge. According to two ethnographic surveys, 65 to 70 percent of hunter-gatherer groups are at war at least every two years, 90 percent engage in war at least once a generation, and virtually all the rest report a cultural memory of war in the past.[67] That means that hunter-gatherers often fight, but they can avoid war for long stretches of time. Figure 2–3 reveals two tribes,

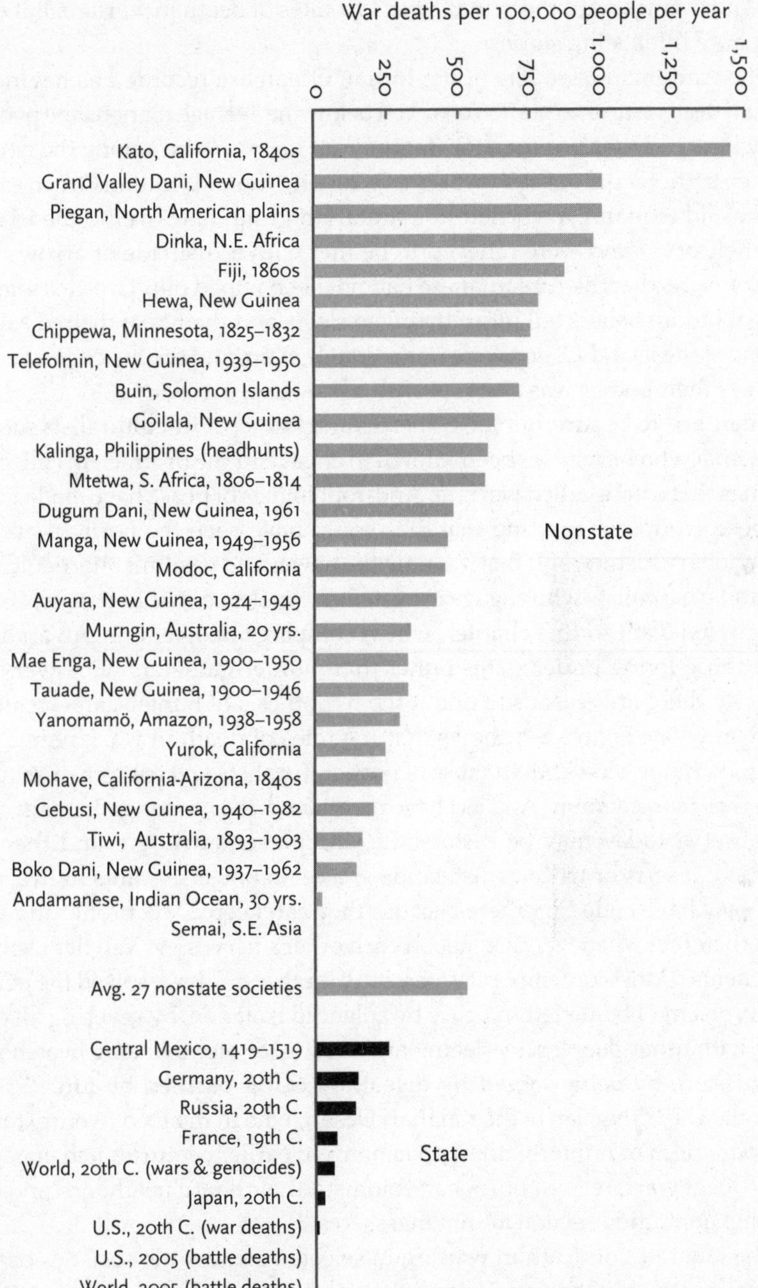

FIGURE 2–3. Rate of death in warfare in nonstate and state societies

Sources: Nonstate: Hewa and Goilala from Gat, 2006; others from Keeley, 1996. Central Mexico, Germany, Russia, France, Japan: Keeley, 1996; see notes 62 and 63. United States in the 20th century: Leland & Oboroceanu, 2010; see note 64. World in 20th century: White, 2011; see note 65. World in 2005: Human Security Report Project, 2008; see notes 57 and 58.

the Andamanese and the Semai, with low rates of death in warfare. But even they have interesting stories.

The Andaman Islanders of the Indian Ocean are recorded as having an annual death rate of 20 per 100,000, well below the average for nonstate peoples (which exceeds 500 per 100,000). But they are known to be among the fiercest hunter-gatherer groups left on earth. Following the 2004 Indian Ocean earthquake and tsunami, a worried humanitarian group flew over to the islands in a helicopter and were relieved to be met with a fusillade of arrows and spears, signs that the Andamanese had not been wiped out. Two years later a pair of Indian fishers fell into a drunken sleep, and their boat drifted ashore on one of the islands. They were immediately slain, and the helicopter sent to retrieve their bodies was also met with a shower of arrows.[68]

There are, to be sure, hunter-gatherers and hunter-horticulturalists such as the Semai who have *never* been known to engage in the protracted, collective killings that can be called warfare. Anthropologists of peace have made much of these groups, suggesting that they could have been the norm in human evolutionary history, and that it is only the newer and wealthier horticulturalists and pastoralists who engage in systematic violence. The hypothesis is not directly relevant to this chapter, which compares people living in anarchy with those living under states rather than hunter-gatherers with everyone else. But there are reasons to doubt the hypothesis of hunter-gatherer innocence anyway. Figure 2–3 shows that the rates of death in warfare in these societies, though lower than those of horticulturalists and tribesmen, overlap with them considerably. And as I have mentioned, the hunter-gatherer groups we observe today may be historically unrepresentative. We find them in parched deserts or frozen wastelands where no one else wants to live, and they may have ended up there because they can keep a low profile and vote with their feet whenever they get on each other's nerves. As Van der Dennen comments, "Most contemporary 'peaceful' foragers . . . have solved the perennial problem of being left in peace by splendid isolation, by severing all contacts with other peoples, by fleeing and hiding, or else by being beaten into submission, by being tamed by defeat, by being pacified by force."[69] For example, the !Kung San of the Kalahari Desert, who in the 1960s were extolled as a paradigm of hunter-gatherer harmony, in earlier centuries had engaged in frequent warfare with European colonists, their Bantu neighbors, and one another, including several all-out massacres.[70]

The low rates of death in warfare in selected small-scale societies can be misleading in another way. Though they may avoid war, they do commit the occasional murder, and their homicide rates can be compared to those of modern state societies. I've plotted them in figure 2–4 on a scale that is fifteen times larger than that of figure 2–3. Let's begin with the right-most gray bar in the nonstate cluster. The Semai are a hunting and horticulturalist tribe who were described in a book called *The Semai: A Nonviolent People of Malaya* and who go out of their way to avoid the use of force. While there aren't many Semai homicides, there aren't many Semai. When the anthropologist Bruce Knauft did the arithmetic, he found that their homicide rate was 30 per 100,000 per year, which puts it in the range of the infamously dangerous American cities in their most violent years and at three times the rate of the United States as a whole in its most violent decade.[71] The same kind of long division has deflated the peaceful reputation of the !Kung, the subject of a book called *The Harmless People*, and of the Central Arctic Inuit (Eskimos), who inspired a book called *Never in Anger*.[72] Not only do these harmless, nonviolent, anger-free people murder each other at rates far greater than Americans or Europeans do, but the murder rate among the !Kung went down by a third after their territory had been brought under the control of the Botswana government, as the Leviathan theory would predict.[73]

The reduction of homicide by government control is so obvious to anthropologists that they seldom document it with numbers. The various "paxes" that one reads about in history books—the Pax Romana, Islamica, Mongolica, Hispanica, Ottomana, Sinica, Britannica, Australiana (in New Guinea), Canadiana (in the Pacific Northwest), and Praetoriana (in South Africa)—refer to the reduction in raiding, feuding, and warfare in the territories brought under the control

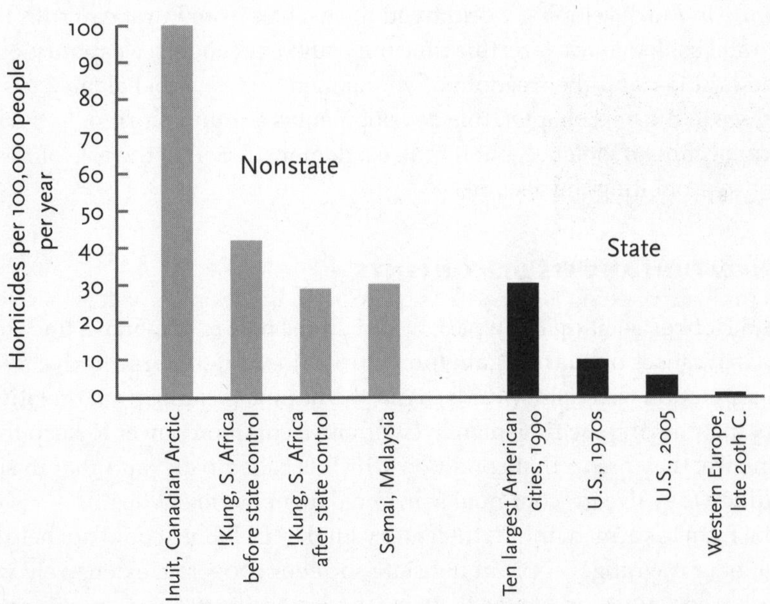

FIGURE 2–4. Homicide rates in the least violent nonstate societies compared to state societies

Sources: !Kung and Central Arctic Inuit: Gat, 2006; Lee, 1982. Semai: Knauft, 1987. Ten largest U.S. cities: Zimring, 2007, p. 140. United States: FBI Uniform Crime Reports; see note 73. Western Europe (approximation): World Health Organization; see note 66 to chap. 3, p. 701.

of an effective government.[74] Though imperial conquest and rule can themselves be brutal, they do reduce endemic violence among the conquered. The Pacification Process is so pervasive that anthropologists often treat it as a methodological nuisance. It goes without saying that peoples that have been brought under the jurisdiction of a government will not fight as much, so they are simply excluded from studies of violence in indigenous societies. The effect is also noticeable to the people themselves. As an Auyana man living in New Guinea under the Pax Australiana put it, "Life was better since the government came" because "a man could now eat without looking over his shoulder and could leave his house in the morning to urinate without fear of being shot."[75]

The anthropologists Karen Ericksen and Heather Horton have quantified the way that the presence of government can move a society away from lethal vengeance. In a survey of 192 traditional studies, they found that one-on-one revenge was common in foraging societies, and kin-against-kin blood feuds were common in tribal societies that had not been pacified by a colonial or national government, particularly if they had an exaggerated culture of manly honor.[76] Adjudication by tribunals and courts, in contrast, was common in societies that had fallen under the control of a centralized government, or that had resource bases and inheritance patterns that gave people more of a stake in social stability.

One of the tragic ironies of the second half of the 20th century is that when colonies in the developing world freed themselves from European rule, they often slid back into warfare, this time intensified by modern weaponry, organized militias, and the freedom of young men to defy tribal elders.[77] As we shall see in the next chapter, this development is a countercurrent to the historical decline of violence, but it is also a demonstration of the role of Leviathans in propelling the decline.

CIVILIZATION AND ITS DISCONTENTS

So did Hobbes get it right? In part, he did. In the nature of man we find three principal causes of quarrel: gain (predatory raids), safety (preemptive raids), and reputation (retaliatory raids). And the numbers confirm that relatively speaking, "during the time men live without a common power to keep them all in awe, they are in that condition which is called war," and that in such condition they live in "continual fear, and danger of violent death."

But from his armchair in 17th-century England, Hobbes could not help but get a lot of it wrong. People in nonstate societies cooperate extensively with their kin and allies, so life for them is far from "solitary," and only intermittently is it nasty and brutish. Even if they are drawn into raids and battles every few years, that leaves a lot of time for foraging, feasting, singing, storytelling, childrearing, tending to the sick, and the other necessities and pleasures of life. In a draft of a previous book, I casually referred to the Yanomamö as "the fierce people," alluding to the title of the famous book by the anthropologist Napoleon Chagnon. An anthropologist colleague wrote in the margin: "Are the babies fierce? Are the old women fierce? Do they eat fiercely?"

As for their lives being "poor," the story is mixed. Certainly societies without an organized state enjoy "no commodious building; no instruments of moving and removing such things as require much force; no knowledge of the face of the earth; no account of time, [and] no letters," since it's hard to develop these things if the warriors from the next village keep waking you up with poisoned arrows, abducting the women, and burning your huts. But the first peoples who gave up hunting and gathering for settled agriculture struck a hard bargain for themselves. Spending your days behind a plow, subsisting on starchy cereal grains, and living cheek by jowl with livestock and thousands of other people can be hazardous to your health. Studies of skeletons by Steckel and his colleagues show that compared to hunter-gatherers, the first city dwellers were anemic, infected, tooth-decayed, and almost two and a half inches shorter.[78] Some biblical scholars believe that the story of the fall from the Garden of Eden was a cultural memory of the transition from foraging to agriculture: "In the sweat of thy face shalt thou eat bread."[79]

So why did our foraging ancestors leave Eden? For many, it was never an explicit choice: they had multiplied themselves into a Malthusian trap in which the fat of the land could no longer support them, and they had to grow their food themselves. The states emerged only later, and the foragers who lived at their frontiers could either be absorbed into them or hold out in their old way of life. For those who had the choice, Eden may have been just too dangerous. A few cavities, the odd abscess, and a couple of inches in height were a small price to pay for a fivefold better chance of not getting speared.[80]

The improved odds of a natural death came with another price, captured by the Roman historian Tacitus: "Formerly we suffered from crimes; now we suffer from laws." The Bible stories we examined in chapter 1 suggest that the first kings kept their subjects in awe with totalistic ideologies and brutal punishments. Just think of the wrathful deity watching people's every move, the regulation of daily life by arbitrary laws, the stonings for blasphemy and nonconformity, the kings with the power to expropriate a woman into their harem or cut a baby in half, the crucifixions of thieves and cult leaders. In these respects the Bible was accurate. Social scientists who study the emergence of states have noted that they began as stratified theocracies in which elites secured their economic privileges by enforcing a brutal peace on their underlings.[81]

Three scholars have analyzed large samples of cultures to quantify the correlation between the political complexity of early societies and their reliance on absolutism and cruelty.[82] The archaeologist Keith Otterbein has shown that societies with more centralized leadership were more likely to kill women in battles (as opposed to abducting them), to keep slaves, and to engage in human sacrifice. The sociologist Steven Spitzer has shown that complex societies are more likely to criminalize victimless activities like sacrilege, sexual deviance, disloyalty, and witchcraft, and to punish offenders by torture, mutilation, enslavement, and execution. And the historian and anthropologist Laura Betzig has shown that complex societies tend to fall under the control of despots: leaders who are guaranteed to get their way in conflicts, who can kill with impunity, and who have large harems of women at their disposal. She found that despotism in this sense emerged among the Babylonians, Israelites, Romans, Samoans, Fijians, Khmer, Aztecs, Incas, Natchez (of the lower Mississippi), Ashanti, and other kingdoms throughout Africa.

When it came to violence, then, the first Leviathans solved one problem but created another. People were less likely to become victims of homicide or casualties of war, but they were now under the thumbs of tyrants, clerics, and kleptocrats. This gives us the more sinister sense of the word *pacification:* not just the bringing about of peace but the imposition of absolute control by a coercive government. Solving this second problem would have to wait another few millennia, and in much of the world it remains unsolved to this day.

3

THE CIVILIZING PROCESS

It is impossible to overlook the extent to which civilization is built upon a renunciation of instinct.

—Sigmund Freud

For as long as I have known how to eat with utensils, I have struggled with the rule of table manners that says that you may not guide food onto your fork with your knife. To be sure, I have the dexterity to capture chunks of food that have enough mass to stay put as I scoot my fork under them. But my feeble cerebellum is no match for finely diced cubes or slippery little spheres that ricochet and roll at the touch of the tines. I chase them around the plate, desperately seeking a ridge or a slope that will give me the needed purchase, hoping they will not reach escape velocity and come to rest on the tablecloth. On occasion I have seized the moment when my dining companion glances away and have placed my knife to block their getaway before she turns back to catch me in this faux pas. Anything to avoid the ignominy, the boorishness, the intolerable uncouthness of using a knife for some purpose other than cutting. Give me a lever long enough, said Archimedes, and a fulcrum on which to place it, and I shall move the world. But if he knew his table manners, *he could not have moved some peas onto his fork with his knife!*

I remember, as a child, questioning this pointless prohibition. What is so terrible, I asked, about using your silverware in an efficient and perfectly sanitary way? It's not as if I were asking to eat mashed potatoes with my hands. I lost the argument, as all children do, when faced with the rejoinder "Because I said so," and for decades I silently grumbled about the unintelligibility of the rules of etiquette. Then one day, while doing research for this book, the scales fell from my eyes, the enigma evaporated, and I forever put aside my resentment of the no-knife rule. I owe this epiphany to the most important thinker you have never heard of, Norbert Elias (1897–1990).

Elias was born in Breslau, Germany (now Wrocław, Poland), and studied sociology and the history of science.[1] He fled Germany in 1933 because he was Jewish, was detained in a British camp in 1940 because he was German, and lost both parents to the Holocaust. On top of these tragedies, Nazism brought one more into his life: his magnum opus, *The Civilizing Process*, was published in Germany in 1939, a time when the very idea seemed like a bad joke. Elias vagabonded from one university to another, mostly teaching night school, and retrained as a psychotherapist before settling down at the University of Leicester, where he taught until his retirement in 1962. He emerged from obscurity in 1969 when *The Civilizing Process* was published in English translation, and he was recognized as a major figure only in the last decade of his life, when an astonishing fact came to light. The discovery was not about the rationale behind table manners but about the history of homicide.

In 1981 the political scientist Ted Robert Gurr, using old court and county records, calculated thirty estimates of homicide rates at various times in English history, combined them with modern records from London, and plotted them on a graph.[2] I've reproduced it in figure 3–1, using a logarithmic scale in which the same vertical distance separates 1 from 10, 10 from 100, and 100 from 1000. The rate is calculated in the same way as in the preceding chapter, namely the number of killings per 100,000 people per year. The log scale is necessary because the homicide rate declined so precipitously. The graph shows that from the 13th century to the 20th, homicide in various parts of England plummeted by a factor of ten, fifty, and in some cases a hundred—for example, from 110 homicides per 100,000 people per year in 14th-century Oxford to less than 1 homicide per 100,000 in mid-20th-century London.

The graph stunned almost everyone who saw it (including me—as

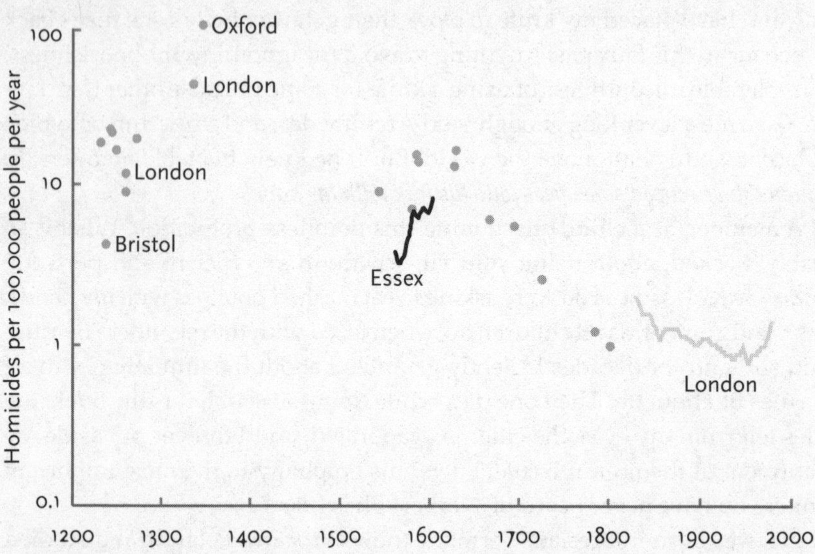

FIGURE 3–1. Homicide rates in England, 1200–2000: Gurr's 1981 estimates

Source: Data from Gurr, 1981, pp. 303–4, 313.

I mentioned in the preface, it was the seed that grew into this book). The discovery confounds every stereotype about the idyllic past and the degenerate present. When I surveyed perceptions of violence in an Internet questionnaire, people guessed that 20th-century England was about 14 percent more violent than 14th-century England. In fact it was 95 percent less violent.[3]

This chapter is about the decline of homicide in Europe from the Middle Ages to the present, and its counterparts and counterexamples in other times and places. I have borrowed the title of the chapter from Elias because he was the only major social thinker with a theory that could explain it.

THE EUROPEAN HOMICIDE DECLINE

Before we try to explain this remarkable development, let's be sure it is real. Following the publication of Gurr's graph, several historical criminologists dug more deeply into the history of homicide.[4] The criminologist Manuel Eisner assembled a much larger set of estimates on homicide in England across the centuries, drawing on coroners' inquests, court cases, and local records.[5] Each dot on the graph in figure 3–2 is an estimate from some town or jurisdiction, plotted once again on a logarithmic scale. By the 19th century the British government was keeping annual records of homicide for the entire country, which are plotted on the graph as a gray line. Another historian, J. S. Cockburn, compiled continuous data from the county of Kent between 1560 and 1985, which Eisner superimposed on his own data as the black line.[6]

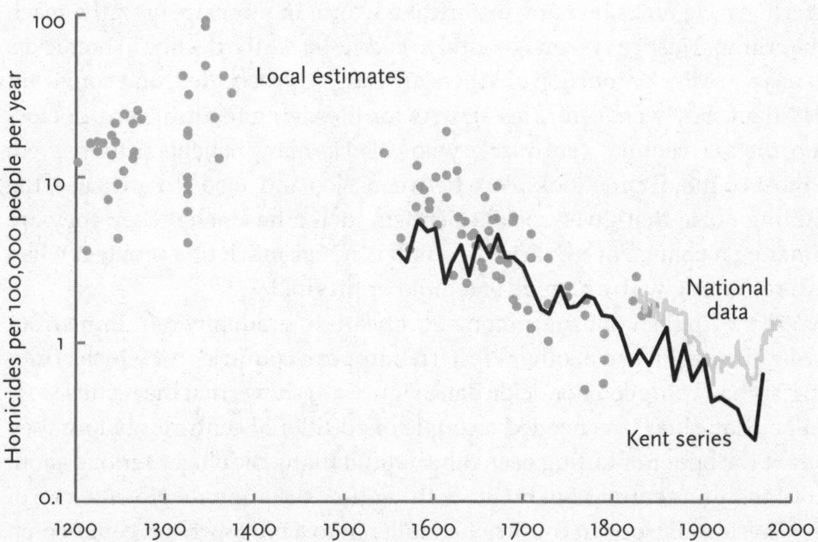

FIGURE 3–2. Homicide rates in England, 1200–2000

Source: Graph from Eisner, 2003.

Once again we see a decline in annual homicide rates, and it is not small: from between 4 and 100 homicides per 100,000 people in the Middle Ages to around 0.8 (eight-tenths of a homicide) per 100,000 in the 1950s. The timing shows that the high medieval murder rates cannot be blamed on the social upheavals that followed the Black Death around 1350, because many of the estimates predated that epidemic.

Eisner has given a lot of thought to how much we should trust these numbers. Homicide is the crime of choice for measurers of violence because regardless of how the people of a distant culture conceptualize crime, a dead body is hard to define away, and it always arouses curiosity about who or what produced it. Records of homicide are therefore a more reliable index of violence than records of robbery, rape, or assault, and they usually (though not always) correlate with them.[7]

Still, it's reasonable to wonder how the people of different eras reacted to these killings. Were they as likely as we are to judge a killing as intentional or accidental, or to prosecute the killing as opposed to letting it pass? Did people in earlier times always kill at the same percentage of the rate that they raped, robbed, and assaulted? How successful were they in saving the lives of victims of assault and thereby preventing them from becoming victims of homicide?

Fortunately, these questions can be addressed. Eisner cites studies showing that when people today are presented with the circumstances of a centuries-old murder and asked whether they think it was intentional, they usually come to the same conclusion as did the people at the time. He has shown that in most periods, the rates of homicide do correlate with the rates of other violent crimes. He notes that any historical advance in forensics or in the reach of the criminal justice system is bound to *underestimate* the decline in homicide, because a greater proportion of killers are caught, prosecuted, and convicted today than they were centuries ago. As for lifesaving medical care, doctors before the 20th century were quacks who killed as many patients as they saved; yet most of the decline took place between 1300 and 1900.[8] In any case, the sampling noise that gives social scientists such a headache when they are estimating a change of a quarter or a half is not as much of a problem when they are dealing with a change of tenfold or fiftyfold.

Were the English unusual among Europeans in gradually refraining from murder? Eisner looked at other Western European countries for which criminologists had compiled homicide data. Figure 3–3 shows that the results were similar. Scandinavians needed a couple of additional centuries before they thought the better of killing each other, and Italians didn't get serious about it until the 19th century. But by the 20th century the annual homicide rate of every Western European country had fallen into a narrow band centered on 1 per 100,000.

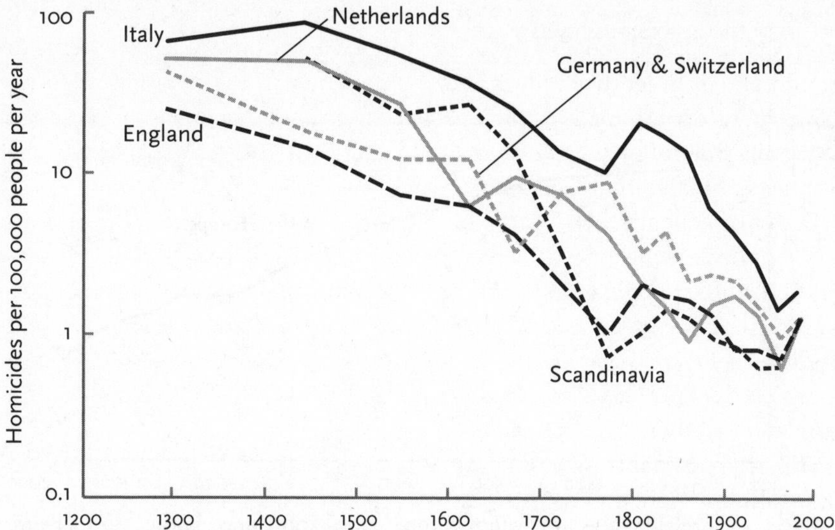

FIGURE 3–3. Homicide rates in five Western European regions, 1300–2000
Source: Data from Eisner, 2003, table 1.

To put the European decline in perspective, let's compare it to the rates for nonstate societies that we encountered in chapter 2. In figure 3–4 I have extended the vertical axis up to 1,000 on the log scale to accommodate the additional order of magnitude required by the nonstate societies. Even in the late Middle Ages, Western Europe was far less violent than the unpacified nonstate societies and the Inuit, and it was comparable to the thinly settled foragers such as the Semai and the !Kung. And from the 14th century on, the European homicide rate sank steadily, with a tiny bounce in the last third of the 20th century.

While Europe was becoming less murderous overall, certain patterns in homicide remained constant.[9] Men were responsible for about 92 percent of the killings (other than infanticide), and they were most likely to kill when they were in their twenties. Until the 1960s uptick, cities were generally safer than the countryside. But other patterns changed. In the earlier centuries the upper and lower social classes engaged in homicide at comparable rates. But as the homicide rate fell, it dropped far more precipitously among the upper classes than among the lower ones, an important social change to which we will return.[10]

Another historical change was that homicides in which one man kills another man who is unrelated to him declined far more rapidly than did the killing of children, parents, spouses, and siblings. This is a common pattern in homicide statistics, sometimes called Verkko's Law: rates of male-on-male

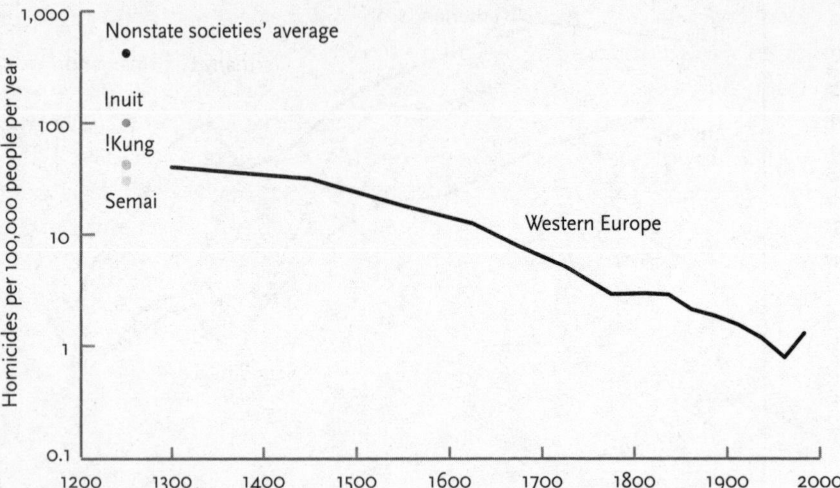

FIGURE 3–4. Homicide rates in Western Europe, 1300–2000, and in nonstate societies
Sources: Nonstate (geometric mean of 26 societies, not including Semai, Inuit, and !Kung): see figure 2–3. Europe: Eisner, 2003, table 1; geometric mean of five regions; missing data interpolated.

violence fluctuate more across different times and places than rates of domestic violence involving women or kin.[11] Martin Daly and Margo Wilson's explanation is that family members get on each other's nerves at similar rates in all times and places because of deeply rooted conflicts of interest that are inherent to the patterns of genetic overlap among kin. Macho violence among male acquaintances, in contrast, is fueled by contests of dominance that are more sensitive to circumstances. How violent a man must be to keep his rank in the pecking order in a given milieu depends on his assessment of how violent the *other* men are, leading to vicious or virtuous circles that can spiral up or down precipitously. I'll explore the psychology of kinship in more detail in chapter 7, and of dominance in chapter 8.

EXPLAINING THE EUROPEAN HOMICIDE DECLINE

Now let's consider the implications of the centuries-long decline in homicide in Europe. Do you think that city living, with its anonymity, crowding, immigrants, and jumble of cultures and classes, is a breeding ground for violence? What about the wrenching social changes brought on by capitalism and the Industrial Revolution? Is it your conviction that small-town life, centered on church, tradition, and fear of God, is our best bulwark against murder and mayhem? Well, think again. As Europe became more urban, cosmopolitan, commercial, industrialized, and secular, it got safer and safer. And that brings us back to the ideas of Norbert Elias, the only theory left standing.

Elias developed the theory of the Civilizing Process not by poring over numbers, which weren't available in his day, but by examining the texture of everyday life in medieval Europe. He examined, for instance, a series of drawings from the 15th-century German manuscript *The Medieval Housebook*, a depiction of daily life as seen through the eyes of a knight.[12]

In the detail shown in figure 3–5, a peasant disembowels a horse as a pig sniffs his exposed buttocks. In a nearby cave a man and a woman sit in the stocks. Above them a man is being led to the gallows, where a corpse is already

FIGURE 3–5. Detail from "Saturn," *Das Mittelalterliche Hausbuch* (*The Medieval Housebook*, 1475–80)

Sources: Reproduced in Elias, 1939/2000, appendix 2; see Graf zu Waldburg Wolfegg, 1988.

hanging, and next to it is a man who has been broken on the wheel, his shattered body pecked by a crow. The wheel and gibbet are not the focal point of the drawing, but a part of the landscape, like the trees and hills.

Figure 3–6 contains a detail from a second drawing, in which knights are attacking a village. In the lower left a peasant is stabbed by a soldier; above him, another peasant is restrained by his shirttail while a woman, hands in the air, cries out. At the lower right, a peasant is being stabbed in a chapel while his possessions are plundered, and nearby another peasant in fetters is cudgeled by a knight. Above them a group of horsemen are setting fire to a farmhouse, while one of them drives off the farmer's cattle and strikes at his wife.

The knights of feudal Europe were what today we would call warlords.

FIGURE 3–6. Detail from "Mars," *Das Mittelalterliche Hausbuch* (*The Medieval Housebook*, 1475–80)

Sources: Reproduced in Elias, 1939/2000, appendix 2; see Graf zu Waldburg Wolfegg, 1988.

States were ineffectual, and the king was merely the most prominent of the noblemen, with no permanent army and little control over the country. Governance was outsourced to the barons, knights, and other noblemen who controlled fiefs of various sizes, exacting crops and military service from the peasants who lived in them. The knights raided one another's territories in a Hobbesian dynamic of conquest, preemptive attack, and vengeance, and as the *Housebook* illustrations suggest, they did not restrict their killing to other knights. In *A Distant Mirror: The Calamitous 14th Century*, the historian Barbara Tuchman describes the way they made a living:

> These private wars were fought by the knights with furious gusto and a single strategy, which consisted in trying to ruin the enemy by killing and maiming as many of his peasants and destroying as many crops, vineyards, tools, barns, and other possessions as possible, thereby reducing his sources of revenue. As a result, the chief victim of the belligerents was their respective peasantry.[13]

As we saw in chapter 1, to maintain the credibility of their deterrent threat, knights engaged in bloody tournaments and other demonstrations of macho prowess, gussied up with words like *honor, valor, chivalry, glory,* and *gallantry,* which made later generations forget they were bloodthirsty marauders.

The private wars and tournaments were the backdrop to a life that was violent in other ways. As we saw, religious values were imparted with bloody crucifixes, threats of eternal torture, and prurient depictions of mutilated saints. Craftsmen applied their ingenuity to sadistic machines of punishment and execution. Brigands made travel a threat to life and limb, and ransoming captives was big business. As Elias noted, "the little people, too—the hatters, the tailors, the shepherds—were all quick to draw their knives."[14] Even clergymen got into the act. The historian Barbara Hanawalt quotes an account from 14th-century England:

> It happened at Ylvertoft on Saturday next before Martinmass in the fifth year of King Edward that a certain William of Wellington, parish chaplain of Ylvertoft, sent John, his clerk, to John Cobbler's house to buy a candle for him for a penny. But John would not send it to him without the money wherefore William became enraged, and, knocking in the door upon him, he struck John in the front part of the head so that his brains flowed forth and he died forthwith.[15]

Violence pervaded their entertainment as well. Tuchman describes two of the popular sports of the time: "Players with hands tied behind them competed to kill a cat nailed to a post by battering it to death with their heads, at the risk of cheeks ripped open or eyes scratched out by the frantic animal's claws. . . .

Or a pig enclosed in a wide pen was chased by men with clubs to the laughter of spectators as he ran squealing from the blows until beaten lifeless."[16]

During my decades in academia I have read thousands of scholarly papers on a vast range of topics, from the grammar of irregular verbs to the physics of multiple universes. But the oddest journal article I have ever read is "Losing Face, Saving Face: Noses and Honour in the Late Medieval Town."[17] Here the historian Valentin Groebner documents dozens of accounts from medieval Europe in which one person cut off the nose of another. Sometimes it was an official punishment for heresy, treason, prostitution, or sodomy, but more often it was an act of private vengeance. In one case in Nuremberg in 1520, Hanns Rigel had an affair with the wife of Hanns von Eyb. A jealous von Eyb cut off the nose of Rigel's innocent wife, a supreme injustice multiplied by the fact that Rigel was sentenced to four weeks of imprisonment for adultery while von Eyb walked away scot-free. These mutilations were so common that, according to Groebner,

> the authors of late-medieval surgical textbooks also devote particular attention to nasal injuries, discussing whether a nose once cut off can grow back, a controversial question that the French royal physician Henri de Mondeville answered in his famous *Chirurgia* with a categorical "No." Other fifteenth-century medical authorities were more optimistic: Heinrich von Pforspundt's 1460 pharmacoepia promised, among other things, a prescription for "making a new nose" for those who had lost theirs.[18]

The practice was the source of our strange idiom *to cut off your nose to spite your face.* In late medieval times, cutting off someone's nose was the prototypical act of spite.

Like other scholars who have peered into medieval life, Elias was taken aback by accounts of the temperament of medieval people, who by our lights seem impetuous, uninhibited, almost childlike:

> Not that people were always going around with fierce looks, drawn brows and martial countenances.... On the contrary, a moment ago they were joking, now they mock each other, one word leads to another, and suddenly from the midst of laughter they find themselves in the fiercest feud. Much of what appears contradictory to us—the intensity of their piety, the violence of their fear of hell, their guilt feelings, their penitence, the immense outbursts of joy and gaiety, the sudden flaring and the uncontrollable force of their hatred and belligerence—all these, like the rapid changes of mood, are in reality symptoms of one and the same structuring of the emotional life. The drives, the emotions were vented more freely, more directly, more openly than later. It is only to us, in whom everything is more subdued, moderate, and calculated, and in whom social taboos are built much more

> deeply into the fabric of our drive-economy as self-restraints, that the unveiled intensity of this piety, belligerence, or cruelty appears to be contradictory.[19]

Tuchman too writes of the "childishness noticeable in medieval behavior, with its marked inability to restrain any kind of impulse."[20] Dorothy Sayers, in the introduction to her translation of *The Song of Roland*, adds, "The idea that a strong man should react to great personal and national calamities by a slight compression of the lips and by silently throwing his cigarette into the fireplace is of very recent origin."[21]

Though the childishness of the medievals was surely exaggerated, there may indeed be differences in degree in the mores of emotional expression in different eras. Elias spends much of *The Civilizing Process* documenting this transition with an unusual database: manuals of etiquette. Today we think of these books, like *Amy Vanderbilt's Everyday Etiquette* and *Miss Manners' Guide to Excruciatingly Correct Behavior*, as sources of handy tips for avoiding embarrassing peccadilloes. But at one time they were serious guides to moral conduct, written by the leading thinkers of the day. In 1530 the great scholar Desiderius Erasmus, one of the founders of modernity, wrote an etiquette manual called *On Civility in Boys* which was a bestseller throughout Europe for two centuries. By laying down rules for what people ought not to do, these manuals give us a snapshot of what they must have been doing.

The people of the Middle Ages were, in a word, gross. A number of the advisories in the etiquette books deal with eliminating bodily effluvia:

> Don't foul the staircases, corridors, closets, or wall hangings with urine or other filth. • Don't relieve yourself in front of ladies, or before doors or windows of court chambers. • Don't slide back and forth on your chair as if you're trying to pass gas. • Don't touch your private parts under your clothes with your bare hands. • Don't greet someone while they are urinating or defecating. • Don't make noise when you pass gas. • Don't undo your clothes in front of other people in preparation for defecating, or do them up afterwards. • When you share a bed with someone in an inn, don't lie so close to him that you touch him, and don't put your legs between his. • If you come across something disgusting in the sheet, don't turn to your companion and point it out to him, or hold up the stinking thing for the other to smell and say "I should like to know how much that stinks."

Others deal with blowing one's nose:

> Don't blow your nose onto the tablecloth, or into your fingers, sleeve, or hat. • Don't offer your used handkerchief to someone else. • Don't carry your handkerchief in your mouth. • "Nor is it seemly, after wiping your nose, to

> spread out your handkerchief and peer into it as if pearls and rubies might have fallen out of your head."[22]

Then there are fine points of spitting:

> Don't spit into the bowl when you are washing your hands. • Do not spit so far that you have to look for the saliva to put your foot on it. • Turn away when spitting, lest your saliva fall on someone. • "If anything purulent falls to the ground, it should be trodden upon, lest it nauseate someone."[23] • If you notice saliva on someone's coat, it is not polite to make it known.

And there are many, many pieces of advice on table manners:

> Don't be the first to take from the dish. • Don't fall on the food like a pig, snorting and smacking your lips. • Don't turn the serving dish around so the biggest piece of meat is near you. • "Don't wolf your food like you're about to be carried off to prison, nor push so much food into your mouth that your cheeks bulge like bellows, nor pull your lips apart so that they make a noise like pigs." • Don't dip your fingers into the sauce in the serving dish. • Don't put a spoon into your mouth and then use it to take food from the serving dish. • Don't gnaw on a bone and put it back in the serving dish. • Don't wipe your utensils on the tablecloth. • Don't put back on your plate what has been in your mouth. • Do not offer anyone a piece of food you have bitten into. • Don't lick your greasy fingers, wipe them on the bread, or wipe them on your coat. • Don't lean over to drink from your soup bowl. • Don't spit bones, pits, eggshells, or rinds into your hand, or throw them on the floor. • Don't pick your nose while eating. • Don't drink from your dish; use a spoon. • Don't slurp from your spoon. • Don't loosen your belt at the table. • Don't clean a dirty plate with your fingers. • Don't stir sauce with your fingers. • Don't lift meat to your nose to smell it. • Don't drink coffee from your saucer.

In the mind of a modern reader, these advisories set off a train of reactions. How inconsiderate, how boorish, how animalistic, how immature those people must have been! These are the kinds of directives you'd expect a parent to give to a three-year-old, not a great philosopher to a literate readership. Yet as Elias points out, the habits of refinement, self-control, and consideration that are second nature to us had to be acquired—that's why we call them *second* nature—and they developed in Europe over the course of its modern history.

The sheer quantity of the advice tells a story. The three-dozen-odd rules are not independent of one another but exemplify a few themes. It's unlikely that each of us today had to be instructed in every rule individually, so that if some mother had been remiss in teaching one of them, her adult son would still be blowing his nose into the tablecloth. The rules in the list (and many more that are not) are deducible from a few principles: Control your appetites; Delay gratification; Consider the sensibilities of others; Don't act like a peasant; Distance yourself from your animal nature. And the penalty for these infractions was assumed to be internal: a sense of shame. Elias notes that the etiquette books rarely mention health and hygiene. Today we recognize that the emotion of disgust evolved as an unconscious defense against biological contamination.[24] But an understanding of microbes and infection did not arrive until well into the 19th century. The only explicit rationales stated in the etiquette books are to avoid acting like a peasant or an animal and to avoid offending others.

In the European Middle Ages, sexual activity too was less discreet. People were publicly naked more often, and couples took only perfunctory measures to keep their coitus private. Prostitutes offered their services openly; in many English towns, the red-light district was called Gropecunt Lane. Men would discuss their sexual exploits with their children, and a man's illegitimate offspring would mix with his legitimate ones. During the transition to modernity, this openness came to be frowned upon as uncouth and then as unacceptable.

The change left its mark in the language. Words for peasantry took on a second meaning as words for turpitude: *boor* (which originally just meant "farmer," as in the German *Bauer* and Dutch *boer*); *villain* (from the French *vilein*, a serf or villager); *churlish* (from English *churl*, a commoner); *vulgar* (common, as in the term *vulgate*); and *ignoble*, not an aristocrat. Many of the words for the fraught actions and substances became taboo. Englishmen used to swear by invoking supernatural beings, as in *My God!* and *Jesus Christ!* At the start of the modern era they began to invoke sexuality and excretion, and the "Anglo-Saxon four-letter words," as we call them today, could no longer be used in polite company.[25] As the historian Geoffrey Hughes has noted, "The days when the dandelion could be called the *pissabed*, a heron could be called a *shitecrow* and the windhover could be called the *windfucker* have passed away with the exuberant phallic advertisement of the codpiece."[26] *Bastard*, *cunt*, *arse*, and *whore* also passed from ordinary to taboo.

As the new etiquette took hold, it also applied to the accoutrements of violence, particularly knives. In the Middle Ages, most people carried a knife and would use it at the dinner table to carve a chunk of meat off the roasted carcass, spear it, and bring it to their mouths. But the menace of a lethal weapon within reach at a communal gathering, and the horrific image of a knife pointed at a face, became increasingly repellent. Elias cites a number of points of etiquette that center on the use of knives:

> Don't pick your teeth with your knife. • Don't hold your knife the entire time you are eating, but only when you are using it. • Don't use the tip of your knife to put food into your mouth. • Don't cut bread; break it. • If you pass

someone a knife, take the point in your hand and offer him the handle. • Don't clutch your knife with your whole hand like a stick, but hold it in your fingers. • Don't use your knife to point at someone.

It was during this transition that the fork came into common use as a table utensil, so that people no longer had to bring their knives to their mouths. Special knives were set at the table so people would not have to unsheathe their own, and they were designed with rounded rather than pointed ends. Certain foods were never to be cut with a knife, such as fish, round objects, and bread—hence the expression *to break bread together.*

Some of the medieval knife taboos remain with us today. Many people will not give a knife as a present unless it is accompanied by a coin, which the recipient gives back, to make the transaction a sale rather than a gift. The ostensible reason is to avoid the symbolism of "severing the friendship," but a more likely reason is to avoid the symbolism of directing an unsolicited knife in the friend's direction. A similar superstition makes it bad luck to hand someone a knife: one is supposed to lay it down on the table and allow the recipient to pick it up. Knives in table settings are rounded at the end and no sharper than needed: steak knives are brought out for tough meat, and blunter knives substituted for fish. And knives may be used only when they are absolutely necessary. It's rude to use a knife to eat a piece of cake, to bring food to your mouth, to mix ingredients ("Stir with a knife, stir up strife"), or to push food onto your fork.

Aha!

Elias's theory, then, attributes the decline in European violence to a larger psychological change (the subtitle of his book is *Sociogenetic and Psychogenetic Investigations*). He proposed that over a span of several centuries, beginning in the 11th or 12th and maturing in the 17th and 18th, Europeans increasingly inhibited their impulses, anticipated the long-term consequences of their actions, and took other people's thoughts and feelings into consideration. A culture of honor—the readiness to take revenge—gave way to a culture of dignity—the readiness to control one's emotions. These ideals originated in explicit instructions that cultural arbiters gave to aristocrats and noblemen, allowing them to differentiate themselves from the villains and boors. But they were then absorbed into the socialization of younger and younger children until they became second nature. The standards also trickled down from the upper classes to the bourgeoisie that strove to emulate them, and from them to the lower classes, eventually becoming a part of the culture as a whole.

Elias helped himself to Freud's structural model of the psyche, in which children acquire a conscience (the superego) by internalizing the injunctions of their parents when they are too young to understand them. At that point the child's ego can apply these injunctions to keep their biological impulses (the id) in check. Elias stayed away from Freud's more exotic claims (such as the primeval parricide, the death instinct, and the oedipal complex), and his psychology is thoroughly modern. In chapter 9 we will look at a faculty of the mind that psychologists call self-control, delay of gratification, and shallow temporal discounting and that laypeople call counting to ten, holding your horses, biting your tongue, saving for a rainy day, and keeping your pecker in your pocket.[27] We will also look at a faculty that psychologists call empathy, intuitive psychology, perspective-taking, and theory of mind and that laypeople call getting into other people's heads, seeing the world from their point of view, walking a mile in their moccasins, and feeling their pain. Elias anticipated the scientific study of both of these better angels.

Critics of Elias have pointed out that all societies have standards of propriety about sexuality and excretion which presumably grow out of innate emotions surrounding purity, disgust, and shame.[28] As we will see, the degree to which societies moralize these emotions is a major dimension of variation across cultures. Though medieval Europe certainly did not lack norms of propriety altogether, it seems to have lain at the far end of the envelope of cultural possibilities.

To his credit, Elias leapfrogged academic fashion in not claiming that early modern Europeans "invented" or "constructed" self-control. He claimed only that they toned up a mental faculty that had always been a part of human nature but which the medievals had underused. He repeatedly drove the point home with the pronouncement "There is no zero point."[29] As we shall see in chapter 9, exactly how people dial their capacity for self-control up or down is an interesting topic in psychology. One possibility is that self-control is like a muscle, so that if you exercise it with table manners it will be stronger across the board and more effective when you have to stop yourself from killing the person who just insulted you. Another possibility is that a particular setting of the self-control dial is a social norm, like how close you can stand to another person or how much of your body has to be covered in public. A third is that self-control can be adjusted adaptively according to its costs and benefits in the local environment. Self-control, after all, is not an unmitigated good. The problem with having too much self-control is that an aggressor can use it to his advantage, anticipating that you may hold back from retaliating because it's too late to do any good. But if he had reason to believe that you would lash out reflexively, consequences be damned, he might treat you with more respect in the first place. In that case people might adjust a self-control slider according to the dangerousness of those around them.

At this point in the story, the theory of the Civilizing Process is incomplete, because it appeals to a process that is endogenous to the phenomenon it is trying to explain. A decline in violent behavior, it says, coincided with a decline

As my grandfather would have put it, "Goyische kopp!"—gentile head. Jews were brought in as moneylenders and middlemen but were just as often persecuted and expelled. The era's economic backwardness was enforced by laws which decreed that prices should be fixed at a "just" level reflecting the cost of the raw material and the value of the labor added to it. "To ensure that no one gained an advantage over anyone else," Tuchman explains, "commercial law prohibited innovation in tools or techniques, underselling below a fixed price, working late by artificial light, employing extra apprentices or wife and under-age children, and advertising of wares or praising them to the detriment of others."[36] This is a recipe for a zero-sum game, and leaves predation as the only way people could add to their wealth.

A *positive*-sum game is a scenario in which agents have choices that can improve the lots of both of them at the same time. A classic positive-sum game in everyday life is the exchange of favors, where each person can confer a large benefit to another at a small cost to himself or herself. Examples include primates who remove ticks from each other's backs, hunters who share meat whenever one of them has felled an animal that is too big for him to consume on the spot, and parents who take turns keeping each other's children out of trouble. As we shall see in chapter 8, a key insight of evolutionary psychology is that human cooperation and the social emotions that support it, such as sympathy, trust, gratitude, guilt, and anger, were selected because they allow people to flourish in positive-sum games.[37]

A classic positive-sum game in economic life is the trading of surpluses. If a farmer has more grain than he can eat, and a herder has more milk than he can drink, both of them come out ahead if they trade some wheat for some milk. As they say, everybody wins. Of course, an exchange at a single moment in time only pays when there is a division of labor. There would be no point in one farmer giving a bushel of wheat to another farmer and receiving a bushel of wheat in return. A fundamental insight of modern economics is that the key to the creation of wealth is a division of labor, in which specialists learn to produce a commodity with increasing cost-effectiveness and have the means to exchange their specialized products efficiently. One infrastructure that allows efficient exchange is transportation, which makes it possible for producers to trade their surpluses even when they are separated by distance. Another is money, interest, and middlemen, which allow producers to exchange many kinds of surpluses with many other producers at many points in time.

Positive-sum games also change the incentives for violence. If you're trading favors or surpluses with someone, your trading partner suddenly becomes more valuable to you alive than dead. You have an incentive, moreover, to anticipate what he wants, the better to supply it to him in exchange for what you want. Though many intellectuals, following in the footsteps of Saints Augustine and Jerome, hold businesspeople in contempt for their selfishness and greed, in fact a free market puts a premium on empathy.[38] A good businessperson has to keep the customers satisfied or a competitor will woo them away, and the more customers he attracts, the richer he will be. This idea, which came to be called *doux commerce* (gentle commerce), was expressed by the economist Samuel Ricard in 1704:

> Commerce attaches [people] to one another through mutual utility. . . . Through commerce, man learns to deliberate, to be honest, to acquire manners, to be prudent and reserved in both talk and action. Sensing the necessity to be wise and honest in order to succeed, he flees vice, or at least his demeanor exhibits decency and seriousness so as not to arouse any adverse judgment on the part of present and future acquaintances.[39]

And this brings us to the second exogenous change. Elias noted that in the late Middle Ages people began to unmire themselves from technological and economic stagnation. Money increasingly replaced barter, aided by the larger national territories in which a currency could be recognized. The building of roads, neglected since Roman times, resumed, allowing the transport of goods to the hinterlands of the country and not just along its coasts and navigable rivers. Horse transport became more efficient with the use of horseshoes that protected hooves from paving stones and yokes that didn't choke the poor horse when it pulled a heavy load. Wheeled carts, compasses, clocks, spinning wheels, treadle looms, windmills, and water mills were also perfected in the later Middle Ages. And the specialized expertise needed to implement these technologies was cultivated in an expanding stratum of craftsmen. The advances encouraged the division of labor, increased surpluses, and lubricated the machinery of exchange. Life presented people with more positive-sum games and reduced the attractiveness of zero-sum plunder. To take advantage of the opportunities, people had to plan for the future, control their impulses, take other people's perspectives, and exercise the other social and cognitive skills needed to prosper in social networks.

The two triggers of the Civilizing Process—the Leviathan and gentle commerce—are related. The positive-sum cooperation of commerce flourishes best inside a big tent presided over by a Leviathan. Not only is a state well suited to provide the public goods that serve as infrastructure for economic cooperation, such as money and roads, but it can put a thumb on the scale on which players weigh the relative payoffs of raiding and trading. Suppose a knight can either plunder ten bushels of grain from his neighbor or, by expending the same amount of time and energy, raise the money to buy five bushels from him. The theft option looks pretty good. But if the knight anticipates that the state will fine him six bushels for the theft, he'd be left with only four, so he's better off with honest toil. Not only do the Leviathan's incentives make commerce more attractive, but commerce makes the job of the Leviathan easier. If the honest alternative of buying the grain hadn't been available, the state would have had to threaten to squeeze ten bushels out of the knight to deter him from plundering, which is harder to enforce than squeezing five bushels out of him. Of course, in reality the state's sanctions may be the threat of physical punishment rather than a fine, but the principle is the same: it's easier to deter people from crime if the lawful alternative is more appealing.

The two civilizing forces, then, reinforce each other, and Elias considered them to be part of a single process. The centralization of state control and its monopolization of violence, the growth of craft guilds and bureaucracies, the replacement of barter with money, the development of technology, the enhancement of trade, the growing webs of dependency among far-flung individuals, all fit into an organic whole. And to prosper within that whole, one had to cultivate faculties of empathy and self-control until they became, as he put it, second nature.

Indeed the "organic" analogy is not far-fetched. The biologists John Maynard Smith and Eörs Szathmáry have argued that an evolutionary dynamic similar to the Civilizing Process drove the major transitions in the history of life. These transitions were the successive emergence of genes, chromosomes, bacteria, cells with nuclei, organisms, sexually reproducing organisms, and animal societies.[40] In each transition, entities with the capacity to be either selfish or cooperative tended toward cooperation when they could be subsumed into a larger whole. They specialized, exchanged benefits, and developed safeguards to prevent one of them from exploiting the rest to the detriment of the whole. The journalist Robert Wright sketches a similar arc in his book *Nonzero*, an allusion to positive-sum games, and extends it to the history of human societies.[41] In the final chapter of this book I will take a closer look at overarching theories of the decline of violence.

The theory of the Civilizing Process passed a stringent test for a scientific hypothesis: it made a surprising prediction that turned out to be true. Back in 1939 Elias had no access to the statistics of homicide; he worked from narrative histories and old books of etiquette. When Gurr, Eisner, Cockburn, and others surprised the world of criminology with their graphs showing a decline in killings, Elias had the only theory that anticipated it. But with everything else we have learned about violence in recent decades, how well does the theory fare?

Elias himself was haunted by the not-so-civilized behavior of his native Germany during World War II, and he labored to explain that "decivilizing process" within the framework of his theory.[42] He discussed the fitful history of German unification and the resulting lack of trust in a legitimate central authority. He documented the persistence of a militaristic culture of honor among its elites, the breakdown of a state monopoly on violence with the rise of communist and fascist militias, and a resulting contraction of empathy for groups perceived to be outsiders, particularly the Jews. It would be a stretch to say that he rescued his theory with these analyses, but perhaps he shouldn't have tried. The horrors of the Nazi era did not consist in an upsurge in feuding among warlords or of citizens stabbing each other over the dinner table, but in violence whose scale, nature, and causes are altogether different. In fact in Germany during the Nazi years the declining trend for one-on-one homicides continued (see, for example, figure 3–19).[43] In chapter 8 we will see how the compartmentalization of the moral sense, and the distribution of belief and enforcement among different sectors of a population, can lead to ideologically driven wars and genocides even in otherwise civilized societies.

Eisner pointed out another complication for the theory of the Civilizing Process: the decline of violence in Europe and the rise of centralized states did not always proceed in lockstep.[44] Belgium and the Netherlands were at the forefront of the decline, yet they lacked strong centralized governments. When Sweden joined the trend, it wasn't on the heels of an expansion in state power either. Conversely, the Italian states were in the rearguard of the decline in violence, yet their governments wielded an enormous bureaucracy and police force. Nor did cruel punishments, the enforcement method of choice among early modern monarchs, reduce violence in the areas where they were carried out with the most relish.

Many criminologists believe that the source of the state's pacifying effect isn't just its brute coercive power but the trust it commands among the populace. After all, no state can post an informant in every pub and farmhouse to monitor breaches of the law, and those that try are totalitarian dictatorships that rule by fear, not civilized societies where people coexist through self-control and empathy. A Leviathan can civilize a society only when the citizens feel that its laws, law enforcement, and other social arrangements are legitimate, so that they don't fall back on their worst impulses as soon as Leviathan's back is turned.[45] This doesn't refute Elias's theory, but it adds a twist. An imposition of the rule of law may end the bloody mayhem of feuding warlords, but reducing rates of violence further, to the levels enjoyed by modern European societies, involves a more nebulous process in which certain populations accede to the rule of law that has been imposed on them.

Libertarians, anarchists, and other skeptics of the Leviathan point out that when communities are left to their own devices, they often develop norms of cooperation that allow them to settle their disputes nonviolently, without laws, police, courts, or the other trappings of government. In *Moby-Dick*, Ishmael explains how American whalers thousands of miles from the reach of the law dealt with disputes over whales that had been injured or killed by one ship and then claimed by another:

> Thus the most vexatious and violent disputes would often arise between the fishermen, were there not some written or unwritten, universal, undisputed law applicable to all cases.

> . . . Though no other nation [but Holland] has ever had any written whaling law, yet the American fishermen have been their own legislators and lawyers in this matter. . . . These laws might be engraven on a Queen Anne's farthing, or the barb of a harpoon, and worn round the neck, so small are they.
>
> I. A Fast-Fish belongs to the party fast to it.
>
> II. A Loose-Fish is fair game for anybody who can soonest catch it.

Informal norms of this kind have emerged among fishers, farmers, and herders in many parts of the world.[46] In *Order Without Law: How Neighbors Settle Disputes*, the legal scholar Robert Ellickson studied a modern American version of the ancient (and frequently violent) confrontation between pastoralists and farmers. In northern California's Shasta County, traditional ranchers are essentially cowboys, grazing their cattle in open country, while modern ranchers raise cattle in irrigated, fenced ranches. Both kinds of ranchers coexist with farmers who grow hay, alfalfa, and other crops. Straying cattle occasionally knock down fences, eat crops, foul streams, and wander onto roads where vehicles can hit them. The county is carved into "open ranges," in which an owner is not legally liable for most kinds of accidental damage his cattle may cause, and "closed ranges," in which he is strictly liable, whether he was negligent or not. Ellickson discovered that victims of harm by cattle were loath to invoke the legal system to settle the damages. In fact, most of the residents—ranchers, farmers, insurance adjustors, even lawyers and judges—held beliefs about the applicable laws that were flat wrong. But the residents got along by adhering to a few tacit norms. Cattle owners were always responsible for the damage their animals caused, whether a range was open or closed; but if the damage was minor and sporadic, property owners were expected to "lump it." People kept rough long-term mental accounts of who owed what, and the debts were settled in kind rather than in cash. (For example, a cattleman whose cow damaged a rancher's fence might at a later time board one of the rancher's stray cattle at no charge.) Deadbeats and violators were punished with gossip and with occasional veiled threats or minor vandalism. In chapter 9 we'll take a closer look at the moral psychology behind such norms, which fall into a category called equality matching.[47]

As important as tacit norms are, it would be a mistake to think that they obviate a role for government. The Shasta County ranchers may not have called in Leviathan when a cow knocked over a fence, but they were living in its shadow and knew it would step in if their informal sanctions escalated or if something bigger were at stake, such as a fight, a killing, or a dispute over women. And as we shall see, their current level of peaceful coexistence is itself the legacy of a local version of the Civilizing Process. In the 1850s, the annual homicide rate of northern California ranchers was around 45 per 100,000, comparable to those of medieval Europe.[48]

I think the theory of the Civilizing Process provides a large part of the explanation for the modern decline of violence not only because it predicted the remarkable plunge in European homicide but because it makes correct predictions about the times and places in the modern era that do not enjoy the blessed 1-per-100,000-per-year rate of modern Europe. Two of these rule-proving exceptions are zones that the Civilizing Process never fully penetrated: the lower strata of the socioeconomic scale, and the inaccessible or inhospitable territories of the globe. And two are zones in which the Civilizing Process went into reverse: the developing world, and the 1960s. Let's visit them in turn.

VIOLENCE AND CLASS

Other than the drop in numbers, the most striking feature of the decline in European homicide is the change in the socioeconomic profile of killing. Centuries ago rich people were as violent as poor people, if not more so.[49] Gentlemen would carry swords and would not hesitate to use them to avenge insults. They often traveled with retainers who doubled as bodyguards, so an affront or a retaliation for an affront could escalate into a bloody street fight between gangs of aristocrats (as in the opening scene of *Romeo and Juliet*). The economist Gregory Clark examined records of deaths of English aristocrats from late medieval times to the Industrial Revolution. I've plotted his data in figure 3–7, which shows that in the 14th and 15th centuries an astonishing 26 percent of

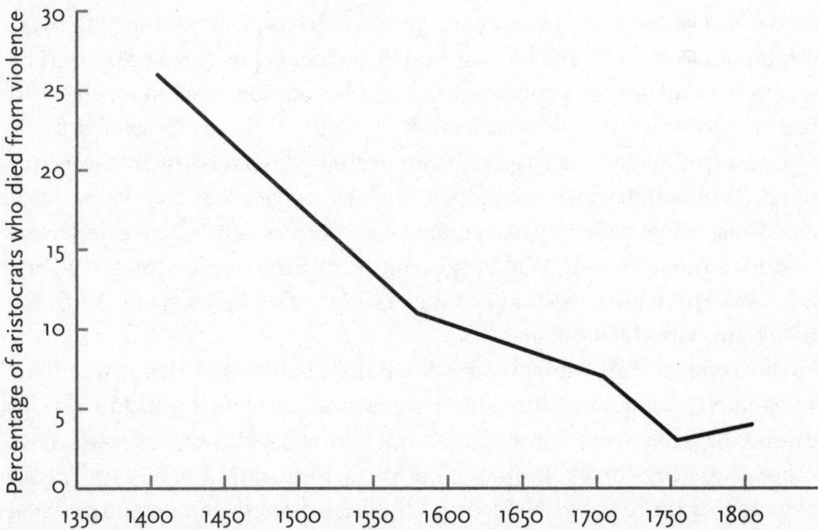

FIGURE 3–7. Percentage of deaths of English male aristocrats from violence, 1330–1829
Source: Data from Clark, 2007a, p. 122; data representing a range of years are plotted at the midpoint of the range.

male aristocrats died from violence—about the same rate that we saw in figure 2–2 as the average for preliterate tribes. The rate fell into the single digits by the turn of the 18th century, and of course today it is essentially zero.

A homicide rate measured in percentage points is still remarkably high, and well into the 18th and 19th centuries violence was a part of the lives of respectable men, such as Alexander Hamilton and Aaron Burr. Boswell quotes Samuel Johnson, who presumably had no trouble defending himself with words, as saying, "I have beat many a fellow, but the rest had the wit to hold their tongues."[50] Members of the upper classes eventually refrained from using force against one another, but with the law watching their backs, they reserved the right to use it against their inferiors. As recently as 1859 the British author of *The Habits of a Good Society* advised:

> There are men whom nothing but a physical punishment will bring to reason, and with these we shall have to deal at some time of our lives. A lady is insulted or annoyed by an unwieldy bargee, or an importunate and dishonest cabman. One well-dealt blow settles the whole matter. . . . A man therefore, whether he aspires to be a gentleman or not, should learn to box. . . . There are but few rules for it, and those are suggested by common sense. Strike out, strike straight, strike suddenly; keep one arm to guard, and punish with the other. Two gentlemen never fight; the art of boxing is brought into use in punishing a stronger and more imprudent man of a class beneath your own.[51]

The European decline of violence was spearheaded by a decline in *elite* violence. Today statistics from every Western country show that the overwhelming majority of homicides and other violent crimes are committed by people in the lowest socioeconomic classes. One obvious reason for the shift is that in medieval times, one *achieved* high status through the use of force. The journalist Steven Sailer recounts an exchange from early-20th-century England: "A hereditary member of the British House of Lords complained that Prime Minister Lloyd George had created new Lords solely because they were self-made millionaires who had only recently acquired large acreages. When asked, 'How did your ancestor become a Lord?' he replied sternly, 'With the battle-ax, sir, with the battle-ax!' "[52]

As the upper classes were putting down their battle-axes, disarming their retinues, and no longer punching out bargees and cabmen, the middle classes followed suit. They were domesticated not by the royal court, of course, but by other civilizing forces. Employment in factories and businesses forced employees to acquire habits of decorum. An increasingly democratic political process allowed them to identify with the institutions of government and society, and it opened up the court system as a way to pursue their grievances.

And then came an institution that was introduced in London in 1828 by Sir Robert Peel and soon named after him, the municipal police, or bobbies.[53]

The main reason that violence correlates with low socioeconomic status today is that the elites and the middle class pursue justice with the legal system while the lower classes resort to what scholars of violence call "self-help." This has nothing to do with *Women Who Love Too Much* or *Chicken Soup for the Soul;* it is another name for vigilantism, frontier justice, taking the law into your own hands, and other forms of violent retaliation by which people secured justice in the absence of intervention by the state.

In an influential article called "Crime as Social Control," the legal scholar Donald Black argued that most of what we call crime is, from the point of view of the perpetrator, the pursuit of justice.[54] Black began with a statistic that has long been known to criminologists: only a minority of homicides (perhaps as few as 10 percent) are committed as a means to a practical end, such as killing a homeowner during a burglary, a policeman during an arrest, or the victim of a robbery or rape because dead people tell no tales.[55] The most common motives for homicide are moralistic: retaliation after an insult, escalation of a domestic quarrel, punishing an unfaithful or deserting romantic partner, and other acts of jealousy, revenge, and self-defense. Black cites some cases from a database in Houston:

> One in which a young man killed his brother during a heated discussion about the latter's sexual advances toward his younger sisters, another in which a man killed his wife after she "dared" him to do so during an argument about which of several bills they should pay, one where a woman killed her husband during a quarrel in which the man struck her daughter (his stepdaughter), one in which a woman killed her 21-year-old son because he had been "fooling around with homosexuals and drugs," and two others in which people died from wounds inflicted during altercations over the parking of an automobile.

Most homicides, Black notes, are really instances of capital punishment, with a private citizen as the judge, jury, and executioner. It's a reminder that the way we conceive of a violent act depends on which of the corners of the violence triangle (see figure 2–1) we stake out as our vantage point. Consider a man who is arrested and tried for wounding his wife's lover. From the point of view of the law, the aggressor is the husband and the victim is society, which is now pursuing justice (an interpretation, recall, captured in the naming of court cases, such as *The People vs. John Doe*). From the point of view of the lover, the aggressor is the husband, and he is the victim; if the husband gets off on an acquittal or mistrial or plea bargain, there is no justice, as the lover is enjoined from pursuing revenge. And from the point of view of the husband, *he* is the victim (of cuckoldry), the lover is the aggressor, and justice has been done—but now he is the victim of a second act of aggression, in which the state is the aggressor and the lover is an accomplice. Black notes:

> Those who commit murder . . . often appear to be resigned to their fate at the hands of the authorities; many wait patiently for the police to arrive; some even call to report their own crimes. . . . In cases of this kind, indeed, the individuals involved might arguably be regarded as martyrs. Not unlike workers who violate a prohibition to strike—knowing they will go to jail—or others who defy the law on grounds of principle, they do what they think is right, and willingly suffer the consequences.[56]

These observations overturn many dogmas about violence. One is that violence is caused by a deficit of morality and justice. On the contrary, violence is often caused by a surfeit of morality and justice, at least as they are conceived in the minds of the perpetrators. Another dogma, cherished among psychologists and public health researchers, is that violence is a kind of disease.[57] But this public health theory of violence flouts the basic definition of a disease, namely a malfunction that causes suffering to the individual.[58] Most violent people insist there is nothing wrong with them; it's the victim and bystanders who think there's a problem. A third dubious belief about violence is that lower-class people engage in it because they are financially needy (for example, stealing food to feed their children) or because they are expressing rage against society. The violence of a lower-class man may indeed express rage, but it is aimed not at society but at the asshole who scraped his car and dissed him in front of a crowd.

In an article inspired by Black called "The Decline of Elite Homicide," the criminologist Mark Cooney shows that many lower-status people—the poor, the uneducated, the unmarried, and members of minority groups—are effectively stateless. Some make a living from illegal activities like drug dealing, gambling, selling stolen goods, and prostitution, so they cannot file lawsuits or call the police to enforce their interests in business disputes. In that regard they share their need for recourse to violence with certain *high*-status people, namely dealers in contraband such as Mafiosi, drug kingpins, and Prohibition rumrunners.

But another reason for their statelessness is that lower-status people and the legal system often live in a condition of mutual hostility. Black and Cooney report that in dealing with low-income African Americans, police "seem to vacillate between indifference and hostility, . . . reluctant to become involved in their affairs but heavy handed when they do so."[59] Judges and prosecutors too "tend to be . . . uninterested in the disputes of low-status people, typically disposing of them quickly and, to the parties involved, with an unsatisfactory penal emphasis."[60] Here is a Harlem police sergeant quoted by the journalist Heather MacDonald:

> Last weekend, a known neighborhood knucklehead hit a kid. In retaliation, the kid's whole family shows up at the perp's apartment. The victim's sisters kick in the apartment door. But the knucklehead's mother beat the shit out of the sisters, leaving them lying on the floor with blood coming from their mouths. The victim's family was looking for a fight: I could charge them with trespass. The perp's mother is eligible for assault three for beating up the opposing family. But all of them were street shit, garbage. They will get justice in their own way. I told them: "We can all go to jail, or we can call it a wash." Otherwise, you'd have six bodies in prison for BS behavior. The district attorney would have been pissed. And none of them would ever show up in court.[61]

Not surprisingly, lower-status people tend not to avail themselves of the law and may be antagonistic to it, preferring the age-old alternative of self-help justice and a code of honor. The police sergeant's compliment about the kind of people he deals with in his precinct was returned by the young African Americans interviewed by the criminologist Deanna Wilkinson:

> *Reggie:* The cops working my neighborhood don't belong working in my neighborhood. How you gonna send white cops to a black neighborhood to protect and serve? You can't do that cause all they gonna see is the black faces that's committing the crimes. They all look the same. The ones that's not committing crimes looks like the niggas that is committing crimes and everybody is getting harassed.
>
> *Dexter:* They make worser cause niggas [the police] was fuckin' niggas [youth] up. They crooked theyself, you know what I mean? Them niggas [the police officers] would run up on the drug spot, take my drugs, they'll sell that shit back on the street, so they could go rush-knock someone else.
>
> *Quentin [speaking of a man who had shot his father]:* There's a chance he could walk, what am I supposed to do? . . . If I lose my father, and they don't catch this guy, I'm gonna get his family. That's the way it works out here. That's the way all this shit out here works. If you can't get him, get them. . . . Everybody grow up with the shit, they want respect, they want to be the man.[62]

The historical Civilizing Process, in other words, did not eliminate violence, but it did relegate it to the socioeconomic margins.

VIOLENCE AROUND THE WORLD

The Civilizing Process spread not only downward along the socioeconomic scale but outward across the geographic scale, from a Western European epicenter. We saw in figure 3–3 that England was the first to pacify itself, followed closely by Germany and the Low Countries. Figure 3–8 plots this outward ripple on maps of Europe in the late 19th and early 21st centuries.

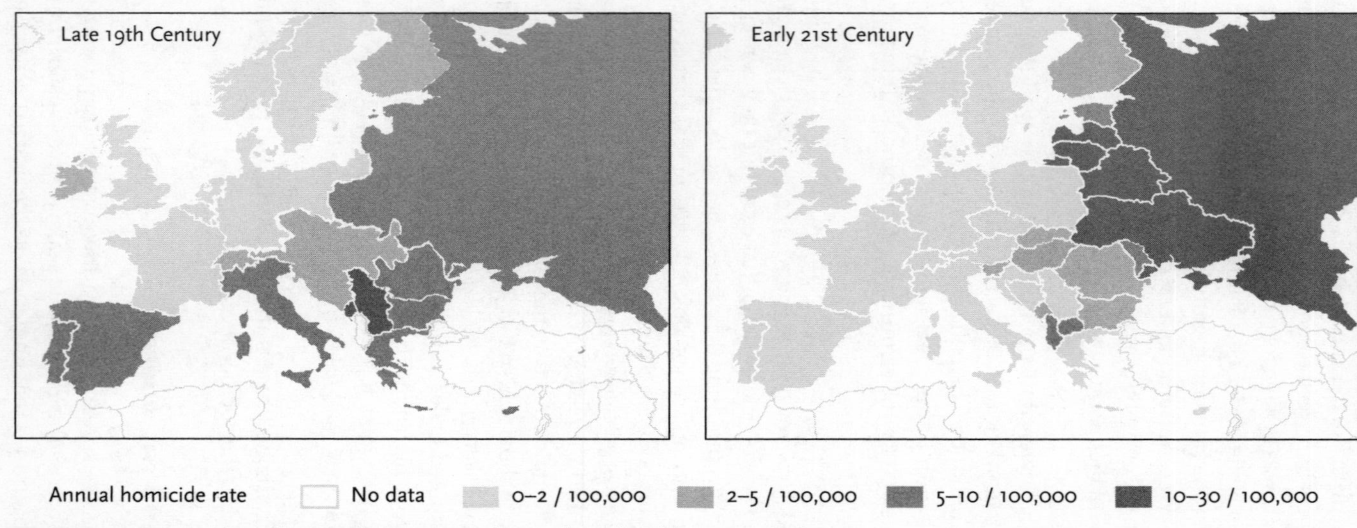

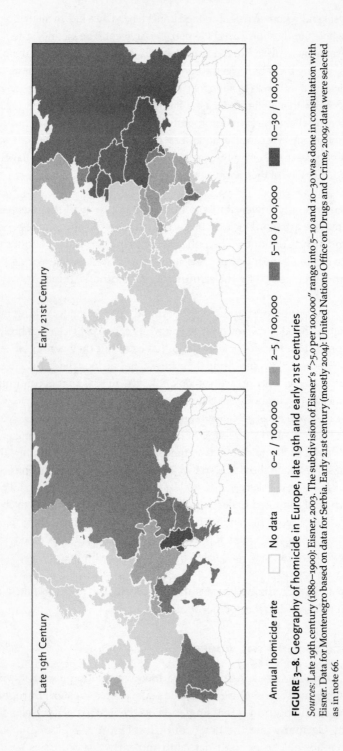

FIGURE 3–8. Geography of homicide in Europe, late 19th and early 21st centuries

Sources: Late 19th century (1880–1900): Eisner, 2003. The subdivision of Eisner's ">5.0 per 100,000" range into 5–10 and 10–30 was done in consultation with Eisner. Data for Montenegro based on data for Serbia. Early 21st century (mostly 2004): United Nations Office on Drugs and Crime, 2009; data were selected as in note 66.

In the late 1800s, Europe had a peaceable bull's-eye in the northern industrialized countries (Great Britain, France, Germany, Denmark, and the Low Countries), bordered by slightly stroppier Ireland, Austria-Hungary, and Finland, surrounded in turn by still more violent Spain, Italy, Greece, and the Slavic countries. Today the peaceable center has swelled to encompass all of Western and Central Europe, but a gradient of lawlessness extending to Eastern Europe and the mountainous Balkans is still visible.

There are gradients within each of these countries as well: the hinterlands and mountains remained violent long after the urbanized and densely farmed centers had calmed down. Clan warfare was endemic to the Scottish highlands until the 18th century, and to Sardinia, Sicily, Montenegro, and other parts of the Balkans until the 20th.[63] It's no coincidence that the two blood-soaked classics with which I began this book—the Hebrew Bible and the Homeric poems—came from peoples that lived in rugged hills and valleys.

What about the rest of the world? Though most European countries have kept statistics on homicide for a century or more, the same cannot be said for the other continents. Even today the police-blotter tallies that departments report to Interpol are often unreliable and sometimes incredible. Many governments feel that their degree of success in keeping their citizens from murdering each other is no one else's business. And in parts of the developing world, warlords dress up their brigandage in the language of political liberation movements, making it hard to draw a line between casualties in a civil war and homicides from organized crime.[64]

With those limitations in mind, let's take a peek at the distribution of homicide in the world today. The most reliable data come from the World Health Organization (WHO), which uses public health records and other sources to estimate the causes of death in as many countries as possible.[65] The UN Office on Drugs and Crime has supplemented these data with high and low estimates for every country in the world. Figure 3–9 plots these numbers for 2004 (the year covered in the office's most recent report) on a map of the world.[66] The good news is that the median national homicide rate among the world's countries in this dataset is 6 per 100,000 per year. The overall homicide rate for the entire world, ignoring the division into countries, was estimated by the WHO in 2000 as 8.8 per 100,000 per year.[67] Both estimates compare favorably to the triple-digit values for pre-state societies and the double-digit values for medieval Europe.

The map shows that Western and Central Europe make up the least violent region in the world today. Among the other states with credible low rates of homicide are those carved out of the British Empire, such as Australia, New Zealand, Fiji, Canada, the Maldives, and Bermuda. Another former British colony defies the pattern of English civility; we will examine this strange country in the next section.

Several Asian countries have low homicide rates as well, particularly those

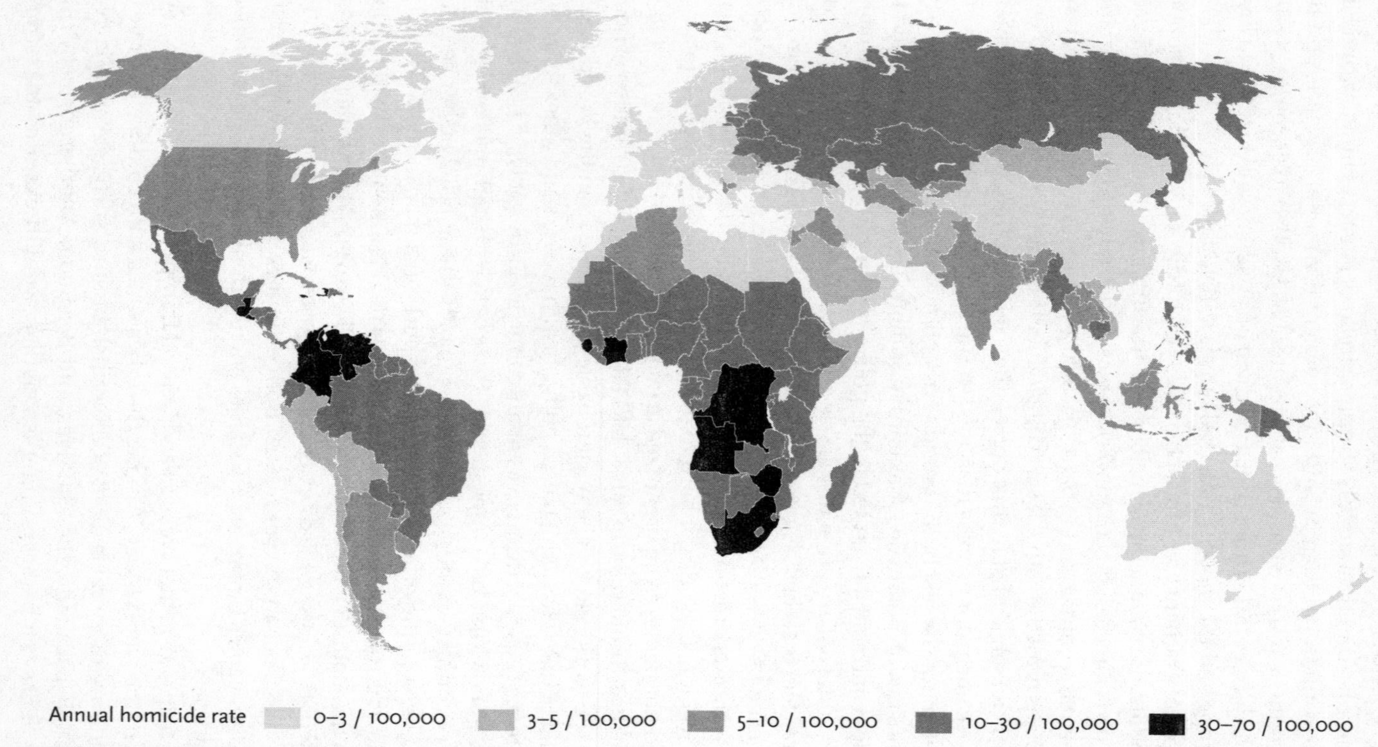

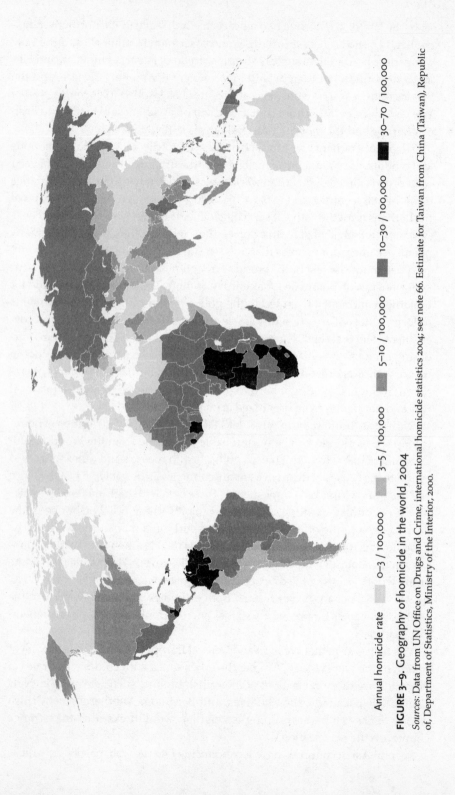

FIGURE 3–9. Geography of homicide in the world, 2004

Sources: Data from UN Office on Drugs and Crime, international homicide statistics 2004; see note 66. Estimate for Taiwan from China (Taiwan), Republic of, Department of Statistics, Ministry of the Interior, 2000.

that have adopted Western models, such as Japan, Singapore, and Hong Kong. Also reporting a low homicide rate is China (2.2 per 100,000). Even if we take the data from this secretive state at face value, in the absence of time-series data we have no way of knowing whether it is best explained by millennia of centralized government or by the authoritarian nature of its current regime. Established autocracies (including many Islamic states) keep close tabs on their citizens and punish them surely and severely when they step out of line; that's why we call them "police states." Not surprisingly, they tend to have low rates of violent crime. But I can't resist an anecdote which suggests that China, like Europe, underwent a civilizing process over the long term. Elias noted that knife taboos, which accompanied the reduction of violence in Europe, have been taken one step further in China. For centuries in China, knives have been reserved for the chef in the kitchen, where he cuts the food into bite-sized pieces. Knives are banned from the dining table altogether. "The Europeans are barbarians," Elias quotes them as saying. "They eat with swords."[68]

What about the other parts of the world? The criminologist Gary LaFree and the sociologist Orlando Patterson have shown that the relationship between crime and democratization is an inverted U. Established democracies are relatively safe places, as are established autocracies, but emerging democracies and semi-democracies (also called anocracies) are often plagued by violent crime and vulnerable to civil war, which sometimes shade into each other.[69] The most crime-prone regions in the world today are Russia, sub-Saharan Africa, and parts of Latin America. Many of them have corrupt police forces and judicial systems which extort bribes out of criminals and victims alike and dole out protection to the highest bidder. Some, like Jamaica (33.7), Mexico (11.1), and Colombia (52.7), are racked by drug-funded militias that operate beyond the reach of the law. Over the past four decades, as drug trafficking has increased, their rates of homicide have soared. Others, like Russia (29.7) and South Africa (69), may have undergone decivilizing processes in the wake of the collapse of their former governments.

The decivilizing process has also racked many of the countries that switched from tribal ways to colonial rule and then suddenly to independence, such as those in sub-Saharan Africa and Papua New Guinea (15.2). In her article "From Spears to M-16s," the anthropologist Polly Wiessner examines the historical trajectory of violence among the Enga, a New Guinean tribal people. She begins with an excerpt from the field notes of an anthropologist who worked in their region in 1939:

> We were now in the heart of the Lai Valley, one of the most beautiful in New Guinea, if not in the world. Everywhere were fine well-laid out garden plots, mostly of sweet potato and groves of casuarinas. Well-cut and graded roads traversed the countryside, and small parks . . . dotted the landscape, which resembled a huge botanical garden.

She compares it to her own diary entry from 2004:

> The Lai Valley is a virtual wasteland—as the Enga say, "cared for by the birds, snakes, and rats." Houses are burned to ash, sweet potato gardens overgrown with weeds, and trees razed to jagged stumps. In the high forest, warfare rages on, fought by "Rambos" with shotguns and high-powered rifles taking the lives of many. By the roadside where markets bustled just a few years before, there is an eerie emptiness.[70]

The Enga were never what you could call a peaceable people. One of their tribes, the Mae Enga, are represented by a bar in figure 2–3: it shows that they killed each other in warfare at an annual rate of about 300 per 100,000, dwarfing the worst rates we have been discussing in this chapter. All the usual Hobbesian dynamics played out: rape and adultery, theft of pigs and land, insults, and of course, revenge, revenge, and more revenge. Still, the Enga were conscious of the waste of war, and some of the tribes took steps, intermittently successful, to contain it. For example, they developed Geneva-like norms that outlawed war crimes such as mutilating bodies or killing negotiators. And though they sometimes were drawn into destructive wars with other villages and tribes, they worked to control the violence within their own communities. Every human society is faced with a conflict of interest between the younger men, who seek dominance (and ultimately mating opportunities) for themselves, and the older men, who seek to minimize internecine damage within their extended families and clans. The Enga elders forced obstreperous young men into "bachelor cults," which encouraged them to control their vengeful impulses with the help of proverbs like "The blood of a man does not wash off easily" and "You live long if you plan the death of a pig, but not if you plan the death of a person."[71] And consonant with the other civilizing elements in their culture, they had norms of propriety and cleanliness, which Wiessner described to me in an e-mail:

> The Enga cover themselves with raincapes when they defecate, so as not to offend anybody, even the sun. For a man to stand by the road, turn his back and pee is unthinkably crude. They wash their hands meticulously before they cook food; they are extremely modest about covering genitals, and so on. Not so great with snot.

Most important, the Enga took well to the Pax Australiana beginning in the late 1930s. Over the span of two decades warfare plummeted, and many of the Enga were relieved to set aside violence to settle their disputes and "fight in courts" instead of on the battlefield.

When Papua New Guinea gained independence in 1975, violence among the Enga shot back up. Government officials doled out land and perks to their clansmen, provoking intimidation and revenge from the clans left in the cold. Young men left the bachelor cults for schools that prepared them for nonexistent jobs, then joined "Raskol" criminal gangs that were unrestrained by elders and the norms they had imposed. They were attracted by alcohol, drugs, nightclubs, gambling, and firearms (including M-16s and AK-47s) and went on rampages of rape, plunder, and arson, not unlike the knights of medieval Europe. The state was weak: its police were untrained and outgunned, and its corrupt bureaucracy was incapable of maintaining order. In short, the governance vacuum left by instant decolonization put the Papuans through a decivilizing process that left them with neither traditional norms nor modern third-party enforcement. Similar degenerations have occurred in other former colonies in the developing world, forming eddies in the global flow toward lower rates of homicide.

It's easy for a Westerner to think that violence in lawless parts of the world is intractable and permanent. But at various times in history communities have gotten so fed up with the bloodshed that they have launched what criminologists call a civilizing offensive.[72] Unlike the unplanned reductions in homicide that came about as a by-product of the consolidation of states and the promotion of commerce, a civilizing offensive is a deliberate effort by sectors of a community (often women, elders, or clergy) to tame the Rambos and Raskols and restore civilized life. Wiessner reports on a civilizing offensive in the Enga province in the 2000s.[73] Church leaders tried to lure young men from the thrill of gang life with exuberant sports, music, and prayer, and to substitute an ethic of forgiveness for the ethic of revenge. Tribal elders, using the cell phones that had been introduced in 2007, developed rapid response units to apprise one another of disputes and rush to the trouble spot before the fighting got out of control. They reined in the most uncontrollable firebrands in their own clans, sometimes with brutal public executions. Community governments were set up to restrict gambling, drinking, and prostitution. And a newer generation was receptive to these efforts, having seen that "the lives of Rambos are short and lead nowhere." Wiessner quantified the results: after having increased for decades, the number of killings declined significantly from the first half of the 2000s to the second. As we shall see, it was not the only time and place in which a civilizing offensive has paid off.

VIOLENCE IN THESE UNITED STATES

Violence is as American as cherry pie.

—H. Rap Brown

The Black Panther spokesman may have mixed up his fruits, but he did express a statistically valid generalization about the United States. Among Western democracies, the United States leaps out of the homicide statistics. Instead of clustering with kindred peoples like Britain, the Netherlands, and Germany, it hangs out with toughs like Albania and Uruguay, close to the median rate

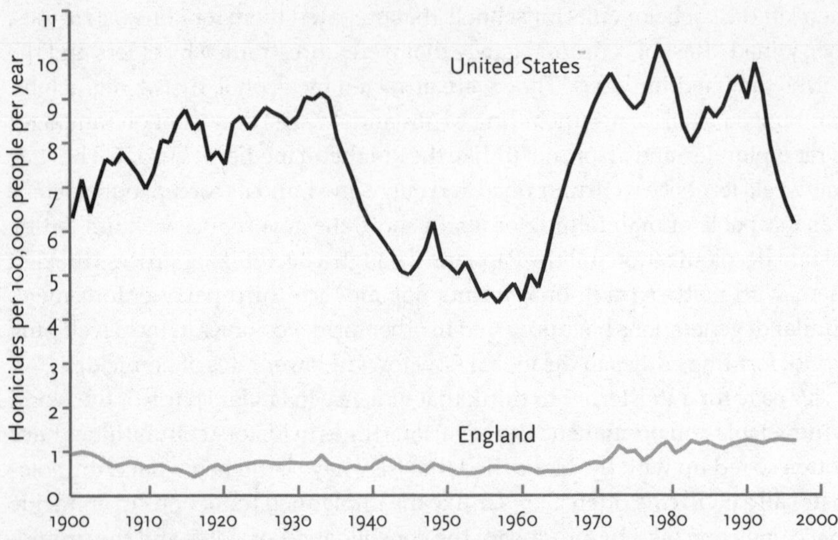

FIGURE 3–10. Homicide rates in the United States and England, 1900–2000

Sources: Graph from Monkkonen, 2001, pp. 171, 185–88; see also Zahn & McCall, 1999, p. 12. Note that Monkkonen's U.S. data differ slightly from the FBI Uniform Crime Reports data plotted in figure 3–18 and cited in this chapter.

for the entire world. Not only has the homicide rate for the United States not wafted down to the levels enjoyed by every European and Commonwealth democracy, but it showed no overall decline during the 20th century, as we see in figure 3–10. (For the 20th-century graphs, I will use a linear rather than a logarithmic scale.)

The American homicide rate crept up until 1933, nose-dived in the 1930s and 1940s, remained low in the 1950s, and then was launched skyward in 1962, bouncing around in the stratosphere in the 1970s and 1980s before returning to earth starting in 1992. The upsurge in the 1960s was shared with every other Western democracy, and I'll return to it in the next section. But why did the United States start the century with homicide rates so much higher than England's, and never close the gap? Could it be a counterexample to the generalization that countries with good governments and good economies enjoy a civilizing process that pushes their rate of violence downward? And if so, what is unusual about the United States? In newspaper commentaries one often reads pseudo-explanations like this: "Why is America more violent? It's our cultural predisposition to violence."[74] How can we find our way out of this logical circle? It's not just that America is gun-happy. Even if you subtract all the killings with firearms and count only the ones with rope, knives, lead pipes, wrenches, candlesticks, and so on, Americans commit murders at a higher rate than Europeans.[75]

Europeans have always thought America is uncivilized, but that is only partly true. A key to understanding American homicide is to remember that

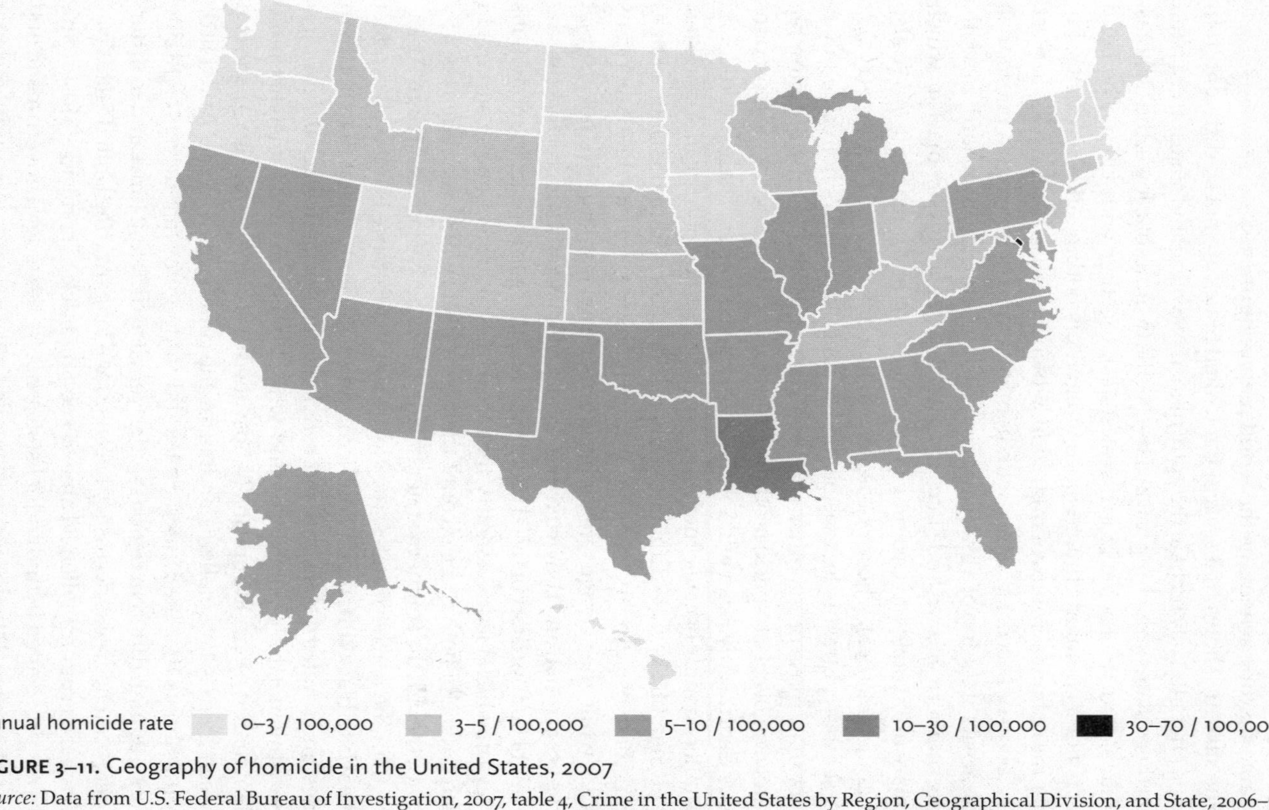

FIGURE 3–11. Geography of homicide in the United States, 2007

Source: Data from U.S. Federal Bureau of Investigation, 2007, table 4, Crime in the United States by Region, Geographical Division, and State, 2006–7.

the United States was originally a plural noun, as in *these United States*. When it comes to violence, the United States is not a country; it's three countries. Figure 3–11 is a map that plots the 2007 homicide rates for the fifty states, using the same shading scheme as the world map in figure 3–9.

The shading shows that *some* of the United States are not so different from Europe after all. They include the aptly named New England states, and a band of northern states stretching toward the Pacific (Minnesota, Iowa, the Dakotas, Montana, and the Pacific Northwest states), together with Utah. The band reflects not a common climate, since Oregon's is nothing like Vermont's, but rather the historical routes of migration, which tended to go from east to west. This ribbon of peaceable states, with homicide rates of less than 3 per 100,000 per year, sits at the top of a gradient of increasing homicide from north to south. At the southern end we find states like Arizona (7.4) and Alabama (8.9), which compare unfavorably to Uruguay (5.3), Jordan (6.9), and Grenada (4.9). We also find Louisiana (14.2), whose rate is close to that of Papua New Guinea (15.2).[76]

A second contrast is less visible on the map. Louisiana's homicide rate is higher than those of the other southern states, and the District of Columbia (a barely visible black speck) is off the scale at 30.8, in the range of the most dangerous Central American and southern African countries. These jurisdictions are outliers mainly because they have a high proportion of African Americans. The current black-white difference in homicide rates within the United States is stark. Between 1976 and 2005 the average homicide rate for white Americans was 4.8, while the average rate for black Americans was 36.9.[77] It's not just that blacks get arrested and convicted more often, which would suggest that the race gap might be an artifact of racial profiling. The same gap appears in anonymous surveys in which victims identify the race of their attackers, and in surveys in which people of both races recount their own history of violent offenses.[78] By the way, though the southern states have a higher percentage of African Americans than the northern states, the North-South difference is not a by-product of the white-black difference. Southern whites are more violent than northern whites, and southern blacks are more violent than northern blacks.[79]

So while northern Americans and white Americans are somewhat more violent than Western Europeans (whose median homicide rate is 1.4), the gap between them is far smaller than it is for the country as a whole. And a little digging shows that the United States did undergo a state-driven civilizing process, though different regions underwent it at different times and to different degrees. Digging is necessary because for a long time the United States was a backwards country when it came to keeping track of homicide. Most homicides are prosecuted by individual states, not by the federal government, and good nationwide statistics weren't compiled until the 1930s. Also, until recently "the United States" was a moving target. The lower forty-eight were not fully assembled until 1912, and many states were periodically infused with a shot of immigrants who changed the demographic profile until they coalesced in

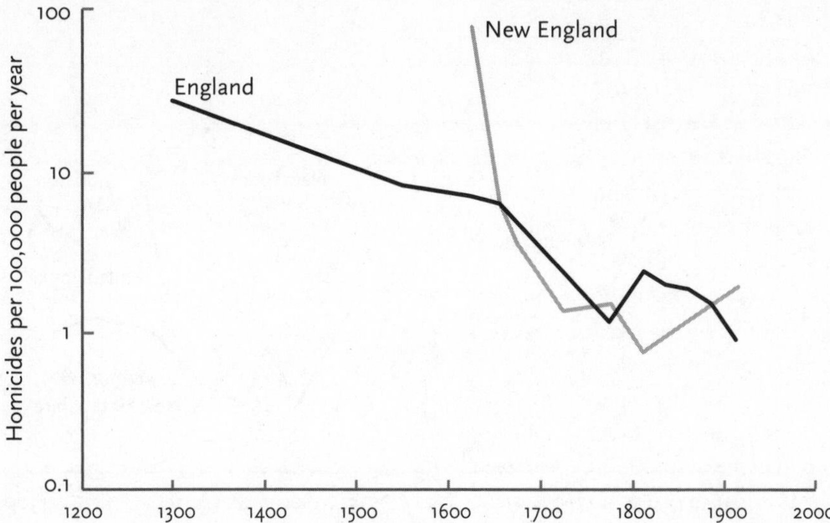

FIGURE 3–12. Homicide rates in England, 1300–1925, and New England, 1630–1914

Sources: Data for England: Eisner, 2003. Data for New England: 1630–37, Roth, 2001, p. 55; 1650–1800: Roth, 2001, p. 56; 1914: Roth, 2009, p. 388. Roth's estimates have been multiplied by 0.65 to convert the rate from per-adults to per-people; see Roth, 2009, p. 495. Data representing a range of years are plotted at the midpoint of the range.

the melting pot. For these reasons, historians of American violence have had to make do with shorter time series from smaller jurisdictions. In *American Homicide* Randolph Roth has recently assembled an enormous number of small-scale datasets for the three centuries of American history before the national statistics were compiled. Though most of the trends are roller coasters rather than toboggan runs, they do show how different parts of the country became civilized as the anarchy of the frontier gave way—in part—to state control.

Figure 3–12 superimposes Roth's data from New England on Eisner's compilation of homicide rates from England. The sky-high point for colonial New England represents Roth's Elias-friendly observation that "the era of frontier violence, during which the homicide rate stood at over 100 per 100,000 adults per year, ended in 1637 when English colonists and their Native American allies established their hegemony over New England." After this consolidation of state control, the curves for old England and New England coincide uncannily.

The rest of the Northeast also saw a plunge from triple-digit and high-double-digit homicide rates to the single digits typical of the world's countries today. The Dutch colony of New Netherland, with settlements from Connecticut to Delaware, saw a sharp decline in its early decades, from 68 to 15 per 100,000 (figure 3–13). But when the data resume in the 19th century, we start to see the United States diverging from the two mother countries. Though the more rural and ethnically homogeneous parts of New England (Vermont and New Hampshire) continue to hover in the peaceful basement beneath

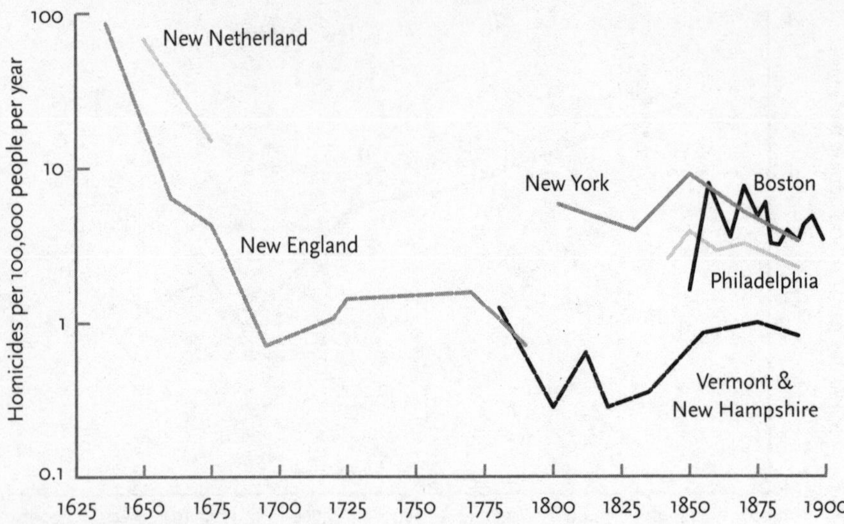

FIGURE 3–13. Homicide rates in the northeastern United States, 1636–1900

Sources: Data from Roth, 2009, whites only. New England: pp. 38, 62. New Netherland: pp. 38, 50. New York: p. 185. New Hampshire and Vermont: p. 184. Philadelphia: p. 185. Data representing a range of years are plotted at the midpoint of the range. Estimates have been multiplied by 0.65 to convert the rate from per-adults to per-people; see Roth, 2009, p. 495. Estimates for "unrelated adults" have been multiplied by 1.1 to make them approximately commensurable with estimates for all adults.

1 in 100,000, the city of Boston became more violent in the middle of the 19th century, overlapping cities in former New Netherland such as New York and Philadelphia.

The zigzags for the northeastern cities show two twists in the American version of the Civilizing Process. The middling altitude of these lines along the homicide scale, down from the ceiling but hovering well above the floor, suggests that the consolidation of a frontier under government control can bring the annual homicide rate down by an order of magnitude or so, from around 100 per 100,000 to around 10. But unlike what happened in Europe, where the momentum continued all the way down to the neighborhood of 1, in America the rate usually got stuck in the 5-to-15 range, where we find it today. Roth suggests that once an effective government has pacified the populace from the 100 to the 10 range, additional reductions depend on the degree to which people accept the legitimacy of the government, its laws, and the social order. Eisner, recall, made a similar observation about the Civilizing Process in Europe.

The other twist on the American version of the Civilizing Process is that in many of Roth's mini-datasets, violence *increased* in the middle decades of the 19th century.[80] The buildup and aftermath of the Civil War disrupted the social balance in many parts of the country, and the northeastern cities saw a wave of immigration from Ireland, which (as we have seen) lagged behind England in its homicide decline. Irish Americans in the 19th century, like African

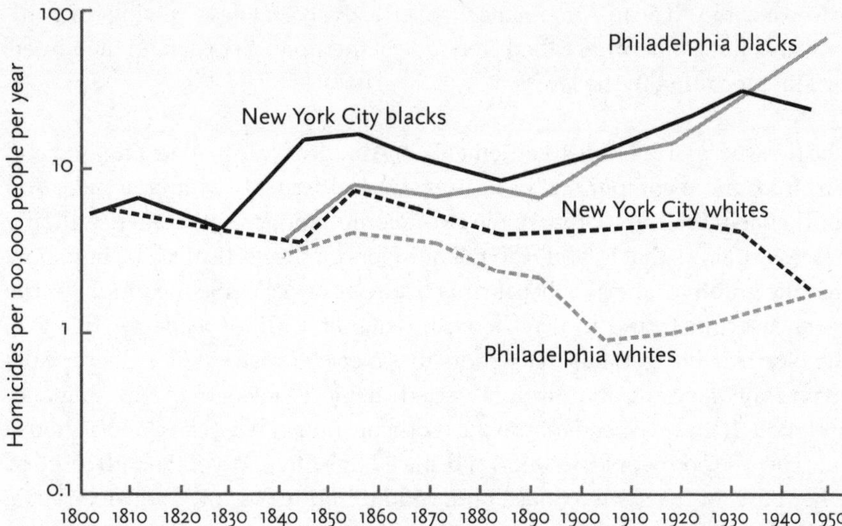

FIGURE 3–14. Homicide rates among blacks and whites in New York and Philadelphia, 1797–1952

Sources: New York 1797–1845: Roth, 2009, p. 195. New York 1856–85: Average of Roth, 2009, p. 195, and Gurr, 1989a, p. 39. New York 1905–53: Gurr, 1989a, p. 39. Philadelphia: 1842–94: Roth, 2009, p. 195. Philadelphia 1907–28: Lane, 1989, p. 72 (15-year averages). Philadelphia, 1950s: Gurr, 1989a, pp. 38–39. Roth's estimates have been multiplied by 0.65 to convert the rate from per-adults to per-people; see Roth, 2009, p. 495. His estimates for Philadelphia were, in addition, multiplied by 1.1 and 1.5 to compensate, respectively, for unrelated versus all victims and indictments versus homicides (Roth, 2009, p. 492). Data representing a range of years are plotted at the midpoint of the range.

Americans in the 20th, were more pugnacious than their neighbors, in large part because they and the police did not take each other seriously.[81] But in the second half of the 19th century police forces in American cities expanded, became more professional, and began to serve the criminal justice system rather than administering their own justice on the streets with their nightsticks. In major northern cities well into the 20th century, homicide rates for white Americans declined.[82]

But the second half of the 19th century also saw a fateful change. The graphs I have shown so far plot the rates for American whites. Figure 3–14 shows the rates for two cities in which black-on-black and white-on-white homicides can be distinguished. The graph reveals that the racial disparity in American homicide has not always been with us. In the northeastern cities, in New England, in the Midwest, and in Virginia, blacks and whites killed at similar rates throughout the first half of the 19th century. Then a gap opened up, and it widened even further in the 20th century, when homicides among African Americans skyrocketed, going from three times the white rate in New York in the 1850s to almost thirteen times the white rate a century later.[83] A probe into the causes, including economic and residential segregation, could fill another book. But one of them, as we have seen, is that communities of lower-income African Americans were effectively stateless, relying on a culture of honor (sometimes called "the code of the streets") to defend their interests rather calling in the law.[84]

The first successful English settlements in America were in New England and Virginia, and a comparison of figure 3–13 and figure 3–15 might make you think that in their first century the two colonies underwent similar civilizing processes. Until, that is, you read the numbers on the vertical axis. They show that the graph for the Northeast runs from 0.1 to 100, while the graph for the Southeast runs from 1 to 1,000, ten times higher. Unlike the black-white gap, the North-South gap has deep roots in American history. The Chesapeake colonies of Maryland and Virginia started out more violent than New England, and though they descended into the moderate range (between 1 and 10 homicides per 100,000 people per year) and stayed there for most of the 19th century, other parts of the settled South bounced around in the low 10-to-100 range, such as the Georgia plantation counties shown on the graph. Many remote and mountainous regions, such as the Georgia backcountry and Tennessee-Kentucky border, continued to float in the uncivilized 100s, some of them well into the 19th century.

Why has the South had such a long history of violence? The most sweeping answer is that the civilizing mission of government never penetrated the

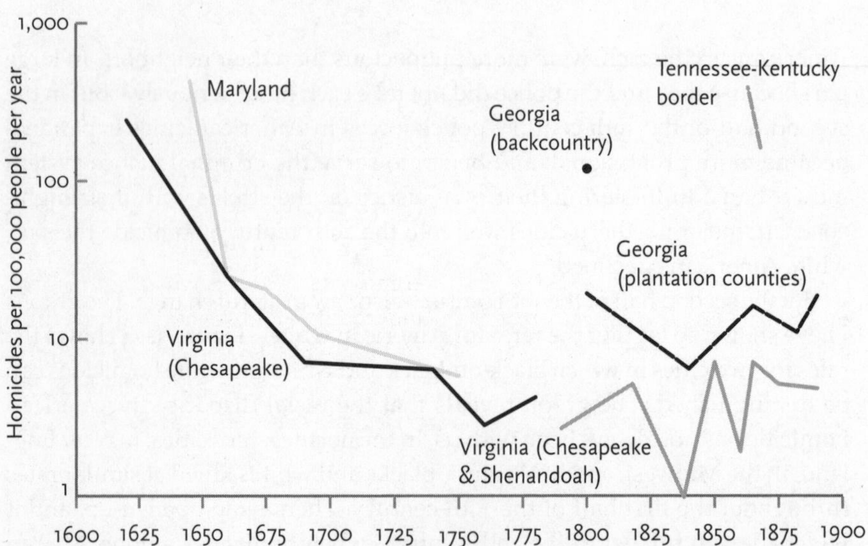

FIGURE 3–15. Homicide rates in the southeastern United States, 1620–1900

Sources: Data from Roth, 2009, whites only. Virginia (Chesapeake): pp. 39, 84. Virginia (Chesapeake and Shenandoah): p. 201. Georgia: p. 162. Tennessee-Kentucky: pp. 336–37. Zero value for Virginia, 1838, plotted as 1 since the log of 0 is undefined. Estimates have been multiplied by 0.65 to convert the rate from per-adults to per-people; see Roth, 2009, p. 495.

American South as deeply as it had the Northeast, to say nothing of Europe. The historian Pieter Spierenburg has provocatively suggested that "democracy came too early" to America.[85] In Europe, first the state disarmed the people and claimed a monopoly on violence, then the people took over the apparatus of the state. In America, the people took over the state before it had forced them to lay down their arms—which, as the Second Amendment famously affirms, they reserve the right to keep and bear. In other words Americans, and especially Americans in the South and West, never fully signed on to a social contract that would vest the government with a monopoly on the legitimate use of force. In much of American history, legitimate force was also wielded by posses, vigilantes, lynch mobs, company police, detective agencies, and Pinkertons, and even more often kept as a prerogative of the individual.

This power sharing, historians have noted, has always been sacred in the South. As Eric Monkkonen puts it, in the 19th century "the South had a deliberately weak state, eschewing things such as penitentiaries in favor of local, personal violence."[86] Homicides were treated lightly if the killing was deemed "reasonable," and "most killings . . . in the rural South were reasonable, in the sense that the victim had not done everything possible to escape from the killer, that the killing resulted from a personal dispute, or because the killer and victim were the kinds of people who kill each other."[87]

The South's reliance on self-help justice has long been a part of its mythology. It was instilled early in life, such as in the maternal advice given to the young Andrew Jackson (the dueling president who claimed to rattle with bullets when he walked): "Never . . . sue anyone for slander or assault or battery; always settle those cases yourself."[88] It was flaunted by pugnacious icons of the mountainous South like Daniel Boone and Davy Crockett, the "King of the Wild Frontier." It fueled the war between the prototypical feuding families, the Hatfields and McCoys of the Kentucky–West Virginia backcountry. And it not only swelled the homicide statistics for as long as they have been recorded, but has left its mark on the southern psyche today.[89]

Self-help justice depends on the credibility of one's prowess and resolve, and to this day the American South is marked by an obsession with credible deterrence, otherwise known as a culture of honor. The essence of a culture of honor is that it does not sanction predatory or instrumental violence, but only retaliation after an insult or other mistreatment. The psychologists Richard Nisbett and Dov Cohen have shown that this mindset continues to pervade southern laws, politics, and attitudes.[90] Southerners do not outkill northerners in homicides carried out during robberies, they found, only in those sparked by quarrels. In surveys, southerners do not endorse the use of violence in the abstract, but only to protect home and family. The laws of the southern states sanction this morality. They give a person wide latitude to kill in defense of self or property, put fewer restrictions on gun purchases, allow corporal punishment ("paddling") in schools, and specify the death penalty for murder, which their judicial systems are happy to carry out. Southern men and women are more likely to serve in the military, to study at military academies, and to take hawkish positions on foreign policy.

In a series of ingenious experiments, Nisbett and Cohen also showed that honor looms large in the behavior of individual southerners. In one study, they sent fake letters inquiring about jobs to companies all over the country. Half of them contained the following confession:

> There is one thing I must explain, because I feel I must be honest and I want no misunderstandings. I have been convicted of a felony, namely manslaughter. You will probably want an explanation for this before you send me an application, so I will provide it. I got into a fight with someone who was having an affair with my fiancée. I lived in a small town, and one night this person confronted me in front of my friends at the bar. He told everyone that he and my fiancée were sleeping together. He laughed at me to my face and asked me to step outside if I was man enough. I was young and didn't want to back down from a challenge in front of everyone. As we went into the alley, he started to attack me. He knocked me down, and he picked up a bottle. I could have run away and the judge said I should have, but my pride wouldn't let me. Instead I picked up a pipe that was laying in the alley and hit him with it. I didn't mean to kill him, but he died a few hours later at the hospital. I realize that what I did was wrong.

The other half contained a similar paragraph in which the applicant confessed to a felony conviction for grand theft auto, which, he said, he had foolishly committed to help support his wife and young children. In response to the letter confessing to the honor killing, companies based in the South and West were more likely than those in the North to send the letter-writer a job application, and their replies were warmer in tone. For example, the owner of one southern store apologized that she had no jobs available at the time and added:

> As for your problem of the past, anyone could probably be in the situation you were in. It was just an unfortunate incident that shouldn't be held against you. Your honesty shows that you are sincere. . . . I wish you the best of luck for your future. You have a positive attitude and a willingness to work. Those are the qualities that businesses look for in an employee. Once you get settled, if you are near here, please stop in and see us.[91]

No such warmth came from companies based in the North, nor from any company when the letter confessed to auto theft. Indeed, northern companies were more forgiving of the auto theft than the honor killing; the southern and western companies were more forgiving of the honor killing than the auto theft.

Nisbett and Cohen also captured the southern culture of honor in the lab. Their subjects were not bubbas from the bayous but affluent students at the University of Michigan who had lived in the South for at least six years. Students were recruited for a psychology experiment on "limited response time conditions on certain facets of human judgment" (a bit of gobbledygook to hide the real purpose of the study). In the hallway on their way to the lab, the students had to pass by an accomplice of the experimenter who was filing papers in a cabinet. In half of the cases, when the student brushed past the accomplice, he slammed the drawer shut and muttered, "Asshole." Then the experimenter (who was kept in the dark as to whether the student had been insulted) welcomed the student into the lab, observed his demeanor, gave him a questionnaire, and drew a blood sample. The students from the northern states, they found, laughed off the insult and behaved no differently from the control group who had entered without incident. But the insulted students from the southern states walked in fuming. They reported lower self-esteem in a questionnaire, and their blood samples showed elevated levels of testosterone and of cortisol, a stress hormone. They behaved more dominantly toward the experimenter and shook his hand more firmly, and when approaching another accomplice in the narrow hallway on their way out, they refused to step aside to let him pass.[92]

Is there an exogenous cause that might explain why the South rather than the North developed a culture of honor? Certainly the brutality needed to maintain a slave economy might have been a factor, but the most violent parts of the South were backcountry regions that never depended on plantation slavery (see figure 3–15). Nisbett and Cohen were influenced by David Hackett Fisher's *Albion's Seed*, a history of the British colonization of the United States, and zeroed in on the origins of the first colonists from different parts of Europe. The northern states were settled by Puritan, Quaker, Dutch, and German farmers, but the interior South was largely settled by Scots-Irish, many of them sheepherders, who hailed from the mountainous periphery of the British Isles beyond the reach of the central government. Herding, Nisbett and Cohen suggest, may have been an exogenous cause of the culture of honor. Not only does a herder's wealth lie in stealable physical assets, but those assets have feet and can be led away in an eyeblink, far more easily than land can be stolen out from under a farmer. Herders all over the world cultivate a hair trigger for violent retaliation. Nisbett and Cohen suggest that the Scots-Irish brought their culture of honor with them and kept it alive when they took up herding in the South's mountainous frontier. Though contemporary southerners are no longer shepherds, cultural mores can persist long after the ecological circumstances that gave rise to them are gone, and to this day southerners behave as if they have to be tough enough to deter livestock rustlers.

The herding hypothesis requires that people cling to an occupational strategy for centuries after it has become dysfunctional, but the more general theory of a culture of honor does not depend on that assumption. People often take up herding in mountainous areas because it's hard to grow crops on mountains, and mountainous areas are often anarchic because they are the hardest regions for a state to conquer, pacify, and administer. The immediate trigger for self-help justice, then, is anarchy, not herding itself. Recall that the ranchers of Shasta County have herded cattle for more than a century, yet when one of them suffers a minor loss of cattle or property, he is expected to "lump it," not lash out with violence to defend his honor. Also, a recent study that compared southern counties in their rates of violence and their suitability for herding found no correlation when other variables were controlled.[93]

So it's sufficient to assume that settlers from the remote parts of Britain ended up in the remote parts of the South, and that both regions were lawless for a long time, fostering a culture of honor. We still have to explain why their culture of honor is so self-sustaining. After all, a functioning criminal justice system has been in place in southern states for some time now. Perhaps honor has staying power because the first man who dares to abjure it would be heaped with contempt for cowardice and treated as an easy mark.

The American West, even more than the American South, was a zone of anarchy until well into the 20th century. The cliché of Hollywood westerns that "the nearest sheriff is ninety miles away" was the reality in millions of square miles of territory, and the result was the other cliché of Hollywood westerns, ever-present violence. Nabokov's Humbert Humbert, drinking in American popular culture during his cross-country escape with Lolita, savors the "ox-stunning fisticuffs" of the cowboy movies:

> There was the mahogany landscape, the florid-faced, blue-eyed roughriders, the prim pretty schoolteacher arriving in Roaring Gulch, the rearing horse, the spectacular stampede, the pistol thrust through the shivered windowpane, the stupendous fist fight, the crashing mountain of dusty old-fashioned furniture, the table used as a weapon, the timely somersault, the pinned hand still groping for the dropped bowie knife, the grunt, the sweet crash of fist against chin, the kick in the belly, the flying tackle; and immediately after a plethora of pain that would have hospitalized a Hercules, nothing to show but the rather becoming bruise on the bronzed cheek of the warmed-up hero embracing his gorgeous frontier bride.[94]

In *Violent Land*, the historian David Courtwright shows that the Hollywood horse operas were accurate in the levels of violence they depicted, if not in their romanticized image of cowboys. The life of a cowboy alternated between dangerous, backbreaking work and payday binges of drinking, gambling, whoring, and brawling. "For the cowboy to become a symbol of the American experience required an act of moral surgery. The cowboy as mounted protector and risk-taker was remembered. The cowboy as dismounted drunk sleeping it off on the manure pile behind the saloon was forgotten."[95]

In the American Wild West, annual homicide rates were fifty to several hundred times higher than those of eastern cities and midwestern farming regions: 50 per 100,000 in Abilene, Kansas, 100 in Dodge City, 229 in Fort Griffin, Texas, and 1,500 in Ellsworth, Kansas.[96] The causes were right out of Hobbes. The criminal justice system was underfunded, inept, and often corrupt. "In 1877," notes Courtwright, "some five thousand men were on the wanted list in Texas alone, not a very encouraging sign of efficiency in law enforcement."[97] Self-help justice was the only way to deter horse thieves, cattle rustlers, highwaymen, and other brigands. The guarantor of its deterrent threat was a reputation for resolve that had to be defended at all costs, epitomized by the epitaph on a Colorado grave marker: "He Called Bill Smith a Liar."[98] One eyewitness described the casus belli of a fight that broke out during a card game in the caboose of a cattle train. One man remarked, "I don't like to play cards with a dirty deck." A cowboy from a rival company thought he said "dirty neck," and when the gunsmoke cleared, one man was dead and three wounded.[99]

It wasn't just cowboy country that developed in Hobbesian anarchy; so did parts of the West settled by miners, railroad workers, loggers, and itinerant laborers. Here is an assertion of property rights found attached to a post during the California Gold Rush of 1849:

> All and everybody, this is my claim, fifty feet on the gulch, cordin to Clear Creek District Law, backed up by shotgun amendments. . . . Any person found trespassing on this claim will be persecuted to the full extent of the law. This is no monkey tale butt I will assert my rites at the pint of the sicks shirter if leagally necessary so taik head and good warning.[100]

Courtwright cites an average annual homicide rate at the time of 83 per 100,000 and points to "an abundance of other evidence that Gold Rush California was a brutal and unforgiving place. Camp Names were mimetic: Gouge Eye, Murderers Bar, Cut-Throat Gulch, Graveyard Flat. There was a Hangtown, a Helltown, a Whiskeytown, and a Gomorrah, though, interestingly, no Sodom."[101] Mining boom towns elsewhere in the West also had annual homicide rates in the upper gallery: 87 per 100,000 in Aurora, Nevada; 105 in Leadville, Colorado; 116 in Bodie, California; and a whopping 24,000 (almost one in four) in Benton, Wyoming.

In figure 3–16 I've plotted the trajectory of western violence, using snapshots of annual homicide rates that Roth provides for a given region at two or more times. The curve for California shows a rise around the 1849 Gold Rush, but then, together with that of the southwestern states, it shows the signature of the Civilizing Process: a greater-than-tenfold decline in homicide rates, from the range of 100 to 200 per 100,000 people to the range of 5 to 15 (though, as in

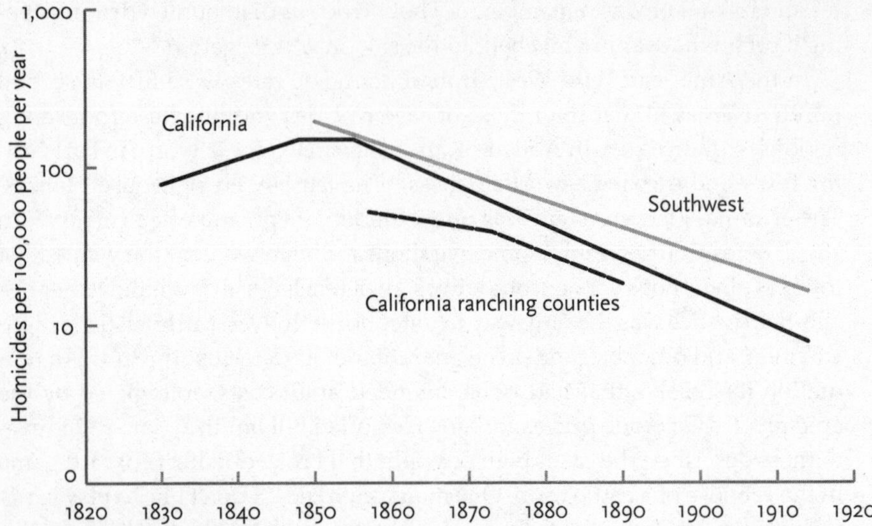

FIGURE 3–16. Homicide rates in the southwestern United States and California, 1830–1914

Sources: Data from Roth, 2009, whites only. California (estimates): pp. 183, 360, 404. California ranching counties: p. 355. Southwest, 1850 (estimate): p. 354. Southwest, 1914 (Arizona, Nevada, and New Mexico): p. 404. Estimates have been multiplied by 0.65 to convert the rate from per-adults to per-people; see Roth, 2009, p. 495.

the South, the rates did not continue to fall into the 1s and 2s of Europe and New England). I've included the decline for the California ranching counties, like those studied by Ellickson, to show how their current norm-governed coexistence came only after a long period of lawless violence.

So at least five of the major regions of the United States—the Northeast, the middle Atlantic states, the coastal South, California, and the Southwest—underwent civilizing processes, but at different times and to different extents. The decline of violence in the American West lagged that in the East by two centuries and spanned the famous 1890 announcement of the closing of the American frontier, which symbolically marked the end of anarchy in the United States.

Anarchy was not the only cause of the mayhem in the Wild West and other violent zones in expanding America such as laborers' camps, hobo villages, and Chinatowns (as in, "Forget it, Jake; it's Chinatown"). Courtwright shows that the wildness was exacerbated by a combination of demography and evolutionary psychology. These regions were peopled by young, single men who had fled impoverished farms and urban ghettos to seek their fortune in the harsh frontier. The one great universal in the study of violence is that most of it is committed by fifteen-to-thirty-year-old men.[102] Not only are males the more competitive sex in most mammalian species, but with *Homo sapiens* a man's position in the pecking order is secured by reputation, an investment with a lifelong payout that must be started early in adulthood.

The violence of men, though, is modulated by a slider: they can allocate their energy along a continuum from competing with other men for access to women to wooing the women themselves and investing in their children, a continuum that biologists sometimes call "cads versus dads."[103] In a social ecosystem populated mainly by men, the optimal allocation for an individual man is at the "cad" end, because attaining alpha status is necessary to beat away the competition and a prerequisite to getting within wooing distance of the scarce women. Also favoring cads is a milieu in which women are more plentiful but some of the men can monopolize them. In these settings it can pay to gamble with one's life because, as Daly and Wilson have noted, "any creature that is recognizably on track toward complete reproductive failure must somehow expend effort, often at risk of death, to try to improve its present life trajectory."[104] The ecosystem that selects for the "dad" setting is one with an equal number of men and women and monogamous matchups between them. In those circumstances, violent competition offers the men no reproductive advantages, but it does threaten them with a big disadvantage: a man cannot support his children if he is dead.

Another biological contribution to frontier violence was neurobiological rather than sociobiological, namely the ubiquity of liquor. Alcohol interferes with synaptic transmission throughout the cerebrum, especially in the prefrontal cortex (see figure 8–3), the region responsible for self-control. An inebriated brain is less inhibited sexually, verbally, and physically, giving us idioms like *beer goggles*, *roaring drunk*, and *Dutch courage*. Many studies have shown that people with a tendency toward violence are more likely to act on it when they are under the influence of alcohol.[105]

The West was eventually tamed not just by flinty-eyed marshals and hanging judges but by an influx of women.[106] The Hollywood westerns' "prim pretty schoolteacher[s] arriving in Roaring Gulch" captures a historical reality. Nature abhors a lopsided sex ratio, and women in eastern cities and farms eventually flowed westward along the sexual concentration gradient. Widows, spinsters, and young single women sought their fortunes in the marriage market, encouraged by the lonely men themselves and by municipal and commercial officials who became increasingly exasperated by the degeneracy of their western hellholes. As the women arrived, they used their bargaining position to transform the West into an environment better suited to their interests. They insisted that the men abandon their brawling and boozing for marriage and family life, encouraged the building of schools and churches, and shut down saloons, brothels, gambling dens, and other rivals for the men's attention. Churches, with their coed membership, Sunday morning discipline, and glorification of norms on temperance, added institutional muscle to the women's civilizing offensive. Today we guffaw at the Women's Christian

Temperance Union (with its ax-wielding tavern terrorist Carrie Nation) and at the Salvation Army, whose anthem, according to the satire, includes the lines "We never eat cookies 'cause cookies have yeast / And one little bite turns a man to a beast." But the early feminists of the temperance movement were responding to the very real catastrophe of alcohol-fueled bloodbaths in male-dominated enclaves.

The idea that young men are civilized by women and marriage may seem as corny as Kansas in August, but it has become a commonplace of modern criminology. A famous study that tracked a thousand low-income Boston teenagers for forty-five years discovered that two factors predicted whether a delinquent would go on to avoid a life of crime: getting a stable job, and marrying a woman he cared about and supporting her and her children. The effect of marriage was substantial: three-quarters of the bachelors, but only a third of the husbands, went on to commit more crimes. This difference alone cannot tell us whether marriage keeps men away from crime or career criminals are less likely to get married, but the sociologists Robert Sampson, John Laub, and Christopher Wimer have shown that marriage really does seem to be a pacifying cause. When they held constant all the factors that *typically* push men into marriage, they found that *actually* getting married made a man less likely to commit crimes immediately thereafter.[107] The causal pathway has been pithily explained by Johnny Cash: Because you're mine, I walk the line.

An appreciation of the Civilizing Process in the American West and rural South helps to make sense of the American political landscape today. Many northern and coastal intellectuals are puzzled by the culture of their red state compatriots, with their embrace of guns, capital punishment, small government, evangelical Christianity, "family values," and sexual propriety. Their opposite numbers are just as baffled by the blue staters' timidity toward criminals and foreign enemies, their trust in government, their intellectualized secularism, and their tolerance of licentiousness. This so-called culture war, I suspect, is the product of a history in which white America took two different paths to civilization. The North is an extension of Europe and continued the court- and commerce-driven Civilizing Process that had been gathering momentum since the Middle Ages. The South and West preserved the culture of honor that sprang up in the anarchic parts of the growing country, balanced by their own civilizing forces of churches, families, and temperance.

DECIVILIZATION IN THE 1960s

But when you talk about destruction, don't you know that you can count me out . . . in.

—John Lennon, "Revolution 1"

For all the lags and mismatches between the historical trajectories of the United States and Europe, they did undergo one trend in synchrony: their rates of violence did a U-turn in the 1960s.[108] Figures 3–1 through 3–4 show that European countries underwent a bounce in homicide rates that brought them back to levels they had said goodbye to a century before. And figure 3–10 shows that in the 1960s the homicide rate in America went through the roof. After a three-decade free fall that spanned the Great Depression, World War II, and the Cold War, Americans multiplied their homicide rate by more than two and a half, from a low of 4.0 in 1957 to a high of 10.2 in 1980.[109] The upsurge included every other category of major crime as well, including rape, assault, robbery, and theft, and lasted (with ups and downs) for three decades. The cities got particularly dangerous, especially New York, which became a symbol of the new criminality. Though the surge in violence affected all the races and both genders, it was most dramatic among black men, whose annual homicide rate had shot up by the mid-1980s to 72 per 100,000.[110]

The flood of violence from the 1960s through the 1980s reshaped American culture, the political scene, and everyday life. Mugger jokes became a staple of comedians, with mentions of Central Park getting an instant laugh as a well-known death trap. New Yorkers imprisoned themselves in their apartments with batteries of latches and deadbolts, including the popular "police lock," a steel bar with one end anchored in the floor and the other propped up against the door. The section of downtown Boston not far from where I now live was called the Combat Zone because of its endemic muggings and stabbings. Urbanites quit other American cities in droves, leaving burned-out cores surrounded by rings of suburbs, exurbs, and gated communities. Books, movies, and television series used intractable urban violence as their backdrop, including *Little Murders, Taxi Driver, The Warriors, Escape from New York, Fort Apache the Bronx, Hill Street Blues*, and *Bonfire of the Vanities*. Women enrolled in self-defense courses to learn how to walk with a defiant gait, to use their keys, pencils, and spike heels as weapons, and to execute karate chops or jujitsu throws to overpower an attacker, role-played by a volunteer in a Michelin-man-tire suit. Red-bereted Guardian Angels patrolled the parks and the transit system, and in 1984 Bernhard Goetz, a mild-mannered engineer, became a folk hero for shooting four young muggers in a New York subway car. A fear of crime helped elect decades of conservative politicians, including Richard Nixon in 1968 with his "Law and Order" platform (overshadowing the Vietnam War as a campaign issue), George H. W. Bush in 1988 with his insinuation that Michael Dukakis, as governor of Massachusetts, had approved a prison furlough program that had released a rapist, and many senators and congressmen who promised to "get tough on crime." Though the popular reaction was overblown—far more people are killed every year in car accidents than in homicides, especially among those who don't get into arguments with young men in bars—the sense that violent crime had multiplied was not a figment of their imaginations.

The rebounding of violence in the 1960s defied every expectation. The decade was a time of unprecedented economic growth, nearly full employment, levels of economic equality for which people today are nostalgic, historic racial progress, and the blossoming of government social programs, not to mention medical advances that made victims more likely to survive being shot or knifed. Social theorists in 1962 would have happily bet that these fortunate conditions would lead to a continuing era of low crime. And they would have lost their shirts.

Why did the Western world embark on a three-decade binge of crime from which it has never fully recovered? This is one of several local reversals of the long-term decline of violence that I will examine in this book. If the analysis is on the right track, then the historical changes I have been invoking to explain the decline should have gone into reverse at the time of the surges.

An obvious place to look is demographics. The 1940s and 1950s, when crime rates hugged the floor, were the great age of marriage. Americans got married in numbers not seen before or since, which removed men from the streets and planted them in suburbs.[111] One consequence was a bust in violence. But the other was a boom in babies. The first baby boomers, born in 1946, entered their crime-prone years in 1961; the ones born in the peak year, 1954, entered in 1969. A natural conclusion is that the crime boom was an echo of the baby boom. Unfortunately, the numbers don't add up. If it were just a matter of there being more teenagers and twenty-somethings who were committing crimes at their usual rates, the increase in crime from 1960 to 1970 would have been 13 percent, not 135 percent.[112] Young men weren't simply more numerous than their predecessors; they were more violent too.

Many criminologists have concluded that the 1960s crime surge cannot be explained by the usual socioeconomic variables but was caused in large part by a change in cultural norms. Of course, to escape the logical circle in which people are said to be violent because they live in a violent culture, it's necessary to identify an exogenous cause for the cultural change. The political scientist James Q. Wilson has argued that demographics were an important trigger after all, not because of the absolute numbers of young people but because of their relative numbers. He makes the point by commenting on a quotation from the demographer Norman Ryder:

> "There is a perennial invasion of barbarians who must somehow be civilized and turned into contributors to fulfillment of the various functions requisite to societal survival." That "invasion" is the coming of age of a new generation of young people. Every society copes with this enormous socialization process more or less successfully, but occasionally that process is literally swamped by a quantitative discontinuity in the number of persons involved. . . . In 1950 and still in 1960 the "invading army" (those aged fourteen to twenty-four) were outnumbered three to one by the size of the "defending army" (those aged twenty-five to sixty-four). By 1970 the ranks

of the former had grown so fast that they were only outnumbered two to one by the latter, a state of affairs that had not existed since 1910.[113]

Subsequent analyses showed that this explanation is not, by itself, satisfactory. Age cohorts that are larger than their predecessors do not, in general, commit more crimes.[114] But I think Wilson was on to something when he linked the 1960s crime boom to a kind of intergenerational decivilizing process. In many ways the new generation tried to push back against the eight-century movement described by Norbert Elias.

The baby boomers were unusual (I know, we baby boomers are always saying we're unusual) in sharing an emboldening sense of solidarity, as if their generation were an ethnic group or a nation. (A decade later it was pretentiously referred to as "Woodstock Nation.") Not only did they outnumber the older generation, but thanks to new electronic media, they felt the strength of their numbers. The baby boomers were the first generation to grow up with television. And television, especially in the three-network era, allowed them to know that other baby boomers were sharing their experiences, and to know that the others knew that they knew. This common knowledge, as economists and logicians call it, gave rise to a horizontal web of solidarity that cut across the vertical ties to parents and authorities that had formerly isolated young people from one another and forced them to kowtow to their elders.[115] Much like a disaffected population that feels its strength only when it assembles at a rally, baby boomers saw other young people like themselves in the audience of *The Ed Sullivan Show* grooving on the Rolling Stones and knew that every other young person in America was grooving at the same time, and knew that the others knew that they knew.

The baby boomers were bonded by another new technology of solidarity, first marketed by an obscure Japanese company called Sony: the transistor radio. The parents of today who complain about the iPods and cell phones that are soldered onto the ears of teenagers forget that their own parents made the same complaint about them and their transistor radios. I can still remember the thrill of tuning in to signals from New York radio stations bouncing off the late-night ionosphere into my bedroom in Montreal, listening to Motown and Dylan and the British invasion and psychedelia and feeling that something was happening here, but Mr. Jones didn't know what it was.

A sense of solidarity among fifteen-to-thirty-year-olds would be a menace to civilized society even in the best of times. But this decivilizing process was magnified by a trend that had been gathering momentum throughout the 20th century. The sociologist Cas Wouters, a translator and intellectual heir of Elias, has argued that after the European Civilizing Process had run its course, it was superseded by an *informalizing process*. The Civilizing Process had been a flow of norms and manners from the upper classes downward.

But as Western countries became more democratic, the upper classes became increasingly discredited as moral paragons, and hierarchies of taste and manners were leveled. The informalization affected the way people dressed, as they abandoned hats, gloves, ties, and dresses for casual sportswear. It affected the language, as people started to address their friends with first names instead of *Mr.* and *Mrs.* and *Miss.* And it could be seen in countless other ways in which speech and demeanor became less mannered and more spontaneous.[116] The stuffy high-society lady, like the Margaret Dumont character in the Marx Brothers movies, became a target of ridicule rather than emulation.

After having been steadily beaten down by the informalizing process, the elites then suffered a second hit to their legitimacy. The civil rights movement had exposed a moral blot on the American establishment, and as critics shone a light on other parts of society, more stains came into view. Among them were the threat of a nuclear holocaust, the pervasiveness of poverty, the mistreatment of Native Americans, the many illiberal military interventions, particularly the Vietnam War, and later the despoliation of the environment and the oppression of women and homosexuals. The stated enemy of the Western establishment, Marxism, gained prestige as it made inroads in third-world "liberation" movements, and it was increasingly embraced by bohemians and fashionable intellectuals. Surveys of popular opinion from the 1960s through the 1990s showed a plummeting of trust in every social institution.[117]

The leveling of hierarchies and the harsh scrutiny of the power structure were unstoppable and in many ways desirable. But one of the side effects was to undermine the prestige of aristocratic and bourgeois lifestyles that had, over the course of several centuries, become less violent than those of the working class and underclass. Instead of values trickling down from the court, they bubbled up from the street, a process that was later called "proletarianization" and "defining deviancy down."[118]

These currents pushed against the civilizing tide in ways that were celebrated in the era's popular culture. The backsliding, to be sure, did not originate in the two prime movers of Elias's Civilizing Process. Government control did not retreat into anarchy, as it had in the American West and in newly independent third-world countries, nor did an economy based on commerce and specialization give way to feudalism and barter. But the next step in Elias's sequence—the psychological change toward greater self-control and interdependence—came under steady assault in the counterculture of the generation that came of age in the 1960s.

A prime target was the inner governor of civilized behavior, self-control. Spontaneity, self-expression, and a defiance of inhibitions became cardinal virtues. "If it feels good, do it," commanded a popular lapel button. *Do It* was the title of a book by the political agitator Jerry Rubin. "Do It 'Til You're Satisfied (Whatever It Is)" was the refrain of a popular song by BT Express. The body was elevated over the mind: Keith Richards boasted, "Rock and roll is music from the neck downwards." And adolescence was elevated over adulthood: "Don't trust anyone over thirty," advised the agitator Abbie Hoffman; "Hope I die before I get old," sang The Who in "My Generation." Sanity was denigrated, and psychosis romanticized, in movies such as *A Fine Madness, One Flew Over the Cuckoo's Nest, King of Hearts*, and *Outrageous*. And then of course there were the drugs.

Another target of the counterculture was the ideal that individuals should be embedded in webs of dependency that obligate them to other people in stable economies and organizations. If you wanted an image that contradicted this ideal as starkly as possible, it might be a rolling stone. Originally from a song by Muddy Waters, the image resonated with the times so well that it lent itself to *three* icons of the culture: the rock group, the magazine, and the famous song by Bob Dylan (in which he taunts an upper-class woman who has become homeless). "Tune in, turn on, drop out," the motto of onetime Harvard psychology instructor Timothy Leary, became a watchword of the psychedelia movement. The idea of coordinating one's interests with others in a job was treated as selling out. As Dylan put it:

> Well, I try my best
> To be just like I am,
> But everybody wants you
> To be just like them.
> They say sing while you slave and I just get bored.
> I ain't gonna work on Maggie's farm no more.

Elias had written that the demands of self-control and the embedding of the self into webs of interdependence were historically reflected in the development of timekeeping devices and a consciousness of time: "This is why tendencies in the individual so often rebel against social time as represented by his or her super-ego, and why so many people come into conflict with themselves when they wish to be punctual."[119] In the opening scene of the 1969 movie *Easy Rider*, Dennis Hopper and Peter Fonda conspicuously toss their wristwatches into the dirt before setting off on their motorcycles to find America. That same year, the first album by the band Chicago (when they were known as the Chicago Transit Authority) contained the lyrics "Does anybody really know what time it is? Does anybody really care? If so I can't imagine why." All this made sense to me when I was sixteen, and so I discarded my own Timex. When my grandmother saw my naked wrist, she was incredulous: "How can you be a mensch without a zager?" She ran to a drawer and pulled out a Seiko she had bought during a visit to the 1970 World's Fair in Osaka. I have it to this day.

Together with self-control and societal connectedness, a third ideal came under attack: marriage and family life, which had done so much to domesticate male violence in the preceding decades. The idea that a man and a woman should devote their energies to a monogamous relationship in which they raise their children in a safe environment became a target of howling ridicule. That life was now the soulless, conformist, consumerist, materialist, ticky-tacky, plastic, white-bread, *Ozzie and Harriet* suburban wasteland.

I don't remember anyone in the 1960s blowing his nose into a tablecloth, but popular culture did celebrate the flouting of standards of cleanliness, propriety, and sexual continence. The hippies were popularly perceived as unwashed and malodorous, which in my experience was a calumny. But there's no disputing that they rejected conventional standards of grooming, and an enduring image from Woodstock was of naked concertgoers frolicking in the mud. One could trace the reversal of conventions of propriety on album covers alone (figure 3–17). There was *The Who Sell Out*, with a sauce-dribbling Roger Daltrey immersed in a bath of baked beans; the Beatles' *Yesterday and Today*, with the lovable moptops adorned with chunks of raw meat and decapitated dolls (quickly recalled); the Rolling Stones' *Beggars Banquet*, with a photo of a filthy public toilet (originally censored); and *Who's Next*, in which the four musicians are shown zipping up their flies while walking away from a urine-spattered wall. The flouting of propriety extended to famous live performances, as when Jimi Hendrix pretended to copulate with his amplifier at the Monterey Pop Festival.

Throwing away your wristwatch or bathing in baked beans is, of course, a far cry from committing actual violence. The 1960s were supposed to be the era of peace and love, and so they were in some respects. But the glorification of dissoluteness shaded into an indulgence of violence and then into violence itself. At the end of every concert, The Who famously smashed their instruments to smithereens, which could be dismissed as harmless theater were it not for the fact that drummer Keith Moon also destroyed dozens of hotel rooms, partly deafened Pete Townshend by detonating his drums onstage, beat up his wife, girlfriend, and daughter, threatened to injure the hands of a keyboardist of the Faces for dating his ex-wife, and accidentally killed his

FIGURE 3–17. Flouting conventions of cleanliness and propriety in the 1960s

bodyguard by running over him with his car before dying himself in 1978 of the customary drug overdose.

Personal violence was sometimes celebrated in song, as if it were just another form of antiestablishment protest. In 1964 Martha Reeves and the Vandellas sang "Summer's here and the time is right for dancing in the street." Four years later the Rolling Stones replied that the time was right for *fighting* in the street. As part of their "satanic majesty" and "sympathy for the devil," the Stones had a theatrical ten-minute song, "Midnight Rambler," which acted out a rape-murder by the Boston Strangler, ending with the lines "I'm gonna smash down on your plate-glass window / Put a fist, put a fist through your steel-plated door / I'll . . . stick . . . my . . . knife . . . right . . . down . . . your . . . throat!" The affectation of rock musicians to treat every thug and serial killer as a dashing "rebel" or "outlaw" was satirized in *This Is Spinal Tap* when the band speaks of their plans to write a rock musical based on the life of Jack the Ripper. (Chorus: "You're a naughty one, Saucy Jack!")

Less than four months after Woodstock, the Rolling Stones held a free concert at the Altamont Speedway in California, for which the organizers had hired the Hell's Angels, romanticized at the time as "outlaw brothers of the counterculture," to provide security. The atmosphere at the concert (and perhaps the 1960s) is captured in this description from Wikipedia:

> A huge circus performer weighing over 350 pounds and hallucinating on LSD stripped naked and ran berserk through the crowd toward the stage, knocking guests in all directions, prompting a group of Angels to leap from the stage and club him unconscious. [*citation needed*]

No citation is needed for what happened next, since it was captured in the documentary *Gimme Shelter*. A Hell's Angel beat up the lead singer of Jefferson Airplane onstage, Mick Jagger ineffectually tried to calm the increasingly obstreperous mob, and a young man in the audience, apparently after pulling a gun, was stabbed to death by another Angel.

When rock music burst onto the scene in the 1950s, politicians and clergymen vilified it for corrupting morals and encouraging lawlessness. (An amusing video reel of fulminating fogies can be seen in Cleveland's Rock and Roll Hall of Fame and Museum.) Do we now have to—gulp—admit they were right? Can we connect the values of 1960s popular culture to the actual rise in violent crimes that accompanied them? Not directly, of course. Correlation is not causation, and a third factor, the pushback against the values of the Civilizing Process, presumably caused both the changes in popular culture and the increase in violent behavior. Also, the overwhelming majority of baby boomers committed no violence whatsoever. Still, attitudes and popular culture surely reinforce each other, and at the margins, where susceptible individuals and subcultures can be buffeted one way or another, there are plausible causal arrows from the decivilizing mindset to the facilitation of actual violence.

One of them was a self-handicapping of the criminal justice Leviathan. Though rock musicians seldom influence public policy directly, writers and intellectuals do, and they got caught up in the zeitgeist and began to rationalize the new licentiousness. Marxism made violent class conflict seem like a route to a better world. Influential thinkers like Herbert Marcuse and Paul Goodman tried to merge Marxism or anarchism with a new interpretation of Freud that connected sexual and emotional repression to political repression and championed a release from inhibitions as part of the revolutionary struggle. Troublemakers were increasingly seen as rebels and nonconformists, or as victims of racism, poverty, and bad parenting. Graffiti vandals were now "artists," thieves were "class warriors," and neighborhood hooligans were "community leaders." Many smart people, intoxicated by radical chic, did incredibly stupid things. Graduates of elite universities built bombs to be set off at army social functions, or drove getaway cars while "radicals" shot guards during armed robberies. New York intellectuals were conned by Marxobabble-spouting psychopaths into lobbying for their release from prison.[120]

In the interval between the onset of the sexual revolution of the early 1960s and the rise of feminism in the 1970s, the control of women's sexuality was seen as a perquisite of sophisticated men. Boasts of sexual coercion and jealous violence appeared in popular novels and films and in the lyrics of rock songs such as the Beatles' "Run for Your Life," Neil Young's "Down by the River," Jimi Hendrix's "Hey Joe," and Ronnie Hawkins's "Who Do You Love?"[121] It was even rationalized in "revolutionary" political writings, such as Eldridge Cleaver's bestselling 1968 memoir *Soul on Ice*, in which the Black Panther leader wrote:

> Rape was an insurrectionary act. It delighted me that I was defying and trampling upon the white man's law, upon his system of values, and that I was defiling his women—and this point, I believe, was the most satisfying to me because I was very resentful over the historical fact of how the white man has used the black woman. I felt I was getting revenge.[122]

Somehow the interests of the women who were defiled in this insurrectionary act never figured into his political principles, nor into the critical reaction to the book (*New York Times*: "Brilliant and revealing"; *The Nation:* "A remarkable book . . . beautifully written"; *Atlantic Monthly:* "An intelligent and turbulent and passionate and eloquent man").[123]

As the rationalizations for criminality caught the attention of judges and legislators, they became increasingly reluctant to put miscreants behind bars. Though the civil liberties reform of the era did not lead to nearly as many vicious criminals "going free on a technicality" as the *Dirty Harry* movies would suggest, law enforcement was indeed retreating as the crime rate was advancing. In the United States from 1962 to 1979, the likelihood that a crime would lead to an arrest dropped from 0.32 to 0.18, the likelihood that an arrest would lead to imprisonment dropped from 0.32 to 0.14, and the likelihood that a crime would lead to imprisonment fell from 0.10 to 0.02, a factor of five.[124]

Even more calamitous than the return of hoodlums to the street was the mutual disengagement between law enforcement and communities, and the resulting deterioration of neighborhood life. Offenses against civil order like vagrancy, loitering, and panhandling were decriminalized, and minor crimes like vandalism, graffiti-spraying, turnstile-jumping, and urinating in public fell off the police radar screens.[125] Thanks to intermittently effective antipsychotic drugs and a change in attitudes toward deviance, the wards of mental hospitals were emptied, which multiplied the ranks of the homeless. Shopkeepers and citizens with a stake in the neighborhood, who otherwise would have kept an eye out for local misbehavior, eventually surrendered to the vandals, panhandlers, and muggers and retreated to the suburbs.

The 1960s decivilizing process affected the choices of individuals as well as policymakers. Many young men decided that they ain't gonna work on Maggie's farm no more and, instead of pursuing a respectable family life, hung out in all-male packs that spawned the familiar cycle of competition for dominance, insult or minor aggression, and violent retaliation. The sexual revolution, which provided men with plentiful sexual opportunities without the responsibilities of marriage, added to this dubious freedom. Some men tried to get a piece of the lucrative trade in contraband drugs, in which self-help justice is the only way to enforce property rights. (The cutthroat market in crack cocaine in the late 1980s had a particularly low barrier for entry because doses of the drug could be sold in small amounts, and the resulting infusion of teenage crack dealers probably contributed to the 25 percent increase in the homicide rate between 1985 and 1991.) On top of the violence that accompanies any market in contraband, the drugs themselves, together with good old-fashioned alcohol, lowered inhibitions and sent sparks onto the tinder.

The decivilizing effects hit African American communities particularly hard. They started out with the historical disadvantages of second-class citizenship, which left many young people teetering between respectable and underclass lifestyles just when the new antiestablishment forces were pushing in the wrong direction. They could count on even less protection from the criminal justice system than white Americans because of the combination of old racism among the police and the new indulgence by the judicial system toward crime, of which they were disproportionately the victims.[126] Mistrust of the criminal justice system turned into cynicism and sometimes paranoia, making self-help justice seem the only alternative.[127]

On top of these strikes came a feature of African American family life first pointed out by the sociologist Daniel Patrick Moynihan in his famous 1965 report, *The Negro Family: The Case for National Action*, for which he was initially vilified but eventually vindicated.[128] A large proportion (today a majority) of black children are born out of wedlock, and many grow up without fathers. This trend, already visible in the early 1960s, may have been multiplied by the sexual revolution and yet again by perverse welfare incentives that encouraged young women to "marry the state" instead of the fathers of their children.[129] Though I am skeptical of theories of parental influence that say that fatherless boys grow up violent because they lack a role model or paternal discipline (Moynihan himself, for example, grew up without a father), widespread fatherlessness can lead to violence for a different reason.[130] All those young men who aren't bringing up their children are hanging out with one another competing for dominance instead. The mixture was as combustible in the inner city as it had been in the cowboy saloons and mining camps of the Wild West, this time not because there were no women around but because the women lacked the bargaining power to force the men into a civilized lifestyle.

RECIVILIZATION IN THE 1990s

It would be a mistake to think of the 1960s crime boom as undoing the decline of violence in the West, or as a sign that historical trends in violence are cyclical, yo-yoing up and down from one era to the next. The annual homicide rate in the United States at its recent worst—10.2 per 100,000 in 1980—was a quarter of the rate for Western Europe in 1450, a tenth of the rate of the traditional Inuit, and a fiftieth of the average rate in nonstate societies (see figure 3–4).

And even that number turned out to be a high-water mark, not a regular occurrence or a sign of things to come. In 1992 a strange thing happened. The homicide rate went down by almost 10 percent from the year before, and it continued to sink for another seven years, hitting 5.7 in 1999, the lowest it had been since 1966.[131] Even more shockingly, the rate stayed put for another seven years and then drooped even further, from 5.7 in 2006 to 4.8 in 2010. The upper line in figure 3–18 plots the American homicide trend since 1950, including the new lowland we have reached in the 21st century.

The graph also shows the trend for Canada since 1961. Canadians kill at less than a third of the rate of Americans, partly because in the 19th century the Mounties got to the western frontier before the settlers and spared them from having to cultivate a violent code of honor. Despite this difference, the ups and downs of the Canadian homicide rate parallel those of their neighbor to the south (with a correlation coefficient between 1961 and 2009 of 0.85), and it sank almost as much in the 1990s: 35 percent, compared to the American decline of 42 percent.[132]

The parallel trajectory of Canada and the United States is one of many surprises in the great crime decline of the 1990s. The two countries differed in their economic trends and in their policies of criminal justice, yet they enjoyed similar drops in violence. So did most of the countries of Western Europe.[133]

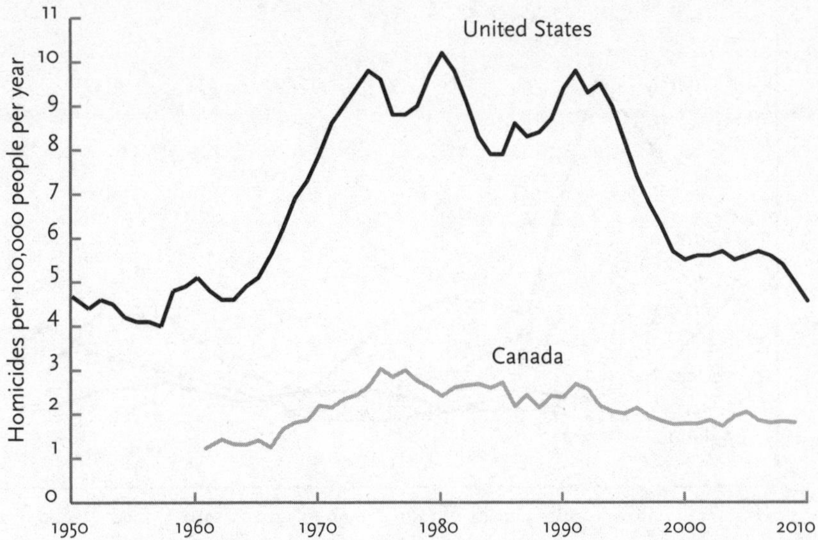

FIGURE 3–18. Homicide rates in the United States, 1950–2010, and Canada, 1961–2009
Sources: Data for United States are from the FBI Uniform Crime Reports 1950–2010: U.S. Bureau of Justice Statistics, 2009; U.S. Federal Bureau of Investigation, 2010b, 2011; Fox & Zawitz, 2007. Data for Canada, 1961–2007: Statistics Canada, 2008. Data for Canada, 2008: Statistics Canada, 2010. Data for Canada, 2009: K. Harris, "Canada's crime rate falls," *Toronto Sun*, Jul. 20, 2010.

Figure 3–19 plots the homicide rates of five major European countries over the past century, showing the historical trajectory we have been tracking: a long-term decline that lasted until the 1960s, an uptick that began in that tumultuous decade, and the recent return to more peaceable rates. Every major Western European country showed a decline, and though it looked for a while as if England and Ireland would be the exceptions, in the 2000s their rates dropped as well.

Not only did people cut down on killing, but they refrained from inflicting other kinds of harm. In the United States the rates of every category of major crime dropped by about half, including rape, robbery, aggravated assault, burglary, larceny, and even auto theft.[134] The effects were visible not just in the statistics but in the fabric of everyday life. Tourists and young urban professionals recolonized American downtowns, and crime receded as a major issue from presidential campaigns.

None of the experts had predicted it. Even as the decline was under way, the standard opinion was that the rise in crime that had begun in the 1960s would even get worse. In a 1995 essay James Q. Wilson wrote:

> Just beyond the horizon, there lurks a cloud that the winds will soon bring over us. The population will start getting younger again. By the end of this decade there will be a million more people between the ages of fourteen and

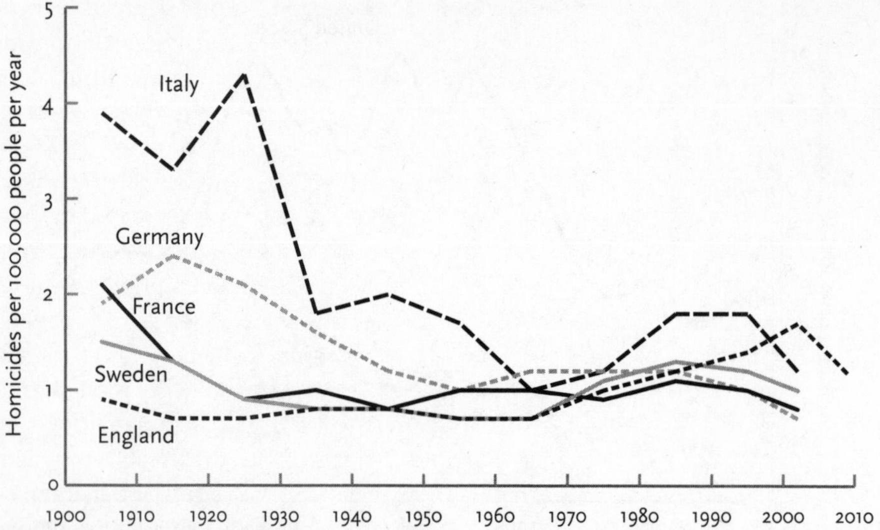

FIGURE 3–19. Homicide rates in five Western European countries, 1900–2009
Sources: Data from Eisner, 2008, except England, 2009, which is from Walker et al., 2009; population estimate from U.K. Office for National Statistics, 2009.

> seventeen than there are now. This extra million will be half male. Six percent of them will become high-rate, repeat offenders—30,000 more young muggers, killers, and thieves than we have now. Get ready.[135]

The cloud beyond the horizon was joined by purple prose from other talking heads on crime. James Alan Fox predicted a "blood bath" by 2005, a crime wave that would "get so bad that it [would] make 1995 look like the good old days."[136] John DiIulio warned of more than a quarter of a million new "super-predators on the streets" by 2010 who would make "the Bloods and the Crips look tame by comparison."[137] In 1991 the former editor of the *Times* of London predicted that "by the year 2000, New York could be a Gotham City without Batman."[138]

As legendary New York mayor Fiorello La Guardia might have said, "When I make a mistake, it's a beaut!" (Wilson was a good sport about it, remarking, "Social scientists should never try to predict the future; they have enough trouble predicting the past.") The mistake of the murder mavens was to have put too much faith in the most recent demographic trends. The crack-fueled violence bubble of the late 1980s involved large numbers of teenagers, and the population of teenagers was set to grow in the 1990s as an echo of the baby boom. But the *overall* crime-prone cohort, which includes twenty-somethings as well as teenagers, actually fell in the 1990s.[139] Even this corrected statistic, though, cannot explain the decline of crime in that decade. The age distribution of a population changes slowly, as each demographic pig makes its way through the population python. But in the 1990s the crime rate lurched downward for seven straight years and promptly parked itself at its new bottom for another nine. As with the takeoff of crime in the 1960s, changes in the *rate* of violence for each age cohort swamped the effect of the *size* of those cohorts.

The other usual suspect in explaining crime trends, the economy, did little better in explaining this one. Though unemployment went down in the United States in the 1990s, it went up in Canada, yet violent crime decreased in Canada as well.[140] France and Germany also saw unemployment go up while violence went down, whereas Ireland and the U.K. saw unemployment go down while violence went up.[141] This is not as surprising as it first appears, since criminologists have long known that unemployment rates don't correlate well with rates of violent crime.[142] (They do correlate somewhat with rates of property crime.) Indeed, in the three years after the financial meltdown of 2008, which caused the worst economic downturn since the Great Depression, the American homicide rate *fell* by another 14 percent, leading the criminologist David Kennedy to explain to a reporter, "The idea that everyone has ingrained into them—that as the economy goes south, crime has to get worse—is wrong. It was never right to begin with."[143]

Among economic measures, inequality is generally a better predictor of violence than unemployment.[144] But the Gini coefficient, the standard index of income inequality, actually *rose* in the United States from 1990 to 2000, while crime was falling, and it had hit its low point in 1968, when crime was soaring.[145] The problem with invoking inequality to explain changes in violence is that while it correlates with violence across states and countries, it does not correlate with violence over time within a state or country, possibly because the real cause of the differences is not inequality per se but stable features of a state's governance or culture that affect both inequality and violence.[146] (For example, in unequal societies, poor neighborhoods are left without police protection and can become zones of violent anarchy.)

Yet another false lead may be found in the kind of punditry that tries to link social trends to the "national mood" following current events. The terrorist attacks of September 11, 2001, led to enormous political, economic, and emotional turmoil, but the homicide rate did not budge in response.

The 1990s crime decline inspired one of the stranger hypotheses in the study of violence. When I told people I was writing a book on the historical decline of violence, I was repeatedly informed that the phenomenon had already been solved. Rates of violence have come down, they explained to me, because after abortion was legalized by the 1973 *Roe v. Wade* U.S. Supreme Court decision, the unwanted children who would ordinarily have grown up to be criminals were not born in the first place, because their begrudging or unfit mothers had had abortions instead. I first heard of this theory in 2001 when it was proposed by the economists John Donohue and Steven Levitt, but it seemed too cute to be true.[147] Any hypothesis that comes out of left field to explain a massive social trend with a single overlooked event will almost certainly turn out to be wrong, even if it has some data supporting it at the time. But Levitt, together with the journalist Stephen Dubner, popularized the theory in their bestseller *Freakonomics*, and now a large proportion of the public believes that crime went down in the 1990s because women aborted their crime-fated fetuses in the 1970s.

To be fair, Levitt went on to argue that *Roe v. Wade* was just one of four causes of the crime decline, and he has presented sophisticated correlational statistics in support of the connection. For example, he showed that the handful of states that legalized abortion before 1973 were the first to see their crime rates go down.[148] But these statistics compare the two ends of a long, hypothetical, and tenuous causal chain—the availability of legal abortion as the first link and the decline in crime two decades later as the last—and ignore all the links in between. The links include the assumptions that legal abortion causes fewer unwanted children, that unwanted children are more likely to become criminals, and that the first abortion-culled generation was the one spearheading the 1990s crime decline. But there are other explanations for the overall correlation (for example, that the large liberal states that first legalized abortion were also the first states to see the rise and fall of the crack epidemic), and the intermediate links have turned out to be fragile or nonexistent.[149]

To begin with, the freakonomics theory assumes that women were just as likely to have conceived unwanted children before and after 1973, and that the only difference was whether the children were born. But once abortion was legalized, couples may have treated it as a backup method of birth control and may have engaged in more unprotected sex. If the women conceived more unwanted children in the first place, the option of aborting more of them could leave the proportion of unwanted children the same. In fact, the proportion of unwanted children could even have *increased* if women were emboldened by the abortion option to have more unprotected sex in the heat of the moment, but then procrastinated or had second thoughts once they were pregnant. That may help explain why in the years since 1973 the proportion of children born to women in the most vulnerable categories—poor, single, teenage, and African American—did not decrease, as the freakonomics theory would predict. It increased, and by a lot.[150]

What about differences among individual women *within* a crime-prone population? Here the freakonomics theory would seem to get things backwards. Among women who are accidentally pregnant and unprepared to raise a child, the ones who terminate their pregnancies are likely to be forward-thinking, realistic, and disciplined, whereas the ones who carry the child to term are more likely to be fatalistic, disorganized, or immaturely focused on the thought of a cute baby rather than an unruly adolescent. Several studies have borne this out.[151] Young pregnant women who opt for abortions get better grades, are less likely to be on welfare, and are more likely to finish school than their counterparts who have miscarriages or carry their pregnancies to term. The availability of abortion thus may have led to a generation that is *more* prone to crime because it weeded out just the children who, whether through genes or environment, were most likely to exercise maturity and self-control.

Also, the freakonomists' theory about the psychological causes of crime comes right out of "Gee, Officer Krupke," when a gang member says of his parents, "They didn't wanna have me, but somehow I was had. Leapin' lizards! That's why I'm so bad!" And it is about as plausible. Though unwanted children may grow up to commit more crimes, it is more likely that women in crime-prone environments have more unwanted children than that unwantedness causes criminal behavior directly. In studies that pit the effects of parenting against the effects of the children's peer environment, holding genes constant, the peer environment almost always wins.[152]

Finally, if easy abortion after 1973 sculpted a more crime-averse generation, the crime decline should have begun with the youngest group and then crept up the age brackets as they got older. The sixteen-year-olds of 1993, for example (who were born in 1977, when abortions were in full swing), should have committed fewer crimes than the sixteen-year-olds of 1983 (who were born in 1967, when abortion was illegal). By similar logic, the twenty-two-year-olds of 1993 should have remained violent, because they were born in pre-*Roe* 1971. Only in the late 1990s, when the first post-*Roe* generation reached their twenties, should the twenty-something age bracket have become less violent. In fact, the opposite happened. When the first post-*Roe* generation came of age in the late 1980s and early 1990s, they did not tug the homicide statistics downward; they indulged in an unprecedented spree of mayhem. The crime decline began when the *older* cohorts, born well before *Roe*, laid down their guns and knives, and from them the lower homicide rates trickled down the age scale.[153]

So how *can* we explain the recent crime decline? Many social scientists have tried, and the best that they can come up with is that the decline had multiple causes, and no one can be certain what they were, because too many things happened at once.[154] Nonetheless, I think two overarching explanations are plausible. The first is that the Leviathan got bigger, smarter, and more effective. The second is that the Civilizing Process, which the counterculture had tried to reverse in the 1960s, was restored to its forward direction. Indeed, it seems to have entered a new phase.

By the early 1990s, Americans had gotten sick of the muggers, vandals, and drive-by shootings, and the country beefed up the criminal justice system in several ways. The most effective was also the crudest: putting more men behind bars for longer stretches of time. The rate of imprisonment in the United States was pretty much flat from the 1920s to the early 1960s, and it even declined a bit until the early 1970s. But then it shot up almost fivefold, and today more than two million Americans are in jail, the highest incarceration rate on the planet.[155] That works out to three-quarters of a percent of *the entire population,* and a much larger percentage of young men, especially African Americans.[156] The American imprisonment binge was set off in the 1980s by several developments. Among them were mandatory sentencing laws (such as California's "Three Strikes and You're Out"), a prison-building boom (in which rural communities that had formerly shouted "Not in my backyard!" now welcomed the economic stimulus), and the War on Drugs (which criminalized possession of small amounts of cocaine and other controlled substances).

Unlike the more gimmicky theories of the crime decline, massive imprisonment is almost certain to lower crime rates because the mechanism by which it operates has so few moving parts. Imprisonment physically removes the most crime-prone individuals from the streets, incapacitating them and subtracting the crimes they would have committed from the statistics. Incarceration is especially effective when a small number of individuals commit a large number of crimes. A classic study of criminal records in Philadelphia, for example, found that 6 percent of the young male population committed more than half the offenses.[157] The people who commit the most crimes expose themselves to the most opportunities to get caught, and so they are the ones most likely to be skimmed off and sent to jail. Moreover, people who commit violent crimes get into trouble in other ways, because they tend to favor instant gratification over long-term benefits. They are more likely to drop out of school, quit work, get into accidents, provoke fights, engage in petty theft and vandalism, and abuse alcohol and drugs.[158] A regime that trawls for drug users or other petty delinquents will net a certain number of violent people as bycatch, further thinning the ranks of the violent people who remain on the streets.

Incarceration can also reduce violence by the familiar but less direct route of deterrence. An ex-convict might think twice about committing another crime once he gets out of jail, and the people who know about him might think twice about following in his footsteps. But proving that incarceration deters people (as opposed to incapacitating them) is easier said than done, because the statistics at any time are inherently stacked against it. The regions with the highest rates of crime will throw the most people in jail, creating the illusion that imprisonment *increases* crime rather than decreasing it. But with suitable ingenuity (for example, correlating increases in imprisonment at one time with decreases in crime at a later time, or seeing if a court order to reduce the prison population leads to a subsequent increase in crime), the deterrence effect can be tested. Analyses by Levitt and other statisticians of crime suggest that deterrence works.[159] Those who prefer real-world experiments to sophisticated statistics may take note of the Montreal police strike of 1969. Within hours of the gendarmes abandoning their posts, that famously safe city was hit with six bank robberies, twelve arsons, a hundred lootings, and two homicides before the Mounties were called in to restore order.[160]

But the case that the incarceration boom led to the crime decline is far from watertight.[161] For one thing, the prison bulge began in the 1980s, but violence did not decline until a decade later. For another, Canada did not go on an imprisonment binge, but its violence rate went down too. These facts don't disprove the theory that imprisonment mattered, but they force it to make additional assumptions, such as that the effect of imprisonment builds up over time, reaches a critical mass, and spills over national borders.

Mass incarceration, even if it does lower violence, introduces problems of its own. Once the most violent individuals have been locked up, imprisoning more of them rapidly reaches a point of diminishing returns, because each additional prisoner become less and less dangerous, and pulling them off the streets makes a smaller and smaller dent in the violence rate.[162] Also, since people tend to get less violent as they get older, keeping men in prison beyond a certain point does little to reduce crime. For all these reasons, there is an optimum rate of incarceration. It's unlikely that the American criminal justice system will find it, because electoral politics keep ratcheting the incarceration rate upward, particularly in jurisdictions in which judges are elected rather than appointed. Any candidate who suggests that too many people are going to jail for too long will be targeted in an opponent's television ads as "soft on crime" and booted out of office. The result is that the United States imprisons far more people than it should, with disproportionate harm falling on African American communities who have been stripped of large numbers of men.

A second way in which Leviathan became more effective in the 1990s was a ballooning of the police.[163] In a stroke of political genius, President Bill Clinton undercut his conservative opponents in 1994 by supporting legislation that promised to add 100,000 officers to the nation's police forces. Additional cops not only nab more criminals but are more noticeable by their presence, deterring people from committing crimes in the first place. And many of the police earned back their old nickname *flatfoots* by walking a beat and keeping an eye on the neighborhood rather than sitting in cars and awaiting a radio call before speeding to a crime scene. In some cities, like Boston, the police were accompanied by parole officers who knew the worst troublemakers individually and had the power to have them rearrested for the slightest infraction.[164] In New York, police headquarters tracked neighborhood crime reports obsessively and held captains' feet to the fire if the crime rate in their precinct started to drift upward.[165] The visibility of the police was multiplied by a mandate to go after nuisance crimes like graffiti, littering, aggressive panhandling, drinking liquor or urinating in public, and extorting cash from drivers at stoplights after a cursory wipe of their windshield with a filthy squeegee. The rationale, originally articulated by James Q. Wilson and George Kelling in their famous Broken Windows theory, was that an orderly environment serves as a reminder that police and residents are dedicated to keeping the peace, whereas a vandalized and unruly one is a signal that no one is in charge.[166]

Did these bigger and smarter police forces actually drive down crime? Research on this question is the usual social science rat's nest of confounded variables, but the big picture suggests that the answer is "yes, in part," even if we can't pinpoint which of the innovations did the trick. Not only do several analyses suggest that *something* in the new policing reduced crime, but the jurisdiction that spent the most effort in perfecting its police, New York City, showed the greatest reduction of all. Once the epitome of urban rot, New York is now one of America's safest cities, having enjoyed a slide in the crime rate that was twice the national average and that continued in the 2000s after the decline in the rest of the country had run out of steam.[167] As the criminologist Franklin Zimring put it in *The Great American Crime Decline*, "If the combination of more cops, more aggressive policing, and management reforms did account for as much as a 35% crime decrease (half the [U.S.] total), it would be by far the biggest crime prevention achievement in the recorded history of metropolitan policing."[168]

What about Broken Windows policing in particular? Most academics hate the Broken Windows theory because it seems to vindicate the view of social conservatives (including former New York mayor Rudy Giuliani) that violence rates are driven by law and order rather than by "root causes" such as poverty and racism. And it has been almost impossible to prove that Broken Windows works with the usual correlational methods because the cities that implemented the policy also hired a lot of police at the same time.[169] But an ingenious set of studies, recently reported in *Science*, has supported the theory using the gold standard of science: an experimental manipulation and a matched control group.

Three Dutch researchers picked an alley in Groningen where Netherlanders park their bicycles and attached an advertising flyer to the handlebars of each one. The commuters had to detach the flyer before they could ride their bikes, but the researchers had removed all the wastebaskets, so they either had to carry the flyer home or toss it on the ground. Above the bicycles was a prominent sign prohibiting graffiti and a wall that the experimenters had either covered in graffiti (the experimental condition) or left clean (the control condition). When the commuters were in the presence of the illegal graffiti, twice as many of them threw the flyer on the ground—exactly what the Broken Windows theory predicted. In other studies, people littered more when they saw unreturned shopping carts strewn about, and when they heard illegal firecrackers being set off in the distance. It wasn't just harmless infractions like littering that were affected. In another experiment, passersby were tempted by an addressed envelope protruding from a mailbox with a five-euro bill visible inside it. When the mailbox was covered in graffiti or surrounded by litter, a quarter of the passersby stole it; when the mailbox was clean, half that many did. The researchers argued that an orderly environment fosters a sense of responsibility not so much by deterrence (since Groningen police rarely penalize litterers) as by the signaling of a social norm: This is the kind of place where people obey the rules.[170]

Ultimately, we must look to a change in norms to understand the 1990s crime bust, just as it was a change in norms that helped explain the boom three decades earlier. Though policing reforms almost certainly contributed to the headlong decline in American violence, particularly in New York, remember that Canada and Western Europe saw declines as well (albeit not by the same amount), and they did not bulk up their prisons or police to nearly the same degree. Even some of the hardest-headed crime statisticians have thrown up their hands and concluded that much of the explanation must lie in difficult-to-quantify cultural and psychological changes.[171]

The Great Crime Decline of the 1990s was part of a change in sensibilities that can fairly be called a recivilizing process. To start with, some of the goofier ideas of the 1960s had lost their appeal. The collapse of communism and a recognition of its economic and humanitarian catastrophes took the romance out of revolutionary violence and cast doubt on the wisdom of redistributing wealth at the point of a gun. A greater awareness of rape and sexual abuse made the ethos "If it feels good, do it" seem repugnant rather than liberating. And the sheer depravity of inner-city violence—toddlers struck by bullets in drive-by shootings, church funerals of teenagers invaded by knife-wielding gangs—could no longer be explained away as an understandable response to racism or poverty.

The result was a wave of civilizing offensives. As we will see in chapter 7, one positive legacy of the 1960s was the revolutions in civil rights, women's rights, children's rights, and gay rights, which began to consolidate power in the 1990s as the baby boomers became the establishment. Their targeting of rape, battering, hate crimes, gay-bashing, and child abuse reframed law-and-order from a reactionary cause to a progressive one, and their efforts to make the home, workplace, schools, and streets safer for vulnerable groups (as in the feminist "Take Back the Night" protests) made these environments safer for everyone.

One of the most impressive civilizing offensives of the 1990s came from African American communities, who set themselves the task of recivilizing their young men. As with the pacifying of the American West a century before, much of the moral energy came from women and the church.[172] In Boston a team of clergymen led by Ray Hammond, Eugene Rivers, and Jeffrey Brown worked in partnership with the police and social service agencies to clamp down on gang violence.[173] They leveraged their knowledge of local communities to identify the most dangerous gang members and put them on notice that the police and community were watching them, sometimes in meetings with their gangs, sometimes in meetings with their mothers and grandmothers. Community leaders also disrupted cycles of revenge by converging on any gang member who had recently been aggrieved and leaning on him to forswear vengeance. The interventions were effective not just because of the threat of an arrest but because the external pressure provided the men with an "out" that allowed them to back off without losing face, much as two brawling men may accede to being pulled apart by weaker interceders. These efforts contributed to the "Boston Miracle" of the 1990s in which the homicide rate dropped fivefold; it has remained low, with some fluctuations, ever since.[174]

The police and courts, for their part, have been redirecting their use of criminal punishment from brute deterrence and incapacitation to the second stage of a civilizing process, enhancing the perceived legitimacy of government force. When a criminal justice system works properly, it's not because rational actors know that Big Brother is watching them 24/7 and will swoop down and impose a cost that will cancel any ill-gotten gain. No democracy has the resources or the will to turn society into that kind of Skinner box. Only a *sample* of criminal behavior can ever be detected and punished, and the sampling should be fair enough that citizens perceive the entire regime to be legitimate. A key legitimator is the perception that the system is set up in such a way that a person, and more importantly the person's adversaries, face a constant chance of being punished if they break the law, so that they all may internalize inhibitions against predation, preemptive attack, and vigilante retribution. But in many American jurisdictions, the punishments had been so capricious as to appear like misfortunes coming out of the blue rather than predictable consequences of proscribed behavior. Offenders skipped probation hearings or failed drug tests with impunity, and they saw their peers get away with it as well, but then one day, in what they experienced as a stroke of bad luck, they were suddenly sent away for years.

But now judges, working with police and community leaders, are broadening their crime-fighting strategy from draconian yet unpredictable punishments for big crimes to small yet reliable punishments for smaller ones—a guarantee, for example, that missing a probation hearing will net the offender a few days in jail.[175] The shift exploits two features of our psychology (which will be explored in the chapter on our better angels). One is that people—especially the people who are likely to get in trouble with the law—steeply discount the future, and respond more to certain and immediate punishments than to hypothetical and delayed ones.[176] The other is that people conceive of their relationships with other people and institutions in moral terms, categorizing them either as contests of raw dominance or as contracts governed by reciprocity and fairness.[177] Steven Alm, a judge who devised a "probation with enforcement" program, summed up the reason for the program's success: "When the system isn't consistent and predictable, when people are punished randomly, they think, My probation officer doesn't like me, or, Someone's prejudiced against me, rather than seeing that everyone who breaks a rule is treated equally, in precisely the same way."[178]

The newer offensive to tamp down violence also aims to enhance the habits of empathy and self-control that are the internal enforcers of the Civilizing Process. The Boston effort was named the TenPoint Coalition after a manifesto with ten stated goals, such as to "Promote and campaign for a cultural shift to help reduce youth violence within the Black community, both physically and verbally, by initiating conversations, introspection and reflection on the thoughts and actions that hold us back as a people, individually and collectively." One of the programs with which it has joined forces, Operation Ceasefire, was explicitly designed by David Kennedy to implement Immanuel Kant's credo that "morality predicated on external pressures alone is never sufficient."[179] The journalist John Seabrook describes one of its empathy-building events:

> At the one I attended, there was a palpable, almost evangelical desire to make the experience transformative for the gangbangers. An older ex–gang member named Arthur Phelps, whom everyone called Pops, wheeled a thirty-seven-year-old woman in a wheelchair to the center of the room. Her name was Margaret Long, and she was paralyzed from the chest down. "Seventeen years ago, I shot this woman," Phelps said, weeping. "And I live with that every day of my life." Then Long cried out, "And I go to the bathroom in a bag," and she snatched out the colostomy bag from inside the pocket of her wheelchair and held it up while the young men stared in horror. When the final speaker, a street worker named Aaron Pullins III, yelled, "Your house is on fire! Your building is burning! You've got to save yourselves! Stand up!," three-quarters of the group jumped to their feet, as if they had been jerked up like puppets on strings.[180]

The 1990s civilizing offensive also sought to glorify the values of responsibility that make a life of violence less appealing. Two highly publicized rallies in the nation's capital, one organized by black men, one by white, affirmed the obligation of men to support their children: Louis Farrakhan's Million Man March, and a march by the Promise Keepers, a conservative Christian movement. Though both movements had unsavory streaks of ethnocentrism, sexism, and religious fundamentalism, their historical significance lay in the larger recivilizing process they exemplified. In *The Great Disruption*, the political scientist Francis Fukuyama notes that as rates of violence went down in the 1990s, so did most other indicators of social pathology, such as divorce, welfare dependency, teenage pregnancy, dropping out of school, sexually transmitted disease, and teenage auto and gun accidents.[181]

The recivilizing process of the past two decades is not just a resumption of the currents that have swept the West since the Middle Ages. For one thing, unlike the original Civilizing Process, which was a by-product of the consolidation of states and the growth of commerce, the recent crime decline has largely come from civilizing offensives that were consciously designed to enhance people's well-being. Also new is a dissociation between the superficial trappings of civilization and the habits of empathy and self-control that we care the most about.

One way in which the 1990s did *not* overturn the decivilization of the 1960s is in popular culture. Many of the popular musicians in recent genres such as punk, metal, goth, grunge, gangsta, and hip-hop make the Rolling Stones look like the Women's Christian Temperance Union. Hollywood movies are bloodier than ever, unlimited pornography is a mouse-click away, and an entirely new form of violent entertainment, video games, has become a major pastime. Yet as these signs of decadence proliferated in the culture, violence went down in real life. The recivilizing process somehow managed to reverse the tide of social dysfunction without turning the cultural clock back to *Ozzie and Harriet*. The other evening I was riding a crowded Boston subway car and saw a fearsome-looking young man clad in black leather, shod in jackboots, painted with tattoos, and pierced by rings and studs. The other passengers were giving him a wide berth when he bellowed, "Isn't anyone going to give up his seat for this old woman? *She could be your grandmother!*"

The cliché about Generation X, who came of age in the 1990s, was that they were media-savvy, ironic, postmodern. They could adopt poses, try on styles, and immerse themselves in seedy cultural genres without taking any of them too seriously. (In this regard they were more sophisticated than the boomers in their youth, who treated the drivel of rock musicians as serious political philosophy.) Today this discernment is exercised by much of Western society. In his 2000 book *Bobos in Paradise*, the journalist David Brooks observed that many members of the middle class have become "bourgeois bohemians" who affect the look of people at the fringes of society while living a thoroughly conventional lifestyle.

Cas Wouters, inspired by conversations with Elias late in his life, suggests that we are living through a new phase in the Civilizing Process. This is the long-term trend of informalization I mentioned earlier, and it leads to what Elias called a "controlled decontrolling of emotional controls" and what Wouters calls third nature.[182] If our first nature consists of the evolved motives that govern life in a state of nature, and our second nature consists of the ingrained habits of a civilized society, then our third nature consists of a conscious reflection on these habits, in which we evaluate which aspects of a culture's norms are worth adhering to and which have outlived their usefulness. Centuries ago our ancestors may have had to squelch all signs of spontaneity and individuality in order to civilize themselves, but now that norms of nonviolence are entrenched, we can let up on particular inhibitions that may be obsolete. In this way of thinking, the fact that women show a lot of skin or that men curse in public is not a sign of cultural decay. On the contrary, it's a sign that they live in a society that is so civilized that they don't have to fear being harassed or assaulted in response. As the novelist Robert Howard put it, "Civilized men are more discourteous than savages because they know they can be impolite without having their skulls split." Maybe the time has even come when I can use a knife to push peas onto my fork.

4

THE HUMANITARIAN REVOLUTION

Those who can make you believe absurdities can make you commit atrocities.

—Voltaire

The world contains a lot of strange museums. There is the Museum of Pez Memorabilia in Burlingame, California, which showcases more than five hundred of the cartoon-headed candy dispensers. Visitors to Paris have long stood in line to see the museum devoted to the city's sewer system. The Devil's Rope Museum in McLean, Texas, "presents every detail and aspect of barbed wire." In Tokyo, the Meguro Museum of Parasitology invites its visitors to "try to think about parasites without a feeling of fear, and take the time to learn about their wonderful world of the Parasites." And then there is the Phallological Museum in Húsavík, "a collection of over one hundred penises and penile parts belonging to almost all the land and sea mammals that can be found in Iceland."

But the museum that I would least like to spend a day in is the Museo della Tortura e di Criminologia Medievale in San Gimignano, Italy.[1] According to a helpful review in www.tripadvisor.com, "The cost is €8,00. Pretty steep for a dozen or so small rooms totalling no more than 100–150 items. If you're into the macabre, though, you should not pass it by. Originals and reproductions of instruments of torture and execution are housed in moodily-lit stone-walled rooms. Each item is accompanied by excellent written descriptions in Italian, French, and English. No details are spared, including which orifice the device was meant for, which limb it was meant to dislocate, who was the usual customer and how the victim would suffer and/or die."

I think even the most atrocity-jaded readers of recent history would find something to shock them in this display of medieval cruelty. There is Judas's Cradle, used in the Spanish Inquisition: the naked victim was bound hand and foot, suspended by an iron belt around the waist, and lowered onto a sharp wedge that penetrated the anus or vagina; when victims relaxed their muscles, the point would stretch and tear their tissues. The Virgin of Nuremberg was a version of the iron maiden, with spikes that were carefully positioned so as not to transfix the victim's vital organs and prematurely end his suffering. A series of engravings show victims hung by the ankles and sawn in half from the crotch down; the display explains that this method of execution was used all over Europe for crimes that included rebellion, witchcraft, and military disobedience. The Pear is a split, spike-tipped wooden knob that was inserted into a mouth, anus, or vagina and spread apart by a screw mechanism to tear the victim open from the inside; it was used to punish sodomy, adultery, incest, heresy, blasphemy, and "sexual union with Satan." The Cat's Paw or Spanish Tickler was a cluster of hooks used to rip and shred a victim's flesh. Masks of Infamy were shaped like the head of a pig or an ass; they subjected a victim both to public humiliation and to the pain of a blade or knob forced into their nose or mouth to prevent them from wailing. The Heretic's Fork had a pair of sharp spikes at each end: one end was propped under the victim's jaw and the other at the base of his neck, so that as his muscles became exhausted he would impale himself in both places.

The devices in the Museo della Tortura are not particularly scarce. Collections of medieval torture instruments may also be found in San Marino, Amsterdam, Munich, Prague, Milan, and the Tower of London. Illustrations of literally hundreds of kinds of torture may be seen in coffee table books like *Inquisition* and *Torment in Art*, some of them reproduced in figure 4–1.[2]

Torture, of course, is not a thing of the past. It has been carried out in modern times by police states, by mobs during ethnic cleansings and genocides, and by democratic governments in interrogations and counterinsurgency operations, most infamously during the administration of George W. Bush following the 9/11 attacks. But the sporadic, clandestine, and universally decried eruptions of torture in recent times cannot be equated with the centuries of institutionalized sadism in medieval Europe. Torture in the Middle Ages was not hidden, denied, or euphemized. It was not just a tactic by which brutal regimes intimidated their political enemies or moderate regimes extracted information from suspected terrorists. It did not erupt from a frenzied crowd stirred up in hatred against a dehumanized enemy. No, torture was woven into the fabric of public life. It was a form of punishment that was cultivated and celebrated, an outlet for artistic and technological creativity. Many of the instruments of torture were beautifully crafted and ornamented. They were designed to inflict not just physical pain, as would a beating, but visceral horrors, such as penetrating sensitive orifices, violating the bodily envelope, displaying the victim in humiliating postures, or putting them in positions where their own flagging stamina would increase their pain and lead to disfigurement or death. Torturers were the era's foremost experts in anatomy and physiology, using their knowledge to maximize agony, avoid nerve damage that might deaden the pain, and prolong consciousness for as long as possible before death. When the victims were female, the sadism was eroticized: the women were stripped naked before being tortured, and their breasts and genitals were often

FIGURE 4–1. Torture in medieval and early modern Europe

Sources: Sawing: Held, 1986, p. 47. Cat's Paw: Held, 1986, p. 107. Impalement: Held, 1986, p. 141. Burning at the stake: Pinker, 2007a. Judas's Cradle: Held, 1986, p. 51. Breaking on the wheel: Puppi, 1990, p. 39.

the targets. Cold jokes made light of the victims' suffering. In France, Judas's Cradle was called "The Nightwatch" for its ability to keep a victim awake. A victim might be roasted alive inside an iron bull so his screams would come out of the bull's mouth, like the bellowing of a beast. A man accused of disturbing the peace might be forced to wear a Noisemaker's Fife, a facsimile of a flute or trumpet with an iron collar that went around his neck and a vise that crushed the bones and joints of his fingers. Many torture devices were shaped like animals and given whimsical names.

Medieval Christendom was a culture of cruelty. Torture was meted out by national and local governments throughout the Continent, and it was codified in laws that prescribed blinding, branding, amputation of hands, ears, noses, and tongues, and other forms of mutilation as punishments for minor crimes. Executions were orgies of sadism, climaxing with ordeals of prolonged killing such as burning at the stake, breaking on the wheel, pulling apart by horses, impalement through the rectum, disembowelment by winding a man's intestines around a spool, and even hanging, which was a slow racking and strangulation rather than a quick breaking of the neck.[3] Sadistic tortures were also inflicted by the Christian church during its inquisitions, witch hunts, and religious wars. Torture had been authorized by the ironically named Pope Innocent IV in 1251, and the order of Dominican monks carried it out with relish. As the *Inquisition* coffee table book notes, under Pope Paul IV (1555–59), the Inquisition was "downright insatiable—Paul, a Dominican and one-time Grand Inquisitor, was himself a fervent and skilled practitioner of torture and atrocious mass murders.[4]

Torture was not just a kind of rough justice, a crude attempt to deter violence with the threat of greater violence. Most of the infractions that sent a person to the rack or the stake were nonviolent, and today many are not even considered legally punishable, such as heresy, blasphemy, apostasy, criticism of the government, gossip, scolding, adultery, and unconventional sexual practices. Both the Christian and secular legal systems, inspired by Roman law, used torture to extract a confession and thereby convict a suspect, in defiance of the obvious fact that a person will say anything to stop the pain. Torture used to secure a confession is thus even more senseless than torture used to deter, terrorize, or extract verifiable information such as the names of accomplices or the location of weapons. Nor were other absurdities allowed to get in the way of the fun. If a victim was burned by fire rather than spared by a miracle, that was taken as proof that he was guilty. A suspected witch would be tied up and thrown into a lake: if she floated, it proved she was a witch and she would then be hanged; if she sank and drowned, it proved she had been innocent.[5]

Far from being hidden in dungeons, torture-executions were forms of popular entertainment, attracting throngs of jubilant spectators who watched the victim struggle and scream. Bodies broken on wheels, hanging from gibbets, or decomposing in iron cages where the victim had been left to die of starvation and exposure were a familiar part of the landscape. (Some of these cages still hang from European public buildings today, such as St. Lambert's Church in Münster.) Torture was often a participatory sport. A victim in the stocks would be tickled, beaten, mutilated, pelted with rocks, or smeared with mud or feces, sometimes leading to suffocation.

Systemic cruelty was far from unique to Europe. Hundreds of methods of torture, applied to millions of victims, have been documented in other civilizations, including the Assyrians, Persians, Seleucids, Romans, Chinese, Hindus, Polynesians, Aztecs, and many African kingdoms and Native American tribes. Brutal killings and punishments were also documented among the Israelites, Greeks, Arabs, and Ottoman Turks. Indeed, as we saw at the end of chapter 2, *all* of the first complex civilizations were absolutist theocracies which punished victimless crimes with torture and mutilation.[6]

This chapter is about the remarkable transformation in history that has left us reacting to these practices with horror. In the modern West and much of the rest of the world, capital and corporal punishments have been effectively eliminated, governments' power to use violence against their subjects has been severely curtailed, slavery has been abolished, and people have lost their thirst for cruelty. All this happened in a narrow slice of history, beginning in the Age of Reason in the 17th century and cresting with the Enlightenment at the end of the 18th.

Some of this progress—and if it isn't progress, I don't know what is—was propelled by ideas: by explicit arguments that institutionalized violence ought to be minimized or abolished. And some of it was propelled by a change in sensibilities. People began to *sympathize* with more of their fellow humans, and were no longer indifferent to their suffering. A new ideology coalesced from these forces, one that placed life and happiness at the center of values, and that used reason and evidence to motivate the design of institutions. The new ideology may be called humanism or human rights, and its sudden impact on Western life in the second half of the 18th century may be called the Humanitarian Revolution.

Today the Enlightenment is often mentioned with a sneer. "Critical theorists" on the left blame it for the disasters of the 20th century; theoconservatives in the Vatican and the American intellectual right long to replace its tolerant secularism with the alleged moral clarity of medieval Catholicism.[7] Even many moderate secular writers disparage the Enlightenment as the revenge of the nerds, the naïve faith that humans are a race of pointy-eared rational actors. This colossal amnesia and ingratitude is possible because of the natural whitewashing of history that we saw in chapter 1, in which the reality behind the atrocities of yesteryear is consigned to the memory hole and is remembered only in bland idioms and icons. If the opening of this chapter has been graphic, it is only to remind you of the realities of the era that the Enlightenment put to an end.

Of course no historical change takes place in a single thunderclap, and humanist currents flowed for centuries before and after the Enlightenment and in parts of the world other than the West.[8] But in *Inventing Human Rights,* the historian Lynn Hunt notes that human rights have been conspicuously affirmed at two moments in history. One was the end of the 18th century, which saw the American Declaration of Independence in 1776 and the French Declaration of the Rights of Man and Citizen in 1789. The other was the mid-point of the 20th century, which saw the Universal Declaration of Human Rights in 1948, followed by a cascade of Rights Revolutions in the ensuing decades (chapter 7).

As we shall see, the declarations were more than feel-good verbiage; the Humanitarian Revolution initiated the abolition of many barbaric practices that had been unexceptionable features of life for most of human history. But the custom that most dramatically illustrates the advance of humanitarian sentiments was eradicated well before that time, and its disappearance is a starting point for understanding the decline of institutionalized violence.

SUPERSTITIOUS KILLING: HUMAN SACRIFICE, WITCHCRAFT, AND BLOOD LIBEL

The most benighted form of institutionalized violence is human sacrifice: the torture and killing of an innocent person to slake a deity's thirst for blood.[9]

The biblical story of the binding of Isaac shows that human sacrifice was far from unthinkable in the 1st millennium BCE. The Israelites boasted that their god was morally superior to those of the neighboring tribes because he demanded only that sheep and cattle be slaughtered on his behalf, not children. But the temptation must have been around, because the Israelites saw fit to outlaw it in Leviticus 18:21: "You shall not give any of your children to devote them by fire to Molech, and so profane the name of your God." For centuries their descendants would have to take measures against people back-sliding into the custom. In the 7th century BCE, King Josiah defiled the sacrificial arena of Tophet so "that no one might burn his son or his daughter as an offering to Molech."[10] After their return from Babylon, the practice of human sacrifice died out among Jews, but it survived as an ideal in one of its break-away sects, which believed that God accepted the torture-sacrifice of an innocent man in exchange for not visiting a worse fate on the rest of humanity. The sect is called Christianity.

Human sacrifice appears in the mythology of all the major civilizations. In addition to the Hebrew and Christian Bibles, it is recounted in the Greek legend in which Agamemnon sacrifices his daughter Iphigenia in hopes of bringing a fair wind for his war fleet; in the episode in Roman history in which four slaves were buried alive to keep Hannibal at bay; in a Druid legend from Wales in which priests killed a child to stop the disappearance of building materials for a fort; and in many legends surrounding the multiarmed Hindu goddess Kali and the feathered Aztec god Quetzalcoatl.

Human sacrifice was more than a riveting myth. Two millennia ago the Roman historian Tacitus left eyewitness accounts of the practice among Germanic tribes. Plutarch described it taking place in Carthage, where tourists today can see the charred remains of the sacrificial children. It has been documented among traditional Hawaiians, Scandinavians, Incas, and Celts (remember Bog Man?). It was a veritable industry among the Aztecs in Mexico, the Khonds in southeast India, and the Ashanti, Benin, and Dahomey kingdoms in western Africa, where victims were sacrificed by the thousands. Matthew White estimates that between the years 1440 and 1524 CE the Aztecs sacrificed about forty people a day, 1.2 million people in all.[11]

Human sacrifice is usually preceded by torture. The Aztecs, for example, lowered their victims into a fire, lifted them out before they died, and cut the beating hearts out of their chests (a spectacle incongruously reenacted in *Indiana Jones and the Temple of Doom* as a sacrifice to Kali in 1930s India). The Dayaks of Borneo inflicted death by a thousand cuts, slowly bleeding the victim to death with bamboo needles and blades. To meet the demand for sacrificial victims, the Aztecs went to war to capture prisoners, and the Khonds raised them for that purpose from childhood.

The killing of innocents was often combined with other superstitious customs. Foundation sacrifices, in which a victim was interred in the foundation of a fort, palace, or temple to mitigate the effrontery of intruding into the gods' lofty realm, were performed in Wales, Germany, India, Japan, and China. Another bright idea that was independently discovered in many kingdoms (including Sumeria, Egypt, China, and Japan) was the burial sacrifice: when a king died, his retinue and harem would be buried with him. The Indian practice of suttee, in which a widow would join her late husband on the funeral pyre, is yet another variation. About 200,000 women suffered these pointless deaths between the Middle Ages and 1829, when the practice was outlawed.[12]

What were these people thinking? Many institutionalized killings, however unforgivable, are at least understandable. People in power kill in order to eliminate enemies, deter troublemakers, or demonstrate their prowess. But sacrificing harmless children, going to war to capture victims, and raising a doomed caste from childhood hardly seem like cost-effective ways to stay in power.

In an insightful book on the history of force, the political scientist James Payne suggests that ancient peoples put a low value on other people's lives because pain and death were so common in their own. This set a low threshold for any practice that had a chance of bringing them an advantage, even if the price was the lives of others. And if the ancients believed in gods, as most people do, then human sacrifice could easily have been seen as offering them that advantage. "Their primitive world was full of dangers, suffering, and nasty surprises, including plagues, famines, and wars. It would be natural for them to ask, 'What kind of god would create such a world?' A plausible answer was: a sadistic god, a god who liked to see people bleed and suffer."[13] So, they might think, if these gods have a minimum daily requirement of human gore, why not be proactive about it? Better him than me.

Human sacrifice was eliminated in some parts of the world by Christian proselytizers, such as Saint Patrick in Ireland, and in others by European colonial powers like the British in Africa and India. Charles Napier, the British army's commander in chief in India, faced with local complaints about the abolition of suttee, replied, "You say that it is your custom to burn widows. Very well. We also have a custom: when men burn a woman alive, we tie a rope around their necks and we hang them. Build your funeral pyre; beside it, my carpenters will build a gallows. You may follow your custom. And then we will follow ours."[14]

In most places, though, human sacrifice died out on its own. It was abandoned by the Israelites around 600 BCE, and by the Greeks, Romans, Chinese, and Japanese a few centuries later. Something about mature, literate states eventually leads them to think the better of human sacrifice. One possibility is that the combination of a literate elite, the rudiments of historical scholarship, and contacts with neighboring societies gives people the means to figure out that the bloodthirsty-god hypothesis is incorrect. They infer that throwing a virgin into a volcano does not, in fact, cure diseases, defeat enemies, or bring them good weather. Another possibility, favored by Payne, is that a more affluent and predictable life erodes people's fatalism and elevates their valuation of other people's lives. Both theories are plausible, but neither is easy to prove, because it's hard to find any scientific or economic advance that coincides with the abandonment of human sacrifice.

The transition away from human sacrifice always has a moral coloring. The people who live through the abolition know they have made progress, and they look with disgust at the unenlightened foreigners who cling to the old ways. One episode in Japan illustrates the expansion of sympathy that must contribute to abolition. When the emperor's brother died in 2 BCE, his entourage was buried with him in a traditional funeral sacrifice. But the victims didn't die for several days, and they "wept and wailed at night," upsetting the emperor and other witnesses. When the emperor's wife died five years later, he changed the custom so that clay images were placed in the tomb instead of live humans. As Payne notes, "The emperor shortchanged the gods because spending human lives had become too dear."[15]

A sanguinary god that hungers for indiscriminate human scapegoats is a rather crude theory of misfortune. When people outgrow it, they are still apt to look to supernatural explanations for bad things that happen to them. The difference is that their explanations become more finely tuned to their particulars. They still feel they have been targeted by supernatural forces, but the forces are wielded by a specific individual rather than a generic god. The name for such an individual is a witch.

Witchcraft is one of the most common motives for revenge among hunter-gatherer and tribal societies. In their theory of causation, there is no such thing as a natural death. Any fatality that cannot be explained by an observable cause is explained by an unobservable one, namely sorcery.[16] It seems incredible to us that so many societies have sanctioned cold-blooded murder for screwball reasons. But certain features of human cognition, combined with certain recurring conflicts of interest, make it a bit more comprehensible. The brain has evolved to ferret out hidden powers in nature, including those that no one can see.[17] Once you start rummaging around in the realm of the unverifiable there is considerable room for creativity, and accusations of sorcery are often blended with self-serving motives. Tribal people, anthropologists have shown, often single out despised in-laws for allegations of witchcraft, a convenient pretext to have them executed. The accusations may also be used to cut a rival down to size (especially one who has boasted that he really does have magical powers), to claim to be holier than everyone else when competing in the local reputational sweepstakes, or to dispose of ornery, eccentric, or burdensome neighbors, especially ones who have no supporting relatives to avenge their deaths.[18]

People may also use allegations of witchcraft to recoup some of the losses from a misfortune by holding another party liable—a bit like American accident victims who trip on a crack or spill hot coffee on themselves and sue everyone in sight. And perhaps the most potent motive is to deter adversaries from plotting against them and covering their tracks: the plotters may be able to disprove any physical connection to the attack, but they can never disprove a nonphysical connection. In Mario Puzo's novel *The Godfather,* Vito Corleone is credited with the principle "Accidents don't happen to people who take accidents as a personal insult." In the movie version, he spells it out to the heads of the other crime families: "I'm a superstitious man. And if some unlucky accident should befall my son, if my son is struck by a bolt of lightning, I will blame some of the people here."

Moralistic accusations can sometimes escalate into denunciations of those who fail to make moralistic accusations, snowballing into extraordinary popular delusions and the madness of crowds.[19] In the 15th century two monks published an exposé of witches called *Malleus Maleficarum,* which the historian Anthony Grafton has called "a strange amalgam of Monty Python and *Mein Kampf.*"[20] Egged on by its revelations, and inspired by the injunction in Exodus 22:18 "Thou shalt not suffer a witch to live," French and German witch-hunters killed between 60,000 and 100,000 accused witches (85 percent of them women) during the next two centuries.[21] The executions, usually by burning at the stake, followed an ordeal of torture in which the women confessed to such crimes as eating babies, wrecking ships, destroying crops, flying on broomsticks on the Sabbath, copulating with devils, transforming their demon lovers into cats and dogs, and making ordinary men impotent by convincing them that they had lost their penises.[22]

The psychology of witchcraft accusations can shade into other blood libels, such as the recurring rumors in medieval Europe that Jews poisoned the wells or killed Christian children during Passover to use their blood for matzo. Thousands of Jews were massacred in England, France, Germany, and the Low Countries during the Middle Ages, emptying entire regions of their Jewish populations.[23]

Witch hunts are always vulnerable to common sense. Objectively speaking, it is impossible for a woman to fly on a broomstick or to turn a man into a cat, and these facts are not too hard to demonstrate if enough people are allowed to compare notes and question popular beliefs. Throughout the Middle Ages there were scattered clerics and politicians who pointed out the obvious, namely that there is no such thing as a witch, and so persecuting someone for witchcraft was a moral abomination. (Unfortunately, some of these skeptics ended up in the torture chambers themselves.)[24] These voices became more prominent during the Age of Reason, and included influential writers such as Erasmus, Montaigne, and Hobbes.

Some officials became infected with the scientific spirit and tested the witchcraft hypothesis for themselves. A Milanese judge killed his mule, accused his servant of committing the misdeed, and had him subjected to torture, whereupon the man confessed to the crime; he even refused to recant on the gallows for fear of being tortured again. (Today this experiment would not be approved by committees for the protection of human subjects in research.) The judge then abolished the use of torture in his court. The writer Daniel Mannix recounts another demonstration:

> The Duke of Brunswick in Germany was so shocked by the methods used by Inquisitors in his duchy that he asked two famous Jesuit scholars to supervise the hearings. After a careful study the Jesuits told the Duke, "The Inquisitors are doing their duty. They are arresting only people who have been implicated by the confession of other witches."
>
> "Come with me to the torture chamber," suggested the Duke. The priests followed him to where a wretched woman was being stretched on the rack. "Let me question her," suggested the Duke. "Now woman, you are a confessed witch. I suspect these two men of being warlocks. What do you say? Another turn of the rack, executioners."
>
> "No, no!" screamed the woman. "You are quite right. I have often seen them at the Sabbat. They can turn themselves into goats, wolves, and other animals."
>
> "What else do you know about them?" demanded the Duke.
>
> "Several witches have had children by them. One woman even had eight children whom these men fathered. The children had heads like toads and legs like spiders."

> The Duke turned to the astonished Jesuits. "Shall I put you to the torture until you confess, my friends?"[25]

One of the Jesuits, Father Friedrich Spee, was so impressed that he wrote a book in 1631 that has been credited with ending witchcraft accusations in much of Germany. The persecution of witches began to subside during the 17th century, when several European states abolished it. The year 1716 was the last time a woman was hanged as a witch in England, and 1749 was the last year a woman was burned as a witch anywhere in Europe.[26]

In most of the world, institutionalized superstitious killing, whether in human sacrifice, blood libel, or witch persecution, has succumbed to two pressures. One is intellectual: the realization that some events, even those with profound personal significance, must be attributed to impersonal physical forces and raw chance rather than the designs of other conscious beings. A great principle of moral advancement, on a par with "Love thy neighbor" and "All men are created equal," is the one on the bumper sticker: "Shit happens."

The other pressure is harder to explain but just as forceful: an increased valuation of human life and happiness. Why are we taken aback by the experiment in which a judge tortured his servant to prove that torture was immoral, harming one to help many? It is because we sympathize with other humans, even if we don't know them, by virtue of the fact that they *are* human, and we parlay that sympathy into bright lines that outlaw the imposition of suffering on an identifiable human being. Even if we have not eliminated the features of human nature that tempt us to blame others for our misfortunes, we have increasingly prevented that temptation from erupting in violence. An increased valuation of the well-being of other people, we shall see, was a common thread in the abandonment of other barbaric practices during the Humanitarian Revolution.

SUPERSTITIOUS KILLING: VIOLENCE AGAINST BLASPHEMERS, HERETICS, AND APOSTATES

Human sacrifice and witch-burnings are just two examples of the harm that can result from people pursuing ends that involve figments of their imagination. Another may be seen in psychotics who kill in pursuit of a delusion, such as Charles Manson's plan to hasten an apocalyptic race war, and John Hinckley's scheme to impress Jodie Foster. But the greatest damage comes from religious beliefs that downgrade the lives of flesh-and-blood people, such as the faith that suffering in this world will be rewarded in the next, or that flying a plane into a skyscraper will earn the pilot seventy-two virgins in heaven. As we saw in chapter 1, the belief that one may escape from an eternity in hell only by accepting Jesus as a savior makes it a moral imperative to coerce people into accepting that belief and to silence anyone who might sow doubt about it.

A broader danger of unverifiable beliefs is the temptation to defend them by violent means. People become wedded to their beliefs, because the validity of those beliefs reflects on their competence, commends them as authorities, and rationalizes their mandate to lead. Challenge a person's beliefs, and you challenge his dignity, standing, and power. And when those beliefs are based on nothing but faith, they are chronically fragile. No one gets upset about the belief that rocks fall down as opposed to up, because all sane people can see it with their own eyes. Not so for the belief that babies are born with original sin or that God exists in three persons or that Ali was the second-most divinely inspired man after Muhammad. When people organize their lives around these beliefs, and then learn of other people who seem to be doing just fine without them—or worse, who credibly rebut them—they are in danger of looking like fools. Since one cannot defend a belief based on faith by persuading skeptics it is true, the faithful are apt to react to unbelief with rage, and may try to eliminate that affront to everything that makes their lives meaningful.

The human toll of the persecution of heretics and nonbelievers in medieval and early modern Christendom beggars the imagination and belies the conventional wisdom that the 20th century was an unusually violent era. Though no one knows exactly how many people were killed in these holy slaughters, we can get a sense from numerical estimates by atrocitologists such as the political scientist R. J. Rummel in his books *Death by Government* and *Statistics of Democide* and the historian Matthew White in his *Great Big Book of Horrible Things* and his "Deaths by Mass Unpleasantness" Web site.[27] They have tried to put numbers on the death tolls of wars and massacres, including those for which conventional statistics are unavailable, by combing the available sources, assessing their credibility with sanity checks and allowances for bias, and selecting a middle value, often the geometric mean of the lowest and the highest credible figures. I'll present Rummel's estimates for this era, which are generally lower than White's.[28]

Between 1095 and 1208 Crusader armies were mobilized to fight a "just war" to retake Jerusalem from Muslim Turks, earning them remission from their sins and a ticket to heaven. They massacred Jewish communities on the way, and after besieging and sacking Nicea, Antioch, Jerusalem, and Constantinople, they slaughtered their Muslim and Jewish populations. Rummel estimates the death toll at 1 million. The world had around 400 million people at the time, about a sixth of the number in the mid-20th century, so the death toll of the Crusader massacres as a proportion of the world population would today come out at around 6 million, equivalent to the Nazis' genocide of the Jews.[29]

In the 13th century the Cathars of southern France embraced the Albigensian heresy, according to which there are two gods, one of good and one of evil. An infuriated papacy, in collusion with the king of France, sent waves of armies to the region, which killed around 200,000 of them. To give you a sense of the armies' tactics, after capturing the city of Bram in 1210 they took a hundred of the defeated soldiers, cut off their noses and upper lips, gouged out the eyes of all but one, and had him lead the others to the city of Cabaret to terrorize its citizens into surrendering.[30] The reason you have never met a Cathar is that the Albigensian Crusade exterminated them. Historians classify this episode as a clear instance of genocide.[31]

Shortly after the suppression of the Albigensian heresy, the Inquisition was set up to root out other heresies in Europe. Between the late 15th and early 18th centuries, the Spanish branch took aim at converts from Judaism and Islam who were suspected of backsliding into their old practices. One transcript from 16th-century Toledo describes the inquisition of a woman who was accused of wearing clean underwear on Saturday, a sign that she was a secret Jew. She was subjected to the rack and the water torture (I'll spare you the details—it was worse than waterboarding), given several days to recover, and tortured again while she desperately tried to figure out what she should confess to.[32] The Vatican today claims that the Inquisition killed only a few thousand people, but it leaves off the books the larger number of victims who were remanded to secular authorities for execution or imprisonment (often a slow death sentence), together with the victims of branch offices in the New World. Rummel estimates the death toll from the Spanish Inquisition at 350,000.[33]

After the Reformation, the Catholic Church had to deal with the vast number of people in northern Europe who became Protestants, often involuntarily after their local prince or king had converted.[34] The Protestants, for their part, had to deal with the breakaway sects that wanted nothing to do with either branch of Christianity, and of course with the Jews. One might think that Protestants, who had been persecuted so viciously for their heresies against Catholic doctrines, would take a dim view of the idea of persecuting heretics, but no. In his 65,000-word treatise *On the Jews and Their Lies*, Martin Luther offered the following advice on what Christians should do with this "rejected and condemned people":

> First, . . . set fire to their synagogues or schools and . . . bury and cover with dirt whatever will not burn, so that no man will ever again see a stone or cinder of them. . . . Second, I advise that their houses also be razed and destroyed. . . . Third, I advise that all their prayer books and Talmudic writings, in which such idolatry, lies, cursing, and blasphemy are taught, be taken from them. . . . Fourth, I advise that their rabbis be forbidden to teach henceforth on pain of loss of life and limb. . . . Fifth, I advise that safe-conduct on the highways be abolished completely for the Jews. . . . Sixth, I advise that usury be prohibited to them, and that all cash and treasure of silver and gold be taken from them and put aside for safekeeping. Seventh, I recommend putting a flail, an ax, a hoe, a spade, a distaff, or a spindle into the hands of young, strong Jews and Jewesses and letting them earn their bread in the sweat of their brow, as was imposed on the children of Adam (Gen. 3[:19]).

> For it is not fitting that they should let us accursed Goyim toil in the sweat of our faces while they, the holy people, idle away their time behind the stove, feasting and farting, and on top of all, boasting blasphemously of their lordship over the Christians by means of our sweat. Let us emulate the common sense of other nations . . . [and] eject them forever from the country.[35]

At least he suffered most of them to live. The Anabaptists (forerunners of today's Amish and Mennonites) got no such mercy. They believed that people should not be baptized at birth but should affirm their faith for themselves, so Luther declared they should be put to death. The other major founder of Protestantism, John Calvin, had a similar view about blasphemy and heresy:

> Some say that because the crime consists only of words there is no cause for such severe punishment. But we muzzle dogs; shall we leave men free to open their mouths and say what they please? . . . God makes it plain that the false prophet is to be stoned without mercy. We are to crush beneath our heels all natural affections when his honour is at stake. The father should not spare his child, nor the husband his wife, nor the friend that friend who is dearer to him than life.[36]

Calvin put his argument into practice by ordering, among other things, that the writer Michael Servetus (who had questioned the trinity) be burned at the stake.[37] The third major rebel against Catholicism was Henry VIII, whose administration burned, on average, 3.25 heretics per year.[38]

With the people who brought us the Crusades and Inquisition on one side, and the people who wanted to kill rabbis, Anabaptists, and Unitarians on the other, it's not surprising that the European Wars of Religion between 1520 and 1648 were nasty, brutish, and long. The wars were fought, to be sure, not just over religion but also over territorial and dynastic power, but the religious differences kept tempers at a fever pitch. According to the classification of the military historian Quincy Wright, the Wars of Religion embrace the French Huguenot Wars (1562–94), the Dutch Wars of Independence, also known as the Eighty Years' War (1568–1648), the Thirty Years' War (1618–48), the English Civil War (1642–48), the wars of Elizabeth I in Ireland, Scotland, and Spain (1586–1603), the War of the Holy League (1508–16), and Charles V's wars in Mexico, Peru, France, and the Ottoman Empire (1521–52).[39] The rates of death in these wars were staggering. During the Thirty Years' War soldiers laid waste to much of present-day Germany, reducing its population by around a third. Rummel puts the death toll at 5.75 million, which as a proportion of the world's population at the time was more than double the death rate of World War I and was in the range of World War II in Europe.[40] The historian Simon Schama estimates that the English Civil War killed almost half a million people, a loss that is proportionally greater than that in World War I.[41]

It wasn't until the second half of the 17th century that Europeans finally began to lose their zeal for killing people with the wrong supernatural beliefs. The Peace of Westphalia, which ended the Thirty Years' War in 1648, confirmed the principle that each local prince could decide whether his state would be Protestant or Catholic and that the minority denomination in each one could more or less live in peace. (Pope Innocent X was not a good sport about this: he declared the Peace "null, void, invalid, unjust, damnable, reprobate, inane, empty of meaning and effect for all time.")[42] The Spanish and Portuguese Inquisitions began to run out of steam in the 17th century, declined further in the 18th, and were shut down in 1834 and 1821, respectively.[43] England put religious killing behind it after the Glorious Revolution of 1688. Though the divisions of Christianity have sporadically continued to skirmish right up to the present (Protestants and Catholics in Northern Ireland, and Catholics and Orthodox Christians in the Balkans), today the disputes are more ethnic and political than theological. Beginning in the 1790s, Jews were granted legal equality in the West, first in the United States, France, and the Netherlands, and then, over the following century, in most of the rest of Europe.

What made Europeans finally decide that it was all right to let their dissenting compatriots risk eternal damnation and, by their bad example, lure others to that fate? Perhaps they were exhausted by the Wars of Religion, but it's not clear why it took thirty years to exhaust them rather than ten or twenty. One gets a sense that people started to place a higher value on human life. Part of this newfound appreciation was an emotional change: a habit of identifying with the pains and pleasures of others. And another part was an intellectual and moral change: a shift from valuing *souls* to valuing *lives*. The doctrine of the sacredness of the soul sounds vaguely uplifting, but in fact is highly malignant. It discounts life on earth as just a temporary phase that people pass through, indeed, an infinitesimal fraction of their existence. Death becomes a mere rite of passage, like puberty or a midlife crisis.

The gradual replacement of lives for souls as the locus of moral value was helped along by the ascendancy of skepticism and reason. No one can deny the difference between life and death or the existence of suffering, but it takes indoctrination to hold beliefs about what becomes of an immortal soul after it has parted company from the body. The 17th century is called the Age of Reason, an age when writers began to insist that beliefs be justified by experience and logic. That undermines dogmas about souls and salvation, and it undermines the policy of forcing people to believe unbelievable things at the point of a sword (or a Judas's Cradle).

Erasmus and other skeptical philosophers noted that human knowledge was inherently fragile. If our eyes can be fooled by a visual illusion (such as an oar that appears to be broken at the water's surface, or a cylindrical tower in the distance that appears to be square), why should we trust our beliefs about more vaporous objects?[44] Calvin's burning of Michael Servetus in 1553 prompted a widespread scrutiny of the very idea of religious persecution.[45] The French scholar Sebastian Castellio led the charge by calling attention to the absurdity of different people being unshakably certain of the truth of their mutually incompatible beliefs. He also noted the horrific moral consequences of acting on these beliefs.

> Calvin says that he is certain, and [other sects] say that they are; Calvin says that they are wrong and wishes to judge them, and so do they. Who shall be judge? Who made Calvin the arbiter of all the sects, that he alone should kill? He has the Word of God and so have they. If the matter is certain, to whom is it so? To Calvin? But then why does he write so many books about manifest truth? . . . In view of the uncertainty we must define the heretic simply as one with whom we disagree. And if then we are going to kill heretics, the logical outcome will be a war of extermination, since each is sure of himself. Calvin would have to invade France and all other nations, wipe out cities, put all the inhabitants to the sword, sparing neither sex nor age, not even babies and the beasts.[46]

The arguments were picked up in the 17th century by, among others, Baruch Spinoza, John Milton (who wrote, "Let truth and falsehood grapple . . . truth is strong"), Isaac Newton, and John Locke. The emergence of modern science proved that deeply held beliefs could be entirely false, and that the world worked by physical laws rather than by divine whims. The Catholic Church did itself no favor by threatening Galileo with torture and committing him to a life sentence of house arrest for espousing what turned out to be correct beliefs about the physical world. And the skeptical mindset, sometimes spiced with humor and common sense, was increasingly allowed to challenge superstition. In *Henry IV, Part 1*, Glendower boasts, "I can call spirits from the vasty deep." Hotspur replies, "Why, so can I, or so can any man; / But will they come when you do call for them?" Francis Bacon, often credited with the principle that beliefs must be grounded in observation, wrote of a man who was taken to a house of worship and shown a painting of sailors who had escaped shipwreck by paying their holy vows. The man was asked whether this didn't prove the power of the gods. "Aye," he answered, "but where are they painted that were drowned after their vows?"[47]

CRUEL AND UNUSUAL PUNISHMENTS

The debunking of superstition and dogma removes one of the pretexts for torture, but leaves it available as a punishment for secular crimes and misdemeanors. People in ancient, medieval, and early modern times thought cruel punishments were perfectly reasonable. The whole point of punishing someone is to make him so unhappy that he and others won't be tempted to engage in the prohibited activity. By that reasoning, the harsher the punishment is, the better it accomplishes what it is designed to do. Also, a state without an effective police and judiciary had to make a little punishment go a long way. It had to make the punishments so memorably brutal that anyone who witnessed them would be terrorized into submission and would spread the word to terrorize others.

But the practical function of cruel punishments was just a part of their appeal. Spectators *enjoyed* cruelty, even when it served no judicial purpose. Torturing animals, for example, was good clean fun. In 16th-century Paris, a popular form of entertainment was cat-burning, in which a cat was hoisted in a sling on a stage and slowly lowered into a fire. According to the historian Norman Davies, "The spectators, including kings and queens, shrieked with laughter as the animals, howling with pain, were singed, roasted, and finally carbonized."[48] Also popular were dogfights, bull runs, cockfights, public executions of "criminal" animals, and bearbaiting, in which a bear would be chained to a post and dogs would tear it apart or be killed in the effort.

Even when they were not actively enjoying torture, people showed a chilling insouciance to it. Samuel Pepys, presumably one of the more refined men of his day, made the following entry in his diary for October 13, 1660:

> Out to Charing Cross, to see Major-general Harrison hanged, drawn, and quartered; which was done there, he looking as cheerful as any man could do in that condition. He was presently cut down, and his head and heart shown to the people, at which there was great shouts of joy. . . . From thence to my Lord's, and took Captain Cuttance and Mr. Sheply to the Sun Tavern, and did give them some oysters.[49]

Pepys's cold joke about Harrison's "looking as cheerful as any man could do in that condition" referred to his being partly strangled, disemboweled, castrated, and shown his organs being burned before being decapitated.

Even the less flamboyant penalties that we remember with the euphemism "corporal punishment" were forms of hideous torture. Today many historical tourist traps have stocks and pillories in which children can pose for pictures. Here is a description of an actual pillorying of two men in 18th-century England:

> One of them being of short stature could not reach the hole made for the admission of the head. The officers of justice nevertheless forced his head through the hole and the poor wretch hung rather than stood. He soon grew black in the face and blood issued from his nostrils, his eyes and his ears. The mob nevertheless attacked him with great fury. The officers opened the pillory and the poor wretch fell down dead on the stand of the instrument.

The other man was so maimed and hurt by what was thrown at him that he lay there without hope of recovery.[50]

Another kind of "corporal punishment" was flogging, the common penalty for insolence or dawdling by British sailors and African American slaves. Whips were engineered in countless models that could flay skin, pulverize flesh into mincemeat, or slice through muscle to the bone. Charles Napier recounted that in the late-18th-century British armed forces, sentences of a thousand lashes were not uncommon:

> I have often seen victims brought out of the hospital three or four times to receive the remainder of the punishment, too severe to be borne without danger of death at one flogging. It was terrible to see the new, tender skin of the scarcely healed back laid bare again to receive the lash. I have seen hundreds of men flogged and have always observed that when the skin is thoroughly cut up or flayed off, the great pain subsides. Men are frequently convulsed and screaming during the time they receive from one lash to three hundred and then they bear the remainder, even to 800 or a thousand without a groan. They will often lie as without life and the drummers appear to be flogging a lump of dead, raw flesh.[51]

The word *keelhaul* is sometimes used to refer to a verbal reprimanding. Its literal sense comes from another punishment in the British navy. A sailor was tied to a rope and pulled around the bottom of the ship's hull. If he didn't drown, he would be slashed to ribbons by the encrusted barnacles.

By the end of the 16th century in England and the Netherlands, imprisonment began to replace torture and mutilation as the penalty for minor crimes. It was not much of an improvement. Prisoners had to pay for food, clothing, and straw, and if they or their families couldn't pay they did without. Sometimes they had to pay for "easement of irons," namely being released from spiked iron collars or from bars that pinned their legs to the floor. Vermin, heat and cold, human waste, and scanty and putrid food not only added to the misery but fostered diseases that made prisons de facto death camps. Many prisons were workhouses in which underfed prisoners were forced to rasp wood, break rocks, or climb moving treadwheels for most of their waking hours.[52]

The 18th century marked a turning point in the use of institutionalized cruelty in the West. In England reformers and committees criticized the "cruelty, barbarity, and extortion" they found in the country's prisons.[53] Graphic reports of torture-executions began to sear the public's conscience. According to a description of the execution of Catherine Hayes in 1726, "As soon as the flames reached her, she tried to push away the faggots with her hands but scattered them. The executioner got hold of the rope around her neck and tried to strangle her but the fire reached his hand and burned it so he had to let it go. More faggots were immediately thrown on the fire and in three or four hours she was reduced to ashes."[54]

The bland phrase *broken on the wheel* cannot come close to capturing the horror of that form of punishment. According to one chronicler, the victim was transformed into a "huge screaming puppet writhing in rivulets of blood, a puppet with four tentacles, like a sea monster, of raw, slimy and shapeless flesh mixed up with splinters of smashed bones."[55] In 1762 a sixty-four-year-old French Protestant named Jean Calas was accused of killing his son to prevent him from converting to Catholicism; in fact, he had tried to conceal the son's suicide.[56] During an interrogation that attempted to draw out the names of his accomplices, he was subjected to the strappado and water torture, then was broken on the wheel. After being left in agony for two hours, Calas was finally strangled in an act of mercy. Witnesses who heard his protestations of innocence as his bones were being broken were moved by the terrible spectacle. Each blow of the iron club "sounded in the bottom of their souls," and "torrents of tears were unleashed, too late, from all the eyes present."[57] Voltaire took up the cause, noting the irony that foreigners judged France by its fine literature and beautiful actresses without realizing that it was a cruel nation that followed "atrocious old customs."[58]

Other prominent writers also began to inveigh against sadistic punishments. Some, like Voltaire, used the language of shaming, calling the practices barbaric, savage, cruel, primitive, cannibalistic, and atrocious. Others, like Montesquieu, pointed out the hypocrisy of Christians' bemoaning their cruel treatment at the hands of Romans, Japanese, and Muslims, yet inflicting the same cruelty themselves.[59] Still others, like the American physician and signer of the Declaration of Independence Benjamin Rush, appealed to the common humanity of readers and the people who were targets of punishment. In 1787 he noted that "the men, or perhaps the women, whose persons we detest, possess souls and bodies composed of the same materials as those of our friends and relations. They are bone of their bone." And, he added, if we consider their misery without emotion or sympathy, then "the principle of sympathy . . . will cease to act altogether; and will soon lose its place in the human breast."[60] The goal of the judicial system should be to rehabilitate wrongdoers rather than harming them, and "the reformation of a criminal can never be effected by a public punishment."[61] The English lawyer William Eden also noted the brutalizing effect of cruel punishments, writing in 1771, "We leave each other to rot like scare-crows in the hedges; and our gibbets are crowded with human carcasses. May it not be doubted, whether a forced familiarity with such objects can have any other effect, than to blunt the sentiments, and destroy the benevolent prejudices of the people?"[62]

Most influential of all was the Milanese economist and social scientist Cesare Beccaria, whose 1764 bestseller *On Crimes and Punishments* influenced every major political thinker in the literate world, including Voltaire, Denis Diderot, Thomas Jefferson, and John Adams.[63] Beccaria began from first principles, namely that the goal of a system of justice is to attain "the greatest happiness of the greatest number" (a phrase later adopted by Jeremy Bentham as the motto of utilitarianism). The only legitimate use of punishment, then, is to deter people from inflicting greater harm on others than the harm inflicted on them. It follows that a punishment should be proportional to the harm of the crime—not to balance some mysterious cosmic scale of justice but to set up the right incentive structure: "If an equal punishment be ordained for two crimes that injure society in different degrees, there is nothing to deter men from committing the greater as often as it is attended with greater advantage." A clearheaded view of criminal justice also entails that the certainty and promptness of a punishment are more important than its severity, that criminal trials should be public and based on evidence, and that the death penalty is unnecessary as a deterrent and not among the powers that should be granted to a state.

Beccaria's essay didn't impress everyone. It was placed on the papal Index of Forbidden Books, and vigorously contested by the legal and religious scholar Pierre-François Muyart de Vouglans. Muyart mocked Beccaria's bleeding-heart sensibility, accused him of recklessly undermining a time-tested system, and argued that strong punishments were needed to counteract man's innate depravity, beginning with his original sin.[64]

But Beccaria's ideas carried the day, and within a few decades punitive torture was abolished in every major Western country, including the newly independent United States in its famous prohibition of "cruel and unusual punishments" in the Eighth Amendment to the Constitution. Though it is impossible to plot the decline of torture precisely (because many countries outlawed different uses at different times), the cumulative graph in figure 4–2 shows when fifteen major European countries, together with the United States, explicitly abolished the major forms of judicial torture practiced there.

I have demarcated the 18th century on this and the other graphs in this chapter to highlight the many humanitarian reforms that were launched in this remarkable slice of history. Another was the prevention of cruelty to animals. In 1789 Jeremy Bentham articulated the rationale for animal rights in a passage that continues to be the watchword of animal protection movements today: "The question is not Can they *reason?* nor Can they *talk?* but, Can they *suffer?*" Beginning in 1800, the first laws against bearbaiting were introduced into Parliament. In 1822 it passed the Ill-Treatment of Cattle Act and in 1835 extended its protections to bulls, bears, dogs, and cats.[65] Like many humanitarian movements that originated in the Enlightenment, opposition to animal cruelty found a second wind during the Rights Revolutions of the second half of the 20th century, culminating in the banning of the last legal blood sport in Britain, the foxhunt, in 2005.

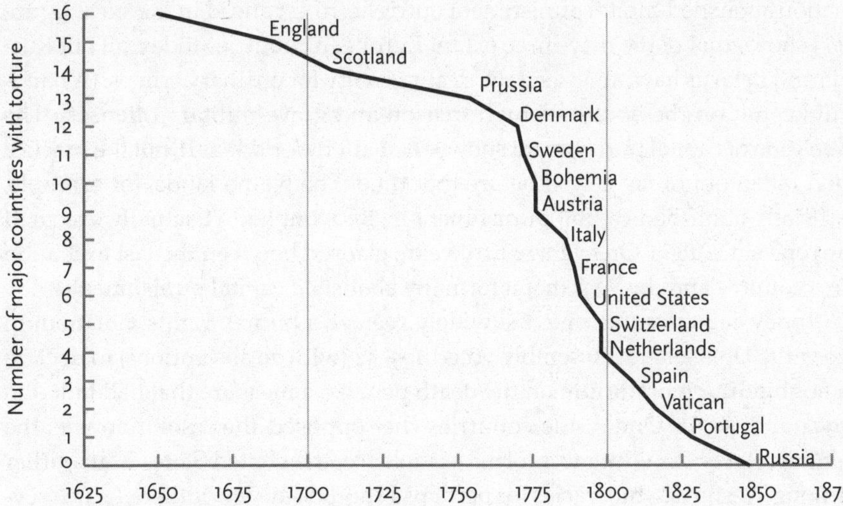

FIGURE 4–2. Time line for the abolition of judicial torture

Sources: Hunt, 2007, pp. 76, 179; Mannix, 1964, pp. 137–38.

CAPITAL PUNISHMENT

When England introduced drop hanging in 1783 and France introduced the guillotine in 1792, it was a moral advance, because an execution that instantly renders the victim unconscious is more humane than one that is designed to prolong his suffering. But execution is still a form of extreme violence, especially when it is applied as frivolously as most states did for most of human history. In biblical, medieval, and early modern times, scores of trivial affronts and infractions were punishable by death, including sodomy, gossiping, stealing cabbages, picking up sticks on the Sabbath, talking back to parents, and criticizing the royal garden.[66] During the last years of the reign of Henry VIII, there were more than ten executions in London *every week*. By 1822 England had 222 capital offenses on the books, including poaching, counterfeiting, robbing a rabbit warren, and cutting down a tree. And with an average trial length at the time of eight and a half minutes, it is certain that many of the people sent to the gallows were innocent.[67] Rummel estimates that between the time of Jesus and the 20th century, 19 million people were executed for trivial offenses.[68]

But as the 18th century came to a close, capital punishment itself was on death row. Public hangings, which had long been rowdy carnivals, were abolished in England in 1783. The display of corpses on gibbets was abolished in 1834, and by 1861 England's 222 capital offenses had been reduced to 4.[69] During the 19th century many European countries stopped executing people for any crime but murder and high treason, and eventually almost every Western nation abolished capital punishment outright. To get ahead in the story, figure 4–3 shows that of the fifty-three extant European countries today, all but Russia and Belarus have abolished the death penalty for ordinary crimes. (A handful keep it on the books for high treason and grave military offenses.) The abolition of capital punishment snowballed after World War II, but the practice had fallen out of favor well before that time. The Netherlands, for example, officially abolished capital punishment in 1982, but hadn't actually executed anyone since 1860. On average fifty years elapsed between the last execution in a country and the year that it formally abolished capital punishment.

Today capital punishment is widely seen as a human rights violation. In 2007 the UN General Assembly voted 105–54 (with 29 abstentions) to declare a nonbinding moratorium on the death penalty, a measure that had failed in 1994 and 1999.[70] One of the countries that opposed the resolution was the United States. As with most forms of violence, the United States is an outlier among Western democracies (or perhaps I should say "*are* outliers," since seventeen states, mostly in the North, have abolished the death penalty as well—four of them within the past two years—and an eighteenth has not carried out an execution in forty-five years).[71] But even the American death penalty, for all its notoriety, is more symbolic than real. Figure 4–4 shows that the rate of executions in the United States as a proportion of its population has plummeted since colonial times, and that the steepest drop was in the 17th and 18th centuries, when so many other forms of institutional violence were being scaled back in the West.

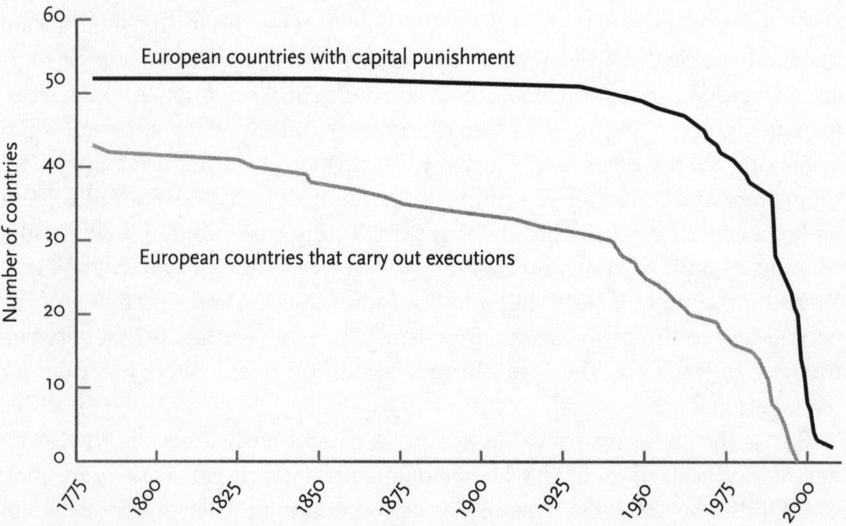

FIGURE 4–3. Time line for the abolition of capital punishment in Europe

Sources: French Ministry of Foreign Affairs, 2007; Capital Punishment U.K., 2004; Amnesty International, 2010.

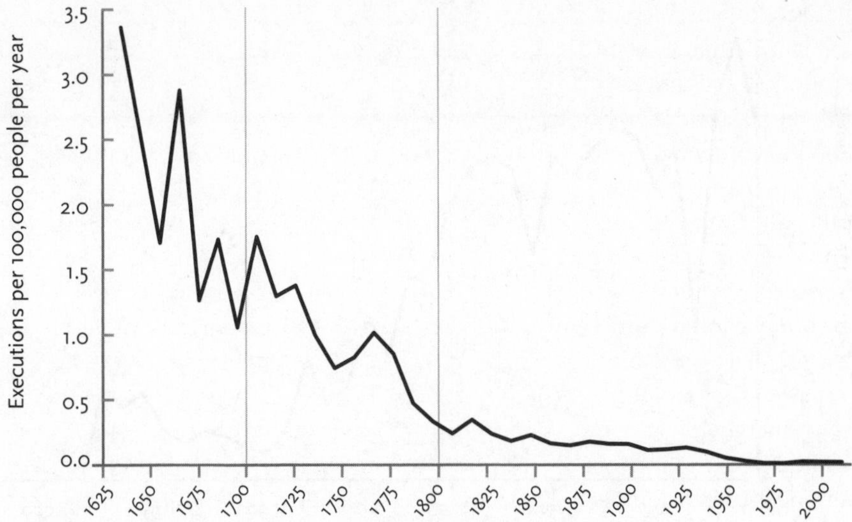

FIGURE 4–4. Execution rate in the United States, 1640–2010
Sources: Payne, 2004, p. 130, based on data from Espy & Smykla, 2002. The figures for the decades ending in 2000 and 2010 are from Death Penalty Information Center, 2010b.

The barely visible swelling in the last two decades reflects the tough-on-crime policies that were a reaction to the homicide boom of the 1960s, 1970s, and 1980s. But in present-day America a "death sentence" is a bit of a fiction, because mandatory legal reviews delay most executions indefinitely, and only a few tenths of a percentage point of the nation's murderers are ever put to death.[72] And the most recent trend points downward: the peak year for executions was 1999, and since then the number of executions per year has been almost halved.[73]

At the same time that the rate of capital punishment went down, so did the number of capital crimes. In earlier centuries people could be executed for theft, sodomy, buggery, bestiality, adultery, witchcraft, arson, concealing birth, burglary, slave revolt, counterfeiting, and horse theft. Figure 4–5 shows the proportion of American executions since colonial times that were for crimes other than homicide. In recent decades the only crime other than murder that has led to an execution is "conspiracy to commit murder." In 2007 the U.S. Supreme Court ruled that the death penalty may not be applied to any crime against an individual "where the victim's life was not taken" (though the death penalty is still available for a few "crimes against the state" such as espionage, treason, and terrorism).[74]

The means of execution has changed as well. Not only has the country long abandoned torture-executions such as burning at the stake, but it has experimented with a succession of "humane" methods, the problem being that the more effectively a method guarantees instant death (say, a few bullets to the

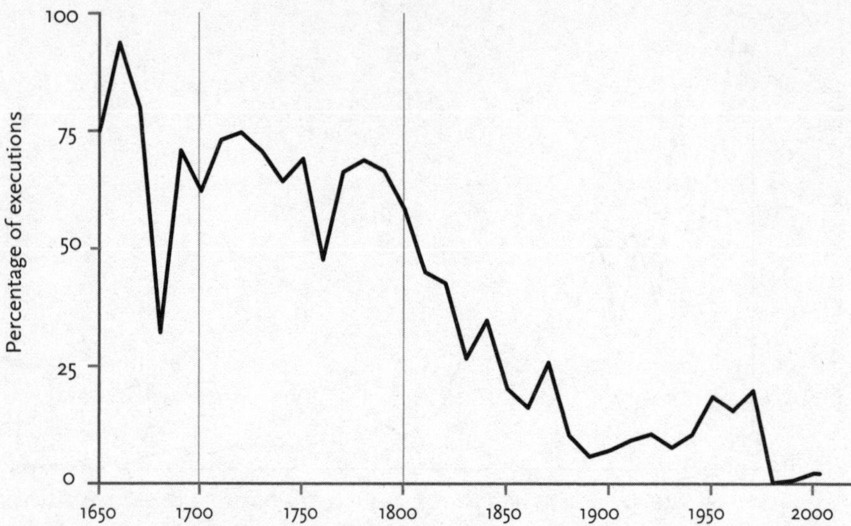

FIGURE 4–5. Executions for crimes other than homicide in the United States, 1650–2002
Sources: Espy & Smykla, 2002; Death Penalty Information Center, 2010a.

brain), the more gruesome it will appear to onlookers, who don't want to be reminded that violence has been applied to kill a living body. Hence the physicality of ropes and bullets gave way to the invisible agents of gas and electricity, which have been replaced by the quasi-medical procedure of lethal injection under general anesthesia—and even that method has been criticized for being too stressful to the dying prisoner. As Payne has noted,

> In reform after reform lawmakers have moderated the death penalty so that it is now but a vestige of its former self. It is not terrifying, it is not swift, and in its present restricted use, it is not certain (only about one murder in two hundred leads to an execution). What does it mean, then, to say that the United States "has" the death penalty? If the United States had the death penalty in robust, traditional form, we would be executing approximately 10,000 prisoners a year, including scores of perfectly innocent people. The victims would be killed in torture-deaths, and these events would be shown on nationwide television to be viewed by all citizens, including children (at 27 executions a day, this would leave little time for any other television fare). That defenders of capital punishment would be appalled by this prospect shows that even they have felt the leavening effects of the increasing respect for human life.[75]

One can imagine that in the 18th century the idea of abolishing capital punishment would have seemed reckless. Undeterred by the fear of a grisly execution, one might have thought, people would not hesitate to murder for profit or revenge. Yet today we know that abolition, far from reversing the centuries-long decline of homicide, proceeded in tandem with it, and that the countries of modern Western Europe, none of which execute people, have the lowest homicide rates in the world. It is one of many cases in which institutionalized violence was once seen as indispensable to the functioning of a society, yet once it was abolished, the society managed to get along perfectly well without it.

SLAVERY

For most of the history of civilization, the practice of slavery was the rule rather than the exception. It was upheld in the Hebrew and Christian Bibles, and was justified by Plato and Aristotle as a natural institution that was essential to civilized society. So-called democratic Athens in the time of Pericles enslaved 35 percent of its population, as did the Roman Republic. Slaves have always been a major booty in wartime, and stateless people of all races were vulnerable to capture.[76] The word *slave* comes from *Slav*, because, as the dictionary informs us, "Slavic peoples were widely captured and enslaved during the Middle Ages." States and armed forces, when they were not used as enslaving devices, were used as enslavement-prevention devices, as we are reminded by the lyric "Rule, Britannia! Britannia rule the waves. Britons never, never, never shall be slaves." Well before Africans were enslaved by Europeans, they were enslaved by other Africans, as well as by Islamic states in North Africa and the Middle East. Some of those states did not abolish legal slavery until recently: Qatar in 1952; Saudi Arabia and Yemen in 1962; Mauritania in 1980.[77]

For captives in war, slavery was often a better fate than the alternative, massacre, and in many societies slavery shaded into milder forms of servitude, employment, military service, and occupational guilds. But violence is inherent to the definition of slavery—if a person did all the work of a slave but had the option of quitting at any time without being physically restrained or punished, we would not call him a slave—and this violence was often a regular part of a slave's life. Exodus 21:20–21 decrees, "When a slave-owner strikes a male or female slave with a rod and the slave dies immediately, the owner shall be punished. But if the slave survives for a day or two, there is no punishment; for the slave is the owner's property." Slaves' lack of ownership of their own bodies left even the better-treated ones vulnerable to vicious exploitation. Women in harems were perpetual rape victims, and the men who guarded them, eunuchs, had their testicles—or in the case of black eunuchs, their entire genitalia—hacked off with a knife and cauterized with boiling butter so they would not bleed to death from the wound.

The African slave trade in particular was among the most brutal chapters in human history. Between the 16th and 19th centuries at least 1.5 million Africans died in transatlantic slave ships, chained together in stifling, filth-ridden holds, and as one observer noted, "those who remain to meet the shore present a picture of wretchedness language cannot express."[78] Millions more perished in forced marches through jungles and deserts to slave markets on the coast or in the Middle East. Slave traders treated their cargo according to the business model of ice merchants, who accept that a certain proportion of their goods will be lost in transport. At least 17 million Africans, and perhaps as many as 65 million, died in the slave trade.[79] The slave trade not only killed people in transit, but by providing a continuous stream of bodies, it encouraged slaveholders to work their slaves to death and replace them with new ones. But even the slaves who were kept in relatively good health lived in the shadow of flogging, rape, mutilation, forced separation from family members, and summary execution.

Slaveholders in many times have manumitted their slaves, often in their wills, as they became personally close to them. In some places, such as Europe in the Middle Ages, slavery gave way to serfdom and sharecropping when it became cheaper to tax people than to keep them in bondage, or when weak states could not enforce a slave owner's property rights. But a mass movement against chattel slavery as an institution arose for the first time in the 18th century and rapidly pushed it to near extinction.

Why did people eventually forswear the ultimate labor-saving device? Historians have long debated the extent to which the abolition of slavery was driven by economics or by humanitarian concerns. At one time the economic explanation seemed compelling. In 1776 Adam Smith reasoned that slavery must be less efficient than paid employment because only the latter was a positive-sum game:

> The work done by slaves, though it appears to cost only their maintenance, is in the end the dearest of any. A person who can acquire no property, can have no other interest but to eat as much, and to labour as little as possible. Whatever work he does beyond what is sufficient to purchase his own maintenance can be squeezed out of him by violence only, and not by any interest of his own.[80]

The political scientist John Mueller points out, "Smith's view garnered adherents, but not, as it happens, among slaveowners. That is, either Smith was wrong, or slaveholders were bad businessmen."[81] Some economists, such as Robert Fogel and Stanley Engerman, have concluded that Smith was at least partly wrong in the case of the antebellum South, which had a reasonably efficient economy for the time.[82] And southern slavery, of course, did not gradually give way to more cost-effective production techniques but had to be obliterated by war and by law.

It took guns and laws to end slavery in much of the rest of the world as well. Britain, once among the most exuberant slave-trading nations, outlawed the slave trade in 1807 and abolished slavery throughout the empire in 1833. By the 1840s it was jawboning other countries to end their participation in the slave trade, backed up by economic sanctions and by almost a quarter of the Royal Navy.[83]

Most historians have concluded that Britain's policing of the abolition of slavery was driven by humanitarian motives.[84] Locke undermined the moral basis for slavery in his 1689 work *Two Treatises on Government*, and though he and many of his intellectual descendants hypocritically profited from the institution, their advocacy of liberty, equality, and the universal rights of man let a genie out of the bottle and made it increasingly awkward for anyone to justify the practice. Many of the Enlightenment writers who inveighed against torture on humanitarian grounds, such as Jacques-Pierre Brisson in France, applied the same logic to oppose slavery. They were joined by Quakers, who founded the influential Society for the Abolition of the Slave Trade in 1787, and by preachers, scholars, free blacks, former slaves, and politicians.[85]

At the same time, many politicians and preachers *defended* slavery, citing the Bible's approval of the practice, the inferiority of the African race, the value of preserving the southern way of life, and a paternalistic concern that freed slaves could not survive on their own. But these rationalizations withered under intellectual and moral scrutiny. The intellectual argument held that it was indefensible to allow one person to own another, arbitrarily excluding him from the community of decision-makers whose interests were negotiated in the social contract. As Jefferson put it, "The mass of mankind has not been born with saddles on their backs, nor a favored few booted and spurred, ready to ride them legitimately."[86] The moral revulsion was stimulated by first-person accounts of what it was like to be a slave. Some were autobiographies, like *The Interesting Narrative of the Life of Olaudah Equiano, the African, Written by Himself* (1789) and *Narrative of the Life of Frederick Douglass, an American Slave* (1845). Even more influential was a work of fiction, Harriet Beecher Stowe's *Uncle Tom's Cabin, or Life Among the Lowly* (1852). The novel depicted a wrenching episode in which mothers were separated from their children, and another in which the kindly Tom was beaten to death for refusing to flog other slaves. The book sold three hundred thousand copies and was a catalyst for the abolitionist movement. According to legend, when Abraham Lincoln met Stowe in 1862, he said, "So you're the little woman who started this great war."

In 1865, after the most destructive war in American history, slavery was abolished by the Thirteenth Amendment to the Constitution. Many countries had abolished it before that time, and France had the dubious distinction of abolishing it twice, first in the wake of the French Revolution in 1794 and again, after Napoleon had restored it in 1802, during the Second Republic in 1848. The rest of the world quickly followed suit. Many encyclopedias provide time lines of the abolition of slavery, which differ slightly in how they delineate territories and what they count as "abolition," but they all show the same pattern: an explosion of abolition proclamations beginning in the late 18th century. Figure 4–6 shows the cumulative number of nations and colonies that have formally abolished slavery since 1575.

Closely related to slavery is the practice of debt bondage. Beginning in biblical and classical times, people who defaulted on their loans could be enslaved, imprisoned, or executed.[87] The word *draconian* comes from the Greek lawgiver Draco, who in 621 BCE codified laws governing the enslavement of debtors. Shylock's right to cut a pound of flesh from Antonio in *The Merchant of Venice* is another reminder of the practice. By the 16th century defaulters were no longer enslaved or executed, but they filled up debtors' prisons by the thousands. Sometimes they were charged for food, despite being broke, and had to survive on what they could beg from passersby through the windows of the jail. In early-19th-century America, thousands of people, including many women, languished in debtors' prisons, half of them for debts of less than ten dollars. In the 1830s a reform movement sprang up which, like the antislavery movement, appealed to both reason and emotion. A congressional committee argued that it ran contrary to the principles of justice "to give the creditor, in any case whatever, power over the body of his debtor." The committee also noted that "if all the victims of oppression were presented to our view in one congregated mass, with all the train of wives, children, and friends, involved in the same ruin, they would exhibit a spectacle at which

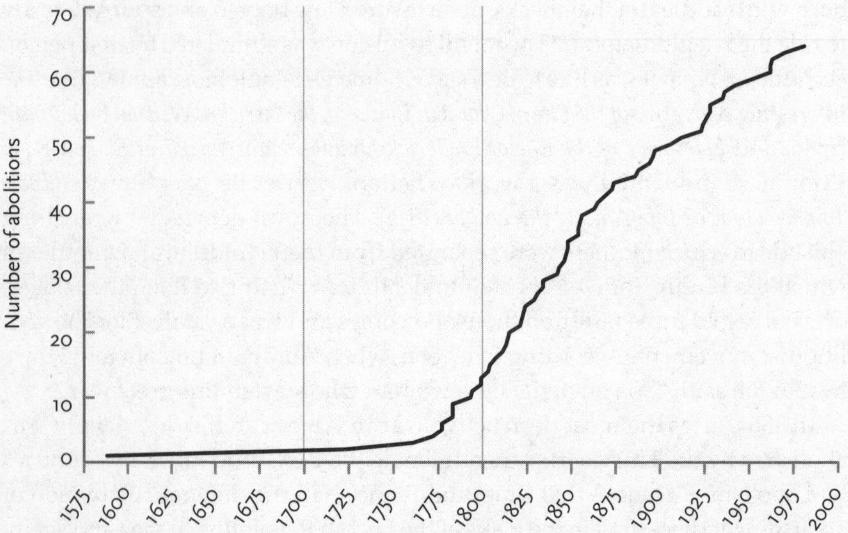

FIGURE 4–6. Time line for the abolition of slavery

Source: The most comprehensive list of abolitions I have found is "Abolition of slavery timeline," Wikipedia, http://en.wikipedia.org/wiki/Abolition_of_slavery_timeline, retrieved Aug. 18, 2009. Included are all entries from "Modern Timeline" that mention formal abolition of slavery in a political jurisdiction.

humanity would shudder."[88] Debt bondage was abolished by almost every American state between 1820 and 1840, and by most European governments in the 1860s and 1870s.

The history of our treatment of debtors, Payne notes, illustrates the mysterious process in which violence has declined in every sphere of life. Western societies have gone from enslaving and executing debtors to imprisoning them and then to seizing their assets to repay the debt. Even the seizure of assets, he points out, is a kind of violence: "When John buys groceries on credit and later refuses to pay for them, he has not used force. If the grocer goes to court and gets the police to seize John's car or bank account, the grocer and police are the ones who are initiating the use of force."[89] And because it is a form of violence, even if people don't usually think of it that way, this practice too has been in decline. The trend in bankruptcy law has been away from punishing debtors or squeezing assets out of them and toward giving them the opportunity of a fresh start. In many states a debtor's house, car, retirement accounts, and spouse's assets are protected, and when a person or company declares bankruptcy, they can write off many debts with impunity. In the old days of debtors' prisons, people might have predicted that this lenience would spell the demise of capitalism, which depends on the repayment of loans. But the commercial ecosystem evolved workarounds for this loss of leverage. Credit checks, credit ratings, loan insurance, and credit cards are just some of the ways that economic life continued after borrowers could no longer be deterred by the threat of legal coercion. An entire category of violence evaporated, and mechanisms that carried out the same function materialized, without anyone realizing that that was what was happening.

Slavery and other forms of bondage, of course, have not been obliterated from the face of the earth. As a result of recent publicity about the trafficking of people for labor and prostitution, one sometimes hears the statistically illiterate and morally obtuse claim that nothing has changed since the 18th century, as if there were no difference between a clandestine practice in a few parts of the world and an authorized practice everywhere in the world. Moreover, modern human trafficking, as heinous as it is, cannot be equated with the horrors of the African slave trade. As David Feingold, who initiated the UNESCO Trafficking Statistics Project in 2003, notes of today's hotbeds of trafficking:

> The identification of trafficking with chattel slavery—in particular, the transatlantic slave trade—is tenuous at best. In the 18th and 19th centuries, African slaves were kidnapped or captured in war. They were shipped to the New World into life-long servitude, from which they or their children could rarely escape. In contrast, although some trafficking victims are kidnapped, for most . . . , trafficking is migration gone terribly wrong. Most leave their homes voluntarily—though sometimes coerced by circumstance—in search

> of a materially better or more exciting life. Along the way, they become enmeshed in a coercive and exploitative situation. However, this situation rarely persists for life; nor . . . do the trafficked become a permanent or hereditary caste.[90]

Feingold also notes that the numbers of trafficking victims reported by activist groups and repeated by journalists and nongovernmental organizations are usually pulled out of thin air and inflated for their advocacy value. Nonetheless, even the activists recognize the fantastic progress that has been made. A statement by Kevin Bales, president of Free the Slaves, though it begins with a dubious statistic, puts the issue in perspective: "While the real number of slaves is the largest there has ever been, it is also probably the smallest proportion of the world population ever in slavery. Today, we don't have to win the legal battle; there's a law against it in every country. We don't have to win the economic argument; no economy is dependent on slavery (unlike in the 19th century, when whole industries could have collapsed). And we don't have to win the moral argument; no one is trying to justify it any more."[91]

The Age of Reason and the Enlightenment brought many violent institutions to a sudden end. Two others had more staying power, and were indulged in large parts of the world for another two centuries: tyranny, and war between major states. Though the first systematic movements to undermine these institutions were nearly strangled in the crib and began to predominate only in our lifetimes, they originated in the grand change in thoughts and sensibilities that make up the Humanitarian Revolution, so I will introduce them here.

DESPOTISM AND POLITICAL VIOLENCE

A government, according to the famous characterization by the sociologist Max Weber, is an institution that holds a monopoly on the legitimate use of violence. Governments, then, are institutions that by their very nature are designed to carry out violence. Ideally this violence is held in reserve as a deterrent to criminals and invaders, but for millennia most governments showed no such restraint and indulged in violence exuberantly.

All of the first complex states were despotisms in the sense of an "exercised right of heads of societies to murder their subjects arbitrarily and with impunity."[92] Evidence for despotism, Laura Betzig has shown, may be found in the records of the Babylonians, Hebrews, Imperial Romans, Samoans, Fijians, Khmer, Natchez, Aztecs, Incas, and nine African kingdoms. Despots put their power to good Darwinian use by living in luxury and enjoying the services of enormous harems. According to a report from the early days of the British colonization of India, "a party given by the Mogul governor of Surat . . . was rudely interrupted when the host fell into a sudden rage and ordered all the dancing girls to be decapitated on the spot, to the stupefaction of the English guests."[93] They could afford to be stupefied only because the mother country had recently put its own despotism behind it. When Henry VIII got into various of his bad moods, he executed two wives, several of their suspected lovers, many of his own advisors (including Thomas More and Thomas Cromwell), the Bible translator William Tyndale, and tens of thousands of others.

The power of despots to kill on a whim is the backdrop to stories told throughout the world. The wise King Solomon proposed to resolve a maternity dispute by butchering the baby in question. The backdrop to the Scheherazade story is a Persian king who murdered a new bride every day. The legendary King Narashimhadev in Orissa, India, demanded that exactly twelve hundred artisans build a temple in exactly twelve years or all would be executed. And in Dr. Seuss's *The Five Hundred Hats of Bartholomew Cubbins*, the protagonist is nearly beheaded for being unable to remove his hat in the presence of the king.

He who lives by the sword dies by the sword, and in most of human history political murder—a challenger killing a leader and taking his place—was the primary mechanism for the transfer of power.[94] A political murderer differs from the modern assassin who tries to make a political statement, wants to go down in the history books, or is stark raving mad. Instead he is typically a member of the political elite, kills a leader to take over his position, and counts on his accession to be recognized as legitimate. Kings Saul, David, and Solomon were all targets or perpetrators of murder plots, and Julius Caesar was one of the thirty-four Roman emperors (out of the total of forty-nine that reigned until the division of the empire) who were killed by guards, high officials, or members of their own families. Manuel Eisner has calculated that between 600 and 1800 CE, about one in eight European monarchs was murdered in office, mostly by noblemen, and that a third of the killers took over the throne.[95]

Political leaders not only kill each other, but commonly commit mass violence against their citizenries. They may torture them, imprison them, execute them, starve them, or work them to death in pharaonic construction projects. Rummel estimates that governments killed 133 million people before the 20th century, and the total may be as high as 625 million.[96] So once raiding and feuding have been brought under control in a society, the greatest opportunity for reducing violence is reducing *government* violence.

By the 17th and 18th centuries, many countries had begun to cut back on tyranny and political murder.[97] Between the early Middle Ages and 1800, Eisner calculates, the European regicide rate declined fivefold, particularly in Western and Northern Europe. A famous example of this change is the fate of the two Stuart kings who locked horns with the English Parliament. In 1649 Charles I was beheaded, but in 1688 his son James II was deposed bloodlessly in the Glorious Revolution. Even after attempting to stage a coup he was merely forced into exile. By 1776 the American revolutionaries had defined "despotism" down to the level of taxing tea and quartering soldiers.

At the same time that governments were gradually becoming less tyrannical, thinkers were seeking a principled way to reel in government violence to the minimum necessary. It began with a conceptual revolution. Instead of taking government for granted as an organic part of the society, or as the local franchise of God's rule over his kingdom, people began to think of a government as a gadget—a piece of technology invented by humans for the purpose of enhancing their collective welfare. Of course, governments had never been deliberately invented, and they had been in place long before history was recorded, so this way of thinking required a considerable leap of the imagination. Thinkers such as Hobbes, Spinoza, Locke, and Rousseau, and later Jefferson, Hamilton, James Madison, and John Adams, fantasized about what life was like in a state of nature, and played out thought experiments about what a group of rational actors would come up with to better their lives. The resulting institutions would clearly bear no resemblance to the theocracies and hereditary monarchies of the day. It's hard to imagine a plausible simulation of rational actors in a state of nature choosing an arrangement that would give them the divine right of kings, "*L'état, c'est moi,*" or inbred ten-year-olds ascending to the throne. Instead, the government would serve at the pleasure of the people it governed. Its power to "keep them all in awe," as Hobbes put it, was not a license to brutalize its citizens in pursuit of its own interests but only a mandate to implement the agreement "that a man be willing, when others are so too . . . to lay down this right to all things; and be contented with so much liberty against other men, as he would allow other men against himself."[98]

It's fair to say that Hobbes himself didn't think through the problem deeply enough. He imagined that somehow people would vest authority in a sovereign or a committee once and for all at the dawn of time, and thereafter it would embody their interests so perfectly that they would never have reason to question it. One only has to think of a typical American congressman or member of the British royal family (to say nothing of a generalissimo or a commissar) to see how this would be a recipe for disaster. Real-life Leviathans are human beings, with all the greed and foolishness we should expect of a specimen of *Homo sapiens*. Locke recognized that people in power would be tempted to "exempt themselves from the obedience to the Laws they make, and suit the Law, both in its making and its execution, to their own private Wish, and thereby come to have a distinct Interest from the rest of the Community, contrary to the end of Society and Government."[99] He called for a separation between the legislative and executive branches of government, and for the citizenry to reserve the power to throw out a government that was no longer carrying out its mandate.

This line of thinking was taken to the next level by the heirs of Hobbes and Locke who hashed out a design for American constitutional government after years of study and debate. They were obsessed with the problem of how a ruling body composed of fallible humans could wield enough force to prevent citizens from preying on each other without arrogating so much that it would become the most destructive predator of all.[100] As Madison wrote, "If men were angels, no government would be necessary. If angels were to govern men, neither external nor internal controls on government would be necessary."[101] And so Locke's ideal of the separation of powers was written into the design of the new government, because "ambition must be made to counteract ambition."[102] The result was the division of government into executive, judicial, and legislative branches, the federalist system in which authority was divided between the states and the national government, and periodic elections to force the government to give some attention to the wishes of the populace and to transfer power in an orderly and peaceable way. Perhaps most important, the government was given a circumscribed mission statement—to secure the life, liberty, and pursuit of happiness of its citizens, with their consent—and, in the form of the Bill of Rights, a set of lines it could not cross in its use of violence against them.

Yet another innovation of the American system was its explicit recognition of the pacifying effects of positive-sum cooperation. The ideal of gentle commerce was implemented in the Commerce, Contract, and Takings clauses of the Constitution, which prevented the government from getting too much in the way of reciprocal exchanges among its citizens.[103]

The forms of democracy that were tried out in the 18th century were what you might expect of the 1.0 release of a complex new technology. The English implementation was weak tea, the French implementation an unmitigated disaster, and the American implementation had a flaw that is best captured in the actor Ice-T's impression of Thomas Jefferson reviewing a draft of the Constitution: "Let's see: freedom of speech; freedom of religion; freedom of the press; you can own niggers . . . Looks good to me!" But the value of the early designs for democracy was their upgradability. Not only did they carve out zones, however restricted, that were free of inquisitions, cruel punishments, and despotic authority, but they contained the means of their own expansion. The statement "We hold these truths to be self-evident, that all men are created equal," however hypocritical at the time, was a built-in rights-widener that could be invoked to end slavery four score and seven years later and other forms of racial coercion a century after that. The idea of democracy, once loosed on the world, would eventually infect larger and larger portions of it, and as we shall see, would turn out to be one of the greatest violence-reduction technologies since the appearance of government itself.

MAJOR WAR

For most of human history, the justification for war was pithily captured by Julius Caesar: "I came. I saw. I conquered." Conquest was what governments did. Empires rose, empires fell, entire populations were annihilated or enslaved, and no one seemed to think there was anything wrong with it. The historical figures who earned the honorific "So-and-So the Great" were not great artists, scholars, doctors, or inventors, people who enhanced human happiness or wisdom. They were dictators who conquered large swaths of territory and the people in them. If Hitler's luck had held out a bit longer, he probably would have gone down in history as Adolf the Great. Even today the standard histories of war teach the reader a great deal about horses and armor and gunpowder but give only the vaguest sense that immense numbers of people were killed and maimed in these extravaganzas.

At the same time, there have always been eyes that zoom in to the scale of the individual women and men affected by war and that have seen its moral dimension. In the 5th century BCE the Chinese philosopher Mozi, the founder of a rival religion to Confucianism and Taoism, noted:

> To kill one man is to be guilty of a capital crime, to kill ten men is to increase the guilt ten-fold, to kill a hundred men is to increase it a hundred-fold. This the rulers of the earth all recognize, and yet when it comes to the greatest crime—waging war on another state—they praise it! . . .
>
> If a man on seeing a little black were to say it is black, but on seeing a lot of black were to say it is white, it would be clear that such a man could not distinguish black and white. . . . So those who recognize a small crime as such, but do not recognize the wickedness of the greatest crime of all—the waging of war on another state—but actually praise it—cannot distinguish right and wrong.[104]

The occasional Western seer too paid homage to the ideal of peace. The prophet Isaiah expressed the hope that "they shall beat their swords into plowshares, and their spears into pruning hooks: nation shall not lift up sword against nation, neither shall they learn war any more."[105] Jesus preached, "Love your enemies, do good to those who hate you, bless those who curse you, pray for those who mistreat you. If someone strikes you on one cheek, turn to him the other also."[106] Though Christianity began as a pacifist movement, things went downhill in 312 CE when the Roman ruler Constantine had a vision of a flaming cross in the sky with the words "In this sign thou shalt conquer" and converted the Roman Empire to this militant version of the faith.

Periodic expressions of pacifism or war-weariness over the next millennium did nothing to stop the nearly constant state of warfare. According to the *Encyclopaedia Britannica*, the premises of international law during the Middle Ages were as follows: "In the absence of an agreed state of truce or peace, war was the basic state of international relations even between independent Christian communities; (2) Unless exceptions were made by means of individual safe conduct or treaty, rulers saw themselves entitled to treat foreigners at their absolute discretion; (3) The high seas were no-man's-land, where anyone might do as he pleased."[107] In the 15th, 16th, and 17th centuries, wars broke out between European countries at a rate of about three new wars a year.[108]

The moral arguments against war are irrefutable. As the musician Edwin Starr put it, "War. Hunh! What is it good for? Absolutely nothing. War means tears to thousands of mothers' eyes, when their sons go to fight and lose their lives." But for most of history this argument has not caught on, for two reasons.

One is the other-guy problem. If a nation decides not to learn war anymore, but its neighbor continues to do so, its pruning hooks will be no match for the neighbor's spears, and it may find itself at the wrong end of an invading army. This was the fate of Carthage against the Romans, India against Muslim invaders, the Cathars against the French and the Catholic Church, and the various countries stuck between Germany and Russia at many times in their history.

Pacifism is also vulnerable to militaristic forces *within* a country. When a country is embroiled in a war or on the verge of one, its leaders have trouble distinguishing a pacifist from a coward or a traitor. The Anabaptists are one of many pacifist sects that have been persecuted throughout history.[109]

To gain traction, antiwar sentiments have to infect many constituencies at the same time. And they have to be grounded in economic and political institutions, so that the war-averse outlook doesn't depend on everyone's deciding to become and stay virtuous. It was in the Age of Reason and the Enlightenment that pacifism evolved from a pious but ineffectual sentiment to a movement with a practicable agenda.

One way to drive home the futility and evil of war is to tap the distancing power of satire. A moralizer can be mocked, a polemicist can be silenced, but a satirist can get the same point across through stealth. By luring an audience into taking the perspective of an outsider—a fool, a foreigner, a traveler—a satirist can make them appreciate the hypocrisy of their own society and the flaws in human nature that foster it. If the audience gets the joke, if the readers or viewers lose themselves in the work, they have tacitly acceded to the author's deconstruction of a norm without anyone having had to rebuff it in so many words. Shakespeare's Falstaff, for example, delivers the finest analysis ever expressed of the concept of honor, the source of so much violence over the course of human history. Prince Hal has urged him into battle, saying "Thou owest God a death." Falstaff muses:

> 'Tis not due yet: I would be loath to pay him before his day. What need I be so forward with him that calls not on me? Well, 'tis no matter; honour pricks me on. Yea, but how if honour prick me off when I come on? How then? Can honour set to a leg? No. Or an arm? No. Or take away the grief of a wound? No. Honour hath no skill in surgery then? No. What is honour? A word. What is that word honour? Air—a trim reckoning! Who hath it? He that died a Wednesday. Doth he feel it? No. Doth he hear it? No. 'Tis insensible then? Yea, to the dead. But will it not live with the living? No. Why? Detraction

will not suffer it. Therefore I'll none of it. Honour is a mere scutcheon—and so ends my catechism.[110]

Detraction will not suffer it! More than a century later, in 1759, Samuel Johnson imagined a Quebec Indian chief commenting on "the art and regularity of European war" in a speech to his people during the Seven Years' War:

> They have a written law among them, of which they boast as derived from him who made the earth and sea, and by which they profess to believe that man will be made happy when life shall forsake him. Why is not this law communicated to us? It is concealed because it is violated. For how can they preach it to an Indian nation, when I am told that one of its first precepts forbids them to do to others what they would not that others should do to them. . . .
>
> The sons of rapacity have now drawn their swords upon each other, and referred their claims to the decision of war; let us look unconcerned upon the slaughter, and remember that the death of every European delivers the country from a tyrant and a robber; for what is the claim of either nation, but the claim of the vulture to the leveret, of the tiger to the fawn?[111]

(A leveret is a young hare.) Jonathan Swift's *Gulliver's Travels* (1726) was the quintessential exercise in the shifting of vantage points, in this case from the Lilliputian to the Brobdingnagian. Swift has Gulliver describe the recent history of his homeland to the King of Brobdingnag:

> He was perfectly astonished with the historical Account I gave him of our Affairs during the last Century, protesting it was only a Heap of Conspiracies, Rebellions, Murders, Massacres, Revolutions, Banishments, the very worst Effects that Avarice, Faction, Hypocrisy, Perfidiousness, Cruelty, Rage, Madness, Hatred, Envy, Lust, Malice, or Ambition could produce. . . .
>
> "As for yourself," (continued the King), "who have spent the greatest Part of your Life in Travelling, I am well disposed to hope you may hitherto have escaped many Vices of your Country. But by what I have gathered from your own Relation, and the Answers I have with much Pain wringed and extorted from you, I cannot but conclude the Bulk of your Natives to be the most pernicious Race of little odious Vermin that Nature ever suffered to crawl upon the Surface of the Earth."[112]

Satires appeared in France as well. In one of his *pensées*, Blaise Pascal (1623–62) imagined the following dialogue: "Why are you killing me for your own benefit? I am unarmed." "Why, do you not live on the other side of the water? My friend, if you lived on this side, I should be a murderer, but since you live on the other side, I am a hero, and it is just."[113] Voltaire's *Candide* (1759) was another novel that slipped scathing antiwar commentary into the mouth of a fictitious character, such as the following definition of war: "A million assassins in uniform, roaming from one end of Europe to the other, murder and pillage with discipline in order to earn their daily bread."

Together with satires suggesting that war was hypocritical and contemptible, the 18th century saw the appearance of theories holding that it was irrational and avoidable. One of the foremost was gentle commerce, the theory that the positive-sum payoff of trade should be more appealing than the zero-sum or negative-sum payoff of war.[114] Though the mathematics of game theory would not be available for another two hundred years, the key idea could be stated easily enough in words: Why spend money and blood to invade a country and plunder its treasure when you can just buy it from them at less expense and sell them some of your own? The Abbé de Saint Pierre (1713), Montesquieu (1748), Adam Smith (1776), George Washington (1788), and Immanuel Kant (1795) were some of the writers who extolled free trade because it yoked the material interests of nations and thus encouraged them to value one another's well-being. As Kant put it, "The spirit of commerce sooner or later takes hold of every people, and it cannot exist side by side with war. . . . Thus states find themselves compelled to promote the noble cause of peace, though not exactly from motives of morality."[115]

As they did with slavery, Quakers founded activist groups that opposed the institution of war. Though the sect's commitment to nonviolence sprang from its religious belief that God speaks through individual human lives, it didn't hurt the cause that they were influential businessmen rather than ascetic Luddites, having founded, among other concerns, Lloyd's of London, Barclays Bank, and the colony of Pennsylvania.[116]

The most remarkable antiwar document of the era was Kant's 1795 essay "Perpetual Peace."[117] Kant was no dreamer; he began the essay with the self-deprecating confession that he took the title of his essay from the caption on an innkeeper's sign with a picture of a burial ground. He then laid out six preliminary steps toward perpetual peace, followed by three sweeping principles. The preliminary steps were that peace treaties should not leave open the option of war; that states should not absorb other states; that standing armies should be abolished; that governments should not borrow to finance wars; that a state should not interfere in the internal governance of another state; and that in war, states should avoid tactics that would undermine confidence in a future peace, such as assassinations, poisonings, and incitements to treason.

More interesting were his "definitive articles." Kant was a strong believer in human nature; elsewhere he had written that "from the crooked timber of humanity no truly straight thing can be made." Thus he began from a Hobbesian premise:

> The state of peace among men living side by side is not the natural state; the natural state is one of war. This does not always mean open hostilities, but

> at least an unceasing threat of war. A state of peace, therefore, must be *established,* for in order to be secured against hostility it is not sufficient that hostilities simply be not committed; and, unless this security is pledged to each by his neighbor (a thing that can occur only in a civil state), each may treat his neighbor, from whom he demands this security, as an enemy.

He then outlined his three conditions for perpetual peace. The first is that states should be democratic. Kant himself preferred the term *republican,* because he associated the word *democracy* with mob rule; what he had in mind was a government dedicated to freedom, equality, and the rule of law. Democracies are unlikely to fight each other, Kant argued, for two reasons. One is that a democracy is a form of government that by design ("having sprung from the pure source of the concept of law") is built around nonviolence. A democratic government wields its power only to safeguard the rights of its citizens. Democracies, Kant reasoned, are apt to externalize this principle to their dealings with other nations, who are no more deserving of domination by force than are their own citizens.

More important, democracies tend to avoid wars because the benefits of war go to a country's leaders whereas the costs are paid by its citizens. In an autocracy "a declaration of war is the easiest thing in the world to decide upon, because war does not require of the ruler, who is the proprietor and not a member of the state, the least sacrifice of the pleasures of his table, the chase, his country houses, his court functions, and the like. He may, therefore, resolve on war as on a pleasure party for the most trivial reasons." But if the citizens are in charge, they will think twice about wasting their own money and blood on a foolish foreign adventure.

Kant's second condition for perpetual peace was that "the law of nations shall be founded on a Federation of Free States"—a "League of Nations," as he also called it. This federation, a kind of international Leviathan, would provide objective, third-party adjudication of disputes, circumventing every nation's tendency to believe that it is always in the right. Just as individuals accede to a social contract in which they surrender some of their freedom to the state to escape the nastiness of anarchy, so it should be with states: "For states in their relation to each other, there cannot be any reasonable way out of the lawless condition which entails only war except that they, like individual men, should give up their savage (lawless) freedom, adjust themselves to the constraints of public law, and thus establish a continuously growing state consisting of various nations which will ultimately include all the nations of the world."

Kant didn't have in mind a world government with a global army. He thought that international laws could be self-enforcing. "The homage which each state pays (at least in words) to the concept of law proves that there is slumbering in man an even greater moral disposition to become master of the evil principle in himself (which he cannot disclaim) and to hope for the same from others." The author of "Perpetual Peace" was, after all, the same man who proposed the Categorical Imperative, which stated that people should act so that the maxim of their action can be universalized. This is all starting to sound a bit starry-eyed, but Kant brought the idea back to earth by tying it to the spread of democracy. Each of two democracies can recognize the validity of the principles that govern the other. That sets them apart from theocracies, which are based on parochial faiths, and from autocracies, which are based on clans, dynasties, or charismatic leaders. In other words, if one state has reason to believe that a neighboring one organizes its political affairs in the same way that it does because both have stumbled upon the same solution to the problem of government, then neither has to worry about the other one attacking, neither will be tempted to attack the other in preemptive self-defense, and so on, freeing everyone from the Hobbesian trap. Today, for example, the Swedes don't stay up at night worrying that their neighbors are hatching plans for Norway Über Alles, or vice versa.

The third condition for perpetual peace is "universal hospitality" or "world citizenship." People from one country should be free to live in safety in others, as long as they don't bring an army in with them. The hope is that communication, trade, and other "peaceable relations" across national boundaries will knit the world's people into a single community, so that a "violation of rights in one place is felt throughout the world."

Obviously the satirists' deglorification of war and Kant's practical ideas on how to reduce it did not catch on widely enough to spare Western civilization the catastrophes of the next century and a half. But as we shall see, they planted the seeds of a movement that would blossom later and turn the world away from war. The new attitudes had an immediate impact as well. Historians have noted a change in the attitudes to war beginning around 1700. Leaders began to profess their love of peace and to claim that war had been forced upon them.[118] As Mueller notes, "No longer was it possible simply and honestly to proclaim like Julius Caesar, 'I came, I saw, I conquered.' Gradually this was changed to 'I came, I saw, he attacked me while I was just standing there looking, I won.' This might be seen as progress."[119]

More tangible progress was seen in the dwindling appeal of imperial power. In the 18th century some of the world's most bellicose nations, such as the Netherlands, Sweden, Spain, Denmark, and Portugal, reacted to military disappointments not by doubling down and plotting a return to glory but by dropping out of the conquest game, leaving war and empire to other countries and becoming commercial nations instead.[120] One of the results, as we shall see in the next chapter, was that wars between great powers became shorter, less frequent, and limited to fewer countries (though the advance of military organization meant that the wars that did occur were more damaging).[121]

And the greatest progress was yet to come. The extraordinary decline of major war in the last sixty years may be a delayed vindication of the ivory-tower theories of Immanuel Kant—if not "perpetual peace," then certainly a "long peace," and one that keeps getting longer. As the great thinkers of the Enlightenment predicted, we owe this peace not just to the belittling of war but to the spread of democracy, the expansion of trade and commerce, and the growth of international organizations.

WHENCE THE HUMANITARIAN REVOLUTION?

We have seen that in the span of just over a century, cruel practices that had been a part of civilization for millennia were suddenly abolished. The killing of witches, the torture of prisoners, the persecution of heretics, the execution of nonconformists, and the enslavement of foreigners—all carried out with stomach-turning cruelty—quickly passed from the unexceptionable to the unthinkable. Payne remarks on how difficult it is to explain these changes:

> The routes whereby uses of force are abandoned are often quite unexpected, even mysterious—so mysterious that one is sometimes tempted to allude to a higher power at work. Time and again one encounters violent practices so rooted and so self-reinforcing that it seems almost magical that they were overcome. One is reduced to pointing to "History" to explain how this immensely beneficial policy—a reduction in the use of force—has been gradually imposed on a human race that has neither consciously sought it nor agreed with it.[122]

One example of this mysterious, unsought progress is the long-term trend away from using force to punish debtors, which most people never realized was a trend. Another is the way that political murder had faded in English-speaking countries well before the principles of democracy had been articulated. In cases like these a nebulous shift in sensibilities may have been a prerequisite to consciously designed reforms. It's hard to imagine how a stable democracy can be implemented until competing factions give up the idea that murder is a good way to allocate power. The recent failure of democracy to take hold in many African and Islamic states is a reminder that a change in the norms surrounding violence has to precede a change in the nuts and bolts of governance.[123]

Still, a gradual shift in sensibilities is often incapable of changing actual practices until the change is implemented by the stroke of a pen. The slave trade, for example, was abolished as a result of moral agitation that persuaded men in power to pass laws and back them up with guns and ships.[124] Blood sports, public hangings, cruel punishments, and debtors' prisons were also shut down by acts of legislators who had been influenced by moral agitators and the public debates they began.

In explaining the Humanitarian Revolution, then, we don't have to decide between unspoken norms and explicit moral argumentation. Each affects the other. As sensibilities change, thinkers who question a practice are more likely to materialize, and their arguments are more likely to get a hearing and then catch on. The arguments may not only persuade the people who wield the levers of power but infiltrate the culture's sensibilities by finding their way into barroom and dinner-table debates where they may shift the consensus one mind at a time. And when a practice has vanished from everyday experience because it was outlawed from the top down, it may fall off the menu of live options in people's imaginations. Just as today smoking in offices and classrooms has passed from commonplace to prohibited to unthinkable, practices like slavery and public hangings, when enough time passed that no one alive could remember them, became so unimaginable that they were no longer brought up for debate.

The most sweeping change in everyday sensibilities left by the Humanitarian Revolution is the reaction to suffering in other living things. People today are far from morally immaculate. They may covet nice objects, fantasize about sex with inappropriate partners, or want to kill someone who has humiliated them in public.[125] But other sinful desires no longer occur to people in the first place. Most people today have no desire to watch a cat burn to death, let alone a man or a woman. In that regard we are different from our ancestors of a few centuries ago, who approved, carried out, and even savored the infliction of unspeakable agony on other living beings. What were these people feeling? And why don't we feel it today?

We won't be equipped to answer this question until we plunge into the psychology of sadism in chapter 8 and empathy in chapter 9. But for now we can look at some historical changes that militated against the indulgence of cruelty. As always, the challenge is to find an exogenous change that precedes the change in sensibilities and behavior so we can avoid the circularity of saying that people stopped doing cruel things because they got less cruel. What changed in people's environment that could have set off the Humanitarian Revolution?

The Civilizing Process is one candidate. Recall that Elias suggested that during the transition to modernity people not only exercised more self-control but also cultivated their sense of empathy. They did so not as an exercise in moral improvement but to hone their ability to get inside the heads of bureaucrats and merchants and prosper in a society that increasingly depended on networks of exchange rather than farming and plunder. Certainly the taste for cruelty clashes with the values of a cooperative society: it must be harder to work with your neighbors if you think they might enjoy seeing you disemboweled. And the reduction in personal violence brought about by the Civilizing Process may have lessened the demand for harsh punishments, just as today demands to "get tough on crime" rise and fall with the crime rate.

Lynn Hunt, the historian of human rights, points to another knock-on effect of the Civilizing Process: the refinements in hygiene and manners, such as eating with utensils, having sex in private, and trying to keep one's effluvia out of view and off one's clothing. The enhanced decorum, she suggests, contributed to the sense that people are *autonomous*—that they own their bodies, which have an inherent integrity and are not a possession of society. Bodily integrity was increasingly seen as worthy of respect, as something that may not be breached at the expense of the person for the benefit of society.

My own sensibilities tend toward the concrete, and I suspect there is a simpler hypothesis about the effect of cleanliness on moral sensibilities: people got less repulsive. Humans have a revulsion to filth and bodily secretions, and just as people today may avoid a homeless person who reeks of feces and urine, people in earlier centuries may have been more callous to their neighbors because those neighbors were more disgusting. Worse, people easily slip from visceral disgust to moralistic disgust and treat unsanitary things as contemptibly defiled and sordid.[126] Scholars of 20th-century atrocities have wondered how brutality can spring up so easily when one group achieves domination over another. The philosopher Jonathan Glover has pointed to a downward spiral of dehumanization. People force a despised minority to live in squalor, which makes them seem animalistic and subhuman, which encourages the dominant group to mistreat them further, which degrades them still further, removing any remaining tug on the oppressors' conscience.[127] Perhaps this spiral of dehumanization runs the movie of the Civilizing Process backwards. It reverses the historical sweep toward greater cleanliness and dignity that led, over the centuries, to greater respect for people's well-being.

Unfortunately the Civilizing Process and the Humanitarian Revolution don't line up in time in a way that would suggest that one caused the other. The rise of government and commerce and the plummeting of homicide that propelled the Civilizing Process had been under way for several centuries without anyone much caring about the barbarity of punishments, the power of kings, or the violent suppression of heresy. Indeed as states became more powerful, they also got crueler. The use of torture to extract confessions (rather than to punish), for example, was reintroduced in the Middle Ages when many states revived Roman law.[128] Something else must have accelerated humanitarian sentiments in the 17th and 18th centuries.

An alternative explanation is that people become more compassionate as their own lives improve. Payne speculates that "when people grow richer, so that they are better fed, healthier, and more comfortable, they come to value their own lives, and the lives of others, more highly."[129] The hypothesis that life used to be cheap but has become dearer loosely fits within the broad sweep of history. Over the millennia the world has moved away from barbaric practices like human sacrifice and sadistic executions, and over the millennia people have been living longer and in greater comfort. Countries that were at the leading edge of the abolition of cruelty, such as 17th-century England and Holland, were also among the more affluent countries of their time. And today it is in the poorer corners of the world that we continue to find backwaters with slavery, superstitious killing, and other barbaric customs.

But the life-was-cheap hypothesis also has some problems. Many of the more affluent states of their day, such as the Roman Empire, were hotbeds of sadism, and today harsh punishments like amputations and stonings may be found among the wealthy oil-exporting nations of the Middle East. A bigger problem is that the timing is off. The history of affluence in the modern West is depicted in figure 4–7, in which the economic historian Gregory Clark plots real income per person (calibrated in terms of how much money would be needed to buy a fixed amount of food) in England from 1200 to 2000.

Affluence began its liftoff only with the advent of the Industrial Revolution in the 19th century. Before 1800 the mathematics of Malthus prevailed: any advance in producing food only bred more mouths to feed, leaving the population as poor as before. This was true not only in England but all over the world. Between 1200 and 1800 measures of economic well-being, such as income, calories per capita, protein per capita, and number of surviving children per woman, showed no upward trend in any European country. Indeed, they were barely above the levels of hunter-gatherer societies. Only when the Industrial Revolution introduced more efficient manufacturing techniques and built an infrastructure of canals and railroads did European economies start to shoot upward and the populace become more affluent. Yet the

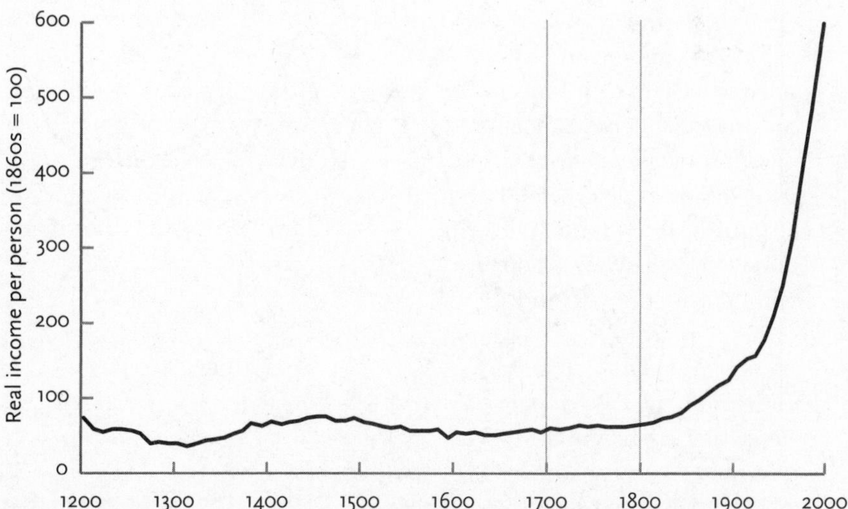

FIGURE 4–7. Real income per person in England, 1200–2000

Source: Graph from Clark, 2007a, p. 195.

humanitarian changes we are trying to explain began in the 17th century and were concentrated in the 18th.

Even if we could show that affluence correlated with humanitarian sensibilities, it would be hard to pinpoint the reasons. Money does not just fill the belly and put a roof over one's head; it also buys better governments, higher rates of literacy, greater mobility, and other goods. Also, it's not completely obvious that poverty and misery should lead people to enjoy torturing others. One could just as easily make the opposite prediction: if you have firsthand experience of pain and deprivation, you should be unwilling to inflict them on others, whereas if you have lived a cushy life, the suffering of others is less real to you. I will return to the life-was-cheap hypothesis in the final chapter, but for now we must seek other candidates for an exogenous change that made people more compassionate.

One technology that did show a precocious increase in productivity before the Industrial Revolution was book production. Before Gutenberg's invention of the printing press in 1452, every copy of a book had to be written out by hand. Not only was the process time-consuming—it took thirty-seven person-days to produce the equivalent of a 250-page book—but it was inefficient in materials and energy. Handwriting is harder to read than type is, and so handwritten books had to be larger, using up more paper and making the book more expensive to bind, store, and ship. In the two centuries after Gutenberg, publishing became a high-tech venture, and productivity in printing and papermaking grew more than twentyfold (figure 4–8), faster than the growth rate of the entire British economy during the Industrial Revolution.[130]

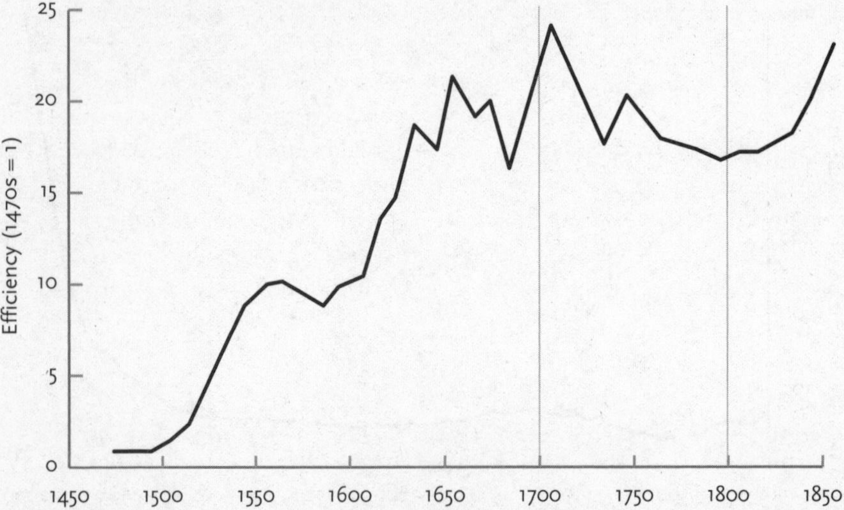

FIGURE 4–8. Efficiency in book production in England, 1470–1860s

Source: Graph from Clark, 2007a, p. 253.

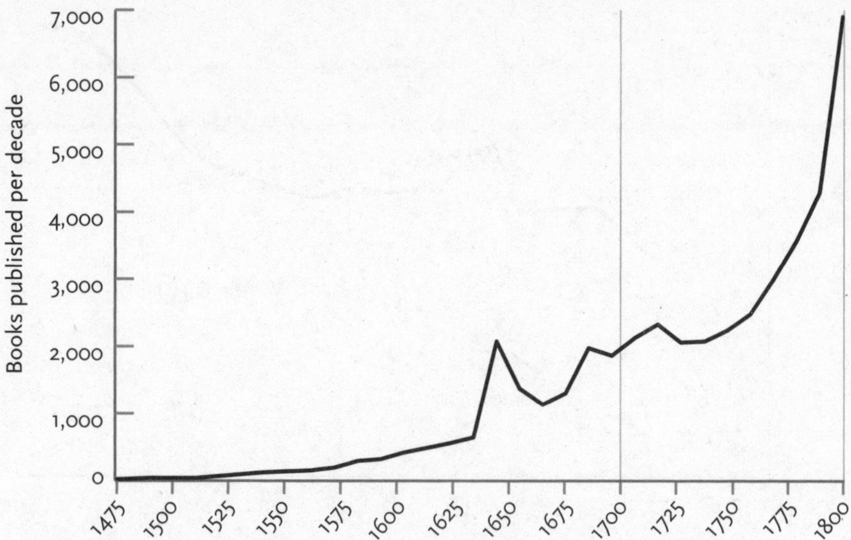

FIGURE 4–9. Number of books in English published per decade, 1475–1800

Sources: Simons, 2001; graph adapted from http://en.wikipedia.org/wiki/File:1477-1799_ESTC_titles_per_decade,_statistics.png.

The newly efficient publishing technology set off an explosion in book publication. Figure 4–9 shows that the number of books published per year rose significantly in the 17th century and shot up toward the end of the 18th.

The books, moreover, were not just playthings for aristocrats and intellectuals. As the literary scholar Suzanne Keen notes, "By the late 18th century, circulating libraries had become widespread in London and provincial towns, and most of what they offered for rent was novels."[131] With more numerous and cheaper books available, people had a greater incentive to read. It's not easy to estimate the level of literacy in periods before the advent of universal schooling and standardized testing, but historians have used clever proxy measures such as the proportion of people who could sign their marriage registers or court declarations. Figure 4–10 presents a pair of time series from Clark which suggest that during the 17th century in England, rates of literacy doubled, and that by the end of the century a majority of Englishmen had learned to read and write.[132]

Literacy was increasing in other parts of Western Europe at the same time. By the late 18th century a majority of French citizens had become literate, and though estimates of literacy don't appear for other countries until later, they suggest that by the early 19th century a majority of men were literate in Denmark, Finland, Germany, Iceland, Scotland, Sweden, and Switzerland as well.[133] Not only were more people reading, but they were reading in different ways, a development the historian Rolf Engelsing has called the Reading Revolution.[134]

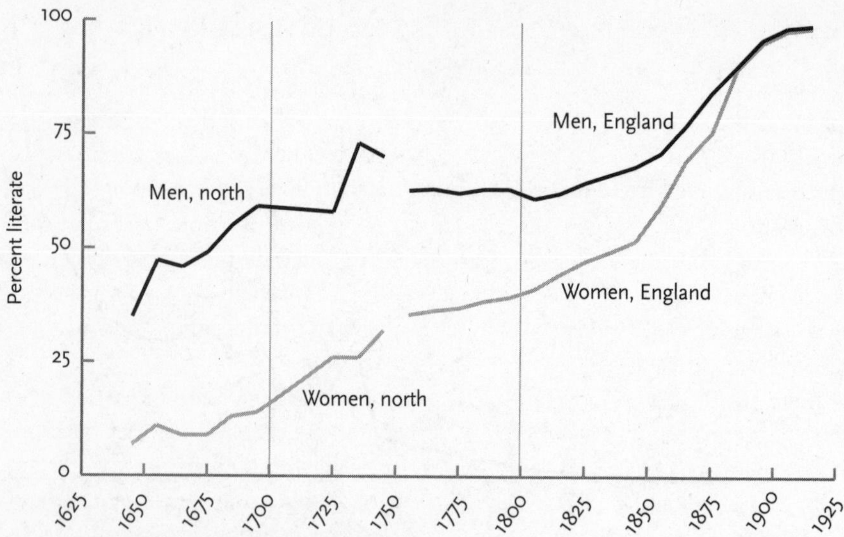

FIGURE 4–10. Literacy rate in England, 1625–1925
Source: Graph adapted from Clark, 2007a, p. 179.

People began to read secular rather than just religious material, to read to themselves instead of in groups, and to read a wide range of topical media, such as pamphlets and periodicals, rather than rereading a few canonical texts like almanacs, devotional works, and the Bible. As the historian Robert Darnton put it, "The late eighteenth century does seem to represent a turning point, a time when more reading matter became available to a wider public, when one can see the emergence of a mass readership that would grow to giant proportions in the nineteenth century with the development of machine-made paper, steam-powered presses, linotype, and nearly universal literacy."[135]

And of course people in the 17th and 18th centuries had more to read about. The Scientific Revolution had revealed that everyday experience is a narrow slice of a vast continuum of scales from the microscopic to the astronomical, and that our own abode is a rock orbiting a star rather than the center of creation. The European exploration of the Americas, Oceania, and Africa, and the discovery of sea routes to India and Asia, had opened up new worlds and revealed the existence of exotic peoples with ways of life very different from the readers' own.

The growth of writing and literacy strikes me as the best candidate for an exogenous change that helped set off the Humanitarian Revolution. The pokey little world of village and clan, accessible through the five senses and informed by a single content provider, the church, gave way to a phantasmagoria of people, places, cultures, and ideas. And for several reasons, the expansion of people's minds could have added a dose of humanitarianism to their emotions and their beliefs.

THE RISE OF EMPATHY AND THE REGARD FOR HUMAN LIFE

The human capacity for compassion is not a reflex that is triggered automatically by the presence of another living thing. As we shall see in chapter 9, though people in all cultures can react sympathetically to kin, friends, and babies, they tend to hold back when it comes to larger circles of neighbors, strangers, foreigners, and other sentient beings. In his book *The Expanding Circle,* the philosopher Peter Singer has argued that over the course of history, people have enlarged the range of beings whose interests they value as they value their own.[136] An interesting question is what inflated the empathy circle. And a good candidate is the expansion of literacy.

Reading is a technology for perspective-taking. When someone else's thoughts are in your head, you are observing the world from that person's vantage point. Not only are you taking in sights and sounds that you could not experience firsthand, but you have stepped inside that person's mind and are temporarily sharing his or her attitudes and reactions. As we shall see, "empathy" in the sense of adopting someone's viewpoint is not the same as "empathy" in the sense of feeling compassion toward the person, but the first can lead to the second by a natural route. Stepping into someone else's vantage point reminds you that the other fellow has a first-person, present-tense, ongoing stream of consciousness that is very much like your own but not the same as your own. It's not a big leap to suppose that the habit of reading other people's words could put one in the habit of entering other people's minds, including their pleasures and pains. Slipping even for a moment into the perspective of someone who is turning black in a pillory or desperately pushing burning faggots away from her body or convulsing under the two hundredth stroke of the lash may give a person second thoughts as to whether these cruelties should ever be visited upon anyone.

Adopting other people's vantage points can alter one's convictions in other ways. Exposure to worlds that can be seen only through the eyes of a foreigner, an explorer, or a historian can turn an unquestioned norm ("That's the way it's done") into an explicit observation ("That's what our tribe happens to do now"). This self-consciousness is the first step toward asking whether the practice could be done in some other way. Also, learning that over the course of history the first can become last and the last can become first may instill the habit of mind that reminds us, "There but for fortune go I."

The power of literacy to lift readers out of their parochial stations is not confined to factual writing. We have already seen how satirical fiction, which transports readers into a hypothetical world from which they can observe the follies of their own, may be an effective way to change people's sensibilities without haranguing or sermonizing.

Realistic fiction, for its part, may expand readers' circle of empathy by seducing them into thinking and feeling like people very different from themselves. Literature students are taught that the 18th century was a turning point in the history of the novel. It became a form of mass entertainment, and by the end of the century almost a hundred new novels were published in England and France every year.[137] And unlike earlier epics which recounted the exploits of heroes, aristocrats, or saints, the novels brought to life the aspirations and losses of ordinary people.

Lynn Hunt points out that the heyday of the Humanitarian Revolution, the late 18th century, was also the heyday of the epistolary novel. In this genre the story unfolds in a character's own words, exposing the character's thoughts and feelings in real time rather than describing them from the distancing perspective of a disembodied narrator. In the middle of the century three melodramatic novels named after female protagonists became unlikely bestsellers: Samuel Richardson's *Pamela* (1740) and *Clarissa* (1748), and Rousseau's *Julie, or the New Hélöise* (1761). Grown men burst into tears while experiencing the forbidden loves, intolerable arranged marriages, and cruel twists of fate in the lives of undistinguished women (including servants) with whom they had nothing in common. A retired military officer, writing to Rousseau, gushed:

> You have driven me crazy about her. Imagine then the tears that her death must have wrung from me. . . . Never have I wept such delicious tears. That reading created such a powerful effect on me that I believe I would have gladly died during that supreme moment.[138]

The philosophes of the Enlightenment extolled the way novels engaged a reader's identification with and sympathetic concern for others. In his eulogy for Richardson, Diderot wrote:

> One takes, despite all precautions, a role in his works, you are thrown into conversation, you approve, you blame, you admire, you become irritated, you feel indignant. How many times did I not surprise myself, as it happens to children who have been taken to the theater for the first time, crying: "Don't believe it, he is deceiving you.". . . His characters are taken from ordinary society . . . the passions he depicts are those I feel in myself.[139]

The clergy, of course, denounced these novels and placed several on the Index of Forbidden Books. One Catholic cleric wrote, "Open these works and you will see in almost all of them the rights of divine and human justice violated, parents' authority over their children scorned, the sacred bonds of marriage and friendship broken."[140]

Hunt suggests a causal chain: reading epistolary novels about characters unlike oneself exercises the ability to put oneself in other people's shoes, which turns one against cruel punishments and other abuses of human rights. As usual, it is hard to rule out alternative explanations for the correlation.

Perhaps people became more empathic for other reasons, which simultaneously made them receptive to epistolary novels and concerned with others' mistreatment.

But the full-strength causal hypothesis may be more than a fantasy of English teachers. The ordering of events is in the right direction: technological advances in publishing, the mass production of books, the expansion of literacy, and the popularity of the novel all preceded the major humanitarian reforms of the 18th century. And in some cases a bestselling novel or memoir demonstrably exposed a wide range of readers to the suffering of a forgotten class of victims and led to a change in policy. Around the same time that *Uncle Tom's Cabin* mobilized abolitionist sentiment in the United States, Charles Dickens's *Oliver Twist* (1838) and *Nicholas Nickleby* (1839) opened people's eyes to the mistreatment of children in British workhouses and orphanages, and Richard Henry Dana's *Two Years Before the Mast: A Personal Narrative of Life at Sea* (1840) and Herman Melville's *White Jacket* helped end the flogging of sailors. In the past century Erich Maria Remarque's *All Quiet on the Western Front*, George Orwell's *1984*, Arthur Koestler's *Darkness at Noon*, Aleksandr Solzhenitsyn's *One Day in the Life of Ivan Denisovich*, Harper Lee's *To Kill a Mockingbird*, Elie Wiesel's *Night*, Kurt Vonnegut's *Slaughterhouse-Five*, Alex Haley's *Roots*, Anchee Min's *Red Azalea*, Azar Nafisi's *Reading Lolita in Tehran*, and Alice Walker's *Possessing the Secret of Joy* (a novel that features female genital mutilation) all raised public awareness of the suffering of people who might otherwise have been ignored.[141] Cinema and television reached even larger audiences and offered experiences that were even more immediate. In chapter 9 we will learn of experiments that confirm that fictional narratives can evoke people's empathy and prick them to action.

Whether or not novels in general, or epistolary novels in particular, were the critical genre in expanding empathy, the explosion of reading may have contributed to the Humanitarian Revolution by getting people into the habit of straying from their parochial vantage points. And it may have contributed in a second way: by creating a hothouse for new ideas about moral values and the social order.

THE REPUBLIC OF LETTERS AND ENLIGHTENMENT HUMANISM

In David Lodge's 1988 novel *Small World*, a professor explains why he believes that the elite university has become obsolete:

> Information is much more portable in the modern world than it used to be. So are people. . . . There are three things which have revolutionized academic life in the last twenty years . . . : jet travel, direct-dialing telephones and the Xerox machine. . . . As long as you have access to a telephone, a Xerox machine, and a conference grant fund, you're OK, you're plugged into the only university that really matters—the global campus.[142]

Morris Zapp had a point, but he overemphasized the technologies of the 1980s. Two decades after his words were written, they have been superseded by e-mail, digital documents, Web sites, blogs, teleconferencing, Skype, and smartphones. And two centuries *before* they were written, the technologies of the day—the sailing ship, the printed book, and the postal service—had already made information and people portable. The result was the same: a global campus, a public sphere, or as it was called in the 17th and 18th centuries, the Republic of Letters.

Any 21st-century reader who dips into intellectual history can't help but be impressed by the blogosphere of the 18th. No sooner did a book appear than it would sell out, get reprinted, get translated into half a dozen languages, and spawn a flurry of commentary in pamphlets, correspondence, and additional books. Thinkers like Locke and Newton exchanged tens of thousands of letters; Voltaire alone wrote more than eighteen thousand, which now fill fifteen volumes.[143] Of course this colloquy unfolded on a scale that by today's standards was glacial—weeks, sometimes even months—but it was rapid enough that ideas could be broached, criticized, amalgamated, refined, and brought to the attention of people in power. A signature example is Beccaria's *On Crimes and Punishments*, which became an instant sensation and the impetus for the abolition of cruel punishments throughout Europe.

Given enough time and purveyors, a marketplace of ideas can not only disseminate ideas but change their composition. No one is smart enough to figure out anything worthwhile from scratch. As Newton (hardly a humble man) conceded in a 1675 letter to fellow scientist Robert Hooke, "If I have seen further it is by standing on the shoulders of giants." The human mind is adept at packaging a complicated idea into a chunk, combining it with other ideas into a more complex assembly, packaging that assembly into a still bigger contrivance, combining it with still other ideas, and so on.[144] But to do so it needs a steady supply of plug-ins and subassemblies, which can come only from a network of other minds.

A global campus increases not only the complexity of ideas but their quality. In hermetic isolation, all kinds of bizarre and toxic ideas can fester. Sunlight is the best disinfectant, and exposing a bad idea to the critical glare of other minds provides at least a chance that it will wither and die. Superstitions, dogmas, and legends ought to have a shorter half-life in a Republic of Letters, together with bad ideas about how to control crime or run a country. Setting fire to a person and seeing whether he burns is a dumb way to determine his guilt. Executing a woman for copulating with devils and turning them into cats is equally inane. And unless you are a hereditary absolutist monarch, you are unlikely to be persuaded that hereditary absolutist monarchy is the optimal form of government.

The jet airplane is the only technology of Lodge's small world of 1988 that has not been made obsolete by the Internet, and that reminds us that sometimes there is no substitute for face-to-face communication. Airplanes can bring people together, but people who live in a city are already together, so cities have long been crucibles of ideas. Cosmopolitan cities can bring together a critical mass of diverse minds, and their nooks and crannies can offer places for mavericks to seek refuge. The Age of Reason and the Enlightenment were also an age of urbanization. London, Paris, and Amsterdam became intellectual bazaars, and thinkers congregated in their salons, coffeehouses, and bookstores to hash out the ideas of the day.

Amsterdam played a special role as an arena of ideas. During the Dutch Golden Age in the 17th century it became a bustling port, open to the flow of goods, ideas, money, and people. It accommodated Catholics, Anabaptists, Protestants of various denominations, and Jews whose ancestors had been expelled from Portugal. It housed many book publishers, who did a brisk business printing controversial books and exporting them to the countries in which they had been banned. One Amsterdammer, Spinoza, subjected the Bible to literary analysis and developed a theory of everything that left no room for an animate God. In 1656 he was excommunicated by his Jewish community, who, with memories of the Inquisition still fresh, were nervous about making waves among the surrounding Christians.[145] It was no tragedy for Spinoza, as it might have been if he had lived in an isolated village, because he just picked up and moved to a new neighborhood and from there to another tolerant Dutch city, Leiden. In both places he was welcomed into the community of writers, thinkers, and artists. John Locke used Amsterdam as a safe haven in 1683 after he had been suspected of taking part in a plot against King Charles II in England. René Descartes also changed addresses frequently, bouncing around Holland and Sweden whenever things got too hot.

The economist Edward Glaeser has credited the rise of cities with the emergence of liberal democracy.[146] Oppressive autocrats can remain in power even when their citizens despise them because of a conundrum that economists call the social dilemma or free-rider problem. In a dictatorship, the autocrat and his henchmen have a strong incentive to stay in power, but no individual citizen has an incentive to depose him, because the rebel would assume all the risks of the dictator's reprisals while the benefits of democracy would flow diffusely to everyone in the country. The crucible of a city, however, can bring together financiers, lawyers, writers, publishers, and well-connected merchants who can collude in pubs and guild halls to challenge the current leadership, dividing the labor and diffusing the risk. Classical Athens, Renaissance Venice, revolutionary Boston and Philadelphia, and the cities of the Low Countries are examples of cities where new democracies were gestated, and today urbanization and democracy tend to go together.

The subversive power of the flow of information and people has never been lost on political and religious tyrants. That is why they suppress speech, writing, and association, and why democracies protect these channels in their bills of rights. Before the rise of cities and literacy, liberating ideas had a harder time being conceived and amalgamated, and so the rise of cosmopolitanism in the 17th and 18th centuries deserves part of the credit for the Humanitarian Revolution.

Bringing people and ideas together, of course, does not determine how those ideas will evolve. The rise of the Republic of Letters and the cosmopolitan city cannot, by themselves, explain why a humanitarian ethics arose in the 18th century, rather than ever-more-ingenious rationales for torture, slavery, despotism, and war.

My own view is that the two developments really are linked. When a large enough community of free, rational agents confers on how a society should run its affairs, steered by logical consistency and feedback from the world, their consensus will veer in certain directions. Just as we don't have to explain why molecular biologists discovered that DNA has four bases—given that they were doing their biology properly, and given that DNA really does have four bases, in the long run they could hardly have discovered anything else—we may not have to explain why enlightened thinkers would eventually argue against African slavery, cruel punishments, despotic monarchs, and the execution of witches and heretics. With enough scrutiny by disinterested, rational, and informed thinkers, these practices cannot be justified indefinitely. The universe of ideas, in which one idea entails others, is itself an exogenous force, and once a community of thinkers enters that universe, they will be forced in certain directions regardless of their material surroundings. I think this process of moral discovery was a significant cause of the Humanitarian Revolution.

I am prepared to take this line of explanation a step further. The reason so many violent institutions succumbed within so short a span of time was that the arguments that slew them belong to a coherent philosophy that emerged during the Age of Reason and the Enlightenment. The ideas of thinkers like Hobbes, Spinoza, Descartes, Locke, David Hume, Mary Astell, Kant, Beccaria, Smith, Mary Wollstonecraft, Madison, Jefferson, Hamilton, and John Stuart Mill coalesced into a worldview that we can call Enlightenment humanism. (It is also sometimes called classical liberalism, though since the 1960s the word *liberalism* has acquired other meanings as well.) Here is a potted account of this philosophy—a rough but more or less coherent composite of the views of these Enlightenment thinkers.

It begins with skepticism.[147] The history of human folly, and our own susceptibility to illusions and fallacies, tell us that men and women are fallible. One therefore ought to seek good *reasons* for believing something. Faith, revelation, tradition, dogma, authority, the ecstatic glow of subjective certainty—all are recipes for error, and should be dismissed as sources of knowledge.

Is there anything we can be certain of? Descartes gave as good an answer as any: our own consciousness. I know that I am conscious, by the very fact of wondering what I can know, and I can also know that my consciousness comprises several kinds of experience. These include the perception of an external world and of other people, and various pleasures and pains, both sensual (such as food, comfort, and sex) and spiritual (such as love, knowledge, and an appreciation of beauty).

We are also committed to reason. If we are asking a question, evaluating possible answers, and trying to persuade others of the value of those answers, then we are reasoning, and therefore have tacitly signed on to the validity of reason. We are also committed to whatever conclusions follow from the careful application of reason, such as the theorems of mathematics and logic.

Though we cannot logically *prove* anything about the physical world, we are entitled to have *confidence* in certain beliefs about it. The application of reason and observation to discover tentative generalizations about the world is what we call science. The progress of science, with its dazzling success at explaining and manipulating the world, shows that knowledge of the universe is possible, albeit always probabilistic and subject to revision. Science is thus a paradigm for how we ought to gain knowledge—not the particular methods or institutions of science but its value system, namely to seek to explain the world, to evaluate candidate explanations objectively, and to be cognizant of the tentativeness and uncertainty of our understanding at any time.

The indispensability of reason does not imply that individual people are always rational or are unswayed by passion and illusion. It only means that people are *capable* of reason, and that a community of people who choose to perfect this faculty and to exercise it openly and fairly can collectively reason their way to sounder conclusions in the long run. As Lincoln observed, you can fool all of the people some of the time, and you can fool some of the people all of the time, but you can't fool all of the people all of the time.

Among the beliefs about the world of which we can be highly confident is that other people are conscious in the same way that we are. Other people are made of the same stuff, seek the same kinds of goals, and react with external signs of pleasure and pain to the kinds of events that cause pain and pleasure in each of us.

By the same reasoning, we can infer that people who are different from us in many superficial ways—their gender, their race, their culture—are like us in fundamental ways. As Shakespeare's Shylock asks:

> Hath not a Jew eyes? hath not a Jew hands, organs, dimensions, senses, affections, passions? fed with the same food, hurt with the same weapons, subject to the same diseases, healed by the same means, warmed and cooled by the same winter and summer, as a Christian is? If you prick us, do we not bleed? if you tickle us, do we not laugh? if you poison us, do we not die? and if you wrong us, shall we not revenge?

The commonality of basic human responses across cultures has profound implications. One is that there is a universal human nature. It encompasses our common pleasures and pains, our common methods of reasoning, and our common vulnerability to folly (not least the desire for revenge). Human nature may be studied, just as anything else in the world may be. And our decisions on how to organize our lives can take the facts of human nature into account—including the discounting of our own intuitions when a scientific understanding casts them in doubt.

The other implication of our psychological commonality is that however much people differ, there can be, in principle, a meeting of the minds. I can appeal to your reason and try to persuade you, applying standards of logic and evidence that both of us are committed to by the very fact that we are both reasoning beings.

The universality of reason is a momentous realization, because it defines a place for morality. If I appeal to you to do something that affects me—to get off my foot, or not to stab me for the fun of it, or to save my child from drowning—then I can't do it in a way that privileges my interests over yours if I want you to take me seriously (say, by retaining my right to stand on your foot, or to stab you, or to let your children drown). I have to state my case in a way that would force me to treat you in kind. I can't act as if my interests are special just because I'm me and you're not, any more than I can persuade you that the spot I am standing on is a special place in the universe just because I happen to be standing on it.[148]

You and I ought to reach this moral understanding not just so we can have a logically consistent conversation but because mutual unselfishness is the only way we can simultaneously pursue our interests. You and I are both better off if we share our surpluses, rescue each other's children when they get into trouble, and refrain from knifing each other than we would be if we hoarded our surpluses while they rotted, let each other's children drown, and feuded incessantly. Granted, I might be a bit better off if I acted selfishly at your expense and you played the sucker, but the same is true for you with me, so if each of us tried for these advantages, we'd both end up worse off. Any neutral observer, and you and I if we could talk it over rationally, would have to conclude that the state we should aim for is the one where we both are unselfish.

Morality, then, is not a set of arbitrary regulations dictated by a vengeful deity and written down in a book; nor is it the custom of a particular culture or tribe. It is a consequence of the interchangeability of perspectives and the opportunity the world provides for positive-sum games. This foundation of morality may be seen in the many versions of the Golden Rule that have been discovered by the world's major religions, and also in Spinoza's Viewpoint of Eternity, Kant's Categorical Imperative, Hobbes and Rousseau's Social Contract, and Locke and Jefferson's self-evident truth that all people are created equal.

From the factual knowledge that there is a universal human nature, and the moral principle that no person has grounds for privileging his or her interests over others', we can deduce a great deal about how we ought to run our affairs. A government is a good thing to have, because in a state of anarchy people's self-interest, self-deception, and fear of these shortcomings in others would lead to constant strife. People are better off abjuring violence, if everyone else agrees to do so, and vesting authority in a disinterested third party. But since that third party will consist of human beings, not angels, their power must be checked by the power of other people, to force them to govern with the consent of the governed. They may not use violence against their citizens beyond the minimum necessary to prevent greater violence. And they should foster arrangements that allow people to flourish from cooperation and voluntary exchange.

This line of reasoning may be called humanism because the value that it recognizes is the flourishing of humans, the only value that cannot be denied. I experience pleasures and pains, and pursue goals in service of them, so I cannot reasonably deny the right of other sentient agents to do the same.

If all this sounds banal and obvious, then you are a child of the Enlightenment, and have absorbed its humanist philosophy. As a matter of historical fact, there is nothing banal or obvious about it. Though not necessarily atheistic (it is compatible with a deism in which God is identified with the nature of the universe), Enlightenment humanism makes no use of scripture, Jesus, ritual, religious law, divine purpose, immortal souls, an afterlife, a messianic age, or a God who responds to individual people. It sweeps aside many secular sources of value as well, if they cannot be shown to be necessary for the enhancement of human flourishing. These include the prestige of the nation, race, or class; fetishized virtues such as manliness, dignity, heroism, glory, and honor; and other mystical forces, quests, destinies, dialectics, and struggles.

I would argue that Enlightenment humanism, whether invoked explicitly or implicitly, underlay the diverse humanitarian reforms of the 18th and 19th centuries. The philosophy was explicitly invoked in the design of the first liberal democracies, most transparently in the "self-evident truths" in the American Declaration of Independence. Later it would spread to other parts of the world, blended with humanistic arguments that had arisen independently in those civilizations.[149] And as we shall see in chapter 7, it regained momentum during the Rights Revolutions of the present era.

For all that, Enlightenment humanism did not, at first, carry the day. Though it helped to eliminate many barbaric practices and established beachheads in the first liberal democracies, its full implications were roundly rejected in much of the world. One objection arose from a tension between the forces of enlightenment we have been exploring in this chapter and the forces of civilization we explored in the previous one—though as we shall see, it is not difficult to reconcile the two. The other objection was more foundational, and its consequences more fateful.

CIVILIZATION AND ENLIGHTENMENT

On the heels of the Enlightenment came the French Revolution: a brief promise of democracy followed by a train of regicides, putsches, fanatics, mobs, terrors, and preemptive wars, culminating in a megalomaniacal emperor and an insane war of conquest. More than a quarter of a million people were killed in the Revolution and its aftermath, and another 2 to 4 million were killed in the Revolutionary and Napoleonic Wars. In reflecting on this catastrophe, it was natural for people to reason, "After this, therefore because of this," and for intellectuals on the right and the left to blame the Enlightenment. This is what you get, they say, when you eat the fruit of the tree of knowledge, steal fire from the gods, and open Pandora's box.

The theory that the Enlightenment was responsible for the Terror and Napoleon is, to put it mildly, dubious. Political murder, massacre, and wars of imperial expansion are as old as civilization, and had long been the everyday stuff of European monarchies, including that of France. Many of the French philosophes from whom the revolutionaries drew their inspiration were intellectual lightweights and did not represent the stream of reasoning that connected Hobbes, Descartes, Spinoza, Locke, Hume, and Kant. The American Revolution, which stuck more closely to the Enlightenment script, gave the world a liberal democracy that has lasted more than two centuries. Toward the end of this book I will argue that the data on the historical decline of violence vindicate Enlightenment humanism and refute its critics on the right and the left. But one of these critics, the Anglo-Irish writer Edmund Burke, deserves our attention, because his argument appeals to the other major explanation for the decline of violence, the civilizing process. The two explanations overlap—both appeal to an expansion of empathy and to the pacifying effects of positive-sum cooperation—but they differ in which aspect of human nature they emphasize.

Burke was the father of intellectual secular conservatism, which is based on what the economist Thomas Sowell has called a tragic vision of human nature.[150] In that vision, human beings are permanently saddled with limitations of knowledge, wisdom, and virtue. People are selfish and shortsighted, and if they are left to their own devices, they will plunge into a Hobbesian war of all against all. The only things that keep people from falling into this abyss are the habits of self-control and social harmony they absorb when they conform to the norms of a civilized society. Social customs, religious traditions, sexual mores, family structures, and long-standing political institutions, even if no one can articulate their rationale, are time-tested work-arounds for the shortcomings of an unchanging human nature and are as indispensable today as when they lifted us out of barbarism.

According to Burke, no mortal is smart enough to design a society from first principles. A society is an organic system that develops spontaneously, governed by myriad interactions and adjustments that no human mind can pretend to understand. Just because we cannot capture its workings in verbal propositions does not mean it should be scrapped and reinvented according to the fashionable theories of the day. Such ham-fisted tinkering will only lead to unintended consequences, culminating in violent chaos.

Burke clearly went too far. It would be mad to say that people should never have agitated against torture, witch hunts, and slavery because these were long-standing traditions and that if they were suddenly abolished society would descend into savagery. The practices themselves were savage, and as we have seen, societies find ways to compensate for the disappearance of violent practices that were once thought to be indispensable. Humanitarianism can be the mother of invention.

But Burke had a point. Unspoken norms of civilized behavior, both in everyday interactions and in the conduct of government, may be a prerequisite to implementing certain reforms successfully. The development of these norms may be the mysterious "historical forces" that Payne remarked on, such as the spontaneous fading of political murder well before the principles of democracy had been articulated, and the sequence in which some abolition movements gave the coup de grâce to practices that were already in decline. They may explain why today it is so hard to impose liberal democracy on countries in the developing world that have not outgrown their superstitions, warlords, and feuding tribes.[151]

Civilization and Enlightenment need not be alternatives in explaining declines of violence. In some periods, tacit norms of empathy, self-control, and cooperation may take the lead, and rationally articulated principles of equality, nonviolence, and human rights may follow. In other periods, it may go in the other direction.

This to-and-fro may explain why the American Revolution was not as calamitous as its French counterpart. The Founders were products not just of the Enlightenment but of the English Civilizing Process, and self-control and cooperation had become second nature to them. "A decent respect to the opinions of mankind requires that they should declare the causes which impel them to the separation," the Declaration politely explains. "Prudence, indeed, will dictate that Governments long established should not be changed for light and transient causes." Prudence, indeed.

But their decency and prudence were more than mindless habits. The Founders consciously deliberated about just those limitations of human nature that made Burke so nervous about conscious deliberation. "What is government itself," asked Madison, "but the greatest of all reflections on human nature?"[152] Democracy, in their vision, had to be designed to counteract the vices of human nature, particularly the temptation in leaders to abuse their power. An acknowledgment of human nature may have been the chief difference between the American revolutionaries and their French confrères, who had the romantic conviction that they were rendering human limitations obsolete. In 1794 Maximilien Robespierre, architect of the Terror, wrote, "The French people seem to have outstripped the rest of humanity by two thousand years; one might be tempted to regard them, living amongst them, as a different species."[153]

In *The Blank Slate* I argued that two extreme visions of human nature—a Tragic vision that is resigned to its flaws, and a Utopian vision that denies it exists—define the great divide between right-wing and left-wing political ideologies.[154] And I suggested that a better understanding of human nature in the light of modern science can point the way to an approach to politics that is more sophisticated than either. The human mind is not a blank slate, and no humane political system should be allowed to deify its leaders or remake its citizens. Yet for all its limitations, human nature includes a recursive, open-ended, combinatorial system for reasoning, which can take cognizance of its own limitations. That is why the engine of Enlightenment humanism, rationality, can never be refuted by some flaw or error in the reasoning of the people in a given era. Reason can always stand back, take note of the flaw, and revise its rules so as not to succumb to it the next time.

BLOOD AND SOIL

A second counter-Enlightenment movement took root in the late 18th and early 19th centuries and was centered not in England but in Germany. The various strands have been explored in an essay by Isaiah Berlin and a book by the philosopher Graeme Garrard.[155] This counter-Enlightenment originated with Rousseau and was developed by theologians, poets, and essayists such as Johann Hamann, Friedrich Jacobi, Johann Herder, and Friedrich Schelling. Its target was not, as it was for Burke, the unintended consequences of Enlightenment reason for social stability, but the foundations of reason itself.

The first mistake, they said, was to start from the consciousness of an individual mind. The disembodied individual reasoner, ripped from his culture and its history, is a figment of the Enlightenment thinker's imagination. A person is not a locus of abstract cogitation—a brain on a stick—but a body with emotions and a part of the fabric of nature.

The second mistake was to posit a universal human nature and a universally valid system of reasoning. People are embedded in a culture and find meaning in its myths, symbols, and epics. Truth does not reside in propositions in the sky, there for everyone to see, but is situated in narratives and archetypes that are particular to the history of a place and give meaning to the lives of its inhabitants.

In this way of thinking, for a rational analyst to criticize traditional beliefs or customs is to miss the point. Only if one enters into the experience of those who live by those beliefs can one truly understand them. The Bible, for example, can be appreciated only by reproducing the experience of ancient shepherds in the Judaean hills. Every culture has a unique *Schwerpunkt*, a center of gravity, and unless we try to occupy it, we cannot comprehend its meaning and value.[156] Cosmopolitanism, far from being a virtue, is a "shedding of all that makes one most human, most oneself."[157] Universality, objectivity, and rationality are out; romanticism, vitalism, intuition, and irrationalism are in. Herder summed up the *Sturm und Drang* (storm and impulse) movement he helped to inspire: "I am not here to think, but to be, feel, live! . . . Heart! Warmth! Blood! Humanity! Life!"[158]

A child of the counter-Enlightenment, then, does not pursue a goal because it is objectively true or virtuous, but because it is a unique product of one's creativity. The wellspring of creativity may be in one's own true self, as the Romantic painters and writers insisted, or it may be in some kind of transcendent entity: a cosmic spirit, a divine flame. Berlin elaborates:

> Others again identified the creative self with a super-personal "organism" of which they saw themselves as elements or members—nation, or church, or culture, or class, or history itself, a mighty force of which they conceived their earthly selves as emanations. Aggressive nationalism, self-identification with the interests of the class, the culture or the race, or the forces of progress—with the wave of the future-directed dynamism of history, something that at once explains and justifies acts which might be abhorred or despised if committed from calculation of selfish advantage or some other mundane motive—this family of political and moral conceptions is so many expressions of a doctrine of self-realization based on defiant rejection of the central theses of the Enlightenment, according to which what is true, or right, or good, or beautiful, can be shown to be valid for all men by the correct application of objective methods of discovery and interpretation, open to anyone to use and verify.[159]

The counter-Enlightenment also rejected the assumption that violence was a problem to be solved. Struggle and bloodshed are inherent in the natural order, and cannot be eliminated without draining life of its vitality and subverting the destiny of mankind. As Herder put it, "Men desire harmony, but nature knows better what is good for the species: it desires strife."[160] The glorification of the struggle in "nature red in tooth and claw" (as Tennyson had put it) was a pervasive theme in 19th-century art and writing. Later it would be retrofitted with a scientific patina in the form of "social Darwinism," though the connection with Darwin is anachronistic and unjust: *The Origin of Species* was published in 1859, long after romantic struggleism had become a popular philosophy, and Darwin himself was a thoroughgoing liberal humanist.[161]

The counter-Enlightenment was the wellspring of a family of romantic movements that gained strength during the 19th century. Some of them influenced the arts and gave us sublime music and poetry. Others became political ideologies and led to horrendous reversals in the trend of declining violence. One of these ideologies was a form of militant nationalism that came to be known as "blood and soil"—the notion that an ethnic group and the land from which it originated form an organic whole with unique moral qualities, and that its grandeur and glory are more precious than the lives and happiness of its individual members. Another was romantic militarism, the idea that (as Mueller has summarized it) "war is noble, uplifting, virtuous, glorious, heroic, exciting, beautiful, holy, thrilling."[162] A third was Marxist socialism, in which history is a glorious struggle between classes, culminating in the subjugation of the bourgeoisie and the supremacy of the proletariat. And a fourth was National Socialism, in which history is a glorious struggle between races, culminating in the subjugation of inferior races and the supremacy of the Aryans.

The Humanitarian Revolution was a milestone in the historical reduction of violence and is one of humanity's proudest achievements. Superstitious killing, cruel punishments, frivolous executions, and chattel slavery may not have been obliterated from the face of the earth, but they have certainly been pushed to the margins. And despotism and major war, which had cast their shadow on humanity since the beginning of civilization, began to show cracks. The philosophy of Enlightenment humanism that united these developments got a toehold in the West and bided its time until more violent ideologies tragically ran their course.

5

THE LONG PEACE

War appears to be as old as mankind, but peace is a modern invention.

—Henry Maine

In the early 1950s, two eminent British scholars reflected on the history of war and ventured predictions on what the world should expect in the years to come. One of them was Arnold Toynbee (1889–1975), perhaps the most famous historian of the 20th century. Toynbee had served in the British Foreign Office during both world wars, had represented the government at the peace conferences following each one, and had been chronicling the rise and fall of twenty-six civilizations in his monumental twelve-volume work *A Study of History*. The patterns of history, as he saw them in 1950, did not leave him optimistic:

> In our recent Western history war has been following war in an ascending order of intensity; and today it is already apparent that the War of 1939–45 was not the climax of this crescendo movement.[1]

Writing in the shadow of World War II and at the dawn of the Cold War and the nuclear age, Toynbee could certainly be forgiven for his bleak prognostication. Many other distinguished commentators were equally pessimistic, and predictions of an imminent doomsday continued for another three decades.[2]

The other scholar's qualifications could not be more different. Lewis Fry Richardson (1881–1953) was a physicist, meteorologist, psychologist, and applied mathematician. His main claim to fame had been devising numerical techniques for predicting the weather, decades before there were computers powerful enough to implement them.[3] Richardson's own prediction about the future came not from erudition about great civilizations but from statistical analysis of a dataset of hundreds of violent conflicts spanning more than a century. Richardson was more circumspect than Toynbee, and more optimistic.

> The occurrence of two world wars in the present century is apt to leave us with the vague belief that the world has become more warlike. But this belief

> needs logical scrutiny. A long future may perhaps be coming without a third world war in it.[4]

Richardson chose statistics over impressions to defy the common understanding that global nuclear war was a certainty. More than half a century later, we know that the eminent historian was wrong and the obscure physicist was right.

This chapter is about the full story behind Richardson's prescience: the trends in war between major nations, culminating in the unexpected good news that the apparent crescendo of war did not continue to a new climax. During the last two decades, the world's attention has shifted to other kinds of conflict, including wars in smaller countries, civil wars, genocides, and terrorism; they will be covered in the following chapter.

STATISTICS AND NARRATIVES

The 20th century would seem to be an insult to the very suggestion that violence has declined over the course of history. Commonly labeled the most violent century in history, its first half saw a cascade of world wars, civil wars, and genocides that Matthew White has called the Hemoclysm, the blood-flood.[5] The Hemoclysm was not just an unfathomable tragedy in its human toll but an upheaval in humanity's understanding of its historical movement. The Enlightenment hope for progress led by science and reason gave way to a sheaf of grim diagnoses: the recrudescence of a death instinct, the trial of modernity, an indictment of Western civilization, man's Faustian bargain with science and technology.[6]

But a century is made up of a hundred years, not fifty. The second half of the 20th century saw a historically unprecedented avoidance of war between the great powers which the historian John Gaddis has called the Long Peace, followed by the equally astonishing fizzling out of the Cold War.[7] How can we make sense of the multiple personalities of this twisted century? And what can we conclude about the prospects for war and peace in the present one?

The competing predictions of Toynbee the historian and Richardson the physicist represent complementary ways of understanding the flow of events in time. Traditional history is a narrative of the past. But if we are to heed George Santayana's advisory to remember the past so as not to repeat it, we need to discern *patterns* in the past, so we can know what to generalize to the predicaments of the present. Inducing generalizable patterns from a finite set of observations is the stock in trade of the scientist, and some of the lessons of pattern extraction in science may be applied to the data of history.

Suppose, for the sake of argument, that World War II was the most destructive event in history. (Or if you prefer, suppose that the entire Hemoclysm deserves that designation, if you consider the two world wars and their associated genocides to be a single protracted historical episode.) What does that tell us about long-term trends in war and peace?

The answer is: nothing. The most destructive event in history had to take place in *some* century, and it could be embedded in any of a large number of very different long-term trends. Toynbee assumed that World War II was a step in an escalating staircase, as in the left panel in figure 5–1. Almost as gloomy is the common suggestion that epochs of war are cyclical, as in the right panel of figure 5–1. Like many depressing prospects, both models have spawned some black humor. I am often asked if I've heard the one about the man who fell off the roof of an office building and shouted to the workers on each floor, "So far so good!" I have also been told (several times) about the turkey who, on the eve of Thanksgiving, remarked on the extraordinary 364-day era of peace between farmers and turkeys he is lucky enough to be living in.[8]

But are the processes of history really as deterministic as the law of gravity or the cycling of the planet? Mathematicians tells us that an infinite number of curves can be drawn through any finite set of points. Figure 5–2 shows two other curves which situate the same episode in very different narratives.

The left panel depicts the radical possibility that World War II was a statistical fluke—that it was neither a step in an escalating series nor a harbinger of things to come, and not part of a trend at all. At first the suggestion seems preposterous. How could a random unfolding of events in time result in so many catastrophes being bunched together in just a decade: the brutal invasions by Hitler, Mussolini, Stalin, and Imperial Japan; the Holocaust; Stalin's purge; the Gulag; and two atomic explosions (to say nothing of World War I and the wars and genocides of the preceding two decades)? Also, the usual wars we find in history books tend to have death tolls in the tens or hundreds of thousands or, very rarely, in the millions. If wars really broke out at random,

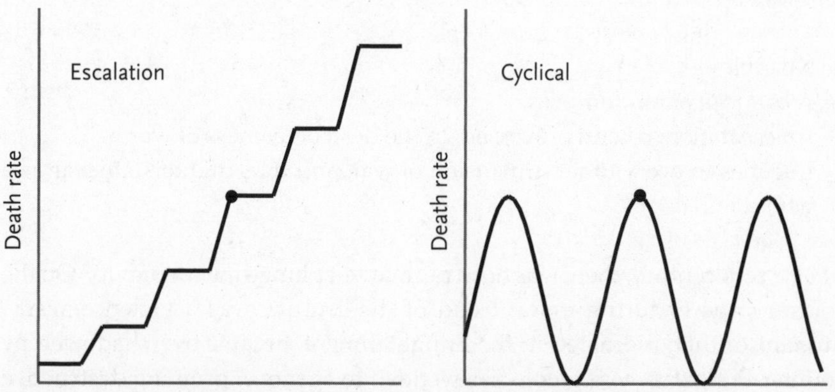

FIGURE 5–1. Two pessimistic possibilities for historical trends in war

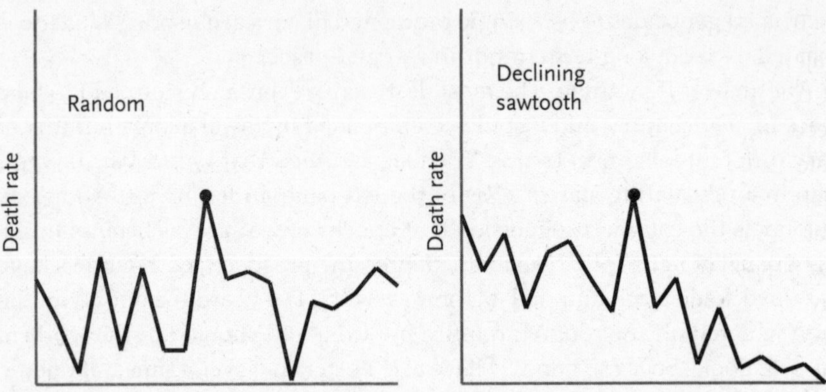

FIGURE 5–2. Two less pessimistic possibilities for historical trends in war

shouldn't a war that led to the deaths of 55 million people be astronomically improbable? Richardson showed that both these intuitions are cognitive illusions. When the iron dice begin to roll (as the German chancellor Theobald von Bethmann-Hollweg put it on the eve of World War I), the unlucky outcomes can be far worse than our primitive imaginations foresee.

The right-hand panel in figure 5–2 places the war in a narrative that is so unpessimistic that it's almost optimistic. Could World War II be an isolated peak in a declining sawtooth—the last gasp in a long slide of major war into historical obsolescence? Again, we will see that this possibility is not as dreamy as it sounds.

The long-term trajectory of war, in reality, is likely to be a superimposition of several trends. We all know that patterns in other complex sequences, such as the weather, are a composite of several curves: the cyclical rhythm of the seasons, the randomness of daily fluctuations, the long-term trend of global warming. The goal of this chapter is to identify the components of the long-term trends in wars between states. I will try to persuade you that they are as follows:

- No cycles.
- A big dose of randomness.
- An escalation, recently reversed, in the destructiveness of war.
- Declines in every other dimension of war, and thus in interstate war as a whole.

The 20th century, then, was not a permanent plunge into depravity. On the contrary, the enduring moral trend of the century was a violence-averse humanism that originated in the Enlightenment, became overshadowed by counter-Enlightenment ideologies wedded to agents of growing destructive power, and regained momentum in the wake of World War II.

To reach these conclusions, I will blend the two ways of understanding the trajectory of war: the statistics of Richardson and his heirs, and the narratives of traditional historians and political scientists. The statistical approach is necessary to avoid Toynbee's fallacy: the all-too-human tendency to hallucinate grand patterns in complex statistical phenomena and confidently extrapolate them into the future. But if narratives without statistics are blind, statistics without narratives are empty. History is not a screen saver with pretty curves generated by equations; the curves are abstractions over real events involving the decisions of people and the effects of their weapons. So we also need to explain how the various staircases, ramps, and sawtooths we see in the graphs emerge from the behavior of leaders, soldiers, bayonets, and bombs. In the course of the chapter, the ingredients of the blend will shift from the statistical to the narrative, but neither is dispensable in understanding something as complex as the long-term trajectory of war.

WAS THE 20th CENTURY REALLY THE WORST?

"The twentieth century was the bloodiest in history" is a cliché that has been used to indict a vast range of demons, including atheism, Darwin, government, science, capitalism, communism, the ideal of progress, and the male gender. But is it true? The claim is rarely backed up by numbers from any century other than the 20th, or by a mention of the hemoclysms of centuries past. The truth is that we will never really know which was the worst century, because it's hard enough to pin down death tolls in the 20th century, let alone earlier ones. But there are two reasons to suspect that the bloodiest-century factoid is an illusion.

The first is that while the 20th century certainly had more violent deaths than earlier ones, it also had more people. The population of the world in 1950 was 2.5 billion, which is about two and a half times the population in 1800, four and a half times that in 1600, seven times that in 1300, and fifteen times that of 1 CE. So the death count of a war in 1600, for instance, would have to be multiplied by 4.5 for us to compare its destructiveness to those in the middle of the 20th century.[9]

The second illusion is *historical myopia:* the closer an era is to our vantage point in the present, the more details we can make out. Historical myopia can afflict both common sense and professional history. The cognitive psychologists Amos Tversky and Daniel Kahneman have shown that people intuitively estimate relative frequency using a shortcut called the availability heuristic: the easier it is to recall examples of an event, the more probable people think it is.[10] People, for example, overestimate the likelihoods of the kinds of accidents that make headlines, such as plane crashes, shark attacks, and terrorist bombings, and they underestimate those that pile up unremarked, like electrocutions, falls, and drownings.[11] When we are judging the density of killings in different centuries, anyone who doesn't consult the numbers is apt to overweight the conflicts that are most recent, most studied, or most sermonized. In a survey of historical memory, I asked a hundred Internet users to write down as many wars as they could remember in five minutes. The responses were heavily weighted toward the world wars, wars fought by the United States, and wars close to the present. Though the earlier centuries, as we shall see, had far more wars, people *remembered* more wars from the recent centuries.

When one corrects for the availability bias and the 20th-century population explosion by rooting around in history books and scaling the death tolls by the world population at the time, one comes across many wars and massacres that could hold their head high among 20th-century atrocities. The table on page 195 is a list from White called "(Possibly) The Twenty (or so) Worst Things People Have Done to Each Other."[12] Each death toll is the median or mode of the figures cited in a large number of histories and encyclopedias. They include not just deaths on the battlefield but indirect deaths of civilians from starvation and disease; they are thus considerably higher than estimates of battlefield casualties, though consistently so for both recent and ancient events. I have added two columns that scale the death tolls and adjust the rankings to what they would be if the world at the time had had the population it did in the middle of the 20th century.

First of all: had you even heard of all of them? (I hadn't.) Second, did you know there were five wars and four atrocities before World War I that killed more people than that war? I suspect many readers will also be surprised to learn that of the twenty-one worst things that people have ever done to each other (that we know of), fourteen were in centuries before the 20th. And all this pertains to absolute numbers. When you scale by population size, only one of the 20th century's atrocities even makes the top ten. The worst atrocity of all time was the An Lushan Revolt and Civil War, an eight-year rebellion during China's Tang Dynasty that, according to censuses, resulted in the loss of two-thirds of the empire's population, a sixth of the world's population at the time.[13]

These figures, of course, cannot all be taken at face value. Some tendentiously blame the entire death toll of a famine or epidemic on a particular war, rebellion, or tyrant. And some came from innumerate cultures that lacked modern techniques for counting and record-keeping. At the same time, narrative history confirms that earlier civilizations were certainly capable of killing in vast numbers. Technological backwardness was no impediment; we know from Rwanda and Cambodia that massive numbers of people can be murdered with low-tech means like machetes and starvation. And in the distant past, implements of killing were not always so low-tech, because military weaponry usually boasted the most advanced technology of the age. The military historian John Keegan notes that by the middle of the 2nd millennium BCE, the chariot allowed nomadic armies to rain death on the civilizations

Rank	Cause	Century	Death toll	Death toll: mid-20th-century equivalent	Adjusted Rank
1	Second World War	20th	55,000,000	55,000,000	9
2	Mao Zedong (mostly government-caused famine)	20th	40,000,000	40,000,000	11
3	Mongol Conquests	13th	40,000,000	278,000,000	2
4	An Lushan Revolt	8th	36,000,000	429,000,000	1
5	Fall of the Ming Dynasty	17th	25,000,000	112,000,000	4
6	Taiping Rebellion	19th	20,000,000	40,000,000	10
7	Annihilation of the American Indians	15th–19th	20,000,000	92,000,000	7
8	Josef Stalin	20th	20,000,000	20,000,000	15
9	Mideast Slave Trade	7th–19th	19,000,000	132,000,000	3
10	Atlantic Slave Trade	15th–19th	18,000,000	83,000,000	8
11	Timur Lenk (Tamerlane)	14th–15th	17,000,000	100,000,000	6
12	British India (mostly preventable famine)	19th	17,000,000	35,000,000	12
13	First World War	20th	15,000,000	15,000,000	16
14	Russian Civil War	20th	9,000,000	9,000,000	20
15	Fall of Rome	3rd–5th	8,000,000	105,000,000	5
16	Congo Free State	19th–20th	8,000,000	12,000,000	18
17	Thirty Years' War	17th	7,000,000	32,000,000	13
18	Russia's Time of Troubles	16th–17th	5,000,000	23,000,000	14
19	Napoleonic Wars	19th	4,000,000	11,000,000	19
20	Chinese Civil War	20th	3,000,000	3,000,000	21
21	French Wars of Religion	16th	3,000,000	14,000,000	17

they invaded. "Circling at a distance of 100 or 200 yards from the herds of unarmored foot soldiers, a chariot crew—one to drive, one to shoot—might have transfixed six men a minute. Ten minutes' work by ten chariots would cause 500 casualties or more, a Battle of the Somme–like toll among the small armies of the period."[14]

High-throughput massacre was also perfected by mounted hordes from the steppes, such as the Scythians, Huns, Mongols, Turks, Magyars, Tatars, Mughals, and Manchus. For two thousand years these warriors deployed meticulously crafted composite bows (made from a glued laminate of wood, tendon, and horn) to run up immense body counts in their sackings and raids. These tribes were responsible for numbers 3, 5, 11, and 15 on the top-twenty-one list, and they take four of the top six slots in the population-adjusted ranking. The Mongol invasions of Islamic lands in the 13th century resulted in the massacre of 1.3 million people in the city of Merv alone, and another 800,000 residents of Baghdad. As the historian of the Mongols J. J. Saunders remarks:

> There is something indescribably revolting in the cold savagery with which the Mongols carried out their massacres. The inhabitants of a doomed town were obliged to assemble in a plain outside the walls, and each Mongol trooper, armed with a battle-axe, was told to kill so many people, ten, twenty or fifty. As proof that orders had been properly obeyed, the killers were sometimes required to cut off an ear from each victim, collect the ears in sacks, and bring them to their officers to be counted. A few days after the massacre, troops were sent back into the ruined city to search for any poor wretches who might be hiding in holes or cellars; these were dragged out and slain.[15]

The Mongols' first leader, Genghis Khan, offered this reflection on the pleasures of life: "The greatest joy a man can know is to conquer his enemies and drive them before him. To ride their horses and take away their possessions. To see the faces of those who were dear to them bedewed with tears, and to clasp their wives and daughters in his arms."[16] Modern genetics has shown this was no idle boast. Today 8 percent of the men who live within the former territory of the Mongol Empire share a Y chromosome that dates to around the time of Genghis, most likely because they descended from him and his sons and the vast number of women they clasped in their arms.[17] These accomplishments set the bar pretty high, but Timur Lenk (aka Tamerlane), a Turk who aimed to restore the Mongol Empire, did his best. He slaughtered tens of thousands of prisoners in each of his conquests of western Asian cities, then marked his accomplishment by building minarets out of their skulls. One Syrian eyewitness counted twenty-eight towers of fifteen hundred heads apiece.[18]

The worst-things list also gives the lie to the conventional wisdom that the 20th century saw a quantum leap in organized violence from a peaceful 19th.

For one thing, the 19th century has to be gerrymandered to show such a leap by chopping off the extremely destructive Napoleonic Wars from its beginning. For another, the lull in war in the remainder of the century applies only to Europe. Elsewhere we find many hemoclysms, including the Taiping Rebellion in China (a religiously inspired revolt that was perhaps the worst civil war in history), the African slave trade, imperial wars throughout Asia, Africa, and the South Pacific, and two major bloodlettings that didn't even make the list: the American Civil War (650,000 deaths) and the reign of Shaka, a Zulu Hitler who killed between 1 and 2 million people during his conquest of southern Africa between 1816 and 1827. Did I leave any continent out? Oh yes, South America. Among its many wars is the War of the Triple Alliance, which may have killed 400,000 people, including more than 60 percent of the population of Paraguay, making it proportionally the most destructive war in modern times.

A list of extreme cases, of course, cannot establish a trend. There were more major wars and massacres before the 20th century, but then there were more centuries before the 20th. Figure 5–3 extends White's list from the top twenty-one to the top hundred, scales them by the population of the world in that era, and shows how they were distributed in time between 500 BCE and 2000 CE.

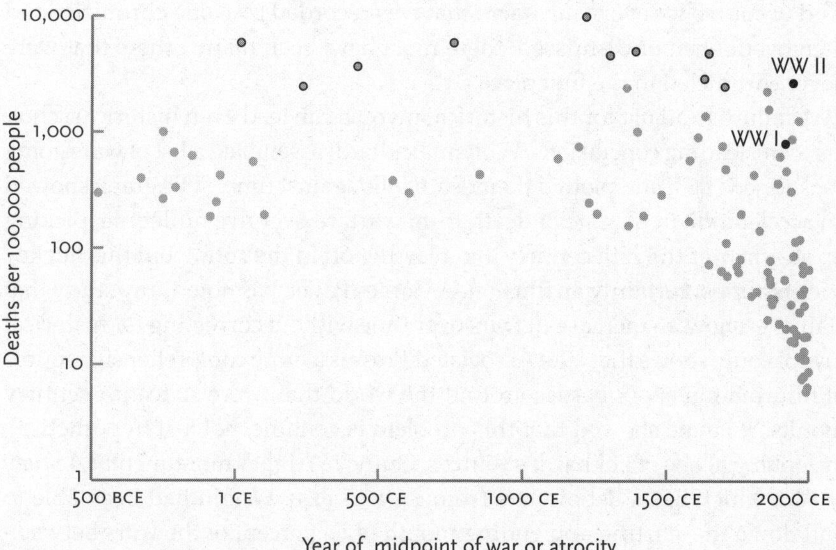

FIGURE 5–3. 100 worst wars and atrocities in human history

Source: Data from White, in press, scaled by world population from McEvedy & Jones, 1978, at the midpoint of the listed range. Note that the estimates are not scaled by the duration of the war or atrocity. Circled dots represent selected events with death rates higher than the 20th-century world wars (from earlier to later): Xin Dynasty, Three Kingdoms, fall of Rome, An Lushan Revolt, Genghis Khan, Mideast slave trade, Timur Lenk, Atlantic slave trade, fall of the Ming Dynasty, and the conquest of the Americas.

Two patterns jump out of the splatter. The first is that the most serious wars and atrocities—those that killed more than a tenth of a percent of the population of the world—are pretty evenly distributed over 2,500 years of history. The other is that the cloud of data tapers rightward and downward into smaller and smaller conflicts for years that are closer to the present. How can we explain this funnel? It's unlikely that our distant ancestors refrained from small massacres and indulged only in large ones. White offers a more likely explanation:

> Maybe the only reason it appears that so many were killed in the past 200 years is because we have more records from that period. I've been researching this for years, and it's been a long time since I found a new, previously unpublicized mass killing from the Twentieth Century; however, it seems like every time I open an old book, I will find another hundred thousand forgotten people killed somewhere in the distant past. Perhaps one chronicler made a note long ago of the number killed, but now that event has faded into the forgotten past. Maybe a few modern historians have revisited the event, but they ignore the body count because it doesn't fit into their perception of the past. They don't believe it was possible to kill that many people without gas chambers and machine guns so they dismiss contrary evidence as unreliable.[19]

And of course for every massacre that was recorded by some chronicler and then overlooked or dismissed, there must have been many others that were never chronicled in the first place.

A failure to adjust for this historical myopia can lead even historical scholars to misleading conclusions. William Eckhardt assembled a list of wars going back to 3000 BCE and plotted their death tolls against time.[20] His graph showed an acceleration in the rate of death from warfare over five millennia, picking up steam after the 16th century and blasting off in the 20th.[21] But this hockey stick is almost certainly an illusion. As James Payne has noted, any study that claims to show an increase in wars over time without correcting for historical myopia only shows that "the Associated Press is a more comprehensive source of information about battles around the world than were sixteenth-century monks."[22] Payne showed that this problem is genuine, not just hypothetical, by looking at one of Eckhardt's sources, Quincy Wright's monumental *A Study of War,* which has a list of wars from 1400 to 1940. Wright had been able to nail down the starting and ending month of 99 percent of the wars between 1875 to 1940, but only 13 percent of the wars between 1480 and 1650, a telltale sign that records of the distant past are far less complete than those of the recent past.[23]

The historian Rein Taagepera quantified the myopia in a different way. He took a historical almanac and stepped through the pages with a ruler, measuring the number of column inches devoted to each century.[24] The range was so great that he had to plot the data on a logarithmic scale (on which an exponential fade looks like a straight line). His graph, reproduced in figure 5–4, shows that as you go back into the past, historical coverage hurtles exponentially downward for two and a half centuries, then falls with a gentler but still exponential decline for the three millennia before.

If it were only a matter of missing a few small wars that escaped the notice of ancient chroniclers, one might be reassured that the body counts were not underestimated, because most of the deaths would be in big wars that no one could fail to notice. But the undercounting may introduce a bias, not just a fuzziness, in the estimates. Keegan writes of a "military horizon."[25] Beneath it are the raids, ambushes, skirmishes, turf battles, feuds, and depredations that historians dismiss as "primitive" warfare. Above it are the organized campaigns for conquest and occupation, including the set-piece battles that war buffs reenact in costume or display with toy soldiers. Remember Tuchman's "private wars" of the 14th century, the ones that knights fought with furious gusto and a single strategy, namely killing as many of another knight's peasants as possible? Many of these massacres were never dubbed The War of Such-and-Such and immortalized in the history books. An undercounting of conflicts below the military horizon could, in theory, throw off the body count for the period as a whole. If more conflicts fell beneath the military horizon in the anarchic feudal societies, frontiers, and tribal lands of the early periods than in the clashes between Leviathans of the later ones, then the earlier periods would appear less violent to us than they really were.

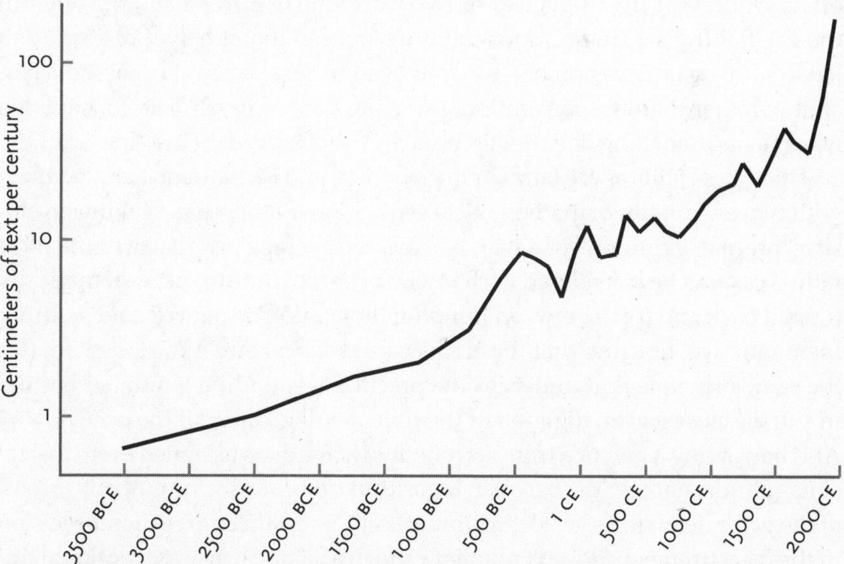

FIGURE 5–4. Historical myopia: Centimeters of text per century in a historical almanac

Source: Data from Taagepera & Colby, 1979, p. 911.

So when one adjusts for population size, the availability bias, and historical myopia, it is far from clear that the 20th century was the bloodiest in history. Sweeping that dogma out of the way is the first step in understanding the historical trajectory of war. The next is to zoom in for a closer look at the distribution of wars over time—which holds even more surprises.

THE STATISTICS OF DEADLY QUARRELS, PART 1: THE TIMING OF WARS

Lewis Richardson wrote that his quest to analyze peace with numbers sprang from two prejudices. As a Quaker, he believed that "the moral evil in war outweighs the moral good, although the latter is conspicuous."[26] As a scientist, he thought there was too much moralizing about war and not enough knowledge: "For indignation is so easy and satisfying a mood that it is apt to prevent one from attending to any facts that oppose it. If the reader should object that I have abandoned ethics for the false doctrine that 'tout comprendre c'est tout pardonner' [to understand all is to forgive all], I can reply that it is only a temporary suspense of ethical judgment, made because 'beaucoup condamner c'est peu comprendre' [to condemn much is to understand little]."[27]

After poring through encyclopedias and histories of different regions of the world, Richardson compiled data on 315 "deadly quarrels" that ended between 1820 and 1952. He faced some daunting problems. One is that most histories are sketchy when it comes to numbers. Another is that it isn't always clear how to count wars, since they tend to split, coalesce, and flicker on and off. Is World War II a single war or two wars, one in Europe and the other in the Pacific? If it's a single war, should we not say that it began in 1937, with Japan's full-scale invasion of China, or even in 1931, when it occupied Manchuria, rather than the conventional starting date of 1939? "The concept of a war as a discrete thing does not fit the facts," he observed. "Thinginess fails."[28]

Thinginess failures are familiar to physicists, and Richardson handled them with two techniques of mathematical estimation. Rather than seeking an elusive "precise definition" of a war, he gave the average priority over the individual case: as he considered each unclear conflict in turn, he systematically flipped back and forth between lumping them into one quarrel and splitting them into two, figuring that the errors would cancel out in the long run. (It's the same principle that underlies the practice of rounding a number ending in 5 to the closest even digit—half the time it will go up, half the time down.) And borrowing a practice from astronomy, Richardson assigned each quarrel a magnitude, namely the base-ten logarithm (roughly, the number of zeroes) of the war's death toll. On a logarithmic scale, a certain degree of imprecision in the measurements doesn't matter as much as it does on a conventional linear scale. For example, uncertainty over whether a war killed 100,000 or 200,000 people translates to an uncertainty in magnitude of only 5 versus 5.3. So

Richardson sorted the magnitudes into logarithmic pigeonholes: 2.5 to 3.5 (that is, between 316 and 3,162 deaths), 3.5 to 4.5 (3,163 to 31,622), and so on. The other advantage of a logarithmic scale is that it allows us to visualize quarrels of a variety of sizes, from turf battles to world wars, on a single scale.

Richardson also faced the problem of what kinds of quarrels to include, which deaths to tally, and how low to go. His criterion for adding a historical event to his database was "malice aforethought," so he included wars of all kinds and sizes, as well as mutinies, insurrections, lethal riots, and genocides; that's why he called his units of analysis "deadly quarrels" instead of haggling over what really deserves the word "war." His magnitude figures included soldiers killed on the battlefield, civilians killed deliberately or as collateral damage, and deaths of soldiers from disease or exposure; he did not count civilian deaths from disease or exposure since these are more properly attributed to negligence than to malice.

Richardson bemoaned an important gap in the historical record: the feuds, raids, and skirmishes that killed between 4 and 315 people apiece (magnitude 0.5 to 2.5), which were too big for criminologists to record but too small for historians. He illustrated the problem of these quarrels beneath the military horizon by quoting from Reginald Coupland's history of the East African slave trade:

> "The main sources of supply were the organized slave-raids in the chosen areas, which shifted steadily inland as tract after tract became 'worked out.' The Arabs might conduct a raid themselves, but more usually they incited a chief to attack another tribe, lending him their own armed slaves and guns to ensure his victory. The result, of course, was an increase in intertribal warfare till 'the whole country was in a flame.' "
>
> How should this abominable custom be classified? Was it all one huge war between Arabs and Negroes which began two thousand years before it ended in 1880? If so it may have caused more deaths than any other war in history. From Coupland's description, however, it would seem more reasonable to regard slave-raiding as a numerous collection of small fatal quarrels each between an Arab caravan and a negro tribe or village, and of magnitudes such as 1, 2, or 3. Detailed statistics are not available.[29]

Nor were they available for 80 revolutions in Latin America, 556 peasant uprisings in Russia, and 477 conflicts in China, which Richardson knew about but was forced to exclude from his tallies.[30]

Richardson did, however, anchor the scale at magnitude 0 by including statistics on homicides, which are quarrels with a death toll of 1 (since $10^0 = 1$). He anticipates an objection by Shakespeare's Portia: "You ought not to mix up murder with war; for murder is an abominable selfish crime, but war is a heroic and patriotic adventure." He replies: "Yet they are both fatal quarrels. Does it never strike you as puzzling that it is wicked to kill one person, but glorious to kill ten thousand?"[31]

Richardson then analyzed the 315 quarrels (without the benefit of a computer) to get a bird's-eye view of human violence and test a variety of hypotheses suggested by historians and his own prejudices.[32] Most of the hypotheses did not survive their confrontation with the data. A common language didn't make two factions less likely to go to war (just think of most civil wars, or the 19th-century wars between South American countries); so much for the "hope" that gave Esperanto its name. Economic indicators predicted little; rich countries, for example, didn't systematically pick on poor countries or vice versa. Wars were not, in general, precipitated by arms races.

But a few generalizations did survive. A long-standing government inhibits fighting: peoples on one side of a national border are less likely to have a civil war than peoples on opposite sides are to have an interstate war. Countries are more likely to fight their neighbors, but great powers are more likely to fight everyone, largely because their far-flung empires make almost everyone their neighbors. Certain cultures, especially those with a militant ideology, are particularly prone to go to war.

But Richardson's most enduring discoveries are about the statistical patterning of wars. Three of his generalizations are robust, profound, and underappreciated. To understand them, we must first take a small detour into a paradox of probability.

Suppose you live in a place that has a constant chance of being struck by lightning at any time throughout the year. Suppose that the strikes are random: every day the chance of a strike is the same, and the rate works out to one strike a month. Your house is hit by lightning today, Monday. What is the most likely day for the *next* bolt to strike your house?

The answer is "tomorrow," Tuesday. That probability, to be sure, is not very high; let's approximate it at 0.03 (about once a month). Now think about the chance that the next strike will be the day after tomorrow, Wednesday. For that to happen, two things have to take place. First lightning has to strike on Wednesday, a probability of 0.03. Second, lightning *can't have struck on Tuesday*, or else Tuesday would have been the day of the next strike, not Wednesday. To calculate that probability, you have to multiply the chance that lightning will not strike on Tuesday (0.97, or 1 minus 0.03) by the chance that lightning will strike on Wednesday (0.03), which is 0.0291, a bit lower than Tuesday's chances. What about Thursday? For that to be the day, lightning can't have struck on Tuesday (0.97) or on Wednesday either (0.97 again) but it must strike on Thursday, so the chances are 0.97 × 0.97 × 0.03, which is 0.0282. What about Friday? It's 0.97 × 0.97 × 0.97 × 0.03, or 0.274. With each day, the odds go down (0.0300 . . . 0.0291 . . . 0.0282 . . . 0.0274), because for a given day to be the next day that lightning strikes, all the previous days have to have been strike-free, and the more of these days there are, the lower the chances are that the streak will continue. To be exact, the probability goes down exponentially, accelerating at an accelerating rate. The chance that the next strike will be thirty days from today is $0.97^{29} \times 0.03$, barely more than 1 percent.

Almost no one gets this right. I gave the question to a hundred Internet users, with the word *next* italicized so they couldn't miss it. Sixty-seven picked the option "every day has the same chance." But that answer, though intuitively compelling, is wrong. If every day were equally likely to be the next one, then a day a thousand years from now would be just as likely as a day a month from now. That would mean that the house would be just as likely to go a thousand years without a strike as to suffer one next month. Of the remaining respondents, nineteen thought that the most likely day was a month from today. Only five of the hundred correctly guessed "tomorrow."

Lightning strikes are an example of what statisticians call a Poisson process (pronounced *pwah-sonh*), named after the 19th-century mathematician and physicist Siméon-Denis Poisson. In a Poisson process, events occur continuously, randomly, and independently of one another. Every instant the lord of the sky, Jupiter, rolls the dice, and if they land snake eyes he hurls a thunderbolt. The next instant he rolls them again, with no memory of what happened the moment before. For reasons we have just seen, in a Poisson process the intervals between events are distributed exponentially: there are lots of short intervals and fewer and fewer of them as they get longer and longer. That implies that events that occur at random will seem to come in clusters, because it would take a *non*random process to space them out.

The human mind has great difficulty appreciating this law of probability. When I was a graduate student, I worked in an auditory perception lab. In one experiment listeners had to press a key as quickly as possible every time they heard a beep. The beeps were timed at random, that is, according to a Poisson process. The listeners, graduate students themselves, knew this, but as soon as the experiment began they would run out of the booth and say, "Your random event generator is broken. The beeps are coming in bursts. They sound like this: "beepbeepbeepbeepbeep . . . beep . . . beepbeep . . . beepitybeepitybeepbeepbeep." They didn't appreciate that that's what randomness sounds like.

This cognitive illusion was first noted in 1968 by the mathematician William Feller in his classic textbook on probability: "To the untrained eye, randomness appears as regularity or tendency to cluster."[33] Here are a few examples of the cluster illusion.

The London Blitz. Feller recounts that during the Blitz in World War II, Londoners noticed that a few sections of the city were hit by German V-2 rockets many times, while others were not hit at all. They were convinced that the rockets were targeting particular kinds of neighborhoods. But when statisticians divided a map of London into small squares and counted the bomb strikes, they found that the strikes followed the distribution of a Poisson process—the bombs, in other words, were falling at random. The episode is depicted in Thomas Pynchon's 1973 novel *Gravity's Rainbow*, in which statistician Roger Mexico has correctly predicted the distribution of bomb strikes, though not their exact locations. Mexico has to deny that he is a psychic and fend off desperate demands for advice on where to hide.

The gambler's fallacy. Many high rollers lose their fortunes because of the gambler's fallacy: the belief that after a run of similar outcomes in a game of chance (red numbers in a roulette wheel, sevens in a game of dice), the next spin or toss is bound to go the other way. Tversky and Kahneman showed that people think that genuine sequences of coin flips (like TTHHTHTTTT) are fixed, because they have more long runs of heads or of tails than their intuitions allow, and they think that sequences that were jiggered to avoid long runs (like HTHTTHTHHT) are fair.[34]

The birthday paradox. Most people are surprised to learn that if there are at least 23 people in a room, the chances that two of them will share a birthday are better than even. With 57 people, the probability rises to 99 percent. In this case the illusory clusters are in the calendar. There are only so many birthdays to go around (366), so a few of the birthdays scattered throughout the year are bound to fall onto the same day, unless there was some mysterious force trying to separate them.

Constellations. My favorite example was discovered by the biologist Stephen Jay Gould when he toured the famous glowworm caves in Waitomo, New Zealand.[35] The worms' pinpricks of light on the dark ceiling made the grotto look like a planetarium, but with one difference: there were no constellations. Gould deduced the reason. Glowworms are gluttonous and will eat anything that comes within snatching distance, so each worm gives the others a wide berth when it stakes out a patch of ceiling. As a result, they are more evenly spaced than stars, which from our vantage point are randomly spattered across the sky. Yet it is the stars that seem to fall into shapes, including the ram, bull, twins, and so on, that for millennia have served as portents to pattern-hungry brains. Gould's colleague, the physicist Ed Purcell, confirmed Gould's intuition by programming a computer to generate two arrays of random dots. The virtual stars were plonked on the page with no constraints. The virtual worms were given a random tiny patch around them in which no other worm could intrude. They are shown in figure 5–5; you can probably guess which is which. The one on the left, with the clumps, strands, voids, and filaments (and perhaps, depending on your obsessions, animals, nudes, or Virgin Marys) is the array that was plotted at random, like stars. The one on the right, which seems to be haphazard, is the array whose positions were nudged apart, like glowworms.

Richardson's data. My last example comes from another physicist, our friend Lewis Fry Richardson. These are real data from a naturally occurring phenomenon. The segments in figure 5–6 represent events of various durations, and they are arranged from left to right in time and from bottom to top in

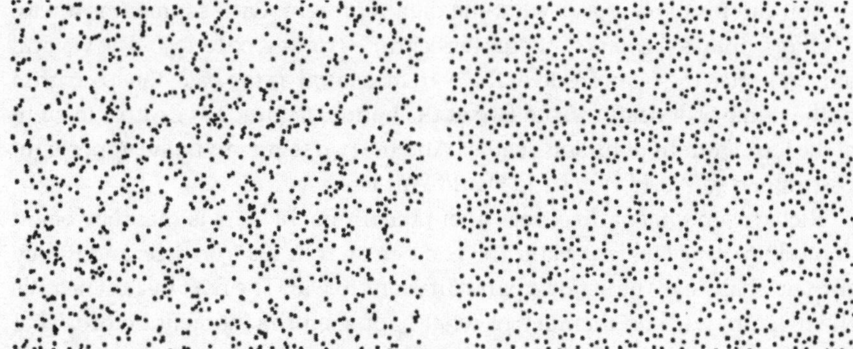

FIGURE 5–5. Random and nonrandom patterns

Sources: Displays generated by Ed Purcell; reproduced from Gould, 1991, pp. 266–67.

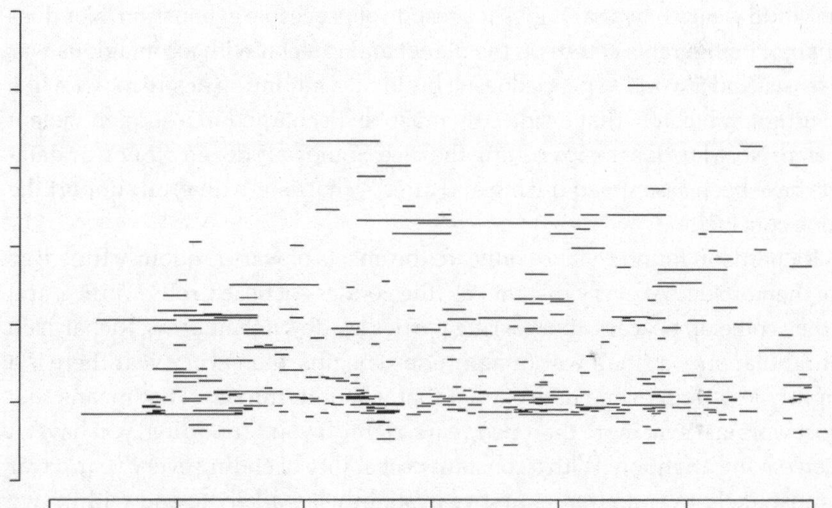

FIGURE 5–6. Richardson's data

Source: Graph from Hayes, 2002, based on data in Richardson, 1960.

magnitude. Richardson showed that the events are governed by a Poisson process: they stop and start at random. Your eye may discern some patterns—for example, a scarcity of segments at the top left, and the two floaters at the top right. But by now you have learned to distrust these apparitions. And indeed Richardson showed that there was no statistically significant trend in the distribution of magnitudes from the beginning of the sequence to the end. Cover up the two outliers with your thumb, and the impression of randomness is total.

You can probably guess what the data represent. Each segment is a war. The horizontal axis marks off quarter-centuries from 1800 to 1950. The vertical axis indicates the magnitude of the war, measured as the base-ten logarithm of the number of deaths, from two at the bottom (a hundred deaths) to eight at the top (a hundred million deaths). And the two segments in the upper right correspond to World War I and World War II.

Richardson's major discovery about the timing of wars is that they begin at random. Every instant Mars, the god of war, rolls his iron dice, and if they turn up snake eyes he sends a pair of nations to war. The next instant he rolls them again, with no memory of what happened the moment before. That would make the distribution of intervals between war onsets exponential, with lots of short intervals and fewer long ones.

The Poisson nature of war undermines historical narratives that see constellations in illusory clusters. It also confounds theories that see grand patterns, cycles, and dialectics in human history. A horrible conflict doesn't make the world weary of war and give it a respite of peaceable exhaustion. Nor does a pair of belligerents cough on the planet and infect it with a contagious war disease. And a world at peace doesn't build up a mounting desire for war, like an unignorable itch, that eventually must be discharged in a sudden violent spasm. No, Mars just keeps rolling the dice. Some half-dozen other war datasets have been assembled during and after Richardson's time; all support the same conclusion.[36]

Richardson found that not only are the onsets of wars randomly timed; so are their offsets. At every instant Pax, the goddess of peace, rolls *her* dice, and if they come up boxcars, the warring parties lay down their arms. Richardson found that once a small war (magnitude 3) begins, then every year there is a slightly less than even chance (0.43) that it will terminate. That means that most wars last a bit more than two years, right? If you're nodding, you haven't been paying attention! With a constant probability of ending every year, a war is most likely to end after its first year, slightly less likely to end within two years, a bit less likely to stretch on to three, and so on. The same is true for larger wars (magnitude 4 to 7), which have a 0.235 chance of coming to an end before another year is up. War durations are distributed exponentially, with the shortest wars being the most common.[37] This tells us that warring nations don't have to "get the aggression out of their system" before they come to their senses, that wars don't have a "momentum" that must be allowed to "play itself out." As soon as a war begins, some combination of antiwar forces—pacifism, fear, rout—puts pressure on it to end.[38]

If wars start and stop at random, is it pointless even to look for historical trends in war? It isn't. The "randomness" in a Poisson process pertains to the relationships among successive events, namely that there is none: the event generator, like the dice, has no memory. But nothing says that the probability has to be constant over long stretches of time. Mars could switch from causing a war whenever the dice land in snake eyes to, say, causing a war whenever they add up to 3, or 6, or 7. Any of these shifts would change the probability of war over time without changing its randomness—the fact that the outbreak of one war doesn't make another war either more or less likely. A Poisson process with a drifting probability is called nonstationary. The possibility that war might decline over some historical period, then, is alive. It would reside in a nonstationary Poisson process with a declining rate parameter.

By the same token, it's mathematically possible for war both to be a Poisson process *and* to display cycles. In theory, Mars could oscillate, causing a war on 3 percent of his throws, then shifting to causing a war on 6 percent, and then going back again. In practice, it isn't easy to distinguish cycles in a nonstationary Poisson process from illusory clusters in a stationary one. A few clusters could fool the eye into thinking that the whole system waxes and wanes (as in the so-called business cycle, which is really a sequence of unpredictable lurches in economic activity rather than a genuine cycle with a constant period). There are good statistical methods that can test for periodicities in time series data, but they work best when the span of time is much longer than the period of the cycles one is looking for, since that provides room for many of the putative cycles to fit. To be confident in the results, it also helps to have a second dataset in which to replicate the analysis, so that one isn't fooled by the possibility of "overfitting" cycles to what are really random clusters in a particular dataset. Richardson examined a number of possible cycles for wars of magnitudes 3, 4, and 5 (the bigger wars were too sparse to allow a test), and found none. Other analysts have looked at longer datasets, and the literature contains sightings of cycles at 5, 15, 20, 24, 30, 50, 60, 120, and 200 years. With so many tenuous candidates, it is safer to conclude that war follows no meaningful cycle at all, and that is the conclusion endorsed by most quantitative historians of war.[39] The sociologist Pitirim Sorokin, another pioneer of the statistical study of war, concluded, "History seems to be neither as monotonous and uninventive as the partisans of the strict periodicities and 'iron laws' and 'universal uniformities' think; nor so dull and mechanical as an engine, making the same number of revolutions in a unit of time."[40]

Could the 20th-century Hemoclysm, then, have been some kind of fluke? Even to think that way seems like monstrous disrespect to the victims. But the statistics of deadly quarrels don't force such an extreme conclusion. Randomness over long stretches of time can coexist with changing probabilities, and certainly some of the probabilities in the 1930s must have been different from those of other decades. The Nazi ideology that justified an invasion of Poland in order to acquire living space for the "racially superior" Aryans was a part of the same ideology that justified the annihilation of the "racially inferior" Jews. Militant nationalism was a common thread that ran through Germany, Italy, and Japan. There was also a common denominator of counter-Enlightenment utopianism behind the ideologies of Nazism and communism. And even if wars are randomly distributed over the long run, there can be an occasional exception. The occurrence of World War I, for example, presumably incremented the probability that a war like World War II in Europe would break out.

But statistical thinking, particularly an awareness of the cluster illusion, suggests that we are apt to *exaggerate* the narrative coherence of this history—to think that what did happen must have happened because of historical forces like cycles, crescendos, and collision courses. Even with all the probabilities in place, highly contingent events, which need not reoccur if we somehow could rewind the tape of history and play it again, may have been necessary to set off the wars with death tolls in the 6s and 7s on the magnitude scale.

Writing in 1999, White repeated a Frequently Asked Question of that year: "Who's the most important person of the Twentieth Century?" His choice: Gavrilo Princip. Who the heck was Gavrilo Princip? He was the nineteen-year-old Serb nationalist who assassinated Archduke Franz Ferdinand of Austria-Hungary during a state visit to Bosnia, after a string of errors and accidents delivered the archduke to within shooting distance. White explains his choice:

> Here's a man who single-handedly sets off a chain reaction which ultimately leads to the deaths of 80 million people.
>
> Top that, Albert Einstein!
>
> With just a couple of bullets, this terrorist starts the First World War, which destroys four monarchies, leading to a power vacuum filled by the Communists in Russia and the Nazis in Germany who then fight it out in a Second World War. . . .
>
> Some people would minimize Princip's importance by saying that a Great Power War was inevitable sooner or later given the tensions of the times, but I say that it was no more inevitable than, say, a war between NATO and the Warsaw Pact. Left unsparked, the Great War could have been avoided, and without it, there would have been no Lenin, no Hitler, no Eisenhower.[41]

Other historians who indulge in counterfactual scenarios, such as Richard Ned Lebow, have made similar arguments.[42] As for World War II, the historian F. H. Hinsley wrote, "Historians are, rightly, nearly unanimous that . . . the causes of the Second World War were the personality and the aims of Adolf Hitler." Keegan agrees: "Only one European really wanted war—Adolf Hitler."[43] The political scientist John Mueller concludes:

> These statements suggest that there was no momentum toward another world war in Europe, that historical conditions in no important way required that contest, and that the major nations of Europe were not on a collision course that was likely to lead to war. That is, had Adolf Hitler gone into art rather than politics, had he been gassed a bit more thoroughly by the British

> in the trenches in 1918, had he, rather than the man marching next to him, been gunned down in the Beer Hall Putsch of 1923, had he failed to survive the automobile crash he experienced in 1930, had he been denied the leadership position in Germany, or had he been removed from office at almost any time before September 1939 (and possibly even before May 1940), Europe's greatest war would most probably never have taken place.[44]

So, too, the Nazi genocide. As we shall see in the next chapter, most historians of genocide agree with the title of a 1984 essay by the sociologist Milton Himmelfarb: "No Hitler, no Holocaust."[45]

Probability is a matter of perspective. Viewed at sufficiently close range, individual events have determinate causes. Even a coin flip can be predicted from the starting conditions and the laws of physics, and a skilled magician can exploit those laws to throw heads every time.[46] Yet when we zoom out to take a wide-angle view of a large number of these events, we are seeing the sum of a vast number of causes that sometimes cancel each other out and sometimes align in the same direction. The physicist and philosopher Henri Poincaré explained that we see the operation of chance in a deterministic world either when a large number of puny causes add up to a formidable effect, or when a small cause that escapes our notice determines a large effect that we cannot miss.[47] In the case of organized violence, someone may want to start a war; he waits for the opportune moment, which may or may not come; his enemy decides to engage or retreat; bullets fly; bombs burst; people die. Every event may be determined by the laws of neuroscience and physics and physiology. But in the aggregate, the many causes that go into this matrix can sometimes be shuffled into extreme combinations. Together with whatever ideological, political, and social currents put the world at risk in the first half of the 20th century, those decades were also hit with a run of extremely bad luck.

Now to the money question: has the probability that a war will break out increased, decreased, or stayed the same over time? Richardson's dataset is biased to show an increase. It begins just after the Napoleonic Wars, slicing off one of the most destructive wars in history at one end, and finishes just after World War II, snagging history's most destructive war at the other. Richardson did not live to see the Long Peace that dominated the subsequent decades, but he was an astute enough mathematician to know that it was statistically possible, and he devised ingenious ways of testing for trends in a time series without being misled by extreme events at either end. The simplest was to separate the wars of different magnitudes and test for trends separately in each range. In none of the five ranges (3 to 7) did he find a significant trend. If anything, he found a slight decline. "There is a suggestion," he wrote, "but not a conclusive proof, that mankind has become less warlike since A.D. 1820. The best available observations show a slight decrease in the number of wars with time. . . . But the distinction is not great enough to show plainly among chance variations."[48] Written at a time when the ashes of Europe and Asia were still warm, this is a testament to a great scientist's willingness to lets facts and reason override casual impressions and conventional wisdom.

As we shall see, analyses of the frequency of war over time from other datasets point to the same conclusion.[49] But the frequency of war is not the whole story; magnitude matters as well. One could be forgiven for pointing out that Richardson's conjecture that mankind was getting less warlike depended on segregating the world wars into a micro-class of two, in which statistics are futile. His other analyses counted all wars alike, with World War II no different from, say, a 1952 revolution in Bolivia with a thousand deaths. Richardson's son pointed out to him that if he divided his data into large and small wars, they seemed to show opposing trends: small wars were becoming considerably less frequent, but larger wars, while fewer in number, were becoming somewhat more frequent. A different way of putting it is that between 1820 and 1953 wars became less frequent but more lethal. Richardson tested the pattern of contrast and found that it was statistically significant.[50] The next section will show that this too was an astute conclusion: other datasets confirm that until 1945, the story of war in Europe and among major nations in general was one of fewer but more damaging wars.

So does that mean that mankind got more warlike or less? There is no single answer, because "warlike" can refer to two different things. It can refer to how likely nations are to go to war, or it can refer to how many people are killed when they do. Imagine two rural counties with the same size population. One of them has a hundred teenage arsonists who delight in setting forest fires. But the forests are in isolated patches, so each fire dies out before doing much damage. The other county has just two arsonists, but its forests are connected, so that a small blaze is likely to spread, as they say, like wildfire. Which county has the worse forest fire problem? One could argue it either way. As far as the amount of reckless depravity is concerned, the first county is worse; as far as the risk of serious damage is concerned, the second is. Nor is it obvious which county will have the greater amount of overall damage, the one with a lot of little fires, or the one with a few big ones. To make sense of these questions, we have to turn from the statistics of time to the statistics of magnitude.

THE STATISTICS OF DEADLY QUARRELS, PART 2: THE MAGNITUDE OF WARS

Richardson made a second major discovery about the statistics of deadly quarrels. It emerged when he counted the number of quarrels of each magnitude—how many with death tolls in the thousands, how many in the tens of thousands, how many in the hundreds of thousands, and so on. It isn't a complete surprise that there were lots of little wars and only a few big ones. What was a surprise was how neat the relationship turned out to be. When

Richardson plotted the log of the number of quarrels of each magnitude against the log of the number of deaths per quarrel (that is, the magnitude itself), he ended up with a graph like figure 5–7.

Scientists are accustomed to seeing data fall into perfect straight lines when they come from hard sciences like physics, such as the volume of a gas plotted against its temperature. But not in their wildest dreams do they expect the messy data from history to be so well behaved. The data we are looking at come from a ragbag of deadly quarrels ranging from the greatest cataclysm in the history of humanity to a coup d'état in a banana republic, and from the dawn of the Industrial Revolution to the dawn of the computer age. The jaw drops when seeing this mélange of data fall onto a perfect diagonal.

Piles of data in which the log of the *frequency* of a certain kind of entity is proportional to the log of the *size* of that entity, so that a plot on log-log paper looks like a straight line, are called power-law distributions.[51] The name comes from the fact that when you put away the logarithms and go back to the original numbers, the probability of an entity showing up in the data is proportional to the size of that entity raised to some power (which translates visually to the slope of the line in the log-log plot), plus a constant. In this case the power is –1.5, which means that with every tenfold jump in the death toll of a war, you can expect to find about a third as many of them. Richardson plotted murders (quarrels of magnitude 0) on the same graph as wars, noting that qualitatively they follow the overall pattern: they are much, much less damaging than the smallest wars and much, much more frequent. But as you can see

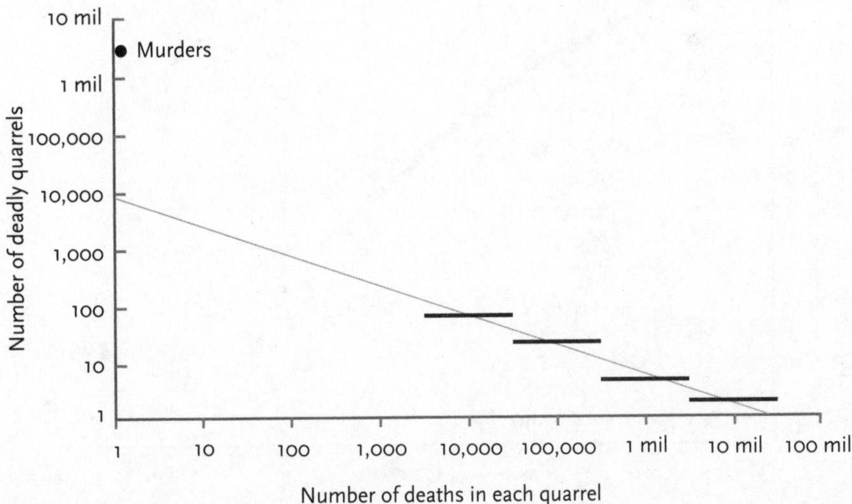

FIGURE 5–7. Number of deadly quarrels of different magnitudes, 1820–1952

Source: Graph adapted from Weiss, 1963, p. 103, based on data from Richardson, 1960, p. 149. The range 1820–1952 refers to the year a war ended.

from their lonely perch atop the vertical axis, high above the point where an extrapolation of the line for the wars would hit it, he was pushing his luck when he said that all deadly quarrels fell along a single continuum. Richardson gamely connected the murder point to the war line with a swoopy curve so that he could interpolate the numbers of quarrels with death tolls in the single digits, the tens, and the hundreds, which are missing from the historical record. (These are the skirmishes beneath the military horizon that fall in the crack between criminology and history.) But for now let's ignore the murders and skirmishes and concentrate on the wars.

Could Richardson just have been lucky with his sample? Fifty years later the political scientist Lars-Erik Cederman plotted a newer set of numbers in a major dataset of battle deaths from the Correlates of War Project, comprising ninety-seven interstate wars between 1820 and 1997 (figure 5–8).[52] They too fall along a straight line in log-log coordinates. (Cederman plotted the data in a slightly different way, but that doesn't matter for our purposes.)[53]

Scientists are intrigued by power-law distributions for two reasons.[54] One is that the distribution keeps turning up in measurements of things that you would think have nothing in common. One of the first power-law distributions was discovered in the 1930s by the linguist G. K. Zipf when he plotted the frequencies of words in the English language.[55] If you count up the instances of each of the words in a large corpus of text, you'll find around a dozen that occur extremely frequently, that is, in more than 1 percent of all word tokens, including *the* (7 percent), *be* (4 percent), *of* (4 percent), *and* (3 percent), and *a*

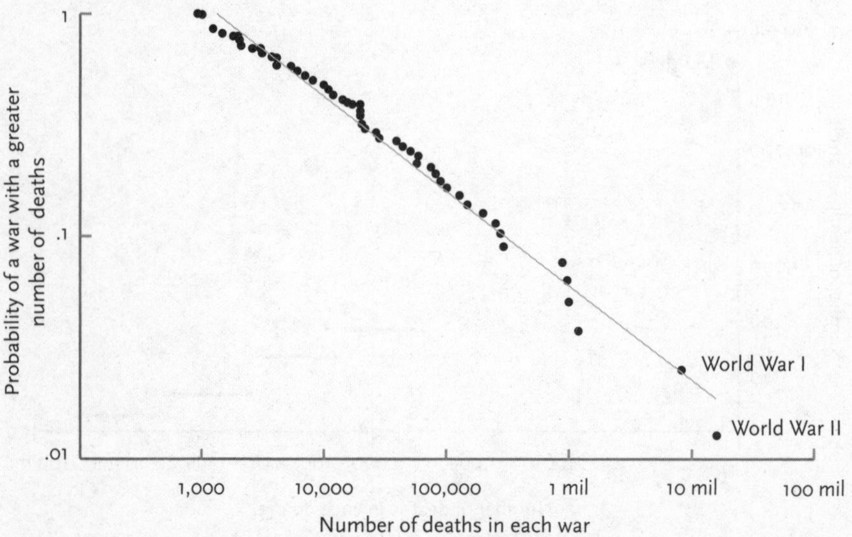

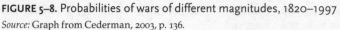

FIGURE 5–8. Probabilities of wars of different magnitudes, 1820–1997

Source: Graph from Cederman, 2003, p. 136.

(2 percent).[56] Around three thousand occur in the medium-frequency range centered on 1 in 10,000, such as *confidence, junior,* and *afraid.* Tens of thousands occur once every million words, including *embitter, memorialize,* and *titular.* And hundreds of thousands have frequencies far less than one in a million, like *kankedort, apotropaic,* and *deliquesce.*

Another example of a power-law distribution was discovered in 1906 by the economist Vilfredo Pareto when he looked at the distribution of incomes in Italy: a handful of people were filthy rich, while a much larger number were dirt-poor. Since these discoveries, power-law distributions have also turned up, among other places, in the populations of cities, the commonness of names, the popularity of Web sites, the number of citations of scientific papers, the sales figures of books and musical recordings, the number of species in biological taxa, and the sizes of moon craters.[57]

The second remarkable thing about power-law distributions is that they look the same over a vast range of values. To understand why this is so striking, let's compare power-law distributions to a more familiar distribution called the normal, Gaussian, or bell curve. With measurements like the heights of men or the speeds of cars on a freeway, most of the numbers pile up around an average, and they tail off in both directions, falling into a curve that looks like a bell.[58] Figure 5–9 shows one for the heights of American males. There are lots of men around 5'10" tall, fewer who are 5'6" or 6'2", not that many who are 5'0" or 6'8", and no one who is shorter than 1'11" or taller than 8'11" (the two extremes in *The Guinness Book of World Records*). The ratio of the tallest man in

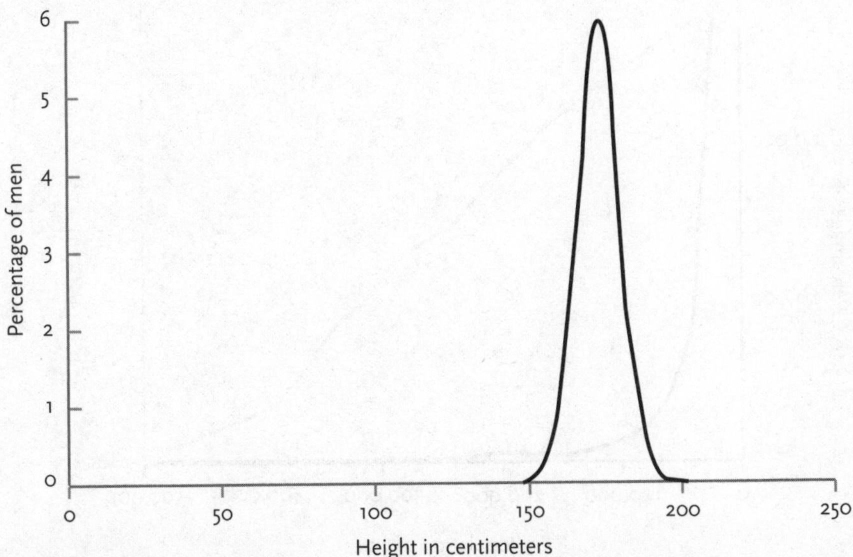

FIGURE 5–9. Heights of males (a normal or bell-curve distribution)
Source: Graph from Newman, 2005, p. 324.

the world to the shortest man in the world is 4.8, and you can bet that you will never meet a man who is 20 feet tall.

But with other kinds of entities, the measurements don't heap up around a typical value, don't fall off symmetrically in both directions, and don't fit within a cozy range. The sizes of towns and cities is a good example. It's hard to answer the question "How big is a typical American municipality?" New York has 8 million people; the smallest municipality that counts as a "town," according to *Guinness*, is Duffield, Virginia, with only 52. The ratio of the largest municipality to the smallest is 150,000, which is very different from the less-than-fivefold variation in the heights of men.

Also, the distribution of sizes of municipalities isn't curved like a bell. As the black line in figure 5–10 shows, it is L-shaped, with a tall spine on the left and a long tail on the right. In this graph, city populations are laid out along a conventional linear scale on the black horizontal axis: cities of 100,000, cities of 200,000, and so on. So are the proportions of cities of each population size on the black vertical axis: three-thousandths (3/1000, or 0.003) of a percent of American municipalities have a population of exactly 20,000, two thousandths of a percent have a population of 30,000, one thousandth of a percent have a population of 40,000, and so on, with smaller and smaller proportions having larger

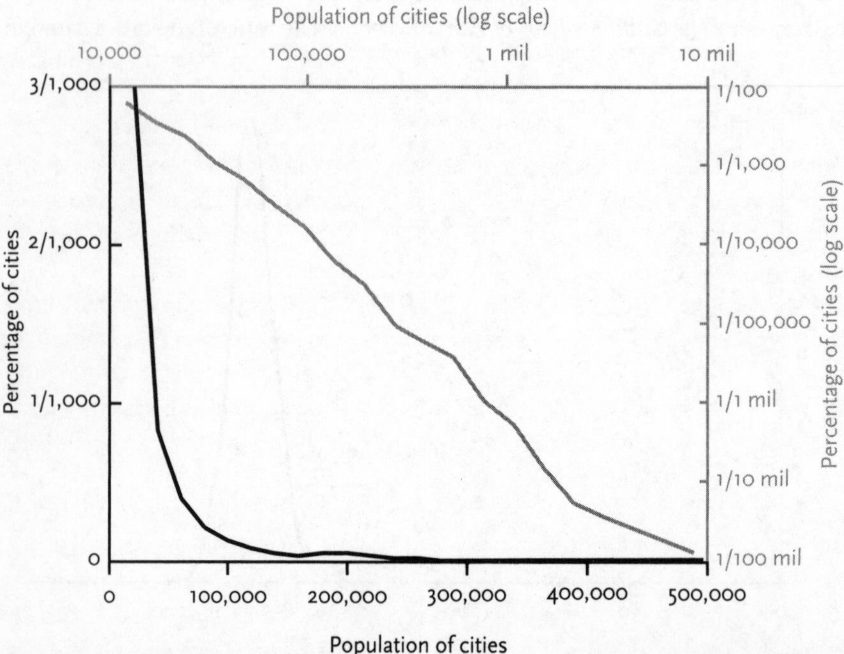

FIGURE 5–10. Populations of cities (a power-law distribution), plotted on linear and log scales
Source: Graph adapted from Newman, 2005, p. 324.

and larger populations.[59] Now the gray axes at the top and the right of the graph stretch out these same numbers on a logarithmic scale, in which *orders of magnitude* (the number of zeroes) are evenly spaced, rather than the values themselves. The tick marks for population sizes are at ten thousand, a hundred thousand, a million, and so on. Likewise the proportions of cities at each population size are arranged along equal order-of-magnitude tick marks: one one-hundredth (1/100, or 0.01) of a percent, one one-thousandth (1/1,000, or 0.001) of a percent, one ten-thousandth, and so on. When the axes are stretched out like this, something interesting happens to the distribution: the L straightens out into a nice line. And that is the signature of a power-law distribution.

Which brings us back to wars. Since wars fall into a power-law distribution, some of the mathematical properties of these distributions may help us understand the nature of wars and the mechanisms that give rise to them. For starters, power-law distributions with the exponent we see for wars do not even have a finite mean. There is no such thing as a "typical war." We should not expect, even on average, that a war will proceed until the casualties pile up to an expected level and then will naturally wind itself down.

Also, power-law distributions are *scale-free*. As you slide up or down the line in the log-log graph, it always looks the same, namely, like a line. The mathematical implication is that as you magnify or shrink the units you are looking at, the distribution looks the same. Suppose that computer files of 2 kilobytes are a quarter as common as files of 1 kilobyte. Then if we stand back and look at files in higher ranges, we find the same thing: files of 2 megabytes are a quarter as common as files of 1 megabyte, and files of 2 terabytes are a quarter as common as files of 1 terabyte. In the case of wars, you can think of it this way. What are the odds of going from a small war, say, with 1,000 deaths, to a medium-size war, with 10,000 deaths? It's the same as the odds of going from a medium-size war of 10,000 deaths to a large war of 100,000 deaths, or from a large war of 100,000 deaths to a historically large war of 1 million deaths, or from a historic war to a world war.

Finally, power-law distributions have "thick tails," meaning that they have a nonnegligible number of extreme values. You will never meet a 20-foot man, or see a car driving down the freeway at 500 miles per hour. But you could conceivably come across a city of 14 million, or a book that was on the bestseller list for 10 years, or a moon crater big enough to see from the earth with the naked eye—or a war that killed 55 million people.

The thick tail of a power-law distribution, which declines gradually rather than precipitously as you rocket up the magnitude scale, means that extreme values are *extremely unlikely* but not *astronomically unlikely*. It's an important difference. The chances of meeting a 20-foot-tall man are astronomically unlikely; you can bet your life it will never happen. But the chances that a city will grow to 20 million, or that a book will stay on the bestseller list for 20 years, is merely extremely unlikely—it probably won't happen, but you could well imagine it happening. I hardly need to point out the implications for war. It is extremely unlikely that the world will see a war that will kill 100 million people, and less likely still that it will have one that will kill a billion. But in an age of nuclear weapons, our terrified imaginations and the mathematics of power-law distributions agree: it is not astronomically unlikely.

So far I've been discussing the causes of war as Platonic abstractions, as if armies were sent into war by equations. What we really need to understand is *why* wars distribute themselves as power laws; that is, what combination of psychology and politics and technology could generate this pattern. At present we can't be sure of the answer. Too many kinds of mechanisms can give rise to power-law distributions, and the data on wars are not precise enough to tell us which is at work.

Still, the scale-free nature of the distribution of deadly quarrels gives us an insight about the drivers of war.[60] Intuitively, it suggests that *size doesn't matter*. The same psychological or game-theoretic dynamics that govern whether quarreling coalitions will threaten, back down, bluff, engage, escalate, fight on, or surrender apply whether the coalitions are street gangs, militias, or armies of great powers. Presumably this is because humans are social animals who aggregate into coalitions, which amalgamate into larger coalitions, and so on. Yet at any scale these coalitions may be sent into battle by a single clique or individual, be it a gang leader, capo, warlord, king, or emperor.

How can the intuition that size doesn't matter be implemented in models of armed conflict that actually generate power-law distributions?[61] The simplest is to assume that the coalitions themselves are power-law-distributed in size, that they fight each other in proportion to their numbers, and that they suffer losses in proportion to their sizes. We know that some human aggregations, namely municipalities, are power-law-distributed, and we know the reason. One of the commonest generators of a power-law distribution is preferential attachment: the bigger something is, the more new members it attracts. Preferential attachment is also known as accumulated advantage, the-rich-get-richer, and the Matthew Effect, after the passage in Matthew 25:29 that Billie Holiday summarized as "Them that's got shall get, them that's not shall lose." Web sites that are popular attract more visitors, making them even more popular; bestselling books are put on bestseller lists, which lure more people into buying them; and cities with lots of people offer more professional and cultural opportunities so more people flock to them. (How are you going to keep them down on the farm after they've seen Paree?)

Richardson considered this simple explanation but found that the numbers didn't add up.[62] If deadly quarrels reflected city sizes, then for every tenfold reduction in the size of a quarrel, there should be ten times as many of them, but in fact there are fewer than four times as many. Also, in recent centuries wars have been fought by states, not cities, and states follow a log-normal distribution (a warped bell curve) rather than a power law.

Another kind of mechanism has been suggested by the science of complex systems, which looks for laws that govern structures that are organized into similar patterns despite being made of different stuff. Many complexity theorists are intrigued by systems that display a pattern called self-organized criticality. You can think of "criticality" as the straw that broke the camel's back: a small input causes a sudden large output. "Self-organized" criticality would be a camel whose back healed right back to the exact strength at which straws of various sizes could break it again. A good example is a trickle of sand falling onto a sandpile, which periodically causes landslides of different sizes; the landslides are distributed according to a power law. An avalanche of sand stops at a point where the slope is just shallow enough to be stable, but the new sand trickling onto it steepens the slope and sets off a new avalanche. Earthquakes and forest fires are other examples. A fire burns a forest, which allows trees to grow back at random, forming clusters that can grow into each other and fuel another fire. Several political scientists have developed computer simulations that model wars on an analogy to forest fires.[63] In these models, countries conquer their neighbors and create larger countries in the same way that patches of trees grow into each other and create larger patches. Just as a cigarette tossed in a forest can set off either a brushfire or a conflagration, a destabilizing event in the simulation of states can set off either a skirmish or a world war.

In these simulations, the destructiveness of a war depends mainly on the territorial size of the combatants and their alliances. But in the real world, variations in destructiveness also depend on the resolve of the two parties to keep a war going, with each hoping that the other will collapse first. Some of the bloodiest conflicts in modern history, such as the American Civil War, World War I, the Vietnam War, and the Iran-Iraq War, were wars of attrition, where both sides kept shoveling men and matériel into the maw of the war machine hoping that the other side would exhaust itself first.

John Maynard Smith, the biologist who first applied game theory to evolution, modeled this kind of standoff as a War of Attrition game.[64] Each of two contestants competes for a valuable resource by trying to outlast the other, steadily accumulating costs as he waits. In the original scenario, they might be heavily armored animals competing for a territory who stare at each other until one of them leaves; the costs are the time and energy the animals waste in the standoff, which they could otherwise use in catching food or pursuing mates. A game of attrition is mathematically equivalent to an auction in which the highest bidder wins the prize and *both* sides have to pay the loser's low bid. And of course it can be analogized to a war in which the expenditure is reckoned in the lives of soldiers.

The War of Attrition is one of those paradoxical scenarios in game theory (like the Prisoner's Dilemma, the Tragedy of the Commons, and the Dollar Auction) in which a set of rational actors pursuing their interests end up worse off than if they had put their heads together and come to a collective and binding agreement. One might think that in an attrition game each side should do what bidders on eBay are advised to do: decide how much the contested resource is worth and bid only up to that limit. The problem is that this strategy can be gamed by another bidder. All he has to do is bid one more dollar (or wait just a bit longer, or commit another surge of soldiers), and he wins. He gets the prize for close to the amount you think it is worth, while you have to forfeit that amount too, without getting anything in return. You would be crazy to let that happen, so you are tempted to use the strategy "Always outbid him by a dollar," which he is tempted to adopt as well. You can see where this leads. Thanks to the perverse logic of an attrition game, in which the loser pays too, the bidders may keep bidding after the point at which the expenditure exceeds the value of the prize. They can no longer win, but each side hopes not to lose as much. The technical term for this outcome in game theory is "a ruinous situation." It is also called a "Pyrrhic victory"; the military analogy is profound.

One strategy that can evolve in a War of Attrition game (where the expenditure, recall, is in time) is for each player to wait a *random* amount of time, with an average wait time that is equivalent in value to what the resource is worth to them. In the long run, each player gets good value for his expenditure, but because the waiting times are random, neither is able to predict the surrender time of the other and reliably outlast him. In other words, they follow the rule: At every instant throw a pair of dice, and if they come up (say) 4, concede; if not, throw them again. This is, of course, like a Poisson process, and by now you know that it leads to an exponential distribution of wait times (since a longer and longer wait depends on a less and less probable run of tosses). Since the contest ends when the first side throws in the towel, the contest durations will also be exponentially distributed. Returning to our model where the expenditures are in soldiers rather than seconds, if real wars of attrition were like the "War of Attrition" modeled in game theory, and if all else were equal, then wars of attrition would fall into an exponential distribution of magnitudes.

Of course, real wars fall into a power-law distribution, which has a thicker tail than an exponential (in this case, a greater number of severe wars). But an exponential can be transformed into a power law if the values are modulated by a second exponential process pushing in the opposite direction. And attrition games have a twist that might do just that. If one side in an attrition game were to leak its intention to concede in the next instant by, say, twitching or blanching or showing some other sign of nervousness, its opponent could capitalize on the "tell" by waiting just a bit longer, and it would win the prize every time. As Richard Dawkins has put it, in a species that often takes part in wars of attrition, one expects the evolution of a poker face.

Now, one also might have guessed that organisms would capitalize on the opposite kind of signal, a sign of continuing resolve rather than impending surrender. If a contestant could adopt some defiant posture that means "I'll stand my ground; I won't back down," that would make it rational for his opposite number to give up and cut its losses rather than escalate to mutual ruin. But there's a reason we call it "posturing." Any coward can cross his arms and glower, but the other side can simply call his bluff. Only if a signal is *costly*—if the defiant party holds his hand over a candle, or cuts his arm with a knife—can he show that he means business. (Of course, paying a self-imposed cost would be worthwhile only if the prize is especially valuable to him, or if he had reason to believe that he could prevail over his opponent if the contest escalated.)

In the case of a war of attrition, one can imagine a leader who has a *changing* willingness to suffer a cost over time, increasing as the conflict proceeds and his resolve toughens. His motto would be: "We fight on so that our boys shall not have died in vain." This mindset, known as loss aversion, the sunk-cost fallacy, and throwing good money after bad, is patently irrational, but it is surprisingly pervasive in human decision-making.[65] People stay in an abusive marriage because of the years they have already put into it, or sit through a bad movie because they have already paid for the ticket, or try to reverse a gambling loss by doubling their next bet, or pour money into a boondoggle because they've already poured so much money into it. Though psychologists don't fully understand why people are suckers for sunk costs, a common explanation is that it signals a public commitment. The person is announcing: "When I make a decision, I'm not so weak, stupid, or indecisive that I can be easily talked out of it." In a contest of resolve like an attrition game, loss aversion could serve as a costly and hence credible signal that the contestant is not about to concede, preempting his opponent's strategy of outlasting him just one more round.

I already mentioned some evidence from Richardson's dataset which suggests that combatants do fight longer when a war is more lethal: small wars show a higher probability of coming to an end with each succeeding year than do large wars.[66] The magnitude numbers in the Correlates of War Dataset also show signs of escalating commitment: wars that are longer in duration are not just costlier in fatalities; they are costlier than one would expect from their durations alone.[67] If we pop back from the statistics of war to the conduct of actual wars, we can see the mechanism at work. Many of the bloodiest wars in history owe their destructiveness to leaders on one or both sides pursuing a blatantly irrational loss-aversion strategy. Hitler fought the last months of World War II with a maniacal fury well past the point when defeat was all but certain, as did Japan. Lyndon Johnson's repeated escalations of the Vietnam War inspired a protest song that has served as a summary of people's understanding of that destructive war: "We were waist-deep in the Big Muddy; The big fool said to push on."

The systems biologist Jean-Baptiste Michel has pointed out to me how escalating commitments in a war of attrition could produce a power-law distribution.

All we need to assume is that leaders keep escalating as a constant proportion of their past commitment—the size of each surge is, say, 10 percent of the number of soldiers that have fought so far. A constant proportional increase would be consistent with the well-known discovery in psychology called Weber's Law: for an increase in intensity to be noticeable, it must be a constant proportion of the existing intensity. (If a room is illuminated by ten lightbulbs, you'll notice a brightening when an eleventh is switched on, but if it is illuminated by a hundred lightbulbs, you won't notice the hundred and first; someone would have to switch on another *ten* bulbs before you noticed the brightening.) Richardson observed that people perceive lost lives in the same way: "Contrast for example the many days of newspaper-sympathy over the loss of the British submarine *Thetis* in time of peace with the terse announcement of similar losses during the war. This contrast may be regarded as an example of the Weber-Fechner doctrine that an increment is judged relative to the previous amount."[68] The psychologist Paul Slovic has recently reviewed several experiments that support this observation.[69] The quotation falsely attributed to Stalin, "One death is a tragedy; a million deaths is a statistic," gets the numbers wrong but captures a real fact about human psychology.

If escalations are proportional to past commitments (and a constant proportion of soldiers sent to the battlefield are killed in battle), then losses will increase exponentially as a war drags on, like compound interest. And if wars are attrition games, their durations will also be distributed exponentially. Recall the mathematical law that a variable will fall into a power-law distribution if it is an exponential function of a second variable that is distributed exponentially.[70] My own guess is that the combination of escalation and attrition is the best explanation for the power-law distribution of war magnitudes.

Though we may not know exactly why wars fall into a power-law distribution, the nature of that distribution—scale-free, thick-tailed—suggests that it involves a set of underlying processes in which size doesn't matter. Armed coalitions can always get a bit larger, wars can always last a bit longer, and losses can always get a bit heavier, with the same likelihood regardless of how large, long, or heavy they were to start with.

The next obvious question about the statistics of deadly quarrels is: What destroys more lives, the large number of small wars or the few big ones? A power-law distribution itself doesn't give the answer. One can imagine a dataset in which the aggregate damage from the wars of each size adds up to the same number of deaths: one war with ten million deaths, ten wars with a million deaths, a hundred wars with a hundred thousand deaths, all the way down to ten million murders with one death apiece. But in fact, distributions with exponents greater than one (which is what we get for wars) will have the numbers skewed toward the tail. A power-law distribution with an exponent in this range is sometimes said to follow the 80:20 rule, also known as the

Pareto Principle, in which, say, the richest 20 percent of the population controls 80 percent of the wealth. The ratio may not be 80:20 exactly, but many power-law distributions have this kind of lopsidedness. For example, the 20 percent most popular Web sites get around two-thirds of the hits.[71]

Richardson added up the total number of deaths from all the deadly quarrels in each magnitude range. The computer scientist Brian Hayes has plotted them in the histogram in figure 5–11. The gray bars, which tally the deaths from the elusive small quarrels (between 3 and 3,162 deaths), don't represent actual data, because they fall in the criminology-history crack and were not available in the sources Richardson consulted. Instead, they show hypothetical numbers that Richardson interpolated with a smooth curve between the murders and the smaller wars.[72] With or without them, the shape of the graph is striking: it has peaks at each end and a sag in the middle. That tells us that the most damaging kinds of lethal violence (at least from 1820 to 1952) were murders and world wars; all the other kinds of quarrels killed far fewer people. That has remained true in the sixty years since. In the United States, 37,000 military personnel died in the Korean War, and 58,000 died in Vietnam; no other war came close. Yet an average of 17,000 people are murdered in the country *every year*, adding up to almost a million deaths since 1950.[73] Likewise, in the world as a whole, homicides outnumber war-related deaths, even if one includes the indirect deaths from hunger and disease.[74]

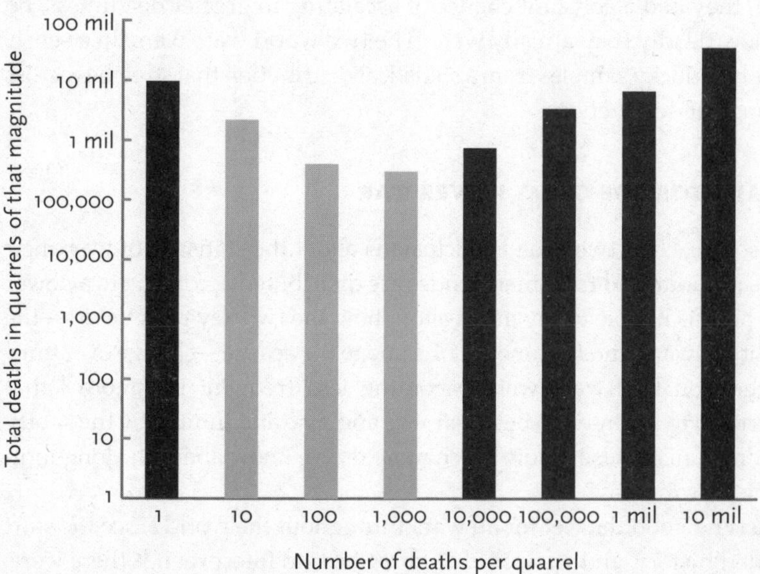

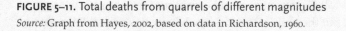

FIGURE 5–11. Total deaths from quarrels of different magnitudes

Source: Graph from Hayes, 2002, based on data in Richardson, 1960.

Richardson also estimated the proportion of deaths that were caused by deadly quarrels of all magnitudes combined, from murders to world wars. The answer was 1.6 percent. He notes: "This is less than one might have guessed from the large amount of attention which quarrels attract. Those who enjoy wars can excuse their taste by saying that wars after all are much less deadly than disease."[75] Again, this continues to be true by a large margin.[76]

That the two world wars killed 77 percent of the people who died in all the wars that took place in a 130-year period is an extraordinary discovery. Wars don't even follow the 80:20 rule that we are accustomed to seeing in power-law distributions. They follow an 80:2 rule: almost 80 percent of the deaths were caused by 2 *percent* of the wars.[77] The lopsided ratio tells us that the global effort to prevent deaths in war should give the highest priority to preventing the largest wars.

The ratio also underscores the difficulty of reconciling our desire for a coherent historical narrative with the statistics of deadly quarrels. In making sense of the 20th century, our desire for a good story arc is amplified by two statistical illusions. One is the tendency to see meaningful clusters in randomly spaced events. Another is the bell-curve mindset that makes extreme values seem astronomically unlikely, so when we come across an extreme event, we reason there must have been extraordinary design behind it. That mindset makes it difficult to accept that the worst two events in recent history, though unlikely, were not astronomically unlikely. Even if the odds had been increased by the tensions of the times, the wars did not have to start. And once they did, they had a constant chance of escalating to greater deadliness, no matter how deadly they already were. The two world wars were, in a sense, horrifically unlucky samples from a statistical distribution that stretches across a vast range of destruction.

THE TRAJECTORY OF GREAT POWER WAR

Richardson reached two broad conclusions about the statistics of war: their timing is random, and their magnitudes are distributed according to a power law. But he was unable to say much about how the two key parameters—the probability of wars, and the amount of damage they cause—change over time. His suggestion that wars were becoming less frequent but more lethal was restricted to the interval between 1820 and 1950 and limited by the spotty list of wars in his dataset. How much more do we know about the long-term trajectory of war today?

There is no good dataset for all wars throughout the world since the start of recorded history, and we wouldn't know how to interpret it if there were. Societies have undergone such radical and uneven changes over the centuries that a single death toll for the entire world would sum over too many different kinds of societies. But the political scientist Jack Levy has assembled a dataset that gives us a clear view of the trajectory of war in a particularly important slice of space and time.

The time span is the era that began in the late 1400s, when gunpowder, ocean navigation, and the printing press are said to have inaugurated the modern age (using one of the many definitions of the word *modern*). That is also the time at which sovereign states began to emerge from the medieval quilt of baronies and duchies.

The countries that Levy focused on are the ones that belong to the *great power system*—the handful of states in a given epoch that can throw their weight around the world. Levy found that at any time a small number of eight-hundred-pound gorillas are responsible for a majority of the mayhem.[78] The great powers participated in about 70 percent of all the wars that Wright included in his half-millennium database for the entire world, and four of them have the dubious honor of having participated in at least a fifth of all European wars.[79] (This remains true today: France, the U.K., the United States, and the USSR/Russia have been involved in more international conflicts since World War II than any other countries.)[80] Countries that slip in or out of the great power league fight far more wars when they are in than when they are out. One more advantage of focusing on great powers is that with footprints that large, it's unlikely that any war they fought would have been missed by the scribblers of the day.

As we might predict from the lopsided power-law distribution of war magnitudes, the wars among great powers (especially the wars that embroiled several great powers at a time) account for a substantial proportion of all recorded war deaths.[81] According to the African proverb (like most African proverbs, attributed to many different tribes), when elephants fight, it is the grass that suffers. And these elephants have a habit of getting into fights with one another because they are not leashed by some larger suzerain but constantly eye each other in a state of nervous Hobbesian anarchy.

Levy set out technical criteria for being a great power and listed the countries that met them between 1495 and 1975. Most of them are large European states: France and England/Great Britain/U.K. for the entire period; the entities ruled by the Habsburg dynasty through 1918; Spain until 1808; the Netherlands and Sweden in the 17th and early 18th centuries; Russia/USSR from 1721 on; Prussia/Germany from 1740 on; and Italy from 1861 to 1943. But the system also includes a few powers outside Europe: the Ottoman Empire until 1699; the United States from 1898 on; Japan from 1905 to 1945; and China from 1949. Levy assembled a dataset of wars that had at least a thousand battle deaths a year (a conventional cutoff for a "war" in many datasets, such as the Correlates of War Project), that had a great power on at least one side, and that had a state on the other side. He excluded colonial wars and civil wars unless a great power was butting into a civil war on the side of the insurgency, which would mean that the war had pitted a great power against a foreign government. Using the Correlates of War Dataset, and in consultation with Levy, I have extended his data through the quarter-century ending in 2000.[82]

Let's start with the clashes of the titans—the wars with at least one great power on each side. Among them are what Levy called "general wars" but which could also be called "world wars," at least in the sense that World War I deserves that name—not that the fighting spanned the globe, but that it embroiled most of the world's great powers. These include the Thirty Years' War (1618–48; six of the seven great powers), the Dutch War of Louis XIV (1672–78; six of seven), the War of the League of Augsburg (1688–97; five of seven), the War of the Spanish Succession (1701–13; five of six), the War of the Austrian Succession (1739–48; six of six), the Seven Years' War (1755–63; six of six), and the French Revolutionary and Napoleonic Wars (1792–1815; six of six), together with the two world wars. There are more than fifty other wars in which two or more great powers faced off.

One indication of the impact of war in different eras is the percentage of time that people had to endure wars between great powers, with their disruptions, sacrifices, and changes in priorities. Figure 5–12 shows the percentage of years in each quarter-century that saw the great powers of the day at war. In two of the early quarter-centuries (1550–75 and 1625–50), the line bumps up against the ceiling: great powers fought each other in all 25 of the 25 years. These periods were saturated with the horrendous European Wars of Religion, including the First Huguenot War and the Thirty Years' War. From there the trend is unmistakably downward. Great powers fought each other for less of

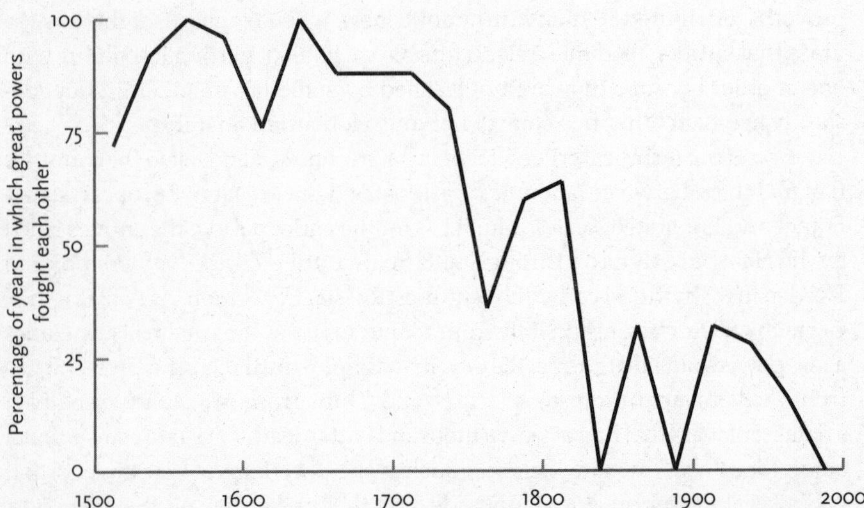

FIGURE 5–12. Percentage of years in which the great powers fought one another, 1500–2000

Source: Graph adapted from Levy & Thompson, 2011. Data are aggregated over 25-year periods.

the time as the centuries proceeded, though with a few partial reversals, including the quarters with the French Revolutionary and Napoleonic Wars and with the two world wars. At the toe of the graph on the right one can see the first signs of the Long Peace. The quarter-century from 1950 to 1975 had one war between the great powers (the Korean War, from 1950 to 1953, with the United States and China on opposite sides), and there has not been once since.

Now let's zoom out and look at a wider view of war: the hundred-plus wars with a great power on one side and any country whatsoever, great or not, on the other.[83] With this larger dataset we can unpack the years-at-war measure from the previous graph into two dimensions. The first is frequency. Figure 5–13 plots how many wars were fought in each quarter-century. Once again we see a decline over the five centuries: the great powers have become less and less likely to fall into wars. During the last quarter of the 20th, only four wars met Levy's criteria: the two wars between China and Vietnam (1979 and 1987), the UN-sanctioned war to reverse Iraq's invasion of Kuwait (1991), and NATO's bombing of Yugoslavia to halt its displacement of ethnic Albanians in Kosovo (1999).

The second dimension is duration. Figure 5–14 shows how long, on average, these wars dragged on. Once again the trend is downward, though with a spike around the middle of the 17th century. This is not a simpleminded consequence of counting the Thirty Years' War as lasting exactly thirty years; following the practice of other historians, Levy divided it into four more circumscribed wars. Even after that slicing, the Wars of Religion in that era were

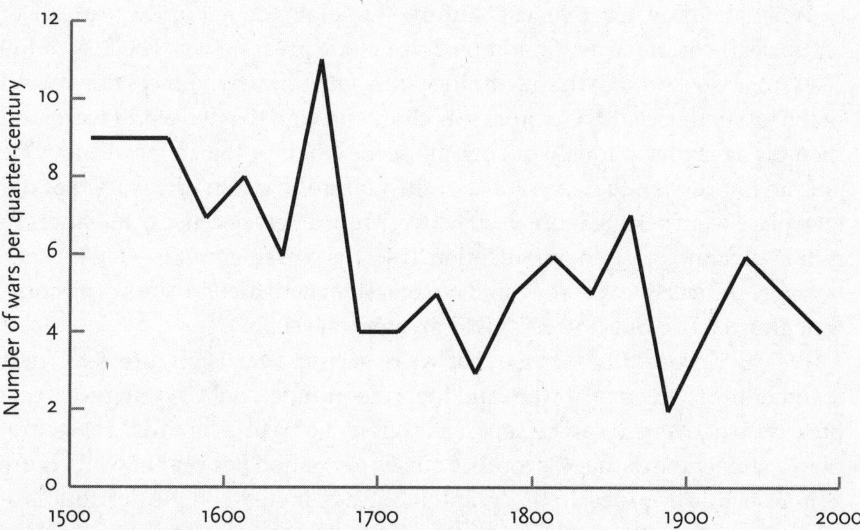

FIGURE 5–13. Frequency of wars involving the great powers, 1500–2000

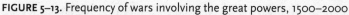

Sources: Graph from Levy, 1983, except the last point, which is based on the Correlates of War Inter-State War Dataset, 1816–1997, Sarkees, 2000, and, for 1997–99, the PRIO Battle Deaths Dataset 1946–2008, Lacina & Gleditsch, 2005. Data are aggregated over 25-year periods.

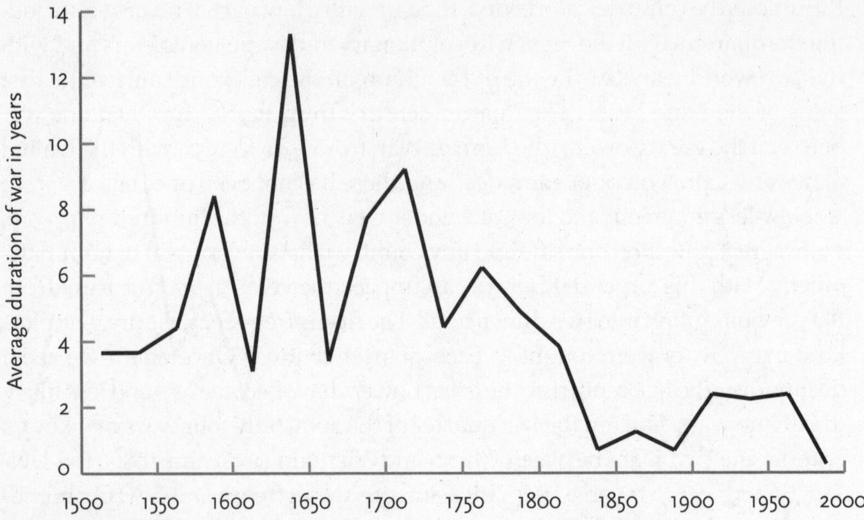

FIGURE 5–14. Duration of wars involving the great powers, 1500–2000

Sources: Graph from Levy, 1983, except the last point, which is based on the Correlates of War Inter-State War Dataset, 1816–1997, Sarkees, 2000, and, for 1997–99, the PRIO Battle Deaths Dataset 1946–2008, Lacina & Gleditsch, 2005. Data are aggregated over 25-year periods.

brutally long. But from then on the great powers sought to end their wars soon after beginning them, culminating in the last quarter of the 20th century, when the four wars involving great powers lasted an average of 97 days.[84]

What about destructiveness? Figure 5–15 plots the log of the number of battle deaths in the wars fought by at least one great power. The loss of life rises from 1500 through the beginning of the 19th century, bounces downward in the rest of that century, resumes its climb through the two world wars, and then plunges precipitously during the second half of the 20th century. One gets an impression that over most of the half-millennium, the wars that did take place were getting more destructive, presumably because of advances in military technology and organization. If so, the crossing trends—fewer wars, but more destructive wars—would be consistent with Richardson's conjecture, though stretched out over a fivefold greater time span.

We can't prove that this is what we're seeing, because figure 5–15 folds together the frequency of wars and their magnitudes, but Levy suggests that pure destructiveness can be separated out in a measure he calls "concentration," namely the damage a conflict causes per nation per year of war. Figure 5–16 plots this measure. In this graph the steady increase in the deadliness of great power wars through World War II is more apparent, because it is not hidden by the paucity of those wars in the later 19th century. What is striking about the latter half of the 20th century is the sudden reversal of the crisscrossing trends of the 450 years preceding it. The late 20th century was unique in

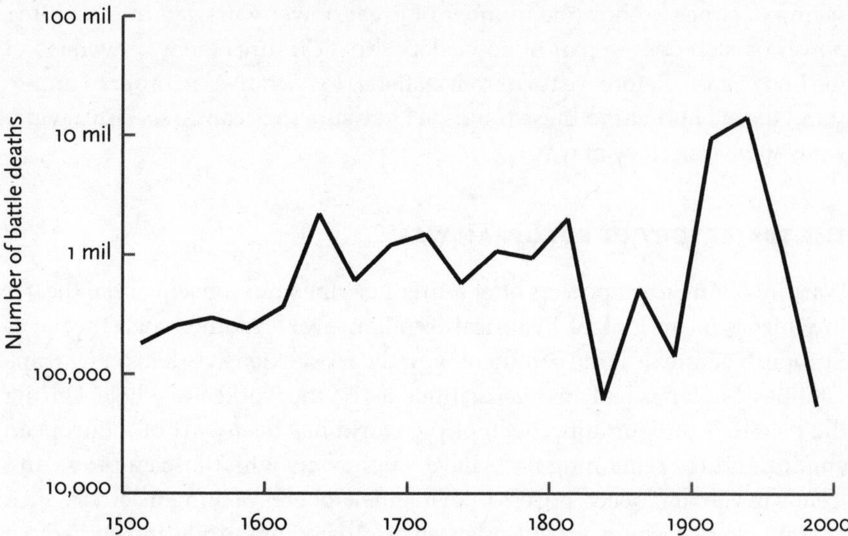

FIGURE 5–15. Deaths in wars involving the great powers, 1500–2000

Sources: Graph from Levy, 1983, except the last point, which is based on the Correlates of War Inter-State War Dataset, 1816–1997, Sarkees, 2000, and, for 1997–99, the PRIO Battle Deaths Dataset 1946–2008, Lacina & Gleditsch, 2005. Data are aggregated over 25-year periods.

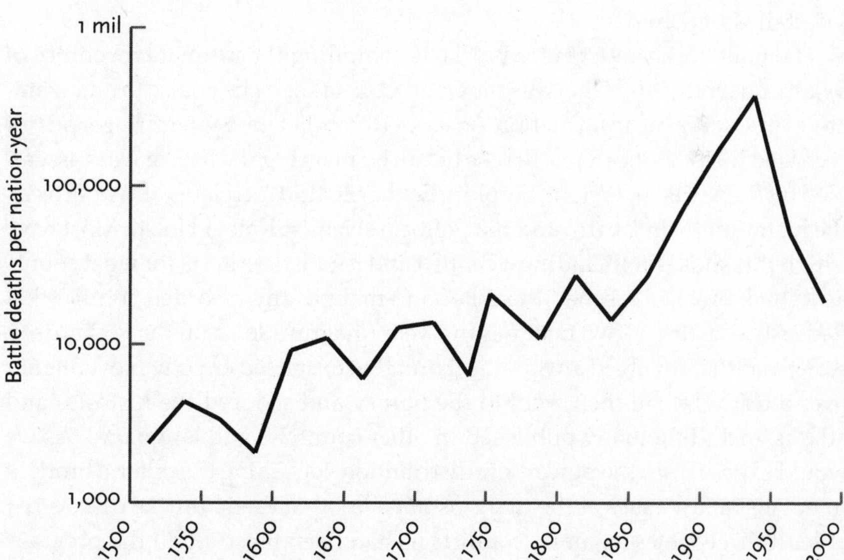

FIGURE 5–16. Concentration of deaths in wars involving the great powers, 1500–2000

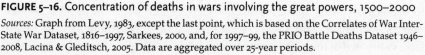

Sources: Graph from Levy, 1983, except the last point, which is based on the Correlates of War Inter-State War Dataset, 1816–1997, Sarkees, 2000, and, for 1997–99, the PRIO Battle Deaths Dataset 1946–2008, Lacina & Gleditsch, 2005. Data are aggregated over 25-year periods.

seeing declines both in the number of great power wars *and* in the killing power of each one—a pair of downslopes that captures the war-aversion of the Long Peace. Before we turn from statistics to narratives in order to understand the events behind these trends, let's be sure they can be seen in a wider view of the trajectory of war.

THE TRAJECTORY OF EUROPEAN WAR

Wars involving great powers offer a circumscribed but consequential theater in which we can look at historical trends in war. Another such theater is Europe. Not only is it the continent with the most extensive data on wartime fatalities, but it has had an outsize influence on the world as a whole. During the past half-millennium, much of the world has been part of a European empire, and the remaining parts have fought wars with those empires. And trends in war and peace, no less than in other spheres of human activity such as technology, fashion, and ideas, often originated in Europe and spilled out to the rest of the world.

The extensive historical data from Europe also give us an opportunity to broaden our view of organized conflict from interstate wars involving the great powers to wars between less powerful nations, conflicts that miss the thousand-death cutoff, civil wars, and genocides, together with deaths of civilians from famine and disease. What kind of picture do we get if we aggregate these other forms of violence—the tall spine of little conflicts as well as the long tail of big ones?

The political scientist Peter Brecke is compiling the ultimate inventory of deadly quarrels, which he calls the Conflict Catalog.[85] His goal is to amalgamate every scrap of information on armed conflict in the entire corpus of recorded history since 1400. Brecke began by merging the lists of wars assembled by Richardson, Wright, Sorokin, Eckhardt, the Correlates of War Project, the historian Evan Luard, and the political scientist Kalevi Holsti. Most have a high threshold for including a conflict and legalistic criteria for what counts as a state. Brecke loosened the criteria to include any recorded conflict that had as few as thirty-two fatalities in a year (magnitude 1.5 on the Richardson scale) and that involved any political unit that exercised effective sovereignty over a territory. He then went to the library and scoured the histories and atlases, including many published in other countries and languages. As we would expect from the power-law distribution, loosening the criteria brought in not just a few cases at the margins but a flood of them: Brecke discovered at least three times as many conflicts as had been listed in all the previous datasets combined. The Conflict Catalog so far contains 4,560 conflicts that took place between 1400 CE and 2000 CE (3,700 of which have been entered into a spreadsheet), and it will eventually contain 6,000. About a third of them have estimates of the number of fatalities, which Brecke divides into military deaths (soldiers killed in battle) and total deaths (which includes the indirect deaths of civilians from war-caused starvation and disease). Brecke kindly provided me with the dataset as it stood in 2010.

Let's start by simply counting the conflicts—not just the wars embroiling great powers, but deadly quarrels great and small. These tallies, plotted in figure 5–17, offer an independent view of the history of war in Europe.

Once again we see a decline in one of the dimensions of armed conflict: how often they break out. When the story begins in 1400, European states were starting conflicts at a rate of more than three a year. That rate has caromed downward to virtually none in Western Europe and to less than one conflict per year in Eastern Europe. Even that bounce is a bit misleading, because half of the conflicts were in countries that are coded in the dataset as "Europe" only because they were once part of the Ottoman or Soviet empire; today they are usually classified as Middle Eastern or Central and South Asian (for example, conflicts in Turkey, Georgia, Azerbaijan, Dagestan, and Armenia).[86] The other Eastern European conflicts were in former republics of Yugoslavia or the Soviet Union. These regions—Yugoslavia, Russia/USSR, and Turkey—were also responsible for the spike of European conflicts in the first quarter of the 20th century.

What about the human toll of the conflicts? Here is where the capaciousness

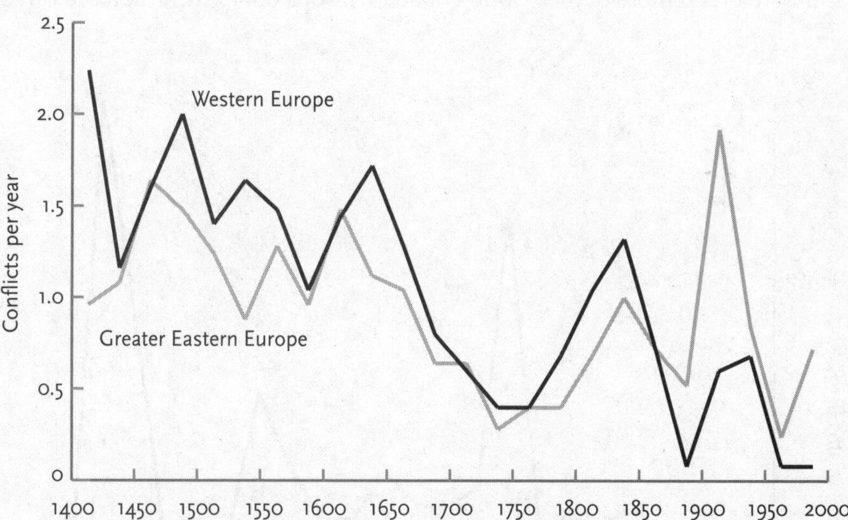

FIGURE 5–17. Conflicts per year in greater Europe, 1400–2000

Sources: Conflict Catalog, Brecke, 1999; Long & Brecke, 2003. The conflicts are aggregated over 25-year periods and include interstate and civil wars, genocides, insurrections, and riots. "Western Europe" includes the territories of the present-day U.K., Ireland, Denmark, Sweden, Norway, France, Belgium, Luxembourg, Netherlands, Germany, Switzerland, Austria, Spain, Portugal, and Italy. "Eastern Europe" includes the territories of the present-day Cyprus, Finland, Poland, Czech Republic, Slovakia, Hungary, Romania, the republics formerly making up Yugoslavia, Albania, Greece, Bulgaria, Turkey (both Europe and Asia), Russia (Europe), Georgia, Armenia, Azerbaijan, and other Caucasus republics.

of the Conflict Catalog comes in handy. The power-law distribution tells us that the biggest of the great power wars should account for the lion's share of the deaths from all wars—at least, from all wars that exceed the thousand-death cutoff, which make up the data I have plotted so far. But Richardson alerted us to the possibility that a large number of smaller conflicts missed by traditional histories and datasets could, in theory, pile up into a substantial number of additional deaths (the gray bars in figure 5–11). The Conflict Catalog is the first long-term dataset that reaches down into that gray area and tries to list the skirmishes, riots, and massacres that fall beneath the traditional military horizon (though of course many more in the earlier centuries may never have been recorded). Unfortunately the catalog is a work in progress, and at present fewer than half the conflicts have fatality figures attached to them. Until it is completed, we can get a crude glimpse of the trajectory of conflict deaths in Europe by filling in the missing values using the median of the death tolls from that quarter-century. Brian Atwood and I have interpolated these values, added up the direct and indirect deaths from conflicts of all types and sizes, divided them by the population of Europe in each period, and plotted them on a linear scale.[87] Figure 5–18 presents this maximalist (albeit tentative) picture of the history of violent conflict in Europe:

The scaling by population size did not eliminate an overall upward trend through 1950, which shows that Europe's ability to kill people outpaced its ability to breed more of them. But what really pops out of the graph are three

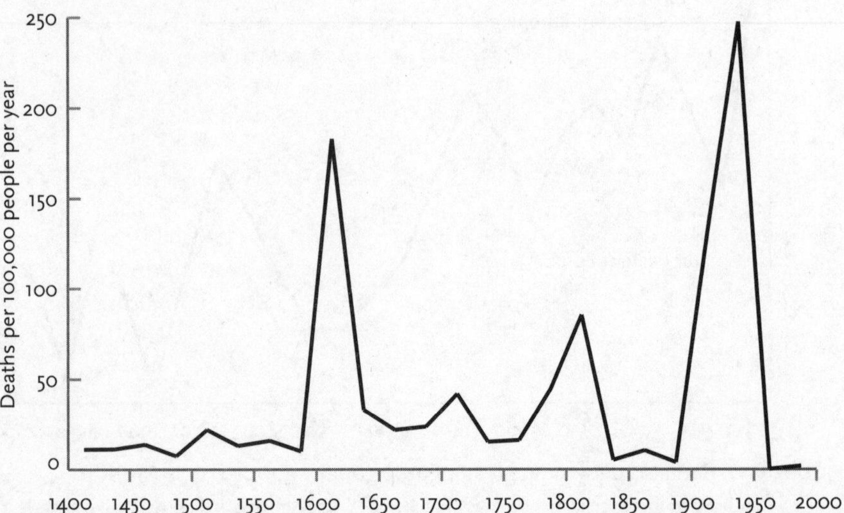

FIGURE 5–18. Rate of death in conflicts in greater Europe, 1400–2000

Sources: Conflict Catalog, Brecke, 1999; Long & Brecke, 2003. Figures are from the "Total Fatalities" column, aggregated over 25-year periods. Redundant entries were eliminated. Missing entries were filled in with the median for that quarter-century. Historical population estimates are from McEvedy & Jones, 1978, taken at the end of the quarter-century. "Europe" is defined as in figure 5–17.

hemoclysms. Other than the quarter-century containing World War II, the most deadly time to have been alive in Europe was during the Wars of Religion in the early 17th century, followed by the quarter with World War I, then the period of the French Revolutionary and Napoleonic Wars.

The career of organized violence in Europe, then, looks something like this. There was a low but steady baseline of conflicts from 1400 to 1600, followed by the bloodbath of the Wars of Religion, a bumpy decline through 1775 followed by the French troubles, a noticeable lull in the middle and late 19th century, and then, after the 20th-century Hemoclysm, the unprecedented ground-hugging levels of the Long Peace.

How can we make sense of the various slow drifts and sudden lurches in violence during the past half-millennium among the great powers and in Europe? We have reached the point at which statistics must hand the baton over to narrative history. In the next sections I'll tell the story behind the graphs by combining the numbers from the conflict-counters with the narratives from historians and political scientists such as David Bell, Niall Ferguson, Azar Gat, Michael Howard, John Keegan, Evan Luard, John Mueller, James Payne, and James Sheehan.

Here is a preview. Think of the zigzags in figure 5–18 as a composite of four currents. Modern Europe began in a Hobbesian state of frequent but small wars. The wars became fewer in number as political units became consolidated into larger states. At the same time the wars that did occur were becoming more lethal, because of a military revolution that created larger and more effective armies. Finally, in different periods European countries veered between totalizing ideologies that subordinated individual people's interests to a utopian vision and an Enlightenment humanism that elevated those interests as the ultimate value.

THE HOBBESIAN BACKGROUND AND THE AGES OF DYNASTIES AND RELIGIONS

The backdrop of European history during most of the past millennium is ever-present warring. Carried over from the knightly raiding and feuding in medieval times, the wars embroiled every kind of political unit that emerged in the ensuing centuries.

The sheer number of European wars is mind-boggling. Brecke has compiled a prequel to the Conflict Catalog which lists 1,148 conflicts from 900 CE to 1400 CE, and the catalog itself lists another 1,166 from 1400 CE to the present—about two new conflicts a year for eleven hundred years.[88] The vast majority of these conflicts, including most of the major wars involving great powers, are outside the consciousness of all but the most assiduous historians. To take some random examples, the Dano-Swedish War (1516–25), the Schmalkaldic War (1546–47), the Franco-Savoian War (1600–1601), the Turkish-Polish War (1673–76), the

War of Julich Succession (1609–10), and the Austria-Sardinia War (1848–49) elicit blank stares from most educated people.[89]

Warring was not just prevalent in practice but accepted in theory. Howard notes that among the ruling classes, "Peace was regarded as a brief interval between wars," and war was "an almost automatic activity, part of the natural order of things."[90] Luard adds that while many battles in the 15th and 16th centuries had moderately low casualty rates, "even when casualties were high, there is little evidence that they weighed heavily with rulers or military commanders. They were seen, for the most part, as the inevitable price of war, which in itself was honourable and glorious."[91]

What were they fighting over? The motives were the "three principal causes of quarrel" identified by Hobbes: predation (primarily of land), preemption of predation by others, and credible deterrence or honor. The principal difference between European wars and the raiding and feuding of tribes, knights, and warlords was that the wars were carried out by organized political units rather than by individuals or clans. Conquest and plunder were the principal means of upward mobility in the centuries when wealth resided in land and resources rather than in commerce and innovation. Nowadays ruling a dominion doesn't strike most of us as an appealing career choice. But the expression "to live like a king" reminds us that centuries ago it was the main route to amenities like plentiful food, comfortable shelter, pretty objects, entertainment on demand, and children who survived their first year of life. The perennial nuisance of royal bastards also reminds us that a lively sex life was a perquisite of European kings no less than of harem-holding sultans, with "serving maids" a euphemism for concubines.[92]

But what the leaders sought was not just material rewards but a spiritual need for dominance, glory, and grandeur—the bliss of contemplating a map and seeing more square inches tinted in the color that represents your dominion than someone else's. Luard notes that even when rulers had little genuine authority over their titular realms, they went to war for "the theoretical right of overlordship: who owed allegiance to whom and for which territories."[93] Many of the wars were pissing contests. Nothing was at stake but the willingness of one leader to pay homage to another in the form of titles, courtesies, and seating arrangements. Wars could be triggered by symbolic affronts such as a refusal to dip a flag, to salute colors, to remove heraldic symbols from a coat of arms, or to follow protocols of ambassadorial precedence.[94]

Though the motive to lead a dominant political bloc was constant through European history, the definition of the blocs changed, and with it the nature and extent of the fighting. In *War in International Society*, the most systematic attempt to combine a dataset of war with narrative history, Luard proposes that the sweep of armed conflict in Europe may be divided into five "ages," each defined by the nature of the blocs that fought for dominance. In fact

Luard's ages are more like overlapping strands in a rope than boxcars on a track, but if we keep that in mind, his scheme helps to organize the major historical shifts in war.

Luard calls the first of his ages, which ran from 1400 to 1559, the Age of Dynasties. In this epoch, royal "houses," or extended coalitions based on kinship, vied for control of European turfs. A little biology shows why the idea of basing leadership on inheritance is a recipe for endless wars of succession.

Rulers always face the dilemma of how to reconcile their thirst for everlasting power with an awareness of their own mortality. A natural solution is to designate an offspring, usually a firstborn son, as a successor. Not only do people think of their genetic progeny as an extension of themselves, but filial affection ought to inhibit any inclination of the successor to hurry things along with a little regicide. This would solve the succession problem in a species in which an organism could bud off an adult clone of himself shortly before he died. But many aspects of the biology of *Homo sapiens* confound the scheme.

First, humans are altricial, with immature newborns and a long childhood. That means that a father can die while a son is too young to rule. Second, character traits are polygenic, and hence obey the statistical law called regression to the mean: however exceptional in courage or wisdom a parent may be, on average his or her children will be less so. (As the critic Rebecca West wrote, 645 years of the Habsburg dynasty produced "no genius, only two rulers of ability . . . , countless dullards, and not a few imbeciles and lunatics.")[95] Third, humans reproduce sexually, which means that every person is the genetic legacy of two lineages, not one, each of which can lay a claim to the person's loyalties when he is alive and to his perquisites when he dies. Fourth, humans are sexually dimorphic, and though the female of the species may, on average, get less emotional gratification from conquest and tyranny than the male, many are capable of cultivating the taste when the opportunity presents itself. Fifth, humans are mildly polygynous, so males are apt to sire bastards, who become rivals to their legitimate heirs. Sixth, humans are multiparous, having several offspring over their reproductive careers. This sets the stage for parent-offspring conflict, in which a son may want to take over a lineage's reproductive franchise before a father is through with it; and sibling rivalry, in which a laterborn may covet the parental investment lavished on a firstborn. Seventh, humans are nepotistic, investing in their siblings' children as well as in their own. Each of these biological realities, and often several at a time, left room for disagreement about who was the appropriate successor of a dead monarch, and the Europeans hashed out these disagreements in countless dynastic wars.[96]

Luard designates 1559 as the inception of the Age of Religions, which lasted until the Treaty of Westphalia ended the Thirty Years' War in 1648. Rival religious coalitions, often aligning with rulers according to the principle *Un roi, une loi, une foi* (One king, one law, one faith), fought for control of cities and states in at least twenty-five international wars and twenty-six civil wars. Usually Protestants warred against Catholics, but during Russia's Time of Troubles (an interregnum between the reign of Boris Godunov and the establishment of the Romanov dynasty), Catholic and Orthodox factions vied for control. The religious fever was not confined to Christendom: Christian countries fought Muslim Turkey, and Sunni and Shiite Muslims fought in four wars between Turkey and Persia.

This is the age that contributed atrocities number 13, 14, and 17 to the population-adjusted top-twenty-one list on page 195, and it is marked by pinnacles of death in figure 5–15 and figure 5–18. The era broke new records for killing partly because of advances in military technology such as muskets, pikes, and artillery. But that could not have been the main cause of the carnage, because in subsequent centuries the technology kept getting deadlier while the death toll came back to earth. Luard singles out religious passion as the cause:

> It was above all the extension of warfare to civilians, who (especially if they worshipped the wrong god) were frequently regarded as expendable, which now increased the brutality of war and the level of casualties. Appalling bloodshed could be attributed to divine wrath. The duke of Alva had the entire male population of Naarden killed after its capture (1572), regarding this as a judgement of God for their hard-necked obstinacy in resisting; just as Cromwell later, having allowed his troops to sack Drogheda with appalling bloodshed (1649), declared that this was a "righteous judgement of God." Thus by a cruel paradox those who fought in the name of their faith were often less likely than any to show humanity to their opponents in war. And this was reflected in the appalling loss of life, from starvation and the destruction of crops as much as from warfare, which occurred in the areas most ravaged by religious conflict in this age.[97]

Names like the "Thirty Years' War" and the "Eighty Years' War," together with the never-equaled spike in war durations shown in figure 5–14, tell us that the Wars of Religion were not just intense but interminable. The historian of diplomacy Garrett Mattingly notes that in this period a major mechanism for ending war was disabled: "As religious issues came to dominate political ones, any negotiations with the enemies of one state looked more and more like heresy and treason. The questions which divided Catholics from Protestants had ceased to be negotiable. Consequently . . . diplomatic contacts diminished."[98] It would not be the last time ideological fervor would act as an accelerant to a military conflagration.

THREE CURRENTS IN THE AGE OF SOVEREIGNTY

Historians consider the Treaty of Westphalia of 1648 not only to have put out the Wars of Religion but to have established the first version of the modern international order. Europe was now partitioned into sovereign states rather than being a crazy quilt of jurisdictions nominally overseen by the Pope and the Holy Roman Emperor. This Age of Sovereignty saw the ascendancy of states that were still linked to dynasties and religions but that really hung their prestige on their governments, territories, and commercial empires. It was this gradual consolidation of sovereign states (culminating a process that began well before 1648) that set off the two opposing trends that have emerged from every statistical study of war we have seen: wars were getting less frequent but more damaging.

A major reason wars declined in number was that the units that could fight each other declined in number. Recall from chapter 3 that the number of political units in Europe shrank from five hundred around the time of the Thirty Years' War to fewer than thirty in the 1950s.[99] Now, you might think that this makes the decline in the frequency of wars just an accounting trick. With the stroke of an eraser, diplomats remove a line on a map that separates warring parties and magically take their conflict out of the "interstate war" books and hide it in the "civil war" books. But in fact the reduction is real. As Richardson showed, when we hold area constant, there are far fewer civil wars within national boundaries than there are interstate wars crossing them. (Just think of England, which hasn't had a true civil war in 350 years, but has fought many interstate wars since then.) It is another illustration of the logic of the Leviathan. As small baronies and duchies coalesced into larger kingdoms, the centralized authorities prevented them from warring with each other for the same reason that they prevented individual citizens from murdering each other (and that farmers prevent their livestock from killing each other): as far as an overlord is concerned, private quarrels within his domain are a dead loss. The reduction in the frequency of war is thus another manifestation of Elias's Civilizing Process.

The greater lethality of the wars that did take place was the result of a development called the military revolution.[100] States got serious about war. This was partly a matter of improved weaponry, especially cannons and guns, but it was more a matter of recruiting greater numbers of people to kill and be killed. In medieval Europe and the Age of Dynasties, rulers were understandably nervous about arming large numbers of their peasants and training them in combat. (One can hear them asking themselves: What could possibly go wrong?)

Instead they assembled ad hoc militias by hiring mercenaries or conscripting miscreants and ne'er-do-wells who could not buy their way out. In his essay "War Making and State Making as Organized Crime," Charles Tilly wrote:

> In times of war . . . , the managers of full-fledged states often commissioned privateers, hired sometime bandits to raid their enemies, and encouraged their regular troops to take booty. In royal service, soldiers and sailors were often expected to provide for themselves by preying on the civilian population: commandeering, raping, looting, taking prizes. When demobilized, they commonly continued the same practices, but without the same royal protection; demobilized ships became pirate vessels, demobilized troops bandits.
>
> It also worked the other way: A king's best source of armed supporters was sometimes the world of outlaws. Robin Hood's conversion to royal archer may be a myth, but the myth records a practice. The distinctions between "legitimate" and "illegitimate" users of violence came clear only very slowly, in the process during which the states' armed forces became relatively unified and permanent.[101]

As armed forces became more unified and permanent, they also became more effective. The thugs who had made up the earlier militias could hurt a lot of civilians, but they were not terribly effective in organized combat because bravery and discipline held no appeal. Mueller explains:

> The motto for the criminal, after all, is not a variation of "Semper fi," "All for one and one for all," "Duty, honor, country," "Banzai," or "Remember Pearl Harbor," but "Take the money and run." Indeed, for a criminal to perish in battle (or in the commission of a bank robbery) is essentially absurd; it is profoundly irrational to die for the thrill of violence and even more so for the procurement of booty, because you can't, after all, take either one with you.[102]

But during the military revolution of the 16th and 17th centuries, states began to form professional standing armies. They conscripted large numbers of men from a cross section of society rather than just from the dregs at the bottom. They used a combination of drill, indoctrination, and brutal punishment to train them for organized combat. And they instilled in them a code of discipline, stoicism, and valor. The result was that when two of these armies clashed, they could rack up high body counts in a hurry.

The military historian Azar Gat has argued that "revolution" is a misnomer for what was really a gradual development.[103] The process of making armies more effective was part of the centuries-long wave of technological and organizational change that made *everything* more effective. Perhaps an even greater advance in battlefield carnage than the original military revolution has been attributed to Napoleon, who replaced set battles in which both sides tried to conserve their soldiers with bold attacks in which a country would deploy every available resource to inflict all-out defeat on its enemy.[104] Yet another "advance" was the tapping of the Industrial Revolution, beginning in the 19th century, to feed and equip ever larger quantities of soldiers and transport them to the battlefront more quickly. The renewable supply of cannon fodder stoked the games of attrition that pushed wars farther out along the tail of the power-law distribution.

During this long run-up in military power, there was a second force (together with the consolidation of states) that drove down the frequency of combat. Many historians have seen the 18th century as a time of respite in the long European history of war. In the preceding chapter I mentioned that imperial powers like Holland, Sweden, Denmark, Portugal, and Spain stopped competing in the great power game and redirected their energies from conquest to commerce. Brecke writes of a "relatively pacific 18th century" (at least from 1713 to 1789), which can be seen as a U in figure 5–17 and as a shallow lopsided W between the peaks for the religious and French wars in figure 5–18. Luard notes that in the Age of Sovereignty from 1648 to 1789, "objectives were often relatively limited; and many wars in any case ended in a draw, from which no country secured its maximum aims. Many wars were lengthy, but the method of fighting was often deliberately restrained and casualties were less heavy than in either the preceding age or subsequent ages." To be sure, the century saw some bloody combat, such as the world war known as the Seven Years' War, but as David Bell notes, "Historians need to be able to make distinctions between shades of horror, and if the eighteenth century did not exactly reduce the slavering dogs of war to 'performing poodles' . . . , its conflicts still ranked among the *least* horrific in European history."[105]

As we saw in chapter 4, this tranquillity was a part of the Humanitarian Revolution connected with the Age of Reason, the Enlightenment, and the dawn of classical liberalism. The calming of religious fervor meant that wars were no longer inflamed with eschatological meaning, so leaders could cut deals rather than fight to the last man. Sovereign states were becoming commercial powers, which tend to favor positive-sum trade over zero-sum conquest. Popular writers were deconstructing honor, equating war with murder, ridiculing Europe's history of violence, and taking the viewpoints of soldiers and conquered peoples. Philosophers were redefining government from a means of implementing the whims of a monarch to a means for enhancing the life, liberty, and happiness of individual people, and tried to think up ways to limit the power of political leaders and incentivize them to avoid war. The ideas trickled upward and infiltrated the attitudes of at least some of the rulers of the day. While their "enlightened absolutism" was still absolutism, it was certainly better than unenlightened absolutism. And liberal democracy (which, as we shall see, appears to be a pacifying force) got its first toeholds in the United States and Great Britain.

COUNTER-ENLIGHTENMENT IDEOLOGIES AND THE AGE OF NATIONALISM

Of course, it all went horribly wrong. The French Revolution and the French Revolutionary and Napoleonic Wars caused as many as 4 million deaths, earning the sequence a spot in the twenty-one worst things people have ever done to each other, and poking up a major peak in the graph of war deaths in figure 5–18.

Luard designates 1789 as the start of the Age of Nationalism. The players in the preceding Age of Sovereignty had been sprawling dynastic empires that were not pinned to a "nation" in the sense of a group of people sharing a homeland, a language, and a culture. This new age was populated by states that were better aligned with nations and that competed with other nation-states for preeminence. Nationalist yearnings set off thirty wars of independence in Europe and led to autonomy for Belgium, Greece, Bulgaria, Albania, and Serbia. They also inspired the wars of national unification of Italy and of Germany. The peoples of Asia and Africa were not yet deemed worthy of national self-expression, so the European nation-states enhanced their own glory by colonizing them.

World War I, in this scheme, was a culmination of these nationalist longings. It was ignited by Serbian nationalism against the Habsburg Empire, inflamed by nationalist loyalties that pitted Germanic peoples against Slavic ones (and soon against the British and the French), and ended with the dismemberment of the multiethnic Habsburg and Ottoman empires into the new nation-states of Central and Eastern Europe.

Luard ends his Age of Nationalism in 1917. That was the year the United States entered the war and rebranded it as a struggle of democracy against autocracy, and in which the Russian Revolution created the first communist state. The world then entered the Age of Ideology, in which democracy and communism fought Nazism in World War II and each other during the Cold War. Writing in 1986, Luard dangled a dash after "1917"; today we might close it with "1989."

The concept of an Age of Nationalism is a bit procrustean. The age begins with the French Revolutionary and Napoleonic Wars because they were inflamed by the national spirit of France, but these wars were just as inflamed by the ideological residue of the French Revolution, well before the so-called Age of Ideology. Also, the age is an unwieldy sandwich, with massively destructive wars at each end and two record-breaking intervals of peace (1815–54 and 1871–1914) in the middle.

A better way to make sense of the past two centuries, Michael Howard has argued, is to see them as a battle for influence among four forces—Enlightenment humanism, conservatism, nationalism, and utopian ideologies—which sometimes joined up in temporary coalitions.[106] Napoleonic France, because it emerged from the French Revolution, became associated in Europe with the French Enlightenment. In fact it is better classified as the first implementation of fascism. Though Napoleon did implement a few rational reforms such as the metric system and codes of civil law (which survive in many French-influenced regions today), in most ways he wrenched the clock back from the humanistic advances of the Enlightenment. He seized power in a coup, stamped out constitutional government, reinstituted slavery, glorified war, had the Pope crown him emperor, restored Catholicism as the state religion, installed three brothers and a brother-in-law on foreign thrones, and waged ruthless campaigns of territorial aggrandizement with a criminal disregard for human life.

Revolutionary and Napoleonic France, Bell has shown, were consumed by a combination of French nationalism *and* utopian ideology.[107] The ideology, like the versions of Christianity that came before it and the fascism and communism that would follow it, was messianic, apocalyptic, expansionist, and certain of its own rectitude. And it viewed its opponents as irredeemably evil: as existential threats that had to be eliminated in pursuit of a holy cause. Bell notes that the militant utopianism was a disfigurement of the Enlightenment ideal of humanitarian progress. To the revolutionaries, Kant's "goal of perpetual peace had value not because it conformed to a fundamental moral law but because it conformed to the historical progress of civilization. . . . And so they opened the door to the idea that in the name of future peace, any and all means might be justified—including even exterminatory war."[108] Kant himself despised this turn, noting that such a war "would allow perpetual peace only upon the graveyard of the whole human race." And the American framers, equally aware of the crooked timber of humanity, were positively phobic about the prospect of imperial or messianic leaders.

After the French ideology had been disseminated across Europe at the point of a bayonet and driven back at enormous cost, it elicited a slew of reactions, which as we saw in chapter 4 are often lumped together as counter-Enlightenments. Howard sees the common denominator as "the view that man is not simply an individual who by the light of reason and observation can formulate laws on the basis of which he can create a just and peaceful society, but rather a member of a community that has moulded him in a fashion he himself cannot fully comprehend, and which has a primary claim on his loyalties."

Recall that there were two counter-Enlightenments, which reacted to the French disruptions in opposite ways. The first was Edmund Burke's conservatism, which held that a society's customs were time-tested implementations of a civilizing process that had tamed humanity's dark side and as such deserved respect alongside the explicit formal propositions of intellectuals and reformers. Burkean conservatism, itself a fine application of reason, represented a small tweaking of Enlightenment humanism. But that ideal was blown to bits in Johann Gottfried von Herder's romantic nationalism, which held that an ethnic group—in the case of Herder, the German *Volk*—had unique qualities that could not be submerged into the supposed universality of humankind, and that were held together by ties of blood and soil rather than by a reasoned social contract.

According to Howard, "this dialectic between Enlightenment and Counter-Enlightenment, between the individual and the tribe, was to pervade, and to a large extent shape, the history of Europe throughout the nineteenth century, and of the world the century after that."[109] During those two centuries Burkean conservatism, Enlightenment liberalism, and romantic nationalism played off one another in shifting alliances (and sometimes became strange bedfellows).

The Congress of Vienna in 1815, when statesmen from the great powers engineered a system of international relations that would last a century, was a triumph of Burkean conservatism, aiming for stability above all else. Nonetheless, Howard observes, its architects "were as much the heirs of the Enlightenment as had been the French revolutionary leaders. They believed neither in the divine right of kings nor the divine authority of the church; but since church and king were necessary tools in the restoration and maintenance of the domestic order that the revolution had so rudely disturbed, their authority had everywhere to be restored and upheld."[110] More important, "they no longer accepted war between major states as an ineluctable element in the international system. The events of the past twenty-five years had shown that it was too dangerous." The great powers took on the responsibility of preserving peace and order (which they pretty much equated), and their Concert of Europe was a forerunner of the League of Nations, the United Nations, and the European Union. This international Leviathan deserves much of the credit for the long intervals of peace in 19th-century Europe.

But the stability was enforced by monarchs who ruled over lumpy amalgams of ethnic groups, which began to clamor for a say in how their affairs were run. The result was a nationalism that, according to Howard, was "based not so much on universal human rights as on the rights of nations to fight their way into existence and to defend themselves once they existed." Peace was not particularly desirable in the short term; it would come about "only when all nations were free. Meanwhile, [nations] claimed the right to use such force as was necessary to free themselves, by fighting precisely the wars of national liberation that the Vienna system had been set up to prevent."[111]

Nationalist sentiments soon intermixed with every other political movement. Once nation-states emerged, they became the new establishment, which the conservatives strove to conserve. As monarchs became icons of their nations, conservatism and nationalism gradually merged.[112] And among many intellectuals, romantic nationalism became entwined with the Hegelian doctrine that history was an inexorable dialectic of progress. As Luard summarized the doctrine, "All history represents the working out of some divine plan; war is the way that sovereign states, through which that plan manifested itself, must resolve their differences, leading to the emergence of superior states (such as the Prussian state), representing the fulfillment of the divine purpose."[113] Eventually the doctrine spawned the messianic, militant, romantic nationalist movements of fascism and Nazism. A similar construction of history as an unstoppable dialectic of violent liberation, but with classes substituted for nations, became the foundation of 20th-century communism.[114]

One might think that the liberal heirs of the British, American, and Kantian Enlightenment would have been opposed to the increasingly militant nationalism. But they found themselves in a pickle: they could hardly defend autocratic monarchies and empires. So liberalism signed on to nationalism in the guise of "self-determination of peoples," which has a vaguely democratic aroma. Unfortunately, the whiff of humanism emanating from that phrase depended on a fatal synecdoche. The term "nation" or "people" came to stand for the individual men, women, and children who made up that nation, and then the political leaders came to stand for the nation. A ruler, a flag, an army, a territory, a language, came to be cognitively equated with millions of flesh-and-blood individuals. The liberal doctrine of self-determination of peoples was enshrined by Woodrow Wilson in a 1916 speech and became the basis for the world order after World War I. One of the people who immediately saw the inherent contradiction in the "self-determination of peoples" was Wilson's own secretary of state, Robert Lansing, who wrote in his diary:

> The phrase is simply loaded with dynamite. It will raise hopes which can never be realized. It will, I fear, cost thousands of lives. In the end, it is bound to be discredited, to be called the dream of an idealist who failed to realize the danger until too late to check those who attempt to put the principle into force. What a calamity that the phrase was ever uttered! What misery it will cause! Think of the feelings of the author when he counts the dead who died because he uttered a phrase![115]

Lansing was wrong about one thing: the cost was not thousands of lives but tens of millions. One of the dangers of "self-determination" is that there is really no such thing as a "nation" in the sense of an ethnocultural group that coincides with a patch of real estate. Unlike features of a landscape like trees and mountains, people have feet. They move to places where the opportunities are best, and they soon invite their friends and relatives to join them. This demographic mixing turns the landscape into a fractal, with minorities inside minorities inside minorities. A government with sovereignty over a territory which claims to embody a "nation" will in fact fail to embody the interests of many of the individuals living within that territory, while taking a proprietary interest in individuals living in other territories. If utopia is a world in which political boundaries coincide with ethnic boundaries, leaders will be tempted to hasten it along with campaigns of ethnic cleansing and irredentism. Also, in the absence of liberal democracy and a robust commitment to human rights, the synecdoche in which a people is equated with its political ruler will turn any international confederation (such as the General Assembly of the United Nations) into a travesty. Tinpot dictators are welcomed into the family of nations and given carte blanche to starve, imprison, and murder their citizens.

Another development of the 19th century that would undo Europe's long interval of peace was romantic militarism: the doctrine that war itself was a salubrious activity, quite apart from its strategic goals. Among liberals and conservatives alike, the notion took hold that war called forth spiritual qualities of heroism, self-sacrifice, and manliness and was needed as a cleansing and invigorating therapy for the effeminacy and materialism of bourgeois society. Nowadays the idea that there could be something inherently admirable about an enterprise that is designed to kill people and destroy things seems barking mad. But in this era, writers gushed about it:

> War almost always enlarges the mind of a people and raises their character.
>
> —Alexis de Tocqueville

> [War is] life itself. . . . We must eat and be eaten so that the world might live. It is only warlike nations which have prospered: a nation dies as soon as it disarms.
>
> —Émile Zola

> The grandeur of war lies in the utter annihilation of puny man in the great conception of the State, and it brings out the full magnificence of the sacrifice of fellow-countrymen for one another . . . the love, the friendliness, and the strength of that mutual sentiment.
>
> —Heinrich von Treitschke

> When I tell you that war is the foundation of all the arts, I mean also that it is the foundation of all the high virtues and faculties of man.
>
> —John Ruskin

> Wars are terrible, but necessary, for they save the State from social petrifaction and stagnation.
>
> —Georg Wilhelm Friedrich Hegel

> [War is] a purging and a liberation.
>
> —Thomas Mann

War is necessary for human progress.
—Igor Stravinsky[116]

Peace, in contrast, was "a dream and not a pleasant one at that," wrote the German military strategist Helmuth von Moltke; "without war, the world would wallow in materialism."[117] Friedrich Nietzsche agreed: "It is mere illusion and pretty sentiment to expect much (even anything at all) from mankind if it forgets how to make war." According to the British historian J. A. Cramb, peace would mean "a world sunk in bovine content . . . a nightmare which shall be realized only when the ice has crept to the heart of the sun, and the stars, left black and trackless, start from their orbits."[118]

Even thinkers who opposed war, such as Kant, Adam Smith, Ralph Waldo Emerson, Oliver Wendell Holmes, H. G. Wells, and William James, had nice things to say about it. The title of James's 1906 essay "The Moral Equivalent of War" referred not to something that was as *bad* as war but to something that would be as *good* as it.[119] He began, to be sure, by satirizing the military romantic's view of war:

> Its "horrors" are a cheap price to pay for rescue from the only alternative supposed, of a world of clerks and teachers, of co-education and zo-ophily, of "consumer's leagues" and "associated charities," of industrialism unlimited, and feminism unabashed. No scorn, no hardness, no valor any more! Fie upon such a cattleyard of a planet!

But then he conceded that "we must make new energies and hardihoods continue the manliness to which the military mind so faithfully clings. Martial virtues must be the enduring cement; intrepidity, contempt of softness, surrender of private interest, obedience to command, must still remain the rock upon which states are built." And so he proposed a program of compulsory national service in which "our gilded youths [would] be drafted off . . . to get the childishness knocked out of them" in coal mines, foundries, fishing vessels, and construction sites.

Romantic nationalism and romantic militarism fed off each other, particularly in Germany, which came late to the party of European states and felt that it deserved an empire too. In England and France, romantic militarism ensured that the prospect of war was not as terrifying as it should have been. On the contrary, Hillaire Belloc wrote, "How I long for the Great War! It will sweep Europe like a broom!"[120] Paul Valéry felt the same way: "I almost desire a monstrous war."[121] Even Sherlock Holmes got into the act; in 1914 Arthur Conan Doyle had him say, "It will be cold and bitter, Watson, and a good many of us may wither before its blast. But it's God's own wind none the less, and a cleaner, better, stronger land will lie in the sunshine when the storm has cleared."[122] Metaphors proliferated: the sweeping broom, the bracing wind, the pruning shears, the cleansing storm, the purifying fire. Shortly before he joined the British navy, the poet Rupert Brooke wrote:

Now, God be thanked Who has matched us with His hour,
And caught our youth, and wakened us from sleeping,
With hand made sure, clear eye, and sharpened power,
To turn, as swimmers into cleanness leaping.

"Of course, the swimmers weren't leaping into clean water but wading into blood." So commented the critic Adam Gopnik in a 2004 review of seven new books that were still, almost a century later, trying to figure out exactly how World War I happened.[123] The carnage was stupefying—8.5 million deaths in combat, and perhaps 15 million deaths overall, in just four years.[124] Romantic militarism by itself cannot explain the orgy of slaughter. Writers had been glorifying war at least since the 18th century, but the post-Napoleonic 19th had had two unprecedented stretches without a great power war. The war was a perfect storm of destructive currents, brought suddenly together by the iron dice of Mars: an ideological background of militarism and nationalism, a sudden contest of honor that threatened the credibility of each of the great powers, a Hobbesian trap that frightened leaders into attacking before they were attacked first, an overconfidence that deluded each of them into thinking that victory would come swiftly, military machines that could deliver massive quantities of men to a front that could mow them down as quickly as they arrived, and a game of attrition that locked the two sides into sinking exponentially greater costs into a ruinous situation—all set off by a Serbian nationalist who had a lucky day.

HUMANISM AND TOTALITARIANISM IN THE AGE OF IDEOLOGY

The Age of Ideology that began in 1917 was an era in which the course of war was determined by the inevitabilist belief systems of the 19th-century counter-Enlightenment. A romantic, militarized nationalism inspired the expansionist programs of Fascist Italy and Imperial Japan, and with an additional dose of racialist pseudoscience, Nazi Germany. The leadership of each of these countries railed against the decadent individualism and universalism of the modern liberal West, and each was driven by the conviction that it was destined to rule over a natural domain: the Mediterranean, the Pacific rim, and the European continent, respectively.[125] World War II began with invasions that were intended to move this destiny along. At the same time a romantic, militarized communism inspired the expansionist programs of the Soviet Union and China, who wanted to give a helping hand to the dialectical process by which the proletariat or peasantry would vanquish the bourgeoisie and establish a dictatorship in country after country. The Cold War was the product of the determination of the United States to contain this movement at something close to its boundaries at the end of World War II.[126]

But this narrative leaves out a major plot that perhaps had the most lasting impact on the 20th century. Mueller, Howard, Payne, and other political historians remind us that the 19th century was host to yet another movement: a continuation of the Enlightenment critique of war.[127] Unlike the strain of liberalism that developed a soft spot for nationalism, this one kept its eye on the individual human being as the entity whose interests are paramount. And it invoked the Kantian principles of democracy, commerce, universal citizenship, and international law as practical means of implementing peace.

The brain trust of the 19th- and early-20th-century antiwar movement included Quakers such as John Bright, abolitionists such as William Lloyd Garrison, advocates of the theory of gentle commerce such as John Stuart Mill and Richard Cobden, pacifist writers such as Leo Tolstoy, Victor Hugo, Mark Twain, and George Bernard Shaw, the philosopher Bertrand Russell, industrialists such as Andrew Carnegie and Alfred Nobel (of Peace Prize fame), many feminists, and the occasional socialist (motto: "A bayonet is a weapon with a worker at each end"). Some of these moral entrepreneurs created new institutions that were designed to preempt or constrain war, such as a court of international arbitration in The Hague and a series of Geneva Conventions on the conduct of war.

Peace first became a popular sensation with the publication of two bestsellers. In 1889 the Austrian novelist Bertha von Suttner published a work of fiction called *Die Waffen nieder!* (Lay Down Your Arms!), a first-person account of the gruesomeness of war. And in 1909 the British journalist Norman Angell published a pamphlet called *Europe's Optical Illusion*, later expanded as *The Great Illusion*, which argued that war was economically futile. Plunder may have been profitable in primitive economies, when wealth lay in finite resources like gold or land or in the handiwork of self-sufficient craftsmen. But in a world in which wealth grows out of exchange, credit, and a division of labor, conquest cannot make a conqueror richer. Minerals don't just jump out of the ground, nor does grain harvest itself, so the conqueror would still have to pay the miners to mine and the farmers to farm. In fact, he would make himself poorer, since the conquest would cost money and lives and would damage the networks of trust and cooperation that allow everyone to enjoy gains in trade. Germany would have nothing to gain by conquering Canada any more than Manitoba would have something to gain by conquering Saskatchewan.

For all its literary popularity, the antiwar movement seemed too idealistic at the time to be taken seriously by the political mainstream. Suttner was called "a gentle perfume of absurdity," and her German Peace Society "a comical sewing bee composed of sentimental aunts of both sexes." Angell's friends told him to "avoid that stuff or you'll be classed with cranks and faddists, with devotees of Higher Thought who go about in sandals and long beards, and live on nuts."[128] H. G. Wells wrote that Shaw was "an elderly adolescent still at play. . . . All through the war we shall have this Shavian accompaniment going on, like an idiot child screaming in a hospital."[129] And though Angell had never claimed that war was obsolete—he argued only that it served no economic purpose, and was terrified that glory-drunk leaders would blunder into it anyway—that was how he was interpreted.[130] After World War I he became a laughingstock, and to this day he remains a symbol for naïve optimism about the impending end of war. While I was writing this book, more than one concerned colleague took me aside to educate me about Norman Angell.

But according to Mueller, Angell deserves the last laugh. World War I put an end not just to romantic militarism in the Western mainstream but to the idea that war was in any way desirable or inevitable. "The First World War," notes Luard, "transformed traditional attitudes toward war. For the first time there was an almost universal sense that the deliberate launching of a war could now no longer be justified."[131] It was not just that Europe was reeling from the loss of lives and resources. As Mueller notes, there had been comparably destructive wars in European history before, and in many cases countries dusted themselves off and, as if having learned nothing, promptly jumped into a new one. Recall that the statistics of deadly quarrels show no signature of war-weariness. Mueller argues that the crucial difference this time was that an articulate antiwar movement had been lurking in the background and could now say "I told you so."

The change could be seen both in the political leadership and in the culture at large. When the destructiveness of the Great War became apparent, it was reframed as "the war to end all wars," and once it was over, world leaders tried to legislate the hope into reality by formally renouncing war and setting up a League of Nations to prevent it. However pathetic these measures may seem in hindsight, at the time they were a radical break from centuries in which war had been regarded as glorious, heroic, honorable, or in the famous words of the military theorist Karl von Clausewitz, "merely the continuation of policy by other means."

World War I has also been called the first "literary war." By the late 1920s, a genre of bitter reflections was making the tragedy and futility of the war common knowledge. Among the great works of the era are the poems and memoirs of Siegfried Sassoon, Robert Graves, and Wilfred Owen, the bestselling novel and popular film *All Quiet on the Western Front*, T. S. Eliot's poem "The Hollow Men," Hemingway's novel *A Farewell to Arms*, R. C. Sherriff's play *Journey's End*, King Vidor's film *The Big Parade*, and Jean Renoir's film *Grand Illusion*—the title adapted from Angell's pamphlet. Like other humanizing works of art, these stories created an illusion of first-person immediacy, encouraging their audiences to empathize with the suffering of others. In an unforgettable scene from *All Quiet on the Western Front*, a young German soldier examines the body of a Frenchman he has just killed:

> No doubt his wife still thinks of him; she does not know what happened. He looks as if he would have often written to her—she will still be getting mail from him—Tomorrow, in a week's time—perhaps even a stray letter a month hence. She will read it, and in it he will be speaking to her. . . .
>
> I speak to him and say to him: ". . . Forgive me, comrade. . . . Why do they never tell us that you are poor devils like us, that your mothers are just as anxious as ours, and that we have the same fear of death, and the same dying and the same agony?" . . .
>
> "I will write to your wife," I say hastily to the dead man. . . . "I will tell her everything I have told you, she shall not suffer, I will help her, and your parents too, and your child—" Irresolutely I take the wallet in my hand. It slips out of my hand and falls open. . . . There are portraits of a woman and a little girl, small amateur photographs taken against an ivy-clad wall. Along with them are letters.[132]

Another soldier asks how wars get started and is told, "Mostly by one country badly offending another." The soldier replies, "A country? I don't follow. A mountain in Germany cannot offend a mountain in France. Or a river, or a wood, or a field of wheat."[133] The upshot of this literature, Mueller notes, was that war was no longer seen as glorious, heroic, holy, thrilling, manly, or cleansing. It was now immoral, repulsive, uncivilized, futile, stupid, wasteful, and cruel.

And perhaps just as important, absurd. The immediate cause of World War I had been a showdown over honor. The leaders of Austria-Hungary had issued a humiliating ultimatum to Serbia demanding that it apologize for the assassination of the archduke and crack down on domestic nationalist movements to their satisfaction. Russia took offense on behalf of its fellow Slavs, Germany took offense at Russia's offense on behalf of its fellow German speakers, and as Britain and France joined in, a contest of face, humiliation, shame, stature, and credibility escalated out of control. A fear of being "reduced to a second-rate power" sent them hurtling toward each other in a dreadful game of chicken.

Contests of honor, of course, had been setting off wars in Europe throughout its bloody history. But honor, as Falstaff noted, is just a word—a social construction, we might say today—and "detraction will not suffer it." Detraction there soon was. Perhaps the best antiwar film of all time is the Marx Brothers' *Duck Soup* (1933). Groucho plays Rufus T. Firefly, the newly appointed leader of Freedonia, and is asked to make peace with the ambassador of neighboring Sylvania:

> I'd be unworthy of the high trust that's been placed in me if I didn't do everything within my power to keep our beloved Freedonia at peace with the

> world. I'd be only too happy to meet Ambassador Trentino and offer him on behalf of my country the right hand of good fellowship. And I feel sure he will accept this gesture in the spirit in which it is offered.
>
> But suppose he doesn't. A fine thing that'll be. I hold out my hand and he refuses to accept it. That'll add a lot to my prestige, won't it? Me, the head of a country, snubbed by a foreign ambassador. Who does he think he is that he can come here and make a sap out of me in front of all my people? Think of it. I hold out my hand. And that hyena refuses to accept it. Why, the cheap, four-flushing swine! He'll never get away with it, I tell you! [The ambassador enters.] So, you refuse to shake hands with me, eh? [He slaps the ambassador.]
>
> *Ambassador:* Mrs. Teasdale, this is the last straw! There's no turning back now! This means war!

Whereupon an outlandish production number breaks out in which the Marx Brothers play xylophone on the pickelhauben of the assembled soldiers and then dodge bullets and bombs while their uniforms keep changing, from Civil War soldier to Boy Scout to British palace guard to frontiersman with coonskin cap. War has been likened to dueling, and recall that dueling was eventually laughed into extinction. War was now undergoing a similar deflation, perhaps fulfilling Oscar Wilde's prophecy that "as long as war is regarded as wicked, it will always have its fascination. When it is looked upon as vulgar, it will cease to be popular."

The butt of the joke was different in the other classic war satire of the era, Charlie Chaplin's *The Great Dictator* (1940). It was no longer the hotheaded leaders of generic Ruritanian countries that were the target, since by now virtually everyone was allergic to a military culture of honor. Instead the buffoons were thinly disguised contemporary dictators who anachronistically embraced that ideal. In one memorable scene, the Hitler and Mussolini characters confer in a barbershop and each tries to dominate the other by raising his chair until both are bumping their heads against the ceiling.

By the 1930s, according to Mueller, Europe's war aversion was prevalent even among the German populace and its military leadership.[134] Though resentment of the terms of the Treaty of Versailles was high, few were willing to start a war of conquest to rectify them. Mueller ran through the set of German leaders who had any chance of becoming chancellor and argued that no one but Hitler showed any desire to subjugate Europe. Even a coup by the German military, according to the historian Henry Turner, would not have led to World War II.[135] Hitler exploited the world's war-weariness, repeatedly professing his love of peace and knowing that no one was willing to stop him while he was still stoppable. Mueller reviews biographies of Hitler to defend the idea, also held by many historians, that one man was mostly responsible for the world's greatest cataclysm:

> After seizing control of the country in 1933, [Hitler] moved quickly and decisively to persuade, browbeat, dominate, outmaneuver, downgrade, and in many instances, murder opponents or would-be opponents. He possessed enormous energy and stamina, exceptional persuasive powers, an excellent memory, strong powers of concentration, an overwhelming craving for power, a fanatical belief in his mission, a monumental self-confidence, a unique daring, a spectacular facility for lying, a mesmerizing oratory style, and an ability to be utterly ruthless to anyone who got in his way or attempted to divert him from his intended course of action. . . .
>
> Hitler needed the chaos and discontent to work with—although he created much of it, too. And surely he needed assistance—colleagues who were worshipfully subservient; a superb army that could be manipulated and whipped into action; a population capable of being mesmerized and led to slaughter; foreign opponents who were confused, disorganized, gullible, myopic, and faint-hearted; neighbors who would rather be prey than fight—although he created much of this as well. Hitler took the conditions of the world as he found them and then shaped and manipulated them to his own ends.[136]

Fifty-five million deaths later (including at least 12 million who died in Japan's own atavistic campaign to dominate East Asia), the world was once again in a position to give peace a chance.

THE LONG PEACE: SOME NUMBERS

I have spent a lot of this chapter on the statistics of war. But now we are ready for the most interesting statistic since 1945: zero. Zero is the number that applies to an astonishing collection of categories of war during the two-thirds of a century that has elapsed since the end of the deadliest war of all time. I'll begin with the most momentous.

• Zero is the number of times that nuclear weapons have been used in conflict. Five great powers possess them, and all of them have waged wars. Yet no nuclear device has been set off in anger. It's not just that the great powers avoided the mutual suicide of an all-out nuclear war. They also avoided using the smaller, "tactical" nuclear weapons, many of them comparable to conventional explosives, on the battlefield or in the bombing of enemy facilities. And the United States refrained from using its nuclear arsenal in the late 1940s when it held a nuclear monopoly and did not have to worry about mutually assured destruction. I've been quantifying violence throughout this book using proportions. If one were to calculate the amount of destruction that nations have actually perpetrated as a proportion of how much they *could* perpetrate, given the destructive capacity available to them, the postwar decades would be many orders of magnitudes more peaceable than any time in history.

None of this was a foregone conclusion. Until the sudden end of the Cold War, many experts (including Albert Einstein, C. P. Snow, Herman Kahn, Carl Sagan, and Jonathan Schell) wrote that thermonuclear doomsday was likely, if not inevitable.[137] The eminent international studies scholar Hans Morgenthau, for example, wrote in 1979, "The world is moving ineluctably towards a third world war—a strategic nuclear war. I do not believe that anything can be done to prevent it."[138] The *Bulletin of the Atomic Scientists,* according to its Web site, aims to "inform the public and influence policy through in-depth analyses, op-eds, and reports on nuclear weapons." Since 1947 it has published the famous Doomsday Clock, a measure of "how close humanity is to catastrophic destruction—the figurative midnight." The clock was unveiled with its minute hand pointing at 7 minutes to midnight, and over the next sixty years it was moved back and forth a number of times between 2 minutes to midnight (in 1953) and 17 minutes to midnight (in 1991). In 2007 the *Bulletin* apparently decided that a clock with a minute hand that moved two minutes in sixty years was due for an adjustment. But rather than tuning the mechanism, they redefined midnight. Doomsday now consists of "damage to ecosystems, flooding, destructive storms, increased drought, and polar ice melt." This is a kind of progress.

• Zero is the number of times that the two Cold War superpowers fought each other on the battlefield. To be sure, they occasionally fought each other's smaller allies and stoked proxy wars among their client states. But when either the United States or the Soviet Union sent troops to a contested region (Berlin, Hungary, Vietnam, Czechoslovakia, Afghanistan), the other stayed out of its way.[139] The distinction matters a great deal because as we have seen, one big war can kill vastly more people than many small wars. In the past, when an enemy of a great power invaded a neutral country, the great power would express its displeasure on the battlefield. In 1979, when the Soviet Union invaded Afghanistan, the United States expressed its displeasure by withdrawing its team from the Moscow Summer Olympics. The Cold War, to everyone's surprise, ended without a shot in the late 1980s shortly after Mikhail Gorbachev ascended to power. It was followed by the peaceful tear-down of the Berlin Wall and then by the mostly peaceful collapse of the Soviet Union.

• Zero is the number of times that any of the great powers have fought each other since 1953 (or perhaps even 1945, since many political scientists don't admit China to the club of great powers until after the Korean War). The war-free interval since 1953 handily breaks the previous two records from the 19th century of 38 and 44 years. In fact, as of May 15, 1984, the major powers of the world had remained at peace with one another for the longest stretch of time since the Roman Empire.[140] Not since the 2nd century BCE, when Teutonic tribes challenged the Romans, has a comparable interval passed without an army crossing the Rhine.[141]

• Zero is the number of interstate wars that have been fought between countries in Western Europe since the end of World War II.[142] It is also the number of interstate wars that have been fought in Europe as a whole since 1956, when the Soviet Union briefly invaded Hungary.[143] Keep in mind that up until that point European states had started around two new armed conflicts *a year* since 1400.

• Zero is the number of interstate wars that have been fought since 1945 between major developed countries (the forty-four with the highest per capita income) anywhere in the world (again, with the exception of the 1956 Hungarian invasion).[144] Today we take it for granted that war is something that happens in smaller, poorer, and more backward countries. But the two world wars, together with the many hyphenated European wars from centuries past (Franco-Prussian, Austro-Prussian, Russo-Swedish, British-Spanish, Anglo-Dutch) remind us that this was not always the way things worked.

• Zero is the number of developed countries that have expanded their territory since the late 1940s by conquering another country. No more Poland getting wiped off the map, or Britain adding India to its empire, or Austria helping itself to the odd Balkan nation. Zero is also the number of times that any country has conquered even *parts* of some other country since 1975, and it is not far from the number of permanent conquests since 1948 (a development we'll soon examine more closely).[145] In fact the process of great power aggrandizement went into reverse. In what has been called "the greatest transfer of power in world history," European countries surrendered vast swaths of territory as they closed down their empires and granted independence to colonies, sometimes peacefully, sometimes because they had lost the will to prevail in colonial wars.[146] As we will see in the next chapter, two entire categories of war—the imperial war to acquire colonies, and the colonial war to keep them—no longer exist.[147]

• Zero is the number of internationally recognized states since World War II that have gone out of existence through conquest.[148] (South Vietnam may be the exception, depending on whether its unification with North Vietnam in 1975 is counted as a conquest or as the end of an internationalized civil war.) During the first half of the 20th century, by comparison, twenty-two states were occupied or absorbed, at a time when the world had far fewer states to begin with.[149] Though scores of nations have gained independence since 1945, and several have broken apart, most of the lines on a world map of 1950 are still present on a world map in 2010. This too is an extraordinary development in a world in which rulers used to treat imperial expansion as part of their job description.

The point of this chapter is that these zeroes—the Long Peace—are a result of one of those psychological retunings that take place now and again over the course of history and cause violence to decline. In this case it is a change within the mainstream of the developed world (and increasingly, the rest of the world) in the shared cognitive categorization of war. For most of human history, influential people who craved power, prestige, or vengeance could count on their political network to ratify those cravings and to turn off their sympathies for the victims of an effort to satisfy them. They believed, in other words, in the legitimacy of war. Though the psychological components of war have not gone away—dominance, vengeance, callousness, tribalism, groupthink, self-deception—since the late 1940s they have been disaggregated in Europe and other developed countries in a way that has driven down the frequency of war.

Some people downplay these stunning developments by pointing out that wars still take place in the developing world, so perhaps violence has only been displaced, not reduced. In the following chapter we will examine armed conflict in the rest of the world, but for now it's worth noting that the objection makes little sense. There is no Law of Conservation of Violence, no hydraulic system in which a compression of violence in one part of the world forces it to bulge out somewhere else. Tribal, civil, private, slave-raiding, imperial, and colonial wars have inflamed the territories of the developing world for millennia. A world in which war continues in some of the poorer countries is still better than a world in which it takes place in both the rich *and* the poor countries, especially given the incalculably greater damage that rich, powerful countries can wreak.

A long peace, to be sure, is not a perpetual peace. No one with a statistical appreciation of history could possibly say that a war between great powers, developed countries, or European states will never happen again. But probabilities can change over spans of time that matter to us. The house odds on the iron dice can decline; the power-law line can sink or tilt. And in much of the world, that appears to have happened.

The same statistical consciousness, though, alerts us to alternative possibilities. Perhaps the odds haven't changed at all, and we're overinterpreting a random run of peaceful years in the same way that we are liable to overinterpret a random cluster of wars or atrocities. Perhaps the pressure for war has been building and the system will blow at any moment.

But probably not. The statistics of deadly quarrels show that war is not a pendulum, a pressure cooker, or a hurtling mass, but a memoryless game of dice, perhaps one with changing odds. And the history of many nations affirms that a peace among them can last indefinitely. As Mueller puts it, if war fever were cyclical, "one would expect the Swiss, Danes, Swedes, Dutch, and Spaniards to be positively *roaring* for a fight by now."[150] Nor are Canadians and Americans losing sleep about an overdue invasion across the world's longest undefended border.

What about the possibility of a run of good luck? Also unlikely. The postwar years are by far the longest period of peace among great powers since they came into being five hundred years ago.[151] The stretch of peace among European states is also the longest in its bellicose history. Just about any statistical test can confirm that the zeroes and near zeroes of the Long Peace are extremely improbable, given the rates of war in the preceding centuries. Taking the frequency of wars between great powers from 1495 to 1945 as a baseline, the chance that there would be a sixty-five-year stretch with only a single great power war (the marginal case of the Korean War) is one in a thousand.[152] Even if we take 1815 as our starting point, which biases the test against us by letting the peaceful post-Napoleonic 19th century dominate the base rate, we find that the probability that the postwar era would have at most four wars involving a great power is less than 0.004, and the probability that it would have at most one war between European states (the Soviet invasion of Hungary in 1956) is 0.0008.[153]

The calculation of probabilities, to be sure, critically depends on how one defines the events. Odds are very different when you estimate them in full knowledge of what happened (a post hoc comparison, also known as "data snooping") and when you lay down your prediction beforehand (a planned or a priori comparison). Recall that the chance that two people in a room of fifty-seven will share a birthday is ninety-nine out of a hundred. In that case we are specifying the exact day only after we identify the pair of people. The chance that someone will share *my* birthday is less than one in seven; in that case we specify the day beforehand. A stock scammer can exploit the distinction by sending out newsletters with every possible prediction about the trajectory of the market. Several months later the fraction of recipients that got the lucky matching run will think he is a genius. A skeptic of the Long Peace could claim that anyone making a big deal of a long run of nonwars at the end of that very run is just as guilty of data snooping.

But in fact there is a paper trail of scholars who, more than two decades ago, noticed that the war-free years were piling up and attributed it to a new mindset that they expected to last. Today we can say that their a priori predictions have been confirmed. The story can be told in titles and dates: Werner Levi's *The Coming End of War* (1981), John Gaddis's "The Long Peace: Elements of Stability in the Postwar International System" (1986), Kalevi Holsti's "The Horsemen of the Apocalypse: At the Gate, Detoured, or Retreating?" (1986), Evan Luard's *The Blunted Sword: The Erosion of Military Power in Modern World Politics* (1988), John Mueller's *Retreat from Doomsday: The Obsolescence of Major War* (1989), Francis Fukuyama's "The End of History?" (1989), James Lee Ray's "The Abolition of Slavery and the End of International War" (1989), and Carl Kaysen's "Is War Obsolete?" (1990).[154] In 1988 the political scientist Robert Jervis captured the phenomenon they were all noticing:

> The most striking characteristic of the postwar period is just that—it can be called "postwar" because the major powers have not fought each other since 1945. Such a lengthy period of peace among the most powerful states is unprecedented.[155]

These scholars were confident that they were not being fooled by a lucky run but were putting their finger on an underlying shift that supported predictions about the future. In early 1990, Kaysen added a last-minute postscript to his review of Mueller's 1989 book in which he wrote:

> It is clear that a profound transformation of the international structure in Europe—and the whole world—is underway. In the past, such changes have regularly been consummated by war. The argument presented in this essay supports the prediction that this time the changes can take place without war (although not necessarily without domestic violence in the states concerned). So far—mid-January—so good. The author and his readers will be eagerly and anxiously testing the prediction each day.[156]

Precocious assessments of the obsolescence of interstate war are especially poignant when they come from military historians. These are the scholars who have spent their lives immersed in the annals of warfare and should be most jaded about the possibility that this time it's different. In his magnum opus *A History of Warfare*, John Keegan (the military historian who is so habitually called "distinguished" that one could be forgiven for thinking it is part of his name) wrote in 1993:

> War, it seems to me, after a lifetime of reading about the subject, mingling with men of war, visiting the sites of war and observing its effects, may well be ceasing to commend itself to human beings as a desirable or productive, let alone rational, means of reconciling their discontents.[157]

The equally distinguished Michael Howard had already written, in 1991:

> [It has become] quite possible that war in the sense of major, organized armed conflict between highly developed societies may not recur, and that a stable framework for international order will become firmly established.[158]

And the no-less-distinguished Evan Luard, our guide to six centuries of war, had written still earlier, in 1986:

> Most startling of all has been the change that has come about in Europe, where there has been a virtual cessation of international warfare. . . . Given the scale and frequency of war during the preceding centuries in Europe, this is a change of spectacular proportions: perhaps the single most striking discontinuity that the history of warfare has anywhere provided.[159]

More than two decades later, none of them would have a reason to change his assessment. In his 2006 book *War in Human Civilization*, a military history that is more sweeping than its predecessors and salted with the Hobbesian realism of evolutionary psychology, Azar Gat wrote:

> Among affluent liberal democracies . . . a true *state* of peace appears to have developed, based on genuine mutual confidence that war between them is practically eliminated even as an option. Nothing like this had ever existed in history.[160]

THE LONG PEACE: ATTITUDES AND EVENTS

The italics in Gat's "true *state* of peace" highlight not just the datum that the number of wars between developed states happens to be zero but a change in the countries' mindsets. The ways that developed countries conceptualize and prepare for war have undergone sweeping changes.

A major feeder of the increasing deadliness of war since 1500 (see figure 5–16) has been conscription, the stocking of national armies with a renewable supply of bodies. By the time of the Napoleonic Wars, most European countries had some form of a draft. Conscientious objection was barely a concept, and recruitment methods were far less polite than the telegram dreaded by young American men in the 1960s that began: "Greetings." The idiom *pressed into service* comes from the institution of press gangs, groups of goons paid by the government to snatch men from the streets and force them into the army or navy. (The Continental Navy during the American Revolutionary War was almost entirely rounded up by press gangs.)[161] Compulsory military service could consume a substantial portion of a man's life—as much as twenty-five years for a serf in 19th-century Russia.

Military conscription represents the application of force squared: people are coerced into servitude, and the servitude exposes them to high odds of being maimed or killed. Other than at times of existential threat, the extent of conscription is a barometer of a country's willingness to sanction the use of force. In the decades after World War II, the world saw a steady reduction in the length of compulsory military service. The United States, Canada, and most European countries have eliminated conscription outright, and in the others it functions more as a citizenship-building exercise than as a training ground for warriors.[162] Payne has compiled statistics on the length of military conscription between 1970 and 2000 in forty-eight long-established nations, which I have updated for 2010 in figure 5–19. They show that conscription was in decline even before the end of the Cold War in the late 1980s. Only 19 percent of these countries did without conscription in 1970. The proportion rose to 35 percent in 2000 and to 50 percent in 2010, and it will soon exceed 50 percent because at least two other countries (Poland and Serbia) plan to abolish the draft in the early 2010s.[163]

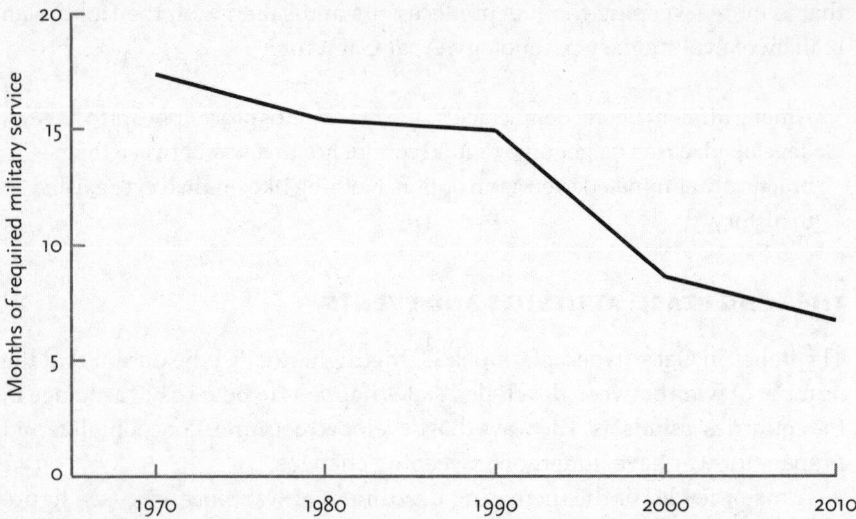

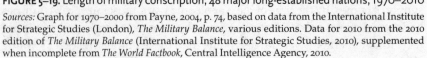

FIGURE 5–19. Length of military conscription, 48 major long-established nations, 1970–2010
Sources: Graph for 1970–2000 from Payne, 2004, p. 74, based on data from the International Institute for Strategic Studies (London), *The Military Balance*, various editions. Data for 2010 from the 2010 edition of *The Military Balance* (International Institute for Strategic Studies, 2010), supplemented when incomplete from *The World Factbook*, Central Intelligence Agency, 2010.

Another indicator of war-friendliness is the size of a nation's military forces as a proportion of its population, whether enlisted by conscription or by television ads promising volunteers that they can be all that they can be. Payne has shown that the proportion of the population that a nation puts in uniform is the best indicator of its ideological embrace of militarism.[164] When the United States demobilized after World War II, it took on a new enemy in the Cold War and never shrank its military back to prewar levels. But figure 5–20 shows that the trend since the mid-1950s has been sharply downward. Europe's disinvestment of human capital in the military sector began even earlier.

Other large countries, including Australia, Brazil, Canada, and China, also shrank their armed forces during this half-century. After the Cold War ended, the trend went global: from a peak of more than 9 military personnel per 100,000 people in 1988, the average across long-established countries plunged to less than 5.5 in 2001.[165] Some of these savings have come from outsourcing noncombat functions like laundry and food services to private contractors, and in the wealthiest countries, from replacing front-line military personnel with robots and drones. But the age of robotic warfare is far in the future, and recent events have shown that the number of available boots on the ground is still a major constraint on the projection of military force. For that matter, the roboticizing of the military is itself a manifestation of the trend we are exploring. Countries

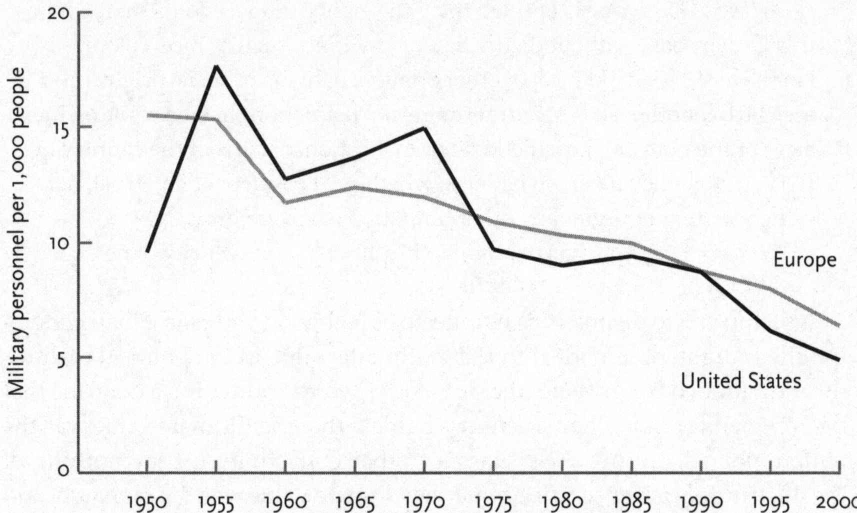

FIGURE 5–20. Military personnel, United States and Europe, 1950–2000

Sources: Correlates of War National Material Capabilities Dataset (1816–2001); http://www.correlatesofwar.org, Sarkees, 2000. Unweighted averages, every five years. "Europe" includes Belgium, Denmark, Finland, France, Greece, Hungary, Ireland, Italy, Luxembourg, the Netherlands, Norway, Poland, Romania, Russia/USSR, Spain, Sweden, Switzerland, Turkey, U.K., Yugoslavia.

have developed these technologies at fantastic expense because the lives of their citizens (and, as we shall see, of foreign citizens) have become dearer.

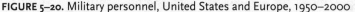

Since wars begin in the minds of men, it is in the minds of men that the defenses of peace must be constructed.

—UNESCO motto

Another indication that the Long Peace is no accident is a set of sanity checks which confirm that the mentality of leaders and populaces has changed. Each component of the war-friendly mindset—nationalism, territorial ambition, an international culture of honor, popular acceptance of war, and indifference to its human costs—went out of fashion in developed countries in the second half of the 20th century.

The first signal event was the 1948 endorsement of the Universal Declaration of Human Rights by forty-eight countries. The declaration begins with these articles:

> *Article 1.* All human beings are born free and equal in dignity and rights. They are endowed with reason and conscience and should act towards one another in a spirit of brotherhood.

Article 2. Everyone is entitled to all the rights and freedoms set forth in this Declaration, without distinction of any kind, such as race, colour, sex, language, religion, political or other opinion, national or social origin, property, birth or other status. Furthermore, no distinction shall be made on the basis of the political, jurisdictional or institutional status of the country or territory to which a person belongs, whether it be independent, trust, non-self-governing or under any other limitation of sovereignty.

Article 3. Everyone has the right to life, liberty, and security of person.

It's tempting to dismiss this manifesto as feel-good verbiage. But in endorsing the Enlightenment ideal that the ultimate value in the political realm is the individual human being, the signatories were repudiating a doctrine that had reigned for more than a century, namely that the ultimate value was the nation, people, culture, *Volk*, class, or other collectivity (to say nothing of the doctrine of earlier centuries that the ultimate value was the monarch, and the people were his or her chattel). The need for an assertion of universal human rights had become evident during the Nuremberg Trials of 1945–46, when some lawyers had argued that Nazis could be prosecuted only for the portion of the genocides they committed in occupied countries like Poland. What they did on their own territory, according to the earlier way of thinking, was none of anyone else's business.

Another sign that the declaration was more than hot air was that the great powers were nervous about signing it. Britain was worried about its colonies, the United States about its Negroes, and the Soviet Union about its puppet states.[166] But after Eleanor Roosevelt shepherded the declaration through eighty-three meetings, it passed without opposition (though pointedly, with eight abstentions from the Soviet bloc).

The era's repudiation of counter-Enlightenment ideology was made explicit forty-five years later by Václav Havel, the playwright who became president of Czechoslovakia after the nonviolent Velvet Revolution had overthrown the communist government. Havel wrote, "The greatness of the idea of European integration on democratic foundations is its capacity to overcome the old Herderian idea of the nation state as the highest expression of national life."[167]

One paradoxical contributor to the Long Peace was the freezing of national borders. The United Nations initiated a norm that existing states and their borders were sacrosanct. By demonizing any attempt to change them by force as "aggression," the new understanding took territorial expansion off the table as a legitimate move in the game of international relations. The borders may have made little sense, the governments within them may not have deserved to govern, but rationalizing the borders by violence was no longer a live option in the minds of statesmen. The grandfathering of boundaries has been, on average, a pacifying development because, as the political scientist John

Vasquez has noted, "of all the issues over which wars could logically be fought, territorial issues seem to be the one most often associated with wars. Few interstate wars are fought without any territorial issue being involved in one way or another."[168]

The political scientist Mark Zacher has quantified the change.[169] Since 1951 there have been only ten invasions that resulted in a major change in national boundaries, all before 1975. Many of them planted flags in sparsely populated hinterlands and islands, and some carved out new political entities (such as Bangladesh) rather than expanding the territory of the conqueror. Ten may sound like a lot, but as figure 5–21 shows, it represents a precipitous drop from the preceding three centuries.

Israel is an exception that proves the rule. The serpentine "green line" where the Israeli and Arab armies stopped in 1949 was not particularly acceptable to anyone at the time, especially the Arab states. But in the ensuing decades it took on an almost mystical status in the international community as Israel's one true correct border. The country has acceded to international pressure to relinquish most of the territory it has occupied in the various wars since then, and within our lifetimes it will probably withdraw from the rest, with some minor swaps of land and perhaps a complicated arrangement regarding Jerusalem, where the norm of immovable borders will clash with the norm of undivided cities. Most other conquests, such as the Indonesian takeover of East Timor, have been reversed as well. The most dramatic recent example

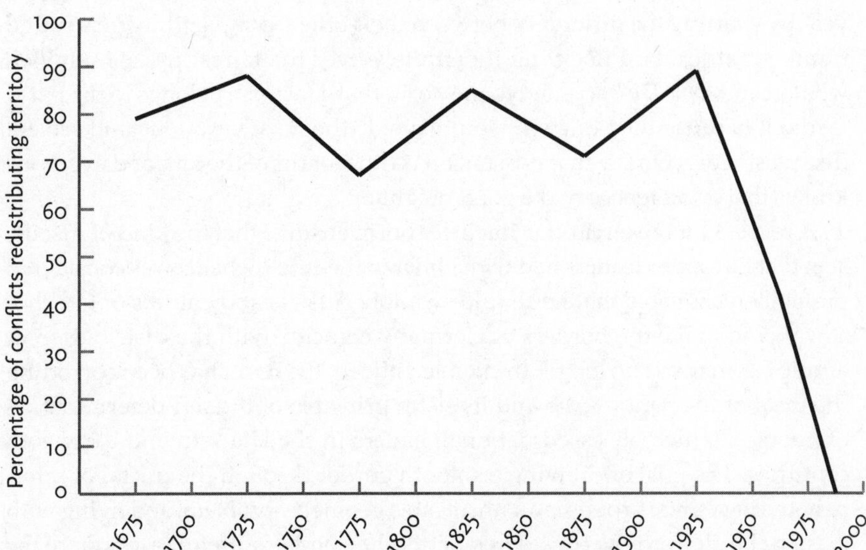

FIGURE 5–21. Percentage of territorial wars resulting in redistribution of territory, 1651–2000

Source: Data from Zacher, 2001, tables 1 and 2; the data point for each half-century is plotted at its midpoint, except for the last half of the 20th century, in which each point represents a quarter-century.

was in 1990, when Saddam Hussein invaded Kuwait (the only time since 1945 that one member of the UN has swallowed another one whole), and an aghast multinational coalition made short work of pushing him out.

The psychology behind the sanctity of national boundaries is not so much empathy or moral reasoning as norms and taboos (a topic that will be explored in chapter 9). Among respectable countries, conquest is no longer a thinkable option. A politician in a democracy today who suggested conquering another country would be met not with counterarguments but with puzzlement, embarrassment, or laughter.

The territorial-integrity norm, Zacher points out, has ruled out not just conquest but other kinds of boundary-tinkering. During decolonization, the borders of newly independent states were the lines that some imperial administrator had drawn on a map decades before, often bisecting ethnic homelands or throwing together enemy tribes. Nonetheless there was no movement to get all the new leaders to sit around a table with a blank map and a pencil and redraw the borders from scratch. The breakup of the Soviet Union and Yugoslavia also resulted in the dashed lines between internal republics and provinces turning into solid lines between sovereign states, without any redrafting.

The sacralization of arbitrary lines on a map may seem illogical, but there is a rationale to the respecting of norms, even arbitrary and unjustifiable ones. The game theorist Thomas Schelling has noted that when a range of compromises would leave two negotiators better off than they would be if they walked away, any salient cognitive landmark can lure them into an agreement that benefits them both.[170] People bargaining over a price, for example, can "get to yes" by splitting the difference between their offers, or by settling on a round number, rather than haggling indefinitely over the fairest price. Melville's whalers in *Moby-Dick* acceded to the norm that a fast-fish belongs to the party fast to it because they knew it would avoid "the most vexatious and violent disputes." Lawyers say that possession is nine tenths of the law, and everyone knows that good fences make good neighbors.

A respect for the territorial-integrity norm ensures that the kind of discussion that European leaders had with Hitler in the 1930s, when it was considered perfectly reasonable that he should swallow Austria and chunks of Czechoslovakia to make the borders of Germany coincide with the distribution of ethnic Germans, is no longer thinkable. Indeed, the norm has been corroding the ideal of the nation-state and its sister principle of the self-determination of peoples, which obsessed national leaders in the late 19th and early 20th centuries. The goal of drawing a smooth border through the fractal of interpenetrating ethnic groups is an unsolvable geometry problem, and living with existing borders is now considered better than endless attempts to square the circle, with its invitations to ethnic cleansing and irredentist conquest.

The territorial-integrity norm brings with it numerous injustices, as ethnic groups may find themselves submerged in political entities that have no benevolent interest in their welfare. The point was not lost on Ishmael, who mused, "What to that redoubted harpooneer, John Bull, is poor Ireland, but a Fast-Fish?" Some of Europe's peaceful borders demarcate countries that were conveniently homogenized by the massive ethnic cleansing of World War II and its aftermath, when millions of ethnic Germans and Slavs were forcibly uprooted from their homes. The developing world is now being held to higher standards, and it is likely, as the sociologist Ann Hironaka has argued, that its civil wars have been prolonged by the insistence that states always be preserved and borders never altered. But on balance, the sacred-border norm appears to have been a good bargain for the world. As we shall see in the next chapter, the death toll from a large number of small civil wars is lower than that from a few big interstate wars, to say nothing of world wars, consistent with the power-law distribution of deadly quarrels. And even civil wars have become fewer in number and less damaging as the modern state evolves from a repository for the national soul to a multiethnic social contract conforming to the principle of human rights.

Together with nationalism and conquest, another ideal has faded in the postwar decades: honor. As Luard understates it, "In general, the value placed on human life today is probably higher, and that placed on national prestige (or 'honor') probably lower, than in earlier times."[171] Nikita Khrushchev, the leader of the Soviet Union during the worst years of the Cold War, captured the new sensibility when he said, "I'm not some czarist officer who has to kill himself if I fart at a masked ball. It's better to back down than to go to war."[172] Many national leaders agree, and have backed down or held their fire in response to provocations that in previous eras would have incited them to war.

In 1979 the United States responded to two affronts in quick succession—the Russian invasion of Afghanistan and the government-indulged takeover of the American embassy in Iran—with little more than an Olympic boycott and a nightly televised vigil. As Jimmy Carter said later, "I could have destroyed Iran with my weaponry, but I felt in the process it was likely that the hostages' lives would be lost, and I didn't want to kill 20,000 Iranians. So I didn't attack."[173] Though American hawks were furious at Carter's wimpiness, their own hero, Ronald Reagan, responded to a 1983 bombing that killed 241 American servicemen in Beirut by withdrawing all American forces from the country, and he sat tight in 1987 when Iraqi jet fighters killed thirty-seven sailors on the USS *Stark*. The 2004 train bombing in Madrid by an Islamist terrorist group, far from whipping the Spanish into an anti-Islamic lather, prompted them to vote out the government that had involved them in the Iraq War, an involvement many felt had brought the attack upon them.

The most consequential discounting of honor in the history of the world was the resolution of the 1962 Cuban Missile Crisis. Though the pursuit of national prestige may have precipitated the crisis, once Khrushchev and Kennedy were in it, they reflected on their mutual need to save face and set that up as a problem for the two of them to solve.[174] Kennedy had read Tuchman's *The Guns of August*, a history of World War I, and knew that an international game of chicken driven by "personal complexes of inferiority and grandeur" could lead to a cataclysm. Robert Kennedy, in a memoir on the crisis, recalled:

> Neither side wanted war over Cuba, we agreed, but it was possible that either side could take a step that—for reasons of "security" or "pride" or "face"—would require a response by the other side, which, in turn, for the same reasons of security, pride, or face, would bring about a counterresponse and eventually an escalation into armed conflict. That was what he wanted to avoid.[175]

Khrushchev's wisecrack about the czarist officer shows that he too was cognizant of the psychology of honor, and he had a similar intuitive sense of game theory. During a tense moment in the crisis, he offered Kennedy this analysis:

> You and I should not now pull on the ends of the rope in which you have tied a knot of war, because the harder you and I pull, the tighter this knot will become. And a time may come when this knot is tied so tight that the person who tied it is no longer capable of untying it, and then the knot will have to be cut.[176]

They untied the knot by making mutual concessions—Khrushchev removed his missiles from Cuba, Kennedy removed his from Turkey, and Kennedy promised not to invade Cuba. Nor was the de-escalation purely a stroke of uncanny good luck. Mueller reviewed the history of superpower confrontations during the Cold War and concluded that the sequence was more like climbing a ladder than stepping onto an escalator. Though several times the leaders began a perilous ascent, with each rung they climbed they became increasingly acrophobic, and always sought a way to gingerly step back down.[177]

And for all the shoe-pounding bluster of the Soviet Union during the Cold War, its leadership spared the world another cataclysm when Mikhail Gorbachev allowed the Soviet bloc, and then the Soviet Union itself, to go out of existence—what the historian Timothy Garton Ash has called a "breathtaking renunciation of the use of force" and a "luminous example of the importance of the individual in history."

This last remark reminds us that historical contingency works both ways. There are parallel universes in which the archduke's driver didn't make a wrong turn in Sarajevo, or in which a policeman aimed differently during the Beer Hall Putsch, and history unfolded with one or two fewer world wars. There are other parallel universes in which an American president listened to his Joint Chiefs of Staff and invaded Cuba, or in which a Soviet leader responded to the breach of the Berlin Wall by calling out the tanks, and history unfolded with one or two more. But given the changing odds set by the prevailing ideas and norms, it is not surprising that in our universe it was the first half of the 20th century that was shaped by a Princip and a Hitler, and the second half by a Kennedy, a Khrushchev, and a Gorbachev.

Yet another historic upheaval in the landscape of 20th-century values was a resistance by the populations of democratic nations to their leaders' plans for war. The late 1950s and early 1960s saw mass demonstrations to Ban the Bomb, whose legacy includes the trident-in-circle peace symbol co-opted by other antiwar movements. By the late 1960s the United States was torn apart by protests against the Vietnam War. Antiwar convictions were no longer confined to sentimental aunts of both sexes, and the idealists who went about in sandals and beards were no longer cranks but a significant proportion of the generation that reached adulthood in the 1960s. Unlike the major artworks deploring World War I, which appeared more than a decade after it was over, popular art in the 1960s condemned the nuclear arms race and the Vietnam War in real time. Antiwar advocacy was woven into prime-time television programs (such as *The Smothers Brothers Comedy Hour* and *M*A*S*H*) and many popular films and songs:

> *Catch-22* • *Fail-Safe* • *Dr. Strangelove* • *Hearts and Minds* • *FTA* • *How I Won the War* • *Johnny Got His Gun* • *King of Hearts* • *M*A*S*H* • *Oh! What a Lovely War* • *Slaughterhouse-Five*

> "Alice's Restaurant" • "Blowin' in the Wind" • "Cruel War" • "Eve of Destruction" • "Feel Like I'm Fixin' to Die Rag" • "Give Peace a Chance" • "Happy Xmas (War Is Over)" • "I Ain't Marchin' Anymore" • "If I Had a Hammer" • "Imagine" • "It's a Hard Rain's a Gonna Fall" • "Last Night I Had the Strangest Dream" • "Machine Gun" • "Masters of War" • "Sky Pilot" • "Three-Five-Zero-Zero" • "Turn! Turn! Turn!" • "Universal Soldier" • "What's Goin' On?" • "With God on Our Side" • "War (What Is It Good For?)" • "Waist-Deep in the Big Muddy" • "Where Have All the Flowers Gone?"

As in the 1700s and the 1930s, artists did not just preach about war to make it seem immoral but satirized it to make it seem ridiculous. During the 1969 Woodstock concert, Country Joe and the Fish sang the jaunty "Feel Like I'm Fixin' to Die Rag," whose chorus was:

> And it's One, Two, Three, what are we fighting for?
> Don't ask me, I don't give a damn; next stop is Vietnam!

And it's Five, Six, Seven, open up the Pearly Gates.
There ain't no time to wonder why; Whoopee! We're all going to die.

In his 1967 monologue "Alice's Restaurant," Arlo Guthrie told of being drafted and sent to an army psychiatrist at the induction center in New York:

> And I went up there, I said, "Shrink, I want to kill. I mean, I wanna, I wanna kill. Kill. I wanna, I wanna see, I wanna see blood and gore and guts and veins in my teeth. Eat dead burnt bodies. I mean kill, Kill, KILL, KILL." And I started jumpin' up and down yelling, "KILL, KILL," and he started jumpin' up and down with me and we was both jumpin' up and down yelling, "KILL, KILL." And the sergeant came over, pinned a medal on me, sent me down the hall, said, "You're our boy."

It's easy to dismiss this cultural moment as baby-boomer nostalgia. As Tom Lehrer satirized it, they won all the battles, but we had the good songs. But in a sense we did win the battles. In the wake of nationwide protests, Lyndon Johnson shocked the country by not seeking his party's nomination in the 1968 presidential election. Though a reaction against the increasingly unruly protests helped elect Richard Nixon in 1968, Nixon shifted the country's war plans from a military victory to a face-saving withdrawal (though not before another twenty thousand Americans and a million Vietnamese had died in the fighting). After a 1973 cease-fire, American troops were withdrawn, and Congress effectively ended the war by prohibiting additional intervention and cutting off funding for the South Vietnamese government.

The United States was then said to have fallen into a "Vietnam Syndrome" in which it shied away from military engagement. By the 1980s it had recovered well enough to fight several small wars and to support anticommunist forces in several proxy wars, but clearly its military policy would never be the same. The phenomenon called "casualty dread," "war aversion," and "the Dover Doctrine" (the imperative to minimize flag-draped coffins returning to Dover Air Force Base) reminded even the more hawkish presidents that the country would not tolerate casualty-intensive military adventures. By the 1990s the only politically acceptable American wars were surgical routs achieved with remote-control technology. They could no longer be wars of attrition that ground up soldiers by the tens of thousands, nor aerial holocausts visited on foreign civilians as in Dresden, Hiroshima, and North Vietnam.

The change is palpable within the American military itself. Military leaders at all levels have become aware that gratuitous killing is a public-relations disaster at home and counterproductive abroad, alienating allies and emboldening enemies.[178] The Marine Corps has instituted a martial-arts program in which leathernecks are indoctrinated in a new code of honor, the Ethical Marine Warrior.[179] The catechism is "The Ethical Warrior is a protector of life.

Whose life? Self and others. Which others? All others." The code is instilled with empathy-expanding allegories such as "The Hunting Story," recounted by Robert Humphrey, a retired officer whose martial bona fides were impeccable, having commanded a rifle platoon on Iwo Jima in World War II.[180] In this story, an American military unit is serving in a poor Asian country, and one day members of the unit go boar hunting as a diversion:

> They took a truck from the motor pool and headed out to the boondocks, stopping at a village to hire some local men to beat the brush and act as guides.
>
> This village was very poor. The huts were made of mud and there was no electricity or running water. The streets were unpaved dirt and the whole village smelled. Flies abounded. The men looked surly and wore dirty clothes. The women covered their faces, and the children had runny noses and were dressed in rags.
>
> It wasn't long before one American in the truck said, "This place stinks." Another said, "These people live just like animals." Finally, a young air force man said, "Yeah, they got nothin' to live for; they may as well be dead."
>
> What could you say? It seemed true enough.
>
> But just then, an old sergeant in the truck spoke up. He was the quiet type who never said much. In fact, except for his uniform, he kind of reminded you of one of the tough men in the village. He looked at the young airman and said, "You think they got nothin' to live for, do you? Well, if you are so sure, why don't you just take my knife, jump down off the back of this truck, and go try to kill one of them?"
>
> There was dead silence in the truck. . . .
>
> The sergeant went on to say, "I don't know either why they value their lives so much. Maybe it's those snotty nosed kids, or the women in the pantaloons. But whatever it is, they care about their lives and the lives of their loved ones, same as we Americans do. And if we don't stop talking bad about them, they will kick us out of this country!"
>
> [A soldier] asked him what we Americans, with all our wealth, could do to prove our respect for the peasants' human equality despite their destitution. The sergeant answered easily, "You got to be brave enough to jump off the back of this truck, knee deep in the mud and sheep dung. You got to have the courage to walk through this village with a smile on your face. And when you see the smelliest, scariest looking peasant, you got to be able to look him in the face and let him know, just with your eyes, that you know he is a man who hurts like you do, and hopes like you do, and wants for his kids just like we all do. It is that way or we lose."

The code of the Ethical Warrior, even as an aspiration, shows that the American armed forces have come a long way from a time when its soldiers referred to Vietnamese peasants as *gooks, slopes,* and *slants* and when the military was slow to investigate atrocities against civilians such as the massacre at My Lai. As former Marine captain Jack Hoban, who helped to implement the Ethical Warrior program, wrote to me, "When I first joined the Marines in the 1970s it was 'Kill, kill, kill.' The probability that there would have been an honor code that trained marines to be 'protectors of all others—including the enemy, if possible' would have been 0 percent."

To be sure, the American-led wars in Afghanistan and Iraq in the first decade of the 21st century show that the country is far from reluctant to go to war. But even they are nothing like the wars of the past. In both conflicts the interstate war phase was quick and (by historical standards) low in battle deaths.[181] Most of the deaths in Iraq were caused by intercommunal violence in the anarchy that followed, and by 2008 the toll of 4,000 American deaths (compare Vietnam's 58,000) helped elect a president who within two years brought the country's combat mission to an end. In Afghanistan, the U.S. Air Force followed a set of humanitarian protocols during the height of the anti-Taliban bombing campaign in 2008 that Human Rights Watch praised for its "very good record of minimizing harm to civilians."[182] The political scientist Joshua Goldstein, in a discussion of how policies of smart targeting had massively reduced civilian deaths in Kosovo and in both Iraq wars, comments on the use of armed drones against Taliban and Al Qaeda targets in Afghanistan and Pakistan in 2009:

> Where an army previously would have blasted its way in to the militants' hideouts, killing and displacing civilians by the tens of thousands as it went, and then ultimately reducing whole towns and villages to rubble with inaccurate artillery and aerial bombing in order to get at a few enemy fighters, now a drone flies in and lets fly a single missile against a single house where militants are gathered. Yes, sometimes such attacks hit the wrong house, but by any historical comparison the rate of civilian deaths has fallen dramatically.
>
> So far has this trend come, and so much do we take it for granted, that a single errant missile that killed ten civilians in Afghanistan was front-page news in February 2010. This event, a terrible tragedy in itself, nonetheless was an exception to a low overall rate of harm to civilians in the middle of a major military offensive, one of the largest in eight years of war. Yet, these ten deaths brought the U.S. military commander in Afghanistan to offer a profuse apology to the president of Afghanistan, and the world news media to play up the event as a major development in the offensive. The point is not that killing ten civilians is OK, but rather that in any previous war, even a few years ago, this kind of civilian death would barely have caused a ripple of attention. Civilian deaths, in sizable numbers, used to be universally considered a necessary and inevitable, if perhaps unfortunate, by-product of war. That we are entering an era when these assumptions no longer apply is good news indeed.[183]

Goldstein's assessment was confirmed in 2011 when *Science* magazine reported data from WikiLeaks documents and from a previously classified civilian casualty database of the American-led military coalition. The documents revealed that around 5,300 civilians had been killed in Afghanistan from 2004 through 2010, the majority (around 80 percent) by Taliban insurgents rather than coalition forces. Even if the estimate is doubled, it would represent an extraordinarily low number of civilian deaths for a major military operation—in the Vietnam War, by comparison, at least 800,000 civilians died in battle.[184]

As big as the change in American attitudes toward war has been, the change in Europe is beyond recognition. As the foreign policy analyst Robert Kagan puts it, "Americans are from Mars, Europeans are from Venus."[185] In February 2003 mass demonstrations in European cities protested the impending American-led invasion of Iraq, drawing a million people each in London, Barcelona, and Rome, and more than half a million in Madrid and Berlin.[186] In London the signs read "No Blood for Oil"; "Stop Mad Cowboy Disease"; "America, the Real Rogue State"; "Make Tea, Not War"; "Down with This Sort of Thing"; and simply "No." Germany and France conspicuously refused to join the United States and Britain, and Spain pulled out soon afterward. Even the war in Afghanistan, which aroused less opposition in Europe, is being fought mainly by American soldiers. Not only do they make up more than half of the forty-four-nation NATO military operation, but the continental forces have acquired a certain reputation when it comes to martial virtues. A Canadian armed forces captain wrote to me from Kabul in 2003:

> During this morning's Kalashnikov concerto, I was waiting for the tower guards in our camp to open fire. I think they were asleep. That's par for the course. Our towers are manned by the Bundeswehr, and they haven't been doing a good job . . . when they're actually there. I qualified that last comment because the Germans have already abandoned the towers several times. The first time was when we got hit by rockets. The remaining instances had something to do with it being cold in the towers. A German Lieutenant with whom I spoke about this lack of honour and basic soldier etiquette replied that it was Canada's responsibility to provide heaters for the towers. I snapped back by mentioning that it was Germany's responsibility to provide warm clothing to its soldiers. I was tempted to mention something about Kabul not being Stalingrad, but I held my tongue.
>
> The German army of today is not what it once was. Or, as I've heard mentioned here several times: "This ain't the Wehrmacht." Given the history of our people, I can make the argument that that's a very good thing indeed. However, since my safety now rests upon the vigilance of the Herrenvolk's progeny, I'm slightly concerned to say the least.[187]

In a book titled *Where Have All the Soldiers Gone? The Transformation of*

Modern Europe (and in Britain, *The Monopoly on Violence: Why Europeans Hate Going to War*), the historian James Sheehan argues that Europeans have changed their very conception of the state. It is no longer the proprietor of a military force that enhances the grandeur and security of the nation, but a provisioner of social security and material well-being. Nonetheless, for all the differences between the American "mad cowboys" and the European "surrender monkeys," the parallel movement of their political culture away from war over the past six decades is more historically significant than their remaining differences.

IS THE LONG PEACE A NUCLEAR PEACE?

What went right? How is it that, in defiance of experts, doomsday clocks, and centuries of European history, World War III never happened? What allowed distinguished military historians to use giddy phrases like "a change of spectacular proportions," "the most striking discontinuity in the history of warfare," and "nothing like this in history"?

To many people, the answer is obvious: the bomb. War had become too dangerous to contemplate, and leaders were scared straight. The balance of nuclear terror deterred them from starting a war that would escalate to a holocaust and put an end to civilization, if not human life itself.[188] As Winston Churchill said in his last major speech to Parliament, "It may well be that we shall by a process of sublime irony have reached a stage in this story where safety will be the sturdy child of terror, and survival the twin brother of annihilation."[189] In the same vein, the foreign policy analyst Kenneth Waltz has suggested that we "thank our nuclear blessings," and Elspeth Rostow proposed that the nuclear bomb be awarded the Nobel Peace Prize.[190]

Let's hope not. If the Long Peace were a nuclear peace, it would be a fool's paradise, because an accident, a miscommunication, or an air force general obsessed with precious bodily fluids could set off an apocalypse. Thankfully, a closer look suggests that the threat of nuclear annihilation deserves little credit for the Long Peace.[191]

For one thing, weapons of mass destruction had never braked the march to war before. The benefactor of the Nobel Peace Prize wrote in the 1860s that his invention of dynamite would "sooner lead to peace than a thousand world conventions, [since] as soon as men will find that in one instant whole armies can be utterly destroyed, they will surely abide in golden peace."[192] Similar predictions have been made about submarines, artillery, smokeless powder, and the machine gun.[193] The 1930s saw a widespread fear that poison gas dropped from airplanes could bring an end to civilization and human life, yet that dread did not come close to ending war either.[194] As Luard puts it, "There is little evidence in history that the existence of supremely destructive weapons alone is capable of deterring war. If the development of bacteriological weapons, poison gas, nerve gases, and other chemical armaments did not deter war in 1939, it is not easy to see why nuclear weapons should do so now."[195]

Also, the theory of the nuclear peace cannot explain why countries without nuclear weapons also forbore war—why, for example, the 1995 squabble over fishing rights between Canada and Spain, or the 1997 dispute between Hungary and Slovakia over damming the Danube, never escalated into war, as crises involving European countries had so often done in the past. During the Long Peace leaders of developed countries never had to calculate which of their counterparts they could get away with attacking (yes for Germany and Italy, no for Britain and France), because they never contemplated a military attack in the first place. Nor were they deterred by nuclear godparents—it wasn't as if the United States had to threaten Canada and Spain with a nuclear spanking if they got too obstreperous in their dispute over flatfish.

As for the superpowers themselves, Mueller points to a simpler explanation for why they avoided fighting each other: they were deterred plenty by the prospect of a conventional war. World War II showed that assembly lines could mass-produce tanks, artillery, and bombers that were capable of killing tens of millions of people and reducing cities to rubble. This was especially obvious in the Soviet Union, which had suffered the greatest losses in the war. It's unlikely that the marginal difference between the unthinkable damage that would be caused by a nuclear war and the thinkable but still staggering damage that would be caused by a conventional war was the main thing that kept the great powers from fighting.

Finally, the nuclear peace theory cannot explain why the wars that did take place often had a nonnuclear force provoking (or failing to surrender to) a nuclear one—exactly the matchup that the nuclear threat ought to have deterred.[196] North Korea, North Vietnam, Iran, Iraq, Panama, and Yugoslavia defied the United States; Afghan and Chechen insurgents defied the Soviet Union; Egypt defied Britain and France; Egypt and Syria defied Israel; Vietnam defied China; and Argentina defied the United Kingdom. For that matter, the Soviet Union established its stranglehold on Eastern Europe during just those years (1945–49) when the United States had nuclear weapons and it did not. The countries that goaded their nuclear superiors were not suicidal. They correctly anticipated that for anything but an existential danger, the implicit threat of a nuclear response was a bluff. The Argentinian junta ordered the invasion of the Falkland Islands in full confidence that Britain would not retaliate by reducing Buenos Aires to a radioactive crater. Nor could Israel have credibly threatened the amassed Egyptian armies in 1967 or 1973, to say nothing of Cairo.

Schelling, and the political scientist Nina Tannenwald, have each written of "a nuclear taboo"—a shared perception that nuclear weapons fall into a uniquely dreadful category.[197] The use of a single tactical nuclear weapon, even one comparable in damage to conventional weaponry, would be seen as a breach in history, a passage into a new world with unimaginable consequences. The obloquy has attached itself to every form of nuclear detonation. The neutron bomb, a weapon that would cause minimal blast damage but would kill soldiers with a transient burst of radiation, fell deadborn from the military lab because of universal loathing, even though, as the political scientist Stanley Hoffman pointed out, it satisfied the moral philosophers' requirements for waging a just war.[198] The half-crazed "Atoms for Peace" schemes of the 1950s and 1960s, in which nuclear explosions would be harnessed to dig canals, excavate harbors, or propel rockets into space, are now the stuff of incredulous reminiscences of a benighted age.

To be sure, the nonuse of nuclear weapons since Nagasaki falls short of an out-and-out taboo.[199] Nuclear bombs don't build themselves, and nations have devoted enormous thought to the design, construction, delivery, and terms of use of these weapons. But this activity has been compartmentalized into a sphere of hypotheticals that barely intersects with the planning of actual wars. And there are telltale signs that the psychology of taboo—a mutual understanding that certain thoughts are evil to think—has been engaged, starting with the word that is most commonly applied to the prospect of nuclear war: unthinkable. In 1964, after Barry Goldwater had mused about how tactical nuclear weapons might be used in Vietnam, Lyndon Johnson's electoral campaign aired the famous "Daisy" television ad, in which footage of a girl counting the petals of a daisy segues into a countdown to a nuclear explosion. The ad has been given some of the credit for Johnson's landslide election victory that year.[200] Religious allusions have surrounded nuclear weapons ever since Robert Oppenheimer quoted the Bhagavad-Gita when he viewed the first atomic test in 1945: "Now I am become Death, the destroyer of worlds." More commonly the language has been biblical: Apocalypse, Armageddon, the End of Days, Judgment Day. Dean Rusk, secretary of state in the Kennedy and Johnson administrations, wrote that if the country had used a nuclear weapon, "we would have worn the mark of Cain for generations to come."[201] The physicist Alvin Weinberg, whose research helped make the bomb possible, asked in 1985:

> Are we witnessing a gradual sanctification of Hiroshima—that is, the elevation of Hiroshima to the status of a profoundly mystical event, an event ultimately of the same religious force as biblical events? I cannot prove it, but I am convinced that the 40th Anniversary of Hiroshima, with its vast outpouring of concern, bears resemblance to the observance of major religious holidays. . . . This sanctification of Hiroshima is one of the most hopeful developments of the nuclear era.[202]

The nuclear taboo emerged only gradually. As we saw in chapter 1, for at least a decade after Hiroshima many Americans thought the A-bomb was adorable. By 1953 John Foster Dulles, secretary of state in the Eisenhower administration, was deploring what he called the "false distinction" and "taboo" surrounding nuclear weapons.[203] During a 1955 crisis involving Taiwan and the People's Republic of China, Eisenhower said, "In any combat where these things can be used on strictly military targets and for strictly military purposes, I see no reason why they shouldn't be used just exactly as you would use a bullet or anything else."[204]

But in the following decade nuclear weapons acquired a stigma that would put such statements beyond the pale. It began to sink in that the weapons' destructive capacity was of a different order from anything in history, that they violated any conception of proportionality in the waging of war, and that plans for civil defense (like backyard fallout shelters and duck-and-cover drills) were a travesty. People became aware that lingering radiation from nuclear fallout could cause chromosome damage and cancer for decades after the actual explosions. The fallout from atmospheric tests had already contaminated rainfall all over the world with strontium 90, a radioactive isotope resembling calcium that is taken up in the bones and teeth of children (inspiring Malvina Reynolds's protest song "What Have They Done to the Rain?").

Though the United States and the USSR continued to develop nuclear technology at a breakneck pace, they began, however hypocritically, to pay homage to nuclear disarmament in conferences and statements. At the same time a grassroots movement began to stigmatize the weapons. Demonstrations and petitions attracted millions of citizens, together with public figures such as Linus Pauling, Bertrand Russell, and Albert Schweitzer. The mounting pressure helped nudge the superpowers to a moratorium and then a ban on atmospheric nuclear testing, and then to a string of arms-control agreements. The Cuban Missile Crisis in 1962 was a tipping point. Lyndon Johnson capitalized on the change to demonize Goldwater in the Daisy ad and called attention to the categorical boundary in a 1964 public statement: "Make no mistake. There is no such thing as a conventional nuclear weapon. For nineteen peril-filled years no nation has loosed the atom against another. To do so now is a political decision of the highest order."[205]

As the world's luck held out, and the two nuclear-free decades grew to three and four and five and six, the taboo fed on itself in the runaway process by which norms become common knowledge. The use of nuclear weapons was unthinkable because everyone knew it was unthinkable, and everyone knew that everyone knew it. The fact that wars both large (Vietnam) and small (Falklands) were not deterred by the increasingly ineffectual nuclear threat was a small price to pay for the indefinite postponement of Armageddon.

A norm that rests only on mutual recognition of that norm is, of course, vulnerable to a sudden unraveling. One might worry—one should worry—that nuclear nations outside the club of great powers, such as India, Pakistan, North Korea, and perhaps soon Iran, may not be party to the common understanding that the use of nuclear weapons is unthinkable. Worse, a terrorist organization that pilfered a stray nuclear weapon could make a point of defying the taboo, since the whole point of international terrorism is to shock the world with the most horrific spectacle imaginable. Once the precedent of a single nuclear explosion was set, one might worry, all restraints would be put aside. A pessimist might argue that even if the Long Peace has not, thus far, depended on nuclear deterrence, it is an ephemeral hiatus. It will surely end as nuclear weapons proliferate, a maniac from the developing world brings the lucky streak to an end, and the taboo comes undone among small and great powers alike.

No judicious person can feel calm about the parlous state of nuclear safety in today's world. But even here, things are not as bad as many people think. In the next chapter, I'll examine the prospect of nuclear terrorism. For now, let's look at nuclear states.

One hopeful sign is that nuclear proliferation has not proceeded at the furious rate that everyone expected. In the 1960 presidential election debates, John F. Kennedy predicted that by 1964 there might be "ten, fifteen, twenty" countries with nuclear weapons.[206] The concern accelerated when China conducted its first nuclear test in 1964, bringing the number of nations in the nuclear club to five in less than twenty years. Tom Lehrer captured popular fears of runaway nuclear proliferation in his song "Who's Next?" which ran through a list of countries that he expected would soon become nuclear powers ("Luxemburg is next to go / And who knows? Maybe Monaco").

But the only country that fulfilled his prophecy is Israel ("'The Lord's my shepherd,' says the Psalm / But just in case—we better get a bomb!"). Contrary to expert predictions that Japan would "unequivocally start on the process of acquiring nuclear weapons" by 1980 and that a reunified Germany "will feel insecure without nuclear weapons," neither country seems interested in developing them.[207] And believe it or not, since 1964 as many countries have *given up* nuclear weapons as have *acquired* them. Say what? While Israel, India, Pakistan, and North Korea currently have a nuclear capability, South Africa dismantled its stash shortly before the collapse of the apartheid regime in 1989, and Kazakhstan, Ukraine, and Belarus said "no thanks" to the arsenals they inherited from the defunct Soviet Union. Also, believe it or not, the number of nonnuclear nations that are pursuing nuclear weapons has plummeted since the 1980s. Figure 5–22, based on a tally by the political scientist Scott Sagan, charts the number of nonnuclear states in each year since 1945 that had programs for developing nuclear weapons.

The downslopes in the curve show that at various times Algeria, Australia, Brazil, Egypt, Iraq, Libya, Romania, South Korea, Switzerland, Sweden, Taiwan, and Yugoslavia have pursued nuclear weapons but then thought the better of it—occasionally through the persuasion of an Israeli air strike, but more often by choice.

How precarious is the nuclear taboo? Will a rogue state inevitably defy the taboo and thereby annul it for the rest of the world? Doesn't history show that

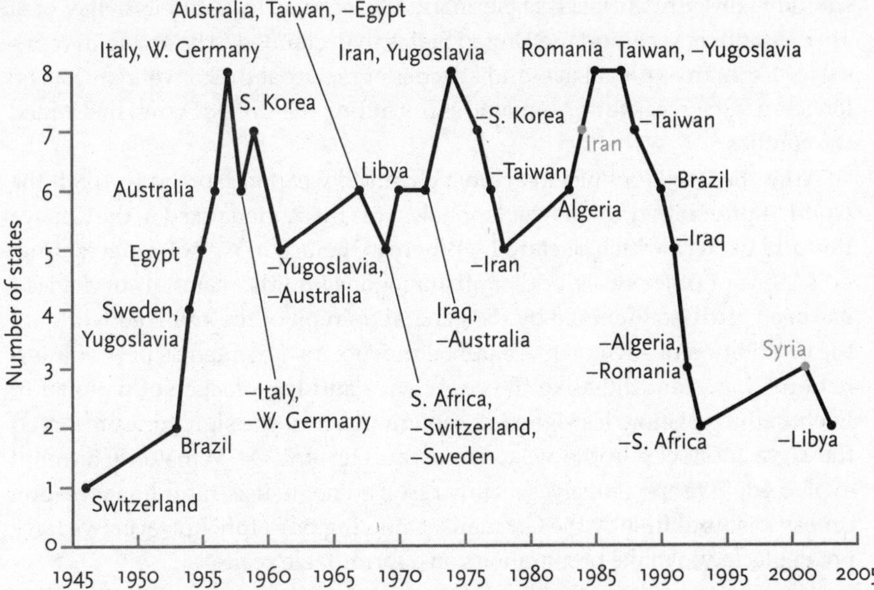

FIGURE 5–22. Nonnuclear states that started and stopped exploring nuclear weapons, 1945–2010

Country names marked with "–" represent the year in which a nuclear program in that country was stopped. The countries labeled in gray were believed to be exploring nuclear weapons in 2010. Though Israel bombed a suspected Syrian nuclear facility in 2007, as of 2010 Syria has refused International Atomic Energy Agency inspections, so it is kept on the list of active states. *Sources:* Graph adapted from Sagan, 2009, with updated information in Sagan, 2010, provided by Scott Sagan and Jane Esberg.

every weapons technology will sooner or later be put to use and then become unexceptionable?

The story of poison gas—the quintessential horror of World War I—is one place to look for an answer. In his book *The Chemical Weapons Taboo*, the political scientist Richard Price recounts how chemical weapons acquired their own stigma during the first half of the 20th century. The Hague Convention of 1899, one of a number of international agreements that aimed to regulate the conduct of war, had banned hollow-point bullets, aerial bombing (from balloons, that is, since the invention of the airplane was four years away), and projectiles that delivered poison gas. Given what was to come, the convention may seem like another candidate for history's dustbin of toothless feel-good manifestos.

But Price shows that even the combatants of World War I felt the need to pay the convention homage. When Germany introduced lethal gas to the battlefield, it claimed that it was retaliating for France's use of tear gas grenades and that anyway, it was conforming to the letter of the law because it didn't deliver the gas in artillery shells but just opened the cylinders and let the wind waft the gas toward the enemy. That these rationalizations were utterly lame shouldn't obscure the fact that Germany felt the need to justify its behavior at all. England, France, and the United States then claimed to be acting in reprisal for Germany's illegal use, and all sides agreed that the convention was no longer in force because nonsignatories (including the United States) had joined the conflict.

After the war, a revulsion against chemical weapons spread through the world. A prohibition with fewer loopholes was institutionalized in the Geneva Protocol of 1925, which declared, "Whereas the use in war of asphyxiating, poisonous or other gases, and of all analogous liquids, materials or devices, has been justly condemned by the general opinion of the civilized world . . . the prohibition of such use . . . shall be universally accepted as part of International Law, binding alike the conscience and the practice of nations."[208] Eventually 133 countries signed it, though many of the signatories reserved the right to stockpile the weapons as a deterrent. As Winston Churchill explained, "We are, ourselves, firmly resolved not to use this odious weapon unless it is used first by the Germans. Knowing our Hun, however, we have not neglected to make preparations on a formidable scale."[209]

Whether or not it was the piece of paper that made the difference, the taboo against the use of poison gas in interstate warfare took hold. Astonishingly, though both sides had tons of the stuff, poison gas was never used on the battlefield during World War II. Each side wanted to avoid the opprobrium of being the first to reintroduce poison gas to the battlefield, especially while the Nazis were hoping that England might accede to their conquest of continental Europe. And each side feared retaliation by the other.

The restraint held even in the face of destabilizing events that might have been expected to trigger an unstoppable escalation. In at least two episodes in Europe, poison gas was accidentally released by Allied forces. Explanations were conveyed to the German commanders, who believed them and did not retaliate.[210] A bit of cognitive compartmentalization helped too. In the 1930s Fascist Italy used poison gas in Abyssinia, and Imperial Japan used it in China. But these events were cordoned off in leaders' minds because they took place in "uncivilized" parts of the world rather than within the family of nations. Neither registered as a breach that would have nullified the taboo.

The only sustained uses of poison gas in war since the 1930s were by Egypt in Yemen in 1967 and by Iraq against Iranian forces (and its own Kurdish citizens) during the war of 1980–88. Defying the taboo may have been Saddam Hussein's undoing. The revulsion against his use of poison gas muted some of the opposition to the United States–led war that deposed him in 2003, and it figured in two of the seven charges against him in the Iraqi trial that led to his execution in 2006.[211] The world's nations formally abolished chemical weapons in 1993, and every known stockpile is in the process of being dismantled.

It's not immediately obvious why, out of all the weapons of war, poison gas was singled out as uniquely abominable—as so uncivilized that even the Nazis kept it off the battlefield. (They clearly had no compunction about using it elsewhere.) It's highly unpleasant to be gassed, but then it's just as unpleasant to be perforated or shredded by pieces of metal. As far as numbers are concerned, gas is far less lethal than bullets and bombs. In World War I fewer than 1 percent of the men who were injured by poison gas died from their injuries, and these fatalities added up to less than 1 percent of the war's death toll.[212] Though chemical warfare is militarily messy—no battlefield commander wants to be at the mercy of which way the wind is blowing—Germany could have used it to devastate the British forces at Dunkirk, and American forces would have found it handy in rooting out the Japanese soldiers hiding in caves in the Pacific Rim. And even if chemical weapons are difficult to deploy, that would hardly make them unique, since most new weapon technologies are ineffective when they are introduced. The first gunpowder weapons, for example, were slow to load, difficult to aim, and apt to blow up in the soldier's face. Nor were chemical weapons the first to be condemned for barbarism: in the era of longbows and pikes, gunpowder weapons were denounced as immoral, unmanly, and cowardly. Why did the taboo against chemical weapons take?

One possibility is that the human mind finds something distinctively repugnant about poison. Whatever suspension of the normal rules of decency allows warriors to do their thing, it seems to license only the sudden and directed application of force against an adversary who has the potential to do the same. Even pacifists may enjoy war movies or video games in which people get shot, stabbed, or blown up, but no one seems to get pleasure from watching a greenish cloud descend on a battlefield and slowly turn men into corpses. The poisoner has long been reviled as a uniquely foul and perfidious killer. Poison is the method of the sorcerer rather than the warrior; of the woman (with her terrifying control of kitchen and medicine chest) rather than the man. In *Venomous Woman*, the literary scholar Margaret Hallissy explains the archetype:

> Poison can never be used as an honorable weapon in a fair duel between worthy opponents, as the sword or gun, male weapons, can. A man who uses such a secret weapon is beneath contempt. Publicly acknowledged rivalry is a kind of bonding in which each worthy opponent gives the other the opportunity to demonstrate prowess. . . . The dueler is open, honest, and strong; the poisoner fraudulent, scheming, and weak. A man with a gun or a sword is a threat, but he declares himself to be so, and his intended victim can arm himself. . . . The poisoner uses superior secret knowledge to compensate for physical inferiority. A weak woman planning a poison is as deadly as a man with a gun, but because she plots in secret, the victim is more disarmed.[213]

Whatever abhorrence of poisoning we might have inherited from our evolutionary or cultural past, it needed a boost from historical contingency to become entrenched as a taboo on the conduct of war. Price conjectures that the critical nonevent was that in World War I, poison gas was never deliberately used against civilians. At least in that application, no taboo-shattering precedent had been set, and the widespread horror in the 1930s about the prospect that gas-dispensing airplanes could annihilate entire cities rallied people into categorically opposing all uses of the weapons.

The analogies between the chemical weapons taboo and the nuclear weapons taboo are clear enough. Today the two are lumped together as "weapons of mass destruction," though nuclear weapons are incomparably more destructive, because each taboo can draw strength from the other by association. The dread of both kinds of weapons is multiplied by the prospect of slow death by sickening and the absence of a boundary between battlefield and civilian life.

The world's experience with chemical weapons offers some morals that are mildly hopeful, at least by the terrifying standards of the nuclear age. Not every lethal technology becomes a permanent part of the military tool kit; some genies can be put back in their bottles; and moral sentiments can sometimes become entrenched as international norms and affect the conduct of war. Those norms, moreover, can be robust enough to withstand an isolated exception, which does not necessarily set off an uncontrollable escalation. That in particular is a hopeful discovery, though it might be good for the world if not too many people were aware of it.

If the world did away with chemical weapons, could it do the same with nuclear weapons? Recently a group of American icons proposed just that in an idealistic manifesto entitled "A World Free of Nuclear Weapons." The icons were not Peter, Paul, and Mary but George Shultz, William Perry, Henry Kissinger, and Sam Nunn.[214] Shultz was secretary of state in the Reagan administration. Perry was secretary of defense under Clinton. Kissinger was national security advisor and secretary of state under Nixon and Ford. Nunn was a chairman of the Senate Armed Services Committee and has long been considered the American lawmaker most knowledgeable about national defense. None could be accused of starry-eyed pacifism.

Supporting them is a dream team of war-hardened statesmen from Democratic and Republican administrations going back to that of John F. Kennedy. They include five former secretaries of state, five former national security advisors, and four former secretaries of defense. In all, three-quarters of the living alumni of those positions signed on to the call for a phased, verified, binding elimination of all nuclear weapons, now sometimes called Global Zero.[215] Barack Obama and Dmitry Medvedev have endorsed it in speeches (one of the reasons Obama was awarded the Nobel Peace Prize in 2009), and several policy think tanks have begun to work out how it might be implemented. The leading road map calls for four phases of negotiation, reduction, and verification, with the last warhead dismantled in 2030.[216]

As one might guess from the résumés of its supporters, Global Zero has some hardheaded realpolitik behind it. Since the end of the Cold War, the nuclear arsenal of the great powers has become an absurdity. It is no longer needed to deter an existential threat from an enemy superpower, and given the nuclear taboo, it serves no other military purpose. The threat of a retaliatory strike cannot deter stateless terrorists, because their bomb would not come with a return address, and if they were religious fanatics there would be nothing on earth that they valued enough to threaten. As praiseworthy as the various nuclear arms reduction agreements have been, they make little difference to global security as long as thousands of weapons remain in existence and the technology to make new ones is not forgotten.

The psychology behind Global Zero is to extend the taboo on *using* nuclear weapons to a taboo on *possessing* them. Taboos depend on a mutual understanding that there are bright lines delineating all-or-none categories, and the line distinguishing zero from more-than-zero is the brightest of all. No country could justify acquiring a nuclear weapon to protect itself against a nuclear-armed neighbor if it had no nuclear-armed neighbors. Nor could it claim that the nuclear legacy nations were hypocritically reserving the right to keep their own weapons. A developing nation could no longer try to look like a grown-up by acquiring a nuclear arsenal if the grown-ups had eschewed the weapons as old-fashioned and repulsive. And any rogue state or terrorist group that flirted with acquiring a nuclear weapon would become a pariah in the eyes of the world—a depraved criminal rather than a redoubtable challenger.

The problem, of course, is how to get there from here. The process of dismantling the weapons would open windows of vulnerability during which one of the remaining nuclear powers could fall under the sway of an expansionist zealot. Nations would be tempted to cheat by retaining a few nukes on the side just in case their adversaries did so. A rogue state might support nuclear terrorists once it was sure that it would never be a target of retaliation. And in a world that lacked nuclear weapons but retained the knowledge of how to build them—and that genie certainly can't be put back in the bottle—a crisis could set off a scramble to rearm, in which the first past the post might be tempted to strike preemptively before its adversary got the upper hand. Some experts on nuclear strategy, including Schelling, John Deutch, and Harold Brown, are skeptical that a nuclear-free world is attainable or even desirable, though others are working out timetables and safeguards designed to answer their objections.[217]

With all these uncertainties, no one should predict that nuclear weapons will go the way of poison gas anytime soon. But it is a sign of the momentum behind the Long Peace that abolition can even be discussed as a foreseeable prospect. If it happens, it would represent the ultimate decline in violence. A nuclear-free world! What realist would have dreamed it?

IS THE LONG PEACE A DEMOCRATIC PEACE?

If the Long Peace is not the sturdy child of terror and the twin brother of annihilation, then whose child is it? Can we identify an exogenous variable—some development that is not part of the peace itself—that blossomed in the postwar years and that we have reason to believe is a generic force against war? Is there a causal story with more explanatory muscle than "Developed countries stopped warring because they got less warlike"?

In chapter 4 we met a two-hundred-year-old theory that offers some predictions. In his essay "Perpetual Peace," Immanuel Kant reasoned that three conditions should reduce the incentives of national leaders to wage war without their having to become any kinder or gentler.

The first is democracy. Democratic government is designed to resolve conflicts among citizens by consensual rule of law, and so democracies should externalize this ethic in dealing with other states. Also, every democracy knows the way every other democracy works, since they're all constructed on the same rational foundations rather than growing out of a cult of personality, a messianic creed, or a chauvinistic mission. The resulting trust among democracies should nip in the bud the Hobbesian cycle in which the fear of a preemptive attack on each side tempts both into launching a preemptive attack. Finally, since democratic leaders are accountable to their people, they should be less likely to initiate stupid wars that enhance their glory at the expense of their citizenries' blood and treasure.

The Democratic Peace, as the theory is now called, has two things going for it as an explanation for the Long Peace. The first is that the trend lines are in the right direction. In most of Europe, democracy has surprisingly shallow roots. The eastern half was dominated by communist dictatorships until 1989, and Spain, Portugal, and Greece were fascist dictatorships until the 1970s. Germany started one world war as a militaristic monarchy, joined by monarchical Austria-Hungary, and another as a Nazi dictatorship, joined by Fascist Italy. Even France needed five tries to get democracy right, interleaved with monarchies, empires, and Vichy regimes. Not so long ago many experts thought that democracy was doomed. In 1975 Daniel Patrick Moynihan lamented that "liberal democracy on the American model increasingly tends to the condition of monarchy in the 19th century: a holdover form of government, one which persists in isolated or peculiar places here and there, and may even serve well enough for special circumstances, but which has simply no relevance to the future. It was where the world was, not where it is going."[218]

Social scientists should never predict the future; it's hard enough to predict the past. Figure 5–23 shows the worldwide fortunes of democracies, autocracies, and anocracies (countries that are neither fully democratic nor fully autocratic) in the decades since World War II. The year in which Moynihan announced the death of democracy was a turning point in the relative fortunes of the different forms of governance, and democracy turned out to be exactly where the world was going, particularly the developed world. Southern Europe became fully democratic in the 1970s, and Eastern Europe by the early 1990s. Currently the only European country classified as an autocracy is Belarus, and all but Russia are full-fledged democracies. Democracies also predominate in the Americas and in major developed countries of the Pacific, such as South Korea and Taiwan.[219] Quite apart from any contribution that democracy might make to international peace, it is a form of government that inflicts the minimum of violence on its own citizens, so the rise of democracy itself must be counted as another milestone in the historical decline of violence.

The second selling point for the Democratic Peace is a factoid that is sometimes elevated to a law of history. Here it is explained by the former U.K. prime minister Tony Blair in a 2008 interview on *The Daily Show* with Jon Stewart:

> *Stewart:* Our president—have you met him? He's a big freedom guy. He believes if everyone was a democracy, there'd be no more fighting.
>
> *Blair:* Well, as a matter of history, no two democracies have gone to war against each other.
>
> *Stewart:* Let me ask you a question. Argentina. Democracy?

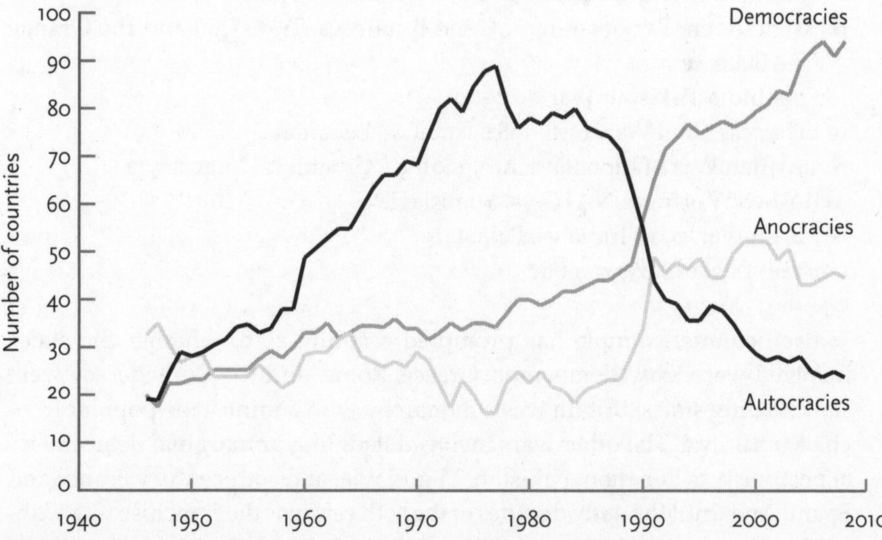

FIGURE 5–23. Democracies, autocracies, and anocracies, 1946–2008

Source: Graph adapted from Marshall & Cole, 2009. Only countries with a 2008 population greater than 500,000 are counted.

Blair: Well, it's a democracy. They elect their president.
Stewart: England. Democracy?
Blair: More or less. It was when I was last there.
Stewart: Uh . . . didn't you guys fight?
Blair: Actually, at the time Argentina was not a democracy.
Stewart: Damn it! I thought I had him.

If developed countries became democratic after World War II, and if democracies never go to war with one another, then we have an explanation for why developed countries stopped going to war after World War II.

As Stewart's skeptical questioning implies, the Democratic Peace theory has come under scrutiny, especially after it provided part of the rationale for Bush and Blair's invasion of Iraq in 2003. History buffs have delighted in coming up with possible counterexamples; here are a few from a collection by White:

- Greek Wars, 5th century BCE: Athens vs. Syracuse
- Punic Wars, 2nd and 3rd centuries BCE: Rome vs. Carthage
- American Revolution, 1775–83: United States vs. Great Britain
- French Revolutionary Wars, 1793–99: France vs. Great Britain, Switzerland, the Netherlands
- War of 1812, 1812–15: United States vs. Great Britain
- Franco-Roman War, 1849: France vs. Roman Republic
- American Civil War, 1861–65: United States vs. Confederate States
- Spanish-American War, 1898: United States vs. Spain
- Anglo-Boer War, 1899–1901: Great Britain vs. Transvaal and the Orange Free State
- First India-Pakistan War, 1947–49
- Lebanese Civil War, 1978, 1982: Israel vs. Lebanon
- Croatian War of Independence, 1991–92: Croatia vs. Yugoslavia
- Kosovo War, 1999: NATO vs. Yugoslavia
- Kargil War, 1999: India vs. Pakistan
- Israel-Lebanon War, 2006[220]

Each counterexample has prompted scrutiny as to whether the states involved were truly democratic. Greece, Rome, and the Confederacy were slaveholding states; Britain was a monarchy with a minuscule popular franchise until 1832. The other wars involved fledgling or marginal democracies at best, such as Lebanon, Pakistan, Yugoslavia, and 19th-century France and Spain. And until the early decades of the 20th century, the franchise was withheld from women, who, as we will see, tend to be more dovish in their voting than men. Most advocates of the Democratic Peace are willing to write off the centuries before the 20th, together with new and unstable democracies, and insist that since then no two mature, stable democracies have fought each other in a war.

Critics of the Democratic Peace theory then point out that if one draws the circle of "democracy" small enough, not that many countries are left in it, so by the laws of probability it's not surprising that we find few wars with a democracy on each side. Other than the great powers, two countries tend to fight only if they share a border, so most of the theoretical matchups are ruled out by geography anyway. We don't need to bring in democracy to explain why New Zealand and Uruguay have never gone to war, or Belgium and Taiwan. If one restricts the database even further by sloughing off early pieces of the time line (restricting it, as some do, to the period after World War II), then a more cynical theory accounts for the Long Peace: since the start of the Cold War, allies of the world's dominant power, the United States, haven't fought each other. Other manifestations of the Long Peace—such as the fact that the great powers never fought each other—were never explained by the Democratic Peace in the first place and, according to the critics, probably came from mutual deterrence, nuclear or conventional.[221]

A final headache for the Democratic Peace theory, at least as it applies to overall war-proneness, is that democracies often don't behave as nicely as Kant said they should. The idea that democracies externalize their law-governed assignment of power and peaceful resolution of conflicts doesn't sit comfortably with the many wars that Britain, France, the Netherlands, and Belgium fought to acquire and defend their colonial empires—at least thirty-three between 1838 and 1920, and a few more extending into the 1950s and even 1960s (such as France in Algeria). Equally disconcerting for Democratic Peaceniks are the American interventions during the Cold War, when the CIA helped overthrow more-or-less democratic governments in Iran (1953), Guatemala (1954), and Chile (1973) which had tilted too far leftward for its liking. The advocates reply that European imperialism, though it did not vanish instantaneously, was plummeting abroad just as democracy was rising at home, and that the American interventions were covert operations hidden from the public rather than wars conducted in full view and thus were exceptions proving the rule.[222]

When a debate devolves into sliding definitions, cherry-picked examples, and ad hoc excuses, it's time to call in the statistics of deadly quarrels. Two political scientists, Bruce Russett and John Oneal, have breathed new life into the Democratic Peace theory by firming up the definitions, controlling the confounding variables, and testing a quantitative version of the theory: not that democracies *never* go to war (in which case every putative counterexample becomes a matter of life or death) but that they go to war *less often* than nondemocracies, all else being equal.[223]

Russett and Oneal untangled the knot with a statistical technique that separates the effects of confounded variables: multiple logistic regression. Say you discover that heavy smokers have more heart attacks, and you want to confirm that the greater risk was caused by the smoking rather than by the lack of exercise that tends to go with smoking. First you try to account for as much of the heart attack data as you can using the nuisance factor, exercise rates. After looking at a large sample of men's health records, you might determine that on average, every additional hour of exercise per week cuts a man's chance of having a heart attack by a certain amount. Still, the correlation is not perfect—some couch potatoes have healthy hearts; some athletes collapse in the gym. The difference between the heart attack rate one would predict, given a certain rate of exercise, and the actual heart attack rate one measures is called a *residual*. The entire set of residuals gives you some numbers to play with in ascertaining the effects of the variable you're really interested in, smoking.

Now you capitalize on a second source of wiggle room. On average, heavy smokers exercise less, but some of them exercise a lot, while some nonsmokers hardly exercise at all. This provides a second set of residuals: the discrepancies between the men's actual rate of smoking and the rate one would predict based on their exercise rate. Finally, you see whether the residuals left over from the smoking-exercise relationship (the degree to which men smoke more or less than you'd predict from their exercise rate) correlate with the residuals left over from the exercise–heart attack relationship (the degree to which men have more or fewer heart attacks than you'd predict from their exercise rate). If the residuals correlate with the residuals, you can conclude that smoking correlates with heart attacks, above and beyond their joint correlation with exercise. And if you measured smoking at an earlier point in the men's lives, and heart attacks at a later point (to rule out the possibility that heart attacks make men smoke, rather than vice versa), you can inch toward the claim that smoking *causes* heart attacks. Multiple regression allows you to do this not just with two tangled predictors, but with any number of them.

A general problem with multiple regression is that the more predictors you want to untangle, the more data you need, because more and more of the variation in the data gets "used up" as each nuisance variable sucks up as much of the variation as it can and the hypothesis you're interested in has to make do with the rest. And fortunately for humanity, but unfortunately for social scientists, interstate wars don't break out all that often. The Correlates of War Project counts only 79 full-fledged interstate wars (killing at least a thousand people a year) between 1823 and 1997, and only 49 since 1900, far too few for statistics. So Russett and Oneal looked at a much larger database that lists militarized interstate disputes—incidents in which a country put its forces on alert, fired a shot across a bow, sent its warplanes scrambling, crossed swords, rattled sabers, or otherwise flexed its military muscles.[224] Assuming that for every war that actually breaks out there are many more disputes that stop short of war but have similar causes, the disputes should be shaped by the same causes as the wars themselves, and thus can serve as a plentiful surrogate for wars. The Correlates of War Project identified more than 2,300 militarized interstate disputes between 1816 and 2001, a number that can satisfy even a data-hungry social scientist.[225]

Russett and Oneal first lined up their units of analysis: pairs of countries in every year from 1886 to 2001 that had at least some risk of going to war, either because they were neighbors or because one of them was a great power. The datum of interest was whether in fact the pair had had a militarized dispute that year. Then they looked at how democratic the *less* democratic member of the pair was the year before, on the assumption that even if a democratic state is war-averse, it still might be dragged into a war by a more belligerent (and perhaps less democratic) adversary. It hardly seems fair to penalize democratic Netherlands in 1940 for getting into a war with its German invaders, so the Netherlands-Germany pair in 1940 would be assigned the rock-bottom democracy score for Germany in 1939.

To circumvent the temptation of data snooping when deciding whether a state was democratic, especially states that call themselves "democracies" on the basis of farcical elections, Russett and Oneal got their numbers from the Polity Project, which assigns each country a democracy score from 0 to 10 based on how competitive its political process is, how openly its leader is chosen, and how many constraints are placed on the leader's power. The researchers also threw into the pot some variables that are expected to affect military disputes through sheer realpolitik: whether a pair of countries were in a formal alliance (since allies are less likely to fight); whether one of them is a great power (since great powers tend to find trouble); and if neither is a great power, whether one is considerably more powerful than the other (because states fight less often when they are mismatched and the outcome would be a foregone conclusion).

So are democracies less likely to get into militarized disputes, all else held constant? The answer was a clear yes. When the less democratic member of a pair was a full autocracy, it doubled the chance that they would have a quarrel compared to an average pair of at-risk countries. When both countries were fully democratic, the chance of a dispute fell by more than half.[226]

In fact, the Democratic Peace theory did even better than its advocates hoped. Not only do democracies avoid disputes with each other, but there is a suggestion that they tend to stay out of disputes across the board.[227] And the reason they don't fight each other is not just that they are birds of a feather: there is no Autocratic Peace, a kind of honor among thieves in which autocracies also avoid disputes with each other.[228] The Democratic Peace held not only over the entire 115 years spanned by the dataset but also in the subspans from 1900 to 1939 and from 1989 to 2001. That shows that the Democratic Peace is not a by-product of a Pax Americana during the Cold War.[229] In fact, there were never any signs of a Pax Americana or a Pax Britannica: the years when one of these countries was the world's dominant military power were no more peaceful than the years in which it was just one power among many.[230] Nor was there any sign that new democracies are stroppy exceptions to the Democratic Peace—just think of the Baltic and Central European countries that embraced democracy after the Soviet empire collapsed, and the South American countries that shook off their military juntas in the 1970s and 1980s, none of which subsequently went to war.[231] Russett and Oneal found only one restriction on the Democratic Peace: it kicked in only around 1900, as one might have expected from the plethora of 19th-century counterexamples.[232]

So the Democratic Peace came out of a tough test in good shape. But that does not mean we should all be freedom guys and try to impose democratic governments on every autocracy we can invade. Democracy is not completely exogenous to a society; it is not a list of procedures for the workings of government from which every other good follows. It is woven into a fabric of civilized attitudes that includes, most prominently, a renunciation of political violence. England and the United States, recall, had prepared the ground for their democracies when their political leaders and their opponents had gotten out of the habit of murdering each other. Without this fabric, democracy brings no guarantee of internal peace. Though new and fragile democracies don't start interstate wars, in the next chapter we will see that they host more than their share of civil wars.

Even when it comes to the aversion of democracies to interstate war, it is premature to anoint democracy as the first cause. Countries with democracy are beneficiaries of the happy end of the Matthew Effect, in which them that's got shall get and them that's not shall lose. Not only are democracies free of despots, but they are richer, healthier, better educated, and more open to international trade and international organizations. To understand the Long Peace, we have to pry these influences apart.

IS THE LONG PEACE A LIBERAL PEACE?

The Democratic Peace is sometimes considered a special case of a Liberal Peace—"liberal" in the sense of classical liberalism, with its emphasis on political and economic freedom, rather than left-liberalism.[233] The theory of the Liberal Peace embraces as well the doctrine of gentle commerce, according to which trade is a form of reciprocal altruism which offers positive-sum benefits for both parties and gives each a selfish stake in the well-being of the other. Robert Wright, who gave reciprocity pride of place in *Nonzero*, his treatise on the expansion of cooperation through history, put it this way: "Among the many reasons I think we shouldn't bomb the Japanese is that they made my minivan."

The vogue word *globalization* reminds us that in recent decades international trade has mushroomed. Many exogenous developments have made trade easier and cheaper. They include transportation technologies such as the jet airplane and the container ship; electronic communication technologies such as the telex, long-distance telephone, fax, satellite, and Internet; trade agreements that have reduced tariffs and regulations; channels of international finance and currency exchange that make it easier for money to flow across borders; and the increased reliance of modern economies on ideas and information rather than on manual labor and physical stuff.

History suggests many examples in which freer trade correlates with greater peace. The 18th century saw both a lull in war and an embrace of commerce, when royal charters and monopolies began to give way to free markets, and when the beggar-thy-neighbor mindset of mercantilism gave way to the everybody-wins mindset of international trade. Countries that withdrew from the great power game and its attendant wars, such as the Netherlands in the 18th century and Germany and Japan in the second half of the 20th, often channeled their national aspirations into becoming commercial powers instead. The protectionist tariffs of the 1930s led to a falloff in international trade and perhaps to a rise in international tensions. The current comity between the United States and China, which have little in common except a river of manufactured goods in one direction and dollars in the other, is a recent reminder of the irenic effects of trade. And rivaling the Democratic Peace theory as a categorical factoid about modern conflict prevention is the Golden Arches theory: no two countries with a McDonald's have ever fought in a war. The only unambiguous Big Mac Attack took place in 1999, when NATO briefly bombed Yugoslavia.[234]

Anecdotes aside, many historians are skeptical that trade, as a general rule, conduces to peace. In 1986, for example, John Gaddis wrote, "These are pleasant things to believe, but there is remarkably little historical evidence to validate them."[235] Certainly, enhancements in the infrastructure supporting trade were not sufficient to yield peace in ancient and medieval times. The technologies that facilitated trade, such as ships and roads, also facilitated plunder, sometimes among the same itinerants, who followed the rule "If there are more of them, trade; if there are more of us, raid."[236] In later centuries, the profits to be gained from trade were so tempting that trade was sometimes imposed with gunboats on colonies and weak countries that resisted it, most infamously in the 19th-century Opium Wars, when Britain fought China to force it to allow British traffickers to sell the addictive drug within its borders. And great power wars often embroiled pairs of countries that had traded with each other a great deal.

Norman Angell inadvertently set back the reputation of the trade-peace connection when he was seen as claiming that free trade had made war obsolete and five years later World War I broke out. Skeptics like to rub it in by pointing out that the prewar years saw unprecedented levels of financial interdependence, including a large volume of trade between England and Germany.[237] And as Angell himself took pains to point out, the economic futility of war is a reason to avoid it only if nations are interested in prosperity in the first place. Many leaders are willing to sacrifice a bit of prosperity (often much more than a bit) to enhance national grandeur, to implement utopian ideologies, or to rectify what they see as historic injustices. Their citizenries, even in democracies, may go along with them.

Russett and Oneal, the number-crunching defenders of the Democratic Peace, also sought to test the theory of the Liberal Peace, and they were skeptical of the skeptics. They noted that though international trade hit a local peak just before World War I, it still was a fraction of the level, relative to gross domestic product, that countries would see after World War II (figure 5–24).

Also, trade may work as a pacifying force only when it is underpinned by international agreements that prevent a nation from suddenly lurching toward protectionism and cutting off the air supply of its trading partners. Gat argues that around the turn of the 20th century, Britain and France were making noises about becoming imperial autarkies that would live off trade within their colonial empires. This sent Germany into a panic and gave its leaders the idea that it needed an empire too.[238]

With examples and counterexamples on both sides, and the many statistical confounds between trade and other good things (democracy, membership in international organizations, membership in alliances, and overall prosperity), it was time once again for multiple regression. For every pair of at-risk nations, Russett and Oneal entered the amount of trade (as a proportion of GDP) for the more trade-dependent member. They found that countries that depended more on trade in a given year were less likely to have a militarized

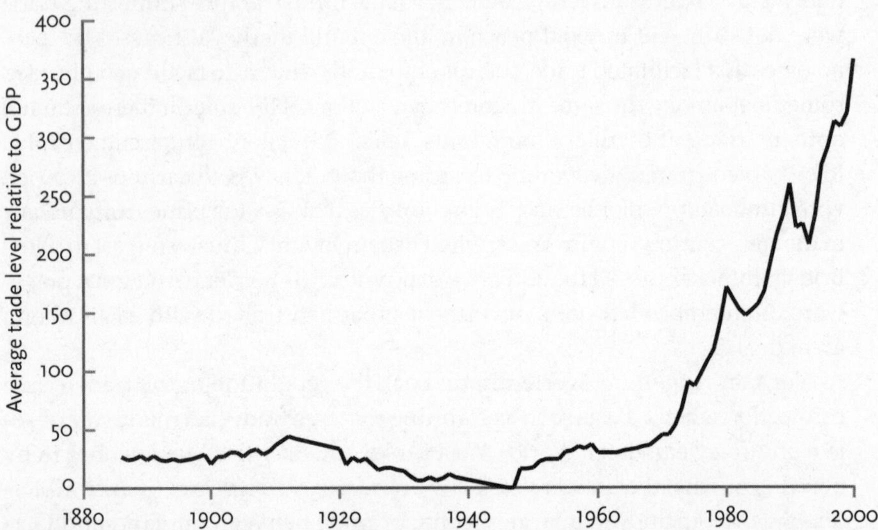

FIGURE 5–24. International trade relative to GDP, 1885–2000

Source: Graph from Russett, 2008, based on data from Gleditsch, 2002.

dispute in the subsequent year, even controlling for democracy, power ratio, great power status, and economic growth.[239] Other studies have shown that the pacifying effects of trade depend on the countries' level of development: those that have access to the financial and technological infrastructure that lowers the cost of trade are most likely to resolve their disputes without displays of military force.[240] This is consistent with the suggestions of Angell and Wright that broad historical changes have tilted financial incentives away from war and toward trade.

Russett and Oneal found that it was not just the level of bilateral trade between the two nations in a pair that contributed to peace, but the dependence of each country on trade across the board: a country that is open to the global economy is less likely to find itself in a militarized dispute.[241] This invites a more expansive version of the theory of gentle commerce. International trade is just one facet of a country's commercial spirit. Others include an openness to foreign investment, the freedom of citizens to enter into enforceable contracts, and their dependence on voluntary financial exchanges as opposed to self-sufficiency, barter, or extortion. The pacifying effects of commerce in this broad sense appear to be even more robust than the pacifying effects of democracy. A democratic peace strongly kicks in only when *both* members of a pair of countries are democratic, but the effects of commerce are demonstrable when *either* member of the pair has a market economy.[242]

Such findings have led some political scientists to entertain a heretical idea called the Capitalist Peace.[243] The word *liberal* in Liberal Peace refers both to the political openness of democracy and to the economic openness of capitalism, and according to the Capitalist Peace heresy, it's the economic openness that does most of the pacifying. In arguments that are sure to leave leftists speechless, advocates claim that many of Kant's arguments about democracy apply just as well to capitalism. Capitalism pertains to an economy that runs by voluntary contracts between citizens rather than government command and control, and that principle can bring some of the same advantages that Kant adduced for democratic republics. The ethic of voluntary negotiation within a country (like the ethic of law-governed transfer of power) is naturally externalized to its relationships with other countries. The transparency and intelligibility of a country with a free market economy can reassure its neighbors that it is not going on a war footing, which can defuse a Hobbesian trap and cramp a leader's freedom to engage in risky bluffing and brinkmanship. And whether or not a leader's power is constrained by the ballot box, in a market economy it is constrained by stakeholders who control the means of production and who might oppose a disruption of international trade that's bad for business. These constraints put a brake on a leader's personal ambition for glory, grandeur, and cosmic justice and on his temptation to respond to a provocation with a reckless escalation.

Democracies tend to be capitalist and vice versa, but the correlation is imperfect: China, for example, is capitalist but autocratic, and India is democratic but until recently was heavily socialist. Several political scientists have exploited this slippage and have pitted democracy and capitalism against each other in analyses of datasets of militarized disputes or other international crises. Like Russett and Oneal, they all find a clear pacifying effect of capitalist variables such as international trade and openness to the global economy. But some of them disagree with the duo about whether democracy also makes a contribution to peace, once its correlation with capitalism is statistically removed.[244] So while the relative contributions of political and economic liberalism are currently mired in regression wonkery, the overarching theory of the Liberal Peace is on solid ground.

The very idea of a Capitalist Peace is a shock to those who remember when capitalists were considered "merchants of death" and "masters of war." The irony was not lost on the eminent peace researcher Nils Petter Gleditsch, who ended his 2008 presidential address to the International Studies Association with an updating of the 1960s peace slogan: "Make money, not war."[245]

IS THE LONG PEACE A KANTIAN PEACE?

In the wake of World War II, leading thinkers were desperate to figure out what had gone wrong and tossed around a number of schemes for preventing a repeat performance. Mueller explains the most popular one:

> Some Western scientists, apparently consumed with guilt over having participated in the development of a weapon that could kill with new efficiency, . . . took time out from their laboratories and studies to consider human affairs. They quickly came to conclusions expressed with an evangelical certainty they would never have used in discussing the physical world. Although he had done his greatest work in physics while a citizen of the sovereign nation of Switzerland, Einstein proved as immune to the Swiss example as everyone else. "As long as there are sovereign nations possessing great power," he declared, "war is inevitable." . . . Fortunately, he and other scientists had managed to discover the one device that could solve the problem. "Only the creation of a world government can prevent the impending self-destruction of mankind."[246]

World government seems like a straightforward extension of the logic of the Leviathan. If a national government with a monopoly on the use of force is the solution to the problem of homicide among individuals and of private and civil wars among factions, isn't a *world* government with a monopoly on the legitimate use of *military* force the solution to the problem of wars among nations? Most intellectuals did not go as far as Bertrand Russell, who in 1948 proposed that the Soviet Union should be given an ultimatum that unless it immediately submitted to world government, the United States would attack it with nuclear weapons.[247] But world government was endorsed by, among others, Einstein, Wendell Willkie, Hubert Humphrey, Norman Cousins, Robert Maynard Hutchins, and William O. Douglas. Many people thought world government would gradually emerge out of the United Nations.

Today the campaign for world government lives on mainly among kooks and science fiction fans. One problem is that a functioning government relies on a degree of mutual trust and shared values among the people it governs which is unlikely to exist across the entire globe. Another is that a world government would have no alternatives from which it could learn better governance, or to which its disgruntled citizens could emigrate, and hence it would have no natural checks against stagnation and arrogance. And the United Nations is unlikely to morph into a government that anyone would want to be governed by. The Security Council is hamstrung by the veto power that the great powers insisted on before ceding it any authority, and the General Assembly is more of a soapbox for despots than a parliament of the world's people.

In "Perpetual Peace," Kant envisioned a "federation of free states" that would fall well short of an international Leviathan. It would be a gradually expanding club of liberal republics rather than a global megagovernment, and it would rely on the soft power of moral legitimacy rather than on a monopoly on the use of force. The modern equivalent is the intergovernmental organization or IGO—a bureaucracy with a limited mandate to coordinate the policies of participating nations in some area in which they have a common interest. The international entity with the best track record for implementing world peace is probably not the United Nations, but the European Coal and Steel Community, an IGO founded in 1950 by France, West Germany, Belgium, the Netherlands, and Italy to oversee a common market and regulate the production of the two most important strategic commodities. The organization was specifically designed as a mechanism for submerging historic rivalries and ambitions—especially West Germany's—in a shared commercial enterprise. The Coal and Steel Community set the stage for the European Economic Community, which in turn begot the European Union.[248]

Many historians believe that these organizations helped keep war out of the collective consciousness of Western Europe. By making national borders porous to people, money, goods, and ideas, they weakened the temptation of nations to fall into militant rivalries, just as the existence of the United States weakens any temptation of, say, Minnesota and Wisconsin to fall into a militant rivalry. By throwing nations into a club whose leaders had to socialize and work together, they enforced certain norms of cooperation. By serving as an impartial judge, they could mediate disputes among member nations. And by holding out the carrot of a vast market, they could entice applicants to give up their empires (in the case of Portugal) or to commit themselves to liberal democracy (in the case of former Soviet satellites and, perhaps soon, Turkey).[249]

Russett and Oneal propose that membership in intergovernmental organizations is the third vertex of a triangle of pacifying forces which they attribute to Kant, the other two being democracy and trade. (Though Kant did not single out trade in "Perpetual Peace," he extolled it elsewhere, so Russett and Oneal felt they could take some license in drawing their triangle.) The international organizations needn't have utopian or even idealistic missions. They can coordinate defense, currency, postal service, tariffs, canal traffic, fishing rights, pollution, tourism, war crimes, weights and measures, road signs, anything—as long as they are voluntary associations of governments. Figure 5–25 shows how membership in these organizations steadily increased during the 20th century, with a bump after World War II.

To verify whether IGO membership made an independent contribution to peace, or just went along for the ride with democracy and trade, Russett and Oneal counted the number of IGOs that every pair of nations jointly belonged to, and they threw it into the regression analysis together with the democracy and trade scores and the realpolitik variables. The researchers concluded that Kant got it right three out of three times: democracy favors peace, trade favors peace, and membership in intergovernmental organizations favors peace. A pair of countries that are in the top tenth of the scale on all three variables are 83 percent less likely than an average pair of countries to have a militarized dispute in a given year, which means the likelihood is very close to zero.[250]

Might Kant have been right in an even grander sense? Russett and Oneal defended the Kantian triangle with sophisticated correlations. But a causal

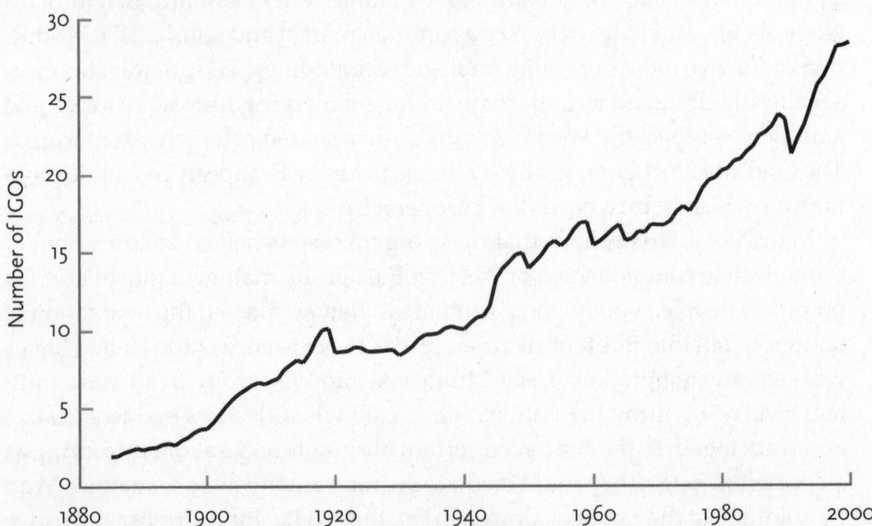

FIGURE 5–25. Average number of IGO memberships shared by a pair of countries, 1885–2000
Source: Graph from Russett, 2008.

story derived from correlational data is always vulnerable to the possibility that some hidden entity is the real cause of both the effect one is trying to explain and the variables one is using to explain it. In the case of the Kantian triangle, each putative pacifying agent may depend on a deeper and even more Kantian cause: a willingness to resolve conflicts by means that are acceptable to all the affected parties, rather than by the stronger party imposing its will on the weaker one. Nations become stable democracies only when their political factions tire of murder as the means of assigning power. They engage in commerce only when they put a greater value on mutual prosperity than on unilateral glory. And they join intergovernmental organizations only when they are willing to cede a bit of sovereignty for a bit of mutual benefit. In other words, by signing on to the Kantian variables, nations and their leaders are increasingly acting in such a way that the principle behind their actions can be made universal. Could the Long Peace represent the ascendancy in the international arena of the Categorical Imperative?[251]

Many scholars in international relations would snort at the very idea. According to an influential theory tendentiously called "realism," the absence of a world government consigns nations to a permanent state of Hobbesian anarchy. That means that leaders must act like psychopaths and consider only the national self-interest, unsoftened by sentimental (and suicidal) thoughts of morality.[252]

Realism is sometimes defended as a consequence of the existence of human nature, where the underlying theory of human nature is that people are self-interested rational animals. But as we shall see in chapters 8 and 9, humans are also *moral* animals: not in the sense that their behavior is moral in the light of disinterested ethical analysis, but in the sense that it is guided by moral intuitions supported by emotions, norms, and taboos. Humans are also cognitive animals, who spin out beliefs and use them to guide their actions. None of these endowments pushes our species toward peace by default. But it is neither sentimental nor unscientific to imagine that particular historical moments can engage the moral and cognitive faculties of leaders and their coalitions in a combination that inclines them toward peaceful coexistence. Perhaps the Long Peace is one of them.

In addition to the three proximate Kantian causes, then, the Long Peace may depend on an ultimate Kantian cause. Norms among the influential constituencies in developed countries may have evolved to incorporate the conviction that war is inherently immoral because of its costs to human well-being, and that it can be justified only on the rare occasions when it is certain to prevent even greater costs to human well-being. If so, interstate war among developed countries would be going the way of customs such as slavery, serfdom, breaking on the wheel, disemboweling, bearbaiting, cat-burning, heretic-burning, witch-drowning, thief-hanging, public executions, the display of rotting corpses on gibbets, dueling, debtors' prisons, flogging, keelhauling, and other practices that passed from unexceptionable to controversial to immoral to unthinkable to not-thought-about during the Humanitarian Revolution.

Can we identify exogenous causes of the new humanitarian aversion to war among developed countries? In chapter 4 I conjectured that the Humanitarian Revolution was accelerated by publishing, literacy, travel, science, and other cosmopolitan forces that broaden people's intellectual and moral horizons. The second half of the 20th century has obvious parallels. It saw the dawn of television, computers, satellites, telecommunications, and jet travel, and an unprecedented expansion of science and higher education. The communications guru Marshall McLuhan called the postwar world a "global village." In a village, the fortunes of other people are immediately felt. If the village is the natural size of our circle of sympathy, then perhaps when the village goes global, the villagers will experience greater concern for their fellow humans than when it embraced just the clan or tribe. A world in which a person can open the morning paper and meet the eyes of a naked, terrified little girl running toward him from a napalm attack nine thousand miles away is not a world in which a writer can opine that war is "the foundation of all the high virtues and faculties of man" or that it "enlarges the mind of a people and raises their character."

The end of the Cold War and the peaceful dissolution of the Soviet empire have also been linked to the easier movement of people and ideas at the end of the 20th century.[253] By the 1970s and 1980s the Soviet Union's attempt to retain its power by totalitarian control of media and travel was becoming a significant handicap. Not only was it becoming ludicrous for a modern economy to do without photocopiers, fax machines, and personal computers (to say nothing of the nascent Internet), but it was impossible for the country's rulers to keep scientists and policy wonks from learning about the ideas in the increasingly prosperous West, or to keep the postwar generation from learning about rock music, blue jeans, and other perquisites of personal freedom. Mikhail Gorbachev was a man of cosmopolitan tastes, and he installed in his administration many analysts who had traveled and studied in the West. The Soviet leadership made a verbal commitment to human rights in the 1975 Helsinki Accords, and a cross-border network of human rights activists were trying to get the populace to hold them to it. Gorbachev's policy of *glasnost* (openness) allowed Aleksandr Solzhenitzyn's *The Gulag Archipelago* to be serialized in 1989, and it allowed debates in the Congress of People's Deputies to be televised, exposing millions of Russians to the brutality of the past Soviet leadership and the ineptitude of the current one.[254] Silicon chips, jet airplanes, and the electromagnetic spectrum were loosing ideas that helped to corrode the Iron Curtain. Though today's authoritarian China may seem to be straining the hypothesis that technology and travel are liberalizing forces, its leadership is incomparably less murderous than Mao's insular regime, as the numbers in the next chapter will show.

There may be another reason why antiwar sentiments finally took. The trajectory of violent deaths in Europe that we saw in figure 5–18 is a craggy landscape in which three pinnacles—the Wars of Religion, the French Revolutionary and Napoleonic Wars, and the two world wars—are followed by extended basins, each at a lower altitude than the preceding one. After each hemoclysm, world leaders tried, with some success, to make a recurrence less likely. Of course their treaties and concerts did not last forever, and an innumerate reading of history may invite the conclusion that the days of the Long Peace are running out and that an even bigger war is waiting to be born. But the Poisson pitter-patter of war shows no periodicity, no cycle of buildup and release. Nothing prevents the world from learning from its mistakes and driving the probability lower each time.

Lars-Erik Cederman went back to Kant's essays and discovered a twist in his prescription for perpetual peace. Kant was under no illusion that national leaders were sagacious enough to deduce the conditions of peace from first principles; he realized they would need to learn them from bitter historical experience. In an essay called "Idea for a Universal History with a Cosmopolitan Purpose," he wrote:

> Wars, tense and unremitting preparations, and the resultant distress which every state must eventually feel within itself, even in the midst of peace—these are the means by which nature drives nations to make initially imperfect attempts, but finally, after many devastations, upheavals and even complete inner exhaustion of their powers, to take the step which reason could have suggested to them even without so many sad experiences—that of abandoning their lawless state of savagery.[255]

Cederman suggests that Kant's theory of peace-through-learning should be combined with his theory of peace-through-democracy. Though all states, including democracies, start off warlike (since many democracies began as great powers), and all states can be blindsided by sudden terrible wars, democracies may be better equipped to learn from their catastrophes, because of their openness to information and the accountability of their leaders.[256]

Cederman plotted the historical trajectory of militarized disputes from 1837 to 1992 within pairs of democracies and other pairs of countries (figure 5–26). The inclined sawtooth for democracies shows that they started out warlike and thereafter underwent periodic shocks that sent their rate of disputes skyward. But after each peak their dispute rate quickly fell back to earth. Cederman also found that the learning curve was steeper for mature democracies than for newer ones. Autocracies too returned to more peaceable levels after the sudden shocks of major wars, but they did so more slowly and erratically. The fuzzy idea that after the 20th-century Hemoclysm an increasingly democratic world "got tired of war" and "learned from its mistakes" may have some truth to it.[257]

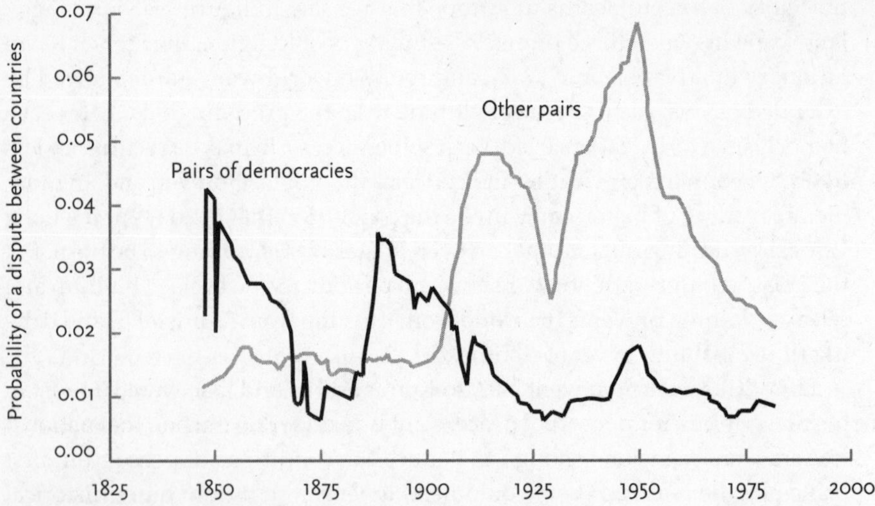

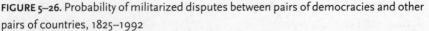

FIGURE 5–26. Probability of militarized disputes between pairs of democracies and other pairs of countries, 1825–1992

Source: Graph from Cederman, 2001. The curves plot 20-year moving averages for at-risk pairs of countries.

A popular theme in the antiwar ballads of the 1960s was that evidence of the folly of war had always been available but that people stubbornly refused to see it. "How many deaths will it take till they learn that too many people have died? The answer, my friend, is blowin' in the wind." "Where have all the soldiers gone? Gone to graveyards, every one. When will they ever learn?" After half a millennium of wars of dynasties, wars of religion, wars of sovereignty, wars of nationalism, and wars of ideology, of the many small wars in the spine of the distribution and a few horrendous ones in the tail, the data suggest that perhaps, at last, we're learning.

6

THE NEW PEACE

Macbeth's self-justifications were feeble—and his conscience devoured him. Yes, even Iago was a little lamb too. The imagination and the spiritual strength of Shakespeare's evildoers stopped short at a dozen corpses. Because they had no *ideology.*

—Aleksandr Solzhenitsyn

You would think that the disappearance of the gravest threat in the history of humanity would bring a sigh of relief among commentators on world affairs. Contrary to expert predictions, there was no invasion of Western Europe by Soviet tanks, no escalation of a crisis in Cuba or Berlin or the Middle East to a nuclear holocaust.[1] The cities of the world were not vaporized; the atmosphere was not poisoned by radioactive fallout or choked with debris that blacked out the sun and sent *Homo sapiens* the way of the dinosaurs. Not only that, but a reunified Germany did not turn into a fourth reich, democracy did not go the way of monarchy, and the great powers and developed nations did not fall into a third world war but rather a long peace, which keeps getting longer. Surely the experts have been acknowledging the improvements in the world's fortunes from a few decades ago.

But no—the pundits are glummer than ever! In 1989 John Gray foresaw "a return to the classical terrain of history, a terrain of great power rivalries . . . and irredentist claims and wars."[2] A *New York Times* editor wrote in 2007 that this return had already taken place: "It did not take long [after 1989] for the gyre to wobble back onto its dependably blood-soaked course, pushed along by fresh gusts of ideological violence and absolutism."[3] The political scientist Stanley Hoffman said that he has been discouraged from teaching his course on international relations because after the end of the Cold War, one heard "about nothing but terrorism, suicide bombings, displaced people, and genocides."[4] The pessimism is bipartisan: in 2007 the conservative writer Norman Podhoretz published a book called *World War IV* (on "the long struggle against Islamofascism"), while the liberal columnist Frank Rich wrote that the world was "a more dangerous place than ever."[5] If Rich is correct, then the world was more dangerous in 2007 than it was during the two world wars, the Berlin crises of 1949 and 1961, the Cuban Missile Crisis, and all the wars in the Middle East. That's pretty dangerous.

Why the gloom? Partly it's the result of market forces in the punditry business, which favor the Cassandras over the Pollyannas. Partly it arises from human temperament: as David Hume observed, "The humour of blaming the present, and admiring the past, is strongly rooted in human nature, and has an influence even on persons endowed with the profoundest judgment and most extensive learning." But mainly, I think, it comes from the innumeracy of our journalistic and intellectual culture. The journalist Michael Kinsley recently wrote, "It is a crushing disappointment that Boomers entered adulthood with Americans killing and dying halfway around the world, and now, as Boomers reach retirement and beyond, our country is doing the same damned thing."[6] This assumes that 5,000 Americans dying is the same damned thing as 58,000 Americans dying, and that a hundred thousand Iraqis being killed is the same damned thing as several million Vietnamese being killed. If we don't keep an eye on the numbers, the programming policy "If it bleeds it leads" will feed the cognitive shortcut "The more memorable, the more frequent," and we will end up with what has been called a false sense of insecurity.[7]

This chapter is about three kinds of organized violence that have stoked the new pessimism. They were given short shrift in the preceding chapter, which concentrated on wars among great powers and developed states. The Long Peace has not seen an end to these other kinds of conflict, leaving the impression that the world is "a more dangerous place than ever."

The first kind of organized violence embraces all the other categories of war, most notably the civil wars and wars between militias, guerrillas, and paramilitaries that plague the developing world. These are the "new wars" or "low-intensity conflicts" that are said to be fueled by "ancient hatreds."[8] Familiar images of African teenagers with Kalashnikovs support the impression that the global burden of war has not declined but has only been displaced from the Northern to the Southern Hemisphere.

The new wars are thought to be especially destructive to civilians because of the hunger and disease they leave in their wake, which are omitted from most counts of war dead. According to a widely repeated statistic, at the beginning of the 20th century 90 percent of war deaths were suffered by soldiers and 10 percent by civilians, but by the end of the century these proportions had reversed. Horrifying estimates of fatalities from famines and epidemics, rivaling the death toll of the Nazi Holocaust, have been reported in war-torn countries such as the Democratic Republic of the Congo.

The second kind of organized violence I will track is the mass killing of ethnic and political groups. The hundred-year period from which we have recently escaped has been called "the age of genocide" and "a century of genocide." Many commentators have written that ethnic cleansing emerged with modernity, was held at bay by the hegemony of the superpowers, returned with a vengeance with the end of the Cold War, and today is as prevalent as ever.

The third is terrorism. Since the September 11, 2001, attacks on the United States, the fear of terrorism has led to a massive new bureaucracy, two foreign wars, and obsessive discussion in the political arena. The threat of terrorism is said to pose an "existential threat" to the United States, having the capacity to "do away with our way of life" or to end "civilization itself."[9]

Each of these scourges, of course, continues to take a toll in human lives. The question I will ask in this chapter is exactly how big a toll, and whether it has increased or decreased in the past few decades. It's only recently that political scientists have tried to measure these kinds of destruction, and now that they have, they have reached a surprising conclusion: *All these kinds of killing are in decline.*[10] The decreases are recent enough—in the past two decades or less—that we cannot count on them lasting, and in recognition of their tentative nature I will call this development the New Peace. Nonetheless the trends are genuine declines of violence and deserve our careful attention. They are substantial in size, opposite in sign to the conventional wisdom, and suggestive of ways we might identify what went right and do more of it in the future.

THE TRAJECTORY OF WAR IN THE REST OF THE WORLD

What was the rest of the world doing during the six hundred years when the great powers and European states went through their Ages of Dynasties, Religions, Sovereignty, Nationalism, and Ideology; were racked by two world wars; and then fell into a long peace? Unfortunately the Eurocentric bias of the historical record makes it impossible to trace out curves with any confidence. Before the advent of colonialism, large swaths of Africa, the Americas, and Asia were host to predation, feuding, and slave-raiding that slunk beneath the military horizon or fell in the forest without any historian hearing them. Colonialism itself was implemented in many imperial wars that the great powers waged to acquire their colonies, suppress revolts, and fend off rivals. Throughout this era there were plenty of wars. For the period from 1400 through 1938, Brecke's Conflict Catalog lists 276 violent conflicts in the Americas, 283 in North Africa and the Middle East, 586 in sub-Saharan Africa, 313 in Central and South Asia, and 657 in East and Southeast Asia.[11] Historical myopia prevents us from plotting trustworthy trends in the frequency or deadliness of the wars, but we saw in the preceding chapter that many were devastating. They included civil and interstate wars that were proportionally (and in some cases absolutely) more lethal than anything taking place in Europe, such as the American Civil War, the Taiping Rebellion in China, the War of the Triple Alliance in South America, and the conquests of Shaka Zulu in southern Africa.

In 1946, just when Europe, the great powers, and the developed world started racking up their peaceful zeroes, the historical record for the world as a whole snaps into focus. That is the first year covered in a meticulous dataset compiled by Bethany Lacina, Nils Petter Gleditsch, and their colleagues at the Peace Research Institute of Oslo called the PRIO Battle Deaths Dataset.[12] The dataset includes every known armed conflict that killed as few as twenty-five people in a year. The conflicts that rise to the level of a thousand deaths a year are promoted to "wars," matching the definition used in the Correlates of War Project, but they are otherwise given no special treatment. (I will continue to use the word *war* in its nontechnical sense to refer to armed conflicts of all sizes.)

The PRIO researchers aim for criteria that are as reliable as possible, so that analysts can compare regions of the world and plot trends over time using a fixed yardstick. Without strict criteria—when analysts use direct battlefield deaths for some wars but include indirect deaths from epidemics and famines in others, or when they count army-against-army wars in some regions but throw in genocides in others—comparisons are meaningless and are too easily used as propaganda for one cause or another. The PRIO analysts comb through histories, media stories, and reports from government and human rights organizations to tally deaths from war as objectively as possible. The counts are conservative; indeed, they are certainly underestimates, because they omit all deaths that are merely conjectured or whose causes cannot be ascertained with confidence. Similar criteria, and overlapping data, are used in other conflict datasets, including those of the Uppsala Conflict Data Project (UCDP), whose data begin in 1989; the Stockholm International Peace Research Institute (SIPRI), which uses adjusted UCDP data; and the Human Security Report Project (HSRP), which draws on both the PRIO and UCDP datasets.[13]

Like Lewis Richardson, the new conflict-counters have to deal with failures of thinginess, and so they divide the conflicts into categories using obsessive-compulsive criteria.[14] The first cut distinguishes three kinds of mass violence that vary in their causes and, just as importantly, in their countability. The concept of "war" (and its milder version, "armed conflict") applies most naturally to multiple killing that is organized and socially legitimated. That invites a definition in which a "war" must have a government on at least one side, and the two sides must be contesting some identifiable resource, usually a territory or the machinery of government. To make this clear, the datasets call wars in this narrow sense "state-based armed conflicts," and they are the only conflicts for which data go all the way back to 1946.

The second category embraces "nonstate" or "intercommunal" conflict, and it pits warlords, militias, or paramilitaries (often aligned with ethnic or religious groups) against each other.

The third category has the clinical name "one-sided violence" and embraces genocides, politicides, and other massacres of unarmed civilians, whether perpetrated by governments or by militias. The exclusion of one-sided violence from the PRIO dataset is in part a tactical choice to divide violence into categories with different causes, but it is also a legacy of historians' long-standing fascination with war at the expense of genocide, which only recently has been recognized as more destructive of human life.[15] Rudolph Rummel, the political scientist Barbara Harff, and the UCDP have collected datasets of genocides, which we will examine in the next section.[16]

The first of the three categories, state-based conflicts, is then subdivided according to whom the government is fighting. The prototypical war is the *interstate* war, which pits two states against each other, such as the Iran-Iraq War of 1980–88. Then there are *extrastate* or *extrasystemic* wars, in which a government wages war on an entity outside its borders that is not a recognized state. These are generally imperial wars, in which a state fights indigenous forces to acquire a colony, or colonial wars, in which it fights to retain one, such as France in Algeria from 1954 to 1962.

Finally there are civil or *intrastate* wars, in which the government fights an insurrection, rebellion, or secessionist movement. These are further subdivided into civil wars that are completely internal (such as the recently concluded war in Sri Lanka between the government and the Tamil Tigers) and the *internationalized intrastate* wars in which a foreign army intervenes, usually to help a government defend itself against the rebels. The wars in Afghanistan and Iraq both began as interstate conflicts (the United States and its allies against Taliban-controlled Afghanistan, and the United States and its allies against Baathist-controlled Iraq), but as soon as the governments were toppled and the invading armies remained in the country to support the new governments against insurgencies, the conflicts were reclassified as internationalized intrastate conflicts.

Now there's the question of which deaths to count. The PRIO and UCDP datasets tally direct or *battle-related deaths*—the people who are shot, stabbed, clubbed, gassed, blown up, drowned, or deliberately starved as part of a contest in which the perpetrators themselves have to worry about getting hurt.[17] The victims may be soldiers, or they may be civilians who were caught in the crossfire or killed in "collateral damage." The battle-related death statistics exclude *indirect deaths* arising from disease, starvation, stress, and the breakdown of infrastructure. When indirect deaths are added to direct deaths to yield the entire toll attributable to the war, the sum may be called *excess deaths*.

Why do the datasets exclude indirect deaths? It's not to write these kinds of suffering out of the history books, but because direct deaths are the only ones that can be counted with confidence. Direct deaths also conform to our basic intuition of what it means for an agent to be responsible for an effect that it causes, namely that the agent foresees the effect, intends for it to happen, and makes it happen via a chain of events that does not have too many uncontrollable intervening links.[18] The problem with estimating indirect deaths is that it requires us to undertake the philosophical exercise of simulating in our imagination the possible world in which the war didn't occur and estimating the number of deaths that took place in that world, which then is used as a baseline. And that requires something close to omniscience. Would a postwar famine have taken place even if the war had not broken out because of the ineptitude of the overthrown government? What if there was a drought that year—should the famine deaths be blamed on the war or on the weather? If the rate of death from hunger was going down in the years before a war, should we assume that it would have declined even further if the war hadn't occurred, or should we freeze it at its level in the last year before the war? If Saddam Hussein had not been deposed, would he have gone on to kill more political enemies than the number of people who died in the intercommunal violence following his defeat? Should we add the 40 to 50 million victims of the 1918 influenza pandemic to the 15 million who were killed in World War I, because the flu virus would not have evolved its virulence if the war hadn't packed so many troops into trenches?[19] Estimating indirect deaths requires answering these sorts of questions in a consistent way for hundreds of conflicts, an impossible undertaking.

Wars, in general, tend to be destructive in many ways at once, and the ones that kill more people on the battlefield also generally lead to more deaths from famine, disease, and the disruption of services. To the extent that they do, trends in battle deaths can serve as a proxy for trends in overall destructiveness. But they don't in every case, and later in the chapter we will ask whether developing nations, with their fragile infrastructure, are more vulnerable to knock-on effects than advanced nations, and whether this ratio has changed over time, making battle deaths a misleading index of trends in the human toll of conflict.

Now that we have the precision instrument of conflict datasets, what do they tell us about the recent trajectory of war in the entire world? Let's begin with the bird's-eye view of the 20th century in figure 6–1. The viewing was arranged by Lacina, Gleditsch, and Russett, who retrofitted numbers from the Correlates of War Project from 1900 to 1945 to the PRIO dataset from 1946 to 2005, and divided the numbers by the size of the world's population, to yield an individual's risk of dying in battle over the century.

The graph reminds us of the freakish destructiveness of the two world wars. They were not steps on a staircase, or swings of a pendulum, but massive spikes poking through a bumpy lowland. The drop-off in the rate of battle deaths after the early 1940s (peaking at 300 per 100,000 people per year) has been precipitous; the world has seen nothing close to that level since.

Eagle-eyed readers will spot a decline within the decline, from some small peaks in the immediate postwar decade to the low-lying flats of today. Let's zoom in on this trend in figure 6–2, while also subdividing the battle deaths according to the type of war that caused them.

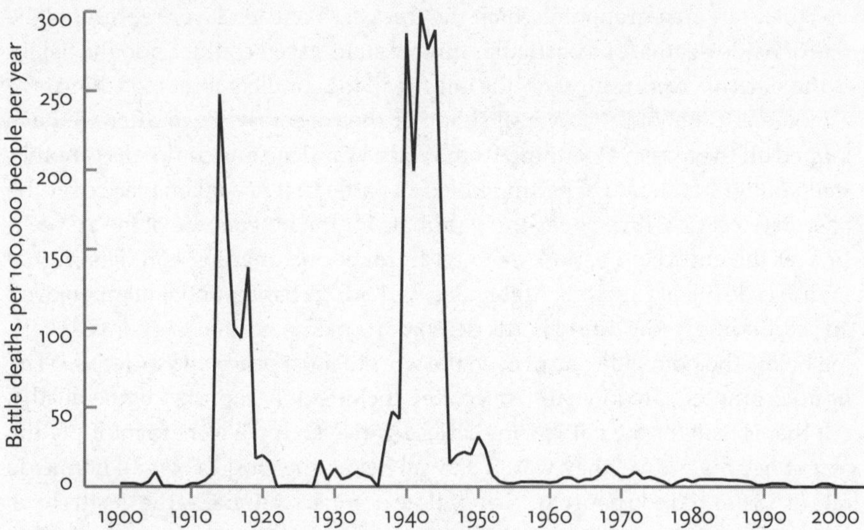

FIGURE 6–1. Rate of battle deaths in state-based armed conflicts, 1900–2005
Source: Graph from Russett, 2008, based on Lacina, Gleditsch, & Russett, 2006.

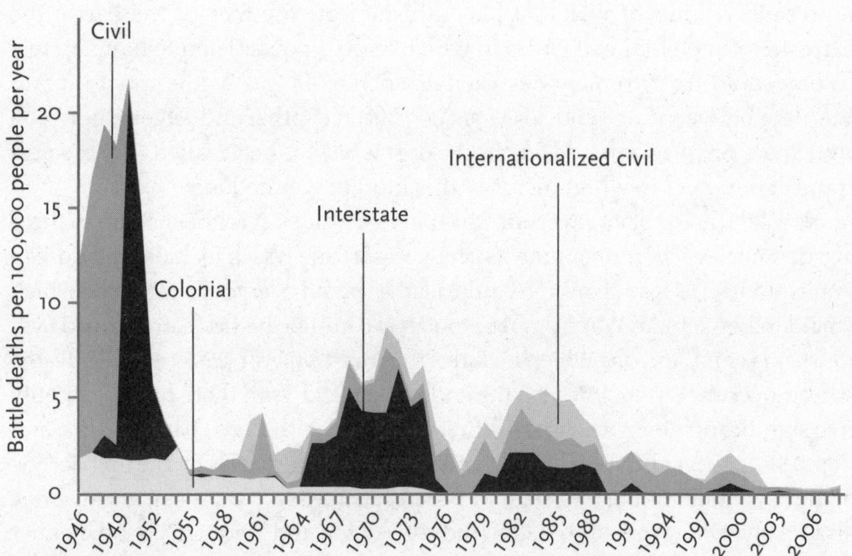

FIGURE 6–2. Rate of battle deaths in state-based armed conflicts, 1946–2008
Civilian and military battle deaths in state-based armed conflicts, divided by world population. *Sources:* UCDP/PRIO Armed Conflict Dataset; see Human Security Report Project, 2007, based on data from Lacina & Gleditsch, 2005, updated in 2010 by Tara Cooper. "Best" estimate used when available; otherwise the geometric mean of the "High" and "Low" estimates is used. World population figures from U.S. Census Bureau, 2010c. Population data for 1946–49 were taken from McEvedy & Jones, 1978, and multiplied by 1.01 to make them commensurable with the rest.

This is an area graph, in which the thickness of each layer represents the rate of battle deaths for a particular kind of state-based conflict, and the height of the stack of layers represents the rate for all the conflicts combined. First take a moment to behold the overall shape of the trajectory. Even after we have lopped off the massive ski-jump from World War II, no one could miss another steep falloff in the rate of getting killed in battle that has taken place over the past sixty years, with a paper-thin laminate for the first decade of the 21st century at the end. This period, even with thirty-one ongoing conflicts in that mid-decade (including Iraq, Afghanistan, Chad, Sri Lanka, and Sudan), enjoyed an astoundingly low rate of battle deaths: around 0.5 per 100,000 per year, falling below the homicide rate of even the world's most peaceable societies.[20] The figures, granted, are lowballs, since they include only reported battle deaths, but that is true for the entire time series. And even if we were to multiply the recent figures by five, they would sit well below the world's overall homicide rate of 8.8 per 100,000 per year.[21] In absolute numbers, annual battle deaths have fallen by more than 90 percent, from around half a million per year in the late 1940s to around thirty thousand a year in the early 2000s. So believe it or not, from a global, historical, and quantitative perspective, the dream of the 1960s folk songs has come true: the world has (almost) put an end to war.

Let's take our jaws off the table and look more closely at what happened category by category. We can start with the pale patch at the bottom left, which represents a kind of war that has vanished off the face of the earth: the extrastate or colonial war. Wars in which a great power tried to hang on to a colony could be extremely destructive, such as France's attempts to retain Vietnam between 1946 and 1954 (375,000 battle deaths) and Algeria between 1954 and 1962 (182,500 battle deaths).[22] After what has been called "the greatest transfer of power in world history," this kind of war no longer exists.

Now look at the black layer, for wars between states. It is bunched up in three large patches, each thinner than its predecessor: one which includes the Korean War from 1950 to 1953 (a million battle deaths spread over four years), one which includes the Vietnam War from 1962 to 1975 (1.6 million battle deaths spread over fourteen years), and one which includes the Iran-Iraq War (645,000 battle deaths spread over nine years).[23] Since the end of the Cold War, there have been only two significant interstate wars: the first Gulf War, with 23,000 battle deaths, and the 1998–2000 war between Eritrea and Ethiopia, with 50,000. By the first decade of the new millennium, interstate wars had become few in number, mostly brief, and relatively low in battle deaths (India-Pakistan and Eritrea-Djibouti, neither of which counts as a "war" in the technical sense of having a thousand deaths a year, and the quick overthrow of the regimes in Afghanistan and Iraq). In 2004, 2005, 2006, 2007, and 2009, there were no interstate conflicts at all.

The Long Peace—an avoidance of major war among great powers and developed states—is spreading to the rest of the world. Aspiring great powers no longer feel the need to establish their greatness by acquiring an empire or picking on weaker countries: China boasts of its "peaceful rise" and Turkey of a policy it calls "zero problems with neighbors"; Brazil's foreign minister recently crowed, "I don't think there are many countries that can boast that they have 10 neighbors and haven't had a war in the last 140 years."[24] And East Asia seems to be catching Europe's distaste for war. Though in the decades after World War II it was the world's bloodiest region, with ruinous wars in China, Korea, and Indochina, from 1980 to 1993 the number of conflicts and their toll in battle deaths plummeted, and they have remained at historically unprecedented lows ever since.[25]

As interstate war was being snuffed out, though, civil wars began to flare up. We see this in the enormous dark gray wedge at the left of figure 6–2, mainly representing the 1.2 million battle deaths in the 1946–50 Chinese Civil War, and a fat lighter gray bulge at the top of the stack in the 1980s, which contains the 435,000 battle deaths in the Soviet Union–bolstered civil war in Afghanistan. And snaking its way through the 1980s and 1990s, we find a continuation of the dark gray layer with a mass of smaller civil wars in countries such as Angola, Bosnia, Chechnya, Croatia, El Salvador, Ethiopia, Guatemala, Iraq, Liberia, Mozambique, Somalia, Sudan, Tajikistan, and Uganda. But even this slice tapers down in the 2000s to a slender layer.

To get a clearer picture of what the numbers here are telling us, it helps to disaggregate the death tolls into the two main dimensions of war: how many there were, and how lethal each kind was. Figure 6–3 shows the raw totals of

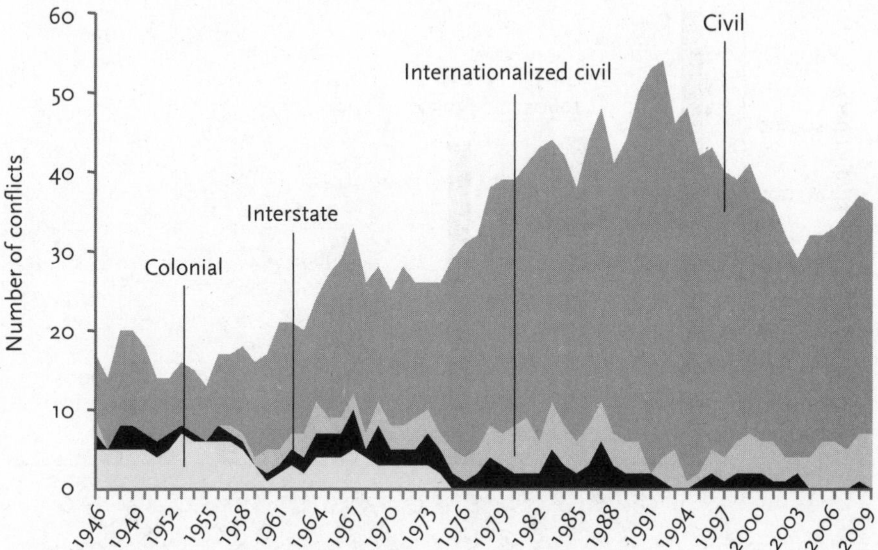

FIGURE 6–3. Number of state-based armed conflicts, 1946–2009

Sources: UCDP/PRIO Armed Conflict Dataset; see Human Security Report Project, 2007, based on data from Lacina & Gleditsch, 2005, updated in 2010 by Tara Cooper.

the conflicts of each kind, disregarding their death tolls, which, recall, can be as low as twenty-five. As colonial wars disappeared and interstate wars were petering out, internationalized civil wars vanished for a brief instant at the end of the Cold War, when the Soviet Union and the United States stopped supporting their client states, and then reappeared with the policing wars in Yugoslavia, Afghanistan, Iraq, and elsewhere. But the big news was an explosion in the number of purely internal civil wars that began around 1960, peaked in the early 1990s, and then declined through 2003, followed by a slight bounce.

Why do the sizes of the patches look so different in the two graphs? It's because of the power-law distribution for wars, in which a small number of wars in the tail of the L-shaped distribution are responsible for a large percentage of the deaths. More than half of the 9.4 million battle deaths in the 260 conflicts between 1946 and 2008 come from just five wars, three of them between states (Korea, Vietnam, Iran-Iraq) and two within states (China and Afghanistan). Most of the downward trend in the death toll came from reeling in that thick tail, leaving fewer of the really destructive wars.

In addition to the differences in the contributions of wars of different *sizes* to the overall death tolls, there are substantial differences in the contributions of the wars of different *kinds*. Figure 6–4 shows the second dimension of war, how many people an average war kills.

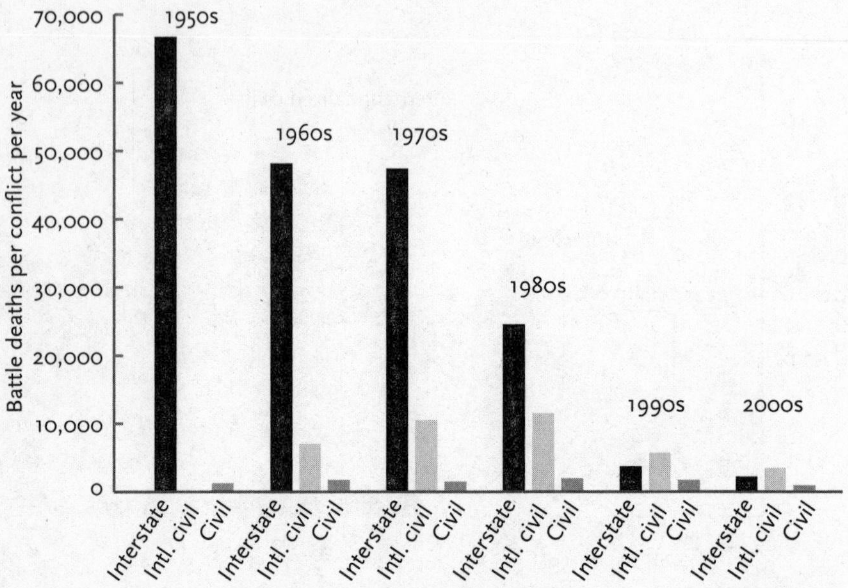

FIGURE 6–4. Deadliness of interstate and civil wars, 1950–2005

Sources: UCDP/PRIO Armed Conflict Dataset, Lacina & Gleditsch, 2005; adapted by the Human Security Report Project; Human Security Centre, 2006.

Until recently the most lethal kind of war *by far* was the interstate war. There is nothing like a pair of Leviathans amassing cannon fodder, lobbing artillery shells, and pulverizing each other's cities to rack up truly impressive body counts. A distant second and third are the wars in which a Leviathan projects its might in some other part of the world to prop up a beleaguered government or keep a grip on its colonies. Pulling up the rear are the internal civil wars, which, at least since the Chinese slaughterhouse in the late 1940s, have been far less deadly. When a gang of Kalashnikov-toting rebels harasses the government in a small country that the great powers don't care about, the damage they do is more limited. And even these fatality rates have decreased over the past quarter-century.[26] In 1950 the average armed conflict (of any kind) killed thirty-three thousand people; in 2007 it killed less than a thousand.[27]

How can we make sense of the juddering trajectory of conflict since the end of World War II, easing into the lull of the New Peace? One major change has been in the theater of armed conflict. Wars today take place mainly in poor countries, mostly in an arc that extends from Central and East Africa through the Middle East, across Southwest Asia and northern India, and down into Southeast Asia. Figure 6–5 shows ongoing conflicts in 2008 as black dots, and shades in the countries containing the "bottom billion," the people with the lowest income. About half of the conflicts take place in the countries with the poorest sixth of the people. In the decades before 2000, conflicts were scattered in other poor parts of the world as well, such as Central America and West Africa. Neither the economic nor the geographic linkage with war is a constant of history. Recall that for half a millennium the wealthy countries of Europe were constantly at each other's throats.

The relation between poverty and war in the world today is smooth but highly nonlinear. Among wealthy countries in the developed world, the risk of civil war is essentially zero. For countries with a per capita gross domestic product of around $1,500 a year (in 2003 U.S. dollars), the probability of a new conflict breaking out within five years rises to around 3 percent. But from there downward the risk shoots up: for countries with a per capita GDP of $750, it is 6 percent; for countries whose people earn $500, it is 8 percent; and for those that subsist on $250, it is 15 percent.[28]

A simplistic interpretation of the correlation is that poverty causes war because poor people have to fight for survival over a meager pool of resources. Though undoubtedly some conflicts are fought over access to water or arable land, the connection is far more tangled than that.[29] For starters, the causal arrow also goes in the other direction. War causes poverty, because it's hard to generate wealth when roads, factories, and granaries are blown up as fast as they are built and when the most skilled workers and managers are constantly being driven from their workplaces or shot. War has been called

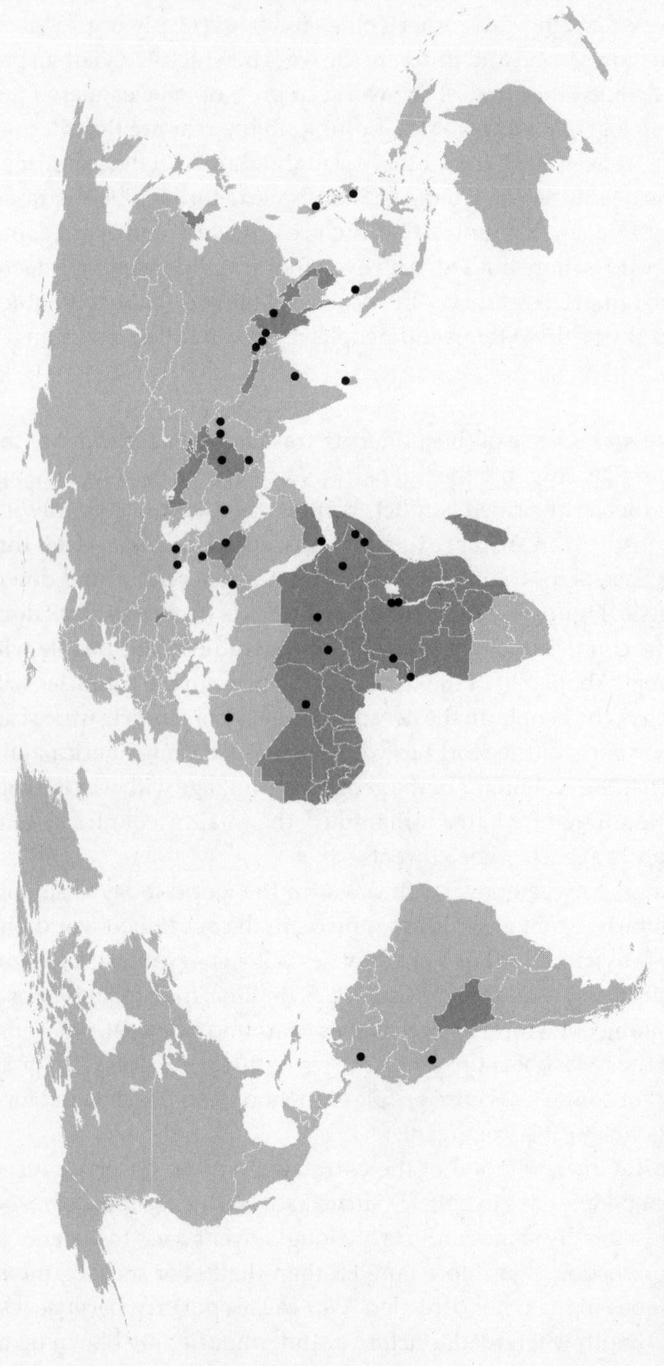

FIGURE 6–5. Geography of armed conflict, 2008

Countries in dark gray contain the "bottom billion" or the world's poorest people. Dots represent sites of armed conflict in 2008. *Sources:* Data from Håvard Strand and Andreas Forø Tollefsen, Peace Research Institute of Oslo (PRIO); adapted from a map by Halvard Buhaug and Siri Rustad in Gleditsch, 2008.

"development in reverse," and the economist Paul Collier has estimated that a typical civil war costs the afflicted country $50 billion.[30]

Also, neither wealth nor peace comes from having valuable stuff in the ground. Many poor and war-torn African countries are overflowing with gold, oil, diamonds, and strategic metals, while affluent and peaceable countries such as Belgium, Singapore, and Hong Kong have no natural resources to speak of. There must be a third variable, presumably the norms and skills of a civilized trading society, that causes both wealth and peace. And even if poverty does cause conflict, it may do so not because of competition over scarce resources but because the most important thing that a little wealth buys a country is an effective police force and army to keep domestic peace. The fruits of economic development flow far more to a government than to a guerrilla force, and that is one of the reasons that the economic tigers of the developing world have come to enjoy a state of relative tranquillity.[31]

Whatever effects poverty may have, measures of it and of other "structural variables," like the youth and maleness of a country's demographics, change too slowly to fully explain the recent rise and fall of civil war in the developing world.[32] Their effects, though, interact with the country's form of governance. The thickening of the civil war wedge in the 1960s had an obvious trigger: decolonization. European governments may have brutalized the natives when conquering a colony and putting down revolts, but they generally had a fairly well-functioning police, judiciary, and public-service infrastructure. And while they often had their pet ethnic groups, their main concern was controlling the colony as a whole, so they enforced law and order fairly broadly and in general did not let one group brutalize another with too much impunity. When the colonial governments departed, they took competent governance with them. A similar semianarchy burst out in parts of Central Asia and the Balkans in the 1990s, when the communist federations that had ruled them for decades suddenly unraveled. One Bosnian Croat explained why ethnic violence erupted only after the breakup of Yugoslavia: "We lived in peace and harmony because every hundred meters we had a policeman to make sure we loved each other very much."[33]

Many of the governments of the newly independent colonies were run by strongmen, kleptocrats, and the occasional psychotic. They left large parts of their countries in anarchy, inviting the predation and gang warfare we saw in Polly Wiessner's account of the decivilizing process in New Guinea in chapter 3. They siphoned tax revenue to themselves and their clans, and their autocracies left the frozen-out groups no hope for change except by coup or insurrection. They responded erratically to minor disorders, letting them build up and then sending death squads to brutalize entire villages, which only inflamed the opposition further.[34] Perhaps an emblem for the era was Jean-Bédel Bokassa of the Central African Empire, the name he gave to the small country formerly called the Central African Republic. Bokassa had seventeen wives, personally carved up (and according to rumors, occasionally ate) his political enemies, had schoolchildren beaten to death when they protested expensive mandatory uniforms bearing his likeness, and crowned himself emperor in a ceremony (complete with a gold throne and diamond-studded crown) that cost one of the world's poorest countries a third of its annual revenue.

During the Cold War many tyrants stayed in office with the blessing of the great powers, who followed the reasoning of Franklin Roosevelt about Nicaragua's Anastasio Somoza: "He may be a son of a bitch, but he's our son of a bitch."[35] The Soviet Union was sympathetic to any regime it saw as advancing the worldwide communist revolution, and the United States was sympathetic to any regime that kept itself out of the Soviet orbit. Other great powers such as France tried to stay on the good side of any regime that would supply them with oil and minerals. The autocrats were armed and financed by one superpower, insurrectionists who fought them were armed by the other, and both patrons were more interested in seeing their client win than in seeing the conflict come to an end. Figure 6–3 reveals a second expansion of civil wars around 1975, when Portugal dismantled its colonial empire and the American defeat in Vietnam emboldened insurrections elsewhere in the world. The number of civil wars peaked at fifty-one in 1991, which, not coincidentally, is the year the Soviet Union went out of existence, taking the Cold War–stoked proxy conflicts with it.

Only a fifth of the decline in conflicts, though, can be attributed to the disappearance of proxy wars.[36] The end of communism removed another source of fuel to world conflict: it was the last of the antihumanist, struggle-glorifying creeds in Luard's Age of Ideologies (we'll look at a new one, Islamism, later in this chapter). Ideologies, whether religious or political, push wars out along the tail of the deadliness distribution because they inflame leaders into trying to outlast their adversaries in destructive wars of attrition, regardless of the human costs. The three deadliest postwar conflicts were fueled by Chinese, Korean, and Vietnamese communist regimes that had a fanatical dedication to outlasting their opponents. Mao Zedong in particular was not embarrassed to say that the lives of his citizens meant nothing to him: "We have so many people. We can afford to lose a few. What difference does it make?"[37] On one occasion he quantified "a few"—300 million people, or half the country's population at the time. He also stated that he was willing to take an equivalent proportion of humanity with him in the cause: "If the worse came to the worst and half of mankind died, the other half would remain while imperialism would be razed to the ground and the whole world would become socialist."[38]

As for China's erstwhile comrades in Vietnam, much has been written, often by the chastened decision-makers themselves, about the American miscalculations in that war. The most fateful was their underestimation of the ability of the North Vietnamese and Vietcong to absorb casualties. As the war unfolded, American strategists like Dean Rusk and Robert McNamara were incredulous that a backward country like North Vietnam could resist the most powerful army on earth, and they were always confident that the next escalation would force it to capitulate. As John Mueller notes:

> If battle death rate as a percentage of pre-war population is calculated for each of the hundreds of countries that have participated in international and colonial wars since 1816, it is apparent that Vietnam was an extreme case. . . . The Communist side accepted battle death rates that were about twice as high as those accepted by the fanatical, often suicidal, Japanese in World War II, for example. Furthermore, the few combatant countries that did experience loss rates as high as that of the Vietnamese Communists were mainly those such as the Germans and Soviets in World War II, who were fighting to the death for their national existence, not for expansion like the North Vietnamese. In Vietnam, it seems, the United States was up against an incredibly well-functioning organization—patient, firmly disciplined, tenaciously led, and largely free from corruption or enervating self-indulgence. Although the communists often experienced massive military setbacks and periods of stress and exhaustion, they were always able to refit themselves, rearm, and come back for more. It may well be that, as one American general put it, "they were in fact the best enemy we have faced in our history."[39]

Ho Chi Minh was correct when he prophesied, "Kill ten of our men and we will kill one of yours. In the end, it is you who will tire." The American democracy was willing to sacrifice a tiny fraction of the lives that the North Vietnamese dictator was willing to forfeit (no one asked the proverbial ten men how they felt about this), and the United States eventually conceded the war of attrition despite having every other advantage. But by the 1980s, as China and Vietnam were changing from ideological to commercial states and easing their reigns of terror over their populations, they were less willing to inflict comparable losses in unnecessary wars.

A world that is less invigorated by honor, glory, and ideology and more tempted by the pleasures of bourgeois life is a world in which fewer people are killed. After Georgia lost a five-day war with Russia in 2008 over control of the tiny territories of Abkhazia and South Ossetia, Georgia's president Mikheil Saakashvili explained to a *New York Times* writer why he decided not to organize an insurgency against the occupation:

> We had a choice here. We could turn this country into Chechnya—we had enough people and equipment to do that—or we had to do nothing and stay a modern European country. Eventually we would have chased them away, but we would have had to go to the mountains and grow beards. That would have been a tremendous national philosophical and emotional burden.[40]

The explanation was melodramatic, even disingenuous—Russia had no intention of occupying Georgia—but it does capture one of the choices in the developing world that lies behind the New Peace: go to the mountains and grow beards, or do nothing and stay a modern country.

Other than the end of the Cold War and the decline of ideology, what led to the mild reduction in the number of civil wars during the past two decades, and the steep reduction in battle deaths of the last one? And why do conflicts persist in the developing world (thirty-six in 2008, all but one of them civil wars) when they have essentially disappeared in the developed world?

A good place to start is the Kantian triangle of democracy, open economies, and engagement with the international community. Russett and Oneal's statistical analyses, described in the preceding chapter, embrace the entire world, but they include only disputes between states. How well does the triad of pacifying factors apply to civil wars within developing countries, where most of today's conflicts take place? Each variable, it turns out, has an important twist.

One might think that if a lot of democracy is a good thing in inhibiting war, then a little democracy is still better than none. But with civil wars it doesn't work that way. Earlier in the chapter (and in chapter 3, when we examined homicide across the world), we came across the concept of anocracy, a form of rule that is neither fully democratic nor fully autocratic.[41] Anocracies are also known among political scientists as semidemocracies, praetorian regimes, and (my favorite, overheard at a conference) crappy governments. These are administrations that don't do anything well. Unlike autocratic police states, they don't intimidate their populations into quiescence, but nor do they have the more-or-less fair systems of law enforcement of a decent democracy. Instead they often respond to local crime with indiscriminate retaliation on entire communities. They retain the kleptocratic habits of the autocracies from which they evolved, doling out tax revenues and patronage jobs to their clansmen, who then extort bribes for police protection, favorable verdicts in court, or access to the endless permits needed to get anything done. A government job is the only ticket out of squalor, and having a clansman in power is the only ticket to a government job. When control of the government is periodically up for grabs in a "democratic election," the stakes are as high as in any contest over precious and indivisible spoils. Clans, tribes, and ethnic groups try to intimidate each other away from the ballot box and then fight to overturn an outcome that doesn't go their way. According to the *Global Report on Conflict, Governance, and State Fragility*, anocracies are "about six times more likely than democracies and two and one-half times as likely as autocracies to experience new outbreaks of societal wars" such as ethnic civil wars, revolutionary wars, and coups d'état.[42]

Figure 5–23 in the preceding chapter shows why the vulnerability of anocracies to violence has become a problem. As the number of autocracies in the world began to decline in the late 1980s, the number of anocracies began to increase. Currently they are distributed in a crescent from Central Africa through the Middle East and West and South Asia that largely coincides with the war zones in figure 6–5.[43]

The vulnerability to civil war of countries in which control of the government is a winner-take-all jackpot is multiplied when the government controls windfalls like oil, gold, diamonds, and strategic minerals. Far from being a blessing, these bonanzas create the so-called resource curse, also known as the paradox of plenty and fool's gold. Countries with an abundance of nonrenewable, easily monopolized resources have slower economic growth, crappier governments, and more violence. As the Venezuelan politician Juan Pérez Alfonzo put it, "Oil is the devil's excrement."[44] A country can be accursed by these resources because they concentrate power and wealth in the hands of whoever monopolizes them, typically a governing elite but sometimes a regional warlord. The leader becomes obsessed with fending off rivals for his cash cow and has no incentive to foster the networks of commerce that enrich a society and knit it together in reciprocal obligations. Collier, together with the economist Dambisa Moyo and other policy analysts, has called attention to a related paradox. Foreign aid, so beloved of crusading celebrities, can be another poisoned chalice, because it can enrich and empower the leaders through whom it is funneled rather than building a sustainable economic infrastructure. Expensive contraband like coca, opium, and diamonds is a third curse, because it opens a niche for cutthroat politicians or warlords to secure the illegal enclaves and distribution channels.

Collier observes that "the countries at the bottom coexist with the 21st century, but their reality is the 14th century: civil war, plague, ignorance."[45] The analogy to that calamitous century, which stood on the verge of the Civilizing Process before the consolidation of effective governments, is apt. In *The Remnants of War,* Mueller notes that most armed conflict in the world today no longer consists of campaigns for territory by professional armies. It consists instead of plunder, intimidation, revenge, and rape by gangs of unemployable young men serving warlords or local politicians, much like the dregs rounded up by medieval barons for their private wars. As Mueller puts it:

> Many of these wars have been labeled "new war," "ethnic conflict," or, most grandly, "clashes of civilizations." But in fact, most, though not all, are more nearly opportunistic predation by packs, often remarkably small ones, of criminals, bandits, and thugs. They engage in armed conflict either as mercenaries hired by desperate governments or as independent or semi-independent warlord or brigand bands. The damage perpetrated by these entrepreneurs of violence, who commonly apply ethnic, nationalist, civilizational, or religious rhetoric, can be extensive, particularly to the citizens who are their chief prey, but it is scarcely differentiable from crime.[46]

Mueller cites eyewitness reports that confirm that the infamous civil wars and genocides of the 1990s were largely perpetrated by gangs of drugged or drunken hooligans, including those in Bosnia, Colombia, Croatia, East Timor, Kosovo, Liberia, Rwanda, Sierra Leone, Somalia, Zimbabwe, and other countries in the African-Asian conflict crescent. Mueller describes some of the "soldiers" in the 1989–96 Liberian Civil War:

> Combatants routinely styled themselves after heroes in violent American action movies like *Rambo, Terminator,* and *Jungle Killer,* and many went under such fanciful noms de guerre as Colonel Action, Captain Mission Impossible, General Murder, Young Colonel Killer, General Jungle King, Colonel Evil Killer, General War Boss III, General Jesus, Major Trouble, General Butt Naked, and, of course, General Rambo. Particularly in the early years, rebels decked themselves out in bizarre, even lunatic attire: women's dresses, wigs, and pantyhose; decorations composed of human bones; painted fingernails; even (perhaps in only one case) headgear made of a flowery toilet seat.[47]

The political scientists James Fearon and David Laitin have backed up such vignettes with data confirming that civil wars today are fought by small numbers of lightly armed men who use their knowledge of the local landscape to elude national forces and intimidate informants and government sympathizers. These insurgencies and rural guerrilla wars may have any number of pretexts, but at heart they are less ethnic, religious, or ideological contests than turf battles between street gangs or Mafiosi. In a regression analysis of 122 civil wars between 1945 and 1999, Fearon and Laitin found that, holding per capita income constant (which they interpret as a proxy for government resources), civil wars were *not* more likely to break out in countries that were ethnically or religiously diverse, that had policies which discriminated against minority religions or languages, or that had high levels of income inequality. Civil wars were more likely to break out in countries that had large populations, mountainous terrain, new or unstable governments, significant oil exports, and (perhaps) a large proportion of young males. Fearon and Laitin conclude, "Our theoretical interpretation is more Hobbesian than economic. Where states are relatively weak and capricious, both fears and opportunities encourage the rise of local would-be rulers who supply a rough justice while arrogating the power to 'tax' for themselves and, often, a larger cause."[48]

Just as the uptick in civil warfare arose from the decivilizing anarchy of decolonization, the recent decline may reflect a recivilizing process in which competent governments have begun to protect and serve their citizens rather than preying on them.[49] Many African nations have traded in their Bokassa-style psychopaths for responsible democrats and, in the case of Nelson Mandela, one of history's greatest statesmen.[50]

The transition required an ideological change as well, not just in the affected countries but in the wider international community. The historian Gérard Prunier has noted that in 1960s Africa, independence from colonial rule became a messianic ideal. New nations made it a priority to adopt the trappings of sovereignty, such as airlines, palaces, and nationally branded institutions. Many were influenced by "dependency theorists" who advocated that third-world governments disengage from the global economy and cultivate self-sufficient industries and agrarian sectors, which most economists today consider a ticket to penury. Often economic nationalism was combined with a romantic militarism that glorified violent revolution, symbolized in two icons of the 1960s, the soft-color portrait of a glowing Mao and the hard-edged graphic of a dashing Che. When dictatorships by glorious revolutionaries lost their cachet, democratic elections became the new elixir. No one found much romance in the frumpy institutions of the Civilizing Process, namely a competent government and police force and a dependable infrastructure for trade and commerce. Yet history suggests that these institutions are necessary for the reduction of chronic violence, which is a prerequisite to every other social good.

During the past two decades the great powers, donor nations, and intergovernmental organizations (such as the African Union) have begun to press the point. They have ostracized, penalized, shamed, and in some cases invaded states that have come under the control of incompetent tyrants.[51] Measures to track and fight government corruption have become more common, as has the identification of barriers that penalize developing nations in global trade. Some combination of these unglamorous measures may have begun to reverse the governmental and social pathologies that had loosed civil wars on the developing world from the 1960s through the early 1990s.

Decent governments tend to be reasonably democratic and market-oriented, and several regression studies have looked at datasets on civil conflict for signs of a Liberal Peace like the one that helps explain the avoidance of wars between developed nations. We have already seen that the first leg of the peace, democracy, does not reduce the *number* of civil conflicts, particularly when it comes in the rickety form of an anocracy. But it does seem to reduce their *severity*. The political scientist Bethany Lacina has found that civil wars in democracies have fewer than half the battle deaths of civil wars in nondemocracies, holding the usual variables constant. In his 2008 survey of the Liberal Peace, Gleditsch concluded that "democracies rarely experience large-scale civil wars."[52] The second leg of the Liberal Peace is even stronger. Openness to the global economy, including trade, foreign investment, aid with strings attached, and access to electronic media, appears to drive down both the likelihood *and* the severity of civil conflict.[53]

The theory of the Kantian Peace places the weight of peace on three legs, the third of which is international organizations. One type of international organization in particular can claim much of the credit for driving down civil wars: international peacekeeping forces.[54] In the postcolonial decades civil wars piled up not so much because they broke out at an increasing rate but because they broke out at a higher rate than they ended (2.2 outbreaks a year compared to 1.8 terminations), and thus began to accumulate.[55] By 1999 an average civil war had been going on for fifteen years! That began to change in the late 1990s and 2000s, when civil wars started to fizzle out faster than new ones took their place. They also tended to end in negotiated settlements, without a clear victor, rather than being fought to the bitter end. Formerly these embers would smolder for a couple of years and then flare up again, but now they were more likely to die out for good.

This burst of peace coincides with a burst of peacekeepers. Figure 6–6 shows that beginning in the late 1980s the international community stepped up its peacekeeping operations and, more importantly, staffed them with increasing numbers of peacekeepers so they could do their job properly. The end of the Cold War was a turning point, because at last the great powers were more interested in seeing a conflict end than in seeing their proxy win.[56] The rise of peacekeeping is also a sign of the humanist times. War is increasingly seen as repugnant, and that includes wars that kill black and brown people.

Peacekeeping is one of the things that the United Nations, for all its foibles, does well. (It doesn't do so well at preventing wars in the first place.) In *Does Peacekeeping Work?* the political scientist Virginia Page Fortna answers the question in her title with "a clear and resounding yes."[57] Fortna assembled a

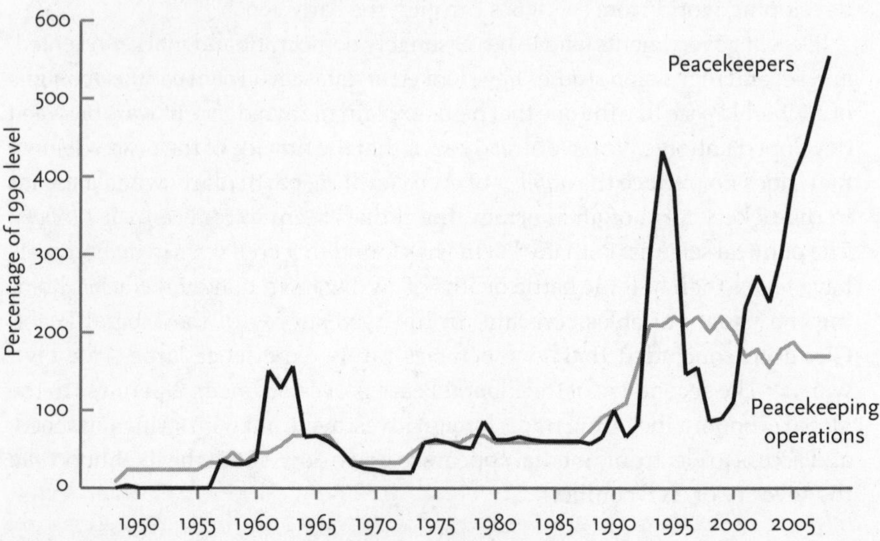

FIGURE 6–6. Growth of peacekeeping, 1948–2008

Source: Graph from Gleditsch, 2008, based on research by Siri Rustad.

dataset of 115 cease-fires in civil wars from 1944 to 1997 and examined whether the presence of a peacekeeping mission lowered the chances that the war would reignite. The dataset included missions by the UN, by permanent organizations such as NATO and the African Union, and by ad hoc coalitions of states. She found that the presence of peacekeepers reduced the risk of recidivism into another war by *80 percent*. This doesn't mean that peacekeeping missions are always successful—the genocides in Bosnia and Rwanda are two conspicuous failures—just that they prevent wars from restarting on average. Peacekeepers need not be substantial armies. Just as scrawny referees can pull apart brawling hockey players, lightly armed and even unarmed missions can get in between militias and induce them to lay down their weapons. And even when they don't succeed at that, they can serve as a tripwire for bringing in the bigger guns. Nor do peacekeepers have to be blue-helmeted soldiers. Functionaries who scrutinize elections, reform the police, monitor human rights, and oversee the functioning of bad governments also make a difference.

Why does peacekeeping work? The first reason comes right out of *Leviathan:* the larger and better-armed missions can retaliate directly against violators of a peace agreement on either side, raising the costs of aggression. The imposed costs and benefits can be reputational as well as material. A member of a mission commented on what led Afonso Dhlakama and his RENAMO rebel force to sign a peace agreement with the government of Mozambique: "For Dhlakama, it meant a great deal to be taken seriously, to go to cocktail parties and be treated with respect. Through the UN he got the government to stop calling RENAMO 'armed bandits.' It felt good to be wooed."[58]

Even small missions can be effective at keeping a peace because they can free the adversaries from a Hobbesian trap in which each side is tempted to attack out of fear of being attacked first. The very act of accepting intrusive peacekeepers is a costly (hence credible) signal that each side is serious about not attacking. Once the peacekeepers are in place, they can reinforce this security by monitoring compliance with the agreement, which allows them to credibly reassure each side that the other is not secretly rearming. They can also assume everyday policing activities, which deter the small acts of violence that can escalate into cycles of revenge. And they can identify the hotheads and spoilers who want to subvert the agreement. Even if a spoiler does launch a provocative attack, the peacekeepers can credibly reassure the target that it was a rogue act rather than the opening shot in a resumption of aggression.

Peacekeeping initiatives have other levers of influence. They can try to stamp out the trade in contraband that finances rebels and warlords, who are often the same people. They can dangle pork-barrel funding as an incentive to leaders who abide by the peace, enhancing their power and electoral popularity. As one Sierra Leonean said of a presidential candidate, "If Kabbah go, white man go, UN go, money go."[59] Also, since third-world soldiers (like premodern soldiers) are often paid in opportunities to plunder, the money can be applied to "demobilization, disarmament, and reintegration" programs that aim to draw General Butt Naked and his comrades back into civil society. With guerrillas who have more of an ideological agenda, the fact that the bribes come from a neutral party rather than a despised enemy allows them to feel they have not sold out. Leverage can also be applied to force political leaders to open their governments to rival political or ethnic groups. As with the financial sweeteners, the fact that the concessions are made to a neutral party rather than to the hated foe provides the conceder with an opportunity to save face. Desmond Malloy, a UN worker in Sierra Leone, observed that "peacekeepers create an atmosphere for negotiations. [Concessions] become a point of pride—it's a human trait. So you need a mechanism that allows negotiations without losing dignity and pride."[60]

For all these encouraging statistics, news readers who are familiar with the carnage in the Democratic Republic of the Congo, Iraq, Sudan, and other deathtraps may not be reassured. The PRIO/UCDP data we have been examining are limited in two ways. They include only state-based conflicts: wars in which at least one of the sides is a government. And they include only battle-related deaths: fatalities caused by battlefield weapons. What happens to the trends when we start looking for the keys that don't fall under these lampposts?

The first exclusion consists of the nonstate conflicts (also called intercommunal violence), in which warlords, militias, mafias, rebel groups, or paramilitaries, often affiliated with ethnic groups, go after each other. These conflicts usually occur in failed states, almost by definition. A war that doesn't even bother to invite the government represents the ultimate failure of the state's monopoly on violence.

The problem with nonstate conflicts is that until recently war buffs just weren't interested in them. No one kept track, so there's nothing to count, and we cannot plot the trends. Even the United Nations, whose mission is to prevent "the scourge of war," refuses to keep statistics on intercommunal violence (or on any other form of armed conflict), because its member states don't want social scientists poking around inside their borders and exposing the violence that their murderous governments cause or their inept governments fail to prevent.[61]

Nonetheless, a broad look at history suggests that nonstate conflicts today must be far fewer than they were in decades and centuries past, when less of the earth's surface was controlled by states. Tribal battles, slave raids, pillagings by raiders and horse tribes, pirate attacks, and private wars by noblemen and warlords, all of them nonstate, were scourges of humanity for millennia. During China's "warlord era" from 1916 to 1928, more than 900,000 people were killed by competing military chieftains in just a dozen years.[62]

It was only in 2002 that nonstate conflicts began to be tabulated. Since then the UCDP has maintained a Non-State Conflict Dataset, and it contains three revelations. First, nonstate conflicts are in some years as numerous as state-based conflicts—which says more about the scarcity of war than about the prevalence of intercommunal combat. Most of them, not surprisingly, are in sub-Saharan Africa, though a growing number are in the Middle East (most prominently, Iraq). Second, nonstate conflicts kill far fewer people than conflicts that involve a government, perhaps a quarter as many. Again, this is not surprising, since governments almost by definition are in the violence business. Third, the trend in the death toll from 2002 to 2008 (the most recent year covered in the dataset) has been mostly downward, despite 2007's being the deadliest year for intercommunal violence in Iraq.[63] So as best as anyone can tell, it seems unlikely that nonstate conflicts kill enough people to stand as a counterexample to the decline in the worldwide toll of armed conflict that constitutes the New Peace.

A more serious challenge is the number of indirect deaths of civilians from the hunger, disease, and lawlessness exacerbated by war. One often reads that a century ago only 10 percent of the deaths in war were suffered by civilians, but that today the figure is 90 percent. Consistent with this claim are new surveys by epidemiologists that reveal horrendous numbers of "excess deaths" (direct and indirect) among civilians. Rather than counting bodies from media reports and nongovernmental organizations, surveyors ask a sample of people whether they know someone who was killed, then extrapolate the proportion to the population as a whole. One of these surveys, published in the medical journal *Lancet* in 2006, estimated that 600,000 people died in the war in Iraq between 2003 and 2006—overwhelmingly more than the 80,000 to 90,000 battle deaths counted for that period by PRIO and by the Iraq Body Count, a respected nongovernmental organization.[64] Another survey in the Democratic Republic of the Congo put the death toll from its civil war at 5.4 million—about thirty-five times the PRIO battle-death estimate, and more than half of the total of *all* the battle deaths it has recorded in all wars since 1946.[65] Even granting that the PRIO figures are intended as lower bounds (because of the stringent requirements that deaths be attributed to a cause), this is quite a discrepancy, and raises doubts about whether, in the big picture, the decline in battle deaths can really be interpreted as an advance in peace.

Casualty figures are always moralized, and it's not surprising that these three numbers, which have been used to indict, respectively, the 20th century, Bush's invasion of Iraq, and the world's indifference to Africa, have been widely disseminated. But an objective look at the sources suggests that the revisionist estimates are not credible (which, needless to say, does not imply that anyone should be indifferent to civilian deaths in wartime).

First off, the commonly cited 10-percent-to-90-percent reversal in civilian casualties turns out to be completely bogus. The political scientists Andrew Mack (of HSRP), Joshua Goldstein, and Adam Roberts have each tried to track down the source of this meme, since they all knew that the data needed to underpin it do not exist.[66] They also knew that the claim fails basic sanity checks. For much of human history, peasants have subsisted on what they could grow, producing little in the way of a surplus. A horde of soldiers living off the land could easily tip a rural population into starvation. The Thirty Years' War in particular saw not only numerous massacres of civilians but the deliberate destruction of homes, crops, livestock, and water supplies, adding up to truly horrendous civilian death tolls. The American Civil War, with its blockades, crop-burnings, and scorched-earth campaigns, caused an enormous number of civilian casualties (the historical reality behind Scarlett O'Hara's vow in *Gone With the Wind:* "As God is my witness, I'll never be hungry again").[67] During World War I the battlefront moved through populated areas, raining artillery shells on towns and villages, and each side tried to starve the other's civilians with blockades. And as I have mentioned, if one includes the victims of the 1918 flu epidemic as indirect deaths from the war, one could multiply the number of civilian casualties many times over. World War II, also in the first half of the 20th century, decimated civilians with a holocaust, a blitz, *Slaughterhouse-Five*–like firebombings of cities in Germany and Japan, and not one but two atomic explosions. It seems unlikely that today's wars, however destructive to civilians, could be substantially worse.

Goldstein, Roberts, and Mack traced the meme to a chain of garbled retellings in which different kinds of casualty estimates were mashed up: battle deaths in one era were compared with battle deaths, indirect deaths, injuries, and refugees in another. Mack and Goldstein estimate that civilians suffer around half of the battle deaths in war, and that the ratio varies from war to war but has not increased over time. Indeed, we shall see that it has recently decreased by a substantial margin.

The most widely noted of the recent epidemiological estimates is the *Lancet* study of deaths in Iraq.[68] A team of eight Iraqi health workers went door to door in eighteen regions and asked people about recent deaths in the family. The epidemiologists subtracted the death rate for the years before the 2003 invasion from the death rate for the years after, figuring that the difference could be attributed to the war, and multiplied that proportion by the size of the population of Iraq. This arithmetic suggested that 655,000 more Iraqis died than if the invasion had never taken place. And 92 percent of these excess deaths, the families indicated, were direct battle deaths from gunshots, airstrikes, and car bombs, not indirect deaths from disease or starvation. If so, the standard body counts would be underestimates by a factor of around seven.

Without meticulous criteria for selecting a sample, though, extrapolations to an entire population can be wildly off. A team of statisticians led by Michael Spagat and Neil Johnson found these estimates incredible and discovered that a disproportionate number of the surveyed families lived on major streets and intersections—just the places where bombings and shootings are most likely.[69] An improved study conducted by the World Health Organization came up with a figure that was a quarter of the *Lancet* number, and even that required inflating an original estimate by a fudge factor of 35 percent to compensate for lying, moves, and memory lapses. Their unadjusted figure, around 110,000, is far closer to the battle-death body counts.[70]

Another team of epidemiologists extrapolated from retrospective surveys of war deaths in thirteen countries to challenge the entire conclusion that battle deaths have declined since the middle of the 20th century.[71] Spagat, Mack, and their collaborators have examined them and shown that the estimates are all over the map and are useless for tracking war deaths over time.[72]

What about the report of 5.4 million deaths (90 percent of them from disease and hunger) in the civil war in the Democratic Republic of the Congo?[73] It also turns out to be inflated. The International Rescue Committee (IRC) got the number by taking an estimate of the prewar death rate that was far too low (because it came from sub-Saharan Africa as a whole, which is better off than the DRC) and subtracting it from an estimate of the rate during the war that was far too high (because it came from areas where the IRC was providing humanitarian assistance, which are just the areas with the highest impact from war). The HSRP, while acknowledging that the indirect death toll in the DRC is high—probably over a million—cautions against accepting estimates of excess deaths from retrospective survey data, since in addition to all of their sampling pitfalls, they require dubious conjectures about what would have happened if a war had not taken place.[74]

Amazingly, the HSRP has collected evidence that death rates from disease and hunger have tended to go *down*, not up, during the wars of the past three decades.[75] It may sound like they are saying that war is healthy for children and other living things after all, but that is not their point. Instead, they document that deaths from malnutrition and hunger in the developing world have been dropping steadily over the years, and that the civil wars of today, which are fought by packs of insurgents in limited regions of a country, have not been destructive enough to reverse the tide. In fact, when medical and food assistance is rushed to a war zone, where it is often administered during humanitarian cease-fires, the progress can accelerate.

How is this possible? Many people are unaware of what UNICEF calls the Child Survival Revolution. (The revolution pertains to adult survival too, though children under five are the most vulnerable population and hence the ones most dramatically helped.) Humanitarian assistance has gotten smarter. Rather than just throwing money at a problem, aid organizations have adapted discoveries from the science of public health about which scourges kill the most people and which weapon against each one is the most cost-effective. Most childhood deaths in the developing world come from four causes: malaria; diarrheal diseases such as cholera and dysentery; respiratory infections such as pneumonia, influenza, and tuberculosis; and measles. Each is preventable or treatable, often remarkably cheaply. Mosquito nets, antimalarial drugs, antibiotics, water purifiers, oral rehydration therapy (a bit of salt and sugar in clean water), vaccinations, and breast-feeding (which reduces diarrheal and respiratory diseases) can save enormous numbers of lives. Over the last three decades, vaccination alone (which in 1974 protected just 5 percent of the world's children and today protects 75 percent) has saved 20 million lives.[76] Ready-to-use therapeutic foods like Plumpy'nut, a peanutbutterish goop in a foil package that children are said to like, can make a big dent in malnutrition and starvation.

Together these measures have slashed the human costs of war and belied the worry that an increase in indirect deaths has canceled or swamped the decrease in battle deaths. The HSRP estimates that during the Korean War about 4.5 percent of the population died from disease and starvation in every year of the four-year conflict. During the DRC civil war, even if we accept the overly pessimistic estimate of 5 million indirect deaths, it would amount to 1 percent of the country's population per year, a reduction of more than fourfold from Korea.[77]

It's not easy to see the bright side in the developing world, where the remnants of war continue to cause tremendous misery. The effort to whittle down the numbers that quantify the misery can seem heartless, especially when the numbers serve as propaganda for raising money and attention. But there is a moral imperative in getting the facts right, and not just to maintain credibility. The discovery that fewer people are dying in wars all over the world can thwart cynicism among compassion-fatigued news readers who might otherwise think that poor countries are irredeemable hellholes. And a better understanding of what drove the numbers down can steer us toward doing things that make people better off rather than congratulating ourselves on how altruistic we are. Among the surprises in the statistics are that some things that sound exciting, like instant independence, natural resources, revolutionary Marxism (when it is effective), and electoral democracy (when it is not) can increase deaths from violence, and some things that sound boring, like effective law enforcement, openness to the world economy, UN peacekeepers, and Plumpy'nut, can decrease them.

THE TRAJECTORY OF GENOCIDE

Of all the varieties of violence of which our sorry species is capable, genocide stands apart, not only as the most heinous but as the hardest to comprehend. We can readily understand why from time to time people enter into deadly quarrels over money, honor, or love, why they punish wrongdoers to excess, and why they take up arms to combat other people who have taken up arms. But that someone should want to slaughter millions of innocents, including women, children, and the elderly, seems to insult any claim we may have to comprehend our kind. Whether it is called genocide (killing people because of their race, religion, ethnicity, or other indelible group membership), politicide (killing people because of their political affiliation), or democide (any mass killing of civilians by a government or militia), killing-by-category targets people for what they *are* rather than what they *do* and thus seems to flout the usual motives of gain, fear, and vengeance.[78]

Genocide also shocks the imagination by the sheer number of its victims. Rummel, who was among the first historians to try to count them all, famously estimated that during the 20th century 169 million people were killed by their governments.[79] The number is, to be sure, a highball estimate, but most atrocitologists agree that in the 20th century more people were killed by democides than by wars.[80] Matthew White, in a comprehensive overview of the published estimates, reckons that 81 million people were killed by democide and another 40 million by man-made famines (mostly by Stalin and Mao), for a total of 121 million. Wars, in comparison, killed 37 million soldiers and 27 million civilians in battle, and another 18 million in the resulting famines, for a total of 82 million deaths.[81] (White adds, though, that about half of the democide deaths took place during wars and may not have been possible without them.)[82]

Killing so many people in so short a time requires methods of mass production of death that add another layer of horror. The Nazis' gas chambers and crematoria will stand forever as the most shocking visual symbols of genocide. But modern chemistry and railroads are by no means necessary for high-throughput killing. When the French revolutionaries suppressed a revolt in the Vendée region in 1793, they hit upon the idea of packing prisoners into barges, sinking them below the water's surface long enough to drown the human cargo, and then floating them up for the next batch.[83] Even during the Holocaust, the gas chambers were not the most efficient means of killing. The Nazis killed more people with their *Einsatzgruppen*, or mobile firing squads, which were foreshadowed by other teams of quick-moving soldiers with projectile weapons such as Assyrians in chariots and Mongols on horses.[84] During the genocide of Hutus by Tutsis in Burundi in 1972 (a predecessor of the reverse genocide in Rwanda twenty-two years later), a perpetrator explained:

> Several techniques, several, several. One can gather two thousand persons in a house—in a prison, let us say. There are some halls which are large. The house is locked. The men are left there for fifteen days without eating, without drinking. Then one opens. One finds cadavers. Not beaten, not anything. Dead.[85]

The bland military term "siege" hides the fact that depriving a city of food and finishing off the weakened survivors is a time-honored and cost-effective form of extermination. As Frank Chalk and Kurt Jonassohn point out in *The*

History and Sociology of Genocide, "The authors of history textbooks hardly ever reported what the razing of an ancient city meant for its inhabitants."[86] One exception is the Book of Deuteronomy, which offers a backdated prophecy that was based on the Assyrian or Babylonian conquest:

> In the desperate straits to which the enemy siege reduces you, you will eat the fruit of your womb, the flesh of your sons and daughters whom the LORD your God has given you. Even the most refined and gentle of men among you will begrudge food to his own brother, to the wife whom he embraces, and to the last of his remaining children, giving to none of them any of the flesh of his children whom he is eating, because nothing else remains to him, in the desperate straits to which the enemy siege will reduce you in all your towns. She who is the most refined and gentle among you, so gentle and refined that she does not venture to set the sole of her foot on the ground, will begrudge food to the husband whom she embraces, to her own son, and to her own daughter, begrudging even the afterbirth that comes out from between her thighs, and the children that she bears, because she will eat them in secret for lack of anything else, in the desperate straits to which the enemy siege will reduce you in your towns.[87]

Apart from numbers and methods, genocides sear the moral imagination by the gratuitous sadism indulged in by the perpetrators. Eyewitness accounts from every continent and decade recount how victims are taunted, tormented, and mutilated before being put to death.[88] In *The Brothers Karamazov,* Dostoevsky commented on Turkish atrocities in Bulgaria during the Russo-Turkish War of 1877–78, when unborn children were ripped from their mothers' wombs and prisoners were nailed by their ears to a fence overnight before being hanged: "People speak sometimes about the 'animal' cruelty of man, but that is terribly unjust and offensive to animals. No animal could ever be so cruel as a man, so artfully, so artistically cruel. A tiger simply gnaws and tears, that is all he can do. It would never occur to him to nail people by their ears overnight, even if he were able to do it."[89] My own reading of histories of genocide has left me with images to disturb sleep for a lifetime. I'll recount two that lodge in the mind not because of any gore (though such accounts are common enough) but because of their cold-bloodedness. Both are taken from the philosopher Jonathan Glover's *Humanity: A Moral History of the Twentieth Century.*

During the Chinese Cultural Revolution of 1966–75, Mao encouraged marauding Red Guards to terrorize "class enemies," including teachers, managers, and the descendants of landlords and "rich peasants," killing perhaps 7 million.[90] In one incident:

> Young men ransacking an old couple's house found boxes of precious French glass. When the old man begged them not to destroy the glass, one of the

group hit him in the mouth with a club, leaving him spitting out blood and teeth. The students smashed the glass and left the couple on their knees crying.[91]

During the Holocaust, Christian Wirth commanded a slave labor compound in Poland, where Jews were worked to death sorting the clothes of their murdered compatriots. Their children had been taken from them and sent to the death camps.

> Wirth allowed one exception. . . . One Jewish boy around ten was given sweets and dressed up as a little SS man. Wirth and he rode among the prisoners, Wirth on a white horse and the boy on a pony, both using machine-guns to kill prisoners (including the boy's mother) at close range.[92]

Glover allows himself a comment: "To this ultimate expression of contempt and mockery, no reaction of disgust and anger is remotely adequate."

How could people do these things? Making sense of killing-by-category, insofar as we can do so at all, must begin with the psychology of categories.[93]

People sort other people into mental pigeonholes according to their affiliations, customs, appearances, and beliefs. Though it's tempting to think of this stereotyping as a kind of mental defect, categorization is indispensable to intelligence. Categories allow us to make inferences from a few observed qualities to a larger number of unobserved ones. If I note the color and shape of a fruit and classify it as a raspberry, I can infer that it will taste sweet, satisfy my hunger, and not poison me. Politically correct sensibilities may bridle at the suggestion that a group of people, like a variety of fruit, may have features in common, but if they didn't, there would be no cultural diversity to celebrate and no ethnic qualities to be proud of. Groups of people cohere because they really do share traits, albeit statistically. So a mind that generalizes about people from their category membership is not ipso facto defective. African Americans today really are more likely to be on welfare than whites, Jews really do have higher average incomes than WASPs, and business students really are more politically conservative than students in the arts—on average.[94]

The problem with categorization is that it often goes beyond the statistics. For one thing, when people are pressured, distracted, or in an emotional state, they forget that a category is an approximation and act as if a stereotype applies to every last man, woman, and child.[95] For another, people tend to *moralize* their categories, assigning praiseworthy traits to their allies and condemnable ones to their enemies. During World War II, for example, Americans thought that Russians had more positive traits than Germans; during the Cold War they thought it was the other way around.[96] Finally, people tend to *essentialize* groups. As children, they tell experimenters that a baby whose parents have been switched at birth will speak the language of her biological rather than her adoptive parents. As they get older, people tend to think that members of particular ethnic and religious groups share a quasi-biological essence, which makes them homogeneous, unchangeable, predictable, and distinct from other groups.[97]

The cognitive habit of treating people as instances of a category gets truly dangerous when people come into conflict. It turns Hobbes's trio of violent motives—gain, fear, and deterrence—from the bones of contention in an individual quarrel to the casus belli in an ethnic war. Historical surveys have shown that genocides are caused by this triad of motives, with, as we shall see, two additional toxins spiked into the brew.[98]

Some genocides begin as matters of convenience. Natives are occupying a desirable territory or are monopolizing a source of water, food, or minerals, and invaders would rather have it for themselves. Eliminating the people is like clearing brush or exterminating pests, and is enabled by nothing fancier in our psychology than the fact that human sympathy can be turned on or off depending on how another person is categorized. Many genocides of indigenous peoples are little more than expedient grabs of land or slaves, with the victims typed as less than human. Such genocides include the numerous expulsions and massacres of Native Americans by settlers or governments in the Americas, the brutalization of African tribes by King Leopold of Belgium in the Congo Free State, the extermination of the Herero by German colonists in South-West Africa, and the attacks on Darfuris by government-encouraged Janjaweed militias in the 2000s.[99]

When conquerors find it expedient to suffer the natives to live so that they can provide tribute and taxes, genocide can have a second down-to-earth function. A reputation for a willingness to commit genocide comes in handy for a conqueror because it allows him to present a city with an ultimatum to surrender or else. To make the threat credible, the invader has to be prepared to carry it out. This was the rationale behind the annihilation of the cities of western Asia by Genghis Khan and his Mongol hordes.

Once the conquerors have absorbed a city or territory into an empire, they may keep it in line with the threat that they will come down on any revolt like a ton of bricks. In 68 CE the governor of Alexandria called in Roman troops to put down a rebellion by the Jews against Roman rule. According to the historian Flavius Josephus, "Once [the Jews] were forced back, they were unmercifully and completely destroyed. Some were caught in the open field, others forced into their houses, which were plundered and then set on fire. The Romans showed no mercy to the infants, had no regard for the aged, and went on in the slaughter of persons of every age, until all the place was overflowed with blood, and 50,000 Jews lay dead."[100] Similar tactics have been used in 20th-century counterinsurgency campaigns, such as the ones by the Soviets in Afghanistan and right-wing military governments in Indonesia and Central America.

When a dehumanized people is in a position to defend itself or turn the tables, it can set a Hobbesian trap of group-against-group fear. Either side may see the other as an existential threat that must be preemptively taken out. After the breakup of Yugoslavia in the 1990s, Serbian nationalists' genocide of Bosnians and Kosovars was partly fueled by fears that they would be the victims of massacres themselves.[101]

If members of a group have seen their comrades victimized, have narrowly escaped victimization themselves, or paranoically worry they have been targeted for victimization, they may stoke themselves into a moralistic fury and seek vengeance on their perceived assailants. Like all forms of revenge, a retaliatory massacre is pointless once it has to be carried out, but a well-advertised and implacable *drive* to carry it out, regardless of its costs at the time, may have been programmed into people's brains by evolution, cultural norms, or both as a way to make the deterrent credible.

These Hobbesian motives don't fully explain why predation, preemption, or revenge should be directed against entire *groups* of people rather than the individuals who get in the way or make trouble. The cognitive habit of pigeonholing may be one reason, and another is explained in *The Godfather: Part II* when the young Vito Corleone's mother begs a Sicilian don to spare the boy's life:

> *Widow:* Don Francesco. You murdered my husband, because he would not bend. And his oldest son Paolo, because he swore revenge. But Vitone is only nine, and dumb-witted. He never speaks.
>
> *Francesco:* I'm not afraid of his words.
>
> *Widow:* He is weak.
>
> *Francesco:* He will grow strong.
>
> *Widow:* The child cannot harm you.
>
> *Francesco:* He will be a man, and then he will come for revenge.

And come for revenge he does. Later in the film the grown Vito returns to Sicily, seeks an audience with the don, whispers his name into the old man's ear, and cuts him open like a sturgeon.

The solidarity among the members of a family, clan, or tribe—in particular, their resolve to avenge killings—makes them all fair game for someone with a bone to pick with any one of them. Though equal-sized groups in frequent contact tend to constrain their revenge to an-eye-for-an-eye reciprocity, repeated violations may turn episodic anger into chronic hatred. As Aristotle wrote, "The angry man wishes the object of his anger to suffer in return; hatred wishes its object not to exist."[102] When one side finds itself with an advantage in numbers or tactics, it may seize the opportunity to impose a final solution. Feuding tribes are well aware of genocide's practical advantages. The anthropologist Rafael Karsten worked with the Jivaro of Amazonian Ecuador (a tribe that contributed one of the long bars to the graph of rates of death in warfare in figure 2–2) and recounts their ways of war:

> Whereas the small feuds within the sub-tribes have the character of a private blood-revenge, based on the principle of just retaliation, the wars between the different tribes are in principle wars of extermination. In these there is no question of weighing life against life; the aim is to completely annihilate the enemy tribe. . . . The victorious party is all the more anxious to leave no single person of the enemy's people, not even small children, alive, as they fear lest these should later appear as avengers against the victors.[103]

Half a world away, the anthropologist Margaret Durham offered a similar vignette from an Albanian tribe that ordinarily abided by norms for measured revenge:

> In February 1912 an amazing case of wholesale justice was reported to me. . . . A certain family of the Fandi bairak [subtribe] had long been notorious for evil-doing—robbing, shooting, and being a pest to the tribe. A gathering of all the heads condemned all the males of the family to death. Men were appointed to lay in wait for them on a certain day and pick them off; and on that day the whole seventeen of them were shot. One was but five and another but twelve years old. I protested against thus killing children who must be innocent and was told: "It was bad blood and must not be further propagated." Such was the belief in heredity that it was proposed to kill an unfortunate woman who was pregnant, lest she should bear a male and so renew the evil.[104]

The essentialist notion of "bad blood" is one of several biological metaphors inspired by a fear of the revenge of the cradle. People anticipate that if they leave even a few of a defeated enemy alive, the remnants will multiply and cause trouble down the line. Human cognition often works by analogy, and the concept of an irksome collection of procreating beings repeatedly calls to mind the concept of vermin.[105] Perpetrators of genocide the world over keep rediscovering the same metaphors to the point of cliché. Despised people are rats, snakes, maggots, lice, flies, parasites, cockroaches, or (in parts of the world where they are pests) monkeys, baboons, and dogs.[106] "Kill the nits and you will have no lice," wrote an English commander in Ireland in 1641, justifying an order to kill thousands of Irish Catholics.[107] "A nit would make a louse," recalled a Californian settler leader in 1856 before slaying 240 Yuki in revenge for their killing of a horse.[108] "Nits make lice," said Colonel John Chivington before the Sand Creek Massacre, which killed hundreds of Cheyenne and Arapaho in 1864.[109] Cankers, cancers, bacilli, and viruses are other insidious biological agents that lend themselves as figures of speech in the poetics of genocide. When it came to the Jews, Hitler mixed his metaphors, but they were always biological: Jews were viruses; Jews were bloodsucking parasites; Jews were a mongrel race; Jews had poisonous blood.[110]

The human mind has evolved a defense against contamination by biological agents: the emotion of disgust.[111] Ordinarily triggered by bodily secretions, animal parts, parasitic insects and worms, and vectors of disease, disgust impels people to eject the polluting substance and anything that looks like it or has been in contact with it. Disgust is easily moralized, defining a continuum in which one pole is identified with spirituality, purity, chastity, and cleansing and the other with animality, defilement, carnality, and contamination.[112] And so we see disgusting agents as not just physically repellent but also morally contemptible. Many metaphors in the English language for a treacherous person use a disease vector as their vehicle—*a rat, a louse, a worm, a cockroach.* The infamous 1990s term for forced displacement and genocide was *ethnic cleansing.*

Metaphorical thinking goes in both directions. Not only do we apply disgust metaphors to morally devalued peoples, but we tend to morally devalue people who are physically disgusting (a phenomenon we encountered in chapter 4 when considering Lynn Hunt's theory that a rise in hygiene in Europe caused a decline in cruel punishments). At one pole of the continuum, white-clad ascetics who undergo rituals of purification are revered as holy men and women. At the other, people living in degradation and filth are reviled as subhuman. The chemist and writer Primo Levi described this spiral during the transport of Jews to the death camps in Germany:

> The SS escort did not hide their amusement at the sight of men and women squatting wherever they could, on the platforms and in the middle of the tracks, and the German passengers openly expressed their disgust: people like this deserve their fate, just look how they behave. These are not *Menschen,* human beings, but animals, it's clear as the light of day.[113]

The emotional pathways to genocide—anger, fear, and disgust—can occur in various combinations. In *Worse than War,* a history of 20th-century genocide, the political scientist Daniel Goldhagen points out that not all genocides have the same causes. He classifies them according to whether the victim group is *dehumanized* (a target of moralized disgust), *demonized* (a target of moralized anger), both, or neither.[114] A dehumanized group may be exterminated like vermin, such as the Hereros in the eyes of German colonists, Armenians in the eyes of Turks, black Darfuris in the eyes of Sudanese Muslims, and many indigenous peoples in the eyes of European settlers. A demonized group, in contrast, is thought to be equipped with the standard human reasoning faculties, which makes them all the more culpable for embracing a heresy or rejecting the one true faith. Among these modern heretics were the victims of communist autocracies, and the victims of their opposite number, the right-wing dictatorships in Chile, Argentina, Indonesia, and El Salvador. Then there are the out-and-out demons—groups that manage to be both repulsively sub-human *and* despicably evil. This is how the Nazis saw the Jews, and how Hutus and Tutsis saw each other. Finally, there may be groups that are not reviled as evil or subhuman but are feared as potential predators and eliminated in preemptive attacks, such as in the Balkan anarchy following the breakup of Yugoslavia.

So far I have tried to explain genocide in the following way. The mind's habit of essentialism can lump people into categories; its moral emotions can be applied to them in their entirety. The combination can transform Hobbesian competition among individuals or armies into Hobbesian competition among peoples. But genocide has another fateful component. As Solzhenitsyn pointed out, to kill by the millions you need an *ideology*.[115] Utopian creeds that sub-merge individuals into moralized categories may take root in powerful regimes and engage their full destructive might. For this reason it is ideologies that generate the outliers in the distribution of genocide death tolls. Divisive ideologies include Christianity during the Crusades and the Wars of Religion (and in an offshoot, the Taiping Rebellion in China); revolutionary romanti-cism during the politicides of the French Revolution; nationalism during the genocides in Ottoman Turkey and the Balkans; Nazism in the Holocaust; and Marxism during the purges, expulsions, and terror-famines in Stalin's Soviet Union, Mao's China, and Pol Pot's Cambodia.

Why should utopian ideologies so often lead to genocide? At first glance it seems to make no sense. Even if an actual utopia is unattainable for all kinds of practical reasons, shouldn't the quest for a perfect world at least leave us with a better one—a world that is 60 percent of the way to perfection, say, or even 15 percent? After all, a man's reach must exceed his grasp. Shouldn't we aim high, dream the impossible dream, imagine things that never were and ask "why not"?

Utopian ideologies invite genocide for two reasons. One is that they set up a pernicious utilitarian calculus. In a utopia, everyone is happy forever, so its moral value is infinite. Most of us agree that it is ethically permissible to divert a runaway trolley that threatens to kill five people onto a side track where it would kill only one. But suppose it were a hundred million lives one could save by diverting the trolley, or a billion, or—projecting into the indefinite future—infinitely many. How many people would it be permissible to sacrifice to attain that infinite good? A few million can seem like a pretty good bargain.

Not only that, but consider the people who learn about the promise of a perfect world yet nonetheless oppose it. They are the only things standing in the way of a plan that could lead to infinite goodness. How evil are they? You do the math.

The second genocidal hazard of a utopia is that it has to conform to a tidy blueprint. In a utopia, everything is there for a reason. What about the people? Well, groups of people are diverse. Some of them stubbornly, perhaps essentially, cling to values that are out of place in a perfect world. They may be entrepreneurial in a world that works by communal sharing, or bookish in a world that works by labor, or brash in a world that works by piety, or clannish in a world that works by unity, or urban and commercial in a world that has returned to its roots in nature. If you are designing the perfect society on a clean sheet of paper, why not write these eyesores out of the plans from the start?

In *Blood and Soil: A World History of Genocide and Extermination from Sparta to Darfur,* the historian Ben Kiernan notes another curious feature of utopian ideologies. Time and again they hark back to a vanished agrarian paradise, which they seek to restore as a healthful substitute for prevailing urban decadence. In chapter 4 we saw that after the Enlightenment had emerged from the intellectual bazaar of cosmopolitan cities, the German counter-Enlightenment romanticized the attachment of a people to their land—the blood and soil of Kiernan's title. The ungovernable metropolis, with its fluid population and ethnic and occupational enclaves, is an affront to a mindset that envisions a world of harmony, purity, and organic wholeness. Many of the nationalisms of the 19th and early 20th centuries were guided by utopian images of ethnic groups flourishing in their native homelands, often based on myths of ancestral tribes who settled the territory at the dawn of time.[116] This agrarian utopianism lay behind Hitler's dual obsessions: his loathing of Jewry, which he associated with commerce and cities, and his deranged plan to depopulate Eastern Europe to provide farmland for German city-dwellers to colonize. Mao's massive agrarian communes and Pol Pot's expulsion of Cambodian city-dwellers to rural killing fields are other examples.

Commercial activities, which tend to be concentrated in cities, can themselves be triggers of moralistic hatred. As we shall see in chapter 9, people's intuitive sense of economics is rooted in tit-for-tat exchanges of concrete goods or services of equivalent value—say, three chickens for one knife. It does not easily grasp the abstract mathematical apparatus of a modern economy, such as money, profit, interest, and rent.[117] In intuitive economics, farmers and craftsmen produce palpable items of value. Merchants and other middlemen, who skim off a profit as they pass goods along without causing new stuff to come into being, are seen as parasites, despite the value they create by enabling transactions between producers and consumers who are unacquainted or separated by distance. Moneylenders, who loan out a sum and then demand additional money in return, are held in even greater contempt, despite the service they render by providing people with money at times in their lives when it can be put to the best use. People tend to be oblivious to the intangible contributions of merchants and moneylenders and view them as bloodsuckers. (Once again the metaphor comes from biology.) Antipathy toward individual middlemen can easily transfer to antipathy to ethnic groups. The capital necessary to prosper in middlemen occupations consists mainly of expertise rather than land or factories, so it is easily shared among kin and friends, and it is highly portable. For these reasons it's common for particular ethnic groups to specialize in the middleman niche and to move to whatever communities currently lack them, where they tend to become prosperous minorities—and targets of envy and resentment.[118] Many victims of discrimination, expulsion, riots, and genocide have been social or ethnic groups that specialize in middlemen niches. They include various bourgeois minorities in the Soviet Union, China, and Cambodia, the Indians in East Africa and Oceania, the Ibos in Nigeria, the Armenians in Turkey, the Chinese in Indonesia, Malaysia, and Vietnam, and the Jews in Europe.[119]

Democides are often scripted into the climax of an eschatological narrative, a final spasm of violence that will usher in millennial bliss. The parallels between the utopian ideologies of the 19th and 20th centuries and the apocalyptic visions of traditional religions have often been noticed by historians of genocide. Daniel Chirot, writing with the social psychologist Clark McCauley, observes:

> Marxist eschatology actually mimicked Christian doctrine. In the beginning, there was a perfect world with no private property, no classes, no exploitation, and no alienation—the Garden of Eden. Then came sin, the discovery of private property, and the creation of exploiters. Humanity was cast from the Garden to suffer inequality and want. Humans then experimented with a series of modes of production, from the slave, to the feudal, to the capitalist mode, always seeking the solution and not finding it. Finally there came a true prophet with a message of salvation, Karl Marx, who preached the truth of Science. He promised redemption but was not heeded, except by his close disciples who carried the truth forward. Eventually, however, the proletariat, the carriers of the true faith, will be converted by the religious elect, the leaders of the party, and join to create a more perfect world. A final, terrible revolution will wipe out capitalism, alienation, exploitation, and inequality. After that, history will end because there will be perfection on earth, and the true believers will have been saved.[120]

Drawing on the work of the historians Joachim Fest and George Mosse, they also comment on Nazi eschatology:

> It was not an accident that Hitler promised a Thousand Year Reich, a millennium of perfection, similar to the thousand-year reign of goodness promised in Revelation before the return of evil, the great battle between good and evil, and the final triumph of God over Satan. The entire imagery of his Nazi Party and regime was deeply mystical, suffused with religious, often

Christian, liturgical symbolism, and it appealed to a higher law, to a mission decreed by fate and entrusted to the prophet Hitler.[121]

Finally, there are the job requirements. Would you want the stress and responsibility of running a perfect world? Utopian leadership selects for monumental narcissism and ruthlessness.[122] Its leaders are possessed of a certainty about the rectitude of their cause and an impatience for incremental reforms or on-the-fly adjustments guided by feedback from the human consequences of their grand schemes. Mao, who had his image plastered all over China and his little red book of sayings issued to every citizen, was described by his doctor and only confidant Li Zhisui as voracious for flattery, demanding of sexual servicing by concubines, and devoid of warmth and compassion.[123] In 1958 he had a revelation that the country could double its steel production in a year if peasant families contributed to the national output by running backyard smelters. On pain of death for failing to meet the quotas, peasants melted down their woks, knives, shovels, and doorknobs into lumps of useless metal. It was also revealed to him that China could grow large quantities of grain on small plots of land, freeing the rest for grasslands and gardens, if farmers planted the seedlings deep and close together so that class solidarity would make them grow strong and thick.[124] Peasants were herded into communes of 50,000 to implement this vision, and anyone who dragged his feet or pointed out the obvious was executed as a class enemy. Impervious to signals from reality informing him that his Great Leap Forward was a great leap backward, Mao masterminded a famine that killed between 20 million and 30 million people.

The motives of leaders are critical in understanding genocide, because the psychological ingredients—the mindset of essentialism; the Hobbesian dynamic of greed, fear, and vengeance; the moralization of emotions like disgust; and the appeal of utopian ideologies—do not overcome an entire population at once and incite them to mass killing. Groups that avoid, distrust, or even despise each other can coexist without genocide indefinitely.[125] Think, for example, of African Americans in the segregated American South, Palestinians in Israel and the occupied territories, and Africans in South Africa under apartheid. Even in Nazi Germany, where anti-Semitism had been entrenched for centuries, there is no indication that anyone but Hitler and a few fanatical henchmen thought it was a good idea for the Jews to be exterminated.[126] When a genocide *is* carried out, only a fraction of the population, usually a police force, military unit, or militia, actually commits the murders.[127]

In the 1st century CE, Tacitus wrote, "A shocking crime was committed on the unscrupulous initiative of a few individuals, with the blessing of more, and amid the passive acquiescence of all." According to the political scientist Benjamin Valentino in *Final Solutions,* that division of labor applies to the genocides of the 20th century as well.[128] A leader or small clique decides that the time for genocide is right. He gives the go-ahead to a relatively small force of armed men, made up a mixture of true believers, conformists, and thugs (often recruited, as in medieval armies, from the ranks of criminals, drifters, and other unemployable young men). They count on the rest of the population not to get in their way, and thanks to features of social psychology that we will explore in chapter 8, they generally don't. The psychological contributors to genocide, such as essentialism, moralization, and utopian ideologies, are engaged to different degrees in each of these constituencies. They consume the minds of the leaders and the true believers but have to tip the others only enough to allow the leaders to make their plans a reality. The indispensability of leaders to 20th-century genocide is made plain by the fact that when the leaders died or were removed by force, the killings stopped.[129]

If this analysis is on the right track, genocides can emerge from toxic reactions among human nature (including essentialism, moralization, and intuitive economics), Hobbesian security dilemmas, millennial ideologies, and the opportunities available to leaders. The question now is: how has this interaction changed over the course of history?

It's not an easy question to answer, because historians have never found genocide particularly interesting. Since antiquity the stacks of libraries have been filled with scholarship on war, but scholarship on genocide is nearly nonexistent, though it killed more people. As Chalk and Jonassohn point out of ancient histories, "We know that empires have disappeared and that cities were destroyed, and we suspect that some wars were genocidal in their results; but we do not know what happened to the bulk of the populations involved in these events. Their fate was simply too unimportant. When they were mentioned at all, they were usually lumped together with the herds of oxen, sheep, and other livestock."[130]

As soon as one realizes that the sackings, razings, and massacres of past centuries are what we would call genocide today, it becomes utterly clear that genocide is not a phenomenon of the 20th century. Those familiar with classical history know that the Athenians destroyed Melos during the 5th-century-BCE Peloponnesian War; according to Thucydides, "the Athenians thereupon put to death all who were of military age and made slaves of the women and children." Another familiar example is the Romans' destruction of Carthage and its population during the Third Punic War in the 3rd century BCE, a war so total that the Romans, it was said, sowed salt into the ground to make it forever unfarmable. Other historical genocides include the real-life bloodbaths that inspired the ones narrated in the *Iliad,* the *Odyssey,* and the Hebrew Bible; the massacres and sackings during the Crusades; the suppression of the Albigensian heresy; the Mongol invasions; the European witch hunts; and the carnage of the European Wars of Religion.

The authors of recent histories of mass killing are adamant that the idea of an unprecedented "century of genocide" (the 20th) is a myth. On their first page Chalk and Jonassohn write, "Genocide has been practiced in all regions of the world and during all periods in history," and add that their eleven case studies of pre-20th-century genocides "are not intended to be either exhaustive or representative."[131] Kiernan agrees: "A major conclusion of this book is that genocide indeed occurred commonly before the twentieth century." One can see what he means with a glance at the first page of his table of contents:

Part One: Early Imperial Expansion

1. Classical Genocide and Early Modern Memory
2. The Spanish Conquest of the New World 1492–1600
3. Guns and Genocide in East Asia 1400–1600
4. Genocidal Massacres in Early Modern Southeast Asia

Part Two. Settler Colonialism

5. The English Conquest of Ireland, 1565–1603
6. Colonial North America, 1600–1776
7. Genocidal Violence in Nineteenth-Century Australia
8. Genocide in the United States
9. Settler Genocides in Africa, 1830–1910[132]

Rummel has fitted a number to his own conclusion that "the mass murder by emperors, kings, sultans, khans, presidents, governors, generals, and other rulers of their own citizens or of those under their protection or control is very much part of our history." He counts 133,147,000 victims of sixteen democides before the 20th century (including ones in India, Iran, the Ottoman Empire, Japan, and Russia) and surmises that there may have been 625,716,000 democide victims in all.[133]

These authors did not compile their lists by indiscriminately piling up every historical episode in which a lot of people died. They are careful to note, for example, that the Native American population was decimated by disease rather than by a program of extermination, while particular incidents *were* blatantly genocidal. In an early example, Puritans in New England exterminated the Pequot nation in 1638, after which the minister Increase Mather asked his congregation to thank God "that on this day we have sent six hundred heathen souls to Hell."[134] This celebration of genocide did not hurt his career. He later became president of Harvard University, and the residential house with which I am currently affiliated is named after him (motto: Increase Mather's Spirit!).

Mather was neither the first nor the last to thank God for genocide. As we saw in chapter 1, Yahweh ordered the Hebrew tribes to carry out dozens of them, and in the 9th century BCE the Moabites returned the favor by massacring the inhabitants of several Hebrew cities in the name of *their* god,

Ashtar-Chemosh.[135] In a passage from the Bhagavad-Gita (written around 400 CE), the Hindu god Krishna upbraids the mortal Arjuna for being reluctant to slay an enemy faction that included his grandfather and tutor: "There is no better engagement for you than fighting on religious principles; and so there is no need for hesitation. . . . The soul can never be cut to pieces by any weapon, nor burned by fire. . . . [Therefore] you are mourning for what is not worthy of grief."[136] Inspired by the conquests of Joshua, Oliver Cromwell massacred every man, woman, and child in an Irish town during the reconquest of Ireland, and explained his actions to Parliament: "It has pleased God to bless our endeavour at Drogheda. The enemy were about 3,000 strong in the town. I believe we put to the sword the whole number."[137] The English Parliament passed a unanimous motion "that the House does approve of the execution done at Drogheda as an act of both justice to them and mercy to others who may be warned of it."[138]

The shocking truth is that until recently most people didn't think there was anything particularly wrong with genocide, as long as it didn't happen to them. One exception was the 16th-century Spanish priest Antonio de Montesinos, who protested the appalling treatment of Native Americans by the Spanish in the Caribbean—and who was, in his own words, "a voice of one crying in the wilderness."[139] There were, to be sure, military codes of honor, some from the Middle Ages, that ineffectually attempted to outlaw the killing of civilians in war, and occasional protests by thinkers of early modernity such as Erasmus and Hugo Grotius. But only in the late 19th century, when citizens began to protest the brutalization of peoples in the American West and the British Empire, did objections to genocide become common.[140] Even then we find Theodore Roosevelt, the future "progressive" president and Nobel Peace laureate, writing in 1886, "I don't go so far as to think that the only good Indians are the dead Indians, but I believe nine out of ten are, and I shouldn't like to inquire too closely in the case of the tenth."[141] The critic John Carey documents that well into the 20th century the British literary intelligentsia viciously dehumanized the teeming masses, whom they considered to be so vulgar and soulless as not to have lives worth living. Genocidal fantasies were not uncommon. In 1908, for example, D. H. Lawrence wrote:

> If I had my way, I would build a lethal chamber as big as the Crystal Palace, with a military band playing softly, and a Cinematograph working brightly; then I'd go out in the back streets and main streets and bring them in, all the sick, the halt, and the maimed; I would lead them gently, and they would smile me a weary thanks; and the band would softly bubble out the "Hallelujah Chorus."[142]

During World War II, when Americans were asked in opinion polls what should be done with the Japanese after an American victory, 10 to 15 percent volunteered the solution of extermination.[143]

The turning point came after the war. The English language did not even have a word for genocide until 1944, when the Polish lawyer Raphael Lemkin coined it in a report on Nazi rule in Europe that would be used a year later to brief the prosecutors at the Nuremberg Trials.[144] In the aftermath of the Nazi destruction of European Jewry, the world was stunned by the enormity of the death toll and by horrific images from the liberated camps: assembly-line gas chambers and crematoria, mountains of shoes and eyeglasses, bodies stacked up like cordwood. In 1948 Lemkin got the UN to approve a Convention on the Prevention and Punishment of the Crime of Genocide, and for the first time in history genocide, regardless of who the victims were, was a crime. James Payne notes a perverse sign of progress. Today's Holocaust deniers at least feel compelled to deny that the Holocaust took place. In earlier centuries the perpetrators of genocide and their sympathizers boasted about it.[145]

No small part in the new awareness of the horrors of genocide was a willingness of Holocaust survivors to tell their stories. Chalk and Jonassohn note that these memoirs are historically unusual.[146] Survivors of earlier genocides had treated them as humiliating defeats and felt that talking about them would only rub in history's harsh verdict. With the new humanitarian sensibilities, genocides became crimes against humanity, and survivors were witnesses for the prosecution. Anne Frank's diary, which recorded her life in hiding in Nazi-occupied Amsterdam before she was deported to her death in Bergen-Belsen, was published by her father shortly after the war. Memoirs of deportations and death camps by Elie Wiesel and Primo Levi were published in the 1960s, and today Frank's *Diary* and Wiesel's *Night* are among the world's most widely read books. In the years that followed, Aleksandr Solzhenitsyn, Anchee Min, and Dith Pran shared their harrowing memories of the communist nightmares in the Soviet Union, China, and Cambodia. Soon other survivors—Armenians, Ukrainians, Gypsies—began to add their stories, joined more recently by Bosnians, Tutsis, and Darfuris. These memoirs are a part of a reorientation of our conception of history. "Throughout most of history," Chalk and Jonassohn note, "only the rulers made news; in the twentieth century, for the first time, it is the ruled who make the news."[147]

Anyone who grew up with Holocaust survivors knows what they had to overcome to tell their stories. For decades after the war they treated their experiences as shameful secrets. On top of the ignominy of victimhood, the desperate straits to which they were reduced could remove the last traces of their humanity in ways they could be forgiven for wanting to forget. At a family occasion in the 1990s, I met a relative by marriage who had spent time in Auschwitz. Within seconds of meeting me he clenched my wrist and recounted this story. A group of men had been eating in silence when one of them slumped over dead. The others fell on his body, still covered in diarrhea, and pried a piece of bread from his fingers. As they divided it, a fierce argument broke out when some of the men felt their share was an imperceptible crumb smaller than the others'. To tell a story of such degradation requires extraordinary courage, backed by a confidence that the hearer will understand it as an accounting of the circumstances and not of the men's characters.

Though the abundance of genocides over the millennia belies the century-of-genocide claim, one still wonders about the trajectory of genocide before, during, and since the 20th century. Rummel was the first political scientist to try to put some numbers together. In his duology *Death by Government* (1994) and *Statistics of Democide* (1997) he analyzed 141 regimes that committed democides in the 20th century through 1987, and a control group of 73 that did not. He collected as many independent estimates of the death tolls as he could find (including ones from pro- and antigovernment sources, whose biases, he assumed, would cancel each other out) and, with the help of sanity checks, chose a defensible value near the middle of the range.[148] His definition of "democide" corresponds roughly to the UCDP's "one-sided violence" and to our everyday concept of "murder" but with a government rather than an individual as the perpetrator: the victims must be unarmed, and the killing deliberate. Democides thus include ethnocides, politicides, purges, terrors, killings of civilians by death squads (including ones committed by private militias to which the government turns a blind eye), deliberate famines from blockades and confiscation of food, deaths in internment camps, and the targeted bombing of civilians such as those in Dresden, Hamburg, Hiroshima, and Nagasaki.[149] Rummel excluded the Great Leap Forward from his 1994 analyses, on the understanding that it was caused by stupidity and callousness rather than malice.[150]

Partly because the phrase "death by government" figured in Rummel's definition of democide and in the title of his book, his conclusion that almost 170 million people were killed by their governments during the 20th century has become a popular meme among anarchists and radical libertarians. But for several reasons, "governments are the main cause of preventable deaths" is not the correct lesson to draw from Rummel's data. For one thing, his definition of "government" is loose, embracing militias, paramilitaries, and warlords, all of which could reasonably be seen as a sign of too little government rather than too much. White examined Rummel's raw data and calculated that the median democide toll by the twenty-four pseudo-governments on his list was around 100,000, whereas the median death toll caused by recognized governments of sovereign states was 33,000. So one could, with more justification, conclude that governments, on average, cause three times *fewer* deaths than alternatives to government.[151] Also, most governments in recent periods do not commit democides at all, and they prevent a far greater number of deaths than the democidal ones cause, by promoting vaccination, sanitation, traffic safety, and policing.[152]

But the main problem with the anarchist interpretation is that it isn't governments in general that kill large numbers of people but a handful of governments of a specific type. To be exact, three-quarters of all the deaths from all 141 democidal regimes were committed by just four governments, which Rummel calls the dekamegamurderers: the Soviet Union with 62 million, the People's Republic of China with 35 million, Nazi Germany with 21 million, and 1928–49 nationalist China with 10 million.[153] Another 11 percent of the total were killed by eleven megamurderers, including Imperial Japan with 6 million, Cambodia with 2 million, and Ottoman Turkey with 1.9 million. The remaining 13 percent of the deaths were spread out over 126 regimes. Genocides don't exactly fall into a power-law distribution, if for no other reason than that the smaller massacres that would go into the tall spine tend not to be counted as "genocides." But the distribution is enormously lopsided, conforming to an 80:4 rule—80 percent of the deaths were caused by 4 percent of the regimes.

Also, deaths from democide were overwhelmingly caused by *totalitarian* governments: the communist, Nazi, fascist, militarist, or Islamist regimes that sought to control every aspect of the societies they ruled. Totalitarian regimes were responsible for 138 million deaths, 82 percent of the total, of which 110 million (65 percent of the total) were caused by the communist regimes.[154] Authoritarian regimes, which are autocracies that tolerate independent social institutions such as businesses and churches, came in second with 28 million deaths. Democracies, which Rummel defines as governments that are open, competitive, elected, and limited in their power, killed 2 million (mainly in their colonial empires, together with food blockades and civilian bombings during the world wars). The skew of the distribution does not just reflect the sheer number of potential victims that totalitarian behemoths like the Soviet Union and China had at their disposal. When Rummel looked at percentages rather than numbers, he found that totalitarian governments of the 20th century racked up a death toll adding up to 4 percent of their populations. Authoritarian governments killed 1 percent. Democracies killed four tenths of 1 percent.[155]

Rummel was one of the first advocates of the Democratic Peace theory, which he argues applied to democides even more than to wars. "At the extremes of Power," Rummel writes, "totalitarian communist governments slaughter their people by the tens of *millions;* in contrast, many democracies can barely bring themselves to execute even serial murderers."[156] Democracies commit fewer democides because their form of governance, by definition, is committed to inclusive and nonviolent means of resolving conflicts. More important, the power of a democratic government is restricted by a tangle of institutional restraints, so a leader can't just mobilize armies and militias on a whim to fan out over the country and start killing massive numbers of citizens. By performing a set of regressions on his dataset of 20th-century regimes, Rummel showed that the occurrence of democide correlates with a lack of democracy, even holding constant the countries' ethnic diversity, wealth, level of development, population density, and culture (African, Asian, Latin American, Muslim, Anglo, and so on).[157] The lessons, he writes, are clear: "The problem is Power. The solution is democracy. The course of action is to foster freedom."[158]

What about the historical trajectory? Rummel tried to break down his 20th-century democides by year, and I've reproduced his data, scaled by world population, in the gray upper line in figure 6–7. Like deaths in wars, deaths in democides were concentrated in a savage burst, the midcentury Hemoclysm.[159] This blood-flood embraced the Nazi Holocaust, Stalin's purges, the Japanese rape of China and Korea, and the wartime firebombings of cities in Europe and Japan. The left slope also includes the Armenian genocide during World War I and the Soviet collectivization campaign, which killed millions of Ukrainians and kulaks, the so-called rich peasants. The right slope embraces the killing of millions of ethnic Germans in newly communized Poland, Czechoslovakia, and Romania, and the victims of forced collectivization in China. It's uncomfortable to say that there's anything good in the trends shown in the graph, but in an important sense there is. The world has seen nothing close

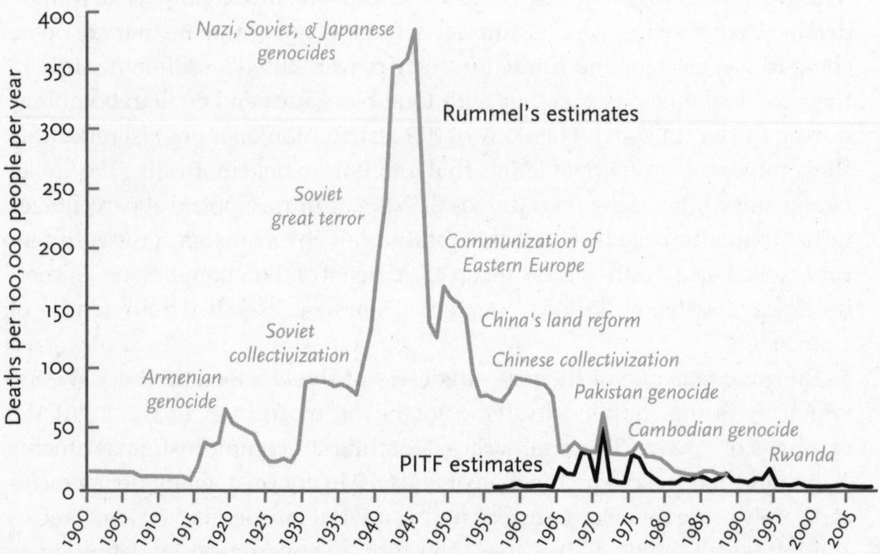

FIGURE 6–7. Rate of deaths in genocides, 1900–2008

Sources: Data for the gray line, 1900–1987, from Rummel, 1997. Data for the black line, 1955–2008, from the Political Instability Task Force (PITF) State Failure Problem Set, 1955–2008, Marshall, Gurr, & Harff, 2009; Center for Systemic Peace, 2010. The death tolls for the latter were geometric means of the ranges in table 8.1 in Harff, 2005, distributed across years according to the proportions in the Excel database. World population figures from U.S. Census Bureau, 2010c. Population figures for the years 1900–1949 were taken from McEvedy & Jones, 1978, and multiplied by 1.01 to make them commensurable with the rest.

to the bloodletting of the 1940s since then; in the four decades that followed, the rate (and number) of deaths from democide went precipitously, if lurchingly, downward. (The smaller bulges represent killings by Pakistani forces during the Bangladesh war of independence in 1971 and by the Khmer Rouge in Cambodia in the late 1970s.) Rummel attributes the falloff in democide since World War II to the decline of totalitarianism and the rise of democracy.[160]

Rummel's dataset ends in 1987, just when things start to get interesting again. Soon communism fell and democracies proliferated—and the world was hit with the unpleasant surprise of genocides in Bosnia and Rwanda. In the impression of many observers, these "new wars" show that we are still living, despite all we should have learned, in an age of genocide.

The historical thread of genocide statistics has recently been extended by the political scientist Barbara Harff. During the Rwanda genocide, some 700,000 Tutsis were killed in just four months by about 10,000 men with machetes, many of them drunkards, addicts, ragpickers, and gang members hastily recruited by the Hutu leadership.[161] Many observers believe that this small pack of *génocidaires* could easily have been stopped by a military intervention by the world's great powers.[162] Bill Clinton in particular was haunted by his own failure to act, and in 1998 he commissioned Harff to analyze the risk factors and warning signs of genocide.[163] She assembled a dataset of 41 genocides and politicides between 1955 (shortly after Stalin died and the process of decolonization began) and 2004. Her criteria were more restrictive than Rummel's and closer to Lemkin's original definition of genocide: episodes of violence in which a state or armed authority intends to destroy, in whole or in part, an identifiable group. Only five of the episodes turned out to be "genocide" in the sense in which people ordinarily understand the term, namely an ethnocide, in which a group is singled out for destruction because of its ethnicity. Most were politicides, or politicides combined with ethnocides, in which members of an ethnic group were thought to be aligned with a targeted political faction.

In figure 6–7 I plotted Harff's PITF data on the same axes with Rummel's. Her figures generally come in well below his, especially in the late 1950s, for which she included far fewer victims of the executions during the Great Leap Forward. But thereafter the curves show similar trends, which are downward from their peak in 1971. Because the genocides from the second half of the 20th century were so much less destructive than those of the Hemoclysm, I've zoomed in on her curve in figure 6–8. The graph also shows the death rates in a third collection, the UCDP One-Sided Violence Dataset, which includes any instance of a government or other armed authority killing at least twenty-five civilians in a year; the perpetrators need not intend to destroy the group per se.[164]

The graph shows that the two decades since the Cold War have *not* seen a recrudescence of genocide. On the contrary, the peak in mass killing (putting

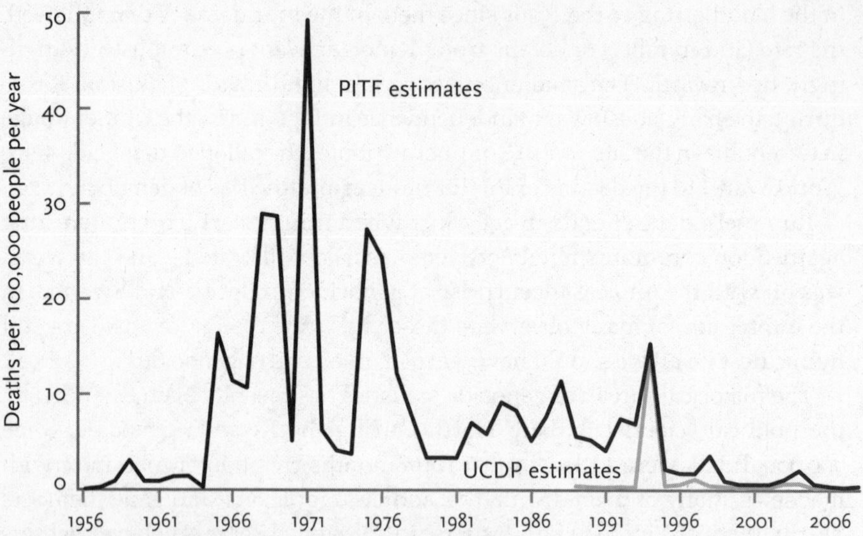

FIGURE 6–8. Rate of deaths in genocides, 1956–2008

Sources: PITF estimates, 1955–2008: same as for figure 6–7. UCDP, 1989–2007: "High Fatality" estimates from http://www.pcr.uu.se/research/ucdp/datasets/ (Kreutz, 2008; Kristine & Hultman, 2007) divided by world population from U.S. Census Bureau, 2010c.

aside China in the 1950s) is located in the mid-1960s to late 1970s. Those fifteen years saw a politicide against communists in Indonesia (1965–66, "the year of living dangerously," with 700,000 deaths), the Chinese Cultural Revolution (1966–75, around 600,000), Tutsis against Hutus in Burundi (1965–73, 140,000), Pakistan's massacre in Bangladesh (1971, around 1.7 million), north-against-south violence in Sudan (1956–72, around 500,000), Idi Amin's regime in Uganda (1972–79, around 150,000), the Cambodian madness (1975–79, 2.5 million), and a decade of massacres in Vietnam culminating in the expulsion of the boat people (1965–75, around half a million).[165] The two decades since the end of the Cold War have been marked by genocides in Bosnia from 1992 to 1995 (225,000 deaths), Rwanda (700,000 deaths), and Darfur (373,000 deaths from 2003 to 2008). These are atrocious numbers, but as the graph shows, they are spikes in a trend that is unmistakably downward. (Recent studies have shown that even some of these figures may be overestimates, but I will stick with the datasets.)[166] The first decade of the new millennium is the most genocide-free of the past fifty years. The UCDP numbers are restricted to a narrower time window and, like all their estimates, are more conservative, but they show a similar pattern: the Rwanda genocide in 1994 leaps out from all the other episodes of one-sided killing, and the world has seen nothing like it since.

Harff was tasked not just with compiling genocides but with identifying their risk factors. She noted that virtually all of them took place in the aftermath of a state failure such as a civil war, revolution, or coup. So she assembled a control group with 93 cases of state failure that did *not* result in genocide, matched as closely as possible to the ones that did, and ran a logistic regression analysis to find out which aspects of the situation the year before made the difference.

Some factors that one might think were important turned out not to be. Measures of ethnic diversity didn't matter, refuting the conventional wisdom that genocides represent the eruption of ancient hatreds that inevitably explode when ethnic groups live side by side. Nor did measures of economic development matter. Poor countries are more likely to have political crises, which are necessary conditions for genocides to take place, but among the countries that did have crises, the poorer ones were no more likely to sink into actual genocide.

Harff did discover six risk factors that distinguished the genocidal from the nongenocidal crises in three-quarters of the cases.[167] One was a country's previous history of genocide, presumably because whatever risk factors were in place the first time did not vanish overnight. The second predictor was the country's immediate history of political instability—to be exact, the number of regime crises and ethnic or revolutionary wars it had suffered in the preceding fifteen years. Governments that feel threatened are tempted to eliminate or take revenge on groups they perceive to be subversive or contaminating, and are more likely to exploit the ongoing chaos to accomplish those goals before opposition can mobilize.[168] A third was a ruling elite that came from an ethnic minority, presumably because that multiplies the leaders' worries about the precariousness of their rule.

The other three predictors are familiar from the theory of the Liberal Peace. Harff vindicated Rummel's insistence that democracy is a key factor in preventing genocides. From 1955 to 2008 autocracies were three and a half times more likely to commit genocides than were full or partial democracies, holding everything else constant. This represents a hat trick for democracy: democracies are less likely to wage interstate wars, to have large-scale civil wars, and to commit genocides. Partial democracies (anocracies) are more likely than autocracies to have violent political crises, as we saw in Fearon and Laitin's analysis of civil wars, but when a crisis does occur, the partial democracies are less likely than autocracies to become genocidal.

Another trifecta was scored by openness to trade. Countries that depend more on international trade, Harff found, are less likely to commit genocides, just as they are less likely to fight wars with other countries and to be riven by civil wars. The inoculating effects of trade against genocide cannot depend, as they do in the case of interstate war, on the positive-sum benefits of trade itself, since the trade we are talking about (imports and exports) does not consist in exchanges with the vulnerable ethnic or political groups. Why, then, should trade matter? One possibility is that Country A might take a communal or moral interest in a group living within the borders of Country B. If B wants to trade with A, it must resist the temptation to exterminate that group. Another is that a desire to engage in trade requires certain peaceable attitudes, including a willingness to abide by international norms and the rule of law, and a mission to enhance the material welfare of its citizens rather than implementing a vision of purity, glory, or perfect justice.

The last predictor of genocide is an exclusionary ideology. Ruling elites that are under the spell of a vision that identifies a certain group as the obstacle to an ideal society, putting it "outside the sanctioned universe of obligation," are far more likely to commit genocide than elites with a more pragmatic or eclectic governing philosophy. Exclusionary ideologies, in Harff's classification, include Marxism, Islamism (in particular, a strict application of Sharia law), militaristic anticommunism, and forms of nationalism that demonize ethnic or religious rivals.

Harff sums up the pathways by which these risk factors erupt into genocide:

> Almost all genocides and politicides of the last half-century were either ideological, exemplified by the Cambodian case, or retributive, as in Iraq [Saddam Hussein's 1988–91 campaign against Iraqi Kurds]. The scenario that leads to *ideological genocide* begins when a new elite comes to power, usually through civil war or revolution, with a transforming vision of a new society purified of unwanted or threatening elements. *Retributive geno-politicides* occur during a protracted internal war . . . when one party, usually the government, seeks to destroy its opponent's support base [or] after a rebel challenge has been militarily defeated.[169]

The decline of genocide over the last third of a century, then, may be traced to the upswing of some of the same factors that drove down interstate and civil wars: stable government, democracy, openness to trade, and humanistic ruling philosophies that elevate the interests of individuals over struggles among groups.

For all the rigor that a logistic regression offers, it is essentially a meat grinder that takes a set of variables as input and extrudes a probability as output. What it hides is the vastly skewed distribution of the human costs of different genocides—the way that a small number of men, under the sway of a smaller number of ideologies, took actions at particular moments in history that caused outsize numbers of deaths. Shifts in the levels of the risk factors certainly pushed around the likelihood of the genocides that racked up thousands, tens of thousands, and even hundreds of thousands of deaths. But the truly monstrous genocides, the ones with tens of millions of victims, depended not so much on gradually shifting political forces as on a few contingent ideas and events.

The appearance of Marxist ideology in particular was a historical tsunami that is breathtaking in its total human impact. It led to the dekamegamurders by Marxist regimes in the Soviet Union and China, and more circuitously, it contributed to the one committed by the Nazi regime in Germany. Hitler read Marx in 1913, and although he detested Marxist socialism, his National Socialism substituted races for classes in its ideology of a dialectical struggle toward utopia, which is why some historians consider the two ideologies "fraternal twins."[170] Marxism also set off reactions that led to politicides by militantly anticommunist regimes in Indonesia and Latin America, and to the destructive civil wars of the 1960s, 1970s, and 1980s stoked by the Cold War superpowers. The point is not that Marxism should be morally blamed for these unintended consequences, just that any historical narrative must acknowledge the sweeping repercussions of this single idea. Valentino notes that no small part of the decline of genocide is the decline of *communism*, and thus "the single most important cause of mass killing in the twentieth century appears to be fading into history."[171] Nor is it likely that it will come back into fashion. During its heyday, violence by Marxist regimes was justified with the saying "You can't make an omelet without breaking eggs."[172] The historian Richard Pipes summarized history's verdict: "Aside from the fact that human beings are not eggs, the trouble is that no omelet has emerged from the slaughter."[173] Valentino concludes that "it may be premature to celebrate 'the end of history,' but if no similarly radical ideas gain the widespread applicability and acceptance of communism, humanity may be able to look forward to considerably less mass killing in the coming century than it experienced in the last."[174]

On top of that singularly destructive ideology were the catastrophic decisions of a few men who took the stage at particular moments in the 20th century. I have already mentioned that many historians have joined the chorus "No Hitler, no Holocaust."[175] But Hitler was not the only tyrant whose obsessions killed tens of millions. The historian Robert Conquest, an authority on Stalin's politicides, concluded that "the nature of the whole Purge depends in the last analysis on the personal and political drives of Stalin."[176] As for China, it is inconceivable that the record-setting famine of the Great Leap Forward would have occurred but for Mao's harebrained schemes, and the historian Harry Harding noted of the country's subsequent politicide that "the principal responsibility for the Cultural Revolution—a movement that affected tens of millions of Chinese—rests with one man. Without a Mao, there could not have been a Cultural Revolution."[177] With such a small number of data points causing such a large share of the devastation, we will never really know how to explain the most calamitous events of the 20th century. The ideologies prepared the ground and attracted the men, the absence of democracy gave them the opportunity, but tens of millions of deaths ultimately depended on the decisions of just three individuals.

THE TRAJECTORY OF TERRORISM

Terrorism is a peculiar category of violence, because it has a cockeyed ratio of fear to harm. Compared to the number of deaths from homicide, war, and genocide, the worldwide toll from terrorism is in the noise: fewer than 400 deaths a year since 1968 from international terrorism (where perpetrators from one country cause damage in another), and about 2,500 a year since 1998 from domestic terrorism.[178] The numbers we have been dealing with in this chapter have been at least two orders of magnitude higher.

But after the September 11, 2001, attacks, terrorism became an obsession. Pundits and politicians turned up the rhetoric to eleven, and the word *existential* (generally modifying *threat* or *crisis*) had not seen as much use since the heyday of Sartre and Camus. Experts proclaimed that terrorism made the United States "vulnerable" and "fragile," and that it threatened to do away with the "ascendancy of the modern state," "our way of life," or "civilization itself."[179] In a 2005 essay in *The Atlantic,* for example, a former White House counterterrorism official confidently prophesied that by the tenth anniversary of the 9/11 attacks the American economy would be shut down by chronic bombings of casinos, subways, and shopping malls, the regular downing of commercial airliners by shoulder-launched missiles, and acts of cataclysmic sabotage at chemical plants.[180] The massive bureaucracy of the Department of Homeland Security was created overnight to reassure the nation with such security theater as color-coded terrorist alerts, advisories to stock up on plastic sheeting and duct tape, obsessive checking of identification cards (despite fakes being so plentiful that George W. Bush's own daughter was arrested for using one to order a margarita), the confiscation of nail clippers at airports, the girding of rural post offices with concrete barriers, and the designation of eighty thousand locations as "potential terrorist targets," including Weeki Wachee Springs, a Florida tourist trap in which comely women dressed as mermaids swim around in large glass tanks.

All this was in response to a threat that has killed a trifling number of Americans. The nearly 3,000 deaths from the 9/11 attacks were literally off the chart—way down in the tail of the power-law distribution into which terrorist attacks fall.[181] According to the Global Terrorism Database of the National Consortium for the Study of Terrorism and Responses to Terrorism (the major publicly available dataset on terrorist attacks), between 1970 and 2007 only one other terrorist attack in the entire world has killed as many as 500 people.[182] In the United States, Timothy McVeigh's bombing of a federal office building in Oklahoma City in 1995 killed 165, a shooting spree by two teenagers at Columbine High School in 1999 killed 17, and no other attack has killed as many as a dozen. Other than 9/11, the number of people killed by terrorists on American soil during these thirty-eight years was 340, and the number killed after 9/11—the date that inaugurated the so-called Age of Terror—was 11.

While some additional plots were foiled by the Department of Homeland Security, many of their claims have turned out to be the proverbial elephant repellent, with every elephant-free day serving as proof of its effectiveness.[183]

Compare the American death toll, with or without 9/11, to other preventable causes of death. Every year more than 40,000 Americans are killed in traffic accidents, 20,000 in falls, 18,000 in homicides, 3,000 by drowning (including 300 in bathtubs), 3,000 in fires, 24,000 from accidental poisoning, 2,500 from complications of surgery, 300 from suffocation in bed, 300 from inhalation of gastric contents, and 17,000 by "other and unspecified nontransport accidents and their sequelae."[184] In fact, in every year but 1995 and 2001, more Americans were killed by lightning, deer, peanut allergies, bee stings, and "ignition or melting of nightwear" than by terrorist attacks.[185] The number of deaths from terrorist attacks is so small that even minor measures to avoid them can *increase* the risk of dying. The cognitive psychologist Gerd Gigerenzer has estimated that in the year after the 9/11 attacks, 1,500 Americans died in car accidents because they chose to drive rather than fly to their destinations out of fear of dying in a hijacked or sabotaged plane, unaware that the risk of death in a plane flight from Boston to Los Angeles is the same as the risk of death in a car trip of twelve miles. In other words the number of people who died by avoiding air travel was six times the number of people who died in the airplanes on September 11.[186] And of course the 9/11 attacks sent the United States into two wars that have taken far more American and British lives than the hijackers did, to say nothing of the lives of Afghans and Iraqis.

The discrepancy between the panic generated by terrorism and the deaths generated by terrorism is no accident. Panic is the whole point of terrorism, as the word itself makes clear. Though definitions vary (as in the cliché "One man's terrorist is another man's freedom fighter"), terrorism is generally understood as premeditated violence perpetrated by a nonstate actor against noncombatants (civilians or off-duty soldiers) in pursuit of a political, religious, or social goal, designed to coerce a government or to intimidate or convey a message to a larger audience. The terrorists may want to extort a government into capitulating to a demand, to sap people's confidence in their government's ability to protect them, or to provoke massive repression that will turn people against their government or bring about violent chaos in which the terrorist faction hopes to prevail. Terrorists are altruistic in the sense of being motivated by a cause rather than by personal profit. They act by surprise and in secrecy; hence the ubiquitous appellation "cowardly." And they are communicators, seeking publicity and attention, which they manufacture through fear.

Terrorism is a form of asymmetrical warfare—a tactic of the weak against the strong—which leverages the psychology of fear to create emotional damage that is disproportionate to its damage in lives or property. Cognitive psychologists such as Tversky, Kahneman, Gigerenzer, and Slovic have shown that the perceived danger of a risk depends on two mental hobgoblins.[187] The first is fathomability: it's better to deal with the devil you know than the devil you don't. People are nervous about risks that are novel, undetectable, delayed in their effects, and poorly understood by the science of the day. The second contributor is dread. People worry about worst-case scenarios, the ones that are uncontrollable, catastrophic, involuntary, and inequitable (the people exposed to the risk are not the ones who benefit from it). The psychologists suggest that the illusions are a legacy of ancient brain circuitry that evolved to protect us against natural risks such as predators, poisons, enemies, and storms. They may have been the best guide to allocating vigilance in the prenumerate societies that predominated in human life until the compilation of statistical databases within the past century. Also, in an era of scientific ignorance these apparent quirks in the psychology of danger may have brought a secondary benefit: people exaggerate threats from enemies to extort compensation from them, to recruit allies against them, or to justify wiping them out preemptively (the superstitious killing discussed in chapter 4).[188]

Fallacies in risk perception are known to distort public policy. Money and laws have been directed at keeping additives out of food and chemical residues out of water supplies which pose infinitesimal risks to health, while measures that demonstrably save lives, such as enforcing lower highway speeds, are resisted.[189] Sometimes a highly publicized accident becomes a prophetic allegory, an ominous portent of an apocalyptic danger. The 1979 accident at the Three Mile Island nuclear power plant killed no one, and probably had no effect on cancer rates, but it halted the development of nuclear power in the United States and thus will contribute to global warming from the burning of fossil fuels for the foreseeable future.

The 9/11 attacks also took on a portentous role in the nation's consciousness. Large-scale terrorist plots were novel, undetectable, catastrophic (compared to what had come before), and inequitable, and thus maximized both unfathomability and dread. The terrorists' ability to gain a large psychological payoff for a small investment in damage was lost on the Department of Homeland Security, which outdid itself in stoking fear and dread, beginning with a mission statement that warned, "Today's terrorists can strike at any place, at any time, and with virtually any weapon." The payoff was not lost on Osama bin Laden, who gloated that "America is full of fear from its north to its south, from its west to its east," and that the $500,000 he spent on the 9/11 attacks cost the country more than half a trillion dollars in economic losses in the immediate aftermath.[190]

Responsible leaders occasionally grasp the arithmetic of terrorism. In an unguarded moment during the 2004 presidential campaign, John Kerry told a *New York Times* interviewer, "We have to get back to the place we were, where terrorists are not the focus of our lives, but they're a nuisance. As a former law-enforcement person, I know we're never going to end prostitution. We're never going to end illegal gambling. But we're going to reduce it, organized crime, to a level where it isn't on the rise. It isn't threatening people's lives every day, and fundamentally, it's something that you continue to fight, but it's not threatening the fabric of your life."[191] Confirming the definition of a *gaffe* in Washington as "something a politician says that is true," George Bush and Dick Cheney pounced on the remark, calling Kerry "unfit to lead," and he quickly backpedaled.

The ups and downs of terrorism, then, are a critical part of the history of violence, not because of its toll in deaths but because of its impact on a society through the psychology of fear. In the future, of course, terrorism really could have a catastrophic death toll if the hypothetical possibility of an attack with nuclear weapons ever becomes a reality. I will discuss nuclear terrorism in the next section but for now will stick to forms of violence that have actually taken place.

Terrorism is not new. After the Roman conquest of Judea two millennia ago, a group of resistance fighters furtively stabbed Roman officials and the Jews who collaborated with them, hoping to force the Romans out. In the 11th century a sect of Shia Muslims perfected an early form of suicide terrorism by getting close to leaders who they thought had strayed from the faith and stabbing them in public, knowing they would immediately be slain by the leader's bodyguards. From the 17th to the 19th century, a cult in India strangled tens of thousands of travelers as a sacrifice to the goddess Kali. These groups did not accomplish any political change, but they left a legacy in their names: the Zealots, the Assassins, and the Thugs.[192] And if you associate the word *anarchist* with a black-clad bomb-thrower, you are recalling a movement around the turn of the 20th century that practiced "propaganda of the deed" by bombing cafés, parliaments, consulates, and banks and by assassinating dozens of political leaders, including Czar Alexander II of Russia, President Sadi Carnot of France, King Umberto I of Italy, and President William McKinley of the United States. The durability of these eponyms and images is a sign of the power of terrorism to lodge in cultural consciousness.

Anyone who thinks that terrorism is a phenomenon of the new millennium has a short memory. The romantic political violence of the 1960s and 1970s included hundreds of bombings, hijackings, and shootings by various armies, leagues, coalitions, brigades, factions, and fronts.[193] The United States had the Black Liberation Army, the Jewish Defense League, the Weather Underground (who took their name from Bob Dylan's lyric "You don't need a weatherman to know which way the wind blows"), the FALN (a Puerto Rican independence group), and of course the Symbionese Liberation Army. The SLA contributed one of the more surreal episodes of the 1970s when they kidnapped newspaper heiress Patty Hearst in 1974 and brainwashed her into joining the group, whereupon she adopted "Tanya" as her nom de guerre, helped them rob a bank, and posed for a photograph in a battle stance with beret and machine gun in front of their seven-headed cobra flag, leaving us one of the iconic images of the decade (together with Richard Nixon's victory salute from the helicopter that would whisk him from the White House for the last time, and the blow-dried Bee Gees in white polyester disco suits).

Europe, during this era, had the Provisional Irish Republican Army and the Ulster Freedom Fighters in the U.K., the Red Brigades in Italy, the Baader-Meinhof Gang in Germany, and the ETA (a Basque separatist group) in Spain, while Japan had the Japanese Red Army and Canada had the Front de Libération du Québec. Terrorism was so much a backdrop to European life that it served as a running joke in Luis Buñuel's 1977 love story *That Obscure Object of Desire*, in which cars and stores blow up at random and the characters barely notice.

Where are they now? In most of the developed world, domestic terrorism has gone the way of the polyester disco suits. It's a little-known fact that most terrorist groups fail, and that all of them die.[194] Lest this seem hard to believe, just reflect on the world around you. Israel continues to exist, Northern Ireland is still a part of the United Kingdom, and Kashmir is a part of India. There are no sovereign states in Kurdistan, Palestine, Quebec, Puerto Rico, Chechnya, Corsica, Tamil Eelam, or Basque Country. The Philippines, Algeria, Egypt, and Uzbekistan are not Islamist theocracies; nor have Japan, the United States, Europe, and Latin America become religious, Marxist, anarchist, or new-age utopias.

The numbers confirm the impressions. In his 2006 article "Why Terrorism Does Not Work," the political scientist Max Abrahms examined the twenty-eight groups designated by the U.S. State Department as foreign terrorist organizations in 2001, most of which had been active for several decades. Putting aside purely tactical victories (such as media attention, new supporters, freed prisoners, and ransom), he found that only 3 of them (7 percent) had attained their goals: Hezbollah expelled multinational peacekeepers and Israeli forces from southern Lebanon in 1984 and 2000, and the Tamil Tigers won control over the northeastern coast of Sri Lanka in 1990. Even that victory was reversed by Sri Lanka's rout of the Tigers in 2009, leaving the terrorist success rate at 2 for 42, less than 5 percent. The success rate is well below that of other forms of political pressure such as economic sanctions, which work about a third of the time. Reviewing its recent history, Abrahms noted that terrorism occasionally succeeds when it has limited territorial goals, like evicting a foreign power from land it had gotten tired of occupying, such as the European powers who in the 1950s and 1960s withdrew from their colonies en masse, terrorism or no terrorism.[195] But it never attains maximalist goals such as imposing an ideology on a state or annihilating it outright. Abrahms also found that the few successes came from campaigns in which the groups targeted military forces rather than civilians and thus were closer to being guerrillas than pure terrorists. Campaigns that primarily targeted civilians always failed.

In her book *How Terrorism Ends*, the political scientist Audrey Cronin examined a larger dataset: 457 terrorist campaigns that had been active since 1968. Like Abrahms, she found that terrorism virtually never works. Terrorist groups die off exponentially over time, lasting, on average, between five and nine years. Cronin points out that "states have a degree of immortality in the international system; groups do not."[196]

Nor do they get what they want. No small terrorist organization has ever taken over a state, and 94 percent fail to achieve *any* of their strategic aims.[197] Terrorist campaigns meet their end when their leaders are killed or captured, when they are rooted out by states, and when they morph into guerrilla or political movements. Many burn out through internal squabbling, a failure of the founders to replace themselves, and the defection of young firebrands to the pleasures of civilian and family life.

Terrorist groups immolate themselves in another way. As they become frustrated by their lack of progress and their audiences start to get bored, they escalate their tactics. They start to target victims who are more newsworthy because they are famous, respected, or simply numerous. That certainly gets people's attention, but not in the way the terrorists intend. Supporters are repulsed by the "senseless violence" and withdraw their money, their safe havens, and their reluctance to cooperate with the police. The Red Brigades in Italy, for example, self-destructed in 1978 when they kidnapped the beloved former prime minister Aldo Moro, kept him in captivity for two months, shot him eleven times, and left his body in the trunk of a car. Earlier the FLQ overplayed its hand during the October Crisis of 1970 when it kidnapped Québec labor minister Pierre Laporte and strangled him with his rosary, also leaving his body in a trunk. McVeigh's killing of 165 people (including 19 children) in the bombing of a federal building in Oklahoma City in 1995 took the stuffing out of the right-wing antigovernment militia movement in the United States. As Cronin puts it, "Violence has an international language, but so does decency."[198]

Attacks on civilians can doom terrorists not just by alienating potential sympathizers but by galvanizing the public into supporting an all-out crackdown. Abrahms tracked public opinion during terrorist campaigns in Israel, Russia, and the United States and found that after a major attack on civilians, attitudes toward the group lurched downward. Any willingness to compromise with the group or to recognize the legitimacy of their grievance evaporated. The public now believed that the terrorists were an existential threat and supported measures that would snuff them out for good. The thing about asymmetric warfare is that one side, by definition, is a lot more powerful than the other. And as the saying goes, the race may not be given to the swift, nor the battle to the strong, but that's the way to bet.

Though terrorist campaigns have a natural arc that bends toward failure, new campaigns can spring up as quickly as old ones fizzle. The world contains an unlimited number of grievances, and as long as the perception that terrorism works stays ahead of the reality, the terrorist meme may continue to infect the aggrieved.

The historical trajectory of terrorism is elusive. Statistics begin only around 1970, when a few agencies began to collect them, and they differ in their recording criteria and their coverage. It can be hard, even in the best of times, to distinguish terrorist attacks from accidents, homicides, and disgruntled individuals going postal, and in war zones the line between terrorism and insurgency can be fuzzy. The statistics are also heavily politicized: various constituencies may try to make the numbers look big, to sow fear of terrorism, or small, to trumpet their success in fighting terrorism. And while the whole world cares about international terrorism, governments often treat domestic terrorism, which kills six to seven times as many people, as no one else's business. The most comprehensive public dataset we have is the Global Terrorism Database, an amalgamation of many of the earlier datasets. Though we can't interpret every zig or zag in the graphs at face value, because some may represent seams and overlaps between databases with different coding criteria, we can try to get a general sense of whether terrorism really has increased in the so-called Age of Terror.[199]

The safest records are those for terrorist attacks on American soil, if for no other reason than that there are so few of them that each can be scrutinized. Figure 6–9 shows all of them since 1970, plotted on a logarithmic scale because

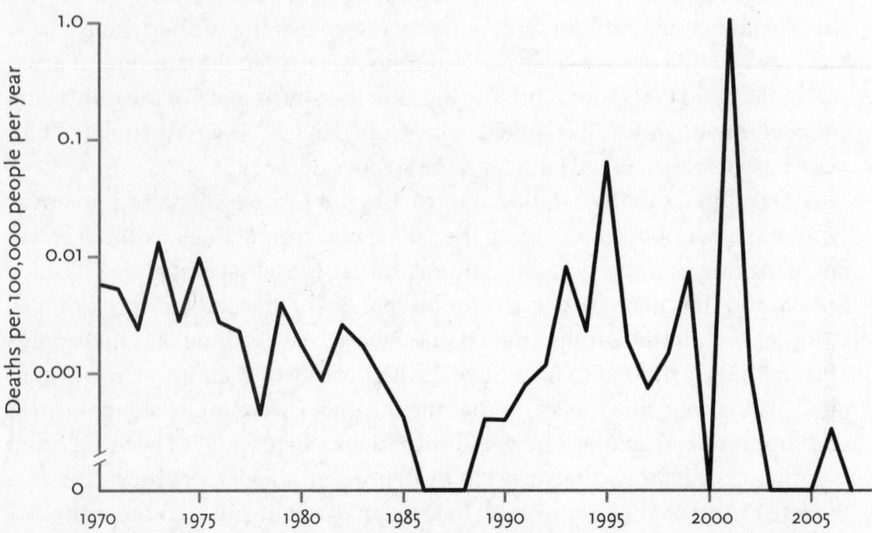

FIGURE 6–9. Rate of deaths from terrorism, United States, 1970–2007

Source: Global Terrorism Database, START (National Consortium for the Study of Terrorism and Responses to Terrorism, 2010, http://www.start.umd.edu/gtd/), accessed on April 6, 2010. The figure for 1993 was taken from the appendix to National Consortium for the Study of Terrorism and Responses to Terrorism, 2009. Since the log of 0 is undefined, years with no deaths are plotted at the arbitrary value 0.0001.

otherwise the line would be a towering spike for 9/11 poking through a barely wrinkled carpet. With the lower altitudes stretched out by the logarithmic scale, we can discern peaks for Oklahoma City in 1995 and Columbine in 1999 (which is a dubious example of "terrorism," but with a single exception, noted below, I never second-guess the datasets when plotting the graphs). Apart from this trio of spikes, the trend since 1970 is, if anything, more downward than upward.

The trajectory of terrorism in Western Europe (figure 6–10) illustrates the point that most terrorist organizations fail and all of them die. Even the spike from the 2004 Madrid train bombings cannot hide the decline from the glory years of the Red Brigades and the Baader-Meinhof Gang.

What about the world as a whole? Though Bush administration statistics released in 2007 seemed to support their warnings about a global increase in terrorism, the HSRP team noticed that their data include civilian deaths from the wars in Iraq and Afghanistan, which would be classified as civil war casualties if they had taken place anywhere else in the world. The picture is different when the criteria are kept consistent and these deaths are excluded. Figure 6–11 shows the worldwide annual rate of death from terrorism (as usual, per 100,000 population) without these deaths. The death tolls for the world as a whole have to be interpreted with caution, because they come from a hybrid dataset and can float up and down with differences in how many news sources

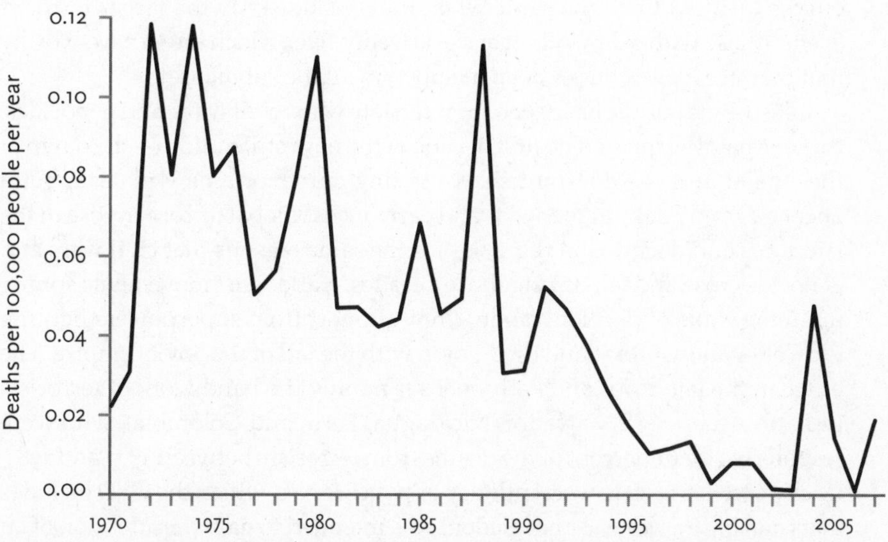

FIGURE 6–10. Rate of deaths from terrorism, Western Europe, 1970–2007

Source: Global Terrorism Database, START (National Consortium for the Study of Terrorism and Responses to Terrorism, 2010, http://www.start.umd.edu/gtd/), accessed on April 6, 2010. Data for 1993 are interpolated. Population figures from UN *World Population Prospects* (United Nations, 2008), accessed April 23, 2010; figures for years not ending in 0 or 5 are interpolated.

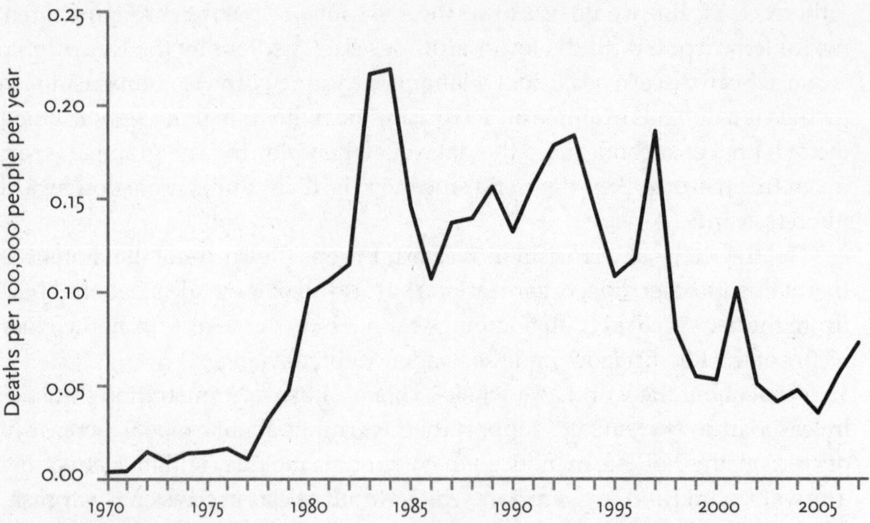

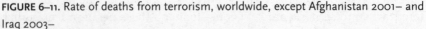

FIGURE 6–11. Rate of deaths from terrorism, worldwide, except Afghanistan 2001– and Iraq 2003–

Source: Global Terrorism Database, START (National Consortium for the Study of Terrorism and Responses to Terrorism, 2010, http://www.start.umd.edu/gtd/), accessed on April 6, 2010. Data for 1993 are interpolated. World population figures from U.S. Census Bureau, 2010c; the population estimate for 2007 is extrapolated.

were consulted in each of the contributing datasets. But the shapes of the curves turn out to be the same when they include only the larger terrorist events (those with death tolls of at least twenty-five), which are so newsworthy that they are likely to have been included in all the subdatasets.

Like the graphs we have seen for interstate wars, civil wars, and genocides, this one has a surprise. The first decade of the new millennium—the dawn of the Age of Terror—does not show a rising curve, or a new plateau, but a decrease from peaks in the 1980s and early 1990s. Global terrorism rose in the late 1970s and declined in the 1990s for the same reasons that civil wars and genocides rose and fell during those decades. Nationalist movements sprang up in the wake of decolonization, drew support from superpowers fighting the Cold War by proxy, and died down with the fall of the Soviet empire. The bulge in the late 1970s and early 1980s is mainly the handiwork of terrorists in Latin America (El Salvador, Nicaragua, Peru, and Colombia), who were responsible for 61 percent of the deaths from terrorism between 1977 and 1984. (Many of these targets were military or police forces, which the GTD includes in its database as long as the incident was intended to gain the attention of an audience rather than to inflict direct damage.)[200] Latin America kept up its contribution in the second rise from 1985 to 1992 (about a third of the deaths), joined by the Tamil Tigers in Sri Lanka (15 percent) and groups in India, the Philippines, and Mozambique. Though some of the terrorist activity in India and the Philippines came from Muslim groups, only a sliver of the deaths occurred in Muslim countries: around 2 percent of them in Lebanon, and 1 percent in Pakistan. The decline of terrorism since 1997 was punctuated by peaks for 9/11 and by a recent uptick in Pakistan, mainly as a spillover from the war in Afghanistan along their nebulous border.

The numbers, then, show that we are not living in a new age of terrorism. If anything, aside from the wars in Iraq and Afghanistan, we are enjoying a *decline* in terrorism from decades in which it was less big a deal in our collective consciousness. Nor, until recently, has terrorism been a particularly Muslim phenomenon.

But isn't it today? Shouldn't we expect the suicide terrorists from Al Qaeda, Hamas, and Hezbollah to be picking up the slack? And what are we hiding by taking the civilian deaths in Iraq and Afghanistan, many of them victims of suicide bombers, out of the tallies? Answering these questions will require a closer look at terrorism, especially suicide terrorism, in the Islamic world.

Though 9/11 did not inaugurate a new age of terror, a case could be made that it foretold an age of Islamist suicide terror. The 9/11 hijackers could not have carried out their attacks had they not been willing to die in the process, and since then the rate of suicide attacks has soared, from fewer than 5 per year in the 1980s and 16 per year in the 1990s to 180 per year between 2001 and 2005. Most of these attacks were carried out by Islamist groups whose expressed motives were at least partly religious.[201] According to the most recent data from the National Counterterrorism Center, in 2008 Sunni Islamic extremists were responsible for almost two-thirds of the deaths from terrorism that could be attributed to a terrorist group.[202]

As a means of killing civilians, suicide terrorism is a tactic of diabolical ingenuity. It combines the ultimate in surgical weapon delivery—the precision manipulators and locomotors called hands and feet, controlled by the human eyes and brain—with the ultimate in stealth—a person who looks just like millions of other people. In technological sophistication, no battle robot comes close. The advantages are not just theoretical. Though suicide terrorism accounts for a minority of terrorist attacks, it is responsible for a majority of the casualties.[203] This bang for the buck can be irresistible to the leaders of a terrorist movement. As one Palestinian official explained, a successful mission requires only "a willing young man . . . nails, gunpowder, a light switch and a short cable, mercury (readily obtainable from thermometers), acetone. . . . The most expensive item is transportation to an Israeli town."[204] The only real technological hurdle is the willingness of the young man. Ordinarily a human being is unwilling to die, the legacy of half a billion years of natural selection. How have terrorist leaders overcome this obstacle?

People have exposed themselves to the risk of dying in wars for as long as there have been wars, but the key term is *risk*. Natural selection works on averages, so a willingness to take a small chance of dying as part of an aggressive coalition that offers a large chance of a big fitness payoff—more land, more women, or more safety—can be favored over the course of evolution.[205] What cannot be favored is a willingness to die with certainty, which would take any genes that allow such willingness along with the dead body. It's not surprising that suicide missions are uncommon in the history of warfare. Foraging bands prefer the safety of raids and ambushes to the hazards of set-piece battles, and even then warriors are not above claiming to have had dreams and omens that conveniently keep them out of risky encounters planned by their comrades.[206]

Modern armies cultivate incentives for soldiers to increase the risk they take on, such as esteem and decorations for bravery, and disincentives for them to reduce the risk, such as the shaming or punishment of cowards and the summary execution of deserters. Sometimes a special class of soldier called file closers trails behind a unit with orders to kill any soldier who fails to advance. The conflicts of interest between war leaders and foot soldiers leads to the well-known hypocrisy of military rhetoric. Here is how a British general waxed about the carnage of World War I: "Not a man shirked going through the extremely heavy barrage, or facing the machine gun and rifle fire that finally wiped them out. . . . I have never seen, indeed could never have imagined, such a magnificent display of gallantry, discipline, and determination." A sergeant described it differently: "We knew it was pointless, even before we went over—crossing open ground like that. But you had to go. You were between the devil and the deep blue sea. If you go forward, you'll likely be shot. If you go back, you'll be court-martialed and shot. What can you do?"[207]

Warriors may accept the risk of death in battle for another reason. The evolutionary biologist J.B.S. Haldane, when asked whether he would lay down his life for his brother, replied, "No, but for two brothers or eight cousins." He was invoking the phenomenon that would later be known as kin selection, inclusive fitness, and nepotistic altruism. Natural selection favors any genes that incline an organism toward making a sacrifice that helps a blood relative, as long as the benefit to the relative, discounted by the degree of relatedness, exceeds the cost to the organism. The reason is that the genes would be helping copies of *themselves* inside the bodies of those relatives and would have a long-term advantage over their narrowly selfish alternatives. Critics who are determined to misunderstand this theory imagine that it requires that organisms consciously calculate their genetic overlap with their kin and anticipate the good it will do their DNA.[208] Of course it requires only that organisms be inclined to pursue goals that help organisms that are statistically likely to be their genetic relatives. In complex organisms such as humans, this inclination is implemented as the emotion of brotherly love.

The small-scale bands in which humans spent much of their evolutionary history were held together by kinship, and people tended to be related to their neighbors. Among the Yanomamö, for example, two individuals picked at random from a village are related almost as closely as first cousins, and people who consider each other relatives are related, on average, even more closely.[209] The genetic overlap tilts the evolutionary payoff toward taking greater risks to life and limb if the risky act might benefit one's fellow warriors. One of the reasons that chimpanzees, unlike other primates, engage in cooperative raiding is that the females, rather than the males, disperse from the troop at sexual maturity, so the males in a troop tend to be related.[210]

As with all aspects of our psychology that have been illuminated by evolutionary theory, what matters is not *actual* genetic relatedness (it's not as if hunter-gatherers, to say nothing of chimpanzees, send off cheek swabs to a genotyping service) but the *perception* of relatedness, as long as the perception was correlated with the reality over long enough spans of time.[211] Among the contributors to the perception of kinship are the experience of having grown up together, having seen one's mother care for the other person, commensal meals, myths of common ancestry, essentialist intuitions of common flesh and blood, the sharing of rituals and ordeals, physical resemblance (often enhanced by hairdressing, tattoos, scarification, and mutilation), and metaphors such as *fraternity, brotherhood, family, fatherland, motherland,* and *blood.*[212] Military leaders use every trick in the book to make their soldiers feel like genetic relatives and take on the biologically predictable risks. Shakespeare made this clear in the most famous motivational speech in the literary history of war, when Henry V addresses his men on St. Crispin's Day:

> And Crispin Crispian shall ne'er go by,
> From this day to the ending of the world,
> But we in it shall be rememberèd—
> We few, we happy few, we band of brothers;
> For he today that sheds his blood with me
> Shall be my brother.

Modern militaries too take pains to group soldiers into bands of brothers—the fire teams, squads, and platoons of half a dozen to several dozen soldiers that serve as a crucible for the primary emotion that moves men to fight in armies, brotherly love. Studies of military psychology have discovered that soldiers fight above all out of loyalty to their platoonmates.[213] The writer William Manchester reminisced about his experience as a Marine in World War II:

> Those men on the line were my family, my home. They were closer to me than I can say, closer than any friends had been or ever would be. They had never let me down, and I couldn't do it to them. . . . I had to be with them, rather than let them die and me live with the knowledge that I might have saved them. Men, I now knew, do not fight for flag or country, for the Marine Corps or glory or any other abstraction. They fight for one another.[214]

Two decades later, another Marine-turned-author, William Broyles, offered a similar reflection on his experience in Vietnam:

> The enduring emotion of war, when everything else has faded, is comradeship. A comrade in war is a man you can trust with anything, because you trust him with your life. . . . Despite its extreme right-wing image, war is the only utopian experience most of us ever have. Individual possessions and advantage count for nothing: the group is everything. What you have is shared with your friends. It isn't a particularly selective process, but a love that needs no reasons, that transcends race and personality and education—all those things that would make a difference in peace.[215]

Though in extremis a man may lay down his life to save a platoon of virtual brothers, it's rarer for him to calmly make plans to commit suicide at some future date on their behalf. The conduct of war would be very different if he did. To avoid panic and rout (at least in the absence of file closers), battle plans are generally engineered so that an individual soldier does not know that he has been singled out for certain death. At a bomber base during World War II, for example, strategists calculated that pilots would have a higher probability of survival if a few of them who drew the short straws in a lottery would fly off to certain death on one-way sorties rather than all of them taking their chances in the fuel-laden planes needed for round trips. But they opted for the higher risk of an unpredictable death over the lower risk of a death that would be preceded by a lengthy period of doom.[216] How do the engineers of suicide terrorism overcome this obstacle?

Certainly an ideology of an afterlife helps, as in the posthumous Playboy Mansion promised to the 9/11 hijackers. (Japanese kamikaze pilots had to make do with the less vivid image of being absorbed into a great realm of the spirit.) But modern suicide terrorism was perfected by the Tamil Tigers, and though the members grew up in Hinduism with its promise of reincarnation, the group's ideology was secular: the usual goulash of nationalism, romantic militarism, Marxism-Leninism, and anti-imperialism that animated 20th-century third-world liberation movements. And in accounts by would-be suicide terrorists of what prompted them to enlist, anticipation of an afterlife, with or without the virgins, seldom figures prominently. So while expectation of a pleasant afterlife may tip the perceived cost-benefit ratio (making it harder to imagine an atheist suicide bomber), it cannot be the only psychological driver.

Using interviews with failed and prospective suicide terrorists, the anthropologist Scott Atran has refuted many common misconceptions about them. Far from being ignorant, impoverished, nihilistic, or mentally ill, suicide terrorists tend to be educated, middle class, morally engaged, and free of obvious psychopathology. Atran concluded that many of the motives may be found in nepotistic altruism.[217]

The case of the Tamil Tigers is relatively easy. They use the terrorist equivalent of file closers, selecting operatives for suicide missions and threatening to kill their families if they withdraw.[218] Only slightly less subtle are the methods of Hamas and other Palestinian terrorist groups, who hold out a carrot rather than a stick to the terrorist's family in the form of generous monthly stipends, lump-sum payments, and massive prestige in the community.[219] Though in general one should not expect extreme behavior to deliver a payoff in biological fitness, the anthropologists Aaron Blackwell and Lawrence Sugiyama have shown that it may do so in the case of Palestinian suicide terrorism. In the West Bank and Gaza many men have trouble finding wives because their families cannot afford a bride-price, they are restricted to marrying parallel cousins, and many women are taken out of the marriage pool by polygynous marriage or by marriage up to more prosperous Arabs in Israel. Blackwell and Sugiyama note that 99 percent of Palestinian suicide terrorists are male, that 86 percent are unmarried, and that 81 percent have at least six siblings, a larger family size than the Palestinian average. When they plugged these and other numbers into a simple demographic model, they found that when a terrorist blows himself up, the financial payoff can buy enough brides for his brothers to make his sacrifice reproductively worthwhile.

Atran has found that suicide terrorists can also be recruited without these direct incentives. Probably the most effective call to martyrdom is the opportunity to join a happy band of brothers. Terrorist cells often begin as gangs of underemployed single young men who come together in cafés, dorms, soccer clubs, barbershops, or Internet chat rooms and suddenly find meaning in their lives by a commitment to the new platoon. Young men in all societies do foolish things to prove their courage and commitment, especially in groups, where individuals may do something they know is foolish because they think that everyone else in the group thinks it is cool.[220] (We will return to this phenomenon in chapter 8.) Commitment to the group is intensified by religion, not just the literal promise of paradise but the feeling of spiritual awe that comes from submerging oneself in a crusade, a calling, a vision quest, or a jihad. Religion may also turn a commitment to the cause into a sacred value—a good that may not be traded off against anything else, including life itself.[221] The commitment can be stoked by the thirst for revenge, which in the case of militant Islamism takes the form of vengeance for the harm and humiliation suffered by any Muslim anywhere on the planet at any time in history, or for symbolic affronts such as the presence of infidel soldiers on sacred Muslim soil. Atran summed up his research in testimony to a U.S. Senate subcommittee:

> When you look at young people like the ones who grew up to blow up trains in Madrid in 2004, carried out the slaughter on the London underground in 2005, hoped to blast airliners out of the sky en route to the United States in 2006 and 2009, and journeyed far to die killing infidels in Iraq, Afghanistan,

> Pakistan, Yemen or Somalia; when you look at whom they idolize, how they organize, what bonds them and what drives them; then you see that what inspires the most lethal terrorists in the world today is not so much the Koran or religious teachings as a thrilling cause and call to action that promises glory and esteem in the eyes of friends, and through friends, eternal respect and remembrance in the wider world that they will never live to enjoy. . . . Jihad is an egalitarian, equal-opportunity employer: . . . fraternal, fast-breaking, thrilling, glorious, and cool. Anyone is welcome to try his hand at slicing off the head of Goliath with a paper cutter.[222]

The local imams are of marginal importance in this radicalization, since young men who want to raise hell rarely look to community elders for guidance. And Al Qaeda has become more a global brand inspiring a diffuse social network than a centralized recruiting organization.

The up-close look at suicide terrorists at first seems pretty depressing, because it suggests we are fighting a multiheaded hydra that cannot be decapitated by killing its leadership or invading its home base. Remember, though, that all terrorist organizations follow an arc toward failure. Are there any signs that Islamist terrorism is beginning to burn out?

The answer is a clear yes. In Israel, sustained attacks on civilians have accomplished what they accomplish everywhere else in the world: erase all sympathy for the group, together with any willingness to compromise with it.[223] After the Second Intifada began, shortly after Yasir Arafat's rejection of the Camp David accords in 2000, the Palestinians' economic and political prospects steadily deteriorated. In the long run, Cronin adds, suicide terrorism is a supremely idiotic tactic because it makes the target nation unwilling to tolerate members of the minority community in their midst, never knowing which among them may be a walking bomb. Though Israel has faced international condemnation for building a security barrier, other countries faced with suicide terrorism, Cronin notes, have taken similar measures.[224] The Palestinian leadership on the West Bank has, more recently, disavowed violence and turned its energies toward competent governance, while Palestinian activist groups have turned to boycotts, civil disobedience, peaceful protests, and other forms of nonviolent resistance.[225] They have even enlisted Rajmohan Gandhi (grandson of Mohandas) and Martin Luther King III for symbolic support. It's too soon to know whether this is a turning point in Palestinian tactics, but a retreat from terrorism would not be historically unprecedented.

The bigger story, though, is the fate of Al Qaeda. Marc Sageman, a former CIA officer who has been keeping tabs on the movement, counted ten serious plots on Western targets in 2004 (many inspired by the invasion of Iraq) but just three in 2008.[226] Not only has Al Qaeda's base in Afghanistan been routed and its leadership decimated (including bin Laden himself in 2011), but in the world of

Muslim opinion its favorables have long been sinking, and its negatives have been rising.[227] In the past six years Muslims have become repulsed by what they increasingly see as nihilistic savagery, consistent with Cronin's remark that decency, not just violence, has an international language. The movement's strategic goals—a pan-Islamic caliphate, the replacement of repressive and theocratic regimes by even more repressive and theocratic regimes, the genocidal killing of infidels—begin to lose their appeal once people start thinking about what they really mean. And Al Qaeda has succumbed to the fatal temptation of all terrorist groups: to stay in the limelight by mounting ever bloodier attacks on ever more sympathetic victims, which in Al Qaeda's case includes tens of thousands of fellow Muslims. Attacks in the mid-2000s on a Bali nightclub, a Jordanian wedding party, an Egyptian resort, the London underground, and cafés in Istanbul and Casablanca massacred Muslims and non-Muslims alike for no discernible purpose. The franchise of the movement known as Al Qaeda in Iraq (AQI) proved to be even more depraved, bombing mosques, marketplaces, hospitals, volleyball games, and funerals, and brutalizing resisters with amputations and beheadings.

The jihad against the jihadis is being fought at many levels. Islamic states such as Saudi Arabia and Indonesia that once indulged Islamist extremists have decided that enough is enough and have begun to crack down. The movement's own gurus have also turned on it. In 2007 one of bin Laden's mentors, the Saudi cleric Salman al-Odah, wrote an open letter accusing him of "fostering a culture of suicide bombings that has caused bloodshed and suffering, and brought ruin to entire Muslim communities and families."[228] He was not afraid to get personal: "My brother Osama, how much blood has been spilt? How many innocent people, children, elderly, and women have been killed . . . in the name of Al Qaeda? Will you be happy to meet God Almighty carrying the burden of these hundreds of thousands or millions on your back?"[229] His indictment struck a chord: two-thirds of the postings on Web sites of Islamist organizations and television networks were favorable, and he has spoken to enthusiastic crowds of young British Muslims.[230] The grand mufti of Saudi Arabia, Abdulaziz al Ash-Sheikh, made it official, issuing a fatwa in 2007 forbidding Saudis to join foreign jihads and condemning bin Laden and his cronies for "transforming our youth into walking bombs to accomplish their own political and military aims."[231] That same year another sage of Al Qaeda, the Egyptian scholar Sayyid Imam Al Sharif (also known as Dr. Fadl), published a book called *Rationalization of Jihad* because, he explained, "Jihad . . . was blemished with grave Sharia violations during recent years. . . . Now there are those who kill hundreds, including women and children, Muslims and non-Muslims in the name of Jihad!"[232]

The Arab street agrees. In a 2008 online Q&A on a jihadist Web site with Ayman al-Zawahiri, Al Qaeda's day-to-day leader, one participant asked, "Excuse me, Mr. Zawahiri, but who is it who is killing, with Your Excellency's blessing, the innocents in Baghdad, Morocco, and Algeria?"[233] Public opinion polls throughout the Islamic world have tapped the outrage. Between 2005 and 2010, the number of respondents in Jordan, Pakistan, Indonesia, Saudi Arabia, and Bangladesh who endorse suicide bombing and other violence against civilians has sunk like a stone, often to around 10 percent. Lest even this figure seem barbarically high, the political scientist Fawaz Gerges (who compiled the data) reminds us that no fewer than 24 percent of Americans tell pollsters that "bombing and other attacks intentionally aimed at civilians are often or sometimes justified."[234]

More important is public opinion in the war zones in which the terrorists rely on the support of the population.[235] In the North-West Frontier Province in Pakistan, support for Al Qaeda plummeted from 70 percent to 4 percent in just five months in late 2007, partly in reaction to the assassination of former prime minister Benazir Bhutto by a suicide bomber. In elections that year Islamists won 2 percent of the national vote—a fivefold decrease since 2002. In a 2007 ABC/BBC poll in Afghanistan, support for jihadist militants nosedived to 1 percent.[236] In Iraq in 2006 a large majority of Sunnis and an overwhelming majority of Kurds and Shias rejected AQI, and by December 2007 the opposition to their attacks on civilians had reached a perfect 100 percent.[237]

Public opinion is one thing, but does it translate into a reduction of violence? Terrorists depend on popular support, so it's highly likely that it does. The year 2007, the turning point in attitudes toward terrorism in the Islamic world, was also a turning point in suicide attacks in Iraq. The Iraq Body Count has documented that vehicle bombs and suicide attacks declined from 21 a day in 2007 to fewer than 8 a day in 2010—still too many, but a sign of progress.[238] Changes in Muslim attitudes do not deserve all the credit; the surge of American soldiers in the first half of 2007 and other military adjustments helped as well. But some of the military developments themselves depended on a shift in attitudes. Muqtada al-Sadr's Mahdi Army, a Shia militia, declared a cease-fire in 2007, and in what has been called the Sunni Awakening tens of thousands of young men have defected from an insurgency against the American-supported government and are participating in the suppression of Al Qaeda in Iraq.[239]

Terrorism is a tactic, not an ideology or a regime, so we will never win the "War on Terror," any more than we will achieve George W. Bush's larger goal (announced in the same post-9/11 speech) to "rid the world of evil." In an age of global media, there will always be an ideologue nursing a grievance somewhere who is tempted by the spectacular return on investment of terrorism—a huge windfall in fear from a trifling outlay in violence—and there will always be bands of brothers willing to risk everything for the comradeship and glory it promises. When terrorism becomes a tactic in a large insurgency, it can do tremendous damage to people and to civil life, and the hypothetical threat of nuclear terrorism (to which I will turn in the final section) gives new meaning to the word *terror.* But in every other circumstance history teaches, and recent events confirm, that terrorist movements carry the seeds of their own destruction.

WHERE ANGELS FEAR TO TREAD

The New Peace is the quantitative decline in war, genocide, and terrorism that has proceeded in fits and starts since the end of the Cold War more than two decades ago. It has not been around for as long as the Long Peace, is not as revolutionary as the Humanitarian Revolution, and has not swept a civilization in the manner of the Civilizing Process. An obvious question is whether it will last. Though I am reasonably confident that during my lifetime France and Germany will not go to war, that cat-burning and the breaking wheel will not make a comeback, and that diners will not routinely stab each other with steak knives or cut off each other's noses, no prudent person could express a similar confidence when it comes to armed conflict in the world as a whole.

I am sometimes asked, "How do you know there won't be a war tomorrow (or a genocide, or an act of terrorism) that will refute your whole thesis?" The question misses the point of this book. The point is not that we have entered an Age of Aquarius in which every last earthling has been pacified forever. It is that substantial reductions in violence *have* taken place, and it is important to understand them. Declines in violence are caused by political, economic, and ideological conditions that take hold in particular cultures at particular times. If the conditions reverse, violence could go right back up.

Also, the world contains a lot of people. The statistics of power-law distributions and the events of the past two centuries agree in telling us that a small number of perpetrators can cause a great deal of damage. If somewhere among the world's six billion people there is a zealot who gets his hands on a stray nuclear bomb, he could single-handedly send the statistics through the roof. But even if he did, we would still need an explanation of why homicide rates fell a hundredfold, why slave markets and debtors' prisons have vanished, and why the Soviets and Americans did not go to war over Cuba, to say nothing of Canada and Spain over flatfish.

The goal of this book is to explain the facts of the past and present, not to augur the hypotheticals of the future. Still, you might ask, isn't it the essence of science to make falsifiable predictions? Shouldn't any claim to understanding the past be evaluated by its ability to extrapolate into the future? Oh, all right. I predict that the chance that a major episode of violence will break out in the next decade—a conflict with 100,000 deaths in a year, or a million deaths overall—is 9.7 percent. How did I come up with that number? Well, it's small enough to capture the intuition "probably not," but not so small that if such an event did occur I would be shown to be flat-out wrong. My point, of course, is that the concept of scientific prediction is meaningless when it comes to a single event—in this case, the eruption of mass violence in the next decade. It would be another thing if we could watch many worlds unfold and tot up the number in which an event happened or did not, but this is the only world we've got.

The truth is, I don't know what will happen across the entire world in the coming decades, and neither does anyone else. Not everyone, though, shares my reticence. A Web search for the text string "the coming war" returns two million hits, with completions like "with Islam," "with Iran," "with China," "with Russia," "in Pakistan," "between Iran and Israel," "between India and Pakistan," "against Saudi Arabia," "on Venezuela," "in America," "within the West," "for Earth's resources," "over climate," "for water," and "with Japan" (the last dating from 1991, which you would think would make everyone a bit more humble about this kind of thing). Books with titles like *The Clash of Civilizations, World on Fire, World War IV,* and (my favorite) *We Are Doomed* boast a similar confidence.

Who knows? Maybe they're right. My aim in the rest of this chapter is to point out that maybe they're wrong. This isn't the first time we've been warned of certain ruin. The experts have predicted civilization-ending aerial gas attacks, global thermonuclear war, a Soviet invasion of Western Europe, a Chinese razing of half of humanity, nuclear powers by the dozen, a revanchist Germany, a rising sun in Japan, cities overrun by teenage superpredators, a world war fought over diminishing oil, nuclear war between India and Pakistan, and weekly 9/11-scale attacks.[240] In this section I'll look at four threats to the New Peace—a civilizational clash with Islam, nuclear terrorism, a nuclear Iran, and climate change—and for each one make the case for "maybe, but maybe not."

The Muslim world, to all appearances, is sitting out the decline of violence. More than two decades of headlines have shocked Westerners with acts of barbarity in the name of Islam. Among them are the 1989 clerical death threat against Salman Rushdie for portraying Muhammad in a novel, the 2002 sentencing of an unmarried pregnant woman in Nigeria to execution by stoning, the fatal stabbing in 2004 of Dutch filmmaker Theo van Gogh for producing Ayaan Hirsi Ali's film about the treatment of women in Islamic countries, the lethal 2005 riots after a Danish newspaper printed editorial cartoons that were disrespectful to the prophet, the jailing and threat of flogging of a British schoolteacher in Sudan who allowed her class to name a teddy bear Muhammad, and of course the 9/11 terrorist attacks, in which nineteen Muslims killed almost three thousand civilians.

The impression that the Muslim world indulges kinds of violence that the West has outgrown is not a symptom of Islamophobia or Orientalism but is borne out by the numbers. Though about a fifth of the world's population is Muslim, and about a quarter of the world's countries have a Muslim majority, more than half of the armed conflicts in 2008 embroiled Muslim countries or insurgencies.[241] Muslim countries force a greater proportion of their citizens into their armies than non-Muslim countries do, holding other factors constant.[242] Muslim groups held two-thirds of the slots on the U.S. State Department's list of foreign terrorist organizations, and (as mentioned) in 2008 Sunni terrorists killed nearly two-thirds of the world's victims of terrorism whose perpetrators could be identified.[243]

In defiance of the rising tide of democracy, only about a quarter of Islamic countries elect their governments, and most of them are only dubiously democratic.[244] Their leaders receive farcically high percentages of the vote, and they exercise the power to jail opponents, outlaw opposition parties, suspend parliament, and cancel elections.[245] It's not just that Islamic countries happen to have risk factors for autocracy, such as being larger, poorer, or richer in oil. Even in a regression analysis that holds these factors constant, countries with larger proportions of Muslims have fewer political rights.[246] Political rights are very much a matter of violence, of course, since they amount to being able to speak, write, and assemble without being dragged off to jail.

The laws and practices of many Muslim countries seem to have missed out on the Humanitarian Revolution. According to Amnesty International, almost three-quarters of Muslim countries execute their criminals, compared to a third of non-Muslim countries, and many use cruel punishments such as stoning, branding, blinding, amputation of tongues or hands, and even crucifixion.[247] More than a hundred million females in Islamic countries have had their genitals mutilated, and many Muslim women have been disfigured with acid or killed outright if they displease their fathers, their brothers, or the husbands who had been forced upon them.[248] Islamic countries were the last to abolish slavery (as recently as 1962 in Saudi Arabia and 1980 in Mauritania), and a majority of the countries in which people continue to be trafficked are Muslim.[249] In many Muslim countries, witchcraft is not just on the books as a crime but is commonly prosecuted. In 2009, for example, Saudi Arabia convicted a man for carrying a phone booklet with characters in an alphabet from his native Eritrea, which the police interpreted as occult symbols. He was lashed three hundred times and imprisoned for more than three years.[250]

Violence is sanctioned in the Islamic world not just by religious superstition but by a hyperdeveloped culture of honor. The political scientists Khaled Fattah and K. M. Fierke have documented how a "discourse of humiliation" runs through the ideology of Islamist organizations.[251] A sweeping litany of affronts—the Crusades, the history of Western colonization, the existence of Israel, the presence of American troops on Arabian soil, the underperformance of Islamic countries—are taken as insults to Islam and used to license indiscriminate vengeance against members of the civilization they hold responsible, together with Muslim leaders of insufficient ideological purity. The radical fringe of Islam harbors an ideology that is classically genocidal: history is seen as a violent struggle that will culminate in the glorious subjugation of an irredeemably evil class of people. Spokesmen for Al Qaeda, Hamas, Hezbollah, and the Iranian regime have demonized enemy groups (Zionists, infidels, crusaders, polytheists), spoken of a millennial cataclysm that would usher in a utopia, and justified the killing of entire categories of people such as Jews, Americans, and those felt to insult Islam.[252]

The historian Bernard Lewis is not the only one who has asked, "What went wrong?" In 2002 a committee of Arab intellectuals under the auspices of the United Nations published the candid *Arab Human Development Report*, said to be "written by Arabs for Arabs."[253] The authors documented that Arab nations were plagued by political repression, economic backwardness, oppression of women, widespread illiteracy, and a self-imposed isolation from the world of ideas. At the time of the report, the entire Arab world exported fewer manufactured goods than the Philippines, had poorer Internet connectivity than sub-Saharan Africa, registered 2 percent as many patents per year as South Korea, and translated about a fifth as many books into Arabic as Greece translates into Greek.[254]

It wasn't always that way. During the Middle Ages, Islamic civilization was unquestionably more refined than Christendom. While Europeans were applying their ingenuity to the design of instruments of torture, Muslims were preserving classical Greek culture, absorbing the knowledge of the civilizations of India and China, and advancing astronomy, architecture, cartography, medicine, chemistry, physics, and mathematics. Among the symbolic legacies of this age are the "Arabic numbers" (adapted from India) and loan words such as *alcohol, algebra, alchemy, alkali, azimuth, alembic,* and *algorithm*. Just as the West had to come from behind to overtake Islam in science, so it was a laggard in human rights. Lewis notes:

> In most tests of tolerance, Islam, both in theory and in practice, compares unfavorably with the Western democracies as they have developed during the last two or three centuries, but very favorably with most other Christian and post-Christian societies and regimes. There is nothing in Islamic history to compare with the emancipation, acceptance, and integration of other-believers and non-believers in the West; but equally, there is nothing in Islamic history to compare with the Spanish expulsion of Jews and Muslims, the Inquisition, the *Auto da fé's*, the wars of religion, not to speak of more recent crimes of commission and acquiescence.[255]

Why did Islam blow its lead and fail to have an Age of Reason, an Enlightenment, and a Humanitarian Revolution? Some historians point to bellicose passages in the Koran, but compared to our own genocidal scriptures, they are nothing that some clever exegesis and evolving norms couldn't spin-doctor away.

Lewis points instead to the historical lack of separation between mosque and state. Muhammad was not just a spiritual leader but a political and military one, and only recently have any Islamic states had the concept of a distinction between the secular and the sacred. With every potential intellectual contribution filtered through religious spectacles, opportunities for absorbing and combining new ideas were lost. Lewis recounts that while works in philosophy and mathematics had been translated from classical Greek into Arabic, works of poetry, drama, and history were not. And while Muslims had a richly developed history of their own civilization, they were incurious about their Asian, African, and European neighbors and about their own pagan ancestors. The Ottoman heirs to classical Islamic civilization resisted the adoption of mechanical clocks, standardized weights and measures, experimental science, modern philosophy, translations of poetry and fiction, the financial instruments of capitalism, and perhaps most importantly, the printing press. (Arabic was the language in which the Koran was written, so printing it was considered an act of desecration.)[256] In chapter 4 I speculated that the Humanitarian Revolution in Europe was catalyzed by a literate cosmopolitanism, which expanded people's circle of empathy and set up a marketplace of ideas from which a liberal humanism could emerge. Perhaps the dead hand of religion impeded the flow of new ideas into the centers of Islamic civilization, locking it into a relatively illiberal stage of development. As if to prove the speculation correct, in 2010 the Iranian government restricted the number of university students who would be admitted to programs in the humanities, because, according to Supreme Leader Ayatollah Ali Khameini, study of the humanities "promotes skepticism and doubt in religious principles and beliefs."[257]

Whatever the historical reasons, a large chasm appears to separate Western and Islamic cultures today. According to a famous theory from the political scientist Samuel Huntington, the chasm has brought us to a new age in the history of the world: the clash of civilizations. "In Eurasia the great historic fault lines between civilizations are once more aflame," he wrote. "This is particularly true along the boundaries of the crescent-shaped Islamic bloc of nations, from the bulge of Africa to Central Asia. Violence also occurs between Muslims, on the one hand, and Orthodox Serbs in the Balkans, Jews in Israel, Hindus in India, Buddhists in Burma and Catholics in the Philippines. Islam has bloody borders."[258]

Though the dramatic notion of a clash of civilizations became popular among pundits, few scholars in international studies take it seriously. Too large a proportion of the world's bloodshed takes place within and between Islamic countries (for example, Iraq's war with Iran in the 1980s, and its invasion of Kuwait in 1990), and too large a proportion takes place within and between non-Islamic countries, for the civilizational fault line to be an accurate summary of violence in the world today. Also, as Nils Petter Gleditsch and

Halvard Buhaug have pointed out, even though an increasing *proportion* of the world's armed conflicts have involved Islamic countries and insurgencies over the past two decades (from 20 to 38 percent), it's not because those conflicts have increased in *number.* As figure 6–12 shows, Islamic conflicts continued at about the same rate while the rest of the world got more peaceful, the phenomenon I have been calling the New Peace.

Most important, the entire concept of "Islamic civilization" does a disservice to the 1.3 billion men and women who call themselves Muslims, living in countries as diverse as Mali, Nigeria, Morocco, Turkey, Saudi Arabia, Bangladesh, and Indonesia. And cutting across the divide of the Islamic world into continents and countries is another divide that is even more critical. Westerners tend to know Muslims through two dubious exemplars: the fanatics who grab headlines with their fatwas and jihads, and the oil-cursed autocrats who rule over them. The beliefs of the hitherto silent (and frequently silenced) majority make less of a contribution to our stereotypes. Can 1.3 billion Muslims really be untouched by the liberalizing tide that has swept the rest of the world in recent decades?

Part of the answer may be found in a massive Gallup poll conducted between 2001 and 2007 on the attitudes of Muslims in thirty-five countries representing 90 percent of the world's Islamic population.[259] The results confirm that most Islamic states will not become secular liberal democracies anytime soon. Majorities of Muslims in Egypt, Pakistan, Jordan, and Bangladesh

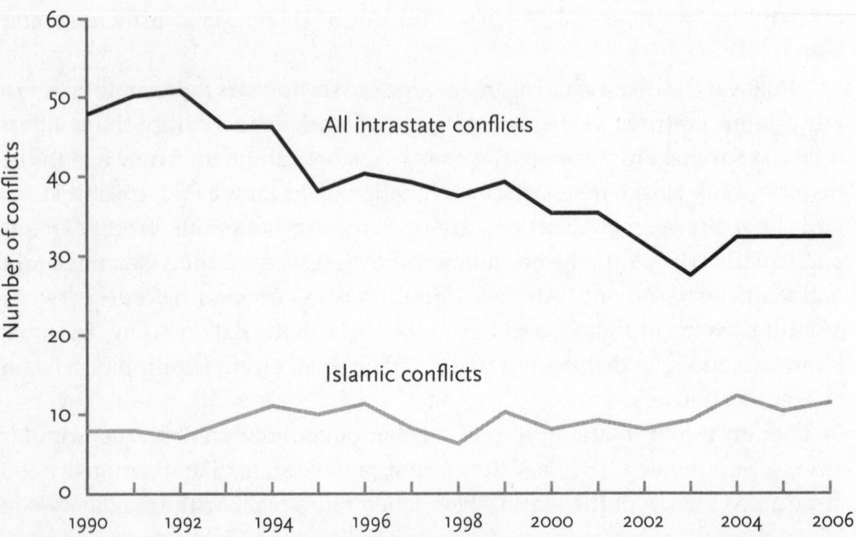

FIGURE 6–12. Islamic and world conflicts, 1990–2006

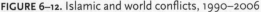

Source: Data from Gleditsch, 2008. "Islamic conflicts" involve Muslim countries or Islamic opposition movements or both. Data assembled by Halvard Buhaug from the UCDP/PRIO conflict dataset and his own coding of Islamic conflicts.

told the pollsters that Sharia, the principles behind Islamic law, should be the only source of legislation in their countries, and majorities in most of the countries said it should be at least one of the sources. On the other hand, a majority of Americans believe that the Bible should be one of the sources of legislation, and presumably they don't mean that people who work on Sunday should be stoned to death. Religion thrives on woolly allegory, emotional commitments to texts that no one reads, and other forms of benign hypocrisy. Like Americans' commitment to the Bible, most Muslims' commitment to Sharia is more a symbolic affiliation with moral attitudes they associate with the best of their culture than a literal desire to see adulteresses stoned to death. In practice, creative and expedient readings of Sharia for liberal ends have often prevailed against the oppressive fundamentalist readings. (The Nigerian woman, for example, was never executed.) Presumably that is why most Muslims see no contradiction between Sharia and democracy. Indeed, despite their professed affection for the idea of Sharia, a large majority believe that religious leaders should have no direct role in drafting their country's constitution.

Though most Muslims distrust the United States, it may not be out of a general animus toward the West or a hostility to democratic principles. Many Muslims feel the United States does *not* want to spread democracy in the Muslim world, and they have a point: the United States, after all, has supported autocratic regimes in Egypt, Jordan, Kuwait, and Saudi Arabia, rejected the election of Hamas in the Palestinian territories, and in 1953 helped overthrow the democratically elected Mossadegh in Iran. France and Germany are viewed more favorably, and between 20 and 40 percent say they admire the "fair political system, respect for human values, liberty, and equality" of Western culture. More than 90 percent would guarantee freedom of speech in their nation's constitution, and large numbers also support freedom of religion and freedom of assembly. Substantial majorities of both sexes in all the major Muslim countries say that women should be allowed to vote without influence from men, to work at any job, to enjoy the same legal rights as men, and to serve in the highest levels of government. And as we have seen, overwhelming majorities of the Muslim world reject the violence of Al Qaeda. Only 7 percent of the Gallup respondents approved the 9/11 attacks, and that was before Al Qaeda's popularity cratered in 2007.

What about mobilization for political violence? A team from the University of Maryland examined the goals of 102 grassroots Muslim organizations in North Africa and the Middle East and found that between 1985 and 2004 the proportion of organizations that endorsed violence dropped from 54 to 14 percent.[260] The proportion committed to nonviolent protests tripled, and the proportion that engaged in electoral politics doubled. These changes helped drive down the terrorism death curve in figure 6–11 and are reflected in the headlines, which feature far less terrorist violence in Egypt and Algeria than we read about a few years ago.

Islamic insularity is also being chipped at by a battery of liberalizing forces: independent news networks such as Al-Jazeera; American university campuses in the Gulf states; the penetration of the Internet, including social networking sites; the temptations of the global economy; and the pressure for women's rights from pent-up internal demand, nongovernmental organizations, and allies in the West. Perhaps conservative ideologues will resist these forces and keep their societies in the Middle Ages forever. But perhaps they won't.

In early 2011, as this book was going to press, a swelling protest movement deposed the leaders of Tunisia and Egypt and was threatening the regimes in Jordan, Bahrain, Libya, Syria, and Yemen. The outcome is unpredictable, but the protesters have been almost entirely nonviolent and non-Islamist, and are animated by a desire for democracy, good governance, and economic vitality rather than global jihad, the restoration of the caliphate, or death to infidels. Even with all these winds of change, it is conceivable that an Islamist tyrant or radical revolutionary group could drag an unwilling populace into a cataclysmic war. But it seems more probable that "the coming war with Islam" will never come. Islamic nations are unlikely to unite and challenge the West: they are too diverse, and they have no civilization-wide animus against us. Some Muslim countries, like Turkey, Indonesia, and Malaysia, are well on the way to becoming fairly liberal democracies. Some will continue to be ruled by SOBs, but they'll be our SOBs. Some will try to muddle through the oxymoron of a Sharia democracy. None is likely to be governed by the ideology of Al Qaeda. This leaves three reasonably foreseeable dangers to the New Peace: nuclear terrorism, the regime in Iran, and climate change.

Though conventional terrorism, as John Kerry gaffed, is a nuisance to be policed rather than a threat to the fabric of life, terrorism with weapons of mass destruction would be something else entirely. The prospect of an attack that would kill millions of people is not just theoretically possible but consistent with the statistics of terrorism. The computer scientists Aaron Clauset and Maxwell Young and the political scientist Kristian Gleditsch plotted the death tolls of eleven thousand terrorist attacks on log-log paper and saw them fall into a neat straight line.[261] Terrorist attacks obey a power-law distribution, which means they are generated by mechanisms that make extreme events unlikely, but not astronomically unlikely.

The trio suggested a simple model that is a bit like the one that Jean-Baptiste Michel and I proposed for wars, invoking nothing fancier than a combination of exponentials. As terrorists invest more time into plotting their attack, the death toll can go up exponentially: a plot that takes twice as long to plan can kill, say, four times as many people. To be concrete, an attack by a single suicide bomber, which usually kills in the single digits, can be planned in a few days or weeks. The 2004 Madrid train bombings, which killed around two hundred, took six months to plan, and 9/11, which killed three thousand, took two years.[262]

But terrorists live on borrowed time: every day that a plot drags on brings the possibility that it will be disrupted, aborted, or executed prematurely. If the probability is constant, the plot durations will be distributed exponentially. (Cronin, recall, showed that terrorist organizations drop like flies over time, falling into an exponential curve.) Combine exponentially growing damage with an exponentially shrinking chance of success, and you get a power law, with its disconcertingly thick tail. Given the presence of weapons of mass destruction in the real world, and religious fanatics willing to wreak untold damage for a higher cause, a lengthy conspiracy producing a horrendous death toll is within the realm of thinkable probabilities.

A statistical model, of course, is not a crystal ball. Even if we could extrapolate the line of existing data points, the massive terrorist attacks in the tail are still extremely (albeit not astronomically) unlikely. More to the point, we *can't* extrapolate it. In practice, as you get to the tail of a power-law distribution, the data points start to misbehave, scattering around the line or warping it downward to very low probabilities. The statistical spectrum of terrorist damage reminds us not to dismiss the worst-case scenarios, but it doesn't tell us how likely they are.

So how likely are they? What do you think the chances are that within the next five years each of the following scenarios will take place? (1) One of the heads of state of a major developed country will be assassinated. (2) A nuclear weapon will be set off in a war or act of terrorism. (3) Venezuela and Cuba will join forces and sponsor Marxist insurrection movements in one or more Latin American countries. (4) Iran will provide nuclear weapons to a terrorist group that will use one of them against Israel or the United States. (5) France will give up its nuclear arsenal.

I gave fifteen of these scenarios to 177 Internet users on a single Web page and asked them to estimate the probability of each. The median estimate that a nuclear bomb would be set off (scenario 2) was 0.20; the median estimate that a nuclear bomb would be set off in the United States or Israel by a terrorist group that obtained it from Iran (scenario 4) was 0.25. About half the respondents judged that the second scenario was more likely than the first. And in doing so, they committed an elementary blunder in the mathematics of probability. The probability of a conjunction of events (A and B both occurring) cannot be greater than the probability of either of them occurring alone. The probability that you will draw a red jack has to be lower than the probability that you will draw a jack, because some jacks you might draw are not red.

Yet Tversky and Kahneman have shown that most people, including statisticians and medical researchers, commonly make the error.[263] Consider the case of Bill, a thirty-four-year-old man who is intelligent but also unimaginative, compulsive, and rather dull. In school he was strong in mathematics but undistinguished in the arts and humanities. What are the chances that Bill plays jazz saxophone? What are the chances that he is an accountant who plays jazz saxophone? Many people give higher odds to the second possibility, but the choice is nonsensical, because there are fewer saxophone-playing accountants than there are saxophone players. In judging probabilities, people rely on the vividness of their imaginations rather than thinking through the laws. Bill fits the stereotype of an accountant but not of a saxophonist, and our intuitions go with the stereotype.

The conjunction fallacy, as psychologists call it, infects many kinds of reasoning. Juries are more likely to believe that a man with shady business dealings killed an employee to prevent him from talking to the police than to believe that he killed the employee. (Trial lawyers thrive on this fallacy, adding conjectural details to a scenario to make it more vivid to a jury, even though every additional detail, mathematically speaking, ought to make it *less* probable.) Professional forecasters give higher odds to an unlikely outcome that is presented with a plausible cause (oil prices will rise, causing oil consumption to fall) than to the same outcome presented naked (oil consumption will fall).[264] And people are willing to pay more for flight insurance against terrorism than for flight insurance against all causes.[265]

You can see where I'm going. The mental movie of an Islamist terrorist group buying a bomb on the black market or obtaining it from a rogue state and then detonating it in a populated area is all too easy to play in our mind's eye. Even if it weren't, the entertainment industry has played it for us in nuclear terrorist dramas like *True Lies, The Sum of All Fears,* and *24*. The narrative is so riveting that we are apt to give it a higher probability than we would if we thought through all the steps that would have to go right for the disaster to happen and multiplied their probabilities. That's why so many of my survey respondents judged an Iran-sponsored nuclear terrorist attack to be more probable than a nuclear attack. The point is not that nuclear terrorism is impossible or even astronomically unlikely. It is just that the probability assigned to it by anyone but a methodical risk analyst is likely to be too high.

What do I mean by "too high"? "With certainty" and "more probable than not" strike me as too high. The physicist Theodore Taylor declared in 1974 that by 1990 it would be too late to prevent terrorists from carrying out a nuclear attack.[266] In 1995 the world's foremost activist on the risks of nuclear terrorism, Graham Allison, wrote that under prevailing circumstances, a nuclear attack on American targets was likely before the decade was out.[267] In 1998 the counterterrorism expert Richard Falkenrath wrote that "it is certain that more and more non-state actors will become capable of nuclear, biological, and chemical weapons acquisition and use."[268] In 2003 UN ambassador John Negroponte judged that there was a "high probability" of an attack with a weapon of mass destruction within two years. And in 2007 the physicist Richard Garwin estimated that the chance of a nuclear terrorist attack was 20 percent per year, or about 50 percent by 2010 and almost 90 percent within a decade.[269]

Like television weather forecasters, the pundits, politicians, and terrorism specialists have every incentive to emphasize the worst-case scenario. It is undoubtedly wise to scare governments into taking extra measures to lock down weapons and fissile material and to monitor and infiltrate groups that might be tempted to acquire them. Overestimating the risk, then, is safer than underestimating it—though only up to a point, as the costly invasion of Iraq in search of nonexistent weapons of mass destruction proves. The professional reputations of experts have proven to be immune to predictions of disasters that never happen, while almost no one wants to take a chance at giving the all-clear and ending up with radioactive egg on his face.[270]

A few brave analysts, such as Mueller, John Parachini, and Michael Levi, have taken the chance by examining the disaster scenarios component by component.[271] For starters, of the four so-called weapons of mass destruction, three are far less massively destructive than good old-fashioned explosives.[272] Radiological or "dirty" bombs, which are conventional explosives wrapped in radioactive material (obtained, for example, from medical waste), would yield only minor and short-lived elevations of radiation, comparable to moving to a city at a higher altitude. Chemical weapons, unless they are released in an enclosed space like a subway (where they would still not do as much damage as conventional explosives), dissipate quickly, drift in the wind, and are broken down by sunlight. (Recall that poison gas was responsible for a tiny fraction of the casualties in World War I.) Biological weapons capable of causing epidemics would be prohibitively expensive to develop and deploy, as well as dangerous to the typically bungling amateur labs that would develop them. It's no wonder that biological and chemical weapons, though far more accessible than nuclear ones, have been used in only three terrorist attacks in thirty years.[273] In 1984 the Rajneeshee religious cult contaminated salad in the restaurants of an Oregon town with salmonella, sickening 751 people and killing none. In 1990 the Tamil Tigers were running low on ammunition while attacking a fort and opened up some chlorine cylinders they found in a nearby paper mill, injuring 60 and killing none before the gas wafted back over them and convinced them never to try it again. The Japanese religious cult Aum Shinrikyo failed in ten attempts to use biological weapons before releasing sarin gas in the Tokyo subways, killing 12. A fourth attack, the 2001 anthrax mailings that killed 5 Americans in media and government offices, turned out to be a spree killing rather than an act of terrorism.

It's really only nuclear weapons that deserve the WMD acronym. Mueller and Parachini have fact-checked the various reports that terrorists got "just this close" to obtaining a nuclear bomb and found that all were apocryphal. Reports of "interest" in procuring weapons on a black market grew into accounts of actual negotiations, generic sketches morphed into detailed blueprints, and flimsy clues (like the aluminum tubes purchased in 2001 by Iraq) were overinterpreted as signs of a development program.

Each of the pathways to nuclear terrorism, when examined carefully, turns out to have gantlets of improbabilities. There may have been a window of vulnerability in the safekeeping of nuclear weapons in Russia, but today most experts agree it has been closed, and that no loose nukes are being peddled in a nuclear bazaar. Stephen Younger, the former director of nuclear weapons research at Los Alamos National Laboratory, has said, "Regardless of what is reported in the news, all nuclear nations take the security of their weapons very seriously."[274] Russia has an intense interest in keeping its weapons out of the hands of Chechen and other ethnic separatist groups, and Pakistan is just as worried about its archenemy Al Qaeda. And contrary to rumor, security experts consider the chance that Pakistan's government and military command will fall under the control of Islamist extremists to be essentially nil.[275] Nuclear weapons have complex interlocks designed to prevent unauthorized deployment, and most of them become "radioactive scrap metal" if they are not maintained.[276] For these reasons, the forty-seven-nation Nuclear Security Summit convened by Barack Obama in 2010 to prevent nuclear terrorism concentrated on the security of fissile material, such as plutonium and highly enriched uranium, rather than on finished weapons.

The dangers of filched fissile material are real, and the measures recommended at the summit are patently wise, responsible, and overdue. Still, one shouldn't get so carried away by the image of garage nukes as to think they are inevitable or even extremely probable. The safeguards that are in place or will be soon will make fissile materials hard to steal or smuggle, and if they went missing, it would trigger an international manhunt. Fashioning a workable nuclear weapon requires precision engineering and fabrication techniques well beyond the capabilities of amateurs. The Gilmore commission, which advises the president and Congress on WMD terrorism, called the challenge "Herculean," and Allison has described the weapons as "large, cumbersome, unsafe, unreliable, unpredictable, and inefficient."[277] Moreover, the path to getting the materials, experts, and facilities in place is mined with hazards of detection, betrayal, stings, blunders, and bad luck. In his book *On Nuclear Terrorism*, Levi laid out all the things that would have to go right for a terrorist nuclear attack to succeed, noting, "Murphy's Law of Nuclear Terrorism: What can go wrong might go wrong."[278] Mueller counts twenty obstacles on the path and notes that even if a terrorist group had a fifty-fifty chance of clearing every one, the aggregate odds of its success would be one in a million. Levi brackets the range from the other end by estimating that even if the path were strewn with only ten obstacles, and the probability that each would be cleared was 80 percent, the aggregate odds of success facing a nuclear terrorist group would be one in ten. Those are not our odds of becoming victims. A terrorist group weighing its options, even with these overly optimistic guesstimates, might well conclude from the long odds that it would better off devoting its resources to projects with a higher chance of success. None of this, to repeat, means that nuclear terrorism is impossible, only that it is not, as so many people insist, imminent, inevitable, or highly probable.

If current pundits are to be believed, then as you are reading these words the New Peace will already have been shattered by a major war, perhaps a nuclear war, with Iran. At the time of this writing, tensions have been rising over the country's nuclear energy program. Iran is currently enriching enough uranium to fashion a nuclear arsenal, and it has defied international demands that it allow inspections and comply with other provisions of the Nuclear Nonproliferation Treaty. The president of Iran, Mahmoud Ahmadinejad, has taunted Western leaders, supported terrorist groups, accused the United States of orchestrating the 9/11 attacks, denied the Holocaust, called for Israel to be "wiped off the map," and prayed for the reappearance of the Twelfth Imam, the Muslim savior who would usher in an age of peace and justice. In some interpretations of Shi'a Islam, this messiah will show up after a worldwide eruption of war and chaos.

All this is, to say the least, disconcerting, and many writers have concluded that Ahmadinejad is another Hitler who will soon develop nuclear weapons and use them on Israel or furnish them to Hezbollah to do so. Even in less dire scenarios, he could blackmail the Middle East into acceding to Iranian hegemony. The prospect might leave Israel or the United States no choice but to bomb its nuclear facilities preemptively, even if it invited years of war and terrorism in response. A 2009 editorial in the *Washington Times* spelled it out: "War with Iran is now inevitable. The only question is: Will it happen sooner or later?"[279]

This chilling scenario of a nuclear attack by Iranian fanatics is certainly possible. But is it *inevitable*, or even highly likely? One can be just as contemptuous of Ahmadinejad, and just as cynical about his motives, while imagining less dire alternatives for the world ahead. John Mueller, Thomas Schelling, and many other foreign affairs analysts have imagined them for us and have concluded that the Iranian nuclear program is not the end of the world.[280]

Iran is a signatory to the Nuclear Nonproliferation Treaty, and Ahmadinejad has repeatedly declared that Iran's nuclear program is intended only for energy and medical research. In 2005 Supreme Leader Khameini (who wields more power than Ahmadinejad) issued a fatwa declaring that nuclear weapons are forbidden under Islam.[281] If the government went ahead and developed the weapons anyway, it would not be the first time in history that national leaders have lied through their teeth. But having painted themselves into this corner, the prospect of forfeiting all credibility in the eyes of the world (including major powers on whom they depend, like Russia, China, Turkey, and Brazil) might at least give them pause.

Ahmadinejad's musings about the return of the Twelfth Imam do not necessarily mean that he plans to hasten it along with a nuclear holocaust. Two of the deadlines by which writers confidently predicted that he would set off the apocalypse (2007 and 2009) have already come and gone.[282] And for what it's worth, here is how he explained his beliefs in a 2009 television interview with NBC correspondent Ann Curry:

> *Curry:* You've said that you believe that his arrival, the apocalypse, would happen in your own lifetime. What do you believe that you should do to hasten his arrival?
>
> *Ahmadinejad:* I have never said such a thing. . . . I was talking about peace. . . . What is being said about an apocalyptic war and—global war, things of that nature. This is what the Zionists are claiming. Imam . . . will come with logic, with culture, with science. He will come so that there is no more war. No more enmity, hatred. No more conflict. He will call on everyone to enter a brotherly love. Of course, he will return with Jesus Christ. The two will come back together. And working together, they would fill this world with love. The stories that have been disseminated around the world about extensive war, apocalyptic wars, so on and so forth, these are false.[283]

As a Jewish atheist, I can't say I find these remarks completely reassuring. But with one obvious change they are not appreciably different from those held by devout Christians; indeed, they are milder, as many Christians do believe in an apocalyptic war and have fantasized about it in bestselling novels. As for the speech containing the phrase that was translated as "wiping Israel off the map," the *New York Times* writer Ethan Bronner consulted Persian translators and analysts of Iranian government rhetoric on the meaning of the phrase in context, and they were unanimous that Ahmadinejad was daydreaming about regime change in the long run, not genocide in the days ahead.[284] The perils of translating foreign bombast bring to mind Khrushchev's boast "We will bury you," which turned out to mean "outlive" rather than "entomb."

There is a parsimonious alternative explanation of Iran's behavior. In 2002 George W. Bush identified Iraq, North Korea, and Iran as the "axis of evil" and proceeded to invade Iraq and depose its leadership. North Korea's leaders saw the writing on the wall and promptly developed a nuclear capability, which (as they no doubt anticipated) has put an end to any musings about the United States invading them too. Shortly afterward Iran put its nuclear program into high gear, aiming to create enough ambiguity as to whether it possesses nuclear weapons, or could assemble them quickly, to squelch any thought of an invasion in the mind of the Great Satan.

If Iran does become a confirmed or suspected nuclear power, the history of the nuclear age suggests that the most likely outcome would be nothing. As we have seen, nuclear weapons have turned out to be useless for anything but deterrence against annihilation, which is why the nuclear powers have repeatedly been defied by their nonnuclear adversaries. The most recent episode of proliferation bears this out. In 2004 it was commonly predicted that if North Korea acquired a nuclear capability, then by the end of the decade it would share it with terrorists and set off a nuclear arms race with South Korea, Japan, and Taiwan.[285] In fact, North Korea did acquire a nuclear capability, the end of the decade has come and gone, and nothing has happened. It's also unlikely that any nation would furnish nuclear ammunition to the loose cannons of a terrorist band, thereby giving up control over how they would be used while being on the hook for the consequences.[286]

In the case of Iran, before it decided to bomb Israel (or license Hezbollah to do so in an incriminating coincidence), with no conceivable benefit to itself, its leaders would have to anticipate a nuclear reprisal by Israeli commanders, who could match them hothead for hothead, together with an invasion by a coalition of powers enraged by the violation of the nuclear taboo. Though the regime is detestable and in many ways irrational, one wonders whether its principals are so indifferent to continuing their hold on power as to choose to annihilate themselves in pursuit of perfect justice in a radioactive Palestine or the arrival of the Twelfth Imam, with or without Jesus at his side. As Thomas Schelling asked in his 2005 Nobel Prize lecture, "What else can Iran accomplish, except possibly the destruction of its own system, with a few nuclear warheads? Nuclear weapons should be too precious to give away or to sell, too precious to waste killing people when they could, held in reserve, make the United States, or Russia, or any other nation, hesitant to consider military action."[287]

Though it may seem dangerous to consider alternatives to the worst-case scenario, the dangers go both ways. In the fall of 2002 George W. Bush warned the nation, "America must not ignore the threat gathering against us. Facing clear evidence of peril, we cannot wait for the final proof—the smoking gun—that could come in the form of a mushroom cloud." The "clear evidence" led to a war that has cost more than a hundred thousand lives and almost a trillion dollars and has left the world no safer. A cocksure certainty that Iran will use nuclear weapons, in defiance of sixty-five years of history in which authoritative predictions of inevitable catastrophes were repeatedly proven wrong, could lead to adventures with even greater costs.

These days one other gloomy scenario is on people's minds. Global temperatures are increasing, which in the decades ahead could lead to a rising sea level, desertification, droughts in some regions, and floods and hurricanes in others. Economies will be disrupted, leading to a competition for resources, and populations will migrate out of distressed regions, leading to friction with their unwelcoming hosts. A 2007 *New York Times* op-ed warned, "Climate stress may well represent a challenge to international security just as dangerous—and more intractable—than the arms race between the United States and the Soviet Union during the Cold War or the proliferation of nuclear weapons among rogue states today."[288] That same year Al Gore and the Intergovernmental Panel on Climate Change were awarded the Nobel Peace Prize for their call to action against global warming because, according to the citation, climate change is a threat to international security. A rising fear lifts all the boats. Calling global warming "a force multiplier for instability," a group of military officers wrote that "climate change will provide the conditions that will extend the war on terror."[289]

Once again it seems to me that the appropriate response is "maybe, but maybe not." Though climate change can cause plenty of misery and deserves to be mitigated for that reason alone, it will not necessarily lead to armed conflict. The political scientists who track war and peace, such as Halvard Buhaug, Idean Salehyan, Ole Theisen, and Nils Gleditsch, are skeptical of the popular idea that people fight wars over scarce resources.[290] Hunger and resource shortages are tragically common in sub-Saharan countries such as Malawi, Zambia, and Tanzania, but wars involving them are not. Hurricanes, floods, droughts, and tsunamis (such as the disastrous one in the Indian Ocean in 2004) do not generally lead to armed conflict. The American dust bowl in the 1930s, to take another example, caused plenty of deprivation but no civil war. And while temperatures have been rising steadily in Africa during the past fifteen years, civil wars and war deaths have been falling. Pressures on access to land and water can certainly cause local skirmishes, but a genuine war requires that hostile forces be organized and armed, and that depends more on the influence of bad governments, closed economies, and militant ideologies than on the sheer availability of land and water. Certainly any connection to terrorism is in the imagination of the terror warriors: terrorists tend to be underemployed lower-middle-class men, not subsistence farmers.[291] As for genocide, the Sudanese government finds it convenient to blame violence in Darfur on desertification, distracting the world from its own role in tolerating or encouraging the ethnic cleansing.

In a regression analysis on armed conflicts from 1980 to 1992, Theisen found that conflict was more likely if a country was poor, populous, politically unstable, and abundant in oil, but not if it had suffered from droughts, water shortages, or mild land degradation. (Severe land degradation did have a small effect.) Reviewing analyses that examined a large number (N) of countries rather than cherry-picking one or two, he concluded, "Those who foresee doom, because of the relationship between resource scarcity and violent internal conflict, have very little support in the large-N literature." Salehyan adds that relatively inexpensive advances in water use and agricultural practices in the developing world can yield massive increases in productivity with a constant or even shrinking amount of land, and that better governance can mitigate the human costs of environmental damage, as it does in developed democracies. Since the state of the environment is at most one ingredient in a mixture that depends far more on political and social organization, resource wars are far from inevitable, even in a climate-changed world.

No reasonable person would prophesy that the New Peace is going to be a long peace, to say nothing of a perpetual peace. There will certainly be wars and terrorist attacks in the decades to come, possibly large ones. On top of the known unknowns—militant Islamism, nuclear terrorists, environmental degradation—there are surely many unknown unknowns. Perhaps new leaders in China will decide to engulf Taiwan once and for all, or Russia will swallow a former Soviet republic or two, provoking an American response. Maybe an aggressive Chavismo will spill out of Venezuela and incite Marxist insurgencies and brutal counterinsurgencies throughout the developing world. Perhaps at this very moment terrorists from some liberation movement no one has heard of are plotting an attack of unprecedented destruction, or an eschatological ideology is fermenting in the mind of a cunning fanatic who will take over a major country and plunge the world back into war. As the *Saturday Night Live* news analyst Roseanne Roseannadanna observed, "It's always something. If it's not one thing, it's another."

But it is just as foolish to let our lurid imaginations determine our sense of the probabilities. It may always be something, but there can be fewer of those things, and the things that happen don't have to be as bad. The numbers tell us that war, genocide, and terrorism have declined over the past two decades—not to zero, but by a lot. A mental model in which the world has a constant allotment of violence, so that every cease-fire is reincarnated somewhere else as a new war, and every interlude of peace is just a time-out in which martial tensions build up and seek release, is factually mistaken. Millions of people are alive today because of the civil wars and genocides that did not take place but that would have taken place if the world had remained as it was in the 1960s, 1970s, and 1980s. The conditions that favored this happy outcome—democracy, prosperity, decent government, peacekeeping, open economies, and the decline of antihuman ideologies—are not, of course, guaranteed to last forever. But nor are they likely to vanish overnight.

Of course we live in a dangerous world. As I have emphasized, a statistical appreciation of history tells us that violent catastrophes may be improbable, but they are not astronomically improbable. Yet that can also be stated in a more hopeful way. Violent catastrophes may not be astronomically improbable, but they are improbable.

7

THE RIGHTS REVOLUTIONS

I have a dream that one day this nation will rise up and live out the true meaning of its creed: "We hold these truths to be self-evident: that all men are created equal."

—Martin Luther King, Jr.

When I was a boy, I was not particularly strong, swift, or agile, and that made organized sports a gantlet of indignities. Basketball meant chucking a series of airballs in the general direction of the backboard. Rope-climbing left me suspended a foot above the floor like a clump of seaweed on a fishing line. Baseball meant long interludes in sun-scorched right field praying that no fly ball would come my way.

But one talent saved me from perpetual pariahhood among my peers: I was not afraid of pain. As long as the blows were delivered fair and square and without ad hominem humiliation, I could mix it up with the best of them. The boy culture that flourished in a parallel universe to that of gym teachers and camp counselors offered many opportunities to redeem myself.

There was pickup hockey and tackle football (sans helmet and pads), where I could check and be checked into the boards, or dive for fumbles in a scrum of bodies. There was murderball, in which one boy clutched a volleyball and counted off the seconds while the others pummeled him until he let go. There was Horse (strictly forbidden by the counselors, doubtless on the orders of lawyers), in which a fat kid ("the pillow") would lean back against a tree, a teammate would bend over and hold him around the waist, and the rest of the team would form a line of backs by holding the waist of the kid in front of him. Each member of the opposing team would then take a running leap and come crashing down on the back of the "horse" until it either collapsed to the ground or supported the riders for three seconds. And during the evening there was Knucks, the outlawed card game in which the loser would be thwhacked on the knuckles with the deck of cards, the number of edge-on and face-on thwacks determined by the point spread and restrained by a complex set of rules about flinching, scraping, and excess force. Mothers would regularly inspect our knuckles for incriminating scabs and bruises.

Nothing organized by grown-ups could compare with these delirious pleasures. The closest they came was dodgeball, with its ecstatic chaos of hiding behind aggressive teammates, ducking projectiles, diving to the floor, and cheating death until the final mortal smack of rubber against skin. It was the only sport in the Orwellianly named "physical education" curriculum that I actually looked forward to.

But now the Boy Gender has lost another battle in its age-old war with camp counselors, phys ed teachers, lawyers, and moms. In school district after school district, *dodgeball has been banned.* A statement by the National Association for Sport and Physical Education, which must have been written by someone who was never a boy, and quite possibly has never met one, explained the reason:

> NASPE believes that dodgeball is not an appropriate activity for K–12 school physical education programs. Some kids may like it—the most skilled, the most confident. But many do not! Certainly not the student who gets hit hard in the stomach, head, or groin. And it is not appropriate to teach our children that you win by hurting others.

Yes, the fate of dodgeball is yet another sign of the historical decline of violence. Recreational violence has a long ancestry in our lineage. Play fighting is common among juvenile primate males, and rough-and-tumble play is one of the most robust sex differences in humans.[1] The channeling of these impulses into extreme sports has been common across cultures and throughout history. Together with Roman gladiatorial combat and medieval jousting tournaments, the bloody history of sports includes recreational fighting with sharp sticks in Renaissance Venice (where noblemen and priests would join in the fun), the Sioux Indian pastime in which boys would try to grab their opponents' hair and knee them in the face, Irish faction fights with stout oak clubs called shillelaghs, the sport of shin-kicking (popular in the 19th-century American South) in which the contestants would lock forearms and kick each other in the shins until one collapsed, and the many forms of bare-knuckle fights whose typical tactics may be inferred from the current rules of boxing (no head-butting, no hitting below the belt, and so on).[2]

But in the past half-century the momentum has been going squarely against boys of all ages. Though people have lost none of their taste for consuming simulated and voluntary violence, they have engineered social life to place the most tempting kinds of real-life violence off-limits. It is part of a current in which Western culture has been extending its distaste for violence farther and farther down the magnitude scale. The postwar revulsion against forms of violence that kill by the millions and thousands, such as war and genocide, has spread to forms that kill by the hundreds, tens, and single digits, such as rioting, lynching, and hate crimes. It has extended from killing to other forms of harm such as rape, assault, battering, and intimidation. It has spread to vulnerable classes of victims that in earlier eras fell outside the circle of protection, such as racial minorities, women, children, homosexuals, and animals. The ban on dodgeball is a weathervane for these winds of change.

The efforts to stigmatize, and in many cases criminalize, temptations to violence have been advanced in a cascade of campaigns for "rights"—civil rights, women's rights, children's rights, gay rights, and animal rights. The movements are tightly bunched in the second half of the 20th century, and I will refer to them as the Rights Revolutions. The contagion of rights in this era may be seen in figure 7–1, which plots the proportion of English-language books (as a percentage of the proportions in 2000) that contain the phrases *civil rights, women's rights, children's rights, gay rights,* and *animal rights* between 1948 (which symbolically inaugurated the era with the signing of the Declaration of Human Rights) and 2000.

As the era begins, the terms *civil rights* and *women's rights* already have a presence, because the ideas had been in the nation's consciousness since the 19th century. *Civil rights* shot up between 1962 and 1969, the era of the most dramatic legal victories of the American civil rights movement. As it began to level off, *women's rights* began its ascent, joined shortly by *children's rights;* then, in the 1970s, *gay rights* appeared on the scene, followed shortly by *animal rights.*

These staggered rises tell a story. Each of the movements took note of the

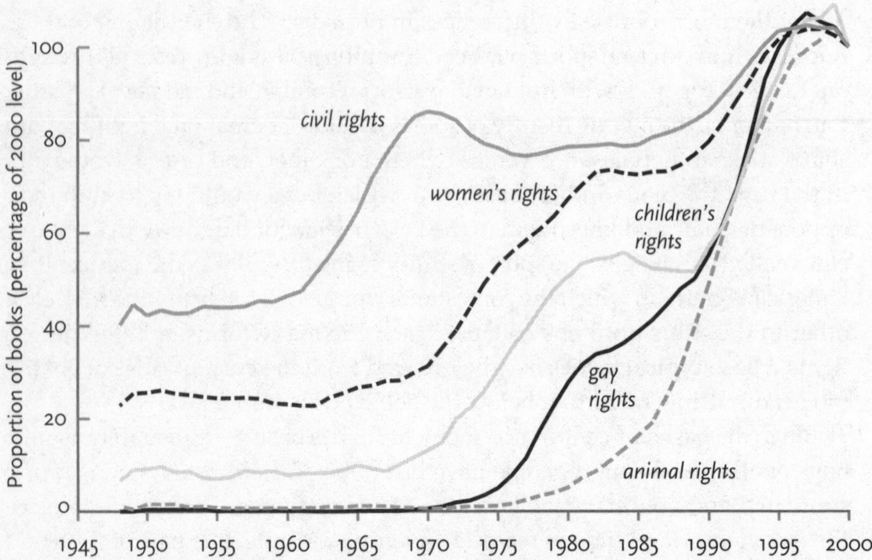

FIGURE 7–1. Use of the terms *civil rights, women's rights, children's rights, gay rights,* and *animal rights* in English-language books, 1948–2000

Source: Five million books digitized by Google Books, analyzed by the Bookworm program, Michel et al., 2011. Bookworm is a more powerful version of the Google Ngram Viewer (ngrams.googlelabs.com), and can analyze the proportion of books, in addition to the proportion of the corpus, in which a search string is found. Plotted as a percentage of the proportion of books containing each term in the year 2000, with a moving average of five years.

success of its predecessors and adopted some of their tactics, rhetoric, and most significantly, moral rationale. During the Humanitarian Revolution two centuries earlier, a cascade of reforms tumbled out in quick succession, instigated by intellectual reflection on entrenched customs, and connected by a humanism that elevated the flourishing and suffering of individual minds over the color, class, or nationality of the bodies that housed them. Then and now the concept of individual rights is not a plateau but an escalator. If a sentient being's right to life, liberty, and the pursuit of happiness may not be compromised because of the color of its skin, then why may it be compromised because of other irrelevant traits such as gender, age, sexual preference, or even species? Dull habit or brute force may prevent people in certain times and places from following this line of argument to each of its logical conclusions, but in an open society the momentum is unstoppable.

The Rights Revolutions replayed some of the themes of the Humanitarian Revolution, but they also replayed one feature of the Civilizing Process. During the transition to modernity, people did not fully appreciate that they were undergoing changes aimed at reducing violence, and once the changes were entrenched, the process was forgotten. When Europeans were mastering norms of self-control, they felt like they were becoming more civilized and courteous, not that they were part of a campaign to drive the homicide statistics downward. Today we give little thought to the rationale behind the customs left behind by that change, such as the revulsion to dinnertime dagger attacks that left us with the condemnation of eating peas with a knife. Likewise the sanctity of religion and "family values" in red-state America is no longer remembered as a tactic to pacify brawling men in cowboy towns and mining camps.

The prohibition of dodgeball represents the overshooting of yet another successful campaign against violence, the century-long movement to prevent the abuse and neglect of children. It reminds us of how a civilizing offensive can leave a culture with a legacy of puzzling customs, peccadilloes, and taboos. The code of etiquette bequeathed by this and the other Rights Revolutions is pervasive enough to have acquired a name. We call it political correctness.

The Rights Revolutions have another curious legacy. Because they are propelled by an escalating sensitivity to new forms of harm, they erase their own tracks and leave us amnesic about their successes. As we shall see, the revolutions have brought us measurable and substantial declines in many categories of violence. But many people resist acknowledging the victories, partly out of ignorance of the statistics, partly because of a mission creep that encourages activists to keep up the pressure by denying that progress has been made. The racial oppression that inspired the first generations of the civil rights movement was played out in lynchings, night raids, antiblack pogroms, and physical intimidation at the ballot box. In a typical battle of today, it may consist of African American drivers being pulled over more often on the highways. (When Clarence Thomas described his successful but contentious 1991

Supreme Court confirmation hearing as a "high-tech lynching," it was the epitome of tastelessness but also a sign of how far we have come.) The oppression of women used to include laws that allowed husbands to rape, beat, and confine their wives; today it is applied to elite universities whose engineering departments do not have a fifty-fifty ratio of male and female professors. The battle for gay rights has progressed from repealing laws that execute, mutilate, or imprison homosexual men to repealing laws that define marriage as a contract between a man and a woman. None of this means we should be satisfied with the status quo or disparage the efforts to combat remaining discrimination and mistreatment. It's just to remind us that the first goal of any rights movement is to protect its beneficiaries from being assaulted or killed. These victories, even if partial, are moments we should acknowledge, savor, and seek to understand.

CIVIL RIGHTS AND THE DECLINE OF LYNCHING AND RACIAL POGROMS

When most people think of the American civil rights movement, they recall a twenty-year run of newsworthy events. It began in 1948, when Harry Truman ended segregation in the U.S. armed forces; accelerated through the 1950s, when the Supreme Court banned segregated schools, Rosa Parks was arrested for refusing to give up her bus seat to a white man, and Martin Luther King organized a boycott in response; climaxed in the early 1960s, when two hundred thousand people marched on Washington and heard King give perhaps the greatest speech in history; and culminated with the passage of the Voting Rights Act of 1965 and the Civil Rights Acts of 1964 and 1968.

Yet these triumphs were presaged by quieter but no less important ones. King began his 1963 speech by noting, "Five score years ago, a great American, in whose symbolic shadow we now stand, signed the Emancipation Proclamation . . . a great beacon-light of hope to millions of negro slaves." Yet "one hundred years later, the negro still is not free." The reason that African Americans did not exercise their rights in the intervening century was that they were intimidated by the threat of violence. Not only did the government use force in administering segregation and discriminatory laws, but African Americans were kept in their place by the category of violence called intercommunal conflict, in which one group of citizens—defined by race, tribe, religion, or language—targets another. In many parts of the United States, African American families were terrorized by organized thugs such as the Ku Klux Klan. And in thousands of incidents, a mob would publicly torture and execute an individual—a lynching—or visit an orgy of vandalism and murder on a community—a racial pogrom, also called a deadly ethnic riot.

In his definitive book on the deadly ethnic riot, the political scientist Donald Horowitz studied reports of 150 episodes of this form of intercommunal violence spanning fifty countries and laid out their common features.[3] An ethnic riot combines aspects of genocide and terrorism with features of its own. Unlike these two other forms of collective violence, it is not planned, has no articulated ideology, and is not masterminded by a leader or implemented by a government or militia, though it does depend on the government sympathizing with the perpetrators and looking the other way. Its psychological roots, though, are the same as those of genocide. One group essentializes the members of another and deems them less than human, inherently evil, or both. A mob forms and strikes against its target, either preemptively, in response to the Hobbesian fear of being targeted first, or retributively, in revenge for a dastardly crime. The inciting threat or crime is typically rumored, embellished, or invented out of whole cloth. The rioters are consumed by their hatred and strike with demonic fury. They burn and destroy assets rather than plundering them, and they kill, rape, torture, and mutilate members of the despised group at random rather than seeking the alleged wrongdoers. Usually they go after their victims with bladed weapons and other hands-on armaments rather than with firearms. The perpetrators (mostly young men, of course) carry out their atrocities in a euphoric frenzy and afterward feel no remorse for what they see as a justifiable response to an intolerable provocation. An ethnic riot doesn't destroy the targeted group, but it kills far greater numbers than terrorism; the death toll averages around a dozen but can range into the hundreds, the thousands, or (as in the nationwide rioting after the partition of India and Pakistan in 1947) the hundreds of thousands. Deadly ethnic riots can be an effective means of ethnic cleansing, sending millions of refugees from their homes in fear of their lives. And like terrorism, deadly riots can exact enormous costs in money and fear, leading to martial law, the abrogation of democracy, coups d'état, and secessionist warfare.[4]

Deadly ethnic riots are by no means an innovation of the 20th century. *Pogrom* is a Russian word that was applied to the frequent anti-Jewish riots in the 19th-century Pale of Settlement, which were just the latest wave in a millennium of intercommunal killings of Jews in Europe. In the 17th and 18th centuries England was swept by hundreds of deadly riots targeting Catholics. One response was a piece of legislation that a magistrate would publicly recite to a mob threatening them with execution if they did not immediately disperse. We remember this crowd-control measure in the expression *to read them the Riot Act*.[5]

The United States also has a long history of intercommunal violence. In the 17th, 18th, and 19th centuries just about every religious group came under assault in deadly riots, including Pilgrims, Puritans, Quakers, Catholics, Mormons, and Jews, together with immigrant communities such as Germans, Poles, Italians, Irish, and Chinese.[6] And as we saw in chapter 6, intercommunal violence against some Native American peoples was so complete that it can be placed in the category of genocide. Though the federal government did not perpetrate any overt genocides, it carried out several ethnic cleansings. The forced expulsion of "five civilized tribes" along the Trail of Tears from their southeastern homelands to present-day Oklahoma resulted in the deaths of tens of thousands from disease, hunger, and exposure. As recently as the 1940s, a hundred thousand Japanese Americans were forced into concentration camps because they were of the same race as the nation the country was fighting.

But the longest-running victims of intercommunal and government-indulged violence were African Americans.[7] Though we tend to think of lynching as a phenomenon of the American South, two of the most atrocious incidents took place in New York City: a 1741 rampage following rumors of a slave revolt in which many African Americans were burned at the stake, and the 1863 draft riots (depicted in the 2002 film *Gangs of New York*) in which at least fifty were lynched. In some years in the postbellum South, thousands of African Americans were killed, and the early 20th century saw race riots killing dozens at a time in more than twenty-five cites.[8]

Rioting of all kinds began to decrease in Europe in the mid-19th century. In the United States deadly rioting began to diminish at the century's end, and by the 1920s it had entered a terminal decline.[9] Using figures from the U.S. Census Bureau, James Payne tabulated the number of lynchings beginning in 1882 and found that they fell precipitously from 1890 to the 1940s (figure 7–2). During these decades, horrific lynchings continued to make the news, and shocking photographs of hanged and burned corpses were published in newspapers and circulated among activists, particularly the National Association

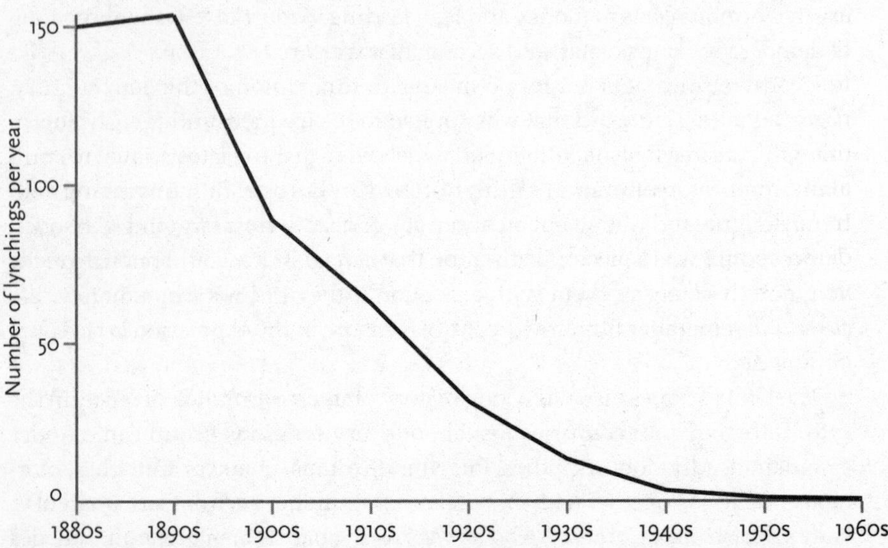

FIGURE 7–2. Lynchings in the United States, 1882–1969

Source: Graph from Payne, 2004, p. 182.

for the Advancement of Colored People. A 1930 photograph of a pair of men hanged in Indiana inspired a schoolteacher named Abel Meeropol to write a poem in protest:

> Southern trees bear strange fruit,
> Blood on the leaves and blood at the root,
> Black body swinging in the Southern breeze,
> Strange fruit hanging from the poplar trees.

(Meeropol and his wife, Anne, would later adopt the orphaned sons of Julius and Ethel Rosenberg, after the couple had been executed for Julius's passing of nuclear secrets to the Soviet Union.) When Meeropol put the poem to music, it became the signature tune of Billie Holiday, and in 1999 *Time* magazine called it the song of the century.[10] Yet in one of those paradoxes of timing that we have often stumbled upon, the conspicuous protest emerged at a time when the crime had already long been in decline. The last famous lynching case came to light in 1955, when fourteen-year-old Emmett Till was kidnapped, beaten, mutilated, and killed in Mississippi after allegedly whistling at a white woman. His murderers were acquitted by an all-white jury in a perfunctory trial.

Fears of a renewal of lynching were raised in the late 1990s, when a vicious murder stunned the nation. In 1998 three racists in Texas abducted an African American man, James Byrd, Jr., beat him senseless, chained him by the ankles to their pickup truck, and dragged him along the pavement for three miles until his body hit a culvert and was torn to pieces. Though the clandestine murder was very different from the lynchings of a century before, in which an entire community would execute a black person in a carnival atmosphere, the word *lynching* was widely applied to the crime. The murder took place a few years after the FBI had begun to gather statistics on so-called hate crimes, namely acts of violence that target a person because of race, religion, or sexual orientation. Since 1996 the FBI has published these statistics in annual reports, allowing us to see whether the Byrd murder was part of a disturbing new trend.[11] Figure 7–3 shows the number of African Americans who were murdered because of their race during the past dozen years. The numbers on the vertical axis do not represent homicides per 100,000 people; they represent the *absolute number* of homicides. Five African Americans were murdered because of their race in 1996, the first year in which records were published, and the number has since gone down to one per year. In a country with 17,000 murders a year, hate-crime murders have fallen into the statistical noise.

Far more common, of course, are the less serious forms of violence, such as aggravated assault (in which the assailant uses a weapon or causes an injury), simple assault, and intimidation (in which the victim is made to feel in danger for his or her personal safety). Though the absolute numbers of racially motivated incidents are alarming—several hundred assaults, several hundred

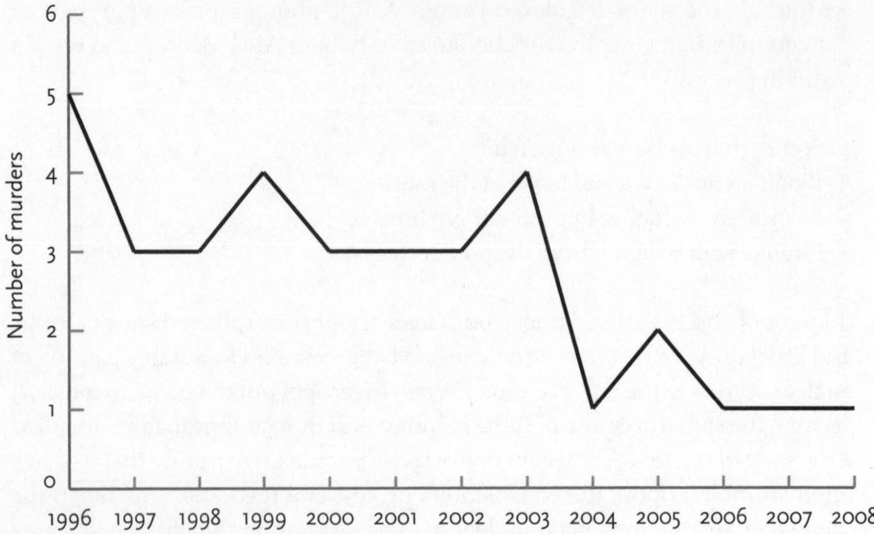

FIGURE 7–3. Hate-crime murders of African Americans, 1996–2008

Source: Data from the annual FBI reports of Hate Crime Statistics (http://www.fbi.gov/hq/cid/civilrights/hate.htm); see U.S. Federal Bureau of Investigation, 2010a.

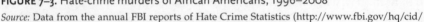

aggravated assaults, and a thousand acts of intimidation a year—they have to be put in the context of American crime numbers during much of that period, which included a *million* aggravated assaults per year. The rate of racially motivated aggravated assaults was about one-half of 1 percent of the rate of all aggravated assaults (322 per 100,000 people per year), and less than the rate that a person of any race would be murdered for any reason. And as figure 7–4 shows, since 1996 all three kinds of hate crime have been in decline.

As lynching died out, so did antiblack pogroms. Horowitz discovered that in the second half of the 20th century in the West, his subject matter, the deadly ethnic riot, ceased to exist.[12] The so-called race riots of the mid-1960s in Los Angeles, Newark, Detroit, and other American cities represented a different phenomenon altogether: African Americans were the rioters rather than the targets, death tolls were low (mostly rioters themselves killed by the police), and virtually all the targets were property rather than people.[13] After 1950 the United States had no riots that singled out a race or ethnic group; nor did other zones of ethnic friction in the West such as Canada, Belgium, Corsica, Catalonia, or the Basque Country.[14]

Some antiblack violence did erupt in the late 1950s and early 1960s, but it took a different form. The attacks are seldom called "terrorism," but that's exactly what they were: they were directed at civilians, low in casualties, high in publicity, intended to intimidate, and directed toward a political goal, namely preventing racial desegregation in the South. And like other terrorist

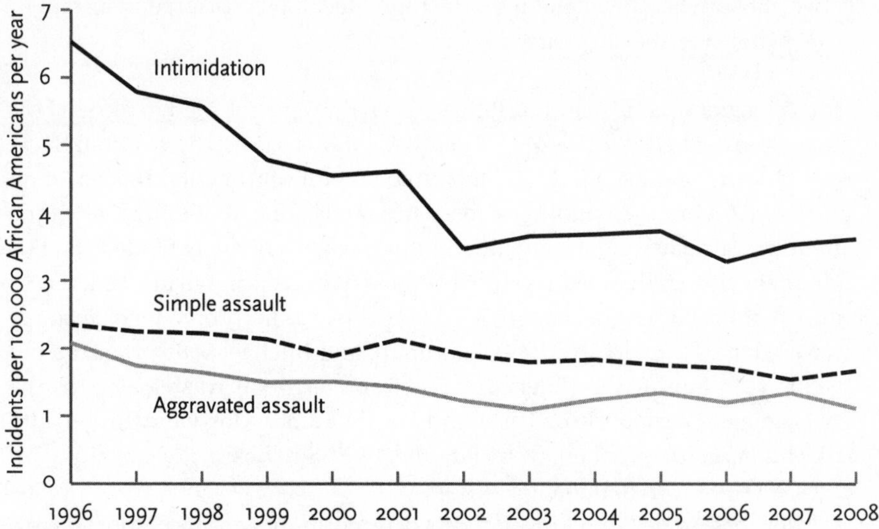

FIGURE 7–4. Nonlethal hate crimes against African Americans, 1996–2008

Source: Data from the annual FBI reports of Hate Crime Statistics (http://www.fbi.gov/hq/cid/civilrights/hate.htm); see U.S. Federal Bureau of Investigation, 2010a. The number of incidents is divided by the population covered by the agencies reporting the statistics multiplied by 0.129, the proportion of African Americans in the population according to the 2000 census.

campaigns, segregationist terrorism sealed its doom when it crossed the line into depravity and turned all public sympathy to its victims. In highly publicized incidents, ugly mobs hurled obscenities and death threats at black children for trying to enroll in all-white schools. One event that left a strong impression in cultural memory was the day six-year-old Ruby Nell Bridges had to be escorted by federal marshals to her first day of school in New Orleans. John Steinbeck, while driving through America to write his memoir *Travels with Charley*, found himself in the Big Easy at the time:

> Four big marshals got out of each car and from somewhere in the automobiles they extracted the littlest negro girl you ever saw, dressed in shining starchy white, with new white shoes on feet so little they were almost round. Her face and little legs were very black against the white.
>
> The big marshals stood her on the curb and a jangle of jeering shrieks went up from behind the barricades. The little girl did not look at the howling crowd, but from the side the whites of her eyes showed like those of a frightened fawn. The men turned her around like a doll and then the strange procession moved up the broad walk toward the school, and the child was even more a mite because the men were so big. Then the girl made a curious hop, and I think I know what it was. I think in her whole life she had not gone ten steps without skipping, but now in the middle of her first step, the

> weight bore her down and her little round feet took measured, reluctant steps between the tall guards.[15]

The incident was also immortalized in a painting published in 1964 in *Look* magazine titled *The Problem We All Live With.* It was painted by Norman Rockwell, the artist whose name is synonymous with sentimental images of an idealized America. In another conscience-jarring incident, four black girls attending Sunday school were killed in 1963 when a bomb exploded at a Birmingham church that had recently been used for civil rights meetings. That same year the civil rights worker Medgar Evers was murdered by Klansmen, as were James Chaney, Andrew Goodman, and Michael Schwerner the following year. Joining the violence by mobs and terrorists was violence by the government. The noble Rosa Parks and Martin Luther King were thrown into jail, and peaceful marchers were assaulted with fire hoses, dogs, whips, and clubs, all shown on national television.

After 1965, opposition to civil rights was moribund, antiblack riots were a distant memory, and terrorism against blacks no longer received support from any significant community. In the 1990s there was a widely publicized report of a string of arson attacks on black churches in the South, but it turned out to be apocryphal.[16] So for all the publicity that hate crimes have received, they have become a blessedly rare phenomenon in modern America.

Lynchings and race riots have declined for other ethnic groups and in other countries as well. The 9/11 attacks and the London and Madrid bombings were just the kind of symbolic provocation that in earlier decades could have led to anti-Muslim riots across the Western world. Yet no riots occurred, and a 2008 review of violence against Muslims by a human rights organization could not turn up a single clear case of a fatality in the West motivated by anti-Muslim hatred.[17]

Horowitz identifies several reasons for the disappearance of deadly ethnic riots in the West. One is governance. For all their abandon in assaulting their victims, rioters are sensitive to their own safety, and know when the police will turn a blind eye. Prompt law enforcement can quell riots and nip cycles of group-against-group revenge in the bud, but the procedures have to be thought out in advance. Since the local police often come from the same ethnic group as the perpetrators and may sympathize with their hatreds, a professionalized national militia is more effective than the neighborhood cops. And since riot police can cause more deaths than they prevent, they must be trained to apply the minimal force needed to disperse a mob.[18]

The other cause of the disappearance of deadly ethnic riots is more nebulous: a rising abhorrence of violence, and of even the slightest trace of a mindset that might lead to it. Recall that the main risk factor of genocides and deadly ethnic riots is an essentialist psychology that categorizes the members of a group as insensate obstacles, as disgusting vermin, or as avaricious, malignant, or heretical villains. These attitudes can be formalized into government policies of the kind that Daniel Goldhagen calls eliminationist and Barbara Harff calls exclusionary. The policies may be implemented as apartheid, forced assimilation, and in extreme cases, deportation or genocide. Ted Robert Gurr has shown that even discriminatory policies that fall short of the extremes are a risk factor for violent ethnic conflicts such as civil wars and deadly riots.[19]

Now imagine policies that are designed to be the diametric opposite of the exclusionary ones. They would not only erase any law in the books that singled out an ethnic minority for unfavorable treatment, but would swing to the opposite pole and mandate *anti*-exclusionary, *un*-eliminationist policies, such as the integration of schools, educational head starts, and racial or ethnic quotas and preferences in government, business, and education. These policies are generally called *remedial discrimination*, though in the United States they go by the name *affirmative action*. Whether or not the policies deserve credit for preventing a backsliding of developed countries into genocide and pogroms, they obviously are designed as the photographic negative of the exclusionary policies that caused or tolerated such violence in the past. And they have been riding a wave of popularity throughout the world.

In a report called "The Decline of Ethnic Political Discrimination 1950–2003," the political scientists Victor Asal and Amy Pate examined a dataset that records the status of 337 ethnic minorities in 124 countries since 1950.[20] (It overlaps with Harff's dataset on genocide, which we examined in chapter 6.) Asal and Pate plotted the percentage of countries with policies that discriminate against an ethnic minority, together with those that discriminate in favor of them. In 1950, as figure 7–5 shows, 44 percent of governments had invidious discriminatory policies; by 2003 only 19 percent did, and they were outnumbered by the governments that had remedial policies.

When Asal and Pate broke down the figures by region, they found that minority groups are doing particularly well in the Americas and Europe, where little official discrimination remains. Minority groups still experience legal discrimination in Asia, North Africa, sub-Saharan Africa, and especially the Middle East, though in each case there have been improvements since the end of the Cold War.[21] The authors conclude, "Everywhere the weight of official discrimination has lifted. While this trend began in Western democracies in the late 1960s, by the 1990s it had reached all parts of the world."[22]

Not only has official discrimination by governments been in decline, but so has the dehumanizing and demonizing mindset in individual people. This claim may seem incredible to the many intellectuals who insist that the United States is racist to the bone. But as we have seen throughout this book, for every moral advance in human history there have been social commentators who insist that we've never had it so bad. In 1968 the political scientist Andrew

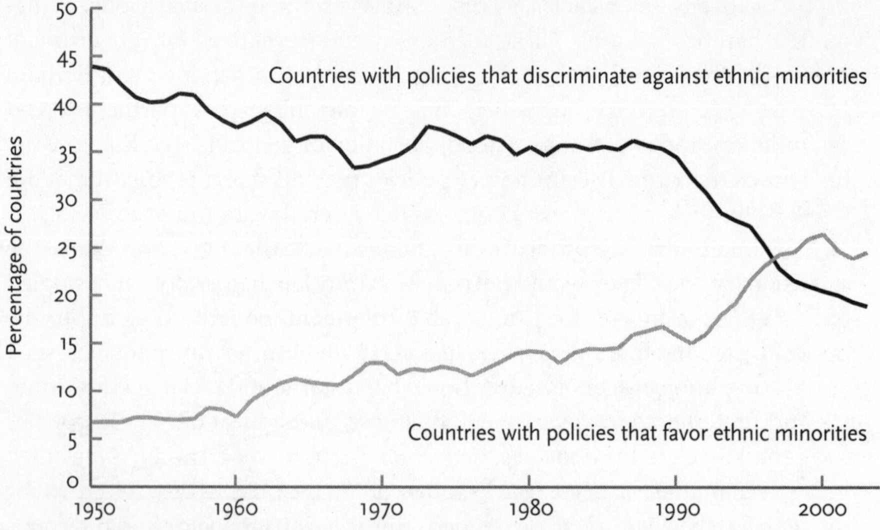

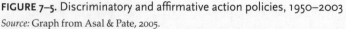

FIGURE 7–5. Discriminatory and affirmative action policies, 1950–2003
Source: Graph from Asal & Pate, 2005.

Hacker predicted that African Americans would soon rise up and engage in "dynamiting of bridges and water mains, firing of buildings, assassination of public officials and private luminaries. And of course there will be occasional rampages."[23] Undeterred by the dearth of dynamitings and the rarity of rampages, he followed up in 1992 with *Two Nations: Black and White, Separate, Hostile, Unequal,* whose message was "A huge racial chasm remains, and there are few signs that the coming century will see it closed."[24] Though the 1990s were a decade in which Oprah Winfrey, Michael Jordan, and Colin Powell were repeatedly named in polls as among the most admired Americans, gloomy assessments on race relations dominated literary life. The legal scholar Derrick Bell, for example, wrote in a 1992 book subtitled *The Permanence of Racism* that "racism is an integral, permanent, and indestructible component of this society."[25]

The sociologist Lawrence Bobo and his colleagues decided to see for themselves by examining the history of white Americans' attitudes toward African Americans.[26] They found that far from being indestructible, overt racism has been steadily disintegrating. Figure 7–6 shows that in the 1940s and early 1950s a majority of Americans said they were opposed to black children attending white schools, and as late as the early 1960s almost half said they would move away if a black family moved in next door. By the 1980s the percentages with these attitudes were in the single digits.

Figure 7–7 tells us that in the late 1950s only 5 percent of white Americans approved of interracial marriage. By the late 1990s two-thirds approved of it, and in 2008 almost 80 percent did. With some questions, like "Should blacks have

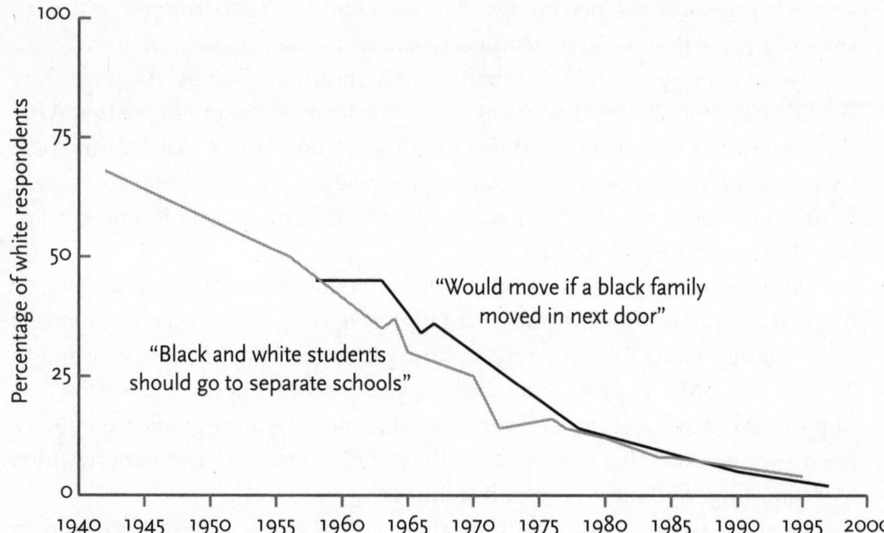

FIGURE 7–6. Segregationist attitudes in the United States, 1942–1997

Sources: "Separate schools": Data from Schuman, Steeh, & Bobo, 1997, originally gathered by the National Opinion Research Center, University of Chicago. "Would move": Data from Schuman, Steeh, & Bobo, 1997, originally gathered by the Gallup Organization.

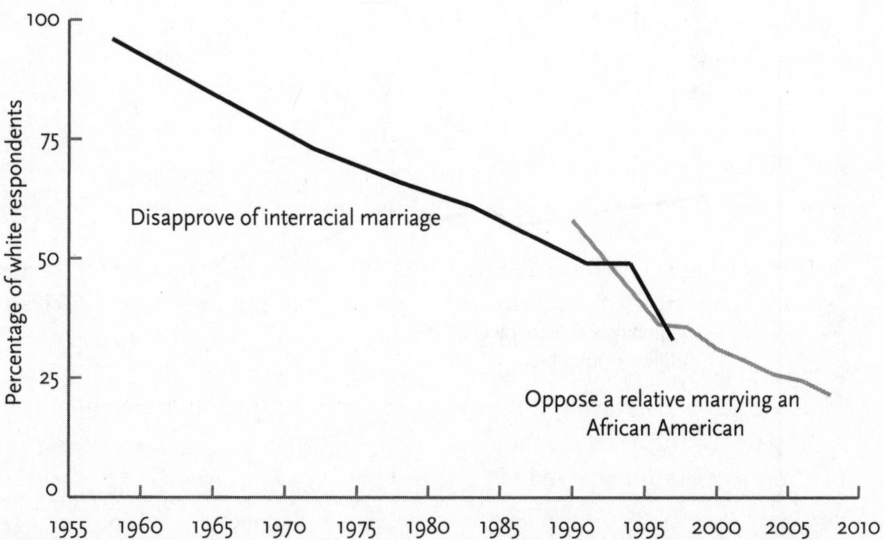

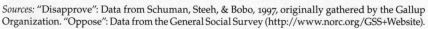

FIGURE 7–7. White attitudes to interracial marriage in the United States, 1958–2008

Sources: "Disapprove": Data from Schuman, Steeh, & Bobo, 1997, originally gathered by the Gallup Organization. "Oppose": Data from the General Social Survey (http://www.norc.org/GSS+Website).

access to any job?" the percentage of racist responses had dropped so low by the early 1970s that pollsters dropped them from their questionnaires.[27]

Also in decline are dehumanizing and demonizing beliefs. Among white Americans, these beliefs historically took the form of the prejudice that African Americans were lazier and less intelligent than whites. But over the past two decades, the proportion of Americans professing these beliefs has been falling, and today the proportion who profess that inequality is a product of low ability is negligible (figure 7–8).

Religious intolerance has been in steady decline as well. In 1924, 91 percent of the students in a middle-American high school agreed with the statement "Christianity is the one true religion and all peoples should be converted to it." By 1980, only 38 percent agreed. In 1996, 62 percent of Protestants and 74 percent of Catholics agreed with the statement "All religions are equally good"—an opinion that would have baffled their ancestors a generation before, to say nothing of those in the 16th century.[28]

The stigmatizing of any attitude that smacks of the dehumanization or demonization of minority groups extends well beyond the polling numbers. It has transformed Western culture, government, sports, and everyday life. For more than fifty years America has been cleansing itself of racist imagery that had accumulated in its popular culture. First to go were demeaning portrayals of African Americans such as blackface musical performances, shows like *Amos 'n' Andy* and *Little Rascals*, films such as Walt Disney's *Song of the*

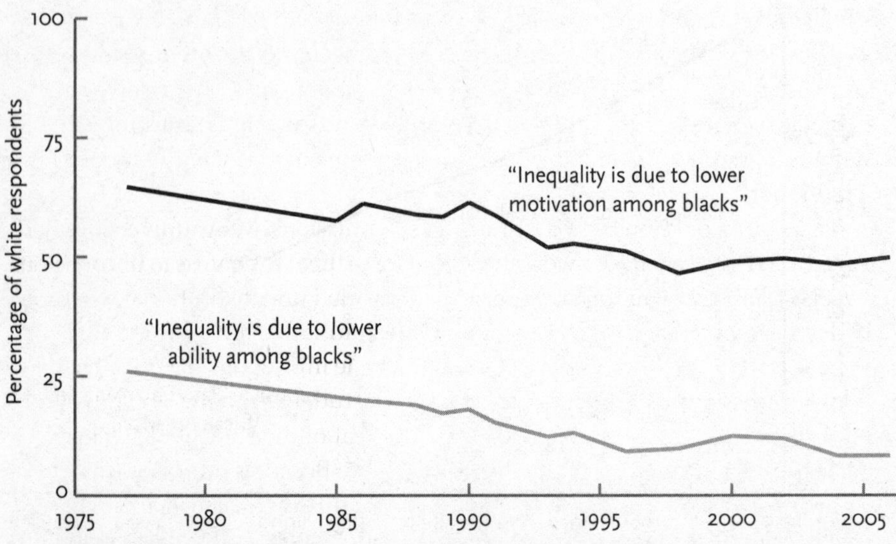

FIGURE 7–8. Unfavorable opinions of African Americans, 1977–2006

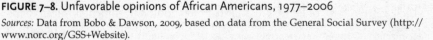

Sources: Data from Bobo & Dawson, 2009, based on data from the General Social Survey (http://www.norc.org/GSS+Website).

South, and many Bugs Bunny cartoons.[29] Caricatures in logos, advertisements, and lawn ornaments have disappeared as well. The peak of the civil rights movement was a turning point, and the taboo was quickly extended to other ethnic groups. I remember as a child the 1964 rollout of a line of powdered drink mixes called Funny Face that came in flavors called Goofy Grape, Loud Mouth Lime, Chinese Cherry, and Injun' Orange, each illustrated with a grotesque caricature. Bad timing. Within two years the latter two were made over into a raceless Choo Choo Cherry and Jolly Olly Orange.[30] We are still living through the rebranding of venerable sports teams that were based on Native American stereotypes, most recently the University of North Dakota Fighting Sioux. Derogatory racial and ethnic jokes, offensive terms for minority groups, and naïve musings about innate racial differences have become taboo in mainstream forums and have ended the careers of several politicians and media figures. Of course, plenty of vicious racism can still be found in the cesspools of the Internet and at the fringes of the political right, but a sharp line divides it from mainstream culture and politics. For instance, in 2002 the Senate Republican minority leader Trent Lott praised the 1948 presidential bid of Strom Thurmond, who at the time was an avowed segregationist. After a firestorm from within his own party, Lott was forced to resign his post.

The campaign to extirpate any precursor to attitudes that could lead to racial violence has defined the bounds of the thinkable and sayable. Racial preferences and set-asides are difficult to justify by rational arguments in a society that professes to judge people not by the color of their skin but by the content of their character. Yet no one in a position of responsibility is willing to eliminate them, because they realize it would decrease the representation of African Americans in professional positions and risk a repolarization of society. So whenever racial preferences are declared illegal or voted out in plebiscites, they are reframed with euphemisms such as "affirmative action" and "diversity" and preserved in workarounds (such as granting university admission to the top percentage of students in every high school rather than to the top percentage statewide).

The race-consciousness continues after admissions. Many universities herd freshmen into sensitivity workshops that force them to confess to unconscious racism, and many more have speech codes (ruled unconstitutional whenever they have been challenged in court) that criminalize any opinion that may cause offense to a minority group.[31] Some of the infractions for "racial harassment" cross over into self-parody, as when a student at an Indiana university was convicted for reading a book on the defeat of the Ku Klux Klan because it featured a Klansman on the cover, and when a Brandeis professor was found guilty for mentioning the term *wetback* in a lecture on racism against Hispanics.[32] Trivial incidents of racial "insensitivity" (such as the 1993 episode in which a University of Pennsylvania student shouted at some late-night revelers to "Shut up, you water buffalo," a slang expression for a rowdy person in his native Hebrew that was construed as a new racial epithet) bring universities to a halt and set off agonized rituals of communal mortification, atonement, and moral cleansing.[33] The only defense of this hypocrisy is that it may be a price worth paying for historically unprecedented levels of racial comity (though it's in the nature of hypocrisy that one cannot say that either).

In *The Blank Slate* I argued that an outsize fear of reintroducing racial hostility has distorted the social sciences by putting a heavy thumb on the nurture side of the nature-nurture scale, even for those aspects of human nature that have nothing to do with racial differences but are universal across the species. The underlying fear is that if *anything* about human nature is innate, then differences among races or ethnic groups might be innate, whereas if the mind is a blank slate at birth, then all minds must start out equally blank. An irony is that a politicized denial of human nature betrays a tacit acceptance of a particularly dark theory of human nature: that human beings are perpetually on the verge of a descent into racial animus, so every resource of the culture must be mobilized against it.

WOMEN'S RIGHTS AND THE DECLINE OF RAPE AND BATTERING

To review the history of violence is to experience repeated bouts of disbelief in learning how categories of violence that we deplore today were perceived in the past. The history of rape provides one of those shocks.

Rape is one of the prime atrocities in the human repertoire. It combines pain, degradation, terror, trauma, the seizure of a woman's means of perpetuating life, and an intrusion into the makeup of her progeny. It is also one of the commonest of atrocities. The anthropologist Donald Brown includes rape in his list of human universals, and it has been chronicled in every age and place. The Hebrew Bible tells of an era in which the brothers of a raped woman could sell her to her rapist, soldiers were entitled by divine decree to ravish nubile captives, and kings acquired concubines by the thousands. Rape, we have seen, was also common in tribal Amazonia, Homeric Greece, medieval Europe, and England during the Hundred Years' War (in Shakespeare's account, Henry V warns a French village to surrender or else their "pure maidens [will] fall into the hand of hot and forcing violation"). Mass rape is a fixture in genocides and pogroms all over the world, including recent rampages in Bosnia, Rwanda, and the Democratic Republic of the Congo. It is also common in the aftermath of military invasions, such as by the Germans in Belgium in World War I, the Japanese in China and the Russians in Eastern Europe in World War II, and the Pakistanis in Bangladesh during its war of independence.[34]

Brown notes that while rape is a human universal, so are proscriptions against rape. Yet one has to look long and hard through history and across cultures to find an acknowledgment of the harm of rape *from the viewpoint of the victim*. "Thou shalt not rape" is not one of the Ten Commandments, though the tenth one does reveal the status of a woman in that world: she is enumerated in a list of her husband's chattels, after his house and before his servants and livestock. Elsewhere in the Bible we learn that a married rape victim was considered guilty of adultery and could be stoned to death, a sentence that was carried over into Sharia law. Rape was seen as an offense not against the woman but against a man—the woman's father, her husband, or in the case of a slave, her owner. Moral and legal systems all over the world codified rape in similar ways.[35] Rape is the theft of a woman's virginity from her father, or of her fidelity from her husband. Rapists can redeem themselves by buying their victim as a wife. Women are culpable for being raped. Rape is a perquisite of a husband, seigneur, slave-owner, or harem-holder. Rape is the legitimate spoils of war.

When medieval European governments began to nationalize criminal justice, rape shifted from a tort against a husband or father to a crime against the state, which ostensibly represented the interests of women and society but in practice tilted the scales well toward the side of the accused. The fact that a false charge of rape is easy to make and hard to defend against was used to put an insuperable burden of proof onto the prosecutrix, as a rape victim was termed in many legal codes. Judges and lawyers sometimes claimed that a woman could not be forced into sex against her will because "you can't thread a moving needle."[36] Police often treated rape as a joke, pressing the victim for pornographic details or dismissing her with wisecracks like "Who'd want to rape you?" or "A rape victim is a prostitute that didn't get paid."[37] In court, the woman often found herself on trial together with the defendant, having to prove that she did not entice, encourage, or consent to her rapist. In many states women were not allowed to serve on juries for sex crimes because they might be "embarrassed" by the testimony.[38]

The prevalence of rape in human history and the invisibility of the victim in the legal treatment of rape are incomprehensible from the vantage point of contemporary moral sensibilities. But they are all too comprehensible from the vantage point of the genetic interests that shaped human desires and sentiments over the course of evolution, before our sensibilities were shaped by Enlightenment humanism. A rape entangles three parties, each with a different set of interests: the rapist, the men who take a proprietary interest in the woman, and the woman herself.[39]

Evolutionary psychologists and many radical feminists agree that rape is governed by the economics of human sexuality. As the feminist writer Andrea Dworkin put it, "A man wants what a woman has—sex. He can steal it (rape), persuade her to give it away (seduction), rent it (prostitution), lease it over the long term (marriage in the United States) or own it outright (marriage in most societies)."[40] What evolutionary psychology adds to this analysis is an explanation of the resource that backs these transactions. In any species in which one sex can reproduce at a faster rate than the other, the participation of the slower-reproducing sex will be a scarce resource over which the faster-reproducing sex competes.[41] In mammals and many birds it is the female who reproduces more slowly, because she is committed to a lengthy period of gestation, and for mammals, lactation. Females are the more discriminating sex, and the males treat restrictions on their access to females as an obstacle to be overcome. Harassment, intimidation, and forced copulation are found in many species, including gorillas, orangutans, and chimpanzees.[42] Among humans, the male may use coercion to get sex when certain risk factors line up: when he is violent, callous, and reckless by temperament; when he is a loser who cannot attract sexual partners by other means; when he is an outcast and has little fear of opprobrium from the community; and when he senses that the risks of punishment are low, such as during conquests and pogroms.[43] Around 5 percent of rapes result in pregnancies, which suggests that rape can bring an evolutionary advantage to the rapist: whatever inclinations sometimes erupt in rape need not have been selected against in our evolutionary history, and may have been selected for.[44] None of this, of course, implies that men are "born to rape," that rapists "can't help it," or that rape is "natural" in the sense of inevitable or excusable. But it does help explain why rape has been a scourge in all human societies.

The second party to a rape is the woman's family, particularly her father, brothers, and husband. The human male is unusual among mammals in that he feeds, protects, and cares for his offspring and for their mother. But this investment is genetically risky. If a man's wife has a secret dalliance, he could be investing in another man's child, which is a form of evolutionary suicide. Any genes that incline him to be indifferent to the risk of cuckoldry will lose out over evolutionary time to genes that incline him to be vigilant. As always, genes don't pull the strings of behavior directly; they exert their influence by shaping the emotional repertoire of the brain, in this case, the emotion of sexual jealousy.[45] Men are enraged at the thought of their partner's infidelity, and they take steps to foreclose that possibility. One step is to threaten her and her prospective partners and to enforce the threat when necessary to keep it credible. Another is to control her movements and her ability to use sexual signals to her advantage. Fathers too can display a proprietariness toward their daughters' sexuality that looks a lot like jealousy. In traditional societies, daughters are sold for a bride-price, and since a virgin is guaranteed not to be bearing another man's child, chastity is a selling point. Fathers, and to some extent brothers and mothers, may try to protect this valuable resource by keeping their girls chaste. The elder generation of women in a society also have an incentive to regulate the sexual competition from the younger one.

Of course, women as well as men are jealous of their partners, as a biologist would predict from the fact that men invest in their offspring. A man's infidelity brings the risk that his investment stream will be alienated by another woman and the children he has with her, and this risk gives his partner an incentive to keep him from straying. But the costs of a partner's infidelity are different for the two sexes, and accordingly a man's jealousy has been found to be more implacable, violent, and tilted toward sexual (rather than emotional) infidelity.[46] In no society are women and in-laws obsessed with the virginity of grooms.

The motives shaped by evolutionary interests do not translate directly into social practices, but they can impel people to lobby for laws and customs that protect those interests. The result is the widespread legal and cultural norms by which men recognize each other's right to control the sexuality of their wives and daughters. The human mind thrives on metaphor, and in the case of women's sexuality the recurring figure of thought is *property*.[47] Property is an elastic concept, and laws in various societies have recognized the ownership of intangibles such as airspace, images, melodies, phrases, electromagnetic bandwidth, and even genes. It's no surprise, then, that the concept of property has also been applied to the ultimate in unpossessability: sentient humans with interests of their own, such as children, slaves, and women.

In their article "The Man Who Mistook His Wife for a Chattel," Margo Wilson and Martin Daly have documented that traditional laws all over the world treat women as the property of their fathers and husbands. Property laws entitle owners to sell, exchange, and dispose of their property without encumbrance, and to expect the community to recognize their right to redress if the property is stolen or damaged by others. With the woman's interests unrepresented in this social contract, rape becomes an offense against the enfranchised men who own her. Rape was conceptualized as a tort for damaged goods, or as the theft of valuable property, as we see in the word *rape* itself, a cognate of *ravage, rapacious,* and *usurp.* It follows that a woman who was not under the protection of a highborn, propertied man was not covered by rape laws, and that the rape of a wife by her husband was an incoherent notion, like stealing one's own property.

Men may also protect their investment by holding the woman strictly liable for any theft or damage of her sexual value. Blaming the victim forecloses any possibility of her explaining away consensual sex as rape, and it incentivizes her to stay out of risky situations and to resist a rapist regardless of the costs to her freedom and safety.

Though the more blatant tropes of the women-as-property metaphor were dismantled in the late Middle Ages, the model has persisted in laws, customs, and emotions into the recent present.[48] Women, but not men, wear engagement rings to signal they are "taken," and many are still "given away" at their weddings by their fathers to their husbands, whereupon they change their surname accordingly. Well into the 1970s marital rape was not a crime in any state, and the legal system underweighted the interests of women in other rapes. Legal scholars who have studied jury proceedings have discovered that jurors must be disabused of the folk theory that women can be negligently liable for their own rapes—a concept not recognized in any contemporary American code of law—or it will creep into their deliberations.[49] And in the realm of the emotions, husbands and boyfriends often find themselves cruelly unsympathetic to their partners after they have been raped, saying things like "Something has been taken from me. I feel cheated. She was all mine before and now she's not." It's not uncommon in the aftermath of a rape for a marriage to unravel.[50]

And finally we get to the third party to the rape: the victim. The same genetic calculus that predicts that men might sometimes be inclined to pressure women into sex, and that the victim's kin may experience rape as an offense against themselves, also predicts that the woman herself should resist and abhor being raped.[51] It is in the nature of sexual reproduction that a female should evolve to exert control over her sexuality. She should choose the time, the terms, and the partner to ensure that her offspring have the fittest, most generous, and most protective father available, and that the offspring are born at the most propitious time. As always, this reproductive spreadsheet is not something that a woman calculates, either consciously or unconsciously; nor is it a chip in her brain that robotically controls her behavior. It is just the backstory of why certain emotions evolved, in this case, the determination of a woman to control her sexuality, and the agony of violation when it has been forcibly wrested from her.[52]

The history of rape, then, is one in which the interests of women had been zeroed out in the implicit negotiations that shaped customs, moral codes, and laws. And our current sensibilities, in which we recognize rape as a heinous crime against the woman, represent a reweighting of those interests, mandated by a humanist mindset that grounds morality in the suffering and flourishing of sentient individuals rather than in power, tradition, or religious practice. The mindset, moreover, has been sharpened into the principle of *autonomy:* that people have an absolute right to their bodies, which may not be treated as a common resource to be negotiated among other interested parties.[53] Our current moral understanding does not seek to balance the interests of a woman not to be raped, the interests of the men who may wish to rape her, and the interests of the husband and fathers who want to monopolize her sexuality. In an upending of the traditional valuation, the woman's ownership of her body counts for everything, and the interests of all other claimants count for nothing. (The only tradeoff we recognize today is the interests of the accused in a criminal proceeding, since his autonomy is at stake too.) The principle of autonomy, recall, was also a linchpin in the abolition of slavery, despotism, debt bondage, and cruel punishments during the Enlightenment.

The idea, seemingly obvious today, that rape is always an atrocity against the rape victim was slow in catching on. In English law there had been some rebalancing toward the interests of victims in the late Middle Ages, but only in the 18th century did the laws settle into a form that is recognizable today.[54]

Not coincidentally, it was also during that era, the age of Enlightenment, that women's rights began to be acknowledged, pretty much for the first time in history. In a 1700 essay Mary Astell took the arguments that had been leveled against despotism and slavery and extended them to the oppression of women:

> If absolute Sovereignty be not necessary in a State how comes it to be so in a Family? or if in a Family why not in a State? since no reason can be alleg'd for the one that will not hold more strongly for the other. . . .
>
> If *all Men are born free*, how is it that all Women are born slaves? As they must be if the being subjected to the *inconstant, uncertain, unknown, arbitrary* Will of Men, be *the perfect Condition of Slavery*?[55]

It took another 150 years for this argument to turn into a movement. The first wave of feminism, bookended in the United States by the Seneca Falls Convention of 1848 and the ratification of the Nineteenth Amendment to the Constitution in 1920, gave women the right to vote, to serve as jurors, to hold property in marriage, to divorce, and to receive an education. But it took the second wave of feminism in the 1970s to revolutionize the treatment of rape.

Much of the credit goes to a 1975 bestseller by the scholar Susan Brownmiller called *Against Our Will*. Brownmiller shone a harsh light on the historical indulgence of rape in religion, law, warfare, slavery, policing, and popular culture. She presented contemporary statistics on rape and first-person accounts of what it is like to be raped and to press charges of rape. And Brownmiller showed how the nonexistence of a female vantage point in society's major institutions had created an atmosphere that made light of rape (as in the common quip "When rape is inevitable you might as well lie back and enjoy it"). She wrote at a time when the decivilizing process of the 1960s had made violence a form of romantic rebellion and the sexual revolution had made lasciviousness a sign of cultural sophistication. The two affectations are more congenial to men than to women, and in combination they made rape almost chic. Brownmiller reproduced discomfitingly heroic portrayals of rapists in middlebrow and highbrow culture, together with cringe-inducing commentaries that assumed that the reader sympathized with them. Stanley Kubrick's 1971 film *A Clockwork Orange*, for example, featured a Beethoven-loving rapscallion who amused himself by beating people senseless and by raping a woman before her husband's eyes. A reviewer from *Newsweek* exclaimed:

> At its most profound level, *A Clockwork Orange* is an odyssey of the human personality, a statement of what it is to be truly human. . . . As a fantasy figure Alex appeals to something dark and primal in all of us. He acts out our desire for instant sexual gratification, for the release of our angers and repressed instincts for revenge, our need for adventure and excitement.[56]

The reviewer seemed to forget, Brownmiller remarked, that there were two sexes who watched the movie: "I am certain no woman believes that the punk with the Pinocchio nose and pair of scissors acted out *her* desire for instant gratification, revenge, or adventure." But the reviewer could not be accused of taking liberties with the intentions of the filmmaker. Kubrick himself used the first person plural to explain its appeal:

> Alex symbolizes man in his natural state, the way he would be if society did not impose its "civilizing" processes upon him. What we respond to subconsciously is Alex's guiltless sense of freedom to kill and rape, and to be our natural savage selves, and it is in this glimpse of the true nature of man that the power of the story derives.[57]

Against Our Will helped put the reform of rape laws and judicial practices onto the national agenda. When the book was published, marital rape was not a crime in any American state; today it has been outlawed in all fifty, and in most of the countries of Western Europe.[58] Rape crisis centers have eased the trauma of reporting and recovering from rape; indeed, on today's campuses one can hardly turn around without seeing an advertisement of their services. Figure 7–9 reproduces a sticker that is pasted above many bathroom sinks at Harvard, offering students no fewer than five agencies they can contact.

Today every level of the criminal justice system has been mandated to take sexual assaults seriously. A recent anecdote conveys the flavor of the change. One of my graduate students was walking in a working-class Boston suburb

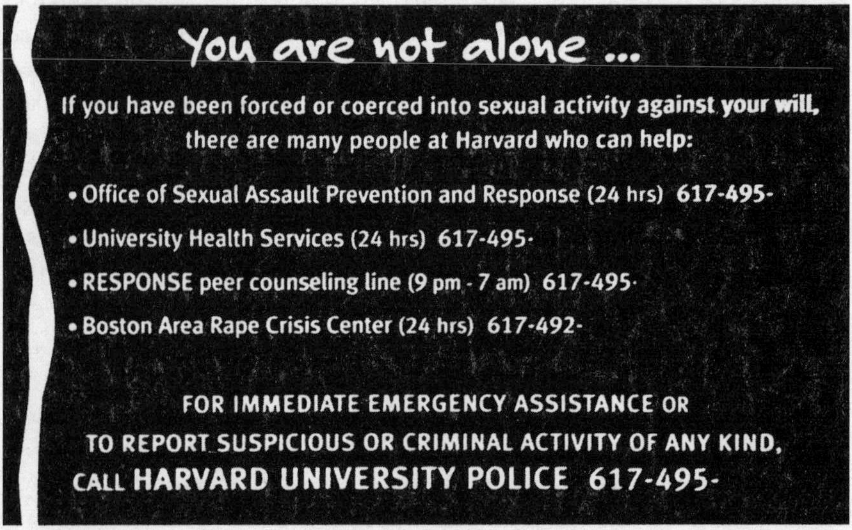

You are not alone ...

If you have been forced or coerced into sexual activity against your will, there are many people at Harvard who can help:

- Office of Sexual Assault Prevention and Response (24 hrs) 617-495-
- University Health Services (24 hrs) 617-495-
- RESPONSE peer counseling line (9 pm - 7 am) 617-495-
- Boston Area Rape Crisis Center (24 hrs) 617-492-

FOR IMMEDIATE EMERGENCY ASSISTANCE OR TO REPORT SUSPICIOUS OR CRIMINAL ACTIVITY OF ANY KIND, CALL HARVARD UNIVERSITY POLICE 617-495-

FIGURE 7–9. Rape prevention and response sticker

and was accosted on a sidewalk by three high school boys, one of whom grabbed her breast and, when she protested, jokingly threatened to hit her. When she reported it to the police, they assigned an undercover officer to conduct a stakeout with her, and the two of them spent three afternoons in an inconspicuous car (a 1978 salmon-colored Cadillac Seville, seized in a drug bust) until she spotted the perpetrator. The assistant district attorney met with her several times and with her consent charged him with second-degree assault, to which he pleaded guilty. Compared to the casual way that even brutal rapes had been treated in earlier decades, this mobilization of the judicial system for a relatively minor offense is a sign of the change in policies.

Also changed beyond recognition is the treatment of rape in popular culture. Today when the film and television industries depict a rape, it is to generate sympathy for the victim and revulsion for her attacker. Popular television series like *Law & Order: Special Victims Unit* drive home the message that sexual attackers at all social stations are contemptible scum and that DNA evidence will inevitably bring them to justice. Most striking of all is the video gaming industry, because it is the medium of the next generation, rivaling cinema and recorded music in revenue. Video gaming is a sprawling anarchy of unregulated content, mostly developed by and for young men. Though the games overflow with violence and gender stereotypes, one activity is conspicuous by its absence. The legal scholar Francis X. Shen has performed a content analysis of video games dating back to the 1980s and discovered a taboo that was close to absolute:

> It seems that rape may be the one thing that you can't put into a video game. . . . Killing scores of people in a game, often brutally, or even destroying entire cities is clearly worse than rape in real life. But in a video game, allowing someone to press the X-button to rape another character is off-limits. The "it's just a game" justification seems to fall flat when it comes to rape. . . . Even in the virtual world of Role Playing Games, rape is taboo.

He uncovered just a handful of exceptions in his worldwide search, and each triggered instant and vehement protest.[59]

But did any of these changes reduce the incidence of rape? The facts of rape are elusive, because rape is notoriously underreported, and at the same time often overreported (as in the highly publicized but ultimately disproven 2006 accusation against three Duke University lacrosse players).[60] Junk statistics from advocacy groups are slung around and become common knowledge, such as the incredible factoid that one in four university students has been raped. (The claim was based on a commodious definition of rape that the alleged victims themselves never accepted; it included, for example, any incident in which a woman consented to sex after having had too much to drink and regretted it afterward.)[61] An imperfect but serviceable dataset is the U.S.

Bureau of Justice Statistics' National Crime Victimization Survey, which since 1973 has methodically interviewed a large and stratified sample of the population to estimate crime rates without the distorting factor of how many victims report a crime to the police.[62] The survey has several features that are designed to minimize underreporting. Ninety percent of the interviewers are women, and after the methodology was improved in 1993, adjustments were made retroactively to the estimates from earlier years to keep the data from all years commensurable. Rape was defined broadly but not too broadly; it included sexual acts coerced by verbal threats as well as by physical force, and it included rapes that were either attempted or completed, of men or of women, homosexual or heterosexual. (In fact, most rapes are man-on-woman.)

Figure 7–10 plots the surveyed annual rate of rape over the past four decades. It shows that in thirty-five years the rate has fallen by an astonishing 80 percent, from 250 per 100,000 people over the age of twelve in 1973 to 50 per 100,000 in 2008. In fact, the decline may be even greater than that, because women have almost certainly been more willing to report being raped in recent years, when rape has been recognized as a serious crime, than they were in earlier years, when rape was often hidden and trivialized.

We learned in chapter 3 that the 1990s saw a decrease in all categories of crime, from homicide to auto theft. One might wonder whether the rape decline is just a special case of the crime decline rather than an accomplishment of the feminist effort to stamp out rape. In figure 7–10 I also plotted the murder rate (from the FBI's Uniform Crime Reports), aligning the two curves at their

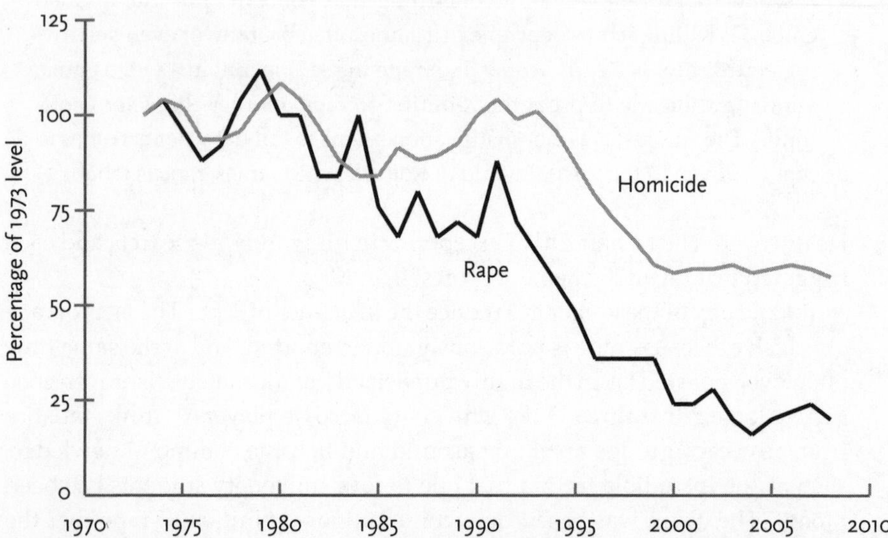

FIGURE 7–10. Rape and homicide rates in the United States, 1973–2008

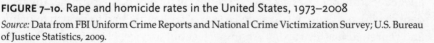

Source: Data from FBI Uniform Crime Reports and National Crime Victimization Survey; U.S. Bureau of Justice Statistics, 2009.

1973 values. The graph shows that the decline of rape is different from the decline of homicide. The murder rate meandered up and down until 1992, fell in the 1990s, and stayed put in the new millennium. The rape rate began to fall around 1979, dropped more steeply during the 1990s, and continued to bounce downward in the new millennium. By 2008 the homicide rate had hit 57 percent of its 1973 level, whereas the rape statistics bottomed out at 20 percent.

If the trend in the survey data is real, the drop in rape is another major decline in violence. Yet it has gone virtually unremarked. Rather than celebrating their success, antirape organizations convey an impression that women are in more danger than ever (as in the university bathroom stickers). And though the thirty-year rape decline needs an explanation that is distinct from the seven-year homicide decline, politicians and criminologists have not jumped into the breach. There is no Broken Windows theory, no Freakonomics theory, that has tried to explain the three-decade plunge.

Probably several causes pushed in the same direction. The portion of the downslope in the 1990s must share some causes with the general crime decline, such as better policing and fewer dangerous men on the streets. Before, during, and after that decline, feminist sensibilities had singled out rape for special attention by the police, courts, and social service agencies. Their effort was enhanced by the Violence Against Women Act of 1994, which added federal funding and oversight to rape prevention, and underwrote the use of rape kits and DNA testing, which put many first-time rapists behind bars rather than waiting for a second or third offense. Indeed, the general crime decline in the 1990s may have been as much a product of the feminist antirape campaign as the other way around. Once the crime binge of the 1960s and 1970s had reached a plateau, it was the feminist campaign against assaults on women that helped to deromanticize street violence, make public safety a right, and spur the recivilizing process of the 1990s.

Though feminist agitation deserves credit for the measures that led to the American rape decline, the country was clearly ready for them. It was not as if anyone argued that women *ought* to be humiliated at police stations and courtrooms, that husbands did have the right to rape their wives, or that rapists should prey on women in apartment stairwells and parking garages. The victories came quickly, did not require boycotts or martyrs, and did not face police dogs or angry mobs. The feminists won the battle against rape partly because there were more women in positions of influence, the legacy of technological changes that loosened the age-old sexual division of labor which had shackled women to hearth and children. But they also won the battle because both sexes had become increasingly feminist.

Despite anecdote-driven claims that women have made no progress because of a "backlash" against feminism, data show that the country's attitudes have become inexorably more progressive. The psychologist Jean Twenge has charted more than a quarter of a century of responses to a standardized questionnaire about attitudes toward women which includes items such as "It is insulting to women to have the 'obey' clause remain in the marriage service," "Women should worry less about their rights and more about becoming good wives and mothers," and "A woman should not expect to go to exactly the same places or to have quite the same freedom of action as a man."[63] Figure 7–11 shows the average of seventy-one studies that probed the attitudes of college-age men and women from 1970 to 1995. Successive generations of students, women and men alike, had increasingly progressive attitudes toward women. In fact, the men of the early 1990s had attitudes that were more feminist than those of the women in the 1970s. Southern students were slightly less feminist than northern ones, but the trends over time were similar, as are attitudes toward women measured in other samples of Americans.

We are all feminists now. Western culture's default point of view has become increasingly unisex. The universalizing of the generic citizen's vantage point, driven by reason and analogy, was an engine of moral progress during the Humanitarian Revolution of the 18th century, and it resumed that impetus during the Rights Revolutions of the 20th. It's no coincidence that the expansion of the rights of women followed on the heels of the expansion of the rights of racial minorities, because if the true meaning of the nation's founding creed is that all men are created equal, then why not all women too? In the case of gender a superficial sign of this universalizing trend is the effort of writers to avoid masculine pronouns such as *he* and *him* to refer to a generic human. A

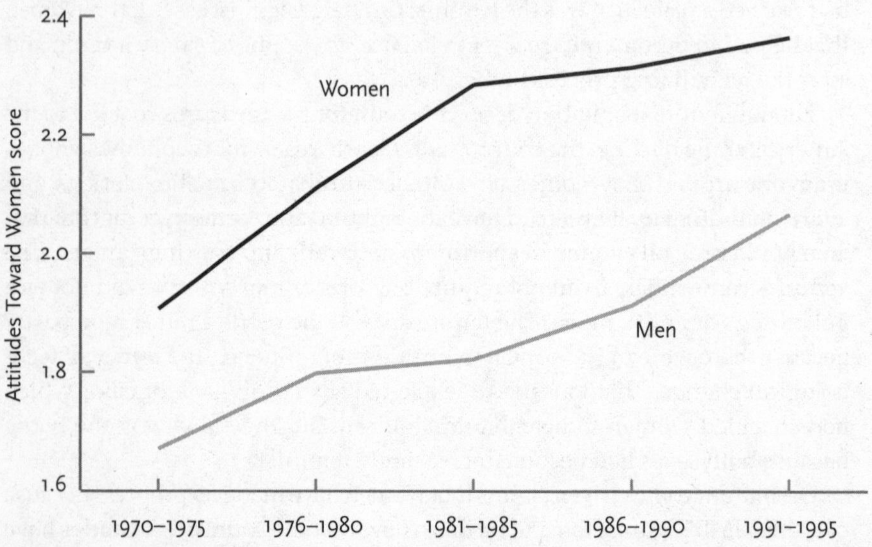

FIGURE 7–11. Attitudes toward women in the United States, 1970–1995
Source: Graph from Twenge, 1997.

deeper sign is the reorienting of moral and legal systems so that they could be justified from a viewpoint that is not specific to men.

Rapists are men; their victims are usually women. The campaign against rape got traction not only because women had muscled their way into power and rebalanced the instruments of government to serve their interests, but also, I suspect, because the presence of women changed the understanding of the men in power. A moral vantage point determines more than who benefits and who pays; it also determines how events are classified as benefits and costs to begin with. And nowhere is this gap in valuation more consequential than in the construal of sexuality by men and by women.

In their book *Warrior Lovers*, an analysis of erotic fiction by women, the psychologist Catherine Salmon and the anthropologist Donald Symons wrote, "To encounter erotica designed to appeal to the other sex is to gaze into the psychological abyss that separates the sexes. . . . The contrasts between romance novels and porn videos are so numerous and profound that they can make one marvel that men and women ever get together at all, much less stay together and successfully rear children."[64] Since the point of erotica is to offer the consumer sexual experiences without having to compromise with the demands of the other sex, it is a window into each sex's unalloyed desires. Pornography for men is visual, anatomical, impulsive, floridly promiscuous, and devoid of context and character. Erotica for women is far more likely to be verbal, psychological, reflective, serially monogamous, and rich in context and character. Men fantasize about copulating with bodies; women fantasize about making love to people.

Rape is not exactly a normal part of male sexuality, but it is made possible by the fact that male desire can be indiscriminate in its choice of a sexual partner and indifferent to the partner's inner life—indeed, "object" can be a more fitting term than "partner." The difference in the sexes' conception of sex translates into a difference in how they perceive the harm of sexual aggression. A survey by the psychologist David Buss shows that men underestimate how upsetting sexual aggression is to a female victim, while women overestimate how upsetting sexual aggression is to a male victim.[65] The sexual abyss offers a complementary explanation of the callous treatment of rape victims in traditional legal and moral codes. It may come from more than the ruthless exercise of power by males over females; it may also come from a parochial inability of men to conceive of a mind unlike theirs, a mind that finds the prospect of abrupt, unsolicited sex with a stranger to be repugnant rather than appealing. A society in which men work side by side with women, and are forced to take their interests into account while justifying their own, is a society in which this thick-headed incuriosity is less likely to remain intact.

The sexual abyss also helps to explain the politically correct ideology of rape. As we have seen, successful campaigns against violence often leave in their wake unexamined codes of etiquette, ideology, and taboo. In the case of rape, the correct belief is that rape has nothing to do with sex and only to do with power. As Brownmiller put it, "From prehistoric times to the present, I believe, rape has played a critical function. It is nothing more or less than a conscious process of intimidation by which *all* men keep *all* women in a state of fear."[66] Rapists, she wrote, are like Myrmidons, the mythical swarm of soldiers descended from ants who fought as mercenaries for Achilles: "Police-blotter rapists in a very real sense perform a myrmidon function for all men in our society."[67] The myrmidon theory, of course, is preposterous. Not only does it elevate rapists to altruistic troopers for a higher cause, and slander all men as beneficiaries of the rape of the women they love, but it assumes that sex is the one thing that no man will ever use violence to attain, and it is contradicted by numerous facts about the statistical distribution of rapists and their victims.[68] Brownmiller wrote that she adapted the theory from the ideas of an old communist professor of hers, and it does fit the Marxist conception that all human behavior is to be explained as a struggle for power between groups.[69] But if I may be permitted an *ad feminam* suggestion, the theory that rape has nothing to do with sex may be more plausible to a gender to whom a desire for impersonal sex with an unwilling stranger is too bizarre to contemplate.

Common sense never gets in the way of a sacred custom that has accompanied a decline of violence, and today rape centers unanimously insist that "rape or sexual assault is not an act of sex or lust—it's about aggression, power, and humiliation, using sex as the weapon. The rapist's goal is domination." (To which the journalist Heather MacDonald replies: "The guys who push themselves on women at keggers are after one thing only, and it's not a reinstatement of the patriarchy.")[70] Because of the sacred belief, rape counselors foist advice on students that no responsible parent would ever give a daughter. When MacDonald asked the associate director of an Office of Sexual Assault Prevention at a major university whether they encouraged students to exercise good judgment with guidelines like "Don't get drunk, don't get into bed with a guy, and don't take off your clothes or allow them to be removed," she replied, "I am uncomfortable with the idea. This indicates that if [female students] are raped it could be their fault—it is never their fault—and how one dresses does not invite rape or violence. . . . I would never allow my staff or myself to send the message it is the victim's fault due to their dress or lack of restraint in any way."

Fortunately, the students whom MacDonald interviewed did not let this sexual correctness get in the way of their own common sense. The party line of the campus rape bureaucracy, however interesting it may be as a topic in the sociology of belief, is a sideshow to a more significant historical development: that in recent decades, a widening of social attitudes and law enforcement to embrace the perspective of women has driven down the incidence of a major category of violence.

The other major category of violence against women has been called wife-beating, battering, spousal abuse, intimate partner violence, and domestic violence. The man uses physical force to intimidate, assault, and in extreme cases kill a current or estranged wife or girlfriend. Usually the violence is motivated by sexual jealousy or a fear that the woman will leave him, though he may also use it to establish dominance in the relationship by punishing her for acts of insubordination, such as challenging his authority or failing to perform a domestic duty.[71]

Domestic violence is the backstop in a set of tactics by which men control the freedom, especially the sexual freedom, of their partners. It may be related to the biological phenomenon of mate-guarding.[72] In many organisms in which males invest in their offspring and females have opportunities to mate with other males, the male will follow the female around, try to keep her away from rival males, and, upon seeing signs that he may have failed, attempt to copulate with her on the spot. Human practices such as veiling, chaperoning, chastity belts, claustration, segregation by sex, and female genital cutting appear to be culturally sanctioned mate-guarding tactics. As an extra layer of protection, men often contract with other men (and sometimes older female kin) to recognize their monopoly over their partners as a legal right. Legal codes in the civilizations of the Fertile Crescent, the Far East, the Americas, Africa, and northern Europe spelled out almost identical corollaries of the equation of women with property.[73] Adultery was a tort against the husband by his romantic rival, entitling him to damages, divorce (with refund of the bride-price), or violent revenge. Adultery was always defined by the marital status of the woman; the man's marital status, and the woman's own preferences in the matter, were immaterial. Well into the first decades of the 20th century, the man of the house was entitled by law to "chastise" his wife.[74]

In Western countries the 1970s saw the repeal of many laws that had treated women as possessions of their husbands. Divorce laws became more symmetrical. A man could no longer claim justifiable provocation when he killed his adulterous wife or her lover. A husband could no longer forcibly confine his wife or prevent her from leaving the house. And a woman's family and friends were no longer guilty of the crime of "harboring" if they gave sanctuary to a fleeing wife.[75] Most parts of the United States now have shelters in which women can escape from an abusive partner, and the legal system has recognized their right to safety by criminalizing domestic violence. Police who used to stay out of "marital spats" are now required by the laws of a majority of states to arrest a spouse if there is probable cause of abuse. In many jurisdictions prosecutors are *obligated* to seek protective orders that keep a potentially abusive spouse away from his home and partner, and then to prosecute him without the option of dropping the case, whether the victim wants it to proceed or not.[76] Originally intended to rescue women who were trapped in a cycle of abuse, apology, forgiveness, and reoffending, the policies have become so intrusive that some legal scholars, such as Jeannie Suk, have argued that they now work against the interests of women by denying them their autonomy.

Attitudes have changed as well. For centuries wife-beating was considered a normal part of marriage, from the quip of the 17th-century playwrights Beaumont and Fletcher that "charity and beating begins at home," to the threat of the 20th-century bus driver Ralph Kramden, "One of these days, Alice . . . POW, right in the kisser." As recently as 1972, the respondents to a survey of the seriousness of various crimes ranked violence toward a spouse 91st in a list of 140. (Respondents in that survey also considered the selling of LSD to be a worse crime than the "forcible rape of a stranger in the park.")[77] Readers who distrust survey data may be interested in an experiment conducted in 1974 by the social psychologists Lance Shotland and Margaret Straw. Students filling out a questionnaire overheard an argument break out between a man and a woman (in reality, actors hired by the experimenters). I will let the authors describe the experimental method:

> After approximately 15 sec of heated discussion, the man physically attacked the woman, violently shaking her while she struggled, resisting and screaming. The screams were loud piercing shrieks, interspersed with pleas to "get away from me." Along with the shrieks, one of two conditions was introduced and then repeated several times. In the Stranger Condition the woman screamed, "I don't know you," and in the Married Condition, "I don't know why I ever married you."[78]

Most of the students ran out of the testing room to see what the commotion was about. In the condition in which the actors played strangers, almost two-thirds of the students intervened, usually by approaching the couple slowly and hoping they would stop. But in the condition in which the actors played husband and wife, fewer than a fifth of the students intervened. Most of them did not even pick up a phone in front of them with a sticker that listed an emergency number for the campus police. When interviewed afterward, they said it was "none of their business." In 1974 violence that was considered unacceptable between strangers was considered acceptable within a marriage.

The experiment almost certainly could not be conducted today because of federal regulations on research with human subjects, another sign of our violence-averse times. But other studies suggest that people today are less likely to think that a violent attack by a man on his wife is none of their business. In a 1995 survey more than 80 percent of the respondents deemed domestic violence a "very important social and legal issue" (more important than children in poverty and the state of the environment), 87 percent believed that intervention is necessary when a man hits his wife even if she is not injured, and 99 percent believe that legal intervention is necessary if a man injures the wife.[79] Surveys that ask the same questions in different decades show striking changes. In 1987 only half of Americans thought it was always wrong for a man to strike his wife with a belt or stick; a decade later 86 percent thought it was always wrong.[80] Figure 7–12 shows the statistically adjusted results of four surveys that asked people whether they approved of a husband slapping a wife. Between 1968 and 1994 the level of approval fell by half, from 20 to 10 percent. Though men are more likely to condone domestic violence than women, the feminist tide carried them along as well, and the men of 1994 were less approving than the women of 1968. The decline was seen in all regions of the country and in samples of whites, blacks, and Hispanics alike.

What about domestic violence itself? Before we look at the trends, we have to consider a surprising claim: that men commit no more domestic violence than women. The sociologist Murray Straus has conducted many confidential and anonymous surveys that asked people in relationships whether they had ever used violence against their partners, and he found no difference between the sexes.[81] In 1978 he wrote, "The old cartoons of the wife chasing her husband with a rolling pin or throwing pots and pans are closer to reality than most (and especially those with feminist sympathies) realize."[82] Some activists have called for greater recognition of the problem of battered men, and for a network of shelters in which men can escape from their violent wives and girlfriends. This would be quite a twist. If women were never the victims of a gendered category of violence called "wife-beating," but rather both sexes have always been equally victimized by "spouse-beating," it would be misleading to ask whether wife-beating has declined over time as a part of the campaign to end violence against women.

To make sense of the survey findings, one has to be careful about what one

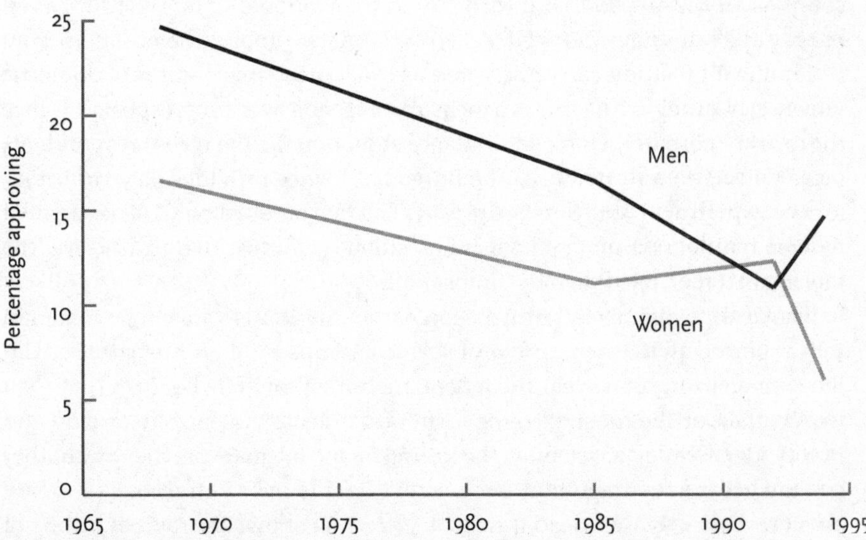

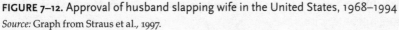

FIGURE 7–12. Approval of husband slapping wife in the United States, 1968–1994
Source: Graph from Straus et al., 1997.

means by domestic violence. It turns out that there really is a distinction between common marital spats that escalate into violence ("the conversations with the flying plates," as Rodgers and Hart put it) and the systematic intimidation and coercion of one partner by another.[83] The sociologist Michael Johnson analyzed data on the interactions between partners in violent relationships and discovered a cluster of controlling tactics that tended to go together. In some couples, one partner threatens the other with force, controls the family finances, restricts the other's movements, redirects anger and violence against the children or pets, and strategically withholds praise and affection. Among couples with a controller, the controllers who used violence were almost exclusively men; the spouses who used violence were almost entirely women, presumably defending themselves or their children. When neither partner was a controller, violence erupted only when an argument got out of hand, and in those couples the men were just a shade more prone to using force than the women. The distinction between controllers and squabblers, then, resolves the mystery of the gender-neutral violence statistics. The numbers in violence surveys are dominated by spats between noncontrolling partners, in which the women give as good as they get. But the numbers from shelter admissions, court records, emergency rooms, and police statistics are dominated by couples with a controller, usually the man intimidating the woman, and occasionally a woman defending herself. The asymmetry is even greater with estranged partners, in which it is the men who do most of the stalking, threatening, and harming. Other studies have confirmed that chronic intimidation, serious violence, and maleness tend to go together.[84]

So has anything changed over time? With the small stuff—the mutual slapping and shoving—perhaps not.[85] But with violence that is severe enough to count as an assault, so that it turns up in the National Crime Victimization Survey, the rates have plunged. As with estimates of rape, the numbers from a victimization study can't be treated as exact measures of rates of domestic violence, but they are useful as a measure of trends over time, especially since the new concern over domestic violence should make the recent respondents more willing to report abuse. The Bureau of Justice provides data from 1993 to 2005, which are plotted in figure 7–13. The rate of reported violence against women by their intimate partners has fallen by almost two-thirds, and the rate against men has fallen by almost half.

The decline almost certainly began earlier. In Straus's surveys, women in 1985 reported twice the number of severe assaults by their husbands as did the women in 1992, the year the federal victimization data begin.[86]

What about the most extreme form of domestic violence, uxoricide and mariticide? To a social scientist, the killing of one intimate partner by another has the great advantage that one needn't haggle over definitions or worry about reporting biases: dead is dead. Figure 7–14 shows the rates of killing of intimate partners from 1976 to 2005, expressed as a ratio per 100,000 people of the same sex.

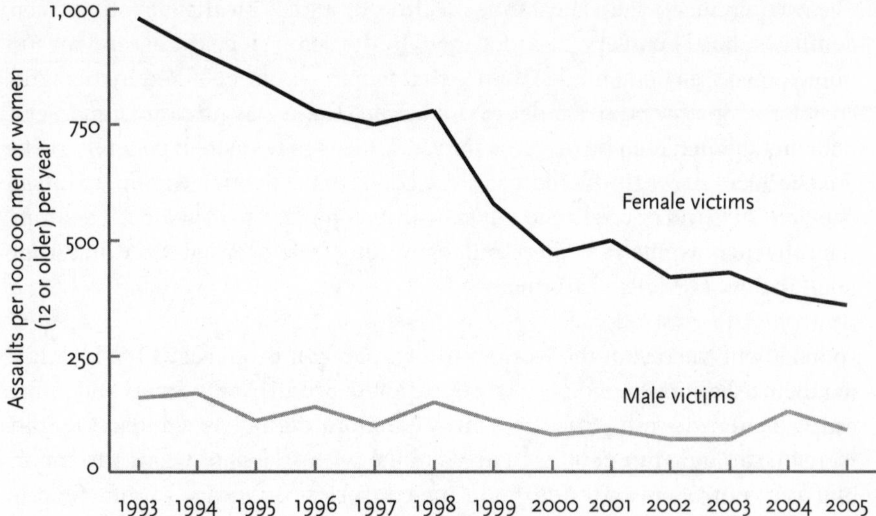

FIGURE 7–13. Assaults by intimate partners in the United States, 1993–2005
Source: Data from U.S. Bureau of Justice Statistics, 2010.

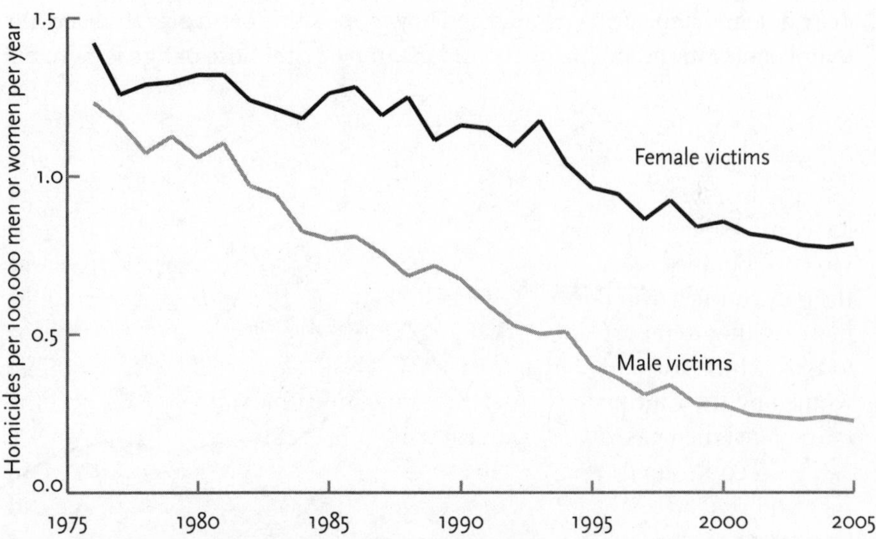

FIGURE 7–14. Homicides of intimate partners in the United States, 1976–2005
Sources: Data from U.S. Bureau of Justice Statistics, 2011, with adjustments by the *Sourcebook of Criminal Justice Statistics Online* (http://www.albany.edu/sourcebook/csv/t31312005.csv). Population figures from the U.S. Census.

Once again, we see a substantial decline, though with an interesting twist: feminism has been very good for men. In the years since the ascendancy of the women's movement, the chance that a man would be killed by his wife, ex-wife, or girlfriend has fallen sixfold. Since there was no campaign to end violence against men during this period, and since women in general are the less homicidal sex, the likeliest explanation is that a woman was apt to kill an abusive husband or boyfriend when he threatened to harm her if she left him. The advent of women's shelters and restraining orders gave women an escape plan that was a bit less extreme.[87]

What about the rest of the world? Unfortunately it is not easy to tell. Unlike homicide, definitions of rape and spousal abuse are all over the map, and police records are misleading because any change in the rate of violence against women can be swamped by changes in the willingness of women to report the abuse to the police. Adding to the confusion, advocacy groups tend to inflate statistics on rates of violence against women and to hide statistics on trends over time. The U.K. Home Office administers a crime victimization survey in England and Wales, but it does not present data on trends in rape or domestic violence.[88] But when the data from separate annual reports are aggregated, as in figure 7–15, they show a dramatic decline in domestic violence, similar to the one seen in the United States. Because of differences in how domestic violence is defined and how population bases are tabulated, the numbers in this graph are not commensurable with those of figure 7–13, but

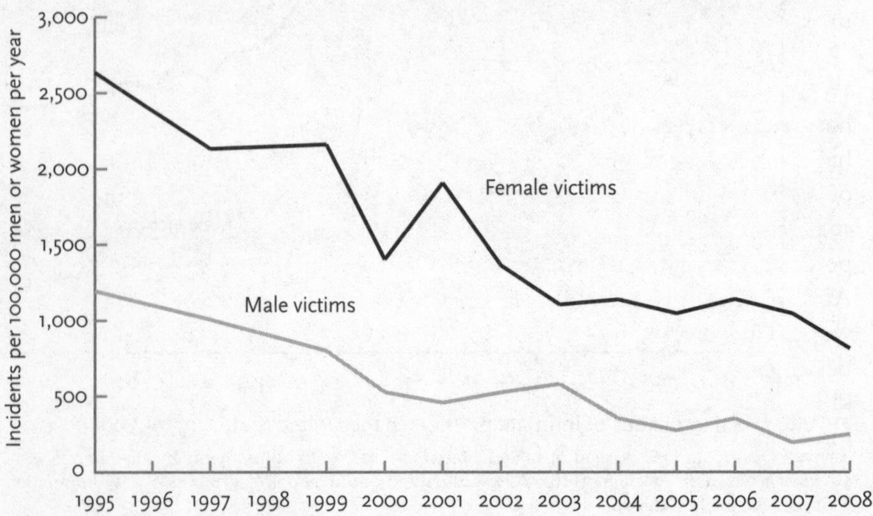

FIGURE 7–15. Domestic violence in England and Wales, 1995–2008

Sources: Data from British Crime Survey, U.K., Home Office, 2010. Data aggregated across years by Dewar Research, 2009. Population estimates from the U.K. Office for National Statistics, 2009.

the trends over time are almost the same. It is safe to guess that similar declines have taken place in other Western democracies, because domestic violence has been a concern in all of them.

Though the United States and other Western nations are often accused of being misogynistic patriarchies, the rest of the world is immensely worse. As I mentioned, surveys of domestic violence in the United States that are broad enough to include minor acts of shoving and slapping show no difference between men and women; the same is true of Canada, Finland, Germany, the U.K., Ireland, Israel, and Poland. But that gender neutrality is a departure from the rest of the world. The psychologist John Archer looked at sex ratios in surveys of domestic violence in sixteen countries and found that in the non-Western ones—India, Jordan, Japan, Korea, Nigeria, and Papua New Guinea—the men do more of the hitting.[89]

The World Health Organization recently published a hodgepodge of rates of serious domestic violence from forty-eight countries.[90] Worldwide, it has been estimated that between a fifth and a half of all women have been victims of domestic violence, and they are far worse off in countries outside Western Europe and the Anglosphere.[91] In the United States, Canada, and Australia, fewer than 3 percent of women report that their partners assaulted them in the previous year, but the reports from other countries are an order of magnitude higher: 27 percent in a Nicaraguan sample, 38 percent in a Korean sample, and 52 percent in a Palestinian sample. Attitudes toward marital violence also show striking differences. About 1 percent of New Zealanders and 4 percent of Singaporeans say that a husband has the right to beat a wife who talks back or disobeys him. But the figures are 78 percent for rural Egyptians, up to 50 percent for Indians in Uttar Pradesh, and 57 percent for Palestinians.

Laws on violence against women also show a lag from the legal reforms of Western democracies.[92] Eighty-four percent of the countries of Western Europe have outlawed or are planning to outlaw domestic violence, and 72 percent have done so for marital rape. Here are the respective figures for other parts of the world: Eastern Europe, 57 and 39 percent; Asia and the Pacific, 51 and 19 percent; Latin America, 94 and 18 percent; sub-Saharan Africa, 35 and 12.5 percent; Arab states, 25 and 0 percent. On top of these injustices, sub-Saharan Africa and South and Southwest Asia are host to systematic atrocities against women that are rare or unheard of in the 21st-century West, including infanticide, genital mutilation, trafficking in child prostitutes and sex slaves, honor killings, attacks on disobedient or under-dowried wives with acid and burning kerosene, and mass rapes during wars, riots, and genocides.[93]

Is the difference in violence against women between the West and the rest just one of the many wholesome factors that are bundled together in the Matthew Effect—democracy, prosperity, economic freedom, education, technology, decent government? Not entirely. Korea and Japan are affluent democracies but have more domestic violence against women, and several Latin American countries that are far less developed appear to have more equal sex ratios and lower absolute rates. This leaves some statistical wiggle room to look for the differences across societies that make women safer, holding affluence constant. Archer found that countries in which women are better represented in government and the professions, and in which they earn a larger proportion of earned income, are less likely to have women at the receiving end of spousal abuse. Also, cultures that are classified as more individualistic, where people feel they are individuals with the right to pursue their own goals, have relatively less domestic violence against women than the cultures classified as collectivist, where people feel they are part of a community whose interests take precedence over their own.[94] These correlations don't prove causation, but they are consistent with the suggestion that the decline of violence against women in the West has been pushed along by a humanist mindset that elevates the rights of individual people over the traditions of the community, and that increasingly embraces the vantage point of women.

Though elsewhere I have been chary about making predictions, I think it's extremely likely that in the coming decades violence against women will decrease throughout the world. The pressure will come both from the top down and from the bottom up. At the top, a consensus has formed within the international community that violence against women is the most pressing human rights problem remaining in the world.[95] There have been symbolic measures such as the International Day for the Elimination of Violence Against Women (November 25) and numerous proclamations from bully pulpits such as the United Nations and its member governments. Though the measures are toothless, the history of denunciations of slavery, whaling, piracy, privateering, chemical weaponry, apartheid, and atmospheric nuclear testing shows that international shaming campaigns can make a difference over the long run.[96] As the head of the UN Development Fund for Women has noted, "There are now more national plans, policies, and laws in place than ever before, and momentum is also growing in the intergovernmental arena."[97]

Among the grassroots, attitudes all over the world will almost certainly ensure that women will gain greater economic and political representation in the coming years. A 2010 survey by the Pew Research Center Global Attitudes Project of twenty-two countries found that in most of them, at least 90 percent of the respondents of both sexes believe that women should have equal rights, including the United States, China, India, Japan, South Korea, Turkey, Lebanon, and countries in Europe and Latin America. Even in Egypt, Jordan, Indonesia, Pakistan, and Kenya, more than 60 percent favor equal rights; only in Nigeria does the proportion fall just short of half.[98] Support for women being allowed to work outside the home is even higher. And recall the global Gallup survey that showed that even in Islamic countries a majority of women believe that women should be able to vote as they please, work at any job, and serve in government, and that in most of the countries, a majority of the men agreed.[99] As this pent-up demand is released, the interests of women are bound to be given greater consideration in their countries' policies and norms. The argument that women should not be assaulted by the men in their lives is irrefutable, and as Victor Hugo noted, "There is nothing more powerful than an idea whose time has come."

CHILDREN'S RIGHTS AND THE DECLINE OF INFANTICIDE, SPANKING, CHILD ABUSE, AND BULLYING

What do Moses, Ishmael, Romulus and Remus, Oedipus, Cyrus the Great, Sargon, Gilgamesh, and Hou Chi (a founder of the Chou Dynasty) have in common? They were all exposed as infants—abandoned by their parents and left to the elements.[100] The image of a helpless baby dying alone of cold, hunger, and predation is a potent tug on the heartstrings, so it is not surprising that a rise from infant exposure to dynastic greatness found its way into the mythologies of Jewish, Muslim, Roman, Greek, Persian, Akkadian, Sumerian, and Chinese civilizations. But the ubiquity of the exposure archetype is not just a lesson in what makes for a good story arc. It is also a lesson on how common infanticide was in human history. From time immemorial, parents have abandoned, smothered, strangled, beaten, drowned, or poisoned many of their newborns.[101]

A survey of cultures by the anthropologist Laila Williamson reveals that infanticide has been practiced on every continent and by every kind of society, from nonstate bands and villages (77 percent of which have an accepted custom of infanticide) to advanced civilizations.[102] Until recently, between 10 and 15 percent of all babies were killed shortly after they were born, and in some societies the rate has been as high as 50 percent.[103] In the words of the historian Lloyd deMause, "All families once practiced infanticide. All states trace their origin to child sacrifice. All religions began with the mutilation and murder of children."[104]

Though infanticide is the most extreme form of maltreatment of children, our cultural heritage tells of many others, including the sacrifice of children to gods; the sale of children into slavery, marriage, and religious servitude; the exploitation of children to clean chimneys and crawl through tunnels in coal mines; and the subjection of children to forms of corporal punishment that verge on or cross over into torture.[105] We have come a long way to arrive at an age in which one-pound preemies are rescued with heroic surgery, children are not expected to be economically productive until their fourth decade, and violence against children has been defined down to dodgeball.

How can we make sense of something that runs as contrary to the continuation of life as killing a newborn? In the concluding chapter of *Hardness of Heart/Hardness of Life*, his magisterial survey of infanticide around the world, the physician Larry Milner makes a confession:

> I began this book with one purpose in mind—to understand, as stated in the Introduction: "How someone can take their own child, and strangle it to death?" When I first raised the question many years ago, I thought the issue to be suggestive of some unique pathologic alteration of Nature's way. It did not seem rational that evolution would maintain an inherited tendency to kill one's offspring when survival was already in such a delicate balance. Darwinian natural selection of genetic material meant that only the survival of the fittest was guaranteed; a tendency toward infanticide must certainly be a sign of unfit behavior that would not pass this reasonable standard. But the answer which has emerged from my research indicates that one of the most "natural" things a human being can do is voluntarily kill its own offspring when faced with a variety of stressful situations.[106]

The solution to Milner's puzzlement lies in the subfield of evolutionary biology called life history theory.[107] The intuition that a mother should treat every offspring as infinitely precious, far from being an implication of the theory of natural selection, is incompatible with it. Selection acts to maximize an organism's expected lifetime reproductive output, and that requires that it negotiate the tradeoff between investing in a new offspring and conserving its resources for current and future offspring. Mammals are extreme among animals in the amount of time, energy, and food they invest in their young, and humans are extreme among mammals. Pregnancy and birth are only the first chapter in a mother's investment career, and a mammalian mother faces an expenditure of more calories in suckling the offspring to maturity than she expended in bearing it.[108] Nature generally abhors the sunk-cost fallacy, and so we expect mothers to assess the offspring and the circumstances to decide whether to commit themselves to the additional investment or to conserve their energy for its born or unborn siblings.[109] If a newborn is sickly, or if the situation is unpromising for its survival, they do not throw good money after bad but cut their losses and favor the healthiest in the litter or wait until times get better and they can try again.

To a biologist, human infanticide is an example of this triage.[110] Until recently, women nursed their children for two to four years before returning to full fertility. Many children died, especially in the perilous first year. Most women saw no more than two or three of their children survive to adulthood, and many did not see any survive. To become a grandmother in the unforgiving environment of our evolutionary ancestors, a woman would have had to make hard choices. The triage theory predicts that a mother would let a newborn die when its prospects for survival to adulthood were poor. The forecast may be based on bad signs in the infant, such as being deformed or unresponsive, or bad signs for successful motherhood, such as being burdened with older children, beset by war or famine, or unable to count on support from relatives or the baby's father. It should also depend on whether she is young enough to have opportunities to try again.

Martin Daly and Margo Wilson tested the triage theory by examining a sample of sixty unrelated societies from a database of ethnographies.[111] Infanticide was documented in a majority of them, and in 112 cases the anthropologists recorded a reason. Eighty-seven percent of the reasons fit the triage theory: the infant was not sired by the woman's husband, the infant was deformed or ill, or the infant had strikes against its chances of surviving to maturity, such as being a twin, having an older sibling close in age, having no father around, or being born into a family that had fallen on hard economic times.

The ubiquity and evolutionary intelligibility of infanticide suggest that for all its apparent inhumanity, it is usually not a form of wanton murder but falls into a special category of violence. Anthropologists who interview these women (or their relatives, since the event may be too painful for the woman to discuss) often recount that the mother saw the death as an unavoidable tragedy and grieved for the lost child. Napoleon Chagnon, for example, wrote of the wife of a Yanomamö headman, "Bahami was pregnant when I began my fieldwork, but she destroyed the infant when it was born—a boy in this case—explaining tearfully that she had no choice. The new baby would have competed with Ariwari, her youngest child, who was still nursing. Rather than expose Ariwari to the dangers and uncertainty of an early weaning, she chose to terminate the newborn instead."[112] Though the Yanomamö are the so-called fierce people, infanticide is not necessarily a manifestation of fierceness across the board. Some warring tribes, particularly in Africa, rarely kill their newborns, while some relatively peaceful ones kill them regularly.[113] The title of Milner's magnum opus comes from a quotation from a 19th-century founder of anthropology, Edward Tylor, who wrote, "Infanticide arises from hardness of life rather than hardness of heart."[114]

The fateful tipping point between keeping and sacrificing a newborn is set both by internal emotions and by cultural norms. In a culture such as ours that reveres birth and takes every step to allow babies to thrive, we tend to think that joyful bonding between mother and newborn is close to reflexive. But in fact it requires overcoming considerable psychological obstacles. In the 1st century CE, Plutarch pointed out an uncomfortable truth:

> There is nothing so imperfect, so helpless, so naked, so shapeless, so foul, as man observed at birth, to whom alone, one might almost say, Nature has given not even a clean passage to the light; but, defiled with blood and covered with filth, and resembling more one just slain than one just born, he is an object for none to touch or lift up or kiss or embrace except for someone who loves with a natural affection.[115]

The "natural affection" is far from automatic. Daly and Wilson, and later the anthropologist Edward Hagen, have proposed that postpartum depression and its milder version, the baby blues, are not a hormonal malfunction but the emotional implementation of the decision period for keeping a child.[116] Mothers with postpartum depression often feel emotionally detached from their newborns and may harbor intrusive thoughts of harming them. Mild depression, psychologists have found, often gives people a more accurate appraisal of their life prospects than the rose-tinted view we normally enjoy. The typical rumination of a depressed new mother—how will I cope with this burden?—has been a legitimate question for mothers throughout history who faced the weighty choice between a definite tragedy now and the possibility of an even greater tragedy later. As the situation becomes manageable and the blues dissipate, many women report falling in love with their baby, coming to see it as a uniquely wonderful individual.

Hagen examined the psychiatric literature on postpartum depression to test five predictions of the theory that it is an evaluation period for investing in a newborn. As predicted, postpartum depression is more common in women who lack social support (they are single, separated, dissatisfied with their marriage, or distant from their parents), who had had a complicated delivery or an unhealthy infant, and who were unemployed or whose husbands were unemployed. He found reports of postpartum depression in a number of non-Western populations which showed the same risk factors (though he could not find enough suitable studies of traditional kin-based societies). Finally, postpartum depression is only loosely tied to measured hormonal imbalances, suggesting that it is not a malfunction but a design feature.

Many cultural traditions work to distance people's emotions from a newborn until its survival seems likely. People may be enjoined from touching, naming, or granting legal personhood to a baby until a danger period is over, and the transition is often marked by a joyful ceremony, as in our own customs of the christening and the bris.[117] Some traditions have a series of milestones, such as traditional Judaism, which grants full legal personhood to a baby only after it has survived thirty days.

If I have tried to make infanticide a bit more comprehensible, it is only to reduce the distance between the vast history in which it was accepted and our contemporary sensibilities in which it is abhorrent. But the chasm that separates them is wide. Even when we acknowledge the harsh evolutionary logic that applies to the hard lives of premodern peoples, many of their infanticides are, by our standards, hard to comprehend and impossible to forgive. Examples from Daly and Wilson's list include the killing of a newborn conceived in adultery, and the killing of all a woman's children from a previous marriage when she takes (or is abducted by) a new husband. And then there are the 14 percent of the infanticidal justifications on the list that, as Daly and Wilson point out, do not easily fall into categories that an evolutionary biologist would have predicted beforehand. They include child sacrifice, an act of spite by a grandfather against his son-in-law, filicides that are committed to eliminate claimants to a throne or to avoid the obligations of kinship customs, and most commonly, the killing of a newborn for no other reason than that she is a girl.

Female infanticide has been put on the world's agenda today by census data revealing a massive shortage of women in the developing world. "A hundred million missing" is the commonly cited statistic for the daughter shortfall, a majority of them in China and India.[118] Many Asian families have a morbid preference for sons. In some countries a pregnant woman can walk into an amniocentesis or ultrasound clinic, and if she learns she is carrying a girl, she can walk next door to an abortion clinic. The technological efficiency of daughter-proofing a pregnancy may make it seem as if the girl shortage is a problem of modernity, but female infanticide has been documented in China and India for more than two thousand years.[119] In China, midwives kept a bucket of water at the bedside to drown the baby if it was a girl. In India there were many methods: "giving a pill of tobacco and bhang to swallow, drowning in milk, smearing the mother's breast with opium or the juice of the poisonous Datura, or covering the child's mouth with a plaster of cow-dung before it drew breath." Then and now, even when daughters are suffered to live, they may not last long. Parents allocate most of the available food to their sons, and as a Chinese doctor explains, "if a boy gets sick, the parents may send him to the hospital at once, but if a girl gets sick, the parents may say to themselves, 'Well, we'll see how she is tomorrow.' "[120]

Female infanticide, also called gendercide and gynecide, is not unique to Asia.[121] The Yanomamö are one of many foraging peoples that kill more newborn daughters than sons. In ancient Greece and Rome, babies were "discarded in rivers, dunghills, or cesspools, placed in jars to starve, or exposed to the elements and beasts in the wild."[122] Infanticide was also common in medieval and Renaissance Europe.[123] In all these places, more girls perished than boys. Often families would kill every daughter born to them until they had a son; subsequent daughters were allowed to live.

Female infanticide is biologically mysterious. Every child has a mother and a father, so if people are concerned about posterity, be it for their genes or their dynasty, culling their own daughters is a form of madness. A basic principle of evolutionary biology is that a fifty-fifty sex ratio at sexual maturity is a stable equilibrium in a population, because if males ever predominated, daughters would be in demand and would have an advantage over sons in attracting partners and contributing children to the next generation. And so it would be for sons if females ever predominated. To the extent that parents can control the sex ratio of their surviving offspring, whether by nature or by nurture, posterity should punish them for favoring sons or daughters across the board.[124]

One naïve hypothesis comes out of the realization that it is the number of females in a population that determines how rapidly it will grow. Perhaps tribes or nations that have multiplied themselves to the Malthusian limit on food or land kill their daughters to achieve zero population growth.[125] One problem for the ZPG theory, however, is that many infanticidal tribes and civilizations were not environmentally stressed. A more serious problem is that it has the fatal flaw of all naïve good-of-the-group theories, namely that the mechanism it proposes is self-undermining. Any family that cheated on the policy and kept its daughters alive would take over the population, stocking it with their grandchildren while the excess bachelor sons of their altruistic neighbors died without issue. The lineages that were inclined to kill their newborn daughters would have died out long ago, and the persistence of female infanticide in any society would be a mystery.

Can evolutionary psychology explain the gender bias? Critics of that approach say that it is merely an exercise in creativity, since one can always come up with an ingenious evolutionary explanation for any phenomenon. But that is an illusion, arising from the fact that so many ingenious evolutionary hypotheses have turned out to be confirmed by the data. Such success is far from guaranteed. One prominent hypothesis that, for all its ingenuity, turned out to be false was the application of the Trivers-Willard theory of sex ratios to female infanticide in humans.[126]

The biologist Robert Trivers and the mathematician Dan Willard reasoned that even though sons and daughters are expected to yield the same number of grandchildren on average, the *maximum* number that each sex can promise is different. A superfit son can outcompete other males and impregnate any number of women and thereby have any number of children, whereas a superfit daughter can have no more than the maximum she can bear and nurture in her reproductive career. On the other hand a daughter is a safer bet—an unfit son will lose the competition with other men and end up childless, whereas an unfit daughter almost never lacks for a willing sex partner. It's not that her fitness is irrelevant—a healthy and desirable daughter will still have more surviving children than an unhealthy and undesirable one—but the difference is not as extreme as it is for boom-or-bust sons. To the extent that parents can predict the fitness of their children (say, by monitoring their own health, nutrition, or territory) and strategically tilt the sex ratio, they should favor sons when they are in better shape than the competition, and favor daughters when they are in worse shape.

The Trivers-Willard theory has been confirmed in many nonhuman species and even, a bit, in *Homo sapiens*. In traditional societies, richer and higher-status people tend to live longer and attract more and better mates, so the theory predicts that higher-status people should favor sons and lower-status people should favor daughters. In some kinds of favoritism (like bequests in wills), that is exactly what happens.[127] But with a very important kind of favoritism—allowing a newborn to live—the theory doesn't work so well. The evolutionary anthropologists Sarah Hrdy and Kristen Hawkes have each shown that the Trivers-Willard theory gets only half of the story right. In India, it's true that the higher castes tend to kill their daughters. Unfortunately, it's not true that the lower castes tend to kill their sons. In fact, it's hard to find a society *anywhere* that kills its sons.[128] The infanticidal cultures of the world are either equal-opportunity baby-killers or they prefer to kill the girls—and with them, the Trivers-Willard explanation for female infanticide in humans.

The ultimate misogyny of female infanticide may suggest a feminist analysis, in which the sexism of a society extends to the right to life itself: being female is a capital offense. But that hypothesis doesn't work either. No matter how sexist these societies were (or are), they did not want a world that was *Frauenfrei*. The men do not live in all-boy treehouses in which no girls are allowed; they depend on women for sex, children, child-rearing, and the gathering or preparation of most of their food. Families that kill their daughters want there to be women around. They just want someone else to raise them. Female infanticide is a kind of social parasitism, a free rider problem, a genealogical tragedy of the commons.[129]

Free rider problems arise when no one owns a common resource, in this case, the pool of potential brides. In a free market in marriages in which parents wielded property rights, sons and daughters would be fungible, and neither sex would be favored across the board. If you really needed a fierce warrior or brawny field hand around the house, it shouldn't matter whether you raised a son for the job or raised a daughter who would bring you a son-in-law. Families with more sons would trade some of them for daughters-in-law and vice versa. True, your son-in-law's parents might prefer him to stay with them, but you could use your bargaining position to force the young man to move in with you if he wanted to have a wife at all. A preference for sons should arise only in a market with distorted property rights, one in which parents, in effect, own their sons but not their daughters.

Hawkes noted that among foraging peoples, female infanticide is more common in patrilocal societies (in which daughters move away to live with their husbands and in-laws) than in matrilocal societies (in which they stay with their parents and their husbands move in with them) or in societies in which the couple goes wherever they want. Patrilocal societies are common in tribes in which neighboring villages are constantly at war, which encourages related men to stay and fight together. They are less common when the enemies are other tribes, and the men have more freedom of movement within their territory. The internally warring societies then fall into a vicious circle in which they kill their newborn girls so their wives can hurry up and bear them more warrior sons, the better to raid other villages, and to defend their own villages against being raided, for a supply of women that had been decimated by their own infanticides. The warring tribes in Homeric Greece were caught in a similar trap.[130]

What about state societies like India and China? In state societies that practice female infanticide, Hawkes noted, parents also own their sons but not their daughters, though for economic rather than military reasons.[131] In stratified societies whose elites have indivisible wealth, the inheritance often goes to a son. In India the caste system was an additional market distorter: lower castes had to pay steep dowries so their daughters could marry higher-caste grooms. In China, parents had a permanent claim on the support of their sons and daughters-in-law extending into their dotage, but not of their daughters and sons-in-law (hence the traditional adage "A daughter is like spilled water.")[132] China's one-child policy, introduced in 1978, made parents' need for a son to support them in their old age all the more acute. In all these cases, sons are an economic asset and daughters a liability, and parents respond to the distorted incentives with the most extreme measures. Today infanticide is illegal in both countries. In China, infanticide is thought to have given way to sex-selective abortions, which are also illegal, though still widely practiced. In India, despite the inroads of the ultrasound-abortion chains, it is thought to remain common.[133] The pressure to reduce these practices will almost certainly increase, if only because governments have finally done the demographic arithmetic and realized that gynecide today means unruly bachelors tomorrow (a phenomenon we will revisit).[134]

Whether they are new mothers in desperate straits, putative fathers doubting their paternity, or parents preferring a son over a daughter, people in the West can no longer kill their newborns with impunity.[135] In 2007 in the United States, 221 infants were murdered out of 4.3 million births. That works out to a rate of 0.00005, or a reduction from the historical average by a factor of two to three thousand. About a quarter of them were killed on their first day of life by their mothers, like the "trash-can moms" who made headlines in the late 1990s by concealing their pregnancies, giving birth in secret (in one case during a high school prom), smothering their newborns, and discarding their bodies in the trash.[136] These women find themselves in similar conditions to those who set the stage for infanticide in human prehistory: they are young, single, give birth alone, and feel they cannot count on the support of their kin. Other infants were killed by fatal abuse, often by a stepfather. Still others perished at the hands of a depressed mother who committed suicide and took her children with her because she could not imagine them living without her. Rarely, a mother with postpartum depression will cross the line into postpartum psychosis and kill her children under the spell of a delusion, like the infamous Andrea Yates, who in 2001 drowned her five children in a bathtub.

What drove down the Western rate of infanticide by more than three orders of magnitude? The first step was to criminalize it. Biblical Judaism prohibited filicide, though it didn't go the whole hog: killing an infant younger than a month did not count as murder, and loopholes were claimed by Abraham,

King Solomon, and Yahweh himself for Plague #10.[137] The prohibition became clearer in Talmudic Judaism and in Christianity, from which it was absorbed into the late Roman Empire. The prohibition came from an ideology that held that lives are owned by God, to be given and taken at his pleasure, so the lives of children no longer belonged to their parents. The upshot was a taboo in Western moral codes and legal systems on taking an identifiable human life: one could not deliberate on the value of the life of an individual in one's midst. (Exceptions were exuberantly made, of course, for heretics, infidels, uncivilized tribes, enemy peoples, and transgressors of any of several hundred laws. And we continue to deliberate on the value of *statistical* lives, as opposed to identifiable lives, every time we send soldiers or police into harm's way, or scrimp on expensive health and safety measures.)

It may seem odd to call the protection of identifiable human life a "taboo," because it seems self-evident. The very act of holding the sacredness of life up to the light to examine it appears to be monstrous. But that reaction is precisely what makes a taboo a taboo, and questioning the identifiable-human-life taboo on intellectual and even moral grounds is certainly possible. In 1911 an English physician, Charles Mercier, presented arguments that infanticide should be considered a less heinous crime than the murder of an older child or an adult:

> The victim's mind is not sufficiently developed to enable it to suffer from the contemplation of approaching suffering or death. It is incapable of feeling fear or terror. Nor is its consciousness sufficiently developed to enable it to suffer pain in appreciable degree. Its loss leaves no gap in any family circle, deprives no children of their breadwinner or their mother, no human being of a friend, helper, or companion.[138]

Today we know that infants do feel pain, but in other ways Mercier's line of reasoning has been taken up by several contemporary philosophers—invariably pilloried when their essays are brought to light—who have probed the shadowy regions of our ethical intuitions in cases of abortion, animal rights, stem cell research, and euthanasia.[139] And while few people would admit to observations like Mercier's, they creep into intuitions that in practice distinguish the killing of a newborn by its mother from other kinds of homicide. Many European legal systems separate the two, defining a separate crime of infanticide or neonaticide, or granting the mother a presumption of temporary insanity.[140] Even in the United States, which makes no such distinction, when a mother kills a newborn prosecutors often don't prosecute, juries rarely convict, and those found guilty often avoid jail.[141] Sometimes, as with the trash-can moms of 1997, a media circus removes any possibility of leniency, but even these young women were paroled after three years in jail.

Like the nuclear taboo, the human life taboo is in general a very good thing. Consider this memoir from a man whose family was migrating with a group of settlers from California to Oregon in 1846. During their journey they came across an abandoned eight-year-old Native American girl, who was starving, naked, and covered with sores.

> A council among the men was held to see what should be done with her. My father wanted to take her along; others wanted to kill her and put her out of her misery. Father said that would be willful murder. A vote was taken and it was decided to do nothing about it, but to leave her where we found her. My mother and my aunt were unwilling to leave the little girl. They stayed behind to do all they could for her. When they finally joined us their eyes were wet with tears. Mother said she had knelt down by the little girl and had asked God to take care of her. One of the young men in charge of the horses felt so badly about leaving her, he went back and put a bullet through her head and put her out of her misery.[142]

Today the story leaves us in shock. But in the moral universe of the settlers, allowing the girl to die and actively ending her life were live options. Though we engage in similar reasoning when we put an aging pet or a horse with a broken leg out of its misery, we place humans in a sacred category. Trumping all calculations based on empathy and mercy is a veto based on human life: an identifiable human's right to live is not negotiable.

The human life taboo was cemented by our reaction to the Nazi Holocaust, which proceeded in stages. It started with the euthanizing of mentally retarded people, psychiatric patients, and children with disabilities, then expanded to homosexuals, inconvenient Slavs, the Roma, and the Jews. Among the masterminds of the Holocaust and the citizens who were complicit with them, each stage may have made the next more thinkable.[143] A bright line at the top of the slippery slope, we now reason, might have prevented people from sliding into depravity. Since the Holocaust a taboo on human manipulations of life and death have put public discussions of infanticide, eugenics, and active euthanasia beyond the pale. But like all taboos, the human life taboo is incompatible with certain features of reality, and fierce debates in bioethics today hinge on how to reconcile it with the fuzziness of the biological boundary that demarcates human life during embryogenesis, comas, and noninstantaneous deaths.[144]

Any taboo that contravenes powerful inclinations in human nature must be fortified with layers of euphemism and hypocrisy, and it may have little practical effect on the proscribed activity. That is what happened with infanticide in most of European history. Perhaps the least contentious claim about human nature is that humans are apt to have sex under a wider range of circumstances than those in which they are capable of bringing up the resulting babies. In the absence of contraception, abortion, or an elaborate system of social welfare, many children will be born without suitable caregivers to bring them to adulthood. Taboo or no taboo, many of those newborns will end up dead.

For almost a millennium and a half the Judeo-Christian prohibition against infanticide coexisted with massive infanticide in practice. According to one historian, exposure of infants during the Middle Ages "was practiced on a gigantic scale with absolute impunity, noticed by writers with most frigid indifference."[145] Milner cites birth records showing an average of 5.1 births among wealthy families, 2.9 among the middle class, and 1.8 among the poor, adding, "There was no evidence that the number of pregnancies followed similar lines."[146] In 1527 a French priest wrote that "the latrines resound with the cries of children who have been plunged into them."[147]

At various points in the late Middle Ages and the early modern period, systems of criminal justice tried to do something about infanticide. The steps they took were a dubious improvement. In some countries, the breasts of unmarried servant women were regularly inspected for signs of milk, and if the woman could not produce a baby, she would be tortured to find out what happened to it.[148] A woman who concealed the birth of a baby who did not survive was presumed guilty of infanticide and put to death, often by being sewn into a sack with a couple of feral cats and thrown into a river. Even with less colorful methods of punishment, the campaign to reduce infanticide by executing young mothers, many of them servants impregnated by the man of the house, began to tug on people's consciences, as they realized they were preserving the sanctity of human life by allowing men to dispose of their inconvenient mistresses.

Various fig leaves were procured. The phenomenon of "overlying," in which a mother would accidentally smother an infant by rolling over it in her sleep, at times became an epidemic. Women were invited to drop off their unwanted babies at foundling homes, some of them equipped with turntables and trapdoors to ensure anonymity. The mortality rates for the inhabitants of these homes ranged from 50 percent to more than 99 percent.[149] Women handed over their infants to wet nurses or "baby farmers" who were known to have similar rates of success. Elixirs of opium, alcohol, and treacle were readily obtainable by mothers and wet nurses to becalm a cranky infant, and at the right dosage it could becalm them very effectively indeed. Many a child who survived infancy was sent to a workhouse, "without the inconvenience of too much food or too much clothing," as Dickens described them in *Oliver Twist*, and where "it did perversely happen in eight and a half cases out of ten, either that it sickened from want and cold, or fell into the fire from neglect, or got half-smothered by accident; in any one of which cases, the miserable little being was usually summoned into another world, and there gathered to the fathers it had never known in this." Even with these contrivances, tiny corpses were a frequent sight in parks, under bridges, and in ditches. According to a

British coroner in 1862, "The police seemed to think no more of finding a dead child than they did of finding a dead cat or a dead dog."[150]

The several-thousandfold reduction in infanticide enjoyed in the Western world today is partly a gift of affluence, which leaves fewer mothers in desperate straits, and partly a gift of technology, in the form of safe and reliable contraception and abortion that has reduced the number of unwanted newborns. But it also reflects a change in the valuation of children. Rather than leaving it a pious aspiration, societies finally made good on the doctrine that the lives of infants are sacred—regardless of who bore them, regardless of how shapeless and foul they were at birth, regardless of how noticeable a gap their loss would leave in a family circle, regardless of how expensive they were to feed and care for. In the 20th century, even before abortions were widely available, a girl who got pregnant was less likely to give birth alone and secretly kill her newborn, because other people had set up alternatives, such as homes for unwed mothers, orphanages that were not death camps, and agencies that found adoptive and foster parents for motherless children. Why did governments, charities, and religions start putting money into these lifesavers? One gets a sense that children became more highly valued, and that our collective circle of concern has widened to embrace their interests, beginning with their interest in staying alive. A look at other aspects of the treatment of children confirms that the recent changes have been sweeping.

Before turning to the bigger picture of the appreciation of children in the West, I must spend a few words on a more jaundiced view of the historical fate of infanticide. According to an alternative history, the major long-term trend in the West is that people have switched from killing children shortly after they are born to killing them shortly after they are conceived.

It is true that in much of the world today, a similar proportion of pregnancies end in abortion as the fraction that in centuries past ended in infanticide.[151] Women in the developed West abort between 12 and 25 percent of their pregnancies; in some of the former communist countries the proportion is greater than half. In 2003 a million fetuses were aborted in the United States, and about 5 million were aborted throughout Europe and the West, with at least another 11 million aborted elsewhere in the world. If abortion counts as a form of violence, the West has made no progress in its treatment of children. Indeed, because effective abortion has become widely available only since the 1970s (especially, in the United States, with the 1973 *Roe v. Wade* Supreme Court decision), the moral state of the West hasn't improved; it has collapsed.

This is not the place to discuss the morality of abortion, but the larger context of trends in violence can provide some insight into how people *conceive* of abortion. Many opponents of legalized abortion predicted that acceptance of the practice would cheapen human life and put society on a slippery slope toward infanticide, euthanasia of the handicapped, a devaluation of the lives of children, and eventually widespread murder and genocide. Today we can say with confidence that that has not happened. Though abortion has been available in most of the Northern Hemisphere for decades, no country has allowed the deadline for abortions during pregnancy to creep steadily forward into legal infanticide, nor has the availability of abortion prepared the ground for euthanasia of disabled children. Between the time when abortion was made widely available and today, the rate of every category of violence has gone down, and, as we shall see, the valuation of the lives of children has shot up.

Opponents of abortion may see the decline in every form of violence but the killing of fetuses as a stunning case of moral hypocrisy. But there is another explanation for the discrepancy. Modern sensibilities have increasingly conceived moral worth in terms of *consciousness*, particularly the ability to suffer and flourish, and have identified consciousness with the activity of the brain. The change is a part of the turning away from religion and custom and toward science and secular philosophy as a source of moral illumination. Just as the legally recognized end of life is now defined by the cessation of brain activity rather than the cessation of a heartbeat, the beginning of life is sensed to depend on the first stirrings of consciousness in the fetus. The current understanding of the neural basis of consciousness ties it to reverberating neural activity between the thalamus and the cerebral cortex, which begins at around twenty-six weeks of gestational age.[152] More to the point, people *conceive* of fetuses as less than fully conscious: the psychologists Heather Gray, Kurt Gray, and Daniel Wegner have shown that people think of fetuses as more capable of experience than robots or corpses, but less capable than animals, babies, children, and adults.[153] The vast majority of abortions are carried out well before the milestone of having a functioning brain, and thus are safely conceptualized, according to this understanding of the worth of human life, as fundamentally different from infanticide and other forms of violence.

At the same time, we might expect a general distaste for the destruction of any kind of living thing to turn people away from abortion even when they don't equate it with murder. And that indeed has happened. It's a little-known fact that rates of abortion are falling throughout the world. Figure 7–16 shows the rates of abortion in the major regions in which data are available (albeit differing widely in quality) in the 1980s, 1996, and 2003.

The decline has been steepest in the countries of the former Soviet bloc, which were said to have had a "culture of abortion." During the communist era abortions were readily available, but contraceptives, like every other consumer good, were allocated by a central commissar rather than by supply and demand, so they were always in short supply. But abortions have also become less common in China, the United States, and the Asian and Islamic countries in which they are legal. Only in India and Western Europe did abortion rates fail to decline, and those are the regions where the rates were lowest to begin with.

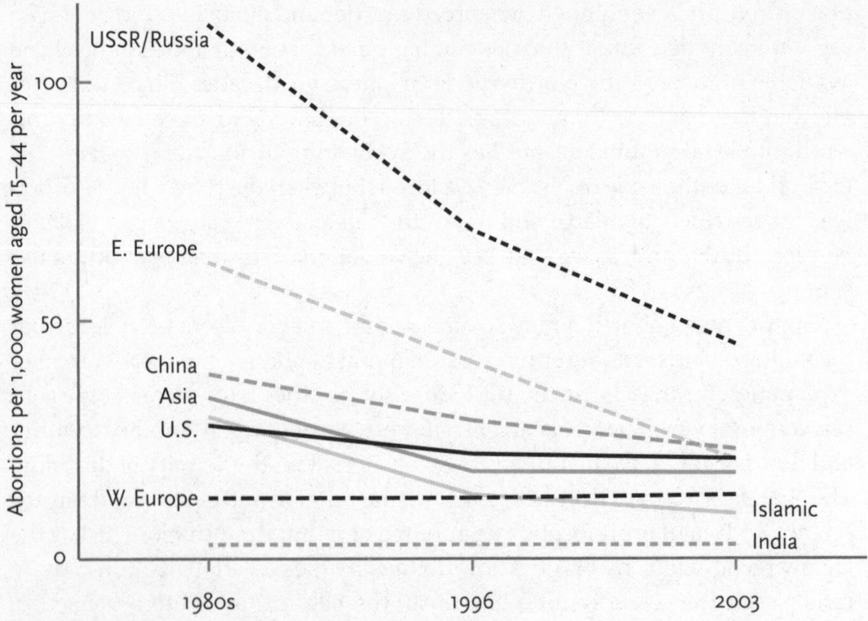

FIGURE 7–16. Abortions in the world, 1980–2003

Sources: 1980s: Henshaw, 1990; 1996 & 2003: Sedgh et al., 2007. "Eastern Europe" comprises Bulgaria, Czechoslovakia/Czech Republic & Slovakia, Hungary, Yugoslavia/Serbia-Montenegro, Romania. "Western Europe" comprises Belgium, Denmark, England and Wales, Finland, the Netherlands, Norway, Scotland, Sweden. "Asia" comprises Singapore, Japan, South Korea (2003 equated with 1996). "Islamic" comprises Tunisia and Turkey.

The causes of much of the decline, to be sure, are practical. Contraception is cheaper and more convenient than abortion, and if it's readily available it will be the first choice of people with the foresight and self-control to use it. But presumably abortion has a moral dimension even among those who undergo it and among their compatriots who want to keep the option safe and legal. Abortion is seen as something to be minimized, even if it is not criminalized. If so, the trends in abortion offer a sliver of common ground in the rancorous debate between the so-called pro-life and pro-choice factions. The countries that allow abortion have not let an indifference to life put them on a slippery slope to infanticide or other forms of violence. But these same countries increasingly act as if abortion is undesirable, and they may be reducing its incidence as part of the move to protect all living things.

During the long, sad history of violence against children, even when infants survived the day of their birth it was only to endure harsh treatment and cruel punishments in the years to come. Though hunter-gatherers tend to use corporal punishment in moderation, the dominant method of child-rearing in every other society comes right out of *Alice in Wonderland:* "Speak roughly to your little boy, and beat him when he sneezes."[154] The reigning theory of child development was that children were innately depraved and could be socialized only by force. The expression "Spare the rod and spoil the child" has been attributed to an advisor to the king of Assyria in the 7th century BCE and may have been the source of Proverbs 13:24, "He that spareth the rod hateth his son: But he that loveth him chasteneth him betimes."[155] A medieval French verse advised, "Better to beat your child when small than to see him hanged when grown." The Puritan minister Cotton Mather (Increase's son) extended the concern for the child's well-being to the hereafter: "Better whipt than Damn'd."[156]

As with all punishments, human ingenuity rose to the technological challenge of delivering experiences that were as unpleasant as possible. DeMause writes of medieval Europe:

> That children with devils in them had to be beaten goes without saying. A panoply of beating instruments existed for that purpose, from cat-o'-nine tails and whips to shovels, canes, iron rods, bundles of sticks, the discipline (a whip made of small chains), the goad (shaped like a cobbler's knife, used to prick the child on the head or hands) and special school instruments like the flapper, which had a pear-shaped end and a round hole to raise blisters. The beatings described in the sources were almost always severe, involved bruising and bloodying of the body, began in infancy, were usually erotically tinged by being inflicted on bare parts of the body near the genitals and were a regular part of the child's daily life.[157]

Severe corporal punishment was common for centuries. One survey found that in the second half of the 18th century, 100 percent of American children were beaten with a stick, whip, or other weapon.[158] Children were also liable to punishment by the legal system. Until the 19th century, British law allowed the death penalty for "strong evidence of malice in a child seven to fourteen years of age," and many teenagers continued to be hanged for petty crimes like arson and burglary until 1908, when the minimum age for execution was raised to sixteen.[159] Even at the turn of the 20th century, German children "were regularly placed on a red-hot iron stove if obstinate, tied to their bedposts for days, thrown into cold water or snow to 'harden' them, [and] forced to kneel for hours every day against the wall on a log while the parents ate and read."[160] During toilet training many children were tormented with enemas, and at school they were "beaten until [their] skin smoked."

The harsh treatment was not unique to Europe. The beating of children has been recorded in ancient Egypt, Sumeria, Babylonia, Persia, Greece, Rome, China, and Aztec Mexico, whose punishments included "sticking the child with thorns, having their hands tied and then being stuck with pointed agave leaves, whippings, and even being held over a fire of dried axi peppers and being made to inhale the acrid smoke."[161] DeMause notes that well into the 20th century,

Japanese children were subjected to "beating and burning of incense on the skin as routine punishments, cruel bowel training with constant enemas, . . . kicking, hanging by the feet, giving cold showers, strangling, driving a needle into the body, cutting off a finger joint."[162] (A psychoanalyst as well as a historian, deMause had plenty of material with which to explain the atrocities of World War II.)

Children were subjected to psychological torture as well. Much of their entertainment was filled with reminders that they might be abandoned by parents, abused by stepparents, or mutilated by ogres and wild animals. Grimm's fairy tales were just a few of the advisories that may be found in children's literature of the misfortunes that can befall a careless or disobedient child. English babies, for example, were soothed to sleep with a lullaby about Napoleon:

Baby, baby, if he hears you,
As he gallops past the house,
Limb from limb at once he'll tear you,
Just as pussy tears a mouse.
And he'll beat you, beat you, beat you,
And he'll beat you all to pap,
And he'll eat you, eat you, eat you,
Every morsel, snap, snap, snap.[163]

A recurring archetype in children's verse is the child who commits a minor slipup or is unjustly blamed for one, whereupon his stepmother butchers him and serves him for dinner to his unwitting father. In a Yiddish version, the victim of one such injustice sings posthumously to his sister:

Murdered by my mother,
Eaten by my father.
And Sheyndele, when they were done
Sucked the marrow from my bones
And threw them out the window.[164]

Why would any parents torture, starve, neglect, and terrify their own children? One might naïvely think that parents would have evolved to nurture their children without stinting, since having viable offspring is the be-all and end-all of natural selection. Children too ought to submit to their parents' guidance without resistance, since it is offered for their own good. The naïve view predicts a harmony between parent and child, since each "wants" the same thing—for the child to grow up healthy and strong enough to have children of its own.

It was Trivers who first noticed that the theory of natural selection predicts no such thing.[165] Some degree of conflict between parent and offspring is rooted in the evolutionary genetics of the family. Parents have to apportion their investment (in resources, time, and risk) across all their children, born and unborn. All things being equal, every offspring is equally valuable, though each benefits from parental investment more when it is young and helpless than when it can fend for itself. The child sees things differently. Though an offspring has an interest in its siblings' welfare, since it shares half its genes with each full sib, it shares *all* of its genes with itself, so it has a disproportionate interest in its *own* welfare. The tension between what a parent wants (an equitable allocation of its worldly efforts to all its children) and what a child wants (a lopsided benefit to itself compared to its siblings) is called parent-offspring conflict. Though the stakes of the conflict are the parents' investment in a child and its siblings, those siblings need not yet exist: a parent must also conserve strength for future children and grandchildren. Indeed, the first dilemma of parenthood—whether to keep a newborn—is just a special case of parent-offspring conflict.

The theory of parent-offspring conflict says nothing about how much investment an offspring should want or how much a parent should be prepared to give. It says only that however much parents are willing to give, the offspring wants a bit more. Children cry when they are in need of help, and parents cannot ignore the cries. But children are expected to cry a bit louder and longer than their objective need calls for. Parents discipline children to keep them out of danger, and socialize them to be effective members of their community. But parents are expected to discipline children a bit more for their own convenience, and to socialize them to be a bit more accommodating to their siblings and kin, than the levels that would be in the interests of the children themselves. As always, the teleological terms in the explanation—"wants," "interests," "for"—don't refer to literal desires in the minds of people, but are shorthand for the evolutionary pressures that shaped those minds.

Parent-offspring conflict explains why child-rearing is always a battle of wills. What it does not explain is why that battle should be fought with rods and birches in one era and lectures and time-outs in another. In retrospect, it's hard to avoid sorrow for the millennia of children who have needlessly suffered at the hands of their caregivers. Unlike the tragedy of war, where each side has to be as fierce as its adversary, the violence of child-rearing is entirely one-sided. The children who were whipped and burned in the past were no naughtier than the children of today, and they ended up no better behaved as adults. On the contrary, we have seen that the rate of impulsive violence of yesterday's adults was far higher than today's. What led the parents of our era to the discovery that they could socialize their children with a fraction of the brute force that was used by their ancestors?

The first nudge was ideological, and like so many other humanitarian reforms it originated in the Age of Reason and the Enlightenment. Children's tactics in parent-offspring conflict have led parents in every era to call them little devils. During the ascendancy of Christianity, that intuition was ratified by a religious belief in innate depravity and original sin. A German preacher in the 1520s, for instance, sermonized that children harbored wishes for "adultery, fornication, impure desires, lewdness, idol worship, belief in magic, hostility, quarreling, passion, anger, strife, dissension, factiousness, hatred, murder, drunkenness, gluttony," and he was just getting started.[166] The expression "beat the devil out of him" was more than a figure of speech! Also, a fatalism about the unfolding of life made child development a matter of fate or divine will rather than the responsibility of parents and teachers.

One paradigm shift came from John Locke's *Some Thoughts Concerning Education*, which was published in 1693 and quickly went viral.[167] Locke suggested that a child was "only as white Paper, or Wax, to be moulded and fashioned as one pleases"—a doctrine also called the tabula rasa (scraped tablet) or blank slate. Locke wrote that the education of children could make "a great difference in mankind," and he encouraged teachers to be sympathetic toward their pupils and to try to take their viewpoints. Tutors should carefully observe the "change in temper" in their students and should help them enjoy their studies. And teachers should not expect young children to show the same "carriage, seriousness, or application" as older ones. On the contrary, "they must be permitted . . . the foolish and childish actions suitable to their years."[168]

The idea that the way children are treated determines the kinds of adults they grow into is conventional wisdom today, but it was news at the time. Several of Locke's contemporaries and successors turned to metaphor to remind people about the formative years of life. John Milton wrote, "The childhood shows the man as morning shows the day." Alexander Pope elevated the correlation to causation: "Just as the twig is bent, the tree's inclined." And William Wordsworth inverted the metaphor of childhood itself: "The child is father of the man." The new understanding required people to rethink the moral and practical implications of the treatment of children. Beating a child was no longer an exorcism of malign forces possessing a child, or even a technique of behavior modification designed to reduce the frequency of bratty behavior in the present. It shaped the kind of person that the child would grow into, so its consequences, foreseen and unforeseen, would alter the makeup of civilization in the future.

Another gestalt shift came from Rousseau, who replaced the Christian notion of original sin with the romantic notion of original innocence. In his 1762 treatise *Émile, or On Education*, Rousseau wrote, "Everything is good as it leaves the hand of the Author of things, and everything degenerates in the hands of man." Foreshadowing the theories of the 20th-century psychologist Jean Piaget, Rousseau divided childhood into a succession of stages centered on Instinct, Sensations, and Ideas. He argued that young children have not yet reached the Age of Ideas, and so should not be expected to reason in the ways of adults. Rather than drilling youngsters in the rules of good and evil, adults should allow children to interact with nature and learn from their experiences. If in the course of exploring the world they damaged things, it was not from an intention to do harm but from their own innocence. "Respect childhood," he implored, and "leave nature to act for a long time before you get involved with acting in its place."[169] The 19th-century Romantic movement inspired by Rousseau saw childhood as a period of wisdom, purity, and creativity, a stage that children should be left to enjoy rather than be disciplined out of. The sensibility is familiar today but was radical at the time.

During the Enlightenment, elite opinion began to incorporate the child-friendly doctrines of the blank slate and original innocence. But historians of childhood place the turning point in the actual treatment of children considerably later, in the decades surrounding the turn of the 20th century.[170] The economist Viviana Zelizer has suggested that the era from the 1870s to the 1930s saw a "sacralization" of childhood among middle- and upper-class parents in the West. That was when children attained the status we now grant them: "economically worthless, emotionally priceless."[171] The era was inaugurated in England when a "baby-farming" scandal led to the formation of the Infant Protection Society in 1870 and to the Infant Life Protection Acts of 1872 and 1897. Around the same time, pasteurization and sterilized bottles meant that fewer infants were outsourced to infanticidal wet nurses. Though the Industrial Revolution originally moved children from backbreaking labor on farms to backbreaking labor in mills and factories, legal reforms increasingly restricted child labor. At the same time, the affluence that flowed from the maturing Industrial Revolution drove rates of infant mortality downward, reduced the need for child labor, and provided a tax stream that could support social services. More children went to school, which soon became compulsory and free. To deal with the packs of urchins, ragamuffins, and artful dodgers who roamed city streets, child welfare agencies founded kindergartens, orphanages, reform schools, fresh-air camps, and boys' and girls' clubs.[172] Stories for children were written to give them pleasure rather than to terrorize them or moralize to them. The Child Study movement aimed for a scientific approach to human development and began to replace the superstition and bunkum of old wives with the superstition and bunkum of child-rearing experts.

We have seen that during periods of humanitarian reform, a recognition of the rights of one group can lead to a recognition of others by analogy, as when the despotism of kings was analogized to the despotism of husbands, and when two centuries later the civil rights movement inspired the women's rights movement. The protection of abused children also benefited from an analogy—in this case, believe it or not, with animals.

In Manhattan in 1874, the neighbors of ten-year-old Mary Ellen McCormack, an orphan being raised by an adoptive mother and her second husband, noticed suspicious cuts and bruises on the girl's body.[173] They reported her to the Department of Public Charities and Correction, which administered the city's jails, poorhouses, orphanages, and insane asylums. Since there were no laws that specifically protected children, the caseworker contacted the American Society for the Protection of Animals. The society's founder saw an analogy between the plight of the girl and the plight of the horses he rescued from violent stable owners. He engaged a lawyer who presented a creative interpretation of habeas corpus to the New York State Supreme Court and petitioned to have her removed from her home. The girl calmly testified:

> Mamma has been in the habit of whipping and beating me almost every day. She used to whip me with a twisted whip—a rawhide. I have now on my head two black-and-blue marks which were made by Mamma with the whip, and a cut on the left side of my forehead which was made by a pair of scissors in Mamma's hand. . . . I never dared speak to anybody, because if I did I would get whipped.

The *New York Times* reprinted the testimony in an article entitled "Inhumane Treatment of a Little Waif," and the girl was removed from the home and eventually adopted by her caseworker. Her lawyer set up the New York Society for the Prevention of Cruelty to Children, the first protective agency for children anywhere in the world. Together with other agencies founded in its wake, it set up shelters for battered children and lobbied for laws that punished their abusive parents. Similarly, in England the first legal case to protect a child against an abusive parent was taken up by the Royal Society for the Prevention of Cruelty to Animals, and out of it grew the National Society for the Prevention of Cruelty to Children.

Though the rollover of the 19th century saw an acceleration in the valuation of children in the West, it was neither an abrupt transition nor a one-shot advance. Expressions of love of children, of grief at their loss, and of dismay at their mistreatment can be found in every period of European history and in every culture.[174] Even many of the parents who treated their children cruelly were often laboring under superstitions that led them to think they were acting in the child's best interests. And as with many declines in violence, it's hard to disentangle all the changes that were happening at once—enlightened ideas, increasing prosperity, reformed laws, changing norms.

But whatever the causes were, they did not stop in the 1930s. Benjamin Spock's perennial bestseller *Baby and Child Care* was considered radical in 1946 because it discouraged mothers from spanking their children, stinting on affection, and regimenting their routines. Though the indulgence of postwar parents was a novelty at the time (widely and spuriously blamed for the excesses of the baby boomers), it was by no means a high-water mark. When the boomers became parents, they were even more solicitous of their children. Locke, Rousseau, and the 19th-century reformers had set in motion an escalator of gentleness in the treatment of children, and in recent decades its rate of ascent has accelerated.

Since 1950, people have become increasingly loath to allow children to become the victims of any kind of violence. The violence people can most easily control, of course, is the violence they inflict themselves, namely by spanking, smacking, slapping, paddling, birching, tanning, hiding, thrashing, and other forms of corporal punishment. Elite opinion on corporal punishment changed dramatically during the 20th century. Other than in fundamentalist Christian groups, it's rare today to hear people say that sparing the rod will spoil the child. Scenes of fathers with belts, mothers with hairbrushes, and teary children tying pillows to their bruised behinds are no longer common in family entertainment.

At least since Dr. Spock, child-care gurus have increasingly advised against spanking.[175] Today every pediatric and psychological association opposes the practice, though not always in language as clear as the title of a recent article by Murray Straus: "Children Should Never, Ever, Be Spanked No Matter What the Circumstances."[176] The expert opinion recommends against spanking for three reasons. One is that spanking has harmful side effects down the line, including aggression, delinquency, a deficit in empathy, and depression. The cause-and-effect theory, in which spanking teaches children that violence is a way to solve problems, is debatable. Equally likely explanations for the correlation between spanking and violence are that innately violent parents have innately violent children, and that cultures and neighborhoods that tolerate spanking also tolerate other kinds of violence.[177] The second reason not to spank a child is that spanking is not particularly effective in reducing misbehavior compared to explaining the infraction to the child and using nonviolent measures like scolding and time-outs. Pain and humiliation distract children from pondering what they did wrong, and if the only reason they have to behave is to avoid these penalties, then as soon as Mom's and Dad's backs are turned they can be as naughty as they like. But perhaps the most compelling reason to avoid spanking is symbolic. Here is Straus's third reason why children should never, ever be spanked: "Spanking contradicts the ideal of nonviolence in the family and society."

Have parents been listening to the experts, or perhaps coming to similar conclusions on their own? Public opinion polls sometimes ask people whether they agree with statements like "It is sometimes necessary to discipline a child with a good, hard spanking" or "There are certain circumstances when it is all right to smack a child." The level of agreement depends on the wording of the question, but in every poll in which the same question has been asked in different years, the trend is downward. Figure 7–17 shows the trends since 1954 from three American datasets, together with surveys from Sweden and New Zealand. Before the early 1980s, around 90 percent of respondents in the English-speaking countries approved of spanking. In less than a generation, the percentage had fallen in some polls to just more than half. The levels of approval

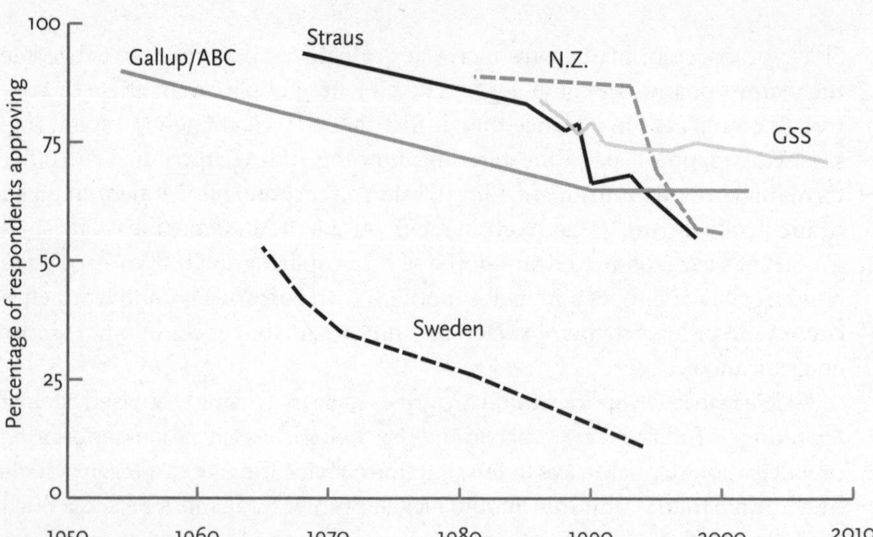

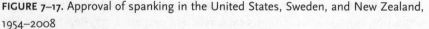

FIGURE 7–17. Approval of spanking in the United States, Sweden, and New Zealand, 1954–2008

Sources: Gallup/ABC: Gallup, 1999; ABC News, 2002. Straus: Straus, 2001, p. 206. General Social Survey: http://www.norc.org/GSS+Website/, weighted means. New Zealand: Carswell, 2001. Sweden: Straus, 2009.

depend on the country and region: Swedes approve of spanking far less than do Americans or Kiwis, and Americans themselves are diverse, as we would expect from the southern culture of honor.[178] In a 2005 survey, spanking approval rates ranged from around 55 percent in northern blue states (those that tend to vote for Democrats), like Massachusetts and Vermont, to more than 85 percent in southern red states (those that tend to vote for Republicans), like Alabama and Arkansas.[179] Across the fifty states, the rate of approval of spanking tracks the homicide rate (the two measures show a correlation of 0.52 on a scale from -1 to 1), which could mean that spanked children grow up to be killers, but more likely that subcultures that encourage the spanking of children also encourage the violent defense of honor among adults.[180] But every region showed a decline, so that by 2006 the southern states disapproved of spanking in the same proportion that the north-central and mid-Atlantic states did in 1986.[181]

What about actual behavior? Many parents will still slap a toddler's hand if the child reaches for a forbidden object, but in the second half of the 20th century every other kind of corporal punishment declined. In the 1930s American parents spanked their children more than 3 times a month, or more than 30 times a year. By 1975 the figure had fallen to 10 times a year and by 1985 to around 7.[182] Even steeper declines were seen in Europe.[183] In the 1950s,

94 percent of Swedes spanked their children, and 33 percent did so every day; by 1995, the figures had plunged to 33 and 4 percent. By 1992, German parents had come a long way from their great-grandparents who had placed their grandparents on hot stoves and tied them to bedposts. But 81 percent still slapped their children on the face, 41 percent spanked them with a stick, and 31 percent beat them to the point of bruising. By 2002, these figures had sunk to 14 percent, 5 percent, and 3 percent.

There remains a lot of variation among countries today. No more than 5 percent of college students in Israel, Hungary, the Netherlands, Belgium, and Sweden recall being hit as a teenager, but more than a quarter of the students in Tanzania and South Africa do.[184] In general, wealthier countries spank their children less, with the exception of developed Asian nations like Taiwan, Singapore, and Hong Kong. The international contrast is replicated among ethnic groups in the United States, where African Americans and Asians spank more than whites.[185] But the level of approval of spanking has declined in all three groups.[186]

In 1979 the government of Sweden outlawed spanking altogether.[187] The other Scandinavian countries soon joined it, followed by several countries of Western Europe. The United Nations and the European Union have called on *all* their member nations to abolish spanking. Several countries have launched public awareness campaigns against the practice, and twenty-four have now made it illegal.

The prohibition of spanking represents a stunning change from millennia in which parents were considered to own their children, and the way they treated them was considered no one else's business. But it is consistent with other intrusions of the state into the family, such as compulsory schooling, mandatory vaccination, the removal of children from abusive homes, the imposition of lifesaving medical care over the objections of religious parents, and the prohibition of female genital cutting by communities of Muslim immigrants in European countries. In one frame of mind, this meddling is a totalitarian imposition of state power into the intimate sphere of the family. But in another, it is part of the historical current toward a recognition of the autonomy of individuals. Children are people, and like adults they have a right to life and limb (and genitalia) that is secured by the social contract that empowers the state. The fact that other individuals—their parents—stake a claim of ownership over them cannot negate that right.

American sentiments tend to weight family over government, and currently no American state prohibits corporal punishment of children by their parents. But when it comes to corporal punishment of children *by* the government, namely in schools, the United States has been turning away from this form of violence. Even in red states, where three-quarters of the people approve of spanking by parents, only 30 percent approve of paddling in schools, and in the blue states the approval rate is less than half that.[188] And since the 1950s the level of approval of corporal punishment in schools has been in decline (figure 7–18). The growing disapproval has been translated into legislation. Figure 7–19 shows the shrinking proportion of American states that still allow corporal punishment in schools.

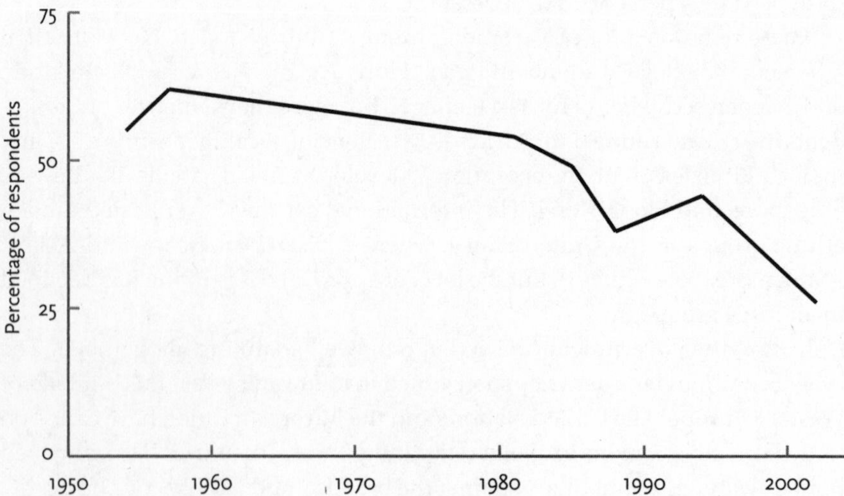

FIGURE 7–18. Approval of corporal punishment in schools in the United States, 1954–2002
Sources: Data for 1954–94 from Gallup, 1999; data for 2002 from ABC News, 2002.

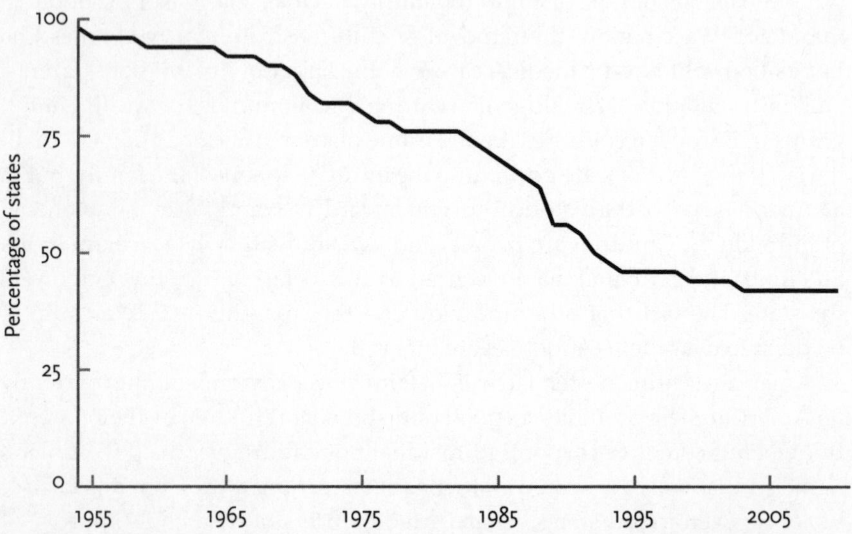

FIGURE 7–19. American states allowing corporal punishment in schools, 1954–2010
Source: Data from Leiter, 2007.

The trend is even more marked in the international arena, where corporal punishment in schools is now seen as a violation of human rights, like other forms of extrajudicial government violence. It has been condemned by the UN Committee on the Rights of the Child, the UN Human Rights Committee, and the UN Committee Against Torture, and has been banned by 106 countries, more than half of the world's total.[189]

Though a majority of Americans still endorse corporal punishment by parents, they draw an increasingly sharp line between mild violence they consider discipline, such as spanking and slapping, and severe violence they consider abuse, such as punching, kicking, whipping, beating, and terrorizing (for example, threatening a child with a knife or gun, or dangling it over a ledge). In his surveys of domestic violence, Straus gave respondents a checklist that included punishments that are now considered abusive. He found that the number of parents admitting to them almost halved between 1975 and 1992, from 20 percent of mothers to a bit more than 10 percent.[190]

A problem in self-reports of violence by perpetrators (as opposed to self-reports by victims) is that a positive response is a confession to a wrongdoing. An apparent decline in parents beating their children may really be a decline in parents owning up to it. At one time a mother who left bruises on her child might consider it within the range of acceptable discipline. But starting in the 1980s, a growing number of opinion leaders, celebrities, and writers of television dramas began to call attention to child abuse, often by portraying abusive parents as reprehensible ogres or grown-up children as permanently scarred. In the wake of this current, a parent who bruised a child in anger might keep her mouth shut when the surveyor called. We do know that child abuse had become more of a stigma in this interval. In 1976, when people were asked, "Is child abuse a serious problem in this country?" 10 percent said yes; when the same question was asked in 1985 and 1999, 90 percent said yes.[191] Straus argued that the downward trend in his violence survey captured both a decline in the acceptance of abuse and a decline in actual abuse; even if much of the decline was in acceptance, he added, that would be something to celebrate. A decreasing tolerance of child abuse led to an expansion in the number of abuse hotlines and child protection officers, and to an expanded mandate among police, social workers, school counselors, and volunteers to look for signs of abuse and take steps that would lead to abusers being punished or counseled and to children being removed from the worst homes.

Have the changes in norms and institutions done any good? The National Child Abuse and Neglect Data System was set up to aggregate data on substantiated cases of child abuse from child protection agencies around the country. The psychologist Lisa Jones and the sociologist David Finkelhor have plotted their data over time and shown that from 1990 to 2007 the rate of physical abuse of children fell by half (figure 7–20).

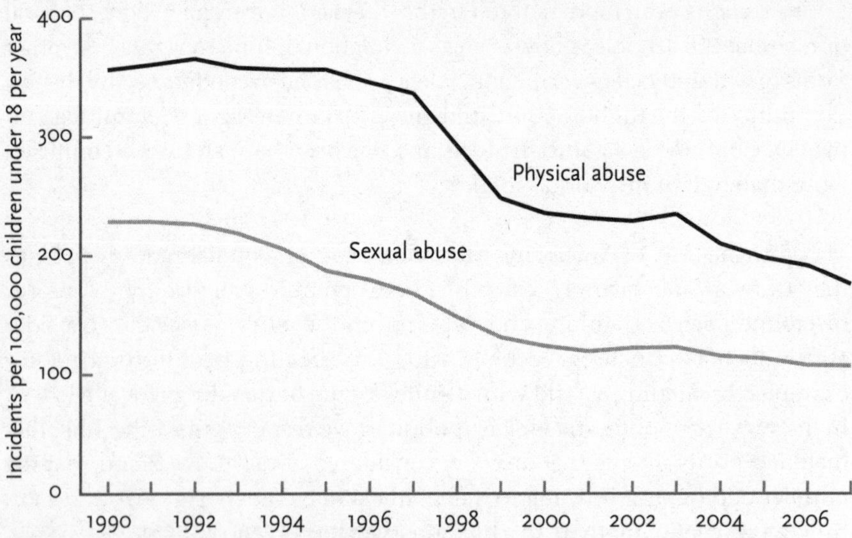

FIGURE 7–20. Child abuse in the United States, 1990–2007
Sources: Data from Jones & Finkelhor, 2007; see also Finkelhor & Jones, 2006.

Jones and Finkelhor also showed that during this time the rate of sexual abuse, and the incidence of violent crimes against children such as assault, robbery, and rape, also fell by a third to two-thirds. They corroborated the declining numbers with sanity checks such as victimization surveys, homicide data, offender confessions, and rates of sexually transmitted diseases, all of which are in decline. In fact over the past two decades the lives of children and adolescents improved in just about every way you can measure. They were also less likely to run away, to get pregnant, to get into trouble with the law, and to kill themselves. England and Wales have also enjoyed a decline in violence against children: a recent report has shown that since the 1970s, the rate of violent deaths of children fell by almost 40 percent.[192]

The decline of child abuse in the 1990s coincided in part with the decline of adult homicide, and its causes are just as hard to pinpoint. Finkelhor and Jones examined the usual suspects. Demography, capital punishment, crack cocaine, guns, abortion, and incarceration can't explain the decline. The prosperity of the 1990s can explain it a little, but can't account for the decline in sexual abuse, nor a second decline of physical abuse in the 2000s, when the economy was in the tank. The hiring of more police and interveners from social service agencies probably helped, and Finkelhor and Jones speculate that another exogenous factor may have made a difference. The early 1990s was the era of Prozac Nation and Running on Ritalin. The massive expansion in the prescription of medication for depression and attention deficit disorder may have lifted many parents out of depression and helped many children control their impulses. Finkelhor and Jones also pointed to nebulous but potentially potent changes in cultural norms. The 1990s, as we saw in chapter 3, hosted a civilizing offensive that reversed some of the licentiousness of the 1960s and made all forms of violence increasingly repugnant. And the Oprahfication of America applied a major stigma to domestic violence, while destigmatizing—indeed beatifying—the victims who brought it to light.

Another kind of violence that torments many children is the violence perpetrated against them by other children. Bullying has probably been around for as long as children have been around, because children, like many juvenile primates, strive for dominance in their social circle by demonstrating their mettle and strength. Many childhood memoirs include tales of cruelty at the hands of other children, and the knuckle-dragging bully is a staple of popular culture. The rogues' gallery includes Butch and Woim in *Our Gang,* Biff Tannen in the *Back to the Future* trilogy, Nelson Muntz in *The Simpsons,* and Moe in *Calvin and Hobbes* (figure 7–21).

Until recently, adults had written off bullying as one of the trials of childhood. "Boys will be boys," they said, figuring that an ability to deal with intimidation in childhood was essential training for the ability to deal with it in adulthood. The victims, for their part, had nowhere to turn, because complaining to a teacher or parent would brand them as snitches and pantywaists and make their lives more hellish than ever.

But in another of those historical gestalt shifts in which a category of violence flips from inevitable to intolerable, bullying has been targeted for elimination. The movement emerged from the ball of confusion surrounding the 1999 Columbine High School massacre, as the media amplified one another's rumors about the causes—Goth culture, jocks, antidepressants, video games, Internet use, violent movies, the rock singer Marilyn Manson—and one of them was bullying. As it turned out, the two assassins were not, as the media endlessly repeated, Goths who had been picked on by jocks.[193] But a popular understanding took hold that the massacre was an act of revenge, and childhood professionals parlayed the urban legend into a campaign against bullying. Fortunately,

FIGURE 7–21. Another form of violence against children

the theory—bully victim today, cafeteria sniper tomorrow—coexisted with more respectable rationales, such as that victims of bullies suffer from depression, impaired performance in school, and an elevated risk of suicide.[194] Currently forty-four states have laws that prohibit bullying in school, and many have mandatory curricula that denounce bullying, encourage empathy, and instruct children in how to resolve their conflicts constructively.[195] Organizations of pediatricians and child psychologists have issued statements calling for prevention efforts, and magazines, television programs, the Oprah Winfrey empire, and even the president of the United States have targeted it as well.[196] In another decade, the facetious treatment of bullying in the *Calvin and Hobbes* cartoon may become as offensive as the spank-the-wife coffee ads from the 1950s are to us today.

Psychological consequences aside, the moral case against bullying is ironclad. As Calvin observed, once you grow up, you can't go beating people up for no reason. We adults protect ourselves with laws, police, workplace regulations, and social norms, and there is no conceivable reason why children should be left more vulnerable, other than laziness or callousness in considering what life is like from their point of view. The increased valuation of children, and the universalizing of moral viewpoints of which it is a part, made the campaign to protect children from violence by their peers inevitable. So too the effort to protect them from other depredations. Children and teenagers have long been victims of petty crimes like the theft of lunch money, vandalism of their possessions, and sexual groping, which fall in the cracks between school regulations and criminal law enforcement. Here too the interests of younger humans are increasingly being recognized.

Has it made a difference? It has begun to do so. In 2004 the U.S. Departments of Justice and Education issued a report on *Indicators of School Crime and Safety* that used victimization surveys and school and police statistics to document trends in violence against students from 1992 to 2003.[197] The survey asked about bullying only in the last three years, but they tracked other kinds of violence for the entire period, and found that fighting, fear at school, and crimes such as theft, sexual assault, robbery, and assault all ramped downward, as figure 7–22 shows.

And contrary to yet another scare that has recently been ginned up by the media, based on widely circulated YouTube videos of female teenagers pummeling one another, the nation's girls have not gone wild. The rates of murder and robbery by girls are at their lowest level in forty years, and rates of weapon possession, fights, assaults, and violent injuries by and toward girls have been declining for a decade.[198] With the popularity of YouTube, we can expect more of these video-driven moral panics (sadist grannies? bloodthirsty toddlers? killer gerbils?) in the years to come.

It is premature to say that the kids are all right, but they are certainly far better off than they used to be. Indeed, in some ways the effort to protect children

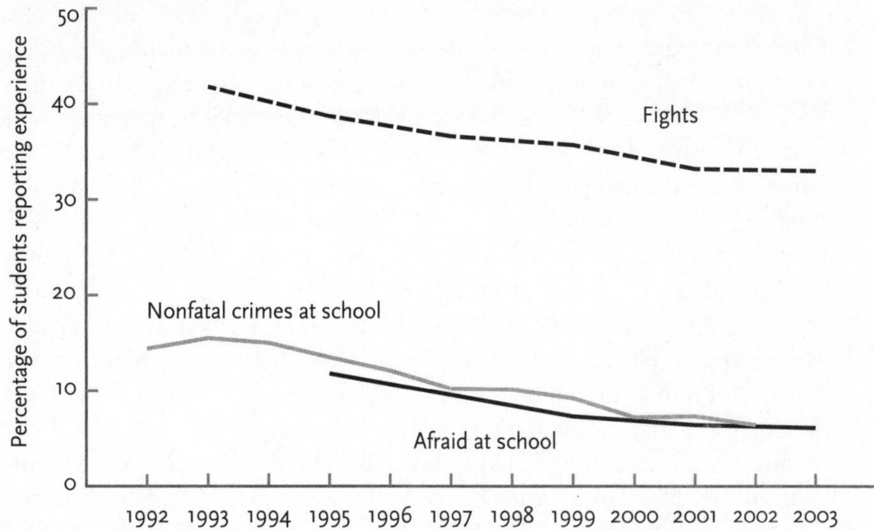

FIGURE 7–22. Violence against youths in the United States, 1992–2003
Source: Data from DeVoe et al., 2004.

against violence has begun to overshoot its target and is veering into the realm of sacrament and taboo.

One of these taboos is what the psychologist Judith Harris calls the Nurture Assumption.[199] Locke and Rousseau helped set off a revolution in the conceptualization of child-rearing by rewriting the role of caregivers from beating bad behavior out of children to shaping the kind of people they would grow into. By the late 20th century, the idea that parents can harm their children by abusing and neglecting them (which is true) grew into the idea that parents can mold their children's intelligence, personalities, social skills, and mental disorders (which is not). Why not? Consider the fact that children of immigrants end up with the accent, values, and norms of their peers, not of their parents. That tells us that children are socialized in their peer group rather than in their families: it takes a village to raise a child. And studies of adopted children have found that they end up with personalities and IQ scores that are correlated with those of their biological siblings but uncorrelated with those of their adopted siblings. That tells us that adult personality and intelligence are shaped by genes, and also by chance (since the correlations are far from perfect, even among identical twins), but are not shaped by parents, at least not by anything they do with all their children. Despite these refutations, the Nurture Assumption developed a stranglehold on professional opinion, and mothers have been advised to turn themselves into round-the-clock parenting machines, charged with stimulating, socializing, and developing the characters of the little blank slates in their care.

Another sacrament is the campaign to quarantine children from the slightest shred of a trace of a hint of a reminder of violence. In Chicago in 2009, after twenty-five students aged eleven to fifteen took part in the age-old sport of a cafeteria food fight, they were rounded up by the police, handcuffed, herded into a paddy wagon, photographed for mug shots, and charged with reckless conduct.[200] Zero-tolerance policies for weapons on school property led to a threat of reform school for a six-year-old Cub Scout who had packed an all-in-one camping utensil in his lunch box, the expulsion of a twelve-year-old girl who had used a utility knife to cut windows out of a paper house for a class project, and the suspension of an Eagle Scout who followed the motto "Be Prepared" by keeping a sleeping bag, drinking water, emergency food, and a two-inch pocketknife in his car.[201] Many schools have hired whistle-wielding "recess coaches" to steer children into constructive organized games, because left on their own they might run into one another, bicker over balls and jump ropes, or monopolize patches of playground.[202]

Adults have increasingly tried to keep depictions of violence out of children's culture. In a climactic sequence of the 1982 movie *E.T.*, Elliott sneaks past a police roadblock with E.T. in the basket of his bicycle. When the film was rereleased in a 20th-anniversary version in 2002, Steven Spielberg had digitally disarmed the officers, using computer-generated imagery to replace their rifles with walkie-talkies.[203] Around Halloween, parents are now instructed to dress their children in "positive costumes" such as historical figures or items of food like carrots or pumpkins rather than zombies, vampires, or characters from slasher films.[204] A memo from a Los Angeles school carried the following costume advisory:

> They should not depict gangs or horror characters, or be scary.
> Masks are allowed only during the parade.
> Costumes may not demean any race, religion, nationality, handicapped condition, or gender.
> No fake fingernails.
> No weapons, even fake ones.

Elsewhere in California, a mother who thought that her children might be frightened by the Halloween tombstones and monsters in a neighbor's yard called the police to report it as a hate crime.[205]

The historical increase in the valuation of children has entered its decadent phase. Now that children are safe from being smothered on the day they are born, starved in foundling homes, poisoned by wet nurses, beaten to death by fathers, cooked in pies by stepmothers, worked to death in mines and mills, felled by infectious diseases, and beaten up by bullies, experts have racked their brains for ways to eke infinitesimal increments of safety from a curve of diminishing or even reversing returns. Children are not allowed to be outside in the middle of the day (skin cancer), to play in the grass (deer ticks), to buy lemonade from a stand (bacteria on lemon peel), or to lick cake batter off spoons (salmonella from uncooked eggs). Lawyer-vetted playgrounds have had their turf padded with rubber, their slides and monkey bars lowered to waist height, and their seesaws removed altogether (so that the kid at the bottom can't jump off and watch the kid at the top come hurtling to the ground—the most fun part of playing on a seesaw). When the producers of *Sesame Street* issued a set of DVDs containing classic programs from the first years of the series (1969–74), they included a warning on the box that the shows were not suitable for children![206] The programs showed kids engaging in dangerous activities like climbing on monkey bars, riding tricycles without helmets, wriggling through pipes, and accepting milk and cookies from kindly strangers. Censored altogether was Monsterpiece Theater, because at the end of each episode the ascotted, smoking-jacketed host, Alistair Cookie (played by Cookie Monster), gobbled down his pipe, which glamorizes the use of tobacco products and depicts a choking hazard.

But nothing has transformed childhood as much as the risk of kidnapping by strangers, a textbook case in the psychology of fear.[207] Since 1979, when six-year-old Etan Patz disappeared on his way to a school bus stop in lower Manhattan, kidnapped children have riveted the nation's attention, thanks to three interest groups that are dedicated to sowing panic among the nation's parents. The grief-stricken parents of murdered children understandably want something good to come out of their tragedies, and several have devoted their lives to raising awareness of child abductions. (One of them, John Walsh, campaigned to have photographs of missing children featured on milk cartons, and hosted a lurid television program, *America's Most Wanted*, which specialized in horrific kidnap-murders.) Politicians, police chiefs, and corporate publicists can smell a no-lose campaign from a mile away—who could be against protecting children from perverts?—and have held ostentatious ceremonies to announce protective measures named after missing children (Code Adam, Amber Alerts, Megan's Law, the National Missing Children Day). The media too can recognize a ratings pump when they see one, and have stoked the fear with round-the-clock vigils, documentaries in constant rotation ("It is every parent's nightmare . . ."), and a *Law and Order* spinoff dedicated to nothing but sex crimes.

Childhood has never been the same. American parents will not let their children out of their sight. Children are chauffeured, chaperoned, and tethered with cell phones, which, far from reducing parents' anxiety, only sends them into a tizzy if a child doesn't answer on the first ring. Making friends in the playground has given way to mother-arranged playdates, a phrase that didn't exist before the 1980s.[208] Forty years ago two-thirds of children walked or biked to school; today 10 percent do. A generation ago 70 percent of children played outside; today the rate is down to 30 percent.[209] In 2008 the nine-year-old son of the journalist Lenore Skenazy begged her to let him go home by himself on the New York subway. She agreed, and he made it home without incident. When she wrote about the vignette in a *New York Sun* column, she found herself at the center of a media frenzy in which she was dubbed "America's Worst Mom." (Sample headline: "Mom Lets 9-Year-Old Take Subway Home Alone: Columnist Stirs Controversy with Experiment in Childhood Independence.") In response she started a movement—Free-Range Children—and proposed National Take Our Children to the Park and Leave Them There Day, intended to get children to learn to play by themselves without constant adult supervision.[210]

Skenazy is not, in fact, America's worst mom. She simply did what no politician, policeman, parent, or producer ever did: she looked up the facts. The overwhelming majority of milk-carton children were not lured into vans by sex perverts, child traffickers, or ransom artists, but were teenagers who ran away from home, or children taken by a divorced parent who was embittered by an unfavorable custody ruling. The annual number of abductions by strangers has ranged from 200 to 300 in the 1990s to about 100 today, around half of whom are murdered. With 50 million children in the United States, that works out to an annual homicide rate of one in a million (0.1 per hundred thousand, to use our usual metric). That's about a twentieth of the risk of drowning and a fortieth of the risk of a fatal car accident. The writer Warwick Cairns calculated that if you *wanted* your child to be kidnapped and held overnight by a stranger, you'd have to leave the child outside and unattended for 750,000 years.[211]

One might reply that the safety of a child is so precious that even if these precautions saved a few lives a year, they would be worth the anxiety and expense. But the reasoning is spurious. People inescapably trade off safety for other good things in life, as when they set aside money for their children's college education rather than installing a sprinkler system in their homes, or drive with their children to a vacation destination rather than letting them play video games in the safety of their bedrooms all summer. The campaign for perfect safety from abductions ignores costs like constricting childhood experience, increasing childhood obesity, instilling chronic anxiety in working women, and scaring young adults away from having children.

And even if minimizing risk *were* the only good in life, the innumerate safety advisories would not accomplish it. Many measures, like the milk-carton wanted posters, are examples of what criminologists call crime-control theater: they advertise that something is being done without actually doing anything.[212] When 300 million people change their lives to reduce a risk to 50 people, they will probably do more harm than good, because of the unforeseen consequences of their adjustments on the vastly *more* than 50 people who are affected by them. To take just two examples, more than twice as many children are hit by cars driven by parents taking their children to school as by other kinds of traffic, so when more parents drive their children to school to prevent them from getting killed by kidnappers, more children get killed.[213] And one form of crime-control theater, electronic highway signs that display the names of missing children to drivers on freeways, may cause slowdowns, distracted drivers, and the inevitable accidents.[214]

The movement over the past two centuries to increase the valuation of children's lives is one of the great moral advances in history. But the movement over the past two decades to increase the valuation to infinity can lead only to absurdities.

GAY RIGHTS, THE DECLINE OF GAY-BASHING, AND THE DECRIMINALIZATION OF HOMOSEXUALITY

It would be an exaggeration to say that the British mathematician Alan Turing explained the nature of logical and mathematical reasoning, invented the digital computer, solved the mind-body problem, and saved Western civilization. But it would not be much of an exaggeration.[215]

In a landmark 1936 paper, Turing laid out a set of simple mechanical operations that was sufficient to compute any mathematical or logical formula that was computable at all.[216] These operations could easily be implemented in a machine—a digital computer—and a decade later Turing designed a practicable version that served as a prototype for the computers we use today. In the interim, he worked for the British decryption unit during World War II and helped to crack the cipher used by the Nazis to communicate with their U-boats, which was instrumental in defeating the German naval blockade and turning around the war. When the war was over, Turing wrote a paper (still widely read today) that equated thinking with computation, thereby offering an explanation of how intelligence could be carried out by a physical system.[217] For good measure, he then tackled one of the hardest problems in science—how the structure of an organism could emerge from a pool of chemicals during embryonic development—and proposed an ingenious solution.

How did Western civilization thank one of the greatest geniuses it ever produced? In 1952 the British government arrested him, withdrew his security clearance, threatened him with prison, and chemically castrated him, driving him to suicide at the age of forty-two.

What did Turing do to earn this stunning display of ingratitude? He had sex with a man. Homosexual acts were illegal in Britain at the time, and he was charged with gross indecency, under the same statute that in the preceding century had broken another genius, Oscar Wilde. Turing's persecution was motivated by a fear that homosexuals were vulnerable to being entrapped by Soviet agents. The fear became risible eight years later when the British war secretary John Profumo was forced to resign because he had had an affair with the mistress of a Soviet spy.

At least since Leviticus 20:13 prescribed the death penalty for a man lying with mankind as he lieth with a woman, many governments have used their monopoly on violence to imprison, torture, mutilate, and kill homosexuals.[218] A gay person who escaped government violence in the form of laws against indecency, sodomy, buggery, unnatural acts, or crimes against nature was vulnerable to violence from his fellow citizens in the form of gay-bashing, homophobic violence, and antigay hate crimes.

Homophobic violence, whether state-sponsored or grassroots, is a mysterious entry in the catalog of human violence, because there is nothing in it for the aggressor. No contested resource is at stake, and since homosexuality is a victimless crime, no peace is gained by deterring it. If anything, one might expect straight men to react to their gay fellows by thinking: "Great! More women for me!" By the same logic, lesbianism should be the most heinous crime imaginable, because it takes women out of the mating pool two at a time. But homophobia has been more prominent in history than lesbophobia.[219] While many legal systems single out male homosexuality for criminalization, no legal system singles out lesbianism, and hate crimes against gay men outnumber hate crimes against gay women by a ratio of almost five to one.[220]

Homophobia is an evolutionary puzzle, as is homosexuality itself.[221] It's not that there's anything mysterious about homosexual *behavior.* Humans are a polymorphously perverse species, and now and again seek sexual gratification from all manner of living and nonliving things that don't contribute to their reproductive output. Men in all-male settings such as ships, prisons, and boarding schools often make do with the available object that resembles a female body more closely than anything else in the vicinity. Pederasty, which offers a softer, smoother, and more docile object, has been institutionalized in a number of societies, including, famously, the elite of ancient Greece. When homosexual behavior is institutionalized, not surprisingly, there is little homophobia as we know it. Women, for their part, are less ardent but more flexible in their sexuality, and many go through phases in life when they are happily celibate, promiscuous, monogamous, or homosexual; hence the phenomenon in American women's colleges of the LUG (lesbian until graduation).[222]

The real puzzle is homosexual *orientation*—why there should be men and women who consistently prefer homosexual mating opportunities to heterosexual ones, or who avoid mating with the opposite sex altogether. At least in men, homosexual orientation appears to be inborn. Gay men generally report that their homosexual attractions began as soon as they felt sexual stirrings shortly before adolescence. And homosexuality is more concordant in identical than in fraternal twins, suggesting that their shared genes play a role. Homosexuality, by the way, is one of the few examples of a nature-nurture debate in which the politically correct position is "nature." If homosexuality is innate, according to the common understanding, then people don't choose to be gay and hence can't be criticized for their lifestyle; nor could they convert the children in their classrooms or Boy Scout troops if they wanted to.

The evolutionary mystery is how any genetic tendency to avoid heterosexual sex can remain in a population for long, since it would have consigned the person to few or no offspring. Perhaps "gay genes" have a compensating advantage, like enhancing fertility when they are carried by women, particularly if they are on the X chromosome, which women have in two copies—the advantage to women would need to be only a bit more than half the disadvantage to men for the gene to spread.[223] Perhaps the putative gay genes lead to homosexuality only in certain environments, which didn't exist while our genes were selected. One ethnographic survey found that in almost 60 percent of preliterate societies, homosexuality was unknown or extremely rare.[224] Or perhaps the genes work indirectly, by making a fetus susceptible to fluctuations in hormones or antibodies which affect its developing brain.

Whatever the explanation, people with a homosexual orientation who grow up in a society that does not cultivate homosexual behavior may find themselves the target of a society-wide hostility. Among traditional societies that take note of homosexuality in their midst, more than twice as many disapprove of it as tolerate it.[225] And in traditional and modern societies alike, the intolerance can erupt in violence. Bullies and toughs may see an easy mark on whom they can prove their machismo to an audience or to one another. And lawmakers may have moralistic convictions about homosexuality that they translate into commandments and statutes. These beliefs may be products of the cross-wiring between disgust and morality that leads people to confuse visceral revulsion with objective sinfulness.[226] That short circuit may convert an impulse to avoid homosexual partners into an impulse to condemn homosexuality. At least since biblical times homophobic sentiments have been translated into laws that punish homosexuals with death or mutilation, especially in Christian and Muslim kingdoms and their former colonies.[227] A chilling 20th-century example was the targeting of homosexuals for elimination during the Holocaust.

During the Enlightenment, the questioning of any moral precept that was based on visceral impulse or religious dogma led to a new look at homosexuality.[228] Montesquieu and Voltaire argued that homosexuality should be decriminalized, though they didn't go so far as to say that it was morally acceptable. In 1785 Jeremy Bentham took the next step. Using utilitarian reasoning, which equates morality with whatever brings the greatest good to the greatest number, Bentham argued that there is nothing immoral about homosexual acts because they make no one worse off. Homosexuality was legalized in France after the Revolution, and in a smattering of other countries in the ensuing decades, as figure 7–23 shows. The movement picked up in the middle of the 20th century and blasted off in the 1970s and 1990s, as the gay rights movement was fueled by the ideal of human rights.

Today homosexuality has been legalized in almost 120 countries, though laws against it remain on the books of another 80, mostly in Africa, the Caribbean, Oceania, and the Islamic world.[229] Worse, homosexuality is punishable

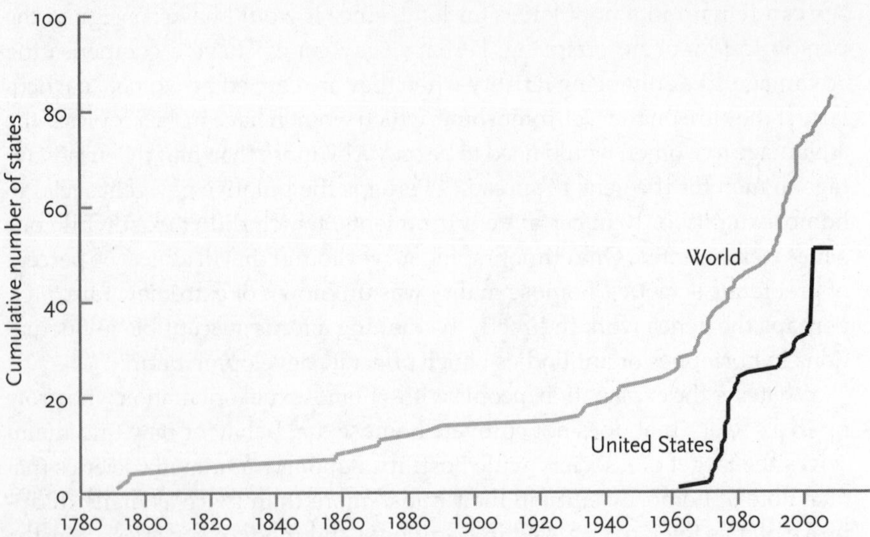

FIGURE 7–23. Time line for the decriminalization of homosexuality, United States and world

Sources: Ottosson, 2006, 2009. Dates for an additional seven countries (Timor-Leste, Surinam, Chad, Belarus, Fiji, Nepal, and Nicaragua) were obtained from "LBGT Rights by Country or Territory," http://en.wikipedia.org/wiki/LGBT_rights. Dates for an additional thirty-six countries that currently allow homosexuality are not listed in either source.

by death in Mauritania, Saudi Arabia, Sudan, Yemen, parts of Nigeria, parts of Somalia, and all of Iran (despite, according to Mahmoud Ahmadinejad, not existing in that country). But the pressure is on. Every human rights organization considers the criminalization of homosexuality to be a human rights violation, and in 2008 in the UN General Assembly, 66 countries endorsed a declaration urging that all such laws be repealed. In a statement endorsing the declaration, Navanethem Pillay, the UN High Commissioner for Human Rights, wrote, "The principle of universality admits no exception. Human rights truly are the birthright of all human beings."[230]

The same graph shows that the decriminalization of homosexuality began later in the United States. As late as 1969, homosexuality was illegal in every state but Illinois, and municipal police would often relieve their boredom on a slow night by raiding a gay hangout and dispersing or arresting the patrons, sometimes with the help of billy clubs. But in 1969 a raid of the Stonewall Inn, a gay dance club in Greenwich Village, set off three days of rioting in protest and galvanized gay communities throughout the country to work to repeal laws that criminalized homosexuality or discriminated against homosexuals. Within a dozen years almost half of American states had decriminalized homosexuality. In 2003, following another burst of decriminalizations, the Supreme Court overturned an antisodomy statute in Texas and ruled that all such laws were unconstitutional. In the majority opinion, Justice Anthony

Kennedy invoked the principle of personal autonomy and the indefensibility of using government power to enforce religious belief and traditional customs:

> Liberty presumes an autonomy of self that includes freedom of thought, belief, expression, and certain intimate conduct. . . . It must be acknowledged, of course, that for centuries there have been powerful voices to condemn homosexual conduct as immoral. The condemnation has been shaped by religious beliefs, conceptions of right and acceptable behavior, and respect for the traditional family. . . . These considerations do not answer the question before us, however. The issue is whether the majority may use the power of the State to enforce these views on the whole society through operation of the criminal law.[231]

Between the first burst of legalization in the 1970s and the collapse of the remaining laws a decade and a half later, Americans' attitudes toward homosexuality underwent a sea change. The rise of AIDS in the 1980s mobilized gay activist groups and led many celebrities to come out of the closet, while others were outed posthumously. They included the actors John Gielgud and Rock Hudson, the singers Elton John and George Michael, the fashion designers Perry Ellis, Roy Halston, and Yves Saint Laurent, the athletes Billie Jean King and Greg Louganis, and the comedians Ellen DeGeneres and Rosie O'Donnell. Popular entertainers such as k.d. lang, Freddie Mercury, and Boy George flaunted gay personas, and playwrights such as Harvey Fierstein and Tony Kushner wrote about AIDS and other gay themes in popular plays and movies. Lovable gay characters began to appear in romantic comedies and in sitcoms such as *Will and Grace* and *Ellen*, and an acceptance of homosexuality among heterosexuals was increasingly depicted as the norm. As Jerry Seinfeld and George Costanza insisted, "*We're not gay!* . . . Not that there's anything wrong with that." As homosexuality was becoming destigmatized, domesticated, and even ennobled, fewer gay people felt the need to keep their sexual orientation hidden. In 1990 my graduate advisor, an eminent psycholinguist and social psychologist who was born in 1925, published an autobiographical essay that began, "When Roger Brown comes out of the closet, the time for courage is past."[232]

Americans increasingly felt that gay people were a part of their real and virtual communities, and that made it harder to keep them outside their circle of sympathy. The changes can be seen in the attitudes they revealed to pollsters. Figure 7–24 shows Americans' opinions on whether homosexuality is morally wrong (from two polling organizations), whether it should be legal, and whether gay people should have equal job opportunities. I've plotted the "yeses" for the last two questions upside down, so that low values for all four questions represent the more tolerant response.

The most gay-friendly opinion, and the first to show a decline, was on equal opportunity. After the civil rights movement, a commitment to fairness had

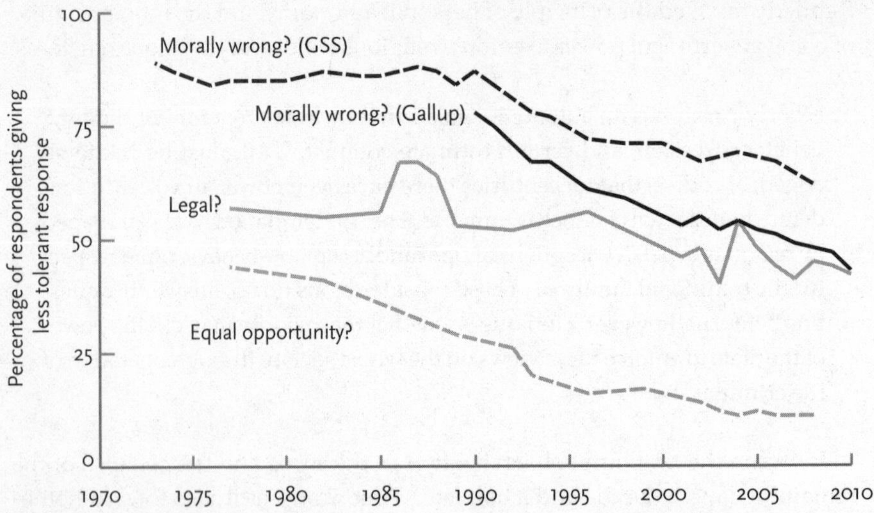

FIGURE 7–24. Intolerance of homosexuality in the United States, 1973–2010

Sources: Morally wrong (GSS): General Social Survey, http://www.norc.org/GSS+Website. All other questions: Gallup, 2001, 2008, 2010. All data represent "yes" responses; data for the "Equal opportunity" and "Legal" questions are subtracted from 100.

become common decency, and Americans were unwilling to accept discrimination against gay people even if they didn't approve of their lifestyle. By the new millennium resistance to equal opportunity had fallen into the zone of crank opinion. Beginning in the late 1980s, the moral judgments began to catch up with the sense of fairness, and more and more Americans were willing to say, "Not that there's anything wrong with that." The headline of a 2008 press release from the Gallup Organization sums up the current national mood: "Americans Evenly Divided on Morality of Homosexuality: However, majority supports legality and acceptance of gay relations."[233]

Liberals are more accepting of homosexuality than conservatives, whites more accepting than blacks, and the secular more tolerant than the religious. But in every sector the trend over time is toward tolerance. Personal familiarity matters: a 2009 Gallup poll showed that the six in ten Americans who have an openly gay friend, relative, or co-worker are more favorable to legalized homosexual relations and to gay marriage than the four in ten who don't. But tolerance is now widespread: even among Americans who have never known a gay person, 62 percent say they would feel comfortable around one.[234]

And in the most significant sector of all, the change has been dramatic. Many people have informed me that younger Americans have become homophobic, based on the observation that they use "That's so gay!" as a putdown. But the numbers say otherwise: the younger the respondents, the more accepting they are of homosexuality.[235] Their acceptance, moreover, is morally deeper. Older tolerant respondents have increasingly come down on the "nature" side of the debate on the causes of homosexuality, and naturists are more tolerant than nurturists because they feel that a person cannot be condemned for a trait he never chose. But teens and twenty-somethings are more sympathetic to the nurture explanation *and* they are more tolerant of homosexuality. The combination suggests that they just find nothing wrong with homosexuality in the first place, so whether gay people can "help it" is beside the point. The attitude is: "Gay? Whatever, dude." Young people, of course, tend to be more liberal than their elders, and it's possible that as they creep up the demographic totem pole they will lose their acceptance of homosexuality. But I doubt it. The acceptance strikes me as a true generational difference, one that this cohort will take with them as they become geriatric. If so, the country will only get increasingly tolerant as their homophobic elders die off.

A populace that accepts homosexuality is likely not just to disempower the police and courts from using force against gay people but to empower them to prevent other citizens from using it. A majority of American states, and more than twenty countries, have hate-crime laws that increase the punishment for violence motivated by a person's sexual orientation, race, religion, or gender. Since the 1990s the federal government has been joining them. The most recent escalation came from the Matthew Shepard and James Byrd, Jr., Hate Crimes Prevention Act of 2009, named after a gay student in Wyoming who in 1998 was beaten, tortured, and tied to a fence overnight to die. (The law's other namesake was the African American man who was murdered that year by being dragged behind a truck.)

So tolerance of homosexuality has gone up, and tolerance of antigay violence has gone down. But have the new attitudes and laws caused a downturn in homophobic violence? The mere fact that gay people have become so much more visible, at least in urban, coastal, and university communities, suggests they feel less menaced by an implicit threat of violence. But it's not easy to show that rates of actual violence have changed. Statistics are available only for the years since 1996, when the FBI started to publish data on hate crimes broken down by the motive, the victim, and the nature of the crime.[236] Even these numbers are iffy, because they depend on the willingness of the victims to report a crime and on the local police to categorize it as a hate crime and report it to the FBI.[237] That isn't as much of a problem with homicides, but unfortunately for social scientists (and fortunately for humanity) not that many people are killed because they are gay. Since 1996 the FBI has recorded fewer than 3 antigay homicides a year from among the 17,000 or so that are committed for every other reason. And as best as we can tell, other antigay hate crimes are uncommon as well. In 2008 the chance that a person would be a victim of aggravated assault because of sexual orientation was 3 per 100,000 gay people, whereas the chance that he would be a victim because he was a human being was more than a hundred times higher.[238]

We don't know whether these odds have gotten smaller over time. Since

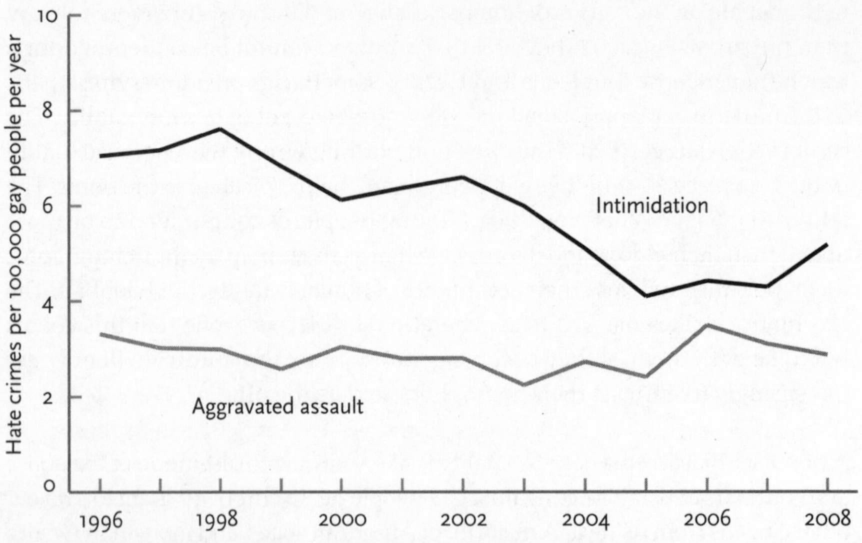

FIGURE 7–25. Antigay hate crimes in the United States, 1996–2008

Source: Data from the annual FBI reports of Hate Crime Statistics (http://www.fbi.gov/hq/cid/civilrights/hate.htm). The number of incidents is divided by the population covered by the agencies reporting the statistics multiplied by 0.03, a common estimate of the incidence of homosexuality in the adult population.

1996 there has been no significant change in the incidence of three of the four major kinds of hate crimes against gay people: aggravated assault, simple assault, or homicide (though the homicides are so rare that trends would be meaningless anyway).[239] In figure 7–25 I've plotted the incidence of the remaining category, which *has* declined, namely intimidation (in which a person is made to feel in danger for his or her personal safety), together with the rate of aggravated assault for comparison.

So while we can't say for sure that gay Americans have become safer from assault, we do know they are safer from intimidation, safer from discrimination and moral condemnation, and perhaps most importantly, completely safe from violence from their own government. For the first time in millennia, the citizens of more than half the countries of the world can enjoy that safety—not enough of them, but a measure of progress from a time in which not even helping to save one's country from defeat in war was enough to keep the government goons away.

ANIMAL RIGHTS AND THE DECLINE OF CRUELTY TO ANIMALS

Let me tell you about the worst thing I have ever done. In 1975, as a twenty-year-old sophomore, I got a summer job as a research assistant in an animal behavior lab. One evening the professor gave me an assignment. Among the rats in the lab was a runt that could not participate in the ongoing studies, so he wanted to use it to try out a new experiment. The first step was to train the rat in what was called a temporal avoidance conditioning procedure. The floor of a Skinner box was hooked up to a shock generator, and a timer that would shock the animal every six seconds unless it pressed a lever, which would give it a ten-second reprieve. Rats catch on quickly and press the lever every eight or nine seconds, postponing the shock indefinitely. All I had to do was throw the rat in the box, start the timers, and go home for the night. When I arrived back at the lab early the next morning, I would find a fully conditioned rat.

But that was not what looked back at me when I opened the box in the morning. The rat had a grotesque crook in its spine and was shivering uncontrollably. Within a few seconds, it jumped with a start. It was nowhere near the lever. I realized that the rat had not learned to press the lever and had spent the night being shocked every six seconds. When I reached in to rescue it, I found it cold to the touch. I rushed it to the veterinarian two floors down, but it was too late, and the rat died an hour later. I had tortured an animal to death.

As the experiment was being explained to me, I had already sensed it was wrong. Even if the procedure had gone perfectly, the rat would have spent twelve hours in constant anxiety, and I had enough experience to know that laboratory procedures don't always go perfectly. My professor was a radical behaviorist, for whom the question "What is it like to be a rat?" was simply incoherent. But I was not, and there was no doubt in my mind that a rat could feel pain. The professor wanted me in his lab; I knew that if I refused, nothing bad would happen. But I carried out the procedure anyway, reassured by the ethically spurious but psychologically reassuring principle that it was standard practice.

The resonance with certain episodes of 20th-century history is too close for comfort, and in the next chapter I will expand on the psychological lesson I learned that day. The reason I bring up this blot on my conscience is to show what *was* standard practice in the treatment of animals at the time. To motivate the animals to work for food, we starved them to 80 percent of their free-feeding weight, which in a small animal means a state of gnawing hunger. In the lab next door, pigeons were shocked through beaded keychains that were fastened around the base of their wings; I saw that the chains had worn right through their skin, exposing the muscle below. In another lab, rats were shocked through safety pins that pierced the skin of their chests. In one experiment on endorphins, animals were given unavoidable shocks described in the paper as "extremely intense, just subtetanizing"—that is, just short of the point where the animal's muscles would seize up in a state of tetanus. The callousness extended outside the testing chambers. One researcher was known to show his anger by picking up the nearest unused rat and throwing it against a wall. Another shared a cold joke with me: a photograph, printed in a scientific journal, of a rat that had learned to avoid shocks by lying on its furry back while pressing the food lever with its forepaw. The caption: "Breakfast in bed."

I'm relieved to say that just five years later, indifference to the welfare of animals among scientists had become unthinkable, indeed illegal. Beginning in the 1980s, any use of an animal for research or teaching had to be approved by an Institutional Animal Care and Use Committee (IACUC), and any scientist will confirm that these committees are not rubber stamps. The size of cages, the amount and quality of food and veterinary care, and the opportunities for exercise and social contact are strictly regulated. Researchers and their assistants must take a training course on the ethics of animal experimentation, attend a series of panel discussions, and pass an exam. Any experiment that would subject an animal to discomfort or distress is placed in a category governed by special regulations and must be justified by its likelihood of providing "a greater benefit to science and human welfare."

Any scientist will also confirm that attitudes among scientists themselves have changed. Recent surveys have shown that animal researchers, virtually without exception, believe that laboratory animals feel pain.[240] Today a scientist who was indifferent to the welfare of laboratory animals would be treated by his or her peers with contempt.

The change in the treatment of laboratory animals is part of yet another rights revolution: the growing conviction that animals should not be subjected to unjustifiable pain, injury, and death. The revolution in animal rights is a uniquely emblematic instance of the decline of violence, and it is fitting that I end my survey of historical declines by recounting it. That is because the change has been driven purely by the ethical principle that one ought not to inflict suffering on a sentient being. Unlike the other Rights Revolutions, the movement for animal rights was not advanced by the affected parties themselves: the rats and pigeons were hardly in a position to press their case. Nor has it been a by-product of commerce, reciprocity, or any other positive-sum negotiation; the animals have nothing to offer us in exchange for our treating them more humanely. And unlike the revolution in children's rights, it does not hold out the promise of an improvement in the makeup of its beneficiaries later in life. The recognition of animal interests was taken forward by human advocates on their behalf, who were moved by empathy, reason, and the inspiration of the other Rights Revolutions. Progress has been uneven, and certainly the animals themselves, if they could be asked, would not allow us to congratulate ourselves too heartily just yet. But the trends are real, and they are touching every aspect of our relationship with our fellow animals.

When we think of indifference to animal welfare, we tend to conjure up images of scientific laboratories and factory farms. But callousness toward animals is by no means modern. In the course of human history it has been the default.[241]

Killing animals to eat their flesh is a part of the human condition. Our ancestors have been hunting, butchering, and probably cooking meat for at least two million years, and our mouths, teeth, and digestive tracts are specialized for a diet that includes meat.[242] The fatty acids and complete protein in meat enabled the evolution of our metabolically expensive brains, and the availability of meat contributed to the evolution of human sociality.[243] The jackpot of a felled animal gave our ancestors something of value to share or trade and set the stage for reciprocity and cooperation, because a lucky hunter with more meat than he could consume on the spot had a reason to share it, with the expectation that he would be the beneficiary when fortunes reversed. And the complementary contributions of hunted meat from men and gathered plants from women created synergies that bonded men and women for reasons other than the obvious ones. Meat also provided men with an efficient way to invest in their offspring, further strengthening family ties.

The ecological importance of meat over evolutionary time left its mark in the psychological importance of meat in human lives. Meat tastes good, and eating it makes people happy. Many traditional cultures have a word for meat hunger, and the arrival of a hunter with a carcass was an occasion for village-wide rejoicing. Successful hunters are esteemed and have better sex lives, sometimes by dint of their prestige, sometimes by explicit exchanges of the carnal for the carnal. And in most cultures, a meal does not count as a feast unless meat is served.[244]

With meat so important in human affairs, it's not surprising that the welfare of the entities whose bodies provide that meat has been low on the list of human priorities. The usual signals that mitigate violence among humans are mostly absent in animals: they are not close kin, they can't trade favors with us, and in most species they don't have faces or expressions that elicit our sympathy. Conservationists are often exasperated that people care only about the charismatic mammals lucky enough to have faces to which humans respond, like grinning dolphins, sad-eyed pandas, and baby-faced juvenile seals. Ugly species are on their own.[245]

The reverence for nature commonly attributed to foraging people in children's books did not prevent them from hunting large animals to extinction or treating captive animals with cruelty. Hopi children, for example, were encouraged to capture birds and play with them by breaking their legs or pulling off their wings.[246] A Web site for Native American cuisine includes the following recipe:

ROAST TURTLE

Ingredients:

One turtle
One campfire

Directions:

Put a turtle on his back on the fire.
When you hear the shell crack, he's done.[247]

The cutting or cooking of live animals by traditional peoples is far from uncommon. The Masai regularly bleed their cattle and mix the blood with milk for a delicious beverage, and Asian nomads cut chunks of fat from the tails of living sheep that they have specially bred for that purpose.[248] Pets too are treated harshly: a recent cross-cultural survey found that half the traditional cultures that keep dogs as pets kill them, usually for food, and more than half abuse them. Among the Mbuti of Africa, for example, "the hunting dogs, valuable as they are, get kicked around mercilessly from the day they are born to the day they die."[249] When I asked an anthropologist friend about the treatment of animals by the hunter-gatherers she had worked with, she replied:

> That is perhaps the hardest part of being an anthropologist. They sensed my weakness and would sell me all kinds of baby animals with descriptions of what they would do to them otherwise. I used to take them far into the desert and release them, they would track them, and bring them back to me for sale again!

The early civilizations that depended on domesticated livestock often had elaborate moral codes on the treatment of animals, but the benefits to the animal were mixed at best. The overriding principle was that animals exist for the benefit of humans. In the Hebrew Bible, God's first words to Adam and Eve in Genesis 1:28 are "Be fruitful, and multiply, and replenish the earth, and subdue it: and have dominion over the fish of the sea, and over the fowl of the air, and over every living thing that moveth upon the earth." Though Adam and Eve were frugivores, after the flood the human diet switched to meat. God told Noah in Genesis 9:2–3: "The fear of you and the dread of you shall be upon every beast of the earth, and upon every fowl of the air, upon all that moveth upon the earth, and upon all the fishes of the sea; into your hand are they delivered. Every moving thing that liveth shall be meat for you; even as the green herb have I given you all things." Until the destruction of the second temple by the Romans in 70 CE, vast numbers of animals were slaughtered by Hebrew priests, not to nourish the people but to indulge the superstition that God had to be periodically placated with a well-done steak. (The smell of charbroiled beef, according to the Bible, is "a soothing aroma" and "a sweet savor" to God.)

Ancient Greece and Rome had a similar view of the place of animals in the scheme of things. Aristotle wrote that "plants are created for the sake of animals, and the animals for the sake of man."[250] Greek scientists put this attitude into practice by dissecting live mammals, including, occasionally, *Homo sapiens*. (According to the Roman medical writer Celsus, physicians in Hellenic Alexandria "procured criminals out of prison by royal permission, and dissecting them alive, contemplated, while they were yet breathing, the parts which nature had before concealed.")[251] The Roman anatomist Galen wrote that he preferred to work with pigs rather than monkeys because of the "unpleasant expression" on the monkeys' faces when he cut into them.[252] His compatriots, of course, delighted in the torture and slaughter of animals in the Colosseum, again not excluding a certain bipedal primate. In Christendom, Saints Augustine and Thomas Aquinas combined biblical with Greek views to ratify the amoral treatment of animals. Aquinas wrote, "By the divine providence [animals] are intended for man's use. . . . Hence it is not wrong for man to make use of them, either by killing or in any other way whatsoever."[253]

When it came to the treatment of animals, modern philosophy got off to a bad start. Descartes wrote that animals were clockwork, so there was no one home to feel pain or pleasure. What sound to us like cries of distress were merely the output of a noisemaker, like a warning buzzer on a machine. Descartes knew that the nervous systems of animals and humans were similar, so from our perspective it's odd that he could grant consciousness to humans while denying it to animals. But Descartes was committed to the existence of the soul, granted to humans by God, and the soul was the locus of consciousness. When he introspected on his own consciousness, he wrote, he could not "distinguish in myself any parts, but apprehend myself to be clearly one and entire. . . . The faculties of willing, feeling, conceiving, etc. cannot be properly speaking said to be its parts, for it is one and the same mind which employs itself in willing and in feeling and understanding."[254] Language too is a faculty of this indivisible thing we call mind or soul. Since animals lack language, they must lack souls; hence they must be without consciousness. A human has a clockwork body and brain, like an animal, but also a soul, which interacts with the brain through a special structure, the pineal gland.

From the standpoint of modern neuroscience, the argument is loopy. Today we know that consciousness depends, down to the last glimmer and itch, on the physiological activity of the brain. We also know that language can be dissociated from the rest of consciousness, most obviously in stroke patients who have lost their ability to speak but have not been turned into insensate robots. But aphasia would not be documented until 1861 (by Descartes' compatriot Paul Broca), and the theory sounded plausible enough at the time. For centuries live animals would be dissected in medical laboratories, encouraged by the church's disapproval of the dissection of human cadavers. Scientists cut the limbs off living animals to see if they would regenerate, drew out their bowels, pulled off their skin, and removed their organs, including their eyes.[255]

Agriculture was no more humane. Practices like gelding, branding, piercing, and the docking of ears and tails have been common in farms for centuries. And cruel practices to fatten animals or tenderize their meat (familiar to us today from protests against foie gras and milk-fed veal) are by no means a modern invention. A history of the British kitchen describes some of the methods of tenderization in the 17th century:

> Poultry, in order to put on flesh after its long journey from the farms, was sewn up by the gut . . . ; turkey were bled to death by hanging them upside down with a small incision in the vein of the mouth; geese were nailed to the floor; salmon and carp were hacked into collops while living to make their flesh firmer; eels were skinned alive, coiled around skewers and fixed through the eye so they could not move. . . . The flesh of the bull, it was believed, was indigestible and unwholesome if the animal was killed without being baited. . . . Calves and pigs were whipped to death with knotted ropes to make the meat more tender, rather than our modern practice of beating the flesh when dead. "Take a red cock that is not too old and beat him to death," begins one . . . recipe.[256]

Factory farming is also not a phenomenon of the 20th century:

> The Elizabethan method of "brawning" or fattening pigs was "to keep them in so close a room that they cannot turn themselves round about . . . whereby they are forced always to lie on their bellies." "They feed in pain," said a contemporary, "lie in pain and sleep in pain." Poultry and game-birds were often fattened in darkness and confinement, sometimes being blinded as well. . . . Geese were thought to put on weight if the webs of their feet were nailed to the floor, and it was the custom of some seventeenth-century housewives to cut the legs off living fowl in the belief that it made their flesh more tender. In 1686 Sir Robert Southwell announced a new invention of "an oxhouse, where the cattle are to eat and drink in the same crib and not to stir until they be fitted for the slaughter." Dorset lambs were specially reared for the Christmas tables of the gentry by being imprisoned in little dark cabins.[257]

Many other millennia-old practices are thoroughly indifferent to animal suffering. Fishhooks and harpoons go back to the stone age, and even fishnets kill by slow suffocation. Bits, whips, spurs, yokes, and heavy loads made life miserable for beasts of burden, especially those who spent their days pushing drive shafts in dark mills and pumping stations. Any reader of *Moby-Dick* knows about the age-old cruelties of whaling. And then there were the blood sports that we saw in chapters 3 and 4, such as head-butting a cat nailed to a post, clubbing a pig, baiting a bear, and watching a cat burn to death.

During this long history of exploitation and cruelty, there had always been forces that pushed for restraint in the treatment of animals. But the driving motive was rarely an empathic concern for their inner lives. Vegetarianism, antivivisectionism, and other pro-animal movements have always had a wide range of rationales.[258] Let's consider a few of them.

I have mentioned in a number of places the mind's tendency to moralize the disgust-purity continuum. The equation holds at both ends of the scale: at one pole, we equate immorality with filth, carnality, hedonism, and dissoluteness; at the other, we equate virtue with purity, chastity, asceticism, and temperance.[259] This cross talk affects our emotions about food. Meat-eating is messy and pleasurable, therefore bad; vegetarianism is clean and abstemious, therefore good.

Also, since the human mind is prone to essentialism, we are apt to take the cliché "you are what you eat" a bit too literally. Incorporating dead flesh into one's body can feel like a kind of contamination, and ingesting a concentrate of animality can threaten to imbue the eater with beastly traits. Even Ivy League university students are vulnerable to the illusion. The psychologist Paul Rozin has shown that students are apt to believe that a tribe that hunts turtles for their meat and wild boar for their bristles are probably good swimmers, whereas a tribe that hunts wild boar for their meat and turtles for their shells are probably tough fighters.[260]

People can be turned against meat by romantic ideologies as well. Prelapsarian, pagan, and blood-and-soil creeds can depict the elaborate process of procuring and preparing animals as a decadent artifice, and vegetarianism as a wholesome living off the land.[261] For similar reasons, a concern over the use of animals in research can feed off an antipathy toward science and intellect in general, as when Wordsworth wrote in "The Tables Turned":

Sweet is the lore which Nature brings;
Our meddling intellect
Mis-shapes the beauteous forms of things:—
We murder to dissect.

Finally, since different subcultures treat animals in different ways, a moralistic concern with how the other guy treats his animals (while ignoring the way we do) can be a form of social one-upmanship. Blood sports in particular offer satisfying opportunities for class warfare, as when the middle class lobbies to outlaw the cockfighting enjoyed by the lower classes and the foxhunts enjoyed by the upper ones.[262] Thomas Macaulay's remark that "the Puritan hated bearbaiting, not because it gave pain to the bear, but because it gave pleasure to the spectators" can mean that campaigns against violence tend to target the mindset of cruelty rather than just the harm to victims. But it also captures the insight that zoophily can shade into misanthropy.

The Jewish dietary laws are an ancient example of the confused motives behind taboos on meat. Leviticus and Deuteronomy present the laws as unadorned diktats, since God is under no obligation to justify his commandments to mere mortals. But according to later rabbinical interpretations, the laws foster a concern with animal welfare, if only by forcing Jews to stop and think about the fact that the source of their meat is a living thing, ultimately belonging to God.[263] Animals must be dispatched by a professional slaughterer who severs the animal's carotid artery, trachea, and esophagus with a clean swipe of a nick-free knife. This indeed may have been the most humane technology of the time, and was certainly better than cutting parts off a living animal or roasting it alive. But it is far from a painless death, and some humane societies today have sought to ban the practice. The commandment not to "seethe a kid in its mother's milk," the basis for the prohibition of mixing meat with dairy products, has also been interpreted as an expression of compassion for animals. But when you think about it, it's really an expression of the sensibilities of the observer. To a kid that is about to be turned into stewing meat, the ingredients of the sauce are the least of its concerns.

Cultures that have gone all the way to vegetarianism are also driven by a mixture of motives.[264] In the 6th century BCE Pythagoras started a cult that did more than measure the sides of triangles: he and his followers avoided meat, largely because they believed in the transmigration of souls from body to body, including those of animals. Before the word *vegetarian* was coined in the 1840s, an abstention from meat and fish was called "the Pythagorean diet." The Hindus too based their vegetarianism on the doctrine of reincarnation, though cynical anthropologists like Marvin Harris have offered a more prosaic explanation: cattle in India were more precious as plow animals and dispensers of milk and dung (used as fuel and fertilizer) than they would have been as the main ingredient in beef curry.[265] The spiritual rationale of Hindu vegetarianism was carried over into Buddhism and Jainism, though with a more explicit concern for animals rooted in a philosophy of nonviolence. Jain monks sweep the ground in front of them so as not to tread on insects, and some wear masks to avoid killing microbes by inhaling them.

But any intuition that vegetarianism and humanitarianism go together was shattered in the 20th-century by the treatment of animals under Nazism.[266] Hitler and many of his henchmen claimed to be vegetarians, not so much out of compassion for animals as from an obsession with purity, a pagan desire to reconnect to the soil, and a reaction to the anthropocentrism and meat rituals of Judaism. In an unsurpassed display of the human capacity for moral compartmentalization, the Nazis, despite their unspeakable experiments on living humans, instituted the strongest laws for the protection of animals in research that Europe had ever seen. Their laws also mandated humane treatment of animals in farms, movie sets, and restaurants, where fish had to be anesthetized and lobsters killed swiftly before they were cooked. Ever since that bizarre chapter in the history of animal rights, advocates of vegetarianism have had to retire one of their oldest arguments: that eating meat makes people aggressive, and abstaining from it makes them peaceful.

Some of the early expressions of a genuinely ethical concern for animals took place in the Renaissance. Europeans had become curious about vegetarianism when reports came back from India of entire nations that lived without meat. Several writers, including Erasmus and Montaigne, condemned the mistreatment of animals in hunting and butchery, and one of them, Leonardo da Vinci, became a vegetarian himself.

But it was in the 18th and 19th centuries that arguments for animal rights began to catch on. Part of the impetus was scientific. Descartes' substance dualism, which considered consciousness a free-floating entity that works separately from the brain, gave way to theories of monism and property dualism that equated, or at least intimately connected, consciousness with brain activity. This early neurobiological thinking had implications for animal welfare. As Voltaire wrote:

> There are barbarians who seize this dog, who so prodigiously surpasses man in friendship, and nail him down to a table, and dissect him alive to show you the mezaraic veins. You discover in him all the same organs of feeling as in yourself. Answer me, Machinist, has Nature really arranged all the springs of feeling in this animal to the end that he might not feel? Has he nerves that he may be incapable of suffering?[267]

And as we saw in chapter 4, Jeremy Bentham's laser-beam analysis of morality led him to pinpoint the issue that should govern our treatment of animals: not whether they can reason or talk, but whether they can suffer. By the early 19th century, the Humanitarian Revolution had been extended from humans to other sentient beings, first targeting the most conspicuous form of sadism toward animals, blood sports, followed by the abuse of beasts of burden, livestock on farms, and laboratory animals. When the first of these measures, a ban on the abuse of horses, was introduced into the British Parliament in 1821, it elicited howls of laughter from MPs who said that it would lead to the protection of dogs and even cats. Within two decades that is exactly what happened.[268] Throughout 19th-century Britain, a blend of humanitarianism and romanticism led to antivivisection leagues, vegetarian movements, and societies for the prevention of cruelty to animals.[269] Biologists' acceptance of the theory of evolution following the publication of *The Origin of Species* in 1859 made it impossible for them to maintain that consciousness was unique to humans, and by the end of the century in Britain, they had acceded to laws banning vivisection.

The campaign to protect animals lost momentum during the middle decades of the 20th century. The austerity from the two world wars had created a meat hunger, and the populace was so grateful for the flood of cheap meat from factory farming that it gave little thought to where the meat came from. Also, beginning in the nineteen-teens, behaviorism took over psychology and philosophy and decreed that the very idea of animal experience was a form of unscientific naïveté: the cardinal sin of anthropomorphism. Around the same time the animal welfare movement, like the pacifist movements of the 19th century, developed an image problem and became associated with do-gooders and health food nuts. Even one of the greatest moral voices of the 20th century, George Orwell, was contemptuous of vegetarians:

> One sometimes gets the impression that the mere words "Socialism" and "Communism" draw towards them with magnetic force every fruit-juice drinker, nudist, sandal-wearer, sex-maniac, Quaker, "Nature Cure" quack, pacifist, and feminist in England. . . . The food-crank is by definition a person willing to cut himself off from human society in hopes of adding five years on to the life of his carcase; that is, a person out of touch with common humanity.[270]

All this changed in the 1970s.[271] The plight of livestock in factory farms was brought to light in Britain in a 1964 book by Ruth Harrison called *Animal Machines*. Other public figures soon took up the cause. Brigid Brophy has been credited with the term *animal rights*, which she deliberately coined by analogy: she wanted to associate "the case for non-human animals with that clutch of egalitarian or libertarian ideas which have sporadically, though quite often with impressively actual political results, come to the rescue of other oppressed classes, such as slaves or homosexuals or women."[272]

The real turning point was the philosopher Peter Singer's 1975 book *Animal Liberation*, the so-called bible of the animal rights movement.[273] The sobriquet is doubly ironic because Singer is a secularist and a utilitarian, and utilitarians have been skeptical of natural rights ever since Bentham called the idea "nonsense on stilts." But following Bentham, Singer laid out a razor-sharp argument for a full consideration of the *interests* of animals, while not necessarily granting them "rights." The argument begins with the realization that it is consciousness rather than intelligence or species membership that makes a being worthy of moral consideration. It follows that we should not inflict avoidable pain on animals any more than we should inflict it on young children or the mentally handicapped. And a corollary is that we should all be vegetarians. Humans can thrive on a modern vegetarian diet, and animals' interests in a life free of pain and premature death surely outweigh the marginal increase in pleasure we get from eating their flesh. The fact that humans "naturally" eat meat, whether by cultural tradition, biological evolution, or both, is morally irrelevant.

Like Brophy, Singer made every effort to analogize the animal welfare movement to the other Rights Revolutions of the 1960s and 1970s. The analogy began with his title, an allusion to colonial liberation, women's liberation, and gay liberation, and it continued with his popularization of the term *speciesism*, a sibling of *racism* and *sexism*. Singer quoted an 18th-century critic of the feminist writer Mary Wollstonecraft who argued that if she was right about women, we would also have to grant rights to "brutes." The critic had intended it as a reductio ad absurdum, but Singer argued that it was a sound deduction. For Singer, these analogies are far more than rhetorical techniques. In another book, *The Expanding Circle*, he advanced a theory of moral progress in which human beings were endowed by natural selection with a kernel of empathy toward kin and allies, and have gradually extended it to wider and wider circles of living things, from family and village to clan, tribe, nation, species, and all of sentient life.[274] The book you are reading owes much to this insight.

Singer's moral arguments were not the only forces that made people sympathetic to animals. In the 1970s it was a *good* thing to be a socialist, fruit-juice drinker, nudist, sandal-wearer, sex-maniac, Quaker, Nature Cure quack, pacifist, and feminist, sometimes all at the same time. The compassion-based argument for vegetarianism was soon fortified with other arguments: that meat was fattening, toxic, and artery-hardening; that growing crops to feed animals rather than people was a waste of land and food; and that the effluvia of farm animals was a major pollutant, particularly methane, the greenhouse gas that comes out of both ends of a cow.

Whether you call it animal liberation, animal rights, animal welfare, or the animal movement, the decades since 1975 in Western culture have seen a growing intolerance of violence toward animals. Changes are visible in at least half a dozen ways.

I've already mentioned the first: the protection of animals in laboratories. Not only are live animals now protected from being hurt, stressed, or killed in the conduct of science, but in high school biology labs the venerable custom of dissecting pickled frogs has gone the way of inkwells and slide rules. (In some schools it has been replaced by V-Frog, a virtual reality dissection program.)[275] And in commercial laboratories the routine use of animals to test cosmetics and household products has come under fire. Since the 1940s, following reports of women being blinded by mascara containing coal tar, many household products have been tested for safety with the infamous Draize procedure, which applies a compound to the eyes of rabbits and looks for signs of damage. Until the 1980s few people had heard of the Draize test, and until the 1990s few would have recognized the term *cruelty-free*, the designation for products that avoid it. Today the term is emblazoned on thousands of consumer goods and has become familiar enough that the label "cruelty-free condoms" does not raise an eyebrow. Animal testing in consumer product labs continues, but has been increasingly regulated and reduced.

Another conspicuous change is the outlawing of blood sports. I have already mentioned that since 2005 the British aristocracy has had to retire its bugles and bloodhounds, and in 2008 Louisiana became the last American state to ban cockfights, a sport that had been popular throughout the world for centuries. Like many prohibited vices, the practice continues, particularly among immigrants from Latin America and Southeast Asia, but it has long been in decline in the United States and has been outlawed in many other countries as well.[276]

Even the proud bullfight has been threatened. In 2004 the city of Barcelona outlawed the deadly contests between matador and beast, and in 2010 the ban was extended to the entire region of Catalonia. The state-run Spanish television network had already ended live coverage of bullfights because they were deemed too violent for children.[277] The European Parliament has considered a continent-wide ban as well. Like formal dueling and other violent customs sanctified by pomp and ceremony, bullfighting may eventually bite the dust, not because compassion condemns it or governments outlaw it, but because detraction will not suffer it. In his 1932 book *Death in the Afternoon*, Ernest Hemingway explained the primal appeal of the bullfight:

> [The matador] must have a spiritual enjoyment of the moment of killing. Killing cleanly and in a way which gives you esthetic pleasure and pride has always been one of the greatest enjoyments of a part of the human race. Once you accept the rule of death, "Thou shalt not kill" is an easily and naturally obeyed commandment. But when a man is still in rebellion against death he has pleasure in taking to himself one of the godlike attributes, that of giving it. This is one of the most profound feelings in those men who enjoy killing. These things are done in pride and pride, of course, is a Christian sin and a pagan virtue. But it is pride which makes the bull-fight and true enjoyment of killing which makes the great matador.

Thirty years later, Tom Lehrer described his experience of a bullfight a bit differently. "There is surely nothing more beautiful in this world," he exclaimed, "than the sight of a lone man facing singlehandedly a half a ton of angry pot roast." In the climactic verse of his ballad, he sang:

> I cheered at the bandilleros' display,
> As they stuck the bull in their own clever way,
> For I hadn't had so much fun since the day
> My brother's dog Rover
> Got run over.

"Rover was killed by a Pontiac," Lehrer added, "and it was done with such grace and artistry that the witnesses awarded the driver both ears and the tail." The reaction of young Spaniards today is closer to Lehrer's than to Hemingway's. Their heroes are not matadors but singers and football players who become famous without the spiritual and aesthetic pride of killing anything. While bullfighting retains a loyal following in Spain, the crowds are middle-aged and older.

Hunting is another pastime that has been in decline. Whether it is from compassion for Bambi or an association with Elmer Fudd, fewer Americans shoot animals for fun. Figure 7–26 shows the declining proportion of Americans in the past three decades who have told the General Social Survey that either they or their spouse hunts. Other statistics show that the average age of hunters is steadily creeping upward.[278]

It's not just that Americans are spending more time behind video screens and less in the great outdoors. According to the U.S. Fish and Wildlife Service, in the decade between 1996 and 2006, while the number of hunters, days of hunting, and dollars spent on hunting declined by about 10 to 15 percent, the number of wildlife watchers, days of wildlife watching, and dollars spent on wildlife watching *increased* by 10 to 20 percent.[279] People still like to commune with animals; they would just rather look at them than shoot them. It remains to be seen whether the decline will be reversed by the locavore craze, in which young urban professionals have taken up hunting to reduce their food miles and harvest their own free-range, grass-fed, sustainable, humanely slaughtered meat.[280]

It's hard to imagine that fishing could ever be considered a humane sport, but anglers are doing their best. Some of them take catch-and-release a step further and release the catch before it even breaks the surface, because exposure to the air is stressful to a fish. Better still is hookless fly-fishing: the angler watches the trout take the fly, feels a little tug on the line, and that's it. One of

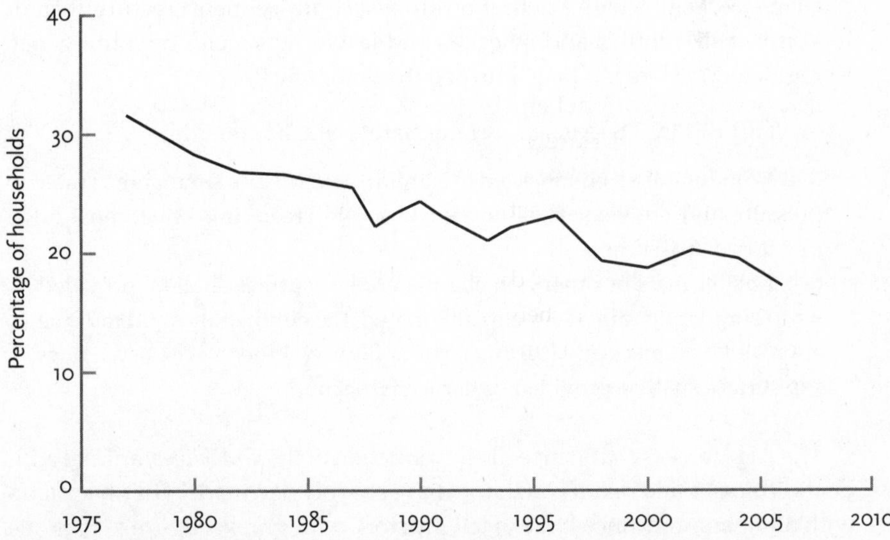

FIGURE 7–26. Percentage of American households with hunters, 1977–2006
Source: General Social Survey, http://www.norc.org/GSS+Website/.

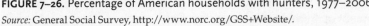

them describes the experience: "I entered the trout's world and got among them in a much more natural way than ever before. I didn't interrupt their feeding rhythms. They took the fly continually, and I still got that little jolt of pleasure you feel when a fish takes your fly. I don't want to harass or harm trout anymore, so now there's a way for me to do that and still keep fishing."[281]

And do you recognize this trope?

No Trees Were Harmed in the Making of This Blog Post.
No Hamsters Were Harmed in the Making of This Book Trailer.
No Polar Bears Were Harmed in the Making of This Commercial.
No Goats Were Harmed in the Writing of This Review.
No Cans of Diet Coke Were Harmed in the Production of This Product.
No Tea Partiers Were Harmed in the Protesting of This Health Care Bill.

It comes from the trademarked certification of the American Humane Association that no animals were harmed in the making of a motion picture, displayed in the rolling credits after the names of the gaffer and key grip.[282] In response to movies that depicted horses plunging over cliffs by actually filming horses plunging over cliffs, the AHA created its film and television unit to develop guidelines for the treatment of animals in films. As the association explains, "Today's consumers, increasingly savvy about animal welfare issues, have forged a partnership with American Humane to demand greater responsibility and accountability from entertainment entities that use animal actors"—a term they insist on, because, they explain, "animals are not props." Their 131-page *Guidelines for the Safe Use of Animals in Filmed Media*, first compiled in 1988, begins with a definition of *animal* ("any sentient creature, including birds, fish, reptiles and insects") and leaves no species or contingency unregulated.[283] Here is a page I turned to at random:

> WATER EFFECTS (Also see Water Safety in Chapter 5.)
>
> 6–2. No animal shall be subjected to extreme, forceful rain simulation. Water pressure and the velocity of any fans used to create this effect must be monitored at all times.
>
> 6–3. Rubber mats or other non-slip material or surface shall be provided when simulating rain. If effects call for mud, the depth of the mud must be approved by American Humane prior to filming. When necessary, a non-slip surface shall be provided underneath the mud.

The AHA boasts that "since the introduction of the *Guidelines*, animal accidents, illnesses and deaths on the set have sharply declined." They back it up with numbers, and since I like to tell my story with graphs, figure 7–27 is one that shows the number of films per year designated as "unacceptable" because of mistreatment of animal actors.

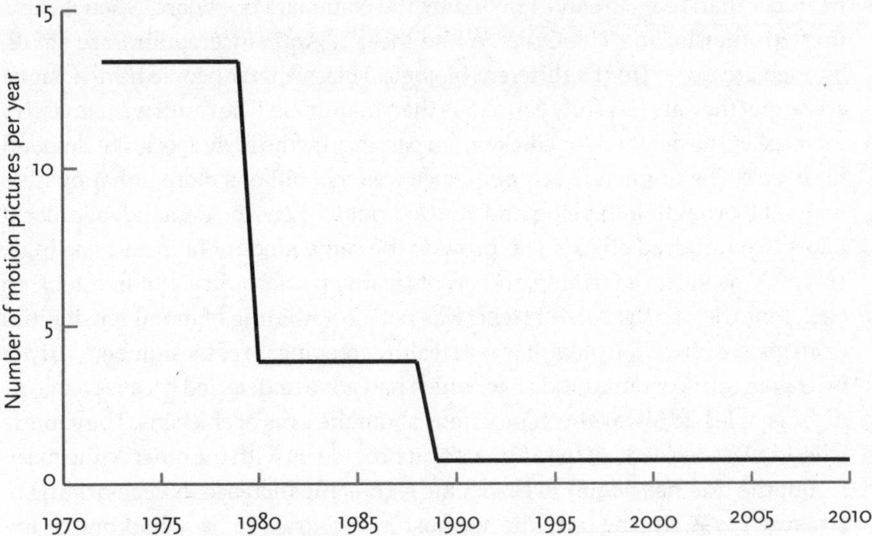

FIGURE 7–27. Number of motion pictures per year in which animals were harmed, 1972–2010
Source: American Humane Association, Film and Television Unit, 2010.

And if that is not enough to convince you that animal rights have been taken to a new level, consider the events of June 16, 2009, as recounted in a *New York Times* article entitled "What's White, Has 132 Rooms, and Flies?" The answer to the riddle is the White House, which had recently become infested with the bugs. During a televised interview a large fly orbited President Obama's head. When the Secret Service did not wrestle it to the ground, the president took matters into his own hands, using one of them to smack the fly on the back of the other. "I got the sucker," boasted the exterminator in chief. The footage became a YouTube sensation but drew a complaint from People for the Ethical Treatment of Animals. They noted on their blog, "It can't be said that President Obama wouldn't hurt a fly," and sent over one of their Katcha Bug Humane Bug Catchers "in the event of future insect incidents."[284]

And finally we get to meat. If someone were to count up every animal that has lived on earth in the past fifty years and tally the harmful acts done to them, he or she might argue that no progress has been made in the treatment of animals. The reason is that the Animal Rights Revolution has been partly canceled out by another development, the Broiler Chicken Revolution.[285] The 1928 campaign slogan "A chicken in every pot" reminds us that chicken was once thought of as a luxury. The market responded by breeding meatier chickens and raising them more efficiently, if less humanely: factory-farmed chickens have spindly legs, live in cramped cages, breathe fetid air, and are handled roughly when transported and slaughtered. In the 1970s consumers became convinced that white meat was healthier than red (a trend exploited by the National Pork Board when it came up with the slogan "The Other White Meat"). And since poultry are small-brained creatures from a different biological class, many people have a vague sense that they are less fully conscious than mammals. The result was a massive increase in the demand for chicken, surpassing, by the early 1990s, the demand for beef.[286] The unintended consequence was that billions more unhappy lives had to be brought into being and snuffed out to meet the demand, because it takes two hundred chickens to provide the same amount of meat as a single cow.[287] Now, factory farming and cruel treatment of poultry and livestock go back centuries, so the baleful trend was not a backsliding of moral sensibilities or an increase in callousness. It was a stealthy creeping up of the numbers, driven by changes in economics and taste, which had gone undetected because a majority of people had always been incurious about the lives of chickens. The same is true, to a lesser extent, of the animals that provide us with the other white meat.

But the tide has begun to turn. One sign is the increase in vegetarianism. I'm sure I was not the only dinner host in the 1990s to have had one of my guests announce as he sat down at the table, "Oh, I forgot to tell you. I don't eat dead animals." Since that era the question "Do you have any food restrictions?" has become a part of the etiquette of a dinner invitation, and participants at conference dinners can now tick a box that will replace a plate of rubber chicken with a plate of sodden eggplant. The trend was noted in 2002, when *Time* magazine ran a cover story entitled: "Should You Be a Vegetarian? Millions of Americans Are Going Meatless."

The food industry has responded with a cornucopia of vegetarian and vegan products. The faux-meat section of my local supermarket offers Soyburgers, Gardenburgers, Seitanburgers, Veggie Burger Meatless Patties, Tofu Pups, Not Dogs, Smart Dogs, Fakin Bacon, Jerquee, Tofurky, Soy Sausage, Soyrizo, Chik Patties, Meatless Buffalo Wings, Celebration Roast, Tempeh Strips, Terkettes, Veggie Protein Slices, Vege-Scallops, and Tuno. The technological and verbal ingenuity is testimony both to the popularity of the new vegetarianism and to the persistence of ancient meat hunger. Those who enjoy a hearty breakfast can serve their Veggie Breakfast Strips with Tofu Scramblers, perhaps in an omelet with Soya Kaas, Soymage, or Veganrella. And for dessert there's Ice Bean, Rice Dream, and Tofutti, perhaps garnished with Hip Whip and a cherry on top. The ultimate replacement for meat would be animal tissue grown in culture, sometimes called meat without feet. The ever-optimistic People for the Ethical Treatment of Animals has offered a million-dollar prize to the first scientist who brings cultured chicken meat to market.[288]

For all the visibility of vegetarianism, pure vegetarians still make up just a few percentage points of the population. It's not easy being green. Vegetarians are surrounded by dead animals and the carnivores who love them, and meat hunger has not been bred out of them. It's not surprising that many fall off the wagon: at any moment there are three times as many lapsed vegetarians as observant ones.[289] Many of those who continue to call themselves vegetarians have convinced themselves that a fish is a vegetable, because they partake of fish and seafood and sometimes even chicken.[290] Others parse their dietary restrictions like Conservadox Jews in a Chinese restaurant, allowing themselves exemptions for narrowly defined categories or for food eaten outside the home. The demographic sector with the largest proportion of vegetarians is teenage girls, and their principal motive may not be compassion for animals. Vegetarianism among teenage girls is highly correlated with eating disorders.[291]

But is vegetarianism at least trending upward? As best we can tell, it is. In the U.K. the Vegetarian Society gathers up the results of every opinion poll it can find and presents the data on its information sheets. In figure 7–28 I've plotted the results of all the questions that ask a national sample of respondents whether they are vegetarians. The best-fitting straight line suggests that over the past two decades, vegetarianism has more than tripled, from about 2 percent of the population to about 7 percent. In the United States, the Vegetarian Resource Group has commissioned polling agencies to ask Americans the more stringent question of whether they eat meat, fish, or fowl, excluding the flexitarians and those with creative Linnaean taxonomies. The numbers are smaller, but the trend is similar, more than tripling in about fifteen years.

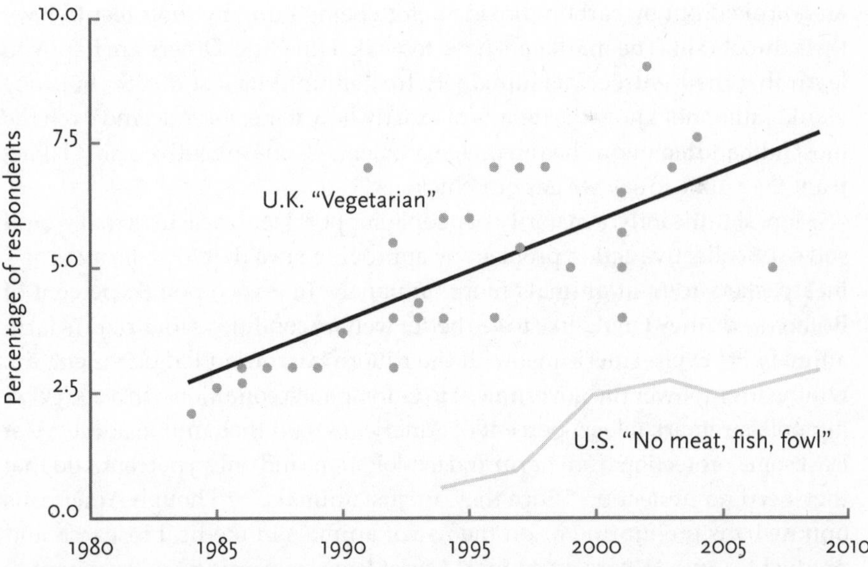

FIGURE 7–28. Vegetarianism in the United States and United Kingdom, 1984–2009

Sources: U.K.: Vegetarian Society, http://www.vegsoc.org/info/. Excluded are polls that ask about households, ask students, or ask about "strict" vegetarians. United States: Vegetarian Resource Group, *Vegetarian Journal.* 2009: http://www.vrg.org/press/2009poll.htm. 2005 and 2003: http://www.vrg.org/journal/vj2006issue4/vj2006issue4poll.htm. 2000: http://www.vrg.org/nutshell/poll2000.htm. 1997: http://www.vrg.org/journal/vj97sep/979poll.htm. 1994: http:www.vrg.org/nutshell/poll.htm.

With all the signs of increasing concern with animal welfare, it may be surprising that the percentage of vegetarians, though rising, is still so low. But it really shouldn't be. Being a vegetarian and being concerned with animal welfare are not the same thing. Not only do vegetarians have motives other than animal welfare—health, taste, ecology, religion, driving their mothers crazy—but people who are concerned with animal welfare may wonder whether the symbolic statement of vegetarianism is the best way to reduce animal suffering. They may feel that the hamburgers they altruistically forgo are unlikely to register amid the noise in the vast national demand for meat or to lead to the sparing of the lives of any cows. And even if they did, the lives of the remaining cows would be no more pleasant. Changing the practices of the food industry is a collective action dilemma, in which individuals are tempted to shirk from private sacrifices that have marginal effects on aggregate welfare.

The increase in vegetarianism, though, is a symbolic indicator of a broader concern for animals that can be seen in other forms. People who don't abstain from meat as a matter of principle may still eat less of it. (American consumption of meat from mammals has declined since 1980.)[292] Restaurants and supermarkets increasingly inform their patrons about what their main course fed on and how freely it ranged while it was still on the hoof or claw. Two of the major poultry processors in the United States announced in 2010 that they were switching to a more humane method of slaughtering, in which the birds are knocked out by carbon dioxide before being hung by their feet to have their throats slit. The marketers have to walk a fine line. Diners are happy to learn that their entrée was humanely treated until its last breath, but they would rather not know the details of exactly how it met its end. And even the most humane technique has an image problem. As one executive said, "I don't want the public to say we gas our chickens."[293]

More significantly, a majority of people support legal measures that would solve the collective action problem by approving laws that force farmers and meatpackers to treat animals more humanely. In a 2000 poll 80 percent of Britons said "they would like to see better welfare conditions for Britain's farm animals."[294] Even Americans, with their more libertarian temperament, are willing to empower the government to enforce such conditions. In a 2003 Gallup poll, a remarkable 96 percent of Americans said that animals deserve at least some protection from harm and exploitation, and only 3 percent said that they need no protection "since they are just animals."[295] Though Americans oppose bans on hunting or on the use of animals in medical research and product testing, 62 percent support "strict laws concerning the treatment of farm animals." And when given the opportunity, they translate their opinions into votes. Livestock rights have been written into the laws of Arizona, Colorado, Florida, Maine, Michigan, Ohio, and Oregon, and in 2008, 63 percent of California voters approved the Prevention of Farm Animal Cruelty Act, which bans veal crates, poultry cages, and sow gestation crates that prevent the animal from moving around.[296] There is a cliché in American politics: as California goes, so goes the country.

And perhaps as Europe goes, so goes California. The European Union has elaborate regulations on animal care "that start with the recognition that animals are sentient beings. The general aim is to ensure that animals need not endure avoidable pain or suffering and obliges the owner/keeper of animals to respect minimum welfare requirements."[297] Not every country has gone so far as Switzerland, which enacted 150 pages of regulations that force dog owners to attend a four-hour "theory" course and legislate how pet owners may house, feed, walk, play with, and dispose of their pets. (No more flushing live goldfish down the toilet.) But even the Swiss balked at a 2010 referendum that would have nationalized a Zurich policy that pays an "animal advocate" to haul offenders into criminal court, including an angler who boasted to a local newspaper that he took ten minutes to land a large pike. (The angler was acquitted; the pike was eaten.)[298] All this may sound like American conservatives' worst nightmare, but they too are willing to allow the government to regulate animal welfare. In the 2003 poll, a majority of Republicans favored passing "strict laws" on the treatment of farm animals.[299]

How far will it go? People often ask me whether I think the moral momentum that carried us from the abolition of slavery and torture to civil rights, women's rights, and gay rights will culminate in the abolition of meat-eating, hunting, and animal experimentation. Will our 22nd-century descendants be as horrified that we ate meat as we are that our ancestors kept slaves?

Maybe, but maybe not. The analogy between oppressed people and oppressed animals has been rhetorically powerful, and insofar as we are all sentient beings, it has a great deal of intellectual warrant. But the analogy is not exact—African Americans, women, children, and gay people are not broiler chickens—and I doubt that the trajectory of animal rights will be a time-lagged copy of the one for human rights. In his book *Some We Love, Some We Hate, Some We Eat,* the psychologist Hal Herzog lays out the many reasons why it's so hard for us to converge on a coherent moral philosophy to govern our dealings with animals. I'll mention a few that have struck me.

One impediment is meat hunger and the social pleasures that go with the consumption of meat. Though traditional Hindus, Buddhists, and Jains prove that a meatless society is possible, the 3 percent market share of vegetarian diets in the United States shows that we are very far from a tipping point. While gathering the data for this chapter, I was excited to stumble upon a 2004 Pew Research poll in which 13 percent of the respondents were vegetarians. Upon reading the fine print, I discovered that it was a poll of supporters of the presidential candidacy of Howard Dean, the left-wing governor of Vermont. That means that even among the crunchiest granolas in Ben-and-Jerry land, 87 percent still eat meat.[300]

But the impediments run deeper than meat hunger. Many interactions between humans and animals will always be zero-sum. Animals eat our houses, our crops, and occasionally our children. They make us itch and bleed. They are vectors for diseases that torment and kill us. They kill each other, including endangered species that we would like to keep around. Without their participation in experiments, medicine would be frozen at its current state, and billions of living and unborn people would suffer and die for the sake of mice. An ethical calculus that gave equal weight to any harm suffered by any sentient being, allowing no chauvinism toward our own species, would prevent us from trading off the well-being of animals for an equivalent well-being of humans—for example, shooting a wild dog to save a little girl. To be sure, the interests of humans could be given some extra points by virtue of our zoological peculiarities, such as that our big brains allow us to savor our lives, reflect on our past and future, dread death, and enmesh our well-being with those of others in dense social networks. But the human life taboo, which among other things protects the lives of mentally incompetent people just because they are people, would have to go. Singer himself unflinchingly accepts this implication of a species-blind morality.[301] But it will not take over Western morality anytime soon.

Ultimately the move toward animal rights will bump against some of the most perplexing enigmas in the space of human thoughts, a place where moral intuitions start to break down. One is the hard problem of consciousness, namely how sentience arises from neural information processing.[302] Descartes was certainly wrong about mammals, and I am pretty sure he was wrong about fish. But was he wrong about oysters? Slugs? Termites? Earthworms? If we wanted ethical certainty in our cooking, gardening, home repair, and recreation, we would need nothing less than a solution to this philosophical conundrum. Another paradox is that human beings are simultaneously rational, moral agents and organisms that are part of nature red in tooth and claw. Something in me objects to the image of a hunter shooting a moose. But why am I not upset by the image of a grizzly bear that renders it just as dead? Why don't I think it's a moral imperative to tempt the bear away with all-soy meatless moose patties? Should we arrange for the gradual extinction of carnivorous species, or even genetically engineer them into herbivores?[303] We recoil from these thought experiments because, rightly or wrongly, we assign some degree of ethical weight to what we feel is "natural." But if the natural carnivory of other species counts for something, why not the natural carnivory of *Homo sapiens*—particularly if we deploy our cognitive and moral faculties to minimize the animals' suffering?

These imponderables, I suspect, prevent the animal rights movement from duplicating the trajectory of the other Rights Revolutions exactly. But for now the location of the finish line is beside the point. There are many opportunities in which enormous suffering by animals can be reduced at a small cost to humans. Given the recent changes in sensibilities, it is certain that the lives of animals will continue to improve.

WHENCE THE RIGHTS REVOLUTIONS?

When I began my research for this chapter, I knew that the decades of the Long Peace and the New Peace were also decades of progress for racial minorities, women, children, gay people, and animals. But I had no idea that in every case, quantifiable measures of violence—hate crimes and rape, wife-beating and child abuse, even the number of motion pictures in which animals were harmed—would all point downward. How can we make sense of all the movements toward nonviolence of the past fifty years?

The trends have a few things in common. In each case they had to swim against powerful currents of human nature. These include the dehumanization and demonization of out-groups; men's sexual rapacity and their proprietary sentiments toward women; manifestations of parent-offspring conflict such as infanticide and corporal punishment; the moralization of sexual disgust in homophobia; and our meat hunger, thrill of the hunt, and boundaries of empathy based on kinship, reciprocity, and charisma.

As if biology didn't make things bad enough, the Abrahamic religions ratified some of our worst instincts with laws and beliefs that have encouraged violence for millennia: the demonization of infidels, the ownership of women, the sinfulness of children, the abomination of homosexuality, the dominion over animals and denial to them of souls. Asian cultures have plenty to be ashamed of too, particularly the mass disowning of daughters that encouraged a holocaust of baby girls. And then there is the entrenchment of norms: beating wives, smacking children, confining calves, and shocking rats were acceptable because everyone had always treated them as acceptable.

Insofar as violence is immoral, the Rights Revolutions show that a moral way of life often requires a decisive rejection of instinct, culture, religion, and standard practice. In their place is an ethics that is inspired by empathy and reason and stated in the language of rights. We force ourselves into the shoes (or paws) of other sentient beings and consider their interests, starting with their interest in not being hurt or killed, and we ignore superficialities that may catch our eye such as race, ethnicity, gender, age, sexual orientation, and to some extent, species.

This conclusion, of course, is the moral vision of the Enlightenment and the strands of humanism and liberalism that have grown out of it. The Rights Revolutions are liberal revolutions. Each has been associated with liberal movements, and each is currently distributed along a gradient that runs, more or less, from Western Europe to the blue American states to the red American states to the democracies of Latin America and Asia and then to the more authoritarian countries, with Africa and most of the Islamic world pulling up the rear. In every case, the movements have left Western cultures with excesses of propriety and taboo that are deservedly ridiculed as political correctness. But the numbers show that the movements have reduced many causes of death and suffering and have made the culture increasingly intolerant of violence in any form.

To hear the liberal punditry talk, one would think that the United States has been hurtling rightward for more than forty years, from Nixon to Reagan to Gingrich to the Bushes and now the angry white men in the Tea Party movement. Yet in every issue touched by the Rights Revolutions—interracial marriage, the empowerment of women, the tolerance of homosexuality, the punishment of children, and the treatment of animals—the attitudes of conservatives have followed the trajectory of liberals, with the result that today's conservatives are more liberal than yesterday's liberals. As the conservative historian George Nash points out, "In practice if not quite in theory American conservatism today stands well to the left of where it stood in 1980."[304] (Maybe that's why the men are so angry.)

What caused the Rights Revolutions? As hard as it was to establish the causes of the Long Peace, New Peace, and 1990s crime decline, it's harder still to pinpoint an exogenous factor that would explain why the Rights Revolutions bunched up when they did. But we can consider the standard candidates.

The postwar years saw an expansion of prosperity, but prosperity has such a diffuse influence on a society that it offers little insight into the revolutions' immediate triggers. Money can buy education, police, social science, social services, media penetration, a professional workforce with more women, and better care of children and animals. It's hard to identify which of these made a difference, and even if we could, it would raise the question of why society chose to distribute its surplus among these various goods in such a way as to reduce harm to vulnerable populations. And though I know of no rigorous statistical analysis, I can discern no correlations between the timing of the various upswings in the consideration of rights from the 1960s to the 2000s and the economic booms and recessions of those decades.

Democratic government obviously played a role. The Rights Revolutions took place in democracies, which are constituted as social contracts among individuals designed to reduce violence among them; as such they contain the seeds of expansion to groups that originally had been overlooked. But the timing remains a puzzle, because democracy is not an entirely exogenous variable. It was the machinery of democracy itself that was at issue during the American civil rights movement, when the political disenfranchisement of blacks was remedied. In the other revolutions too, new groups were invited or argued their way into full partnership in the social contract. Only then could the government be empowered to police violence (or cease its own violence) against members of the affected groups.

During the Rights Revolutions, networks of reciprocity and trade had expanded with the shift from an economy based on stuff to one based on information. Women were less enslaved by domestic chores, and institutions sought talent from a wide pool of human capital rather than just from the local labor supply or old boys' club. As women and members of minority groups were drawn into the wheelings and dealings of government and commerce, they ensured that their interests were factored into their workings. We have seen some evidence for this mechanism: countries with more women in government and the professions have less domestic violence against women, and people who know gay people personally are less likely to disapprove of homosexuality. But as with democracy, the inclusiveness of institutions is not a completely exogenous process. The hidden hand of an information economy may have made institutions more receptive to women, minorities, and gays, but it still took government muscle in the form of antidiscrimination laws to integrate them fully. And in the case of children and animals, there was no market for reciprocal exchanges at all: the beneficence went in one direction.

If I were to put my money on the single most important exogenous cause of the Rights Revolutions, it would be the technologies that made ideas and people increasingly mobile. The decades of the Rights Revolutions were the decades of the electronics revolutions: television, transistor radios, cable, satellite, long-distance telephones, photocopiers, fax machines, the Internet, cell phones, text messaging, Web video. They were the decades of the interstate highway, high-speed rail, and the jet airplane. They were the decades of the unprecedented growth in higher education and in the endless frontier of scientific research. Less well known is that they were also the decades of an explosion in book publishing. From 1960 to 2000, the annual number of books published in the United States increased almost fivefold.[305]

I've mentioned the connection before. The Humanitarian Revolution came out of the Republic of Letters, and the Long Peace and New Peace were children of the Global Village. And remember what went wrong in the Islamic world: it may have been a rejection of the printing press and a resistance to the importation of books and the ideas they contain.

Why should the spread of ideas and people result in reforms that lower violence? There are several pathways. The most obvious is a debunking of ignorance and superstition. A connected and educated populace, at least in aggregate and over the long run, is bound to be disabused of poisonous beliefs, such as that members of other races and ethnicities are innately avaricious or perfidious; that economic and military misfortunes are caused by the treachery of ethnic minorities; that women don't mind being raped; that children must be beaten to be socialized; that people choose to be homosexual as part of a morally degenerate lifestyle; that animals are incapable of feeling pain. The recent debunking of beliefs that invite or tolerate violence calls to mind Voltaire's quip that those who can make you believe absurdities can make you commit atrocities.

Another causal pathway is an increase in invitations to adopt the viewpoints of people unlike oneself. The Humanitarian Revolution had its *Clarissa, Pamela*, and *Julie*, its *Uncle Tom's Cabin* and *Oliver Twist*, its eyewitness reports of people being broken, burned, or flogged. During the electronic age, these empathy technologies were even more pervasive and engaging. African Americans and gay people appeared as entertainers on variety shows, then as guests on talk shows and as sympathetic characters on sitcoms and dramas. Their struggles were depicted in real-time footage of fire hoses and police dogs, and in bestselling books and plays like *Travels with Charley, A Raisin in the Sun,* and *To Kill a Mockingbird.* Telegenic feminists made their case on talk shows, and their views came out of the mouths of characters in soap operas and sitcoms.

And as we will see in chapter 9, it is not just the virtual reality experience of seeing the world through another person's eyes that expands empathy and concern. It is also an intellectual agility—literally a kind of intelligence—which encourages one to step outside the parochial constraints of one's birth and station, to consider hypothetical worlds, and to reflect back on the habits, impulses, and institutions that govern one's beliefs and values. This reflective mindset may be a product of enhanced education, and it may also be a product of electronic media. As Paul Simon marveled:

> These are the days of miracle and wonder,
> This is the long distance call,
> The way the camera follows us in slo-mo
> The way we look to us all.

There is a third way that a flow of information can fertilize moral growth. Scholars who have puzzled over the trajectory of material progress in different parts of the world, such as the economist Thomas Sowell in his *Culture* trilogy and the physiologist Jared Diamond in *Guns, Germs, and Steel,* have concluded that the key to material success is being situated in a large catchment area of innovations.[306] No one is smart enough to invent anything in isolation that anyone else would want to use. Successful innovators not only stand on the shoulders of giants; they engage in massive intellectual property theft, skimming ideas from a vast watershed of tributaries flowing their way. The civilizations of Europe and western Asia conquered the world because their migration and shipping routes allowed traders and conquerors to leave behind inventions that had originated anywhere in the vast Eurasian landmass: cereal crops and alphabetic writing from the Middle East, gunpowder and paper from China, domesticated horses from Ukraine, oceangoing navigation from Portugal, and much else. There is a reason that the literal meaning of *cosmopolitan* is "citizen of the world," and the literal meaning of *insular* is "of an island." Societies that are marooned on islands or in impassable highlands tend to be technologically backward. And morally backward too. We have seen that cultures of honor, whose overriding ethic is tribal loyalty and blood revenge, can survive in mountainous regions long after their lowland neighbors have undergone a civilizing process.

What's true of technological progress may also be true of moral progress. Individuals or civilizations that are situated in a vast *informational* catchment area can compile a moral know-how that is more sustainable and expandable than even the most righteous prophet could devise in isolation. Let me illustrate this point with a potted history of the Rights Revolutions.

In his 1963 essay "Pilgrimage to Nonviolence," Martin Luther King recounted the intellectual threads that he wove into his political philosophy.[307] As a graduate student in theology in the late 1940s and early 1950s, he was, of course, conversant with the Bible and orthodox theology. But he also read renegade theologians such as Walter Rauschenbusch, who criticized the historical accuracy of the Bible and the dogma that Jesus died for people's sins.

King then embarked on "a serious study of the social and ethical theories of the great philosophers, from Plato and Aristotle down to Rousseau, Hobbes, Bentham, Mill, and Locke. All of these masters stimulated my thinking—such as it was—and, while finding things to question in each of them, I nevertheless learned a great deal from their study." He carefully read (and rejected) Nietzsche and Marx, inoculating himself against the autocratic and communist ideologies that would be so seductive to other liberation movements. He also rejected the "anti-rationalism of the continental theologian Karl Barth," while admiring Reinhold Niebuhr's "extraordinary insight into human nature, especially the behavior of nations and social groups. . . . These elements in Niebuhr's thinking helped me to recognize the illusions of a superficial optimism concerning human nature and the dangers of a false idealism."

King's thinking was irrevocably changed one day when he traveled to Philadelphia to hear a lecture by Mordecai Johnson, the president of Howard University. Johnson had recently returned from a trip to India and spoke about Mohandas Gandhi, whose influence had recently culminated in national independence. "His message was so profound and electrifying," King wrote, "that I left the meeting and bought a half-a-dozen books on Gandhi's life and works."

King immediately appreciated that Gandhi's theory of nonviolent resistance was not a moralistic affirmation of love, as nonviolence had been in the teachings of Jesus. Instead it was a set of hardheaded tactics to prevail over an adversary by outwitting him rather than trying to annihilate him. A taboo on violence, King inferred, prevents a movement from being corrupted by thugs and firebrands who are drawn to adventure and mayhem. It preserves morale and focus among followers when the movement suffers early defeats. By removing any pretext for legitimate retaliation by the enemy, it stays on the positive side of the moral ledger in the eyes of third parties, while luring the enemy onto the negative side. For the same reason, it divides the enemy, paring away supporters who find it increasingly uncomfortable to identify themselves with one-sided violence. All the while it can press its agenda by making a nuisance of itself with sit-ins, strikes, and demonstrations. The tactic obviously won't work with all enemies, but it can work with some.

King's historic speech to the March on Washington in 1963 was an ingenious recombination of the intellectual components he had collected during his peripatetic pilgrimage: imagery and language from the Hebrew prophets, the valorization of suffering from Christianity, the ideal of individual rights from the European Enlightenment, cadences and rhetorical tropes from the African American church, and a strategic plan from an Indian who had been steeped in Jain, Hindu, and British culture.

It is not too lazy to say that the rest is history. The moral contrivance assembled by King was thrown back into the idea pool, there to be adapted by the entrepreneurs of the other rights movements. They consciously appropriated its name, its moral rationale, and significantly, many of its tactics.

By the standards of history, a striking feature of the late-20th-century Rights Revolutions is how little violence they employed or even provoked. King himself was a martyr of the civil rights movement, as were the handful of victims of segregationist terrorism. But the urban riots that we associate with the 1960s were not a part of the civil rights movement and erupted after most of its milestones were in place. The other revolutions had hardly any violence at all: there was the nonlethal Stonewall riot, some terrorism from the fringes of the animal rights movement, and that's about it. Their entrepreneurs wrote books, gave speeches, held marches, lobbied legislators, and gathered signatures for plebiscites. They had only to nudge a populace that had become receptive to an ethic based on the rights of individuals and were increasingly repelled by violence in any form. Compare this record to that of earlier movements which ended despotism, slavery, and colonial empires only after bloodbaths that killed people by the hundreds of thousands or millions.

FROM HISTORY TO PSYCHOLOGY

We have come to the end of six chapters that have documented the historical decline of violence. In them we have seen graph after graph that locates the first decade of the new millennium at the bottom of a slope representing the use of force over time. For all the violence that remains in the world, we are living in an extraordinary age. Perhaps it is a snapshot in a progression to an even greater peace. Perhaps it is a bottoming out to a new normal, with the easy reductions all plucked and additional ones harder and harder to reach. Perhaps it is a lucky confluence of good fortune that will soon unravel. But regardless of how the trends extrapolate into the future, something remarkable has brought us to the present.

One of Martin Luther King's most famous quotations was adapted from an 1852 essay by the abolitionist Unitarian minister Theodore Parker:

> I do not pretend to understand the moral universe; the arc is a long one, my eye reaches but little ways; I cannot calculate the curve and complete the

> figure by the experience of sight; I can divine it by conscience. And from what I see I am sure it bends towards justice.[308]

A century and a half later our eyes can see that the arc has bent toward justice in ways that Parker could not have imagined. I do not pretend to understand the moral universe either; nor can I divine it by conscience. But in the next two chapters, let's see how much of it we can understand with science.

8

INNER DEMONS

But man, proud man,
Drest in a little brief authority,
Most ignorant of what he's most assur'd,
His glassy essence, like an angry ape,
Plays such fantastic tricks before high heaven
As makes the angels weep.

—William Shakespeare, *Measure for Measure*

Two aspects of the decline of violence have profound implications for our understanding of human nature: (1) the violence; (2) the decline. The last six chapters have shown that human history is a cavalcade of bloodshed. We have seen tribal raiding and feuding that kills a majority of males, the disposal of newborns that kills a majority of females, the staging of torture for vengeance and pleasure, and killings of enough kinds of victims to fill a page of a rhyming dictionary: homicide, democide, genocide, ethnocide, politicide, regicide, infanticide, neonaticide, filicide, siblicide, gynecide, uxoricide, mariticide, and terrorism by suicide. Violence is found throughout the history and prehistory of our species, and shows no signs of having been invented in one place and spread to the others.

At the same time, those chapters contain five dozen graphs that plot violence over time and display a line that meanders from the top left to the bottom right. Not a single category of violence has been pinned to a fixed rate over the course of history. Whatever causes violence, it is not a perennial urge like hunger, sex, or the need to sleep.

The decline of violence thereby allows us to dispatch a dichotomy that has stood in the way of understanding the roots of violence for millennia: whether humankind is basically bad or basically good, an ape or an angel, a hawk or a dove, the nasty brute of textbook Hobbes or the noble savage of textbook Rousseau. Left to their own devices, humans will not fall into a state of peaceful cooperation, but nor do they have a thirst for blood that must regularly be slaked. There must be at least a grain of truth in conceptions of the human mind that grant it more than one part—theories like faculty psychology, multiple intelligences, mental organs, modularity, domain-specificity, and the metaphor of the mind as a Swiss army knife. Human nature accommodates motives that impel us to violence, like predation, dominance, and vengeance, but also motives that—under the right circumstances—impel us toward peace, like compassion, fairness, self-control, and reason. This chapter and its successor will explore these motives and the circumstances that engage them.

THE DARK SIDE

Before exploring our inner demons, I need to make the case that they exist, because there is a resistance in modern intellectual life to the idea that human nature embraces any motives that incline us toward violence at all.[1] Though the ideas that we evolved from hippie chimps and that primitive people had no concept of violence have been refuted by the facts of anthropology, one still sometimes reads that violence is perpetrated by a few bad apples who do all the damage and that everyone else is peaceful at heart.

It is certainly true that the lives of most people in most societies do not end in violence. The numbers on the vertical axes of the graphs in the preceding chapters have been graduated in single digits, tens, or at most hundreds of killings per 100,000 people per year; only rarely, as in tribal warfare or an unfolding genocide, are the rates in the thousands. It is also true that in most hostile encounters, the antagonists, whether humans or other animals, usually back down before either of them can do serious damage to the other. Even in wartime, many soldiers do not fire their weapons and are racked by post-traumatic stress disorder when they do. Some writers conclude that the vast majority of humans are constitutionally averse to violence and that the high body counts are merely signs of how much harm a few psychopaths can do.

So let me begin by convincing you that most of us—including you, dear reader—are wired for violence, even if in all likelihood we will never have an occasion to use it. We can begin with our younger selves. The psychologist Richard Tremblay has measured rates of violence over the course of the life span and shown that the most violent stage of life is not adolescence or even young adulthood but the aptly named terrible twos.[2] A typical toddler at least sometimes kicks, bites, hits, and gets into fights, and the rate of physical aggression then goes steadily down over the course of childhood. Tremblay remarks, "Babies do not kill each other, because we do not give them access to knives and guns. The question . . . we've been trying to answer for the past 30 years is how do children learn to aggress. [But] that's the wrong question. The right question is how do they learn not to aggress."[3]

Now let's turn to our inner selves. Have you ever fantasized about killing someone you don't like? In separate studies, the psychologists Douglas Kenrick and David Buss have posed this question to a demographic that is known to have exceptionally low rates of violence—university students—and were stunned at the outcome.[4] Between 70 and 90 percent of the men, and between 50 and 80 percent of the women, admitted to having at least one homicidal fantasy in the preceding year. When I described these studies in a lecture, a student shouted, "Yeah, and the others are lying!" At the very least, they may sympathize with Clarence Darrow when he said, "I have never killed a man, but I have read many obituaries with great pleasure."

The motives for imaginary homicides overlap with those on police blotters: a lover's quarrel, a response to a threat, vengeance for an act of humiliation or betrayal, and family conflict, proportionally more often with stepparents than with biological parents. Often the reveries are played out before the mind's eye in theatrical detail, like the jealous revenge fantasy entertained by Rex Harrison while conducting a symphony orchestra in *Unfaithfully Yours*. One young man in Buss's survey estimated that he came "eighty percent of the way" toward killing a former friend who had lied to the man's fiancée that he had been unfaithful to her and then made a move on the fiancée himself:

> First, I would break every bone in his body, starting with his fingers and toes, slowly making my way to larger ones. Then I would puncture his lungs and maybe a few other organs. Basically give him as much pain as possible before killing him.[5]

A woman said she had gone 60 percent of the way toward killing an ex-boyfriend who wanted to get back together and had threatened to send a video of the two of them having sex to her new boyfriend and her fellow students:

> I actually did this. I invited him over for dinner. And as he was in the kitchen, looking stupid peeling the carrots to make a salad, I came to him laughing, gently, so he wouldn't suspect anything. I thought about grabbing a knife quickly and stabbing him in the chest repeatedly until he was dead. I actually did the first thing, but he saw my intentions, and ran away.

Many actual homicides are preceded by lengthy ruminations just like this. The small number of premeditated murders that are actually carried out must be the cusp of a colossal iceberg of homicidal desires submerged in a sea of inhibitions. As the forensic psychiatrist Robert Simon put it in a book title (paraphrasing Freud paraphrasing Plato), *Bad Men Do What Good Men Dream*.

Even people who don't daydream about killing get intense pleasure from vicarious experiences of doing it or seeing it done. People spend large amounts of their time and income immersing themselves in any of a number of genres of bloody virtual reality: Bible stories, Homeric sagas, martyrologies, portrayals of hell, hero myths, Gilgamesh, Greek tragedies, Beowulf, the Bayeux Tapestry, Shakespearean dramas, Grimm's fairy tales, Punch and Judy, opera, murder mysteries, penny dreadfuls, pulp fiction, dime novels, Grand Guignol, murder ballads, films noirs, Westerns, horror comics, superhero comics, the Three Stooges, Tom and Jerry, the Road Runner, video games, and movies starring a certain ex-governor of California. In *Savage Pastimes: A Cultural History of Violent Entertainment*, the literary scholar Harold Schechter shows that today's splatter films are mild stuff compared to the simulated torture and mutilation that have titillated audiences for centuries. Long before computer-generated imagery, theater directors would apply their ingenuity to grisly special effects, such as "phony heads that could be decapitated from dummies and impaled on pikes; fake skin that could be flayed from an actor's torso; concealed bladders filled with animal blood that could produce a satisfying spurt of gore when punctured."[6]

The vast mismatch between the number of violent acts that run through people's imaginations and the number they carry out in the world tells us something about the design of the mind. Statistics on violence underestimate the importance of violence in the human condition. The human brain runs on the Latin adage "If you want peace, prepare for war." Even in peaceable societies, people are fascinated by the logic of bluff and threat, the psychology of alliance and betrayal, the vulnerabilities of a human body and how they can be exploited or shielded. The universal pleasure that people take in violent entertainment, always in the teeth of censorship and moralistic denunciation, suggests that the mind craves information on the conduct of violence.[7] A likely explanation is that in evolutionary history, violence was not so improbable that people could afford not to understand how it works.[8]

The anthropologist Donald Symons has noticed a similar mismatch in the other major content of naughty reverie and entertainment, sex.[9] People fantasize about and make art out of illicit sex vastly more often than they engage in it. Like adultery, violence may be improbable, but when an opportunity arises, the potential consequences for Darwinian fitness are gargantuan. Symons suggests that higher consciousness itself is designed for low-frequency, high-impact events. We seldom muse about daily necessities like grasping, walking, or speaking, let alone pay money to see them dramatized. What grabs our mental spotlight is illicit sex, violent death, and Walter Mittyish leaps of status.

Now to our brains. The human brain is a swollen and warped version of the brains of other mammals. All the major parts may be found in our furry cousins, where they do pretty much the same things, such as process information from the senses, control muscles and glands, and store and retrieve memories. Among these parts is a network of regions that has been called the Rage circuit. The neuroscientist Jaak Panksepp describes what happened when he sent an electrical current through a part of the Rage circuit of a cat:

> Within the first few seconds of the electrical brain stimulation the peaceful animal was emotionally transformed. It leaped viciously toward me with

> claws unsheathed, fangs bared, hissing and spitting. It could have pounced in many different directions, but its arousal was directed right at my head. Fortunately, a Plexiglas wall separated me from the enraged beast. Within a fraction of a minute after terminating the stimulation, the cat was again relaxed and peaceful, and could be petted without further retribution.[10]

The Rage circuit in the cat brain has a counterpart in the human brain, and it too can be stimulated by an electrical current—not in an experiment, of course, but during neurosurgery. A surgeon describes what follows:

> The most significant (and the most dramatic) effect of stimulation has been the eliciting of a range of aggressive responses, from coherent, appropriately directed verbal responses (speaking to surgeon, "I feel I could get up and bite you") to uncontrolled swearing and physically destructive behaviour. . . . On one occasion the patient was asked, 30 sec after cessation of the stimulus, if he had felt angry. He agreed that he had been angry, but that he no longer was, and he sounded very surprised.[11]

Cats hiss; humans swear. The fact that the Rage circuit can activate speech suggests that it is not an inert vestige but has functioning connections with the rest of the human brain.[12] The Rage circuit is one of several circuits that control aggression in nonhuman mammals, and as we shall see, they help make sense of the varieties of aggression in humans as well.

If violence is stamped into our childhoods, our fantasy lives, our art, and our brains, then how is it possible that soldiers are reluctant to fire their guns in combat, when that is what they are there to do? A famous study of World War II veterans claimed that no more than 15 to 25 percent of them were able to discharge their weapons in battle; other studies have found that most of the bullets that are fired miss their targets.[13] Now, the first claim is based on a dubious study, and the second is a red herring—most shots are fired in war not to pick off individual soldiers but to deter any of them from advancing.[14] Nor is it surprising that when a soldier targets an enemy in combat conditions, it isn't easy to score a direct hit. But let's grant that anxiety on the battlefield is high and that many soldiers are paralyzed when the time comes to pull the trigger.

A nervousness about the use of deadly force may also be seen in street fights and barroom brawls. Most confrontations between macho ruffians are nothing like the stupendous fistfights in Hollywood westerns that so impressed Nabokov's Humbert, with "the sweet crash of fist against chin, the kick in the belly, the flying tackle." The sociologist Randall Collins has scrutinized photographs, videotapes, and eyewitness accounts of real fights and found that they are closer to a two-minute penalty for roughing in a boring hockey game than an action-packed brawl in Roaring Gulch.[15] Two men glower, talk trash, swing and miss, clutch each other, sometimes fall to the ground. Occasionally a fist will emerge from the mutual embrace and land a couple of blows, but more often the men will separate, trade angry bluster and face-saving verbiage, and walk away with their egos more bruised than their bodies.

It's true, then, that when men confront each other in face-to-face conflict, they often exercise restraint. But this reticence is not a sign that humans are gentle and compassionate. On the contrary, it's just what one would expect from the analyses of violence by Hobbes and Darwin. Recall from chapter 2 that any tendency toward violence must have evolved in a world in which everyone else was evolving the same tendency. (As Richard Dawkins put it, a living thing differs from a rock or a river because it is inclined to hit back.) That means that the first move toward harming a fellow human simultaneously accomplishes two things:

1. It increases the chance that the target will come to harm.
2. It gives the target an overriding goal of harming you before you harm him.

Even if you prevail by killing him, you will have given his kin the goal of killing you in revenge. It stands to reason that initiating serious aggression in a symmetrical standoff is something a Darwinian creature must consider very, very carefully—a reticence experienced as anxiety or paralysis. Discretion is the better part of valor; compassion has nothing to do with it.

When an opportunity does arise to eliminate a hated opponent with little danger of reprisal, a Darwinian creature will seize on it. We saw this in chimpanzee raiding. When a group of males patrolling a territory encounters a male from another community who has been isolated from his fellows, they will take advantage of the strength in numbers and tear him limb from limb. Pre-state peoples too decimate their enemies not in pitched battles but in stealthy ambushes and raids. Much of human violence is cowardly violence: sucker punches, unfair fights, preemptive strikes, predawn raids, mafia hits, drive-by shootings.

Collins also documents a recurring syndrome that he calls *forward panic,* though a more familiar term would be *rampage.* When an aggressive coalition has stalked or faced off against an opponent in a prolonged state of apprehension and fear, then catches the opponent in a moment of vulnerability, fear turns to rage, and the men will explode in a savage frenzy. A seemingly unstoppable fury drives them to beat the enemy senseless, torture and mutilate the men, rape the women, and destroy their property. A forward panic is violence at its ugliest. It is the state of mind that causes genocides, massacres, deadly ethnic riots, and battles in which no prisoners are taken. It also lies behind episodes of police brutality, such as the savage beating of Rodney King in 1991 after he had been apprehended in a high-speed car chase and had violently resisted his arrest. As the butchery gains momentum, rage may give way to ecstasy, and the rampagers may laugh and whoop in a carnival of barbarity.[16]

No one has to be trained to carry out a rampage, and when they erupt in armies or police squads the commanders are often taken by surprise and have to take steps to quell them, since the overkill and atrocities serve no military or law-enforcement purpose. A rampage may be a primitive adaptation to seize a fleeting opportunity to decisively rout a dangerous enemy before it can remobilize and retaliate. The resemblance to lethal raiding among chimpanzees is uncanny, including the common trigger: an isolated member of the enemy who is outnumbered by a cluster of three or four allies.[17] The instinct behind rampages suggests that the human behavioral repertoire includes scripts for violence that lie quiescent and may be cued by propitious circumstances, rather than building up over time like hunger or thirst.

THE MORALIZATION GAP AND THE MYTH OF PURE EVIL

In *The Blank Slate* I argued that the modern denial of the dark side of human nature—the doctrine of the Noble Savage—was a reaction against the romantic militarism, hydraulic theories of aggression, and glorification of struggle and strife that had been popular in the late 19th and early 20th centuries. Scientists and scholars who question the modern doctrine have been accused of *justifying* violence and have been subjected to vilification, blood libel, and physical assault.[18] The Noble Savage myth appears to be another instance of an antiviolence movement leaving a cultural legacy of propriety and taboo.

But I am now convinced that a denial of the human capacity for evil runs even deeper, and may itself be a feature of human nature, thanks to a brilliant analysis by the social psychologist Roy Baumeister in his book *Evil*.[19] Baumeister was moved to study the commonsense understanding of evil when he noticed that the people who perpetrate destructive acts, from everyday peccadilloes to serial murders and genocides, never think they are doing anything wrong. How can there be so much evil in the world with so few evil people doing it?

When psychologists are confronted with a timeless mystery, they run an experiment. Baumeister and his collaborators Arlene Stillwell and Sara Wotman couldn't very well get people to commit atrocities in the lab, but they reasoned that everyday life has its share of smaller hurts that they could put under the microscope.[20] They asked people to describe one incident in which someone angered them, and one incident in which they angered someone. The order of the two questions was randomly flipped from one participant to the next, and they were separated by a busywork task so the participants wouldn't answer them in quick succession. Most people get angry at least once a week, and nearly everyone gets angry at least once a month, so there was no shortage of material.[21] Both perpetrators and victims recounted plenty of lies, broken promises, violated rules and obligations, betrayed secrets, unfair acts, and conflicts over money.

But that was all that the perpetrators and victims agreed on. The psychologists pored over the narratives and coded features such as the time span of the events, the culpability of each side, the perpetrator's motive, and the aftermath of the harm. If one were to weave composite narratives out of their tallies, they might look something like this:

> *The Perpetrator's Narrative:* The story begins with the harmful act. At the time I had good reasons for doing it. Perhaps I was responding to an immediate provocation. Or I was just reacting to the situation in a way that any reasonable person would. I had a perfect right to do what I did, and it's unfair to blame me for it. The harm was minor, and easily repaired, and I apologized. It's time to get over it, put it behind us, let bygones be bygones.
>
> *The Victim's Narrative:* The story begins long before the harmful act, which was just the latest incident in a long history of mistreatment. The perpetrator's actions were incoherent, senseless, incomprehensible. Either that or he was an abnormal sadist, motivated only by a desire to see me suffer, though I was completely innocent. The harm he did is grievous and irreparable, with effects that will last forever. None of us should ever forget it.

They can't both be right—or more to the point, neither of them can be right all of the time, since the same participants provided a story in which they were the victim and a story in which they were the perpetrator. Something in human psychology distorts our interpretation and memory of harmful events.

This raises an obvious question. Does our inner perpetrator whitewash our crimes in a campaign to exonerate ourselves? Or does our inner victim nurse our grievances in a campaign to claim the world's sympathy? Since the psychologists were not flies on the wall at the time of the actual incidents, they had no way of knowing whose retrospective accounts should be trusted.

In an ingenious follow-up, Stillwell and Baumeister *controlled* the event by writing an ambiguous story in which one college roommate offers to help another with some coursework but reneges for a number of reasons, which leads the student to receive a low grade for the course, change his or her major, and switch to another university.[22] The participants (students themselves) simply had to read the story and then retell it as accurately as possible in the first person, half of them taking the perspective of the perpetrator and half the perspective of the victim. A third group was asked to retell the story in the third person; the details they provided or omitted serve as a baseline for ordinary distortions of human memory that are unaffected by self-serving biases. The psychologists coded the narratives for missing or embellished details that would make either the perpetrator or the victim look better.

The answer to the question "Who should we believe?" turned out to be: neither. Compared to the benchmark of the story itself, and to the recall of the disinterested third-person narrators, both victims and perpetrators distorted the stories to the same extent but in opposite directions, each omitting or embellishing details in a way that made the actions of their character look more reasonable and the other's less reasonable. Remarkably, nothing was at stake in the exercise. Not only had the participants not taken part in the events, but they were not asked to sympathize with the character or to justify anyone's behavior, just to read and remember the story from a first-person perspective. That was all it took to recruit their cognitive processes to the cause of self-serving propaganda.

The diverging narratives of a harmful event in the eyes of the aggressor, the victim, and a neutral party are a psychological overlay on the violence triangle in figure 2–1. Let's call it the Moralization Gap.

The Moralization Gap is part of a larger phenomenon called self-serving biases. People try to look good. "Good" can mean effective, potent, desirable, and competent, or it can mean virtuous, honest, generous, and altruistic. The drive to present the self in a positive light was one of the major findings of 20th-century social psychology. An early exposé was the sociologist Erving Goffman's *The Presentation of Self in Everyday Life,* and recent summaries include Carol Tavris and Elliot Aronson's *Mistakes Were Made (but Not by Me*), Robert Trivers's *Deceit and Self-Deception,* and Robert Kurzban's *Why Everyone (Else) Is a Hypocrite.*[23] Among the signature phenomena are cognitive dissonance, in which people change their evaluation of something they have been manipulated into doing to preserve the impression that they are in control of their actions, and the Lake Wobegon Effect (named after Garrison Keillor's fictitious town in which all the children are above average), in which a majority of people rate themselves above average in every desirable talent or trait.[24]

Self-serving biases are part of the evolutionary price we pay for being social animals. People congregate in groups not because they are robots who are magnetically attracted to one another but because they have social and moral emotions. They feel warmth and sympathy, gratitude and trust, loneliness and guilt, jealousy and anger. The emotions are internal regulators that ensure that people reap the benefits of social life—reciprocal exchange and cooperative action—without suffering the costs, namely exploitation by cheaters and social parasites.[25] We sympathize with, trust, and feel grateful to those who are likely to cooperate with us, rewarding them with our own cooperation. And we get angry at or ostracize those who are likely to cheat, withdrawing cooperation or meting out punishment. A person's own level of virtue is a tradeoff between the esteem that comes from cultivating a reputation as a cooperator and the ill-gotten gains of stealthy cheating. A social group is a marketplace of cooperators of differing degrees of generosity and trustworthiness, and people advertise themselves as being as generous and trustworthy as they can get away with, which may be a bit more generous and trustworthy than they are.

The Moralization Gap consists of complementary bargaining tactics in the negotiation for recompense between a victim and a perpetrator. Like opposing counsel in a lawsuit over a tort, the social plaintiff will emphasize the deliberateness, or at least the depraved indifference, of the defendant's action, together with the pain and suffering the plaintiff endures. The social defendant will emphasize the reasonableness or unavoidability of the action, and will minimize the plaintiff's pain and suffering. The competing framings shape the negotiations over amends, and also play to the gallery in a competition for their sympathy and for a reputation as a responsible reciprocator.[26]

Trivers, the first to propose that the moral emotions are adaptations to cooperation, also identified an important twist. The problem with trying to convey an exaggerated impression of kindness and skill is that other people are bound to develop the ability to see through it, setting in motion a psychological arms race between better liars and better lie detection. Lies can be spotted through internal contradictions (as in the Yiddish proverb "A liar must have a good memory"), or through tells such as hesitations, twitches, blushes, and sweats. Trivers ventured that natural selection may have favored a degree of *self*-deception so as to suppress the tells at the source. We lie to ourselves so that we're more believable when we lie to others.[27] At the same time, an unconscious part of the mind registers the truth about our abilities so that we don't get too far out of touch with reality. Trivers credits George Orwell with an earlier formulation of the idea: "The secret of rulership is to combine a belief in one's own infallibility with a power to learn from past mistakes."[28]

Self-deception is an exotic theory, because it makes the paradoxical claim that something called "the self" can be both deceiver and deceived. It's easy enough to show that people are liable to self-serving *biases*, like a butcher's scale that has been miscalibrated in the butcher's favor. But it's not so easy to show that people are liable to self-*deception*, the psychological equivalent of the dual books kept by shady businesses in which a public ledger is made available to prying eyes and a private ledger with the correct information is used to run the business.[29]

A pair of social psychologists, Piercarlo Valdesolo and David DeSteno, have devised an ingenious experiment that catches people in the act of true, dual-book self-deception.[30] They asked the participants to cooperate with them in planning and evaluating a study in which half of them would get a pleasant and easy task, namely looking through photographs for ten minutes, and half would get a tedious and difficult one, namely solving math problems for forty-five minutes. They told the participants that they were being run in pairs, but that the experimenters had not yet settled on the best way to decide who got which task. So they allowed each participant to choose one of two methods to decide who would get the pleasant task and who would get the unpleasant one. The participants could just choose the easy task for themselves, or they could use a random number generator to decide who got which. Human selfishness being what it is, almost everyone kept the pleasant task for themselves. Later they were given an anonymous questionnaire to evaluate the experiment which unobtrusively slipped in a question about whether the participants thought that their decision had been fair. Human hypocrisy being what it is, most of them said it was. Then the experimenters described the selfish choice to another group of participants and asked them how fairly the selfish subject acted. Not surprisingly, they didn't think it was fair at all. The difference between the way people judge other people's behavior and the way they judge their own behavior is a classic instance of a self-serving bias.

But now comes the key question. Did the self-servers *really*, deep down, believe that they were acting fairly? Or did the conscious spin doctor in their brains just say that, while the unconscious reality-checker registered the truth? To find out, the psychologists *tied up* the conscious mind by forcing a group of participants to keep seven digits in memory while they evaluated the experiment, including the judgment about whether they (or others) had acted fairly. With the conscious mind distracted, the terrible truth came out: the participants judged themselves as harshly as they judged other people. This vindicates Trivers's theory that the truth was in there all along.

I was happy to discover the result, not just because the theory of self-deception is so elegant that it deserves to be true, but because it offers a glimmer of hope for humanity. Though acknowledging a compromising truth about ourselves is among our most painful experiences—Freud posited an armamentarium of defense mechanisms to postpone that dreadful day, such as denial, repression, projection, and reaction formation—it is, at least in principle, possible. It may take ridicule, it may take argument, it may take time, it may take being distracted, but people have the means to recognize that they are not always in the right. Still, we shouldn't deceive ourselves about self-deception. In the absence of these puncturings, the overwhelming tendency is for people to misjudge the harmful acts they have perpetrated or experienced.

Once you become aware of this fateful quirk in our psychology, social life begins to look different, and so do history and current events. It's not just that there are two sides to every dispute. It's that each side *sincerely* believes its version of the story, namely that it is an innocent and long-suffering victim and the other side a malevolent and treacherous sadist. And each side has assembled a historical narrative and database of facts consistent with its sincere belief.[31] For example:

• The Crusades were an upwelling of religious idealism that were marked by a few excesses but left the world with the fruits of cultural exchange. The Crusades were a series of vicious pogroms against Jewish communities that were part of a long history of European anti-Semitism. The Crusades were a brutal invasion of Muslim lands and the start of a long history of humiliation of Islam by Christendom.

• The American Civil War was necessary to abolish the evil institution of slavery and preserve a nation conceived in liberty and equality. The American Civil War was a power grab by a centralized tyranny intended to destroy the way of life of the traditional South.

• The Soviet occupation of Eastern Europe was the act of an evil empire drawing an iron curtain across the continent. The Warsaw Pact was a defensive alliance to protect the Soviet Union and its allies from a repeat of the horrendous losses it had suffered from two German invasions.

• The Six-Day War was a struggle for national survival. It began when Egypt expelled UN peacekeepers and blockaded the Straits of Tiran, the first step in its plan to push the Jews into the sea, and it ended when Israel reunified a divided city and secured defensible borders. The Six-Day War was a campaign of aggression and conquest. It began when Israel invaded its neighbors and ended when it expropriated their land and instituted an apartheid regime.

Adversaries are divided not just by their competitive spin-doctoring but by the calendars with which they measure history and the importance they put on remembrance. The victims of a conflict are assiduous historians and cultivators of memory. The perpetrators are pragmatists, firmly planted in the present. Ordinarily we tend to think of historical memory as a good thing, but when the events being remembered are lingering wounds that call for redress, it can be a call to violence. The slogans "Remember the Alamo!" "Remember the *Maine*!" "Remember the *Lusitania*!" "Remember Pearl Harbor!" and "Remember 9/11!" were not advisories to brush up your history but battle cries that led to Americans' engaging in wars. It is often said that the Balkans are a region that is cursed with too much history per square mile. The Serbs, who in the 1990s perpetrated ethnic cleansings in Croatia, Bosnia, and Kosovo, are also among the world's most aggrieved people.[32] They were inflamed by memories of depredations by the Nazi puppet state in Croatia in World War II, the Austro-Hungarian Empire in World War I, and the Ottoman Turks going back to the Battle of Kosovo in 1389. On the six hundredth anniversary of that battle, President Slobodan Milošević delivered a bellicose speech that presaged the Balkan wars of the 1990s.

In the late 1970s the newly elected separatist government of Québec rediscovered the thrills of 19th-century nationalism, and among other trappings of Québecois patriotism replaced the license-plate motto "La Belle Province" (the beautiful province) with "Je Me Souviens" (I remember). It was never made clear exactly what was being remembered, but most people interpreted it as nostalgia for New France, which had been vanquished by Britain during the Seven Years' War in 1763. All this remembering made Anglophone

Quebeckers a bit nervous and set off an exodus of my generation to Toronto. Fortunately, late-20th-century European pacifism prevailed over 19th-century Gallic nationalism, and Québec today is an unusually cosmopolitan and peaceable part of the world.

The counterpart of too much memory on the part of victims is too little memory on the part of perpetrators. On a visit to Japan in 1992, I bought a tourist guide that included a helpful time line of Japanese history. There was an entry for the period of the Taishō democracy from 1912 to 1926, and then there was an entry for the Osaka World's Fair in 1970. I guess nothing interesting happened in Japan in the years in between.

It's disconcerting to realize that all sides to a conflict, from roommates squabbling over a term paper to nations waging world wars, are convinced of their rectitude and can back up their convictions with the historical record. That record may include some whoppers, but it may just be biased by the omission of facts we consider significant and the sacralization of facts we consider ancient history. The realization is disconcerting because it suggests that in a given disagreement, the other guy might have a point, we may not be as pure as we think, the two sides will come to blows each convinced that it is in the right, and no one will think the better of it because everyone's self-deception is invisible to them.

For example, few Americans today would second-guess the participation of "the greatest generation" in the epitome of a just war, World War II. Yet it's unsettling to reread Franklin Roosevelt's historic speech following Japan's 1941 attack on Pearl Harbor and see that it is a textbook case of a victim narrative. All the coding categories of the Baumeister experiment can be filled in: the fetishization of memory ("a date which will live in infamy"), the innocence of the victim ("The United States was at peace with that nation"), the senselessness and malice of the aggression ("this unprovoked and dastardly attack"), the magnitude of the harm ("The attack yesterday on the Hawaiian Islands has caused severe damage to American naval and military forces. Very many American lives have been lost"), and the justness of retaliation ("the American people in their righteous might will win"). Historians today point out that each of these ringing assertions was, at best, truthy. The United States had imposed a hostile embargo of oil and machinery on Japan, had anticipated possible attacks, had sustained relatively minor military damage, eventually sacrificed 100,000 American lives in response to the 2,500 lost in the attack, forced innocent Japanese Americans into concentration camps, and attained victory with incendiary and nuclear strikes on Japanese civilians that could be considered among history's greatest war crimes.[33]

Even in matters when no reasonable third party can doubt who's right and who's wrong, we have to be prepared, when putting on psychological spectacles, to see that evildoers always think they are acting morally. The spectacles are a painful fit.[34] Just monitor your blood pressure as you read the sentence "Try to see it from Hitler's point of view." (Or Osama bin Laden's, or Kim Jong-il's.) Yet Hitler, like all sentient beings, *had* a point of view, and historians tell us that it was a highly moralistic one. He experienced Germany's sudden and unexpected defeat in World War I and concluded that it could be explained only by the treachery of an internal enemy. He was aggrieved by the Allies' murderous postwar food blockade and their vindictive reparations. He lived through the economic chaos and street violence of the 1920s. And Hitler was an idealist: he had a moral vision in which heroic sacrifices would bring about a thousand-year utopia.[35]

At the smaller scale of interpersonal violence, the most brutal serial killers minimize and even justify their crimes in ways that would be comical if their actions were not so horrific. In 1994 the police quoted a spree killer as saying, "Other than the two we killed, the two we wounded, the woman we pistol-whipped, and the light bulbs we stuck in people's mouths, we didn't really hurt anybody."[36] A serial rapist-murderer interviewed by the sociologist Diana Scully claimed to be "kind and gentle" to the women he captured at gunpoint, and that they enjoyed the experience of being raped. As further proof of this kindness, he noted that when he stabbed his victims "the killing was always sudden, so they wouldn't know it was coming."[37] John Wayne Gacy, who kidnapped, raped, and murdered thirty-three boys, said, "I see myself more as a victim than as a perpetrator," adding, without irony, "I was cheated out of my childhood." His victimization continued into adulthood, when the media inexplicably tried to make him into "an asshole and a scapegoat."[38]

Smaller-time criminals rationalize just as readily. Anyone who has worked with prisoners knows that today's penitentiaries are filled to a man with innocent victims—not just those who were framed by sloppy police work but those whose violence was a form of self-help justice. Remember Donald Black's theory of crime as social control (chapter 3), which seeks to explain why a majority of violent crimes don't bring the perpetrator a tangible benefit.[39] The offender is genuinely provoked by an affront or betrayal; the reprisal that we deem to be excessive—striking a sharp-tongued wife during an argument, killing a swaggering stranger over a parking spot—is from his point of view a natural response to a provocation and the administration of rough justice.

The unease with which we read these rationalizations tells us something about the very act of donning psychological spectacles. Baumeister notes that in the attempt to understand harm-doing, the viewpoint of the scientist or scholar overlaps with the viewpoint of the perpetrator.[40] Both take a detached, amoral stance toward the harmful act. Both are contextualizers, always attentive to the complexities of the situation and how they contributed to the causation of the harm. And both believe that the harm is ultimately explicable. The viewpoint of the moralist, in contrast, is the viewpoint of the victim. The harm is treated with reverence and awe. It continues to evoke sadness and anger long after it was perpetrated. And for all the feeble ratiocination we mortals throw at it, it remains a cosmic mystery, a manifestation of the irreducible and inexplicable existence of evil in the universe. Many chroniclers of the Holocaust consider it immoral even to try to explain it.[41]

Baumeister, with psychological spectacles still affixed, calls this the myth of pure evil. The mindset that we adopt when we don moral spectacles is the mindset of the victim. Evil is the intentional and gratuitous infliction of harm for its own sake, perpetrated by a villain who is malevolent to the bone, inflicted on a victim who is innocent and good. The reason that this is a myth (when seen through psychological spectacles) is that evil in fact is perpetrated by people who are mostly ordinary, and who respond to their circumstances, including provocations by the victim, in ways they feel are reasonable and just.

The myth of pure evil gives rise to an archetype that is common in religions, horror movies, children's literature, nationalist mythologies, and sensationalist news coverage. In many religions evil is personified as the Devil—Hades, Satan, Beelzebub, Lucifer, Mephistopheles—or as the antithesis to a benevolent God in a bilateral Manichean struggle. In popular fiction evil takes the form of the slasher, the serial killer, the bogeyman, the ogre, the Joker, the James Bond villain, or depending on the cinematic decade, the Nazi officer, Soviet spy, Italian gangster, Arab terrorist, inner-city predator, Mexican druglord, galactic emperor, or corporate executive. The evildoer may enjoy money and power, but these motives are vague and ill formed; what he really craves is the infliction of chaos and suffering on innocent victims. The evildoer is an adversary—the enemy of good—and the evildoer is often foreign. Hollywood villains, even if they are stateless, speak with a generic foreign accent.

The myth of pure evil bedevils our attempt to understand real evil. Because the standpoint of the scientist resembles the standpoint of the perpetrator, while the standpoint of the moralizer resembles the standpoint of the victim, the scientist is bound to be seen as "making excuses" or "blaming the victim," or as trying to vindicate the amoral doctrine that "to understand all is to forgive all." (Recall Lewis Richardson's reply that to condemn much is to understand little.) The accusation of relativizing evil is particularly likely when the motive the analyst imputes to the perpetrator appears to be venial, like jealousy, status, or retaliation, rather than grandiose, like the persistence of suffering in the world or the perpetuation of race, class, or gender oppression. It is also likely when the analyst ascribes the motive to every human being rather than to a few psychopaths or to the agents of a malignant political system (hence the popularity of the doctrine of the Noble Savage). The scholar Hannah Arendt, in her writings on the trial of Adolf Eichmann for his role in organizing the logistics of the Holocaust, coined the expression "the banality of evil" to capture what she saw as the ordinariness of the man and the ordinariness of his motives.[42] Whether or not she was right about Eichmann (and historians have shown that he was more of an ideological anti-Semite than Arendt allowed), she was prescient in deconstructing the myth of pure evil.[43] As we shall see, four decades of research in social psychology—some of it inspired by Arendt herself—have underscored the banality of most of the motives that lead to harmful consequences.[44]

In the rest of this chapter I'll lay out the brain systems and motives that incline us toward violence, while trying to identify the inputs that ramp them up or down and thereby offer insight into the historical decline of violence. Appearing to take the perspective of the perpetrator is just one of the dangers that attends this effort. Another is the assumption that nature organized the brain into systems that are morally meaningful to us, such as ones that lead to evil and ones that lead to good. As we shall see, some of the dividing lines between the inner demons of this chapter and the better angels of the next were guided as much by expository convenience as by neurobiological reality, because certain brain systems can cause both the best and the worst in human behavior.

ORGANS OF VIOLENCE

One of the symptoms of the myth of pure evil is to identify violence as an animalistic impulse, as we see in words like *beastly, bestial, brutish, inhuman,* and *wild,* and in depictions of the devil with horns and a tail. But while violence is certainly common in the animal kingdom, to think of it as arising from a single impulse is to see the world through a victim's eyes. Consider all the destructive things that members of our species do to ants. We eat them, poison them, accidentally trample them, and deliberately squish them. Each category of formicide is driven by an utterly distinct motive. But if you were an ant, you might not care about these fine distinctions. We *are* humans, so we tend to think that the terrible things that humans do to other humans come from a single, animalistic motive. But biologists have long noted that the mammalian brain has distinct circuits that underlie very different kinds of aggression.

The most obvious form of aggression in the animal kingdom is predation. Hunters such as hawks, eagles, wolves, lions, tigers, and bears adorn the jerseys of athletes and the coats of arms of nations, and many writers have blamed human violence, as William James did, on "the carnivore within." Yet biologically speaking, predation for food could not be more different from aggression against rivals and threats. Cat people are well aware of the distinction. When their animal companion sets its sights on a beetle on the floorboards, it is crouched, silent, and intently focused. But when one alley cat faces off against another, the cat stands tall, fur erect, hissing and yowling. We saw how neuroscientists can implant an electrode into the Rage circuit of a cat, press a button, and set the animal on attack mode. With the electrode implanted in a different circuit, they can set it on hunting mode and watch in amazement as the cat quietly stalks a hallucinatory mouse.[45]

Like many systems in the brain, the circuits that control aggression are organized in a hierarchy. Subroutines that control the muscles in basic actions are encapsulated in the hindbrain, which sits on top of the spinal cord. But the emotional states that trigger them, such as the Rage circuit, are distributed higher up in the midbrain and forebrain. In cats, for example, stimulating the hindbrain can activate what neuroscientists call sham rage. The cat hisses, bristles, and extends its fangs, but it can be petted without it attacking the petter. If, in contrast, they stimulate the Rage circuit higher up, the resulting emotional state is no sham: the cat is mad as hell and lunges for the experimenter's head.[46] Evolution takes advantage of this modularity. Different mammals use different body parts as offensive weapons, including jaws, fangs, antlers, and in the case of primates, hands. While the hindbrain circuits that drive these peripherals can be reprogrammed or swapped out as a lineage evolves, the central programs that control their emotional states are remarkably conserved.[47] That includes the lineage leading to humans, as neurosurgeons discovered when they found a counterpart to the Rage circuit in the brains of their patients.

Figure 8–1 is a computer-generated model of the brain of a rat, facing left. A rat is a sniffy little animal that depends on its sense of smell, and so it has enormous olfactory bulbs, which have been amputated from the left-hand side of the model to leave room for the rest of the brain in the picture. And like all quadrupeds, the rat is a horizontal creature, so what we think of as the "higher"

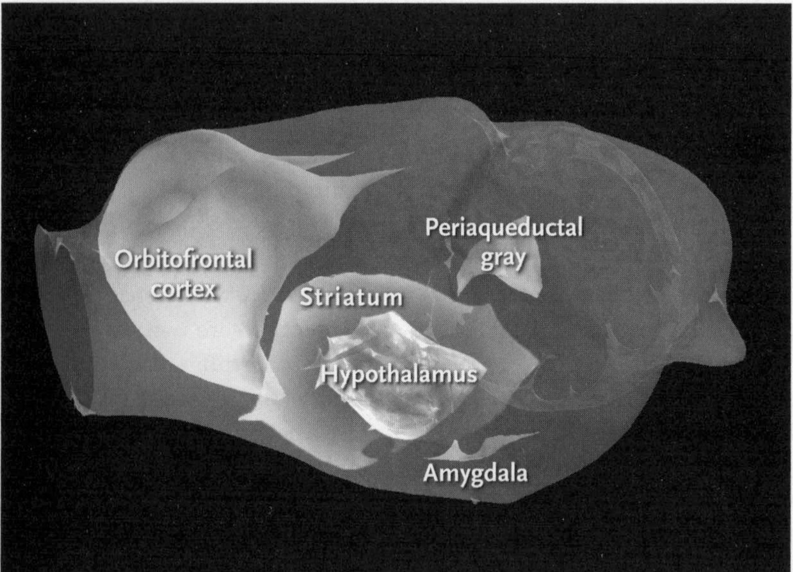

FIGURE 8–1. Rat brain, showing the major structures involved in aggression
Source: Image derived from the Allen Mouse Brain Atlas, http://mouse.brain-map.org.

and "lower" levels of the nervous system are really laid out front to back, with the rat's high-level cogitation, such as it is, located at the front (left) end of the model and the control of the body at the rear (right), extending into the spinal cord, which would spill out of the right edge of the picture if it were shown.

The Rage circuit is a pathway that connects three major structures in the lower parts of the brain.[48] In the midbrain there is a collar of tissue called the periaqueductal gray—"gray" because it consists of gray matter (a tangle of neurons, lacking the white sheaths that insulate output fibers), "periaqueductal" because it surrounds the aqueduct, a fluid-filled canal that runs the length of the central nervous system from the spinal cord up to large cavities in the brain. The periaqueductal gray contains circuits that control the sensorimotor components of rage. They get inputs from parts of the brain that register pain, balance, hunger, blood pressure, heart rate, temperature, and hearing (particularly the shrieks of a fellow rat), all of which can make the animal irritated, frustrated, or enraged. Their outputs feed the motor programs that make the rat lunge, kick, and bite.[49] One of the oldest discoveries in the biology of violence is the link between pain or frustration and aggression. When an animal is shocked, or access to food is taken away, it will attack the nearest fellow animal, or bite an inanimate object if no living target is available.[50]

The periaqueductal gray is partly under the control of the hypothalamus, a cluster of nuclei that regulate the animal's emotional, motivational, and physiological state, including hunger, thirst, and lust. The hypothalamus monitors the temperature, pressure, and chemistry of the bloodstream and sits on top of the pituitary gland, which pumps hormones into the bloodstream that regulate, among other things, the release of cortisol from the adrenal glands and the release of testosterone and estrogen from the gonads. Two of its nuclei, the medial and ventrolateral, are parts of the Rage circuit. "Ventral" refers to the belly side of the animal, as opposed to its "dorsal" or back side. The terms were grandfathered over to the human brain as it evolved its perpendicular perch atop a vertical body, so in the human brain "ventral" points to our feet and "dorsal" to the top of our scalp.

Modulating the hypothalamus is the amygdala, Latin for "almond," the shape it takes in the human brain. The amygdala is a small, multipart organ connected to brain systems for memory and motivation. It applies the emotional coloring to our thoughts and memories, particularly fear. When an animal has been trained to expect a shock after a tone, the amygdala helps to store the connections that give the tone its aura of anxiety and dread. The amygdala also lights up at the sight of a dangerous predator or of a threatening display from a member of the same species. In the case of humans, for example, the amygdala responds to an angry face.

And sitting on top of the entire Rage circuit is the cerebral cortex—the thin layer of gray matter on the outer surface of the cerebral hemispheres where the computations behind perception, thinking, planning, and decision-making are carried out. Each cerebral hemisphere is divided into lobes, and the one at the front, the frontal lobe, computes decisions relevant to how to behave. One of the major patches of the frontal lobes sits on top of the eye sockets in the skull, also known as orbits, so it is called the orbitofrontal cortex, orbital cortex for short.[51] The orbital cortex is densely connected to the amygdala and other emotional circuits, and it helps integrate emotions and memories into decisions about what to do next. When the animal modulates its readiness to attack in response to the circumstances, including its emotional state and any lessons it has learned in the past, it is this part of the brain, behind the eyeballs, that is responsible. By the way, though I have described the control of rage as a top-down chain of command—orbital cortex to amygdala to hypothalamus to periaqueductal gray to motor programs—the connections are all two-way: there is considerable feedback and cross talk among these components and with other parts of the brain.

As I mentioned, predation and rage play out very differently in the behavioral repertoire of a carnivorous mammal and are triggered by electrical stimulation of different parts of the brain. Predation involves a circuit that is part of what Panksepp calls the Seeking system.[52] A major part of the Seeking system runs from a part of the midbrain (not shown in figure 8–1) via a bundle of fibers in the middle of the brain (the medial forebrain bundle) to the lateral hypothalamus, and from there up to the ventral striatum, a major part of the so-called reptilian brain. The striatum is composed of many parallel tracts (giving it a striated appearance), and it is buried deep in the cerebral hemispheres and densely connected to the frontal lobes.

The Seeking system was discovered when the psychologists James Olds and Peter Milner implanted an electrode into the middle of a rat brain, hooked it up to a lever in a Skinner box, and found that the rat would press the lever to stimulate its own brain until it dropped of exhaustion.[53] Originally they thought they had found the pleasure center in the brain, but neuroscientists today believe that the system underlies wanting or craving rather than actual pleasure. (The major realization of adulthood, that you should be careful about what you want because when you get it you may not enjoy it, has a basis in the anatomy of the brain.) The Seeking system is held together not just by wiring but by chemistry. Its neurons signal to each other with a neurotransmitter called dopamine. Drugs that make dopamine more plentiful, like cocaine and amphetamines, jazz the animal up, while drugs that decrease it, like antipsychotic medications, leave the animal apathetic. (The ventral striatum also contains circuits that respond to a different family of transmitters, the endorphins or endogenous opiates. These circuits are more closely related to enjoying a reward once it arrives than to craving it in anticipation.)

The Seeking system identifies goals for the animal to pursue, like access to a lever that it may press to receive food. In more natural settings, the Seeking system motivates a carnivorous animal to hunt. The animal stalks its quarry in what we can imagine is a state of pleasant anticipation. If successful, it dispatches the prey in a quiet bite that is completely unlike the snarling attack of rage.

Animals can attack both in offense and in defense.[54] The simplest trigger of an offensive attack is sudden pain or frustration, the latter delivered as a signal from the Seeking system. The reflex may be seen in some of the primitive responses of a human being. Babies react with rage when their arms are suddenly pinned at their sides, and adults may lash out by swearing or breaking things when they hit their thumb with a hammer or are surprised by not getting what they expect (as in the technique of computer repair called percussive maintenance). Defensive attacks, which in the rat consist of lunging at the head of an adversary rather than kicking and biting its flank, are triggered by yet another brain system, the one that underlies fear. The Fear system, like the Rage system, consists of a circuit that runs from the periaqueductal gray through the hypothalamus to the amygdala. The Fear and Rage circuits are distinct, connecting different nuclei in each of these organs, but their physical proximity reflects the ease with which they interact.[55] Mild fear can trigger freezing or flight, but extreme fear, combined with other stimuli, can trigger an enraged defensive attack. Forward panic or rampage in humans may involve a similar handoff from the Fear system to the Rage system.

Panksepp identifies a fourth motivational system in the mammalian brain that can trigger violence; he calls it the Intermale Aggression or Dominance system.[56] Like Fear and Rage, it runs from the periaqueductal gray through the hypothalamus to the amygdala, connecting yet another trio of nuclei along the way. Each of these nuclei has receptors for testosterone. As Panskepp notes, "In virtually all mammals, male sexuality requires an assertive attitude, so that male sexuality and aggressiveness normally go together. Indeed, these tendencies are intertwined throughout the neuroaxis, and to the best of our limited knowledge, the circuitry for this type of aggression is located near, and probably interacts strongly with, both Rage and Seeking circuits."[57] To psychologize the anatomy, the Seeking system leads a male to willingly, even eagerly, seek out an aggressive challenge with another male, but when the battle is joined and one of them is in danger of defeat or death, focused fighting may give way to blind rage. Panksepp notes that the two kinds of aggression, though they interact with each other, are neurobiologically distinct. When certain parts of the medial hypothalamus or striatum are damaged, the animal is more likely to attack a prey animal or an unwitting experimenter, but less likely to attack another male. And as we shall see, giving an animal (or a man) testosterone does not make him testy across the board. On the contrary, it makes him feel great, while putting a chip on his shoulder when he is faced with a rival male.[58]

One look at a human brain and you know you are dealing with a very unusual mammal. Figure 8–2, with its transparent cortex, shows that all the parts of

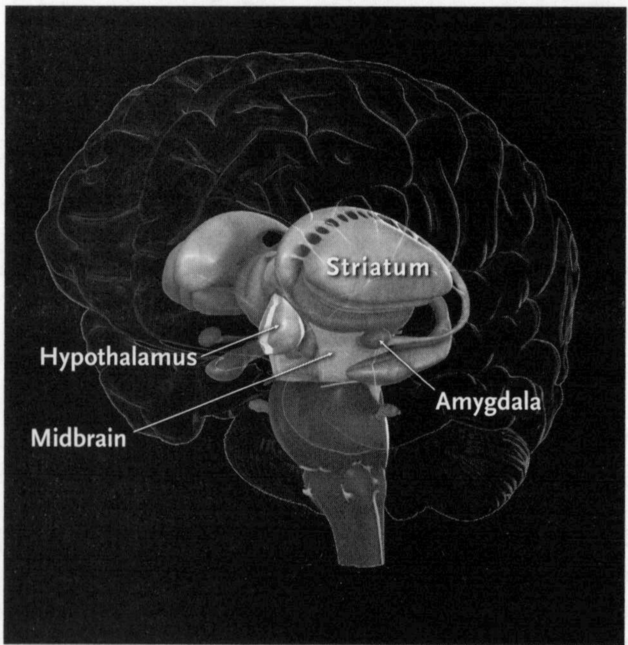

FIGURE 8–2. Human brain, showing the major subcortical structures involved in aggression
Source: 3D Brain Illustration by AXS Biomedical Animation Studio, created for Dolan DNA Learning Center.

the rat brain have been carried over to the human brain, including the organs that house the circuits for rage, fear, and dominance: the amygdala, the hypothalamus, and the periaqueductal gray (which is found inside the midbrain, lining the cerebrospinal canal running through it). The dopamine-fueled striatum, whose ventral portion helps set goals for the whole brain to seek, is also prominent.

But while these structures take up a large proportion of the rat brain, in the human brain they are enveloped by a bloated cerebrum. As figure 8–3 shows, the outsize cerebral cortex has been wrinkled like a wad of newspaper to get it to fit inside the skull. A large part of the cerebrum is taken up by the frontal lobes, which in this view of the brain extend about three-quarters of the way back. The neuroanatomy suggests that in *Homo sapiens* primitive impulses of rage, fear, and craving must contend with the cerebral restraints of prudence, moralization, and self-control—though as in all attempts at taming the wild, it's not always clear who has the upper hand.

Within the frontal lobes, one can readily see how the orbital cortex got its name: it is a big spherical dent that accommodates the bony socket of the eye. Scientists have known that the orbital cortex is involved with regulating the emotions since 1848, when a railroad foreman named Phineas Gage tamped down some blasting powder in a hole in a rock and ignited an explosion that

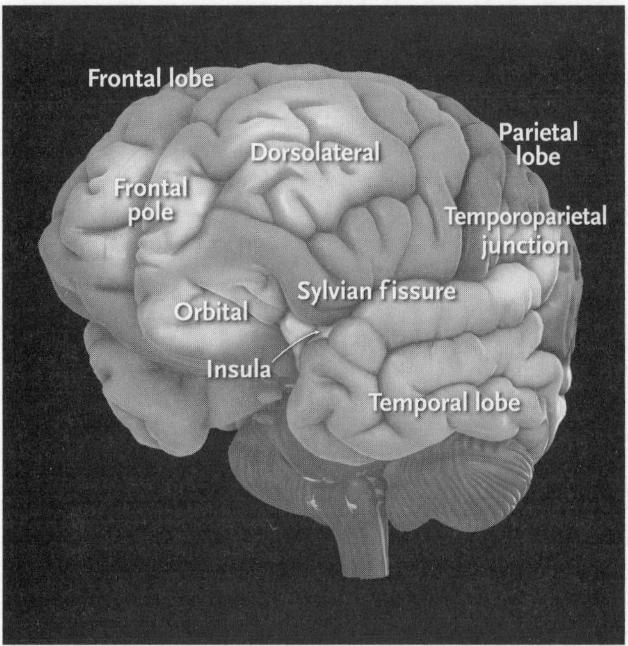

FIGURE 8–3. Human brain, showing the major cortical regions that regulate aggression
Source: 3D Brain Illustration by AXS Biomedical Animation Studio, created for Dolan DNA Learning Center.

sent the tamping iron up through his cheekbone and out the top of his skull.[59] A 20th-century computer reconstruction based on the holes in the skull suggest that the spike tore up his left orbital cortex, together with the ventromedial cortex on the inside wall of the cerebrum. (It is visible in the medial view of the brain in figure 8–4.) The orbital and ventromedial cortex are continuous, wrapping around the bottom edge of the frontal lobe, and neuroscientists often use either term to refer to the combination of the two of them.

Though Gage's senses, memories, and movement were intact, it soon became clear that the damaged parts of his brain had been doing something important. Here is how his physician described the change:

> The equilibrium or balance, so to speak, between his intellectual faculties and animal propensities, seems to have been destroyed. He is fitful, irreverent, indulging at times in the grossest profanity (which was not previously his custom), manifesting but little deference for his fellows, impatient of restraint or advice when it conflicts with his desires, at times pertinaciously obstinate, yet capricious and vacillating, devising many plans of future operations, which are no sooner arranged than they are abandoned in turn for others appearing more feasible. A child in his intellectual capacity and manifestations, he has the animal passions of a strong man. Previous to his

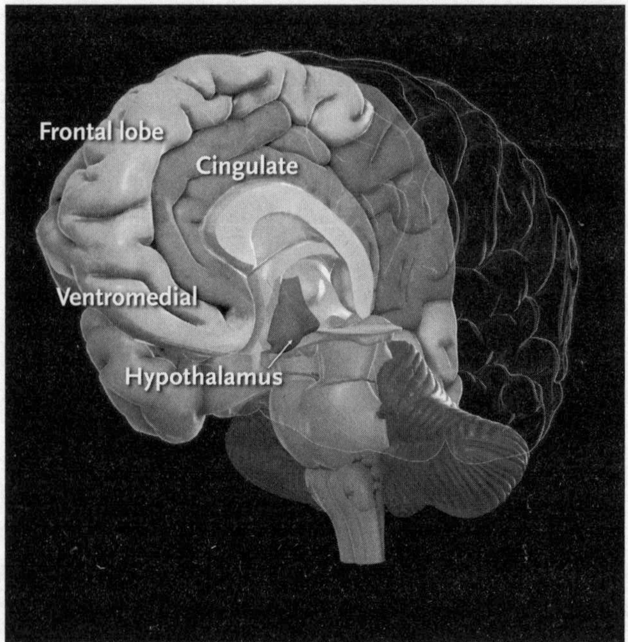

FIGURE 8–4. Human brain, medial view

Source: 3D Brain Illustration by AXS Biomedical Animation Studio, created for Dolan DNA Learning Center.

> injury, although untrained in the schools, he possessed a well-balanced mind, and was looked upon by those who knew him as a shrewd, smart businessman, very energetic and persistent in executing all his plans of operation. In this regard his mind was radically changed, so decidedly that his friends and acquaintances said he was "no longer Gage."[60]

Though Gage eventually recovered much of his equipoise, and the story has been embellished and sometimes garbled in generations of retelling to introductory psychology students, today our understanding of the function of the orbital cortex is broadly consistent with the doctor's description.

The orbital cortex is strongly connected to the amygdala, hypothalamus, and other parts of the brain involved with emotion.[61] It is suffused with neurons that use dopamine as their neurotransmitter and that are connected to the Seeking system in the striatum. It is adjacent to an island of cortex called the insula, the front of which just barely peeks out from beneath the Sylvian fissure in figure 8–3; the rest of the insula extends back under that fissure, obscured by overhanging flaps of the frontal and temporal lobes. The insula registers our physical gut feelings, including the sensation of a distended stomach and other inner states like nausea, warmth, a full bladder, and a pounding heart. The brain takes metaphors like "that makes my blood boil" and "his behavior disgusts me" literally. The cognitive neuroscientist Jonathan Cohen and his team found that when a person feels he is getting a raw deal from another person who is dividing a windfall between them, the insula is fired up. When the stingy allocation is thought to come from a computer, so there's no one to get mad at, the insula remains dark.[62]

The orbital cortex resting on the eyeballs (figure 8–3) and the ventromedial cortex facing inward (figure 8–4) are, as mentioned, adjacent, and it's not easy to distinguish what they do, which is why neuroscientists often lump them together. The orbital cortex seems to be more involved with determining whether an experience is pleasant or unpleasant (befitting its position next to the insula, with its input from the viscera), while the ventromedial cortex is more involved with determining whether you are getting what you want and avoiding what you don't want (befitting its position along the midline of the brain, where the Seeking circuit extends).[63] The distinction may carry over to a difference in the moral realm between an emotional reaction to a harm, and judgment and reflection on it. But the dividing line is fuzzy, and I'll continue to use "orbital" to refer to both parts of the brain.

The inputs to the orbital cortex—gut feelings, objects of desire, and emotional impulses, together with sensations and memories from other parts of the cortex—allow it to serve as the regulator of emotional life. Visceral feelings of anger, warmth, fear, and disgust are combined with the person's goals, and modulating signals are computed and sent back down to the emotional structures from which they originated. Signals are also sent upward to regions of the cortex that carry out cool deliberation and executive control.

This flowchart suggested by the neuroanatomy corresponds fairly well with what psychologists see in the clinic and lab. Making allowances for the difference between the flowery language of medical reports in the 19th century and the clinical jargon of the 21st, today's descriptions of patients with damage to their orbital cortex could have applied to Phineas Gage: "Disinhibited, socially inappropriate, susceptible to misinterpreting others' moods, impulsive, unconcerned with the consequences of their actions, irresponsible in everyday life, lacking in insight into the seriousness of their condition, and prone to weak initiative."[64]

The psychologists Angela Scarpa and Adrian Raine offer a similar list, but with an additional symptom at the end that is relevant to our discussion: "Argumentativeness, lack of concern for consequences of behavior, loss of social graces, impulsivity, distractibility, shallowness, lability, violence."[65] The extra noun came from Raine's own studies, which, instead of selecting patients with orbital brain damage and then examining their personalities, selected people prone to violence and then examined their brains. He focused on people with antisocial personality disorder, defined by the American Psychiatric Association as "a pervasive pattern of disregard for, and violation of, the rights of others," including lawbreaking, deceit, aggressiveness, recklessness, and lack of remorse. People with antisocial personality disorder make up a large proportion of violent felons, and a subset of them, who possess glibness, narcissism, grandiosity, and a superficial charm, are called psychopaths (or sometimes sociopaths). Raine scanned the brains of violence-prone people with antisocial personality disorder and found that the orbital regions were shrunken and less metabolically active, as were other parts of the emotional brain, including the amygdala.[66] In one experiment, Raine compared the brains of prisoners who had committed an impulsive murder with those who had killed with premeditation. Only the impulsive murderers showed a malfunction in their orbital cortex, suggesting that the self-control implemented by this part of the brain is a major inhibitor of violence.

But another part of its job description may come into play as well. Monkeys with lesions in the orbital cortex have trouble fitting into dominance hierarchies, and they get into more fights.[67] Not coincidentally, humans with orbital damage are insensitive to social faux pas. When they hear a story about a woman who inadvertently disparaged a gift she received from a friend, or accidentally divulged that the friend had been excluded from a party list, the patients fail to recognize that anyone had said anything wrong, and don't realize that the friend might have been hurt.[68] Raine found that when people with antisocial personality disorder were asked to compose and deliver a speech about their own faults, which for ordinary people is a nerve-racking ordeal accompanied by embarrassment, shame, and guilt, their nervous systems were unresponsive.[69]

The orbital cortex, then (together with its ventromedial neighbor), is involved in several of the pacifying faculties of the human mind, including self-control, sympathy to others, and sensitivity to norms and conventions. For all that, the orbital cortex is a fairly primitive part of the cerebrum. We saw it in the lowly rat, and its inputs are literally and figuratively from the gut. The more deliberative and intellectual modulators of violence rely on other parts of the brain.

Consider the process of deciding whether to punish someone who has caused a harm. Our sense of justice tells us that the perpetrator's culpability depends not just on the harm done but on his or her mental state—the *mens rea*, or guilty mind, that is necessary for an act to be considered a crime in most legal systems. Suppose a woman kills her husband by putting rat poison in his tea. Our decision as to whether to send her to the electric chair very much depends on whether the container she spooned it out of was mislabeled DOMINO SUGAR or correctly labeled D-CON: KILLS RATS—that is, whether she knew she was poisoning him and wanted him dead, or it was all a tragic accident. A brute emotional reflex to the *actus reus*, the bad act ("She killed her husband! Shame!"), could trigger an urge for retribution regardless of her intention. The crucial role played by the perpetrator's mental state in our assignment of blame is what makes the Moralization Gap possible. Victims insist that the perpetrator deliberately and knowingly wanted to harm them, while perpetrators insist that the harm was unintended.

The psychologists Liane Young and Rebecca Saxe put people in an fMRI scanner and had them read stories involving deliberate and accidental harms.[70] They found that the ability to exculpate harm-doers in the light of their mental state depends on the part of the brain at the junction between the temporal and parietal lobes, which is illuminated in figure 8–3 (though it's actually the counterpart of this region in the right hemisphere that lit up in the study). The temporoparietal junction sits at a crossroads for many kinds of information, including the perception of the position of one's own body, and the perception of the bodies and actions of other people. Saxe had previously shown that the region is necessary for the mental faculty that has been called mentalizing, intuitive psychology, and theory of mind, namely the ability to understand the beliefs and desires of another person.[71]

There is another kind of moral deliberation that goes beyond the gut: weighing the consequences of different courses of action. Consider the old chestnut from moral philosophy: a family is hiding from the Nazis in a cellar. Should they smother their baby to prevent it from crying and giving away their location, which would result in the deaths of everyone in the family, baby included? How about throwing a fat man in front of a runaway trolley so that his bulk will stop it before it slams into five workers on the track? A utilitarian calculus would say that both killings are permissible, because they would sacrifice one life to save five. Yet many people would balk at smothering the baby or heaving the fat man, presumably because they have a visceral reaction against harming an innocent person with their bare hands. In a logically equivalent dilemma, a bystander to the runaway trolley could save the five workers by diverting it onto a side track, where it would kill just one. In this version, everyone agrees that it's permissible to throw the switch and save five lives at a cost of one, presumably because it doesn't *feel like* you're really killing anyone; you're just failing to prevent the trolley from doing it.[72]

The philosopher Joshua Greene, working with Cohen and others, has shown that the visceral reaction against smothering the baby or throwing the man in front of the train comes from the amygdala and orbital cortex, whereas the utilitarian thinking that would save the greatest number of lives is computed in a part of the frontal lobe called the dorsolateral prefrontal cortex, also illuminated in figure 8–3.[73] The dorsolateral cortex is the part of the brain that is most involved in intellectual, abstract problem-solving—it lights up, for example, when people do the problems on an IQ test.[74] When people consider the case of the crying baby in the cellar, both their orbital cortex (which reacts to the horror of smothering the baby) and their dorsolateral cortex (which calculates lives saved and lost) light up, together with a third part of the brain that deals with conflicting impulses—the anterior cingulate cortex in the medial wall of the brain, shown in figure 8–4. The people who deduce that it's all right to smother the baby show greater activation in the dorsolateral cortex.

The temporoparietal junction and dorsolateral prefrontal cortex, which grew tremendously over the course of evolution, give us the wherewithal to perform cool calculations that deem certain kinds of violence justifiable. Our ambivalence about the outputs of those calculations—whether smothering the baby should be thought of as an act of violence or an act that prevents violence—shows that the quintessentially cerebral parts of the cerebrum are neither inner demons nor better angels. They are cognitive tools that can both foster violence and inhibit it, and as we shall see, both powers are exuberantly employed in the distinctively human forms of violence.

My brief tour of the neurobiology of violence barely does justice to our scientific understanding, and our scientific understanding barely does justice to the phenomena themselves. But I hope it has persuaded you that violence does not have a single psychological root but a number of them, working by different principles. To understand them, we need to look not just at the hardware of the brain but also at its software—that is, at the *reasons* people engage in violence. Those reasons are implemented as intricate patterns in the microcircuitry of brain tissue; we cannot read them directly from the neurons, any more than we can understand a movie by putting a DVD under a microscope. So the remainder of the chapter will shift to the bird's-eye view of psychology, while connecting the psychological phenomena to the neuroanatomy.

There are many taxonomies of violence, and they tend to make similar distinctions. I will adapt a four-part scheme from Baumeister, splitting one of his categories in two.[75]

The first category of violence may be called practical, instrumental, exploitative, or predatory. It is the simplest kind of violence: the use of force as a means to an end. The violence is deployed in pursuit of a goal such as greed, lust, or ambition, which is set up by the Seeking system, and it is guided by the entirety of the person's intelligence, for which the dorsolateral prefrontal cortex is a convenient symbol.

The second root of violence is dominance—the drive for supremacy over one's rivals (Baumeister calls it "egotism"). This drive may be tied to the testosterone-fueled Dominance or Intermale Aggression system, though it is by no means confined to males, or even to individual people. As we shall see, groups compete for dominance too.

The third root of violence is revenge—the drive to pay back a harm in kind. Its immediate engine is the Rage system, but it can recruit the Seeking system to its cause as well.

The fourth root is sadism, the joy of hurting. This motive, puzzling and horrifying in equal measure, may be a by-product of several quirks of our psychology, particularly the Seeking system.

The fifth and most consequential cause of violence is ideology, in which true believers weave a collection of motives into a creed and recruit other people to carry out its destructive goals. An ideology cannot be identified with a part of the brain or even with a whole brain, because it is distributed across the brains of many people.

PREDATION

The first category of violence is not really a category at all, because its perpetrators have no destructive motive like hate or anger. They simply take the shortest path to something they want, and a living thing happens to be in the way. At best it is a category by exclusion: the absence of any inhibiting factor like sympathy or moral concern. When Immanuel Kant stated the second formulation of his Categorical Imperative—that an act is moral if it treats a person as an end in itself and not as a means to an end—he was in effect defining morality as the avoidance of this kind of violence.

Predation may also be called exploitative, instrumental, or practical violence.[76] It coincides with Hobbes's first cause of quarrel: to invade for gain. It is Dawkins's survival machine treating another survival machine as a part of its environment, like a rock or a river or a lump of food. It is the interpersonal equivalent of Clausewitz's dictum that war is merely the continuation of policy by other means. It is Willie Sutton's rumored answer to the question of why he robbed banks: "Because that's where the money is." It lies beneath the advice of a farmer to increase a horse's efficiency by castrating it with two bricks. When asked, "Doesn't that hurt," the farmer replies, "Not if you keep your thumbs out of the way."[77]

Because predatory violence is just a means to a goal, it comes in as many varieties as there are human goals. The paradigm case is literal predation—hunting for food or sport—because it involves no animosity toward the victim. Far from hating their quarry, hunters valorize and totemize them, from Paleolithic cave paintings to trophies above the mantels in gentlemen's clubs. Hunters may even empathize with their prey—proof that empathy alone is not a bar to violence. The ecologist Louis Liebenberg studied the remarkable ability of the !Kung San to infer the whereabouts and physical condition of their game from a few faint tracks as they pursue them across the Kalahari Desert.[78] They do it with empathy—putting themselves in the hooves of the animal and imagining what it is feeling and where it is tempted to flee. There may even be an element of love. One night after the ninth inning of a baseball game, I was too comatose to get off the couch or even change the channel and passively watched the following program on the cable sports network. It was a show about fishing, and consisted entirely of footage of a middle-aged man in an aluminum boat on a nondescript stretch of water pulling in one large bass after another. With each catch he brought the fish close to his face and stroked it, making little kissy noises and cooing, "Ooh, aren't you a beauty! You're a pretty one! Yes, you are!"

The chasm between the perpetrator's perspective—amoral, pragmatic, even frivolous—and the victim's is nowhere wider than in our predation of animals. It's safe to say that the bass, if given the chance, would not reciprocate the fisherman's affection, and most people would not want to know the opinion of a broiler chicken or a live lobster on whether the mild pleasure we get from eating their flesh rather than a plate of eggplant justifies the sacrifice they will make. The same incuriosity enables coldhearted predatory violence against humans.

Here are a few examples: Romans suppressing provincial rebellions; Mongols razing cities that resist their conquest; free companies of demobilized soldiers plundering and raping; colonial settlers expelling or massacring indigenous peoples; gangsters whacking a rival, an informant, or an uncooperative official; rulers assassinating a political opponent or vice versa; governments jailing or executing dissidents; warring nations bombing enemy cities; hoodlums injuring a victim who resists a robbery or carjacking; criminals killing an eyewitness to a crime; mothers smothering a newborn they feel they cannot raise. Defensive and preemptive violence—doing it to them before they do it to you—is also a form of instrumental violence.

Predatory violence may be the most extraordinary and perplexing phenomenon in the human moral landscape precisely because it is so mundane and explicable. We read of an atrocity—say, rebel soldiers encamped on a rooftop in Uganda who passed the time by kidnapping women, tying them up, raping them, and throwing them to their deaths—shake our heads, and ask, "How could people do these things?"[79] We refuse to accept obvious answers, like boredom, lust, or sport, because the suffering of the victim is so obscenely disproportionate to the benefit to the perpetrator. We take the victim's point of view and advert to a conception of pure evil. Yet to understand these outrages, we might be better off asking not why they happen but why they don't happen more often.

With the possible exception of Jain priests, all of us engage in predatory violence, if only against insects. In most cases the temptation to prey on humans is inhibited by emotional and cognitive restraints, but in a minority of individuals these restraints are absent. Psychopaths make up 1 to 3 percent of the male population, depending on whether one uses the broad definition of antisocial personality disorder, which embraces many kinds of callous troublemakers, or a narrower definition that picks out the more cunning manipulators.[80] Psychopaths are liars and bullies from the time they are children, show no capacity for sympathy or remorse, make up 20 to 30 percent of violent criminals, and commit half the serious crimes.[81] They also perpetrate nonviolent crimes like bilking elderly couples out of their life savings and running a business with ruthless disregard for the welfare of the workforce or stakeholders. As we saw, the regions of the brain that handle social emotions, especially the amygdala and orbital cortex, are relatively shrunken or unresponsive in psychopaths, though they may show no other signs of pathology.[82] In some people, signs of psychopathy develop after damage to these regions from disease or an accident, but the condition is also partly heritable. Psychopathy may have evolved as a minority strategy that exploits a large population of trusting cooperators.[83] Though no society can stock its militias and armies exclusively with psychopaths, such men are bound to be disproportionately attracted to these adventures, with their prospect of plunder and rape. As we saw in chapter 6, genocides and civil wars often involve a division of labor between the ideologues or warlords who run them and the shock troops, including some number of psychopaths, who are happy to carry them out.[84]

The psychology of predatory violence consists in the human capacity for means-end reasoning and the fact that our faculties of moral restraint do not kick in automatically in our dealings with every living thing. But there are two psychological twists in the way that predatory violence is carried out.

Though predatory violence is purely practical, the human mind does not stick to abstract reasoning for long. It tends to backslide into evolutionarily prepared and emotionally charged categories.[85] As soon as the objects being preyed upon take protective measures in response, emotions are likely to run high. The human prey may hide and regroup, or they may fight back, perhaps even threatening to destroy the predator preemptively, a kind of instrumental violence of their own that gives rise to a security dilemma or Hobbesian trap. In these cases the predator's state of mind may shift from dispassionate means-ends analysis to disgust, hatred, and anger.[86] As we have seen, perpetrators commonly analogize their victims to vermin and treat them with moralized disgust. Or they may see them as existential threats and treat them with hatred, the emotion that, as Aristotle noted, consists of a desire not to punish an adversary but to end its existence. When extermination is not feasible and perpetrators have to continue to deal with their victims, either directly or with the participation of third parties, they may treat them with anger. The predators may respond to the defensive reprisals of their prey as if *they* were the ones under attack, and experience a moralized wrath and a thirst for revenge. Thanks to the Moralization Gap, they will minimize their own first strike as necessary and trivial while magnifying the reprisal as unprovoked and devastating. Each side will count the wrongs differently—the perpetrator tallying an even number of strikes and the victim an odd number—and the difference in arithmetic can stoke a spiral of revenge, a dynamic we will explore in a later section.

There is a second way self-serving biases can fan a small flame of predatory violence into an inferno. People exaggerate not just their moral rectitude but their power and prospects, a subtype of self-serving bias called positive illusions.[87] Hundreds of studies have shown that people overrate their health, leadership ability, intelligence, professional competence, sporting prowess, and managerial skills. People also hold the nonsensical belief that they are inherently lucky. Most people think they are more likely than the average person to attain a good first job, to have gifted children, and to live to a ripe old age. They also think that they are *less* likely than the average person to be the victim of an accident, crime, disease, depression, unwanted pregnancy, or earthquake.

Why should people be so deluded? Positive illusions make people happier, more confident, and mentally healthier, but that cannot be the explanation for why they exist, because it only begs the question of *why* our brains should be designed so that only unrealistic assessments make us happy and confident, as opposed to calibrating our contentment against reality. The most plausible explanation is that positive illusions are a bargaining tactic, a credible bluff. In recruiting an ally to support you in a risky venture, in bargaining for the best deal, or in intimidating an adversary into backing down, you stand to gain if you credibly exaggerate your strengths. Believing your own exaggeration is better than cynically lying about it, because the arms race between lying and lie detection has equipped your audience with the means of seeing through barefaced lies.[88] As long as your exaggerations are not laughable, your audience cannot afford to ignore your self-assessment altogether, because you have more information about yourself than anyone else does, and you have a built-in incentive not to distort your assessment *too* much or you would constantly blunder into disasters. It would be better for the species if no one exaggerated, but our brains were not selected for the benefit of the species, and no individual can afford to be the only honest one in a community of self-enhancers.[89]

Overconfidence makes the tragedy of predation even worse. If people were completely rational, they would launch an act of predatory aggression only if they were likely to succeed and only if the spoils of the success exceeded the losses they would incur in the fighting. By the same token, the weaker party should concede as soon as the outcome was a foregone conclusion. A world with rational actors might see plenty of exploitation, but it should not see many fights or wars. Violence would come about only if the two parties were so closely matched that a fight was the only way to determine who was stronger.

But in a world with positive illusions, an aggressor may be emboldened to attack, and a defender emboldened to resist, well out of proportion to their odds of success. As Winston Churchill noted, "Always remember, however sure you are that you can easily win, that there would not be a war if the other man did not think he also had a chance."[90] The result can be wars of attrition (in both the game-theoretic and military sense), which, as we saw in chapter 5, are among the most destructive events in history, plumping out the tail of high-magnitude wars in the power-law distribution of deadly quarrels.

Military historians have long noted that leaders make decisions in war that are reckless to the point of delusion.[91] The invasions of Russia by Napoleon and, more than a century later, by Hitler are infamous examples. Over the past five centuries, countries that initiated wars have ended up losing them between a quarter and a half of the time, and when they won the victories were often Pyrrhic.[92] Richard Wrangham, inspired by Barbara Tuchman's *The March of Folly: From Troy to Vietnam* and by Robert Trivers's theory of self-deception, suggested that military incompetence is often a matter not of insufficient data or mistakes in strategy but of overconfidence.[93] Leaders overestimate their prospects of winning. Their bravado may rally the troops and intimidate weaker adversaries, but also may put them on a collision course with an enemy who is not as weak as they think and who may be under the spell of an overconfidence of its own.

The political scientist Dominic Johnson, working with Wrangham and others, conducted an experiment to test the idea that mutual overconfidence could lead to war.[94] They ran a moderately complicated war game in which pairs of participants pretended to be national leaders who had opportunities to negotiate with, threaten, or mount a costly attack on each other in competition for diamonds in a disputed border region. The winner of the contest was the player who had more money, if their nation survived at all, at the end of several rounds of play. The players interacted with each other by computer and could not see each other, so the men didn't know whether they were playing with another man or with a woman, and vice versa. Before they began, participants were asked to predict how well they would do relative to everyone else playing the game. The experimenters got a nice Lake Wobegon Effect: a majority thought they would do better than average. Now, in any Lake Wobegon Effect, it's possible that not many people really *are* self-deceived. Suppose 70 percent of people say they are better than average. Since half of any population really is above average, perhaps only 20 percent think too well of themselves. That was not the case in the war game. The more confident a player was, the *worse* he or she did. Confident players launched more unprovoked attacks, especially when playing each other, which triggered mutually destructive retaliation in subsequent rounds. It will come as no surprise to women that the overconfident and mutually destructive pairs of players were almost exclusively men.

To evaluate the overconfidence theory in the real world, it's not enough to notice in hindsight that certain military leaders proved to be mistaken. It has to be shown that at the time of making a fateful decision, a leader had access to information that would have convinced a disinterested party that the venture would probably fail.

In *Overconfidence and War: The Havoc and Glory of Positive Illusions*, Johnson vindicated Wrangham's hypothesis by looking at the predictions made by leaders on the verge of war and showing that they were unrealistically optimistic and contradicted by information available to them at the time. In the weeks preceding World War I, for example, the leaders of England, France, and Russia on one side and of Germany, Austria-Hungary, and the Ottoman

Empire on the other all predicted that the war would be a rout and their victorious troops would be home by Christmas. Ecstatic crowds of young men on both sides poured out of their homes to enlist, not because they were altruists eager to die for their country but because they didn't think they were going to die. They couldn't all be right, and they weren't. In Vietnam, three American administrations escalated the war despite ample intelligence telling them that victory at an acceptable cost was unlikely.

Destructive wars of attrition, Johnson points out, needn't require that both sides be certain or even highly confident of prevailing. All it takes is that the subjective probabilities of the adversaries sum to a value greater than one. In modern conflicts, he notes, where the fog of war is particularly thick and the leadership removed from the facts on the ground, overconfidence can survive longer than it would have in the small-scale battles in which our positive illusions evolved. Another modern danger is that the leadership of nations is likely to go to men who are at the right tail of the distribution of confidence, well into the region of overconfidence.

Johnson expected that wars stoked by overconfidence should be less common in democracies, where the flow of information is more likely to expose the illusions of leaders to cold splashes of reality. But he found that it was the flow of information itself, rather than just the existence of a democratic system, that made the difference. Johnson published his book in 2004, and the choice of an image for the cover was a no-brainer: the famous 2003 photograph of a flight-suited George W. Bush on the deck of an aircraft carrier festooned with the banner "Mission Accomplished." Overconfidence did not undermine the conduct of the Iraq War itself (other than for Saddam Hussein, of course), but it was fatal to the postwar goal of bringing stable democracy to Iraq, which the Bush administration catastrophically failed to plan for. The political scientist Karen Alter conducted an analysis *before* the war broke out showing that the Bush administration was unusually closed in its decision-making process.[95] In a textbook illustration of the phenomenon of groupthink, the prewar policy team believed in its own infallibility and virtue, shut out contradictory assessments, enforced consensus, and self-censored private doubts.[96]

Just before the Iraq War, Defense Secretary Donald Rumsfeld observed,

> There are known knowns; there are things we know we know. We also know there are known unknowns; that is to say we know there are some things we do not know. But there are also unknown unknowns—the ones we don't know we don't know.

Johnson, following a remark by the philosopher Slavoj Žižek, notes that Rumsfeld omitted a crucial fourth category, the unknown knowns—things that are known, or at least could be known, but are ignored or suppressed. It was the unknown knowns that allowed a moderate amount of instrumental violence (a few weeks of shock and awe) to unleash an open-ended exchange of every other kind of violence.

DOMINANCE

The colorful idioms *chest-thumping, having a chip on his shoulder, drawing a line in the sand, throwing down the gauntlet,* and *pissing contest* all denote an action that is inherently meaningless but provokes a contest for dominance. That is a sign that we are dealing with a category that is very different from predatory, practical, or instrumental violence. Even though nothing tangible is at stake in contests for dominance, they are among the deadliest forms of human quarrel. At one end of the magnitude scale, we have seen that many wars in the Ages of Dynasties, Sovereignty, and Nationalism were fought over nebulous claims to national preeminence, including World War I. At the other end of the scale, the single largest motive for homicide is "altercations of relatively trivial origin; insult, curse, jostling, etc."

In their book on homicide, Martin Daly and Margo Wilson advise that "the participants in these 'trivial altercations' behave as if a great deal more is at issue than small change or access to a pool table, and their evaluations of what is at stake deserve our respectful consideration."[97] Contests of dominance are not as ridiculous as they seem. In any zone of anarchy, an agent can protect its interests only by cultivating a reputation for a willingness and an ability to defend itself against depredations. Though this mettle can be demonstrated in retaliation after the fact, it's better to flaunt it proactively, before any damage is done. To prove that one's implicit threats are not hot air, it may be necessary to seek theaters in which one's resolve and retaliatory capacity can be displayed: a way to broadcast the message "Don't fuck with me." Everyone has an interest in knowing the relative fighting abilities of the agents in their midst, because all parties have an interest in preempting any fight whose outcome is a foregone conclusion and that would needlessly bloody both fighters if carried out.[98] When the relative prowesses of the members of a community are stable and widely known, we call it a dominance hierarchy. Dominance hierarchies are based on more than brute strength. Since not even the baddest primate can win a fight of one against three, dominance depends on the ability to recruit allies—who, in turn, don't choose their teammates at random but join up with the stronger and shrewder ones.[99]

The commodity that is immediately at stake in contests of dominance is information, and that feature differentiates dominance from predation in several ways. One is that while contests of dominance can escalate into lethal clashes, especially when the contestants are closely matched and intoxicated with positive illusions, most of the time (in humans and animals alike) they are settled with displays. The antagonists flaunt their strength, brandish their weapons, and play games of brinkmanship; the contest ends when one side backs down.[100] With predation, in contrast, the only point is to obtain an object of desire.

Another implication of the informational stakes in contests of dominance is that the violence is interwoven with exchanges of data. Reputation is a social construction that is built on what logicians call common knowledge. To avert a fight, a pair of rivals must not only know who is stronger, but each must know that the other knows, and must know that the other knows that he knows, and so on.[101] Common knowledge may be undermined by a contrary opinion, and so contests of dominance are fought in arenas of public information. They may be sparked by an insult, particularly in cultures of honor, and in those that sanction formal dueling. The insult is treated like a physical injury or theft, and it sets off an urge for violent revenge (which can make the psychology of dominance blend into the psychology of revenge, discussed in the next section). Studies of American street violence have found that the young men who endorse a code of honor are the ones most likely to commit an act of serious violence in the following year.[102] They also have found that the presence of an audience doubles the likelihood that an argument between two men will escalate to violence.[103]

When dominance is reckoned within a closed group, it is a zero-sum game: if someone's rank goes up, another's has to go down. Dominance tends to erupt in violence within small groups like gangs and isolated workplaces, where a person's rank within the clique determines the entirety of his social worth. If people belong to many groups and can switch in and out of them, they are more likely to find one in which they are esteemed, and an insult or slight is less consequential.[104]

Since the only commodity at stake in contests of dominance is information, once the point has been made about who's the boss, the violence can come to an end without setting off rounds of vendetta. The primatologist Frans de Waal discovered that in most primate species, after two animals have fought, they will reconcile.[105] They may touch hands, kiss, embrace, and in the case of bonobos, have sex. This makes one wonder why they bother to fight in the first place if they were just going to make up afterward, and why they make up afterward if they had reason to fight. The reason is that reconciliation occurs only among primates whose long-term interests are bound together. The ties that bind may be genetic relatedness, collective defense against predators, being in cahoots against a third party, or, in an experiment, getting fed only if they work together.[106] The overlap of interests is not perfect, so they still have reason to fight for dominance or retaliation within the group, but it is not zero, so they cannot afford to smack each other around indefinitely, let alone kill each other. Among primates whose interests are not bound up in any of these ways, adversaries are unforgiving, and violence is likelier to escalate. Chimpanzees, for example, reconcile after a fight within their community, but they never reconcile after a battle or raid with members of a different community.[107]

As we shall see in the next chapter, reconciliation among humans is also governed by the perception of common interests.

The metaphor of competitive distance urination suggests that the gender with the equipment best suited to competing is the gender that is most likely to participate in contests of dominance. Though in many primate species, including humans, both sexes jockey for preeminence, usually against members of their own sex, it seems to loom larger in the minds of men than in the minds of women, taking on a mystical status as a priceless commodity worth almost any sacrifice. Surveys of personal values in men and women find that the men assign a lopsided value to professional status compared to all the other pleasures of life.[108] They take greater risks, and they show more confidence and more overconfidence.[109] Most labor economists consider these sex differences to be a contributor to the gender gap in earnings and professional success.[110]

And men are, of course, by far the more violent sex. Though the exact ratios vary, in every society it is the males more than the females who play-fight, bully, fight for real, carry weapons, enjoy violent entertainment, fantasize about killing, kill for real, rape, start wars, and fight in wars.[111] Not only is the direction of the sex difference universal, but the first domino is almost certainly biological. The difference is found in most other primates, emerges in toddlerhood, and may be seen in boys who (because of anomalous genitalia) are secretly raised as girls.[112]

We have already seen why the sex difference evolved: mammalian males can reproduce more quickly than females, so they compete for sexual opportunities, while females tilt their priorities toward ensuring the survival of themselves and their offspring. Men have more to gain in violent competition, and also less to lose, because fatherless children are more likely to survive than motherless ones. That does not mean that women avoid violence altogether—Chuck Berry speculated that Venus de Milo lost both her arms in a wrestling match over a brown-eyed handsome man—but they find it less appealing. Women's competitive tactics consist in less physically perilous relational aggression such as gossip and ostracism.[113]

In theory, violent competition for mates and violent competition for dominance needn't go together. One doesn't have to invoke dominance to explain why Genghis Khan inseminated so many women that his Y chromosome is common in Central Asia today; it's enough to observe that he killed the women's fathers and husbands. But given that social primates regulate violence by deferring to dominant individuals, dominance and mating success in practice went hand in hand during most of our species' history. In nonstate societies, dominant men have more wives, more girlfriends, and more affairs with other men's wives.[114] In the six earliest empires, the correlation between status and mating success can be quantified precisely. Laura Betzig found that emperors often had thousands of wives and concubines, princes had hundreds, noblemen had dozens, upper-class men had up to a dozen, and middle-class men had three or four.[115] (It follows mathematically that many lower-class men had none—and thus a strong incentive to fight their way out of the lower class.) Recently, with the advent of reliable contraception and the demographic transition, the correlation has been weakened. But wealth, power, and professional success still increase a man's sex appeal, and the most visible clue to physical dominance—height—still gives a man an edge in economic, political, and romantic competition.[116]

Whereas instrumental violence deploys the seeking and calculating parts of the brain, dominance deploys the system that Panksepp calls Intermale Aggression. It really should be called Intrasexual Competition, because it is found in women too, and the human habit of male parental investment means that women as well as men have an evolutionary incentive to compete for mates. Still, at least one part of the circuit, a nucleus in the anterior preoptic portion of the hypothalamus, is twice as large in men as it is in women.[117] And the entire system is studded with receptors for testosterone, which is about five to ten times more plentiful in the bloodstream of men than of women. The hypothalamus, recall, controls the pituitary gland, which can secrete a hormone that tells the testes or the adrenal glands to produce more testosterone.

Though testosterone is often identified in the popular imagination as the cause of male pugnacity—"the substance that drives men to behave with quintessential guyness, to posture, push, yelp, belch, punch and play air-guitar," as the journalist Natalie Angier put it—biologists have been nervous about blaming it for male aggression itself.[118] Raising testosterone undoubtedly makes most birds and mammals more obstreperous, and lowering it makes them less so, as the owner of any neutered dog or cat is aware. But in humans the effects are less easily measured, for a number of boring biochemical reasons, and they are less directly tied to aggression, for an interesting psychological reason.

Testosterone, according to scientists' best guess, does not make men more aggressive across the board, but prepares them for a challenge of dominance.[119] In chimpanzees, testosterone goes up in the presence of a sexually receptive female, and it is correlated with the male's dominance rank, which in turn is correlated with his aggressiveness. In men, testosterone levels rise in the presence of an attractive female and in anticipation of competition with other men, such as in sports. Once a match has begun, testosterone rises even more, and when the match has been decided, testosterone continues to rise in the winner but not in the loser. Men who have higher levels of testosterone play more aggressively, have angrier faces during competition, smile less often, and have firmer handshakes. In experiments they are more likely to lock their gaze onto an angry face, and to perceive a neutral face as angry. It's not just fun and games that pump up the hormone: recall that the southern men who were insulted in Richard Nisbett's experiment on the psychology of honor responded with a rise in testosterone, and that they looked angrier, shook hands more firmly, and walked out of the lab with more of a swagger. At the tail end of the belligerence spectrum, prisoners with higher levels of testosterone have been found to commit more acts of violence.

Testosterone rises in adolescence and young adulthood, and declines in middle age. It also declines when men get married, have children, and spend time with their children. The hormone, then, is an internal regulator of the fundamental tradeoff between parenting effort and mating effort, where mating effort consists both in wooing the opposite sex and in fending off rivals of the same sex.[120] Testosterone may be the knob that turns men into dads or cads.

The rise and fall of testosterone over the life span correlates, more or less, with the rise and fall of male pugnacity. Incidentally, the first law of violence—it's something that young men do—is easier to document than to explain. Though it's clear why men should have evolved to be more violent than women, it's not so clear why young men should be more violent than old men. After all, young men have more years ahead of them, so when they take up a violent challenge, they are gambling with a greater proportion of their unlived lives. On mathematical grounds one might expect the opposite: that as men's days are numbered, they can afford to become increasingly reckless, and a really old man might go on one last spree of rape and murder until a SWAT team cuts him down.[121] One reason this does not happen is that men always have the option of investing in their children, grandchildren, nieces, and nephews, so older men, who are physically weaker but socially and economically stronger, have more to gain in providing for and protecting their families than in siring more offspring.[122] The other is that dominance in humans is a matter of reputation, which can be self-sustaining with a long payout. Everyone loves a winner, and nothing succeeds like success. So it is in the earliest rounds of competition that the reputational stakes are highest.

Testosterone, then, prepares men (and to a lesser extent women) for contests of dominance. It doesn't cause violence directly, because many kinds of violence have nothing to do with dominance, and because many contests of dominance are settled by displays and brinkmanship rather than violence itself. But to the extent that the problem of violence is a problem of young, unmarried, lawless men competing for dominance, whether directly or on behalf of a leader, then violence really is a problem of there being too much testosterone in the world.

The socially constructed nature of dominance can help explain which individuals are most likely to take risks to defend it. Perhaps the most extraordinary popular delusion about violence of the past quarter-century is that it is caused by low self-esteem. That theory has been endorsed by dozens of prominent experts, has inspired school programs designed to get kids to feel better about themselves, and in the late 1980s led the California legislature to form a Task Force to Promote Self-Esteem. Yet Baumeister has shown that the theory could not be more spectacularly, hilariously, achingly wrong. Violence is a problem not of too little self-esteem but of too much, particularly when it is unearned.[123] Self-esteem can be measured, and surveys show that it is the psychopaths, street toughs, bullies, abusive husbands, serial rapists, and hate-crime perpetrators who are off the scale. Diana Scully interviewed many rapists in their prison cells who bragged to her that they were "multitalented superachievers."[124] Psychopaths and other violent people are narcissistic: they think well of themselves not in proportion to their accomplishments but out of a congenital sense of entitlement. When reality intrudes, as it inevitably will, they treat the bad news as a personal affront, and its bearer, who is endangering their fragile reputation, as a malicious slanderer.

Violence-prone personality traits are even more consequential when they infect political rulers, because their hang-ups can affect hundreds of millions of people rather than just the unlucky few who live with them or cross their paths. Unimaginable amounts of suffering have been caused by tyrants who callously presided over the immiseration of their peoples or launched destructive wars of conquest. In chapters 5 and 6 we saw that the tail-thickening wars and dekamegamurders of the 20th century can be attributed in part to the personalities of just three men. Tin-pot tyrants like Saddam Hussein, Mobutu Sese Seko, Moammar Khaddafi, Robert Mugabe, Idi Amin, Jean-Bédel Bokassa, and Kim Jong-il have immiserated their people on a scale that is smaller but still tragic.

The study of the psychology of political leaders, to be sure, has a deservedly poor reputation. It's impossible to test the object of investigation directly, and all too tempting to pathologize people who are morally contemptible. Psychohistory also has a legacy of fanciful psychoanalytic conjectures about what made Hitler Hitler: he had a Jewish grandfather, he had only one testicle, he was a repressed homosexual, he was asexual, he was a sexual fetishist. As the journalist Ron Rosenbaum wrote in *Explaining Hitler*, "The search for Hitler has apprehended not one coherent, consensus image of Hitler but rather many different Hitlers, competing Hitlers, conflicting embodiments of competing visions. Hitlers who might not recognize each other well enough to say '*Heil*' if they came face to face in Hell."[125]

For all that, the more modest field of personality classification, which pigeonholes rather than explains people, has something to say about the psychology of modern tyrants. The *Diagnostic and Statistical Manual of Mental Disorders* (DSM) of the American Psychiatric Association defines narcissistic personality disorder as "a pervasive pattern of grandiosity, need for admiration, and a lack of empathy."[126] Like all psychiatric diagnoses, narcissism is a fuzzy category, and overlaps with psychopathy ("a pervasive pattern of disregard for, and violation of, the rights of others") and with borderline personality disorder ("instability in mood; black and white thinking; chaotic and unstable interpersonal relationships, self-image, identity, and behavior"). But the trio of symptoms at narcissism's core—grandiosity, need for admiration, and lack of empathy—fits tyrants to a T.[127] It is most obvious in their vainglorious monuments, hagiographic iconography, and obsequious mass rallies. And with armies and police forces at their disposal, narcissistic rulers leave their mark in more than statuary; they can authorize vast outlays of violence. As with garden-variety bullies and toughs, the unearned self-regard of tyrants is eternally vulnerable to being popped, so any opposition to their rule is treated not as a criticism but as a heinous crime. At the same time, their lack of empathy imposes no brake on the punishment they mete out to real or imagined opponents. Nor does its allow any consideration of the human costs of another of their DSM symptoms: their "fantasies of unlimited success, power, brilliance, beauty, or ideal love," which may be realized in rapacious conquest, pharaonic construction projects, or utopian master plans. And we have already seen what overconfidence can do in the waging of war.

All leaders, of course, must have a generous dose of confidence to have become leaders, and in this age of psychology, pundits often diagnose leaders they don't like with narcissistic personality disorder. But it's important not to trivialize the distinction between a politician with good teeth and the psychopaths who run their countries into the ground and take large parts of the world with them. Among the pacifying features of democracies is that their leadership-selection procedure penalizes an utter lack of empathy, and their checks and balances limit the damage that a grandiose leader can do. Even within autocracies, the personality of a leader—a Gorbachev as opposed to a Stalin—can have an enormous impact on the statistics of violence.

The damage done by the drive for dominance can be multiplied in a second way. This multiplier depends on a feature of the social mind that can be introduced with a benign anecdote. Every December my heart is warmed by a local tradition: the province of Nova Scotia sends a towering spruce to the city of Boston for its Christmas tree in gratitude for the humanitarian aid that Boston organizations extended to Haligonians after the horrendous 1917 explosion of a munitions-laden ship in Halifax harbor. As a Canadian expatriate in New England, I get to feel good twice: once in gratitude for the generous help that had been extended to my fellow Canucks, once in appreciation for the thoughtful gift that is being returned to my Boston brethren. Yet this entire ritual is, when you think about it, rather odd. I was not a party to either act of generosity, and so I neither earned nor should express gratitude. The people who find, fell, and send the tree never met the original victims and helpers; nor did the people who erect and decorate it. For all I know, not a single person touched by the tragedy is alive today. Yet we all feel the emotions that would be appropriate to a transaction of sympathy and gratitude between a pair of individual people. Everyone's mind contains a representation called "Nova Scotia" and a representation called "Boston" which are granted the full suite of moral emotions and valuations, and individual men and women act out their roles in the social behavior that follows from them.

A part of an individual's personal identity is melded with the identity of the groups that he or she affiliates with.[128] Each group occupies a slot in their minds that is very much like the slot occupied by an individual person, complete with beliefs, desires, and praiseworthy or blameworthy traits. This social identity appears to be an adaptation to the reality of groups in the welfare of individuals. Our fitness depends not just on our own fortunes but on the fortunes of the bands, villages, and tribes we find ourselves in, which are bound together by real or fictive kinship, networks of reciprocity, and a commitment to public goods, including group defense. Within the group, some people help to police the provision of public goods by punishing any parasite who doesn't contribute a fair share, and they are rewarded by the group's esteem. These and other contributions to the group's welfare are psychologically implemented by a partial loss of boundaries between the group and the self. On behalf of our group, we can feel sympathetic, grateful, angry, guilty, trustful, or mistrustful with regard to some other group, and we spread these emotions over the members of that group regardless of what they have done as individuals to deserve them.

Loyalty to groups in competition, such as sports teams or political parties, encourages us to play out our instinct for dominance vicariously. Jerry Seinfeld once remarked that today's athletes churn through the rosters of sports teams so rapidly that a fan can no longer support a group of players. He is reduced to rooting for their team logo and uniforms: "You are standing and cheering and yelling for your clothes to beat the clothes from another city." But stand and cheer we do: the mood of a sports fan rises and falls with the fortunes of his team.[129] The loss of boundaries can literally be assayed in the biochemistry lab. Men's testosterone level rises when their team defeats a rival in a game, just as it rises when they personally defeat a rival in a wrestling match or in singles tennis.[130] It also rises or falls when a favored political candidate wins or loses an election.[131]

The dark side of our communal feelings is a desire for our own group to dominate another group, no matter how we feel about its members as individuals. In a set of famous experiments, the psychologist Henri Tajfel told participants that they belonged to one of two groups defined by some trivial difference, such as whether they preferred the paintings of Paul Klee or Wassily Kandinsky.[132] He then gave them an opportunity to distribute money between a member of their group and a member of the other group; the members were identified only by number, and the participants themselves had nothing to gain or lose from their choice. Not only did they allocate more money to their instant groupmates, but they preferred to penalize a member of the other group (for example, seven cents for a fellow Klee fan, one cent for a Kandinsky fan) than to benefit both individuals at the expense of the experimenter (nineteen cents for a fellow Klee fan, twenty-five cents for a Kandinsky fan). A preference for one's group emerges early in life and seems to be something that must be unlearned, not learned. Developmental psychologists have shown that preschoolers profess racist attitudes that would appall their liberal parents, and that even babies prefer to interact with people of the same race and accent.[133]

The psychologists Jim Sidanius and Felicia Pratto have proposed that people, to varying degrees, harbor a motive they call social dominance, though a more intuitive term is tribalism: the desire that social groups be organized into a hierarchy, generally with one's own group dominant over the others.[134] A social dominance orientation, they show, inclines people to a sweeping array of opinions and values, including patriotism, racism, fate, karma, caste, national destiny, militarism, toughness on crime, and defensiveness of existing arrangements of authority and inequality. An orientation away from social dominance, in contrast, inclines people to humanism, socialism, feminism, universal rights, political progressivism, and the egalitarian and pacifist themes in the Christian Bible.

The theory of social dominance implies that race, the focus of so much discussion on prejudice, is psychologically unimportant. As Tajfel's experiments showed, people can divide the world into in-groups and out-groups based on any ascribed similarity, including tastes in expressionist painters. The psychologists Robert Kurzban, John Tooby, and Leda Cosmides point out that in human evolutionary history members of different races were separated by oceans, deserts, and mountain ranges (which is why racial differences evolved in the first place) and seldom met each other face to face. One's adversaries were villages, clans, and tribes of the same race. What looms large in people's minds is not race but *coalition;* it just so happens that nowadays many coalitions (neighborhoods, gangs, countries) coincide with races. Any invidious treatment that people display toward other races can be just as readily elicited by members of other coalitions.[135] Experiments by the psychologists G. Richard Tucker, Wallace Lambert, and later Katherine Kinzler have shown that one of the most vivid delineators of prejudice is speech: people distrust people who speak with an unfamiliar accent.[136] The effect goes back to the charming story of the origin of the word *shibboleth* in Judges 12:5–6:

> And the Gileadites took the passages of Jordan before the Ephraimites: and it was so, that when those Ephraimites which were escaped said, Let me go over; that the men of Gilead said unto him, Art thou an Ephraimite? If he said, Nay; Then said they unto him, Say now Shibboleth: and he said Sibboleth: for he could not frame to pronounce it right. Then they took him, and slew him at the passages of the Jordan: and there fell at that time of the Ephraimites forty and two thousand.

The phenomenon of *nationalism* can be understood as an interaction between psychology and history. It is the welding together of three things: the emotional impulse behind tribalism; a cognitive conception of the "group" as a people sharing a language, territory, and ancestry; and the political apparatus of government.

Nationalism, Einstein said, is "the measles of the human race." That isn't always true—sometimes it's just a head cold—but nationalism can get virulent when it is comorbid with the group equivalent of narcissism in the psychiatric sense, namely a big but fragile ego with an unearned claim to preeminence. Recall that narcissism can trigger violence when the narcissist is enraged by an insolent signal from reality. Combine narcissism with nationalism, and you get a deadly phenomenon that political scientists call *ressentiment* (French for *resentment*): the conviction that one's nation or civilization has a historical right to greatness despite its lowly status, which can only be explained by the malevolence of an internal or external foe.[137]

Ressentiment whips up the emotions of thwarted dominance—humiliation, envy, and rage—to which narcissists are prone. Historians such as Liah Greenfield and Daniel Chirot have attributed the major wars and genocides in the early decades of the 20th century to ressentiment in Germany and Russia. Both nations felt they were realizing their rightful claims to preeminence, which perfidious enemies had denied them.[138] It has not escaped the notice of observers of the contemporary scene that Russia and the Islamic world both nurse resentments about their undeserved lack of greatness, and that these emotions are nonnegligible threats to peace.[139]

Heading in the other direction are European countries like Holland, Sweden, and Denmark that stopped playing the preeminence game in the 18th century and pegged their self-esteem to more tangible if less heart-pounding achievements like making money and giving their citizens a pleasant lifestyle.[140] Together with countries that never cared about being magnificent in the first place, like Canada, Singapore, and New Zealand, their national pride, though considerable, is commensurate with their achievements, and in the arena of interstate relations they don't make trouble.

Group-level ambition also determines the fate of ethnic neighbors. Experts on ethnicity dismiss the conventional wisdom that ancient hatreds inevitably keep neighboring peoples at each other's throats.[141] After all, there are some six thousand languages spoken on the planet, at least six hundred of which have substantial numbers of speakers.[142] By any reckoning, the number of deadly ethnic conflicts that actually break out is a tiny fraction of the number that could break out. In 1996 James Fearon and David Laitin carried out one such reckoning. They focused on two parts of the world that each housed a combustible mixture of ethnic groups: the republics of the recently dissolved Soviet Union in the early 1990s, which had 45 of them, and newly decolonized Africa from 1960 to 1979, which had at least 160, probably many more. Fearon and Laitin counted the number of civil wars and incidents of intercommunal violence (such as deadly riots) as a proportion of the number of pairs of neighboring ethnic groups. They found that in the former Soviet Union, violence broke out in about 4.4 percent of the opportunities, and in Africa it broke out in fewer than 1 percent. Developed countries with mixtures of ethnic groups, such as New Zealand, Malaysia, Canada, Belgium, and recently the United States, have even better track records of ethnic nonviolence.[143] The groups may get on each other's nerves, but they don't kill each other. Nor should this be surprising. Even if ethnic groups are like people and constantly jockey for status, remember that most of the time people don't come to blows either.

Several things determine whether ethnic groups can coexist without bloodshed. As Fearon and Laitin point out, one important emollient is the way a group treats a loose cannon who attacks a member of the other group.[144] If the malefactor is reeled in and punished by his own community, the victimized group can classify the incident as a one-on-one crime rather than as the first strike in a group-against-group war. (Recall that one reason international peacekeepers are effective is that they can chasten spoilers on one side to the satisfaction of the other.) The political scientist Stephen van Evera suggests that an even bigger factor is ideology. Things get ugly when intermingled ethnic groups long for states of their own, hope to unite with their diasporas in other countries, keep long memories of harms committed by their neighbors' ancestors while being unrepentant for harms committed by their own, and live under crappy governments that mythologize one group's glorious history while excluding others from the social contract.

Many peaceable countries today are in the process of redefining the nation-state by purging it of tribalist psychology. The government no longer defines itself as a crystallization of the yearning of the soul of a particular ethnic group, but as a compact that embraces all the people and groups that happen to find themselves on a contiguous plot of land. The machinery of government is often rubegoldbergian, with complex arrangements of devolution and special status and power-sharing and affirmative action, and the contraption is held together by a few national symbols such as a rugby team.[145] People root for clothing instead of blood and soil. It is a messiness appropriate to the messiness of people's divided selves, with coexisting identities as individuals and as members of overlapping groups.[146]

Social dominance is a guy thing. It's not surprising that men, the more dominance-obsessed gender, have stronger tribalist feelings than women, including racism, militarism, and comfort with inequality.[147] But men are more likely to find themselves at the receiving end of racism too. Contrary to the common assumption that racism and sexism are twin prejudices propping up a white male power structure, with African American women in double jeopardy, Sidanius and Pratto found that minority women are far *less* likely to be the target of racist treatment than minority men. Men's attitudes toward women may be paternalistic or exploitative, but they are not combative, as they tend to be with other men. Sidanius and Pratto explain the difference with reference to the evolution of these invidious attitudes. Sexism ultimately arises from the genetic incentive of men to control the behavior, especially the sexual behavior, of women. Tribalism arises from the incentive of groups of men to compete with other groups for access to resources and mates.

The gender gaps in overconfidence, personal violence, and group-against-group hostility raise a frequently asked question: Would the world be more peaceful if women were in charge? The question is just as interesting if the tense and mood are changed. Has the world become more peaceful because women are more in charge? And will the world become more peaceful when women are even more in charge?

The answer to all three, I think, is a qualified yes. Qualified, because the link between sex and violence is more complicated than just "men are from Mars." In *War and Gender* the political scientist Joshua Goldstein reviewed the intersection of those two categories and discovered that throughout history and in every society men have overwhelmingly made up and commanded the armies.[148] (The archetype of the Amazons and other women warriors owes more to men being turned on by the image of strapping young women in battle gear, like Lara Croft and Xena, than to historical reality.) Even in the feminist 21st century, 97 percent of the world's soldiers, and 99.9 percent of the world's *combat* soldiers, are male. (In Israel, which famously drafts both sexes, women warriors spend most of their time in clinics or behind desks.) Men can also boast about occupying all the top slots in history's list of conquering maniacs, bloodthirsty tyrants, and genocidal thugs.

But women have not been conscientious objectors through all of this bloodshed. On various occasions they have led armed forces or served in combat, and they frequently egg their men into battle or provide logistical support, whether as camp followers in earlier centuries or industrial riveters in the 20th. Many queens and empresses, including Isabella of Spain, Mary and Elizabeth I of England, and Catherine the Great of Russia, acquitted themselves well in internal oppression and external conquest, and several 20th-century heads of state, such as Margaret Thatcher, Golda Meir, Indira Gandhi, and Chandrika Kumaratunga, led their nations in war.[149]

The discrepancy between what women are capable of doing in war and what they typically do is no paradox. In traditional societies women had to worry about abduction, rape, and infanticide by the enemy, so it's not surprising that they should want their men to be on the winning side of a war. In societies with standing armies, differences between the sexes (including upper body strength, the willingness to plunder and kill, and the ability to bear and raise children), combined with the nuisance of mixed-sex armies (such as romantic intrigue between the sexes and dominance contests within them) have always militated toward a division of labor by sex, with the men providing the cannon fodder. As for leadership, women in any era who find themselves in positions of power will obviously carry out their job responsibilities, which in many eras have included the waging of war. A queen in an age of competing dynasties and empires could hardly have afforded to be the world's only pacifist even if she were so inclined. And of course the two sexes' traits overlap considerably, even in those for which the averages might differ, so with any trait relevant to military leadership or combat, many women will be more capable than most men.

But over the long sweep of history, women have been, and will be, a pacifying force. Traditional war is a man's game: tribal women never band together and raid neighboring villages to abduct grooms.[150] This sex difference set the stage for Aristophanes' *Lysistrata,* in which the women of Greece go on a sex strike to pressure their men to end the Peloponnesian War. In the 19th century, feminism often overlapped with pacifism and other antiviolence movements such as abolitionism and animal rights.[151] In the 20th, women's groups have been active, and intermittently effective, in protesting nuclear tests, the Vietnam War, and violent strife in Argentina, Northern Ireland, and the former Soviet Union and Yugoslavia. In a review of almost three hundred American public opinion polls between the 1930s and 1980s, men were found to support the "more violent or forceful option" in 87 percent of the questions, the others being tied.[152] For example, they were more supportive of military confrontation with Germany in 1939, Japan in 1940, Russia in 1960, and Vietnam in 1968. In every American presidential election since 1980, women have cast more votes for the Democratic candidate than men have, and in 2000 and 2004 majorities of women reversed the preference of men and voted against George W. Bush.[153]

Though women are slightly more peace-loving than their menfolk, the men and women of a given society have correlated opinions.[154] In 1961 Americans were asked whether the country should "fight an all-out nuclear war rather than live under communist rule." Eighty-seven percent of the men said yes, while "only" 75 percent of the women felt that way—proof that women are pacifist only in comparison to men of the same time and society. Gender gaps are larger when an issue divides the country (as in the Vietnam War), smaller when there is greater agreement (as in World War II), and nonexistent when the issue obsesses the entire society (as in the attitudes among Israelis and Arabs toward a resolution of the Arab-Israeli conflict).

But women's position in society can affect its fondness for war even if the women themselves are not opposing war. A recognition of women's rights and an opposition to war go together. In Middle Eastern countries, the poll respondents who were more favorable to gender equality were also more favorable to nonviolent solutions to the Arab-Israeli conflict.[155] Several ethnographic surveys of traditional cultures have found that the better a society treats its women, the less it embraces war.[156] The same is true for modern countries, with the usual continuum running from Western Europe to blue American states to red American states to Islamic countries such as Afghanistan and Pakistan.[157] As we shall see in chapter 10, societies that empower their women are less likely to end up with large cohorts of rootless young men, with their penchant for making trouble.[158] And of course the decades of the Long Peace and the New Peace have been the decades of the revolution in women's rights. We don't know what causes what, but biology and history suggest that all else being equal, a world in which women have more influence will be a world with fewer wars.

Dominance is an adaptation to anarchy, and it serves no purpose in a society that has undergone a civilizing process or in an international system regulated by agreements and norms. Anything that deflates the concept of dominance is likely to drive down the frequency of fights between individuals and wars between groups. That doesn't mean that the emotions behind dominance will go away—they are very much a part of our biology, especially in a certain gender—but they can be marginalized.

The mid- and late 20th century saw a deconstruction of the concept of dominance and related virtues like manliness, honor, prestige, and glory. Part of the deflation came from the informalization process, as in the Marx Brothers' burlesque of jingoism in *Duck Soup*. Partly it has come from women's inroads into professional life. Women have the psychological distance to see contests of dominance as boys making noise, so as they have become more influential, dominance has lost some of its aura. (Anyone who has worked in a mixed-sex environment is familiar with a woman belittling the wasteful posturing of her male colleagues as "typical male behavior.") Partly it has come from cosmopolitanism, which exposes us to exaggerated cultures of honor in other countries and thereby gives us a perspective on our own. The word *macho*, recently borrowed from Spanish, has a disdainful air, connoting self-indulgent swagger rather than manly heroism. The Village People's campy "Macho Man" and other homoerotic iconography has further undermined the trappings of masculine dominance.

Another deflationary force, I think, is the progress of biological science and its influence on literate culture. People have increasingly understood the drive for dominance as a vestige of the evolutionary process. A quantitative analysis of Google Books shows recent leaps in the popularity of the biological jargon behind dominance, including *testosterone* beginning in the 1940s, *pecking order* and *dominance hierarchy* beginning in the 1960s, and *alpha male* in the 1990s.[159] Joining them in the 1980s was the facetious pseudo-medical term *testosterone poisoning*. Each of these phrases belittles the stakes in contests for dominance. They imply that the glory men seek may be a figment of their primate imaginations—the symptom of a chemical in their bloodstream, the acting out of instincts that make us laugh when we see them in roosters and baboons. Compare the distancing power of these biological terms to older words like *glorious* and *honorable,* which objectify the prize in a contest of dominance, presupposing that certain accomplishments just *are* glorious or honorable in the very nature of things. The frequency of both terms has been steadily falling in English-language books for a century and a half.[160] An ability to hold our instincts up to the light, rather than naïvely accepting their products in our consciousness as just the way things are, is the first step in discounting them when they lead to harmful ends.

REVENGE

The determination to hurt someone who has hurt you has long been exalted in purple prose. The Hebrew Bible is obsessed with revenge, giving us pithy expressions like "Whoso sheddeth blood will have his blood shed," "An eye for an eye," and "Vengeance is mine." Homer's Achilles describes it as sweeter than flowing honey welling up like smoke from the breasts of men. Shylock cites it as the climax in his listing of human universals, and when asked what he will do with his pound of flesh, replies, "To bait fish withal: if it will feed nothing else, it will feed my revenge."

People in other cultures also wax poetic about the settling of scores. Milovan Djilas, born into a feuding clan of Montenegrins and later a vice president of communist Yugoslavia, called vengeance "the glow in our eyes, the flame in our cheeks, the pounding in our temples, the word that turned to stone in our throats on our hearing that our blood had been shed."[161] A New Guinean man, upon hearing that the killer of his uncle had been paralyzed by an arrow, said, "I feel as if I am developing wings, I feel as if I am about to fly off, and I am very happy."[162] The Apache chief Geronimo, savoring his massacre of four Mexican army companies, wrote:

> Still covered with the blood of my enemies, still holding my conquering weapon, still hot with the joy of battle, victory, and vengeance, I was surrounded by the Apache braves and made war chief of all the Apaches. Then I gave the orders for scalping the slain.
>
> I could not call back my loved ones, I could not bring back the dead Apaches, but I could rejoice in this revenge.

Daly and Wilson comment: "Rejoice? Geronimo wrote these words in a prison cell, his Apache nation broken and nearly extinct. The urge for vengeance seems so futile: There's no use crying over spilt milk, and spilt blood is equally irrevocable."[163]

Yet for all its futility, the urge for vengeance is a major cause of violence. Blood revenge is explicitly endorsed in 95 percent of the world's cultures, and wherever tribal warfare is found, it is one of the major motives.[164] Revenge is the motive of 10 to 20 percent of homicides worldwide and a large percentage of school shootings and private bombings.[165] When directed against groups rather than individuals, it is a major motive of urban riots, terrorist attacks, retaliation against terrorist attacks, and wars.[166] Historians who examine the decisions that led to war in reprisal for an attack note that it is often befogged in a red mist of anger.[167] After Pearl Harbor, for example, the American people were said to react "with a mind-staggering mixture of surprise, awe, mystification, grief, humiliation, and above all, cataclysmic fury."[168] No alternative to war (such as containment or harassment) was ever considered; the very thought would have been tantamount to treason. The reactions to the 9/11 attack were similar: the U.S. invasion of Afghanistan the following month was motivated as much by a sense that something had to be done in reprisal as by a strategic decision that it was the most effective long-term measure against terrorism.[169] The three thousand killings on September 11 had themselves been motivated by revenge, as Osama bin Laden explained in his "Letter to America":

> Allah, the almighty, legislated the permission and the option to take revenge. Thus, if we are attacked, then we have the right to attack back. Whoever has destroyed our villages and towns, then we have the right to destroy their villages and towns. Whoever has stolen our wealth, then we have the right to destroy their economy. And whoever has killed our civilians, then we have the right to kill theirs.[170]

Revenge is not confined to political and tribal hotheads but is an easily pushed button in everyone's brains. The homicidal fantasies to which a large majority of university students confess are almost entirely *revenge* fantasies.[171] And in laboratory studies, students can easily be induced to avenge a humiliation. The students write an essay and are provided with an insulting evaluation written by a fellow student (who is a confederate of the experimenters or entirely fictitious). At that point, Allah smiles: the student is asked to participate in a study that just happens to give him the opportunity to punish his critic by shocking him, blasting him with an air horn, or (in more recent experiments, vetted by violence-averse human subjects committees) forcing him to drink hot sauce in a bogus experiment on taste. It works like a charm.[172]

Revenge is, quite literally, an urge. In one of these experiments, just as a participant was about to deliver the retaliatory shock, the apparatus broke down (thanks to a bit of subterfuge by the experimenter), so he or she was unable to consummate the revenge. All participants then took part in a bogus wine-tasting study. The ones who had never gotten the chance to shock their insulters sampled a lot more of the wine, as if to drown their sorrows.[173]

The neurobiology of revenge begins with the Rage circuit in the midbrain-hypothalamus-amygdala pathway, which inclines an animal who has been hurt or frustrated to lash out at the nearest likely perpetrator.[174] In humans the system is fed by information originating from anywhere in the brain, including the temporoparietal junction, which indicates whether the harm was intended or accidental. The Rage circuit then activates the insular cortex, which gives rise to sensations of pain, disgust, and anger. (Recall that the insula lights up when people feel they have been shortchanged by another person.)[175] None of this is enjoyable, and we know that animals will work to turn off electrical stimulation of the Rage system.

But then the brain can slip into a different mode of information processing. Proverbs like "Revenge is sweet," "Don't get mad; get even," and "Revenge is a dish best served cold" are hypotheses in affective neuroscience. They predict that patterns of activity in the brain can shift from an aversive anger to a cool and pleasurable seeking, the kind that guides the pursuit of delectable food. And as so often happens, the folk neuroscience is correct. Dominique de Quervain and his collaborators gave a sample of men the opportunity to entrust a sum of money to another participant who would invest it for a profit, and then either share the total with the investor or keep it for himself.[176] (The scenario is sometimes called a Trust game.) Participants who had been cheated out of their money were then given the chance to levy a punitive fine on the faithless trustee, though sometimes they would have to pay for the privilege. As they were pondering the opportunity, their brains were scanned, and the scientists found that a part of the striatum (the core of the Seeking system) lit up—the same region that lights up when a person craves nicotine, cocaine, or chocolate. Revenge is sweet, indeed. The more a person's striatum lit up, the more he was willing to pay to punish the crooked trustee, which shows that the activation reflected a genuine desire, something that the person would pay to have consummated. When the participant did choose to pay, his orbital and ventromedial frontal cortex lit up—the part of the brain that weighs the pleasure and pain of different courses of action, in this case presumably the cost of the revenge and the satisfaction it afforded.

Revenge requires the disabling of empathy, and that too can be seen in the brain. Tania Singer and her collaborators ran a similar experiment in which men and women had their trust rewarded or betrayed by a fellow participant.[177] Then they either experienced a mild shock to their fingers, watched a trustworthy partner get shocked, or watched their double-crosser get shocked. When a trustworthy partner got shocked, the participants literally felt their pain: the same part of the insula that lit up when *they* were shocked lit up when they saw the nice guy (or gal) get shocked. When the double-crosser got shocked, the women could not turn off their empathy: their insula still lit up in sympathy. But the men hardened their hearts: their own insula stayed dark, while their striatum and orbital cortex lit up, a sign of a goal sought and consummated. Indeed, those circuits lit up in proportion to the men's stated desire for revenge. The results are in line with the claim by difference feminists such as Carol Gilligan that men are more inclined toward retributive justice and women more toward mercy.[178] The authors of the study, though, caution that the women may have recoiled from the physical nature of the punishment and might have been just as retributive if it had taken the form of a fine, criticism, or ostracism.[179]

There is no gainsaying the cool, sweet pleasure of revenge. A villain getting his comeuppance is a recurring archetype in fiction, and it's not just Dirty Harry Callahan whose day is made when a bad guy is brought to violent justice. One of my most enjoyable moments as a moviegoer was a scene in Peter Weir's award-winning *Witness*. Harrison Ford plays an undercover detective who is assigned to live with an Amish family in rural Pennsylvania. One day, in full Amish drag, he accompanies them to town in their horse-drawn buggy, and they are stopped and harassed by some rural punks. True to their pacifism, the family turns the other cheek, even as one of the punks taunts and bullies their dignified father. The straw-hatted Ford does a slow burn, turns toward the punk, and to the astonishment of the gang and the delight of the arthouse audience, coldcocks him a good one.

What is this madness called revenge? Though our psychotherapeutic culture portrays vengeance as a disease and forgiveness as the cure, the drive for revenge has a thoroughly intelligible function: deterrence.[180] As Daly and Wilson explain, "Effective deterrence is a matter of convincing our rivals that any attempt to advance their interests at our expense will lead to such severe penalties that the competitive gambit will end up a net loss which should never have been undertaken."[181] The necessity of vengeful punishment as a deterrent is not a just-so story but has been demonstrated repeatedly in mathematical and computer models of the evolution of cooperation.[182]

Some forms of cooperation are easy to explain: two people are related, or married, or are teammates or bosom buddies with the same interests, so what is good for one is good for the other, and a kind of symbiotic cooperation comes naturally. Harder to explain is cooperation when people's interests at least partly diverge, and each may be tempted to exploit the other's willingness to cooperate. The simplest way to model this quandary is a positive-sum game called the Prisoner's Dilemma. Imagine a *Law and Order* episode in which two partners in crime are held in separate jail cells and the evidence against them is marginal, so each is offered a deal by the assistant district attorney. If he testifies against his partner ("defects" against him) while the partner stays true ("cooperates"—with the partner, that is), he will go free while his partner is sent away for ten years. If each of them defects and testifies against the other, they will both go to jail, but their sentences will be reduced to six years. If each stays loyal to the other, the prosecutor can only convict them of a lesser crime, and they will be free in six months. Figure 8–5 shows the payoff matrix for their dilemma; the choices and payoffs for the first prisoner (Lefty) are printed in black; those for his partner (Brutus) are printed in gray.

		Brutus's choices	
		Cooperate	Defect
Lefty's choices	Cooperate	6 months (reward) 6 months (reward)	10 years (sucker's payoff) Go free (temptation)
	Defect	Go free (temptation) 10 years (sucker's payoff)	6 years (punishment) 6 years (punishment)

FIGURE 8–5. The Prisoner's Dilemma

Their tragedy is that both ought to cooperate and settle for the reward of a six-month sentence, which gives the game its positive sum. But each will defect, figuring that he's better off either way: if the partner cooperates, he goes free; if the partner defects, he only gets six years rather than the ten he would get if he had cooperated. So he defects, and his partner, following the same reasoning, also defects, and they end up serving six years rather than the six months that they could have served if they had only acted altruistically rather than selfishly.

The Prisoner's Dilemma has been called one of the great ideas of the 20th century, because it distills the tragedy of social life into such a succinct formula.[183] The dilemma arises in any situation in which the best individual payoff is to defect while the partner cooperates, the worst individual payoff is to cooperate while the other defects, the highest total payoff is when both cooperate, and the lowest total payoff is when both defect. Many of life's predicaments have this structure, not least predatory violence, where being the aggressor against a pacifist provides all the benefits of exploitation, but being an aggressor against a fellow aggressor bloodies the two of you, so you should both be pacifists, and you would be but for the fear that the other will be an aggressor. We have come across related tragedies, such as the War of Attrition, the Public Goods game, and the Trust game, wherein individual selfishness is tempting but mutual selfishness is ruinous.

Though a one-shot Prisoner's Dilemma is tragic, an *Iterated* Prisoner's Dilemma, in which players interact repeatedly and accumulate a payoff over many rounds, is truer to life. It can even be a good model of the evolution of cooperation, if the payoffs are doled out, not in years of jail time averted or in dollars and cents, but in number of descendants. Virtual organisms play rounds of the Prisoner's Dilemma, which may be interpreted as opportunities to help each other, say, via mutual grooming, or to refrain from helping; the benefits in health and costs in time translate into the number of surviving offspring. Repeated runs of the game are like generations of organisms evolving by natural selection, and an observer can ask which of several competing strategies will eventually take over the population with its descendants. The combinatorial possibilities are too numerous to allow for a mathematical proof, but strategies can be written into little computer apps that compete in round-robin tournaments, and the theorists can see how they fare in the virtual evolutionary struggle.

In the first such tournament, hosted by the political scientist Robert Axelrod, the winner was a simple strategy of Tit for Tat: cooperate on the first move, then continue to cooperate if your partner cooperates, but defect if he defects.[184] Since cooperation is rewarded and defection punished, defectors will switch to cooperation, and in the long run everyone wins. The idea is identical to Robert Trivers's theory of the evolution of reciprocal altruism, which he had proposed a few years earlier without the mathematical paraphernalia.[185] The positive-sum reward arises from gains in exchange (each can confer a large benefit to the other at a small cost to itself), and the temptation is to exploit the other, taking the benefit without paying the cost. Trivers's theory that the moral emotions are adaptations to cooperation can be translated directly into the Tit for Tat algorithm. Sympathy is cooperating on the first move. Gratitude is cooperating with a cooperator. And anger is defecting against a defector—in other words, punishing in revenge. The punishment can consist in refusing to help, but it can also consist in causing a harm. Vengeance is no disease: it is necessary for cooperation, preventing a nice guy from being exploited.

Hundreds of Iterated Prisoner's Dilemma tournaments have been studied since then, and a few new lessons have emerged.[186] One is that Tit for Tat, simple though it is, can be dissected into features that account for its success and can be recombined into other strategies. These features have been named for personality traits, and the labels may be more than mnemonics; the dynamics of cooperation may explain why we evolved those traits. The first feature behind Tit for Tat's success is that it is *nice:* it cooperates on the first move, thereby tapping opportunities for mutually beneficial cooperation, and it does not defect unless defected upon. The second is that it is *clear:* if a strategy's rules of engagement are so complicated that the other players cannot discern how it is reacting to what they do, then its moves are effectively arbitrary, and if they are arbitrary, the best response is the strategy Always Defect. Tit for Tat is easy for other strategies to cotton on to, and they can adjust their choices in response. Third, Tit for Tat is *retaliatory:* it responds to defection with defection, the simplest form of revenge. And it is *forgiving:* it leaves the gates of repentance open, so if its adversary switches to cooperation after a history of defection, Tit for Tat immediately cooperates in return.[187]

The last feature, forgivingness, turns out to be more important than everyone first appreciated. A weakness of Tit for Tat is that it is vulnerable to error and misunderstanding. Suppose one of the players intends to cooperate but defects by mistake. Or suppose it misperceives another player's cooperation as defection, and defects in retaliation. Then its opponent will defect in retaliation, which forces it to retaliate in turn, and so on, dooming the players to an endless cycle of defection—the software equivalent of a feud. In a noisy world in which misunderstanding and error are possible, Tit for Tat is bested by an even more forgiving strategy called Generous Tit for Tat. Every once in a while Generous Tit for Tat will randomly grant forgiveness to a defector and resume cooperating. The act of unconditional forgiveness can flick a duo that has been trapped in a cycle of mutual defection back onto the path of cooperation.

A problem for overly forgiving strategies, though, is that they can be undone if the population contains a few psychopaths who play Always Defect and a few suckers who play Always Cooperate. The psychopaths proliferate by preying on the suckers, and then become numerous enough to exploit everyone else. One successful contender in such a world is Contrite Tit for Tat, which is more discriminating in its forgiveness. It remembers its *own* behavior, and if a round of mutual defection had been its fault because of a random error or misunderstanding, it allows its opponent one free defection and then switches to cooperation. But if the defection had been triggered by its opponent, it shows no mercy and retaliates. If the opponent is a Contrite Tit for Tatter as well, then it will excuse the justified retaliation, and the pair will settle back into cooperation. So not just vengeance, but also forgiveness and contrition, are necessary for social organisms to reap the benefits of cooperation.

The evolution of cooperation critically depends on the possibility of repeated encounters. It cannot evolve in a one-shot Prisoner's Dilemma, and it collapses even in an Iterated Prisoner's Dilemma if the players know they are playing a limited number of rounds, because as the end of the game approaches, each is tempted to defect without fear of retribution. For similar reasons, subsets of players who are stuck with playing against each other—say because they are neighbors who cannot move—tend to be more forgiving than ones who can pick up and choose another neighborhood in which to find partners. Cliques, organizations, and other social networks are virtual neighborhoods because they force groups of people to interact repeatedly, and they too tilt people toward forgiveness, because mutual defection would be ruinous to everyone.

Human cooperation has another twist. Because we have language, we don't have to deal with people directly to learn whether they are cooperators or defectors. We can ask around, and find out through the grapevine how the person has behaved in the past. This indirect reciprocity, as game theorists call it, puts a tangible premium on reputation and gossip.[188]

Potential cooperators have to balance selfishness against mutual benefit not just in dealing with each other in pairs but when acting collectively in groups. Game theorists have explored a multiplayer version of the Prisoner's Dilemma called the Public Goods game.[189] Each player can contribute money toward a common pool, which then is doubled and divided evenly among the players. (One can imagine a group of fishermen chipping in for harbor improvements such as a lighthouse, or merchants in a block of stores pooling contributions for a security guard.) The best outcome for the group is for everyone to contribute the maximum amount. But the best outcome for an *individual* is to stint on his own contribution and be a free rider on the profits from everyone else's. The tragedy is that contributions will dwindle to zero and everyone ends up worse off. (The biologist Garrett Hardin proposed an identical scenario called the Tragedy of the Commons. Each farmer cannot resist grazing his own cow on the town commons, stripping it bare to everyone's loss. Pollution, overfishing, and carbon emissions are equivalent real-life examples.)[190] But if players have the opportunity to punish free riders, as if in revenge for their exploitation of the group, then the players have an incentive to contribute, and everyone profits.

The modeling of the evolution of cooperation has become increasingly byzantine, because so many worlds can be simulated so cheaply. But in the most plausible of these worlds, we see the evolution of the all-too-human phenomena of exploitation, revenge, forgiveness, contrition, reputation, gossip, cliques, and neighborliness.

So does revenge pay in the real world? Does the credible threat of punishment induce fear in the heart of potential exploiters and deter them from exploiting? The answer from the lab is yes.[191] When people actually act out Prisoner's Dilemma games in experiments, they tend toward Tit for Tat–like strategies and end up enjoying the fruits of cooperation. When they play the Trust game (another version of Prisoner's Dilemma, which was the game used in the neuroimaging experiments on revenge), the ability of an investor to punish a faithless trustee puts enough fear into the trustee to return a fair share of the appreciated investment. In Public Goods games, when people are given the opportunity to punish free riders, people don't free-ride. And remember the studies in which participants' essays were savaged and they had an opportunity to shock their critics in revenge? If they knew that the critic would then get a turn to shock them back—to take revenge for the revenge—they held back on the intensity of the shocks.[192]

Revenge can work as a deterrent only if the avenger has a *reputation* for being willing to avenge and a willingness to carry it out even when it is costly. That helps explain why the urge for revenge can be so implacable, consuming, and sometimes self-defeating (as with pursuers of self-help justice who slay an unfaithful spouse or an insulting stranger).[193] Moreover, it is most effective when the target knows that the punishment came from the avenger so he can recalibrate his behavior toward the avenger in the future.[194] That explains why an avenger's thirst is consummated only when the target knows he has been singled out for the punishment.[195] These impulses implement what judicial theorists call specific deterrence: a punishment is targeted at a wrongdoer to prevent him from repeating a crime.

The psychology of revenge also implements what judicial theorists call general deterrence: a publicly decreed punishment that is designed to scare third parties away from the temptations of crime. The psychological equivalent of general deterrence is the cultivation of a reputation for being the kind of person who cannot be pushed around. (You don't tug on Superman's cape; you don't spit into the wind; you don't pull the mask off the old Lone Ranger; and you don't mess around with Jim.) Experiments have shown that people punish more severely, even at a price that is greater than the amount out of which they have been cheated, when they think an audience is watching.[196] And as we saw, men are twice as likely to escalate an argument into a fight when spectators are around.[197]

The effectiveness of revenge as a deterrent can explain actions that are otherwise puzzling. The rational actor theory, popular in economics and political science, has long been embarrassed by people's behavior in yet another game, the Ultimatum game.[198] One participant, the proposer, gets a sum of money to divide between himself and another participant, the acceptor, who can take it or leave it. If he leaves it, neither side gets anything. A rational proposer would keep the lion's share; a rational respondent would accept the remaining crumbs, no matter how small, because part of a loaf is better than none. In actual experiments the proposer tends to offer almost half of the jackpot, and the respondent doesn't settle for much less than half, even though turning down a smaller share is an act of spite that punishes both participants. Why do actors in these experiments behave so irrationally? The rational actor theory had neglected the psychology of revenge. When a proposal is too stingy, the respondent gets angry—indeed, the neuroimaging study I mentioned earlier, in which the insula lit up in anger, used the Ultimatum game to elicit it.[199] The anger impels the respondent to punish the proposer in revenge. Most proposers anticipate the anger, so they make an offer that is just generous enough to be accepted. When they don't have to worry about revenge, because the rules of the game are changed and the acceptor has to accept the split no matter what (a variation called the Dictator game), the offer is stingier.

We still have a puzzle. If revenge evolved as a deterrent, then why is it used so often in the real world? Why doesn't revenge work like the nuclear arsenals in the Cold War, creating a balance of terror that keeps everyone in line? Why should there ever be cycles of vendetta, with vengeance begetting vengeance?

A major reason is the Moralization Gap. People consider the harms they inflict to be justified and forgettable, and the harms they suffer to be unprovoked and grievous. This bookkeeping makes the two sides in an escalating fight count the number of strikes differently and weigh the inflicted harm differently as well.[200] As the psychologist Daniel Gilbert has put it, the two combatants in a long-running war often sound like a pair of boys in the backseat of a car making their respective briefs to their parents: "He hit me first!" "He hit me harder!"[201]

A simple analogy to the way that misperception can lead to escalation may be found in an experiment by Sukhwinder Shergill, Paul Bays, Chris Frith, and Daniel Wolpert, in which participants placed their finger beneath a bar that could press down on it with a precise amount of force.[202] Their instruction was to press down on the finger of a second participant for three seconds with the same amount of force they were feeling. Then the second participant got the same instructions. The two took turns, each matching the amount of force he or she had just received. After eight turns the second participant was pressing down with about *eighteen times* as much force as was applied in the round that got it started. The reason for the spiral is that people underestimate how much force they apply compared to how much force they feel, so they escalated the pressure by about 40 percent with each turn. In real-world disputes the misperception comes not from an illusion of the sense of touch but from an illusion of the moral sense, but in both cases the result is a spiral of painful escalation.

In many parts of this book I have credited the Leviathan—a government with a monopoly on the legitimate use of force—as a major reducer of violence. Feuding and anarchy go together. We can now appreciate the psychology behind the effectiveness of a Leviathan. The law may be an ass, but it is a disinterested ass, and it can weigh harms without the self-serving distortions of the perpetrator or the victim. Though it is guaranteed that one side will disagree with every decision, the government's monopoly on force prevents the loser from doing anything about it, and it gives him less reason to *want* to do something about it, because he is not conceding weakness to his adversary and has less incentive to carry on the fight to restore his honor. The fashion accessories of Justitia, the Roman goddess of justice, express the logic succinctly: (1) scales; (2) blindfold; (3) sword.

A Leviathan that implements justice at the point of a sword is still using a sword. As we have seen, government vengeance itself can go to excess, as in the cruel punishments and profligate executions before the Humanitarian Revolution and the excessive incarceration in the United States today. Criminal punishment is often harsher than what would be needed as a finely tuned incentive designed to minimize the society's sum of harm. Part of this is by design. The rationale for criminal punishment is not just specific deterrence, general deterrence, and incapacitation. It also embraces just deserts, which is basically citizens' impulse for revenge.[203] Even if we were certain that the perpetrator of a heinous crime would never offend again, nor set an example for anyone else, most people would feel that "justice must be done" and that he should incur some harm to balance the harm he has caused. The psychological impulse behind just deserts is thoroughly intelligible. As Daly and Wilson observe:

> From the perspective of evolutionary psychology, this almost mystical and seemingly irreducible sort of moral imperative is the output of a mental mechanism with a straightforward adaptive function: to reckon justice and administer punishment by a calculus which ensures that violators reap no advantage from their misdeeds. The enormous volume of mystico-religious bafflegab about atonement and penance and divine justice and the like is the attribution to higher, detached authority of what is actually a mundane, pragmatic matter: discouraging self-interested competitive acts by reducing their profitability to nil.[204]

But since it *is* an irreducible imperative, whose evolutionary rationale is invisible to us when we are in the throes of it, the justice that people mete out in practice may be only loosely related to its incentive structure.

The psychologists Kevin Carlsmith, John Darley, and Paul Robinson devised hypothetical cases designed to tease apart deterrence from just deserts.[205] Just deserts is sensitive to the moral worth of the perpetrator's motive. For instance, an embezzler who used his ill-gotten gains to support a lavish lifestyle would seem to deserve a harsher punishment than one who redirected them to the company's underpaid workers in the developing world. Deterrence, in contrast, is sensitive to the incentive structure of the punishment regime. Assuming that malefactors reckon the utility of a misdeed as the probability they will get caught multiplied by the penalty they will incur if they do get caught, then a crime that is hard to detect should get a harsher punishment than one that is easy to detect. For similar reasons, a crime that gets a lot of publicity should be punished more harshly than one that is unpublicized, because the publicized one will leverage the value of the punishment as a general deterrent. When people are asked to mete out sentences to fictitious malefactors in these scenarios, their decisions are affected only by just deserts, not by deterrence. Evil motives draw harsher sentences, but difficult-to-detect or highly publicized infractions do not.

The reforms advocated by the utilitarian economist Cesare Beccaria during the Humanitarian Revolution, which led to the abolition of cruel punishments, were designed to reorient criminal justice away from the raw impulse to make a bad person suffer and toward the practical goal of deterrence. The Carlsmith experiment suggests that people today have not gone all the way into thinking of criminal justice in purely utilitarian terms. But in *The Blank Slate* I argued that even the elements of our judicial practices that seem to be motivated by just deserts may ultimately serve a deterrent function, because if a system ever became too narrowly utilitarian, malefactors would learn to game it. Just deserts can close off that option.[206]

Even the fairest system of criminal justice cannot monitor its citizens wherever they may be and around the clock. It has to count on them internalizing norms of fairness and damping their vengeance before it escalates. In chapter 3 we saw how the ranchers and farmers of Shasta County resolved their grievances without tattling to the police, thanks to reciprocity, gossip, occasional vandalism, and for minor harms, "lumping it."[207] Why do the people in some societies lump it while others experience a glow in their eyes, a flame in their cheeks, and a pounding in their temples? Norbert Elias's theory of the Civilizing Process suggests that government-administered justice can have knock-on effects that lead its citizens to internalize norms of self-restraint and quash their impulses for retribution rather than act on them. We saw many examples in chapters 2 and 3 of how pacification by a government has a whopping effect on lethal vengeance, and in the next chapter we will review experiments showing that self-control in one context can spread into others.

Chapter 3 also introduced the finding that the sheer presence of government brings rates of violence down only so far—from the hundreds of homicides per 100,000 people per year to the tens. A further drop into the single digits may depend on something hazier, such as people's acceptance of the legitimacy of the government and social contract. A recent experiment may have caught a wisp of this phenomenon in the lab. The economists Benedikt Herrmann, Christian Thöni, and Simon Gächter had university students in sixteen countries play Public Goods games (the game in which players contribute money to a pot which is then doubled and redistributed among them), with and without the possibility of punishing one another.[208] The researchers discovered to their horror that in some countries many players punished *generous* contributors to the common good rather than stingy ones. These acts of spite had predictably terrible effects on the group's welfare, because they only reinforced every player's worst instinct to free-ride on the contributions of the others. The contributions soon petered out, and everyone lost. The antisocial punishers seem to have been motivated by an excess of revenge. When they themselves had been punished for a low contribution, rather than being chastened and increasing their contribution on the next round (which is what participants in the original studies conducted in the United States and Western Europe had done), they punished their punishers, who tended to be the altruistic contributors.

What distinguishes the countries in which the targets of punishment repent, such as the United States, Australia, China, and those of Western Europe, from those in which they spitefully retaliate, like Russia, Ukraine, Greece, Saudi Arabia, and Oman? The investigators ran a set of multiple regressions using a dozen traits of the different countries, taken from economic statistics and the results of international surveys. A major predictor of excess revenge turned out to be civic norms: a measure of the degree to which people think it is all right to cheat on their income taxes, claim government benefits to which they are not entitled, and dodge fare-collectors on the subway. (Social scientists believe that civic norms make up a large part of the *social capital* of a country, which is more important to its prosperity than its physical resources.) Where might the civic norms themselves have come from? The World Bank assigns countries a score called the Rule of Law, which reflects how well private contracts can be enforced in courts, whether the legal system is perceived as being fair, the importance of the black market and organized crime, the quality of the police, and the likelihood of crime and violence. In the experiment, the Rule of Law of a country significantly predicted the degree to which its citizens indulged in antisocial revenge: the people in countries with an iffy Rule of Law were more destructively vengeful. With the usual spaghetti of variables, it's impossible to be certain what caused what, but the results are consistent with the idea that the disinterested justice of a decent Leviathan induces citizens to curb their impulse for revenge before it spirals into a destructive cycle.

Revenge, for all its tendency to escalate, must come with a dimmer switch. If it didn't, the Moralization Gap would inflate every affront into an escalating feud, like the experimental subjects who mashed down on each other's fingers harder and harder with every round. Not only does revenge not always escalate, especially in civil societies with the rule of law, but we shouldn't expect it to. The models of the evolution of cooperation showed that the most successful agents dial back their tit-for-tatting with contrition and forgiveness, especially when trapped in the same boat with other agents.

In *Beyond Revenge: The Evolution of the Forgiveness Instinct*, the psychologist Michael McCullough shows that we do have this dimmer switch for revenge.[209] As we have seen, several species of primate can kiss and make up after a fight, at least if their interests are bound by kinship, shared goals, or common enemies.[210] McCullough shows that the human forgiveness instinct is activated under similar circumstances.

The desire for revenge is most easily modulated when the perpetrator falls within our natural circle of empathy. We are apt to forgive our kin and close friends for trespasses that would be unforgivable in others. And when our circle of empathy expands (a process we will examine in the next chapter), our circle of forgivability expands with it.

A second circumstance that cranks down revenge is a relationship with the perpetrator that is too valuable to sever. We may not like them, but we're stuck with them, so we had better learn to live with them. During the presidential primary season, rivals for a party's nomination can spend months slinging each other with mud or worse, and their body language during televised debates makes it clear that they can't stand each other. But when the winner is decided, they bite their lips, swallow their pride, and unite against their common adversary in the other party. In many cases the winner even invites the loser onto the ticket or into the cabinet. The power of a shared goal to induce erstwhile enemies to reconcile was dramatically demonstrated in a famous 1950s experiment in which boys at a summer camp called Robbers Cave were divided into teams and on their own initiative waged war on each other for weeks, with raids and retaliations and dangerous weapons like rocks in socks.[211] But when the psychologists arranged some "accidents" that left the boys no choice but to work together to restore the camp's water supply and to pull a bus out of the muck, they fell into a truce, overcame their enmity, and even made some friendships across team lines.

The third modulator of revenge kicks in when we are assured that the perpetrator has become harmless. For all the warmth and fuzziness of forgiveness, you can't afford to disarm if the person who harmed you is likely to do it again. So if a harm-doer wants to avoid your wrath and get back on your good side, he has to persuade you that he no longer harbors any motive to harm you. He may start out by claiming that his harm was an unfortunate result of a unique set of circumstances that will never be repeated—that is, that the action was unintentional or unavoidable or that the harm it did was unforeseen. Not coincidentally, these are the excuses that harm-doers believe about every harm they do, which is one side of the Moralization Gap. If that doesn't work, he can accept your side of the story by acknowledging that he did something wrong, sympathize with your suffering, cancel the harm with restitution, and commit his credibility to an assurance that he will not repeat it. In other words, he can apologize. All of these tactics, studies have shown, can mollify a rankled victim.

The problem with an apology, of course, is that it can be cheap talk. An insincere apology can be more enraging than none at all, because it compounds the first harm with a second one, namely a cynical ploy to avert revenge. The aggrieved party needs to peer into the perpetrator's soul and see that any intention to harm again has been exorcised. The devices that implement this born-again harmlessness are the self-conscious emotions of shame, guilt, and embarrassment.[212] The problem for the perpetrator is how to make those emotions visible. As with all signaling problems, the way to make a signal credible is to make it costly. When a subordinate primate wants to appease a dominant one, he will make himself small, avert his gaze, and expose vulnerable body parts. The equivalent gestures in humans are called cringing, groveling, or bowing and scraping. We may also hand over control of the conspicuous parts of our bodies to our autonomic nervous system, the involuntary circuitry that controls blood flow, muscle tone, and the activity of the glands. An apology that is certified by blushing, stammering, and tears is more credible than one that is cool, calm, and collected. Crying and blushing are particularly affecting because they are felt from the inside as well as displayed on the outside and hence generate common knowledge. The emoter knows that onlookers know his emotional state, the onlookers know he knows it, and so on. Common knowledge obliterates self-deception: the guilty party can no longer deny the uncomfortable truth.[213]

McCullough notes that our revenge modulators offer a route to public conflict reduction that can supplement the criminal justice system. The potential payoff can be enormous because the court system is expensive, inefficient, unresponsive to the victim's needs, and in its own way violent, since it forcibly incarcerates a guilty perpetrator. Many communities now have programs of *restorative justice*, sometimes supplementing a criminal trial, sometimes replacing it. The perpetrator and victim, often accompanied by family and friends, sit down together with a facilitator, who gives the victim an opportunity to express his or her suffering and anger, and the perpetrator an opportunity to convey sincere remorse, together with restitution for the harm. It sounds like daytime TV, but it can set at least some repentant perpetrators on the straight and narrow, while satisfying their victims and keeping the whole dispute out of the slowly grinding wheels of the criminal justice system.

On the international scene, the last two decades have seen an explosion of apologies by political leaders for crimes committed by their governments. The political scientist Graham Dodds has compiled "a fairly comprehensive chronological listing of major political apologies" through the centuries. His list begins in the year 1077, when "Holy Roman Emperor Henry IV apologized to Pope Gregory VII for church-state conflicts by standing barefoot in the snow for three days."[214] History had to wait more than six hundred years for the next one, when Massachusetts apologized in 1711 to the families of the victims of the Salem witch trials. The first apology of the 20th century, Germany's admission to having started World War I in the Treaty of Versailles in 1919, is perhaps not the best advertisement for the genre. But the spate of apologies in the last two decades bespeak a new era in the self-presentation of states. For the first time in history, the leaders of nations have elevated the ideals of historical truth and international reconciliation above self-serving claims of national infallibility and rectitude. In 1984 Japan sort of apologized for occupying Korea when Emperor Hirohito told the visiting South Korean president, "It is regrettable that there was an unfortunate period in this century." But subsequent decades saw a string of ever-more-forthcoming apologies from other Japanese leaders. In the ensuing decades Germany apologized for the Holocaust, the United States apologized for interning Japanese Americans, the Soviet Union apologized for murdering Polish prisoners during World War II, Britain apologized to the Irish, Indians, and Maori, and the Vatican apologized for its role in the Wars of Religion, the persecution of Jews, the slave trade, and the oppression of women. Figure 8–6 shows how political apologies are a sign of our times.

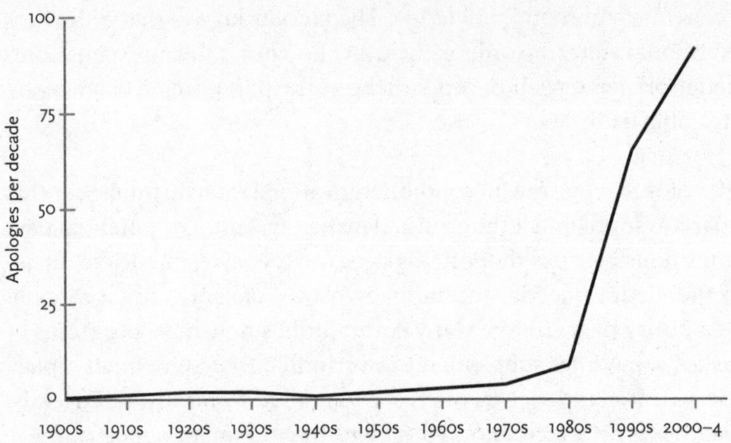

FIGURE 8–6. Apologies by political and religious leaders, 1900–2004
Sources: Data from Dodds, 2003b, and Dodds, 2005.

Do apologies and other conciliatory gestures in the human social repertoire actually avert cycles of revenge? The political scientists William Long and Peter Brecke took up the question in their 2003 book *War and Reconciliation: Reason and Emotion in Conflict Resolution*. Brecke is the scholar who assembled the Conflict Catalog which I relied upon in chapter 5, and he and Long addressed the question with numbers. They selected 114 pairs of countries that fought an interstate war from 1888 to 1991, together with 430 civil wars. They then looked for reconciliation events—ceremonies or rituals that brought the leaders of the warring factions together—and compared the number of militarized disputes (incidents of saber-rattling or fighting) over several decades before and after the event to see if the rituals made any difference. They generated hypotheses and interpreted their findings using both rational actor theory and evolutionary psychology.

When it came to international disputes, emotional gestures made little difference. Long and Brecke identified 21 international reconciliation events and compared the ones that clearly cooled down the belligerents with the ones that left them as disputatious as ever. The successes depended not on symbolic gestures but on costly signaling. The leader of one or both countries made a novel, voluntary, risky, vulnerable, and irrevocable move toward peace that reassured his adversary that he was unlikely to resume hostilities. Anwar Sadat's 1977 speech to the Israeli parliament is the prototype. The gesture was a shocker, and it was unmistakably expensive, later costing Sadat his life. But it led to a peace treaty that has lasted to this day. There were few touchy-feely rituals, and today the two countries are hardly on good terms, but they are at peace. Long and Brecke note that sometimes pairs of countries that looked daggers at each for centuries can turn into good buddies—England and France, England and the United States, Germany and Poland, Germany and France—but the amity comes after decades of coexistence rather than as the immediate outcome of conciliatory gestures.

The psychology of forgiveness, recall, works best when the perpetrator and victim are already bound by kinship, friendship, alliance, or mutual dependence. It is not surprising, then, that conciliatory gestures are more effective in ending civil wars than international ones. The adversaries in a civil war are, at the very least, stuck with each other inside national boundaries, and they have a flag and a soccer team that put them in a fictive coalition. Often the ties run deeper. They may share a language or religion, may work together, and may be related by webs of marriage. In many rebellions and warlord conflicts the fighters may literally be sons, nephews, and neighborhood kids, and communities may have to welcome back the perpetrators of horrible atrocities against them if they are ever to knit their communities together. These and other ties that bind can prepare the way for gestures of apology and reconciliation. These gestures are more effective than the mechanism that leads to peace *between* states, namely the costly signaling of benevolent intentions, because in civil conflicts the two sides are not cleanly separated entities, and so cannot each speak with one voice, exchange messages in safety, and resume the status quo if an initiative fails.

Long and Brecke studied 11 reconciliation events since 1957 that symbolically terminated a civil conflict. With 7 of them (64 percent) there was no return to violence. That figure is impressive: among conflicts that did *not* have a reconciliation event, only 9 percent saw a cessation of violence. The common denominator to the success stories, they found, was a set of conciliation rituals that implemented a symbolic and incomplete justice rather than perfect justice or none at all. Just as a microphone near a loudspeaker can amplify its own output and create an earsplitting howl, retributive justice that visits new harm on the perpetrators can stoke the desire for retaliation in a spiral of competitive victimhood. Conversely, just as the feedback from a microphone can be squelched if the gain is turned down, cycles of communal violence can be squelched if the severity of retributive justice is modulated. A damping of the desire for justice is particularly indispensable after civil conflicts, in which the institutions of justice like the police and prison system are not only fragile but may themselves have been among the main perpetrators of the harm.

The prototype for reconciliation after a civil conflict is South Africa. Invoking the Xhosa concept of *ubuntu* or brotherhood, Nelson Mandela and Desmond Tutu instituted a system of restorative rather than retributive justice to heal the country after decades of violent repression and rebellion under the apartheid regime. As with the tactics of the Rights Revolutions, Mandela and

Tutu's restorative justice both sampled from and contributed to the pool of ideas for nonviolent conflict resolution. Similar programs, Long and Brecke discovered, have cemented civil peace in Mozambique, Argentina, Chile, Uruguay, and El Salvador. They identify four ingredients of the successful elixir.

The first is a round of uncompromised truth-telling and acknowledgment of harm. It may take the form of truth and reconciliation commissions, in which perpetrators publicly own up to the harms they did, or of national fact-finding committees, whose reports are widely publicized and officially endorsed. These mechanisms take direct aim at the self-serving psychology that stokes the Moralization Gap. Though truth-telling sheds no blood, it requires a painful emotional sacrifice on the part of the confessors in the form of shame, guilt, and a unilateral disarmament of their chief moral weapon, the claim to innocence. There is a vast psychological difference between a crime that everyone privately knows about but not everyone acknowledges and one that is "out there" as common knowledge. Just as blushing and tears make an apology more effective, the public acknowledgment of a wrong can rewrite the rules of a relationship between groups.

A second theme in successful reconciliations is an explicit rewriting of people's social identities. People redefine the groups with which they identify. The perpetual victims of a society may take responsibility for running it. Rebels become politicians, bureaucrats, or businesspeople. The military surrenders its claim to embody the nation and demotes itself to their security guards.

The third theme appears to be the most important: incomplete justice. Rather than settling every score, a society has to draw a line under past violations and grant massive amnesty while prosecuting only the blatant ringleaders and some of the more depraved foot soldiers. Even then the punishments take the form of hits to their reputation, prestige, and privileges rather than blood for blood. There may, in addition, be reparations, but their restorative value is registered more on an emotional balance sheet than a financial one. Long and Brecke comment:

> In every instance of successful reconciliation save Mozambique justice was meted out, but never in full measure. This fact may be lamentable, even tragic, from certain legal or moral perspectives, yet it is consistent with the requisites of restoring social order postulated in the forgiveness hypothesis. In all cases of successful reconciliation, retributive justice could neither be ignored nor fully achieved. . . . Disturbing as it may be, people appear to be able to tolerate a substantial amount of injustice wrought by amnesty in the name of social peace.[215]

In other words, peel off the bumper sticker that says "If you want peace, work for justice." Replace it with the one recommended by Joshua Goldstein: "If you want peace, work for peace."[216]

Finally, the belligerents have to signal their commitment to a new relationship with a burst of verbal and nonverbal gestures. As Long and Brecke observe, "Legislatures passed solemn resolutions, peace accords were signed and embraces exchanged by heads of formerly rival groups, statues and monuments to the tragedy were erected, textbooks were rewritten, and a thousand other actions, large and small, were undertaken to underscore the notion that the past was different and the future more hopeful."[217]

The conflict between Israelis and Palestinians stands in many people's minds today as the nastiest ongoing cycle of deadly revenge. Not even Pollyanna would claim to have the key to solving it. But the applied psychology of reconciliation bears out the vision of the Israeli novelist Amos Oz on what a solution will have to look like:

> Tragedies can be resolved in one of two ways: there is the Shakespearean resolution and there is the Chekhovian one. At the end of a Shakespearean tragedy, the stage is strewn with dead bodies and maybe there's some justice hovering high above. A Chekhov tragedy, on the other hand, ends with everybody disillusioned, embittered, heartbroken, disappointed, absolutely shattered, but still alive. And I want a Chekhovian resolution, not a Shakespearean one, for the Israeli/Palestinian tragedy.[218]

SADISM

It's hard to single out the most heinous form of human depravity—there are so many to choose from—but if genocide is the worst by quantity, sadism might be the worst by quality. The deliberate infliction of pain for no purpose but to enjoy a person's suffering is not just morally monstrous but intellectually baffling, because in exchange for the agony of the victim the torturer receives no apparent personal or evolutionary benefit. And unlike many other sins, pure sadism is not a guilty pleasure that most people indulge in their fantasy lives; few of us daydream about watching cats burn to death. Yet torture is a recurring disfigurement in human history and current events, appearing in at least five circumstances.

Sadism can grow out of instrumental violence. The threat of torture can terrify political opponents, and it must at least occasionally be used to make the threat real. Torture may also be used to extract information from a criminal suspect or political enemy. Many police and national security forces engage in mild torture under euphemisms like "the third degree," "moderate physical pressure," and "enhanced interrogation," and these tactics may sometimes be effective.[219] And as moral philosophers since Jeremy Bentham have pointed out, in theory torture can even be justifiable, most famously in the ticking-bomb scenario in which a criminal knows the location of an explosive that will kill and maim many innocent people and only torture would force him to disclose its location.[220]

Yet among the many arguments against the use of torture is that it seldom stays instrumental for long. Torturers get carried away. They inflict so much suffering on their victims that the victims will say anything to make it stop, or become so delirious with agony as to be incapable of responding.[221] Often the victims die, which makes the extraction of information moot. And in cases like the abuse of Iraqi prisoners by American soldiers at Abu Ghraib, the use of torture, far from serving a useful purpose, was a strategic catastrophe for the country that allowed it to happen, inflaming enemies and alienating friends.

A second occasion for torture is in criminal and religious punishment. Here again there is a granule of instrumental motivation, namely to deter wrongdoers with the prospect of pain that would cancel out their gain. Yet as Beccaria and other Enlightenment reformers pointed out, any calculus of deterrence can achieve the same goals with punishments that are less severe but more reliable. And surely the death penalty, if it is applied at all, is a sufficient disincentive to capital crimes without needing the then-customary practice of preceding it with prolonged gruesome torture. In practice, corporal punishment and excruciating capital punishment escalate into orgies of cruelty for its own sake.

Entertainment itself can be a motive for torture, as at the Roman Colosseum and in blood sports like bearbaiting and cat-burning. Tuchman notes that towns in medieval France would sometimes purchase a condemned criminal from another town so they could entertain their citizens with a public execution.[222]

Hideous tortures and mutilations can accompany a rampage by soldiers, rioters, or militiamen, especially when they have been released from apprehension and fear, the phenomenon that Randall Collins calls forward panic. These are the atrocities that accompany pogroms, genocides, police brutality, and military routs, including those in tribal warfare.

Finally there are serial killers, the sickos who stalk, kidnap, torture, mutilate, and kill their victims for sexual gratification. Serial killers like Ted Bundy, John Wayne Gacy, and Jeffrey Dahmer are not the same as garden-variety mass murderers.[223] Mass murderers include men who run amok, like the enraged postal workers who avenge a humiliation and prove their potency by taking as many people as they can with them in a final suicidal outburst. They also include spree killers, like the Washington, D.C., sniper John Muhammad, who stretch out their vengeance and dominance over several weeks. With serial killers, in contrast, the motive is sadism. They are aroused by the prospect of tormenting, disfiguring, dismembering, eviscerating, and slowly draining the life out of victims with their bare hands. Even the most jaded consumer of human atrocities will find something to shock them in Harold Schechter's authoritative compendium *The Serial Killer Files*.

For all its notoriety in rock songs, made-for-TV movies, and Hollywood blockbusters, serial killing is a rare phenomenon. The criminologists James Alan Fox and Jack Levin note that "there may actually be more scholars studying serial murder than there are offenders committing it."[224] And even that small number (like every other tabulation of violence we have examined in this book) has been in decline. In the 1980s, when serial killers were a pop sensation, there were 200 known perpetrators in all, and they killed around 70 victims a year. In the 1990s, there were 141, and in the 2000s, only 61.[225] Those figures may be underestimates (because many serial killers prey on runaways, prostitutes, the homeless, and others whose disappearance may not have been reported as murders), but by any reckoning, no more than two or three dozen serial killers can be at large in the United States at any time, and they are collectively responsible for a tiny fraction of the 17,000 homicides that take place every year.[226]

Serial killing is nothing new. Schechter shows that contrary to a common view that serial killers are the product of our sick society, they have splattered the pages of history for millennia. Caligula, Nero, Bluebeard (probably based on the 15th-century knight Gilles de Rais), Vlad the Impaler, and Jack the Ripper are celebrity examples, and scholars have speculated that legends of werewolves, robber bridegrooms, and demon barbers may have been based on widely retold tales of actual serial killers. All that is new in sadistic killing is the name for the motive, which comes to us courtesy of the most famous serial torturer of all, Donatien Alphonse François, also known as the Marquis de Sade. In earlier centuries serial killers were called murder fiends, bloodthirsty monsters, devils in human shape, or the morally insane.

Though the florid sadism of serial killers is historically rare, the sadism of inquisitors, rampagers, public execution spectators, blood sport fans, and Colosseum audiences is not. And even serial killers don't end up with their avocation because of any gene, brain lesion, or childhood experience we can identify.[227] (They do tend to be victims of childhood sexual and physical abuse, but so are millions of people who don't grow up to be serial killers.) So it's conceivable that the pathway to serial killing can shed light on the pathway that leads ordinary people into sadism as well. How can we make sense of this most senseless variety of violence?

The development of sadism requires two things: motives to enjoy the suffering of others, and a removal of the restraints that ordinarily inhibit people from acting on them.

Though it's painful to admit, human nature comes equipped with at least four motives to take satisfaction in the pain of others. One is a morbid fascination with the vulnerability of living things, a phenomenon perhaps best captured by the word *macabre*. This is what leads boys to pull the legs off grasshoppers and to fry ants with a magnifying glass. It leads adults to rubberneck at the scene of automobile accidents—a vice that can tie up traffic for miles—and to fork over their disposable income to read and watch gory entertainment. The ultimate motive may be mastery over the living world, including our own safety. The implicit lesson of macabre voyeurism may be "But for the swerve of a steering wheel or an unlocked front door, that could have happened to me."[228]

Another appeal of feeling someone's pain is dominance. It can be enjoyable to see how the mighty have fallen, especially if they have been among your tormenters. And when one is looking downward instead of upward, it's reassuring to know that you can exercise the power to dominate others should the need arise. The ultimate form of power over someone is the power to cause them pain at will.[229]

Nowadays neuroscientists will slide people into a magnet to look at just about any human experience. Though no one, to my knowledge, has studied sadism in the scanner, a recent experiment looked at the diluted version, schadenfreude.[230] Male Japanese students lay in an MRI machine and were asked to put themselves in the shoes of a schlemiel who longs for a job in a multinational information technology company but earns mediocre grades, flubs his job interview, warms the bench in his baseball club, ends up with a low-paying job in a retail store, lives in a tiny apartment, and has no girlfriend. At his college reunion he meets a classmate who works for a multinational corporation, lives in a luxury condo, owns a fancy car, dines at French restaurants, collects watches, jet-sets to weekend vacation spots, and "has many opportunities to meet girls after work." The participant also imagines meeting two other classmates, one successful, one unsuccessful, whom the Japanese researchers assumed—correctly, as it turned out—would arouse no envy in the participant because they were women. The participant, still imagining himself the loser, then reads about a string of misfortunes that befall his envied but increasingly Job-like classmate: the classmate is falsely accused of cheating on an exam, he becomes the victim of ugly rumors, his girlfriend has an affair, his company gets into financial trouble, his bonus is small, his car breaks down, his watches are stolen, his apartment building is sprayed with graffiti, he gets food poisoning at the French restaurant, and his vacation is canceled because of a typhoon. The researchers could literally read the gloating off the participants' brains. As the participants read of the misfortunes of their virtual better (though not of the nonthreatening women), their striatum, the part of the Seeking circuit that underlies wanting and liking, lit up like a Tokyo boulevard. The results were the same when women contemplated the downfall of an enviable female rival.

A third occasion for sadism is revenge, or the sanitized third-party version we call justice. The whole point of moralistic punishment is that the wrongdoer suffers for his sins, and we have already seen that revenge can be sweet. Revenge literally turns off the empathic response in the brain (at least among men), and it is consummated only when the avenger knows that the target knows that his suffering is payback for his misdeeds.[231] What better way for the avenger to be certain in that knowledge than to inflict the suffering himself?

Finally, there is sexual sadism. Sadism itself is not a common perversion—among people who indulge in S&M, far more of them are into the M than the S—but milder forms of domination and degradation are not uncommon in pornography, and they may be a by-product of the fact that males are the more ardent and females the more discriminating gender.[232] The circuits for sexuality and aggression are intertwined in the limbic system, and both respond to testosterone.[233]

Male aggression has a sexual component. In interviews, many soldiers describe battlefield routs in explicitly erotic terms. One Vietnam veteran said, "To some people, carrying a gun was like having a permanent hard-on. It was a pure sexual trip every time you got to pull the trigger."[234] Another agreed: "There is . . . just this incredible sense of power in killing five people. . . . The only way I can equate it is to ejaculation. Just an incredible sense of relief, you know, that I did this."[235] Institutionalized torture is often sexualized as well. Female Christian martyrs were depicted as having been sexually mutilated, and when the tables were turned in medieval Christendom, the instruments of torture were often directed at women's erogenous zones.[236] As with the martyrologies, later genres of macabre entertainment such as pulp fiction, Grand Guignol, and "true crime" tabloids often put female protagonists in peril of sexual torture and mutilation.[237] And government torturers in police states have often been reported to be aroused by their atrocities. Lloyd deMause recounts testimony from a survivor of the Holocaust:

> The SS camp commander stood close to the whipping post throughout the flogging. . . . His whole face was already red with lascivious excitement. His hands were plunged deep in his trouser pockets, and it was quite clear that he was masturbating throughout. . . . On more than thirty occasions, I myself have witnessed SS camp commanders masturbating during floggings.[238]

If serial killers represent the taste for rough sex taken to an extreme, the gender difference among serial killers, who come in both sexes, is instructive. Schechter is skeptical of self-anointed "profilers" and "mind hunters," like the Jack Crawford character in *Silence of the Lambs*, but he allows for one kind of inference from the modus operandi of a serial killer to a characteristic trait: "When police discover a corpse with its throat slit, its torso cut open, its viscera removed, and its genitals excised, they are justified in making one basic assumption: the perpetrator was a man."[239] It's not that girls can never grow up to be serial killers; Schechter recounts the stories of several black widows and angels of death. But they go about their pastimes differently. Schechter explains:

> There are unmistakable parallels between [male serial killers'] kind of violence—phallic-aggressive, penetrative, rapacious, and (insofar as it commonly gratifies itself upon the bodies of strangers) undiscriminating—and the typical pattern of male sexual behavior. For this reason, it is possible to see sadistic mutilation-murder as a grotesque distortion . . . of normal male sexuality. . . .
>
> Female psychopaths are no less depraved than their male counterparts. As a rule, however, brutal penetration is not what turns them on. Their excitement comes—not from violating the bodies of strangers with phallic objects—but from a grotesque, sadistic travesty of intimacy and love: from spooning poisoned medicine into the mouth of a trusting patient, for example, or smothering a sleeping child in its bed. In short, from tenderly turning a friend, family member, or dependent into a corpse—from nurturing them to death.[240]

With so many sources for sadism, why are there so few sadists? Obviously the mind must be equipped with safety catches against hurting others, and sadism erupts when they are disabled.

The first that comes to mind is empathy. If people feel each other's pain, then hurting someone else will be felt as hurting oneself. That is why sadism is more thinkable when the victims are demonized or dehumanized beings that lie outside one's circle of empathy. But as I have mentioned (and as we shall explore in the next chapter), for empathy to be a brake on aggression it has to be more than the habit of inhabiting another person's mind. After all, sadists often exercise a perverted ingenuity for intuiting how best to torment their victims. An empathic response must specifically include an alignment of one's own happiness with that of another being, a faculty that is better called sympathy or compassion than empathy. Baumeister points out that an additional emotion has to kick in for sympathy to inhibit behavior: guilt. Guilt, he notes, does not just operate after the fact. Much of our guilt is anticipatory—we refrain from actions that would make us feel bad if we carried them out.[241]

Another brake on sadism is a cultural taboo: the conviction that deliberate infliction of pain is not a thinkable option, regardless of whether it engages one's sympathetic inhibitions. Today torture has been explicitly prohibited by the Universal Declaration of Human Rights and by the 1949 Geneva Conventions.[242] Unlike ancient, medieval, and early modern times when torture was a form of popular entertainment, today the infliction of torture by governments is almost entirely clandestine, showing that the taboo is widely acknowledged—though like most taboos, it is at times hypocritically flouted. In 2001 the legal scholar Alan Dershowitz addressed this hypocrisy by proposing a legal mechanism designed to eliminate sub rosa torture in democracies.[243] The police in a ticking-bomb scenario would have to get a warrant from a disinterested judge before torturing the lifesaving information out of a suspect; all other forms of coercive interrogation would be flatly prohibited. The most common response was outrage. By the very act of examining the taboo on torture, Dershowitz had violated the taboo, and he was widely misunderstood as *advocating* torture rather than seeking to *minimize* it.[244] Some of the more measured critics argued that the taboo in fact serves a useful function. Better, they said, to deal with a ticking-bomb scenario, should one ever occur, on an ad hoc basis, and perhaps even put up with some clandestine torture, than to place torture on the table as a live option, from which it could swell from ticking bombs to a wider range of real or imagined threats.[245]

But perhaps the most powerful inhibition against sadism is more elemental: a visceral revulsion against hurting another person. Most primates find the screams of pain of a fellow animal to be aversive, and they will abstain from food if it is accompanied by the sound and sight of a fellow primate being shocked.[246] The distress is an expression not of the monkey's moral scruples but of its dread of making a fellow animal mad as hell. (It also may be a response to whatever external threat would have caused a fellow animal to issue an alarm call.)[247] The participants in Stanley Milgram's famous experiment, who obeyed instructions to deliver shocks to a bogus fellow participant, were visibly distraught as they heard the shrieks of pain they were inflicting.[248] Even in moral philosophers' hypothetical scenarios like the Trolley Problem, survey-takers recoil from the thought of throwing the fat man in front of the trolley, though they know it would save five innocent lives.[249]

Testimony on the commission of hands-on violence in the real world is consistent with the results of laboratory studies. As we saw, humans don't readily consummate mano a mano fisticuffs, and soldiers on the battlefield may be petrified about pulling the trigger.[250] The historian Christopher Browning's interviews with Nazi reservists who were ordered to shoot Jews at close range showed that their initial reaction was a physical revulsion to what they were doing.[251] The reservists did not recollect the trauma of their first murders in the morally colored ways we might expect—neither with guilt at what they were doing, nor with retroactive excuses to mitigate their culpability. Instead they recalled how viscerally upset they were made by the screams, the gore, and the raw feeling of killing people at close range. As Baumeister sums up their testimony, "The first day of mass murder did not prompt them to engage in spiritual soul-searching so much as it made them literally want to vomit."[252]

There are barriers to sadism, then, but there must also be workarounds, or sadism would not exist. The crudest workaround is evident during rampages, when a window of opportunity to rout the enemy opens up and any revulsion against hands-on harm is suspended. The most sophisticated workaround may be the willing suspension of disbelief that allows us to immerse ourselves in fictional worlds. One part of the brain allows us to lose ourselves in the story and perhaps indulge in a touch of virtual sadism. The other reminds us that it's all make-believe, so our inhibitions don't spoil the pleasure.[253]

Psychopathy is a lifelong disabling of the inhibitions against sadism. Psychopaths have a blunted response in their amygdala and orbital cortex to signs of distress, together with a marked lack of sympathy with the interests of other people.[254] All serial killers are psychopaths, and survivors of brutal government interrogation and punishment often report that some of the guards stood out from others in their sadism, presumably the psychopaths.[255] Yet most psychopaths are not serial killers or even sadists, and in some environments, such as public spectacles of cruelty in medieval Europe, nearly everyone indulged in sadism. That means we need to identify the pathway that leads people, some more easily than others, toward the infliction of pain for pleasure.

Sadism is literally an acquired taste.[256] Government torturers such as police interrogators and prison guards follow a counterintuitive career trajectory. It's not the rookies who are overly exuberant and the veterans who fine-tune the pain to extract the maximum amount of actionable information. Instead it's the veterans who torture prisoners beyond any conceivable purpose. They come to enjoy their work. Other forms of sadism also must be cultivated. Most sexual sadists start out wielding the whips and collars as a favor to the more numerous masochists; only gradually do they start to enjoy it. Serial killers too carry out their first murder with trepidation, distaste, and in its wake, disappointment: the experience had not been as arousing as it had been in their imaginations. But as time passes and their appetite is rewhetted, they find the next one easier and more gratifying, and then they escalate the cruelty to feed what turns into an addiction. One can imagine that when tortures and executions are public and common, as in the European Middle Ages, the acclimatization process can inure an entire population.

It's often said that people can become desensitized to violence, but that is not what happens when people acquire a taste for torture. They are not oblivious to the suffering of others in the way that the neighbors of a fish-processing plant stop noticing the noisome smell. Sadists take pleasure in the suffering of victims, or, in the case of serial killers, positively crave it.[257]

Baumeister explains the acquisition of sadism with the help of a theory of motivation proposed by the psychologist Richard Solomon, based on an analogy with color vision.[258] Emotions come in pairs, Solomon suggested, like complementary colors. The world as seen through rose-tinted goggles eventually returns to neutral, but when the goggles are removed, it looks greenish for a while. That is because our sense of neutral white or gray reflects the present status of a tug-of-war between circuits for the color red (more accurately, longer wavelengths) and circuits for the color green (medium wavelengths). When red-sensitive neurons are overactivated for a protracted period, they habituate and relax their tug, and the rosy tint in our consciousness fades out.

Then when the goggles are removed, the red- and green-sensitive neurons are equally stimulated, but the red ones have been desensitized while the green ones are ready and rested. So the green side predominates in the tug-of-war, and greenness is what we experience.

Solomon suggested that our emotional state, like our perception of the color of the world, is kept in equilibrium by a balance of opposing circuits. Fear is in balance with reassurance, euphoria with depression, hunger with satiety. The main difference between opposing emotions and complementary colors is in how they change with experience. With the emotions, a person's initial reaction gets weaker over time, and the balancing impulse gets stronger. As an experience is repeated, the emotional rebound is more keenly felt than the emotion itself. The first leap in a bungee jump is terrifying, and the sudden yoiiiiing of deceleration exhilarating, followed by an interlude of tranquil euphoria. But with repeated jumps the reassurance component strengthens, which makes the fear subside more quickly and the pleasure arrive earlier. If the most concentrated moment of pleasure is the sudden reversal of panic by reassurance, then the weakening of the panic response over time may require the jumper to try increasingly dangerous jumps to get the same degree of exhilaration. The action-reaction dynamic may be seen with positive initial experiences as well. The first hit of heroin is euphoric, and the withdrawal mild. But as the person turns into a junkie, the pleasure lessens and the withdrawal symptoms come earlier and are more unpleasant, until the compulsion is less to attain the euphoria than to avoid the withdrawal.

According to Baumeister, sadism follows a similar trajectory.[259] An aggressor experiences a revulsion to hurting his victim, but the discomfort cannot last forever, and eventually a reassuring, energizing counteremotion resets his equilibrium to neutral. With repeated bouts of brutality, the reenergizing process gets stronger and turns off the revulsion earlier. Eventually it predominates and tilts the entire process toward enjoyment, exhilaration, and then craving. As Baumeister puts it, the pleasure is in the backwash.

By itself the opponent-process theory is a bit too crude, predicting, for example, that people would hit themselves over the head because it feels so good when they stop. Clearly not all experiences are governed by the same tension between reaction and counterreaction, nor by the same gradual weakening of the first and strengthening of the second. There must be a subset of aversive experiences that especially lend themselves to being overcome. The psychologist Paul Rozin has identified a syndrome of acquired tastes he calls benign masochism.[260] These paradoxical pleasures include consuming hot chili peppers, strong cheese, and dry wine, and partaking in extreme experiences like saunas, skydiving, car racing, and rock climbing. All of them are adult tastes, in which a neophyte must overcome a first reaction of pain, disgust, or fear on the way to becoming a connoisseur. And all are acquired by controlling one's exposure to the stressor in gradually increasing doses. What they have in common is a coupling of high potential gains (nutrition, medicinal benefits, speed, knowledge of new environments) with high potential dangers (poisoning, exposure, accidents). The pleasure in acquiring one of these tastes is the pleasure of pushing the outside of the envelope: of probing, in calibrated steps, how high, hot, strong, fast, or far one can go without bringing on disaster. The ultimate advantage is to open up beneficial regions in the space of local experiences that are closed off by default by innate fears and cautions. Benign masochism is an overshooting of this motive of mastery, and as Solomon and Baumeister point out, the revulsion-overcoming process can overshoot so far as to result in craving and addiction. In the case of sadism, the potential benefits are dominance, revenge, and sexual access, and the potential dangers are reprisals from the victim or victim's allies. Sadists do become connoisseurs—the instruments of torture in medieval Europe, police interrogation centers, and the lairs of serial killers can be gruesomely sophisticated—and sometimes they can become addicts.

The fact that sadism is an acquired taste is both frightening and hopeful. As a pathway prepared by the motivational systems of the brain, sadism is an ever-present danger to individuals, security forces, or subcultures who take the first step and can proceed to greater depravity in secrecy. Yet it does have to be acquired, and if those first steps are blocked and the rest of the pathway bathed in sunlight, the path to sadism can be foreclosed.

IDEOLOGY

Individual people have no shortage of selfish motives for violence. But the really big body counts in history pile up when a large number of people carry out a motive that transcends any one of them: an ideology. Like predatory or instrumental violence, ideological violence is a means to an end. But with an ideology, the end is idealistic: a conception of the greater good.[261]

Yet for all that idealism, it's ideology that drove many of the worst things that people have ever done to each other. They include the Crusades, the European Wars of Religion, the French Revolutionary and Napoleonic Wars, the Russian and Chinese civil wars, the Vietnam War, the Holocaust, and the genocides of Stalin, Mao, and Pol Pot. An ideology can be dangerous for several reasons. The infinite good it promises prevents its true believers from cutting a deal. It allows any number of eggs to be broken to make the utopian omelet. And it renders opponents of the ideology infinitely evil and hence deserving of infinite punishment.

We have already seen the psychological ingredients of a murderous ideology. The cognitive prerequisite is our ability to think through long chains of means-ends reasoning, which encourage us to carry out unpleasant means as a way to bring about desirable ends. After all, in some spheres of life the ends really do justify the means, such as the bitter drugs and painful procedures we undergo as part of a medical treatment. Means-ends reasoning becomes dangerous when the means to a glorious end include harming human beings. The design of the mind can encourage the train of theorization to go in that direction because of our drives for dominance and revenge, our habit of essentializing other groups, particularly as demons or vermin, our elastic circle of sympathy, and the self-serving biases that exaggerate our wisdom and virtue. An ideology can provide a satisfying narrative that explains chaotic events and collective misfortunes in a way that flatters the virtue and competence of believers, while being vague or conspiratorial enough to withstand skeptical scrutiny.[262] Let these ingredients brew in the mind of a narcissist with a lack of empathy, a need for admiration, and fantasies of unlimited success, power, brilliance, and goodness, and the result can be a drive to implement a belief system that results in the deaths of millions.

But the puzzle in understanding ideological violence is not so much psychological as epidemiological: how a toxic ideology can spread from a small number of narcissistic zealots to an entire population willing to carry out its designs. Many ideological beliefs, in addition to being evil, are patently ludicrous—ideas that no sane person would ever countenance on his or her own. Examples include the burning of witches because they sank ships and turned men into cats, the extermination of every last Jew in Europe because their blood would pollute the Aryan race, and the execution of Cambodians who wore eyeglasses because it proved they were intellectuals and hence class enemies. How can we explain extraordinary popular delusions and the madness of crowds?

Groups can breed a number of pathologies of thought. One of them is polarization. Throw a bunch of people with roughly similar opinions into a group to hash them out, and the opinions will become more similar to one another, and more extreme as well.[263] The liberal groups become more liberal; the conservative groups more conservative. Another group pathology is obtuseness, a dynamic that the psychologist Irving Janis called groupthink.[264] Groups are apt to tell their leaders what they want to hear, to suppress dissent, to censor private doubts, and to filter out evidence that contradicts an emerging consensus. A third is animosity between groups.[265] Imagine being locked in a room for a few hours with a person whose opinions you dislike—say, you're a liberal and he or she is a conservative or vice versa, or you sympathize with Israel and the other person sympathizes with the Palestinians or vice versa. Chances are the conversation between the two of you would be civil, and it might even be warm. But now imagine that there are six on your side and six on the other. There would probably be a lot of hollering and red faces and perhaps a small riot. The overall problem is that groups take on an identity of their own in people's minds, and individuals' desire to be accepted within a group, and to promote its standing in comparison to other groups, can override their better judgment.

Even when people are not identifying with a well-defined group, they are enormously influenced by the people around them. One of the great lessons of Stanley Milgram's experiments on obedience to authority, widely appreciated by psychologists, is the degree to which the behavior of the participants depended on the immediate social milieu.[266] Before he ran the experiment, Milgram polled his colleagues, students, and a sample of psychiatrists on how far they thought the participants would go when an experimenter instructed them to shock a fellow participant. The respondents unanimously predicted that few would exceed 150 volts (the level at which the victim demands to be freed), that just 4 percent would go up to 300 volts (the setting that bore the warning "Danger: Severe Shock"), and that only a handful of psychopaths would go all the way to the highest shock the machine could deliver (the setting labeled "450 Volts—XXX"). In fact, *65 percent* of the participants went all the way to the maximum shock, long past the point when the victim's agonized protests had turned to an eerie silence. And they might have kept on shocking the presumably comatose subject (or his corpse) had the experimenter not brought the proceedings to a halt. The percentage barely budged with the sex, age, or occupation of the participants, and it varied only a small amount with their personalities. What did matter was the physical proximity of other people and how they behaved. When the experimenter was absent and his instructions were delivered over the telephone or in a recorded message, obedience fell. When the victim was in the same room instead of an adjacent booth, obedience fell. And when the participant had to work in tandem with a second participant (a confederate of the experimenter), then if the confederate refused to comply, so did the participant. But when the confederate complied, more than 90 percent of the time the participant did too.

People take their cues on how to behave from other people. This is a major conclusion of the golden age of social psychology, when experiments were a kind of guerrilla theater designed to raise consciousness about the dangers of mindless conformity. Following a 1964 news report—almost entirely apocryphal—that dozens of New Yorkers watched impassively as a woman named Kitty Genovese was raped and stabbed to death in their apartment courtyard, the psychologists John Darley and Bibb Latané conducted a set of ingenious studies on so-called bystander apathy.[267] The psychologists suspected that groups of people might fail to respond to an emergency that would send an isolated person leaping to action because in a group, everyone assumes that if no one else is doing anything, the situation couldn't be all that dire. In one experiment, as a participant was filling out a questionnaire, he or she heard a loud crash and a voice calling out from behind a partition: "Oh . . . my foot . . . I . . . can't move it; oh . . . my ankle . . . I can't get this thing off me." Believe it or not, if the participant was sitting with a confederate who continued to fill out the questionnaire as if nothing was happening, 80 percent of the time the participant did nothing too. When the participants were alone, only 30 percent failed to respond.

People don't even need to witness other people behaving callously to behave in uncharacteristically callous ways. It is enough to place them in a fictive group that is defined as being dominant over another one. In another classic psychology-experiment-cum-morality-play (conducted in 1971, before committees for the protection of human subjects put the kibosh on the genre), Philip Zimbardo set up a mock prison in the basement of the Stanford psychology department, divided the participants at random into "prisoners" and "guards," and even got the Palo Alto police to arrest the prisoners and haul them to the campus hoosegow.[268] Acting as the prison superintendent, Zimbardo suggested to the guards that they could flaunt their power and instill fear in the prisoners, and he reinforced the atmosphere of group dominance by outfitting the guards with uniforms, batons, and mirrored sunglasses while dressing the prisoners in humiliating smocks and stocking caps. Within two days some of the guards took their roles too seriously and began to brutalize the prisoners, forcing them to strip naked, clean toilets with their bare hands, do push-ups with the guards standing on their backs, or simulate sodomy. After six days Zimbardo had to call off the experiment for the prisoners' safety. Decades later Zimbardo wrote a book that analogized the unplanned abuses in his own faux prison to the unplanned abuses at the Abu Ghraib prison in Iraq, arguing that a situation in which a group of people is given authority over another group can bring out barbaric behavior in individuals who might never display it in other circumstances.

Many historians of genocide, like Christopher Browning and Benjamin Valentino, have invoked the experiments of Milgram, Darley, Zimbardo, and other social psychologists to make sense of the puzzling participation, or at least acquiescence, of ordinary people in unspeakable atrocities. Bystanders often get caught up in the frenzy around them and join in the looting, gang rapes, and massacres. During the Holocaust, soldiers and policemen rounded up unarmed civilians, lined them up in front of pits, and shot them to death, not out of animus to the victims or a commitment to Nazi ideology but so that they would not shirk their responsibilities or let down their brothers-in-arms. Most of them were not even coerced by a threat of punishment for insubordination. (My own experience in carrying out instructions to shock a laboratory rat against my better judgment makes this disturbing claim utterly believable to me.) Historians have found few if any cases in which a German policeman, soldier, or guard suffered a penalty for refusing to carry out the Nazis' orders.[269] As we shall see in the next chapter, people even *moralize* conformity and obedience. One component of the human moral sense, amplified in many cultures, is the elevation of conformity and obedience to praiseworthy virtues.

Milgram ran his experiments in the 1960s and early 1970s, and as we have seen, many attitudes have changed since then. It's natural to wonder whether Westerners today would still obey the instructions of an authority figure to brutalize a stranger. The Stanford Prison Experiment is too bizarre to replicate exactly today, but thirty-three years after the last of the obedience studies, the social psychologist Jerry Burger figured out a way to carry out a new one that would pass ethical muster in the world of 2008.[270] He noticed that in Milgram's original studies, the 150-volt mark, when the victim first cries out in pain and protest, was a point of no return. If a participant didn't disobey the experimenter then, 80 percent of the time he or she would continue to the highest shock on the board. So Burger ran Milgram's procedure but broke off the experiment at the 150-volt mark, immediately explaining the study to the participants and preempting the awful progression in which so many people tortured a stranger over their own misgivings. The question is: after four decades of fashionable rebellion, bumper stickers that advise the reader to Question Authority, and a growing historical consciousness that ridicules the excuse "I was only following orders," do people still follow the orders of an authority to inflict pain on a stranger? The answer is that they do. Seventy percent of the participants went all the way to 150 volts and so, we have reason to believe, would have continued to fatal levels if the experimenter had permitted it. On the bright side, almost twice as many people disobeyed the experimenter in the 2000s as did in the 1960s (30 percent as compared to 17.5 percent), and the figure might have been even higher if the diverse demographics of the recent study pool had been replaced by the white-bread homogeneity of the earlier ones.[271] But a majority of people will still hurt a stranger against their own inclinations if they see it as part of a legitimate project in their society.

Why do people so often impersonate sheep? It's not that conformity is inherently irrational.[272] Many heads are better than one, and it's usually wiser to trust the hard-won wisdom of millions of people in one's culture than to think that one is a genius who can figure everything out from scratch. Also, conformity can be a virtue in what game theorists call coordination games, where individuals have no rational reason to choose a particular option other than the fact that everyone else has chosen it. Driving on the right or the left side of the road is a classic example: here is a case in which you really don't want to march to the beat of a different drummer. Paper currency, Internet protocols, and the language of one's community are other examples.

But sometimes the advantage of conformity to each individual can lead to pathologies in the group as a whole. A famous example is the way an early technological standard can gain a toehold among a critical mass of users, who use it because so many other people are using it, and thereby lock out superior competitors. According to some theories, these "network externalities" explain the success of English spelling, the QWERTY keyboard, VHS videocassettes, and Microsoft software (though there are doubters in each case). Another example is the unpredictable fortunes of bestsellers, fashions, top-forty singles, and Hollywood blockbusters. The mathematician Duncan Watts set up two versions of a Web site in which users could download garage-band rock music.[273] In one version users could not see how many times a song had already been downloaded. The differences in popularity among songs were slight, and they tended to be stable from one run of the study to another. But in the other version people could see how popular a song had been. These users tended to download the popular songs, making them more popular still, in a runaway positive feedback loop. The amplification of small initial differences led to large chasms between a few smash hits and many duds—and the hits and duds often changed places when the study was rerun.

Whether you call it herd behavior, the cultural echo chamber, the rich get richer, or the Matthew Effect, our tendency to go with the crowd can lead to an outcome that is collectively undesirable. But the cultural products in these examples—buggy software, mediocre novels, 1970s fashion—are fairly innocuous. Can the propagation of conformity through social networks actually lead people to sign on to ideologies they don't find compelling and carry out acts they think are downright wrong? Ever since the rise of Hitler, a debate has raged between two positions that seem equally unacceptable: that Hitler single-handedly duped an innocent nation, and that the Germans would have carried out the Holocaust without him. Careful analyses of social dynamics show that neither explanation is exactly right, but that it's easier for a fanatical ideology to take over a population than common sense would allow.

There is a maddening phenomenon of social dynamics variously called pluralistic ignorance, the spiral of silence, and the Abilene paradox, after an anecdote in which a Texan family takes an unpleasant trip to Abilene one hot afternoon because each member thinks the others want to go.[274] People may endorse a practice or opinion they deplore because they mistakenly think that everyone else favors it. A classic example is the value that college students place on drinking till they puke. In many surveys it turns out that every student, questioned privately, thinks that binge drinking is a terrible idea, but each is convinced that his peers think it's cool. Other surveys have suggested that gay-bashing by young toughs, racial segregation in the American South, honor killings of unchaste women in Islamic societies, and tolerance of the terrorist group ETA among Basque citizens of France and Spain may owe their longevity to spirals of silence.[275] The supporters of each of these forms of group violence did not think it was a good idea so much as they thought that everyone else thought it was a good idea.

Can pluralistic ignorance explain how extreme ideologies may take root among people who ought to know better? Social psychologists have long known that it can happen with simple judgments of fact. In another hall-of-fame experiment, Solomon Asch placed his participants in a dilemma right out of the movie *Gaslight*.[276] Seated around a table with seven other participants (as usual, stooges), they were asked to indicate which of three very different lines had the same length as a target line, an easy call. The six stooges who answered before the participant each gave a patently wrong answer. When their turn came, three-quarters of the real participants defied their own eyeballs and went with the crowd.

But it takes more than the public endorsement of a private falsehood to set off the madness of crowds. Pluralistic ignorance is a house of cards. As the story of the Emperor's New Clothes makes clear, all it takes is one little boy to break the spiral of silence, and a false consensus will implode. Once the emperor's nakedness became common knowledge, pluralistic ignorance was no longer possible. The sociologist Michael Macy suggests that for pluralistic ignorance to be robust against little boys and other truth-tellers, it needs an additional ingredient: enforcement.[277] People not only avow a preposterous belief that they think everyone else avows, but they punish those who fail to avow it, largely out of the belief—also false—that everyone else wants it enforced. Macy and his colleagues speculate that false conformity and false enforcement can reinforce each other, creating a vicious circle that can entrap a population into an ideology that few of them accept individually.

Why would someone punish a heretic who disavows a belief that the person himself or herself rejects? Macy et al. speculate that it's to prove their sincerity—to show other enforcers that they are not endorsing a party line out of expedience but believe it in their hearts. That shields them from punishments by their fellows—who may, paradoxically, only be punishing heretics out of fear that they will be punished if they don't.

The suggestion that unsupportable ideologies can levitate in midair by vicious circles of punishment of those who fail to punish has some history behind it. During witch hunts and purges, people get caught up in cycles of preemptive denunciation. Everyone tries to out a hidden heretic before the heretic outs him. Signs of heartfelt conviction become a precious commodity. Solzhenitsyn recounted a party conference in Moscow that ended with a tribute to Stalin. Everyone stood and clapped wildly for three minutes, then four, then five . . . and then no one dared to be the first to stop. After eleven minutes of increasingly stinging palms, a factory director on the platform finally sat down, followed by the rest of the grateful assembly. He was arrested that evening and sent to the gulag for ten years.[278] People in totalitarian regimes have to cultivate thoroughgoing thought control lest their true feelings betray them. Jung Chang, a former Red Guard and then a historian and memoirist of life under Mao, wrote that on seeing a poster that praised Mao's mother for giving money to the poor, she found herself quashing the heretical thought that the great leader's parents had been rich peasants, the kind of people now denounced as class enemies. Years later, when she heard a public announcement that Mao had died, she had to muster every ounce of thespian ability to pretend to cry.[279]

To show that a spiral of insincere enforcement can ensconce an unpopular belief, Macy, together with his collaborators Damon Centola and Robb Willer, first had to show that the theory was not just plausible but mathematically sound. It's easy to prove that pluralistic ignorance, once it is in place, is a stable equilibrium, because no one has an incentive to be the only deviant in a population of enforcers. The trick is to show how a society can get there from here. Hans Christian Andersen had his readers suspend disbelief in his whimsical premise that an emperor could be hoodwinked into parading around naked; Asch paid his stooges to lie. But how could a false consensus entrench itself in a more realistic world?

The three sociologists simulated a little society in a computer consisting of two kinds of agents.[280] There were true believers, who always comply with a norm and denounce noncompliant neighbors if they grow too numerous. And there were private but pusillanimous skeptics, who comply with a norm if a few of their neighbors are enforcing it, and enforce the norm themselves if a lot of their neighbors are enforcing it. If these skeptics aren't bullied into conforming, they can go the other way and enforce skepticism among their conforming neighbors. Macy and his collaborators found that unpopular norms can become entrenched in some, but not all, patterns of social connectedness. If the true believers are scattered throughout the population and everyone can interact with everyone else, the population is immune to being taken over by an unpopular belief. But if the true believers are clustered within a neighborhood, they can enforce the norm among their more skeptical neighbors, who, overestimating the degree of compliance around them and eager to prove that they do not deserve to be sanctioned, enforce the norm against each other and against *their* neighbors. This can set off cascades of false compliance and false enforcement that saturate the entire society.

The analogy to real societies is not far-fetched. James Payne documented a common sequence in the takeover of Germany, Italy, and Japan by fascist ideologies in the 20th century. In each case a small group of fanatics embraced a "naïve, vigorous ideology that justifies extreme measures, including violence," recruited gangs of thugs willing to carry out the violence, and intimidated growing segments of the rest of the populations into acquiescence.[281]

Macy and his collaborators played with another phenomenon that was first discovered by Milgram: the fact that every member of a large population is connected to everyone else by a short chain of mutual acquaintances—six degrees of separation, according to the popular meme.[282] They laced their virtual society with a few random long-distance connections, which allowed agents to be in touch with other agents with fewer degrees of separation. Agents could thereby sample the compliance of agents in other neighborhoods, disabuse themselves of a false consensus, and resist the pressure to comply or enforce. The opening up of neighborhoods by long-distance channels dissipated the enforcement of the fanatics and prevented them from intimidating enough conformists into setting off a wave that could swamp the society. One is tempted toward the moral that open societies with freedom of speech and movement and well-developed channels of communication are less likely to fall under the sway of delusional ideologies.

Macy, Willer, and Ko Kuwabara then wanted to show the false-consensus effect in real people—that is, to see if people could be cowed into criticizing other people whom they actually agreed with if they feared that everyone else would look down on them for expressing their true beliefs.[283] The sociologists mischievously chose two domains where they suspected that opinions are shaped more by a terror of appearing unsophisticated than by standards of objective merit: wine-tasting and academic scholarship.

In the wine-tasting study, Macy et al. first whipped their participants into a self-conscious lather by telling them they were part of a group that had been selected for its sophistication in appreciating fine art. The group would now take part in the "centuries-old tradition" (in fact, concocted by the experimenters) called a Dutch Round. A circle of wine enthusiasts first evaluate a set of wines, and then evaluate one another's wine-judging abilities. Each participant was given three cups of wine and asked to grade them on bouquet, flavor, aftertaste, robustness, and overall quality. In fact, the three cups had been poured from the same bottle, and one was spiked with vinegar. As in the Asch experiment, the participants, before being asked for their own judgments, witnessed the judgments of four stooges, who rated the vinegary sample higher than one of the unadulterated samples, and rated the other one best of all. Not surprisingly, about half the participants defied their own taste buds and went with the consensus.

Then a sixth participant, also a stooge, rated the wines accurately. Now it was time for the participants to evaluate one another, which some did confidentially and others did publicly. The participants who gave their ratings confidentially respected the accuracy of the honest stooge and gave him high marks, even if they themselves had been browbeaten into conforming. But those who had to offer their ratings publicly compounded their hypocrisy by downgrading the honest rater.

The experiment on academic writing was similar, but with an additional measure at the end. The participants, all undergraduates, were told they had been selected as part of an elite group of promising scholars. They had been assembled, they learned, to take part in the venerable tradition called the Bloomsbury Literary Roundtable, in which readers publicly evaluate a text and then evaluate each other's evaluation skills. They were given a short passage to read by Robert Nelson, Ph.D., a MacArthur "genius grant" recipient and Albert W. Newcombe Professor of Philosophy at Harvard University. (There is no such professor or professorship.) The passage, called "Differential Topology and Homology," had been excerpted from Alan Sokal's "Transgressing the Boundaries: Towards a Transformative Hermeneutics of Quantum Gravity." The essay was in fact the centerpiece of the famous Sokal Hoax, in which the physicist had written a mass of gobbledygook and, confirming his worst suspicions about scholarly standards in the postmodernist humanities, got it published in the prestigious journal *Social Text*.[284]

The participants, to their credit, were not impressed by the essay when they rated it in private. But when they rated it in public after seeing four stooges give it glowing evaluations, they gave it high evaluations too. And when they then rated their fellow raters, including an honest sixth one who gave the essay the low rating it deserved, they gave him high marks in private but low marks in public. Once again the sociologists had demonstrated that people not only endorse an opinion they do not hold if they mistakenly believe everyone else holds it, but they falsely condemn someone else who fails to endorse the opinion. The extra step in this experiment was that Macy et al. got a new group of participants to rate whether the first batch of participants had sincerely believed that the nonsensical essay was good. The new raters judged that the ones who condemned the honest rater were more sincere in their misguided belief than the ones who chose not to condemn him. It confirms Macy's suspicion that enforcement of a belief is perceived as a sign of sincerity, which in turn supports the idea that people enforce beliefs they don't personally hold to make themselves look sincere. And that, in turn, supports their model of pluralistic ignorance, in which a society can be taken over by a belief system that the majority of its members do not hold individually.

It's one thing to say that a sour wine has an excellent bouquet or that academic balderdash is logically coherent. It's quite another to confiscate the last bit of flour from a starving Ukrainian peasant or to line up Jews at the edge of a pit and shoot them. How could ordinary people, even if they were acquiescing to what they thought was a popular ideology, overcome their own consciences and perpetrate such atrocities?

The answer harks back to the Moralization Gap. Perpetrators always have at their disposal a set of self-exculpatory stratagems that they can use to reframe their actions as provoked, justified, involuntary, or inconsequential. In the examples I mentioned in introducing the Moralization Gap, perpetrators rationalize a harm they committed out of self-interested motives (reneging on a promise, robbing or raping a victim). But people also rationalize harms they have been pressured into committing in the service of someone *else's* motives. They can edit their beliefs to make the action seem justifiable to themselves, the better to justify it to others. This process is called cognitive dissonance reduction, and it is a major tactic of self-deception.[285] Social psychologists like Milgram, Zimbardo, Baumeister, Leon Festinger, Albert Bandura, and Herbert Kelman have documented that people have many ways of reducing the dissonance between the regrettable things they sometimes do and their ideal of themselves as moral agents.[286]

One of them is euphemism—the reframing of a harm in words that somehow make it feel less immoral. In his 1946 essay "Politics and the English Language," George Orwell famously exposed the way governments could cloak atrocities in bureaucratese:

> In our time, political speech and writing are largely the defense of the indefensible. Things like the continuance of British rule in India, the Russian purges and deportations, the dropping of the atom bombs on Japan, can indeed be defended, but only by arguments which are too brutal for most people to face, and which do not square with the professed aims of the political parties. Thus political language has to consist largely of euphemism, question-begging and sheer cloudy vagueness. Defenseless villages are bombarded from the air, the inhabitants driven out into the countryside, the cattle machine-gunned, the huts set on fire with incendiary bullets: this is called *pacification*. Millions of peasants are robbed of their farms and sent trudging along the roads with no more than they can carry: this is called *transfer of population* or *rectification of frontiers*. People are imprisoned for years without trial, or shot in the back of the neck or sent to die of scurvy in Arctic lumber camps: this is called *elimination of unreliable elements*. Such phraseology is needed if one wants to name things without calling up mental pictures of them.[287]

Orwell was wrong about one thing: that political euphemism was a phenomenon of his time. A century and a half before Orwell, Edmund Burke complained about the euphemisms emanating from revolutionary France:

> The whole compass of the language is tried to find sinonimies and circumlocutions for massacres and murder. Things are never called by their common names. Massacre is sometimes called *agitation*, sometimes *effervescence*, sometimes *excess;* sometimes *too continued an exercise of a revolutionary power.*[288]

Recent decades have seen, to take just a few examples, *collateral damage* (from the 1970s), *ethnic cleansing* (from the 1990s), and *extraordinary rendition* (from the 2000s).

Euphemisms can be effective for several reasons. Words that are literal synonyms may contrast in their emotional coloring, like *slender* and *skinny*, *fat* and *Rubenesque*, or an obscene word and its genteel synonym. In *The Stuff of Thought* I argued that most euphemisms work more insidiously: not by triggering reactions to the words themselves but by engaging different conceptual interpretations of the state of the world.[289] For example, a euphemism can confer plausible deniability on what is essentially a lie. A listener unfamiliar with the facts could understand *transfer of population* to imply moving vans and train tickets. A choice of words can also imply different motives and hence different ethical valuations. *Collateral damage* implies that a harm was an unintended by-product rather than a desired end, and that makes a legitimate moral difference. One could almost use *collateral damage* with a straight face to describe the hapless worker on the side track who was sacrificed to prevent the runaway trolley from killing five workers on the main one. All of these phenomena—emotional connotation, plausible deniability, and the ascription of motives—can be exploited to alter the way an action is construed.

A second mechanism of moral disengagement is gradualism. People can slide into barbarities a baby step at a time that they would never undertake in a single plunge, because at no point does it feel like they are doing anything terribly different from the current norm.[290] An infamous historical example is the Nazis' euthanizing of the handicapped and mentally retarded and their disenfranchisement, harassment, ghettoization, and deportation of the Jews, which culminated in the events referred to by the ultimate euphemism, *the Final Solution*. Another example is the phasing of decisions in the conduct of war. Material assistance to an ally can morph into military advisors and then into escalating numbers of soldiers, particularly in a war of attrition. The bombing of factories can shade into the bombing of factories near neighborhoods, which can shade into the bombing of neighborhoods. It's unlikely that any participant in the Milgram experiment would have zapped the victim with a 450-volt shock on the first trial; the participants were led to that level in an escalating series, starting with a mild buzz. Milgram's experiment was what game theorists call an Escalation game, which is similar to a War of Attrition.[291] If the participant withdraws from the experiment as the shocks get more severe, he forfeits whatever satisfaction he might have enjoyed from carrying out his responsibilities and advancing the cause of science and thus would have nothing to show for the anxiety he has suffered and the pain he has caused the victim. At each increment, it always seems to pay to stick it out one trial longer and hope that the experimenter will announce that the study is complete.

A third disengagement mechanism is the displacement or diffusion of responsibility. Milgram's mock experimenter always insisted to the participants that he bore full responsibility for whatever happened. When the patter was rewritten and the participant was told that he or she was responsible, compliance plummeted. We have already seen that a second willing participant emboldens the first; Bandura showed that the diffusion of responsibility is a critical factor.[292] When participants in a Milgram-like experiment think that the voltage they have chosen will be averaged with the levels chosen by two other participants, they give stronger shocks. The historical parallels are obvious. "I was only following orders" is the clichéd defense of accused war criminals. And murderous leaders deliberately organize armies, killing squads, and the bureaucracies behind them in such a way that no single person can feel that his actions are necessary or sufficient for the killings to occur.[293]

A fourth way of disabling the usual mechanisms of moral judgment is distancing. We have seen that unless people are in the throes of a rampage or have sunk into sadism, they don't like harming innocent people directly and up close.[294] In the Milgram studies, bringing the victim into the same room as the participant reduced the proportion of participants who delivered the maximum shock by a third. And requiring the participant to force the victim's hand down onto an electrode plate reduced it by more than half. It's safe to say that the pilot of the *Enola Gay* who dropped the atomic bomb over Hiroshima would not have agreed to immolate a hundred thousand people with a flamethrower one at a time. And as we saw in chapter 5, Paul Slovic has confirmed the observation attributed to Stalin that one death is a tragedy but a million deaths is a statistic.[295] People cannot wrap their minds around large (or even small) numbers of people in peril, but will readily mobilize to save the life of a single person with a name and a face.

A fifth means of jiggering the moral sense is to derogate the victim. We have seen that demonizing and dehumanizing a group can pave the way toward harming its members. Bandura confirmed this sequence by allowing some of his participants to overhear the experimenter making an offhand derogatory remark about the ethnicity of a group of people who (they thought) were taking part in the study.[296] The participants who overheard the remark upped the voltage of the shocks they gave to those people. The causal arrow can go in the other direction as well. If people are manipulated into harming someone, they can retroactively downgrade their opinion of the people they have harmed. Bandura found that about half of the participants who had shocked a victim explicitly justified their action. Many did so by blaming the victim (a phenomenon Milgram had noticed as well), writing things like "Poor performance is indicative of laziness and a willingness to test the supervisor."

Social psychologists have identified other gimmicks of moral disengagement, and Bandura's participants rediscovered most of them. They include minimizing the harm ("It would not hurt them too bad"), relativizing the harm ("Everyone is punished for something every day"), and falling back on the requirements of the task ("If doing my job as a supervisor means I must be a son of a bitch, so be it"). The only one they seem to have missed was a tactic called advantageous comparison: "Other people do even worse things."[297]

There is no cure for ideology, because it emerges from many of the cognitive faculties that make us smart. We envision long, abstract chains of causation. We acquire knowledge from other people. We coordinate our behavior with them, sometimes by adhering to common norms. We work in teams, accomplishing feats we could not accomplish on our own. We entertain abstractions, without lingering over every concrete detail. We construe an action in multiple ways, differing in means and ends, goals and by-products.

Dangerous ideologies erupt when these faculties fall into toxic combinations. Someone theorizes that infinite good can be attained by eliminating a demonized or dehumanized group. A kernel of like-minded believers spreads the idea by punishing disbelievers. Clusters of people are swayed or intimidated into endorsing it. Skeptics are silenced or isolated. Self-serving rationalizations allow people to carry out the scheme against what should be their better judgment.

Though nothing can guarantee that virulent ideologies will not infect a country, one vaccine is an open society in which people and ideas move freely and no one is punished for airing dissenting views, including those that seem heretical to polite consensus. The relative immunity of modern cosmopolitan democracies to genocide and ideological civil war is a bit of support for this proposition. The recrudescence of censorship and insularity in regimes that are prone to large-scale violence is the other side of that coin.

PURE EVIL, INNER DEMONS, AND THE DECLINE OF VIOLENCE

At the beginning of this chapter I introduced Baumeister's theory of the myth of pure evil. When people moralize, they take a victim's perspective and assume that all perpetrators of harm are sadists and psychopaths. Moralizers are thereby apt to see historical declines of violence as the outcome of a heroic struggle of justice over evil. The greatest generation defeated the fascists; the civil rights movement defeated the racists; Ronald Reagan's buildup of arms in the 1980s forced the collapse of communism. Now, there surely are evil people in the world—sadistic psychopaths and narcissistic despots obviously qualify—and there surely are heroes. Yet much of the decline in violence seems to have come from changes in the times. Despots died and weren't replaced by new despots; oppressive regimes went out of existence without fighting to the bitter end.

The alternative to the myth of pure evil is that most of the harm that people visit on one another comes from motives that are found in every normal person. And the corollary is that much of the decline of violence comes from people exercising these motives less often, less fully, or in fewer circumstances. The better angels that subdue these demons are the topic of the next chapter. Yet the mere process of identifying our inner demons may be a first step to bringing them under control.

The second half of the 20th century was an age of psychology. Academic research increasingly became a part of the conventional wisdom, including dominance hierarchies, the Milgram and Asch experiments, and the theory of cognitive dissonance. But it wasn't just scientific psychology that filtered into public awareness; it was the general habit of seeing human affairs through a psychological lens. This half-century saw the growth of a species-wide self-consciousness, encouraged by literacy, mobility, and technology: the way the camera follows us in slow-mo, the way we look to us all. Increasingly we see our affairs from two vantage points: from inside our skulls, where the things we experience just *are*, and from a scientist's-eye view, where the things we experience consist of patterns of activity in an evolved brain, with all its illusions and fallacies.

Neither academic psychology nor conventional wisdom is anywhere close to a complete understanding of what makes us tick. But a little psychology can go a long way. It seems to me that a small number of quirks in our cognitive and emotional makeup give rise to a substantial proportion of avoidable human misery.[298] It also seems to me that a shared inkling of these quirks has made some dents in the toll of violence and has the potential to chip away at much more. Each of our five inner demons comes with a design feature that we have begun to notice and would be wise to acknowledge further.

People, especially men, are overconfident in their prospects for success; when they fight each other, the outcome is likely to be bloodier than any of them thought. People, especially men, strive for dominance for themselves and their groups; when contests of dominance are joined, they are unlikely to sort the parties by merit and are likely to be a net loss for everyone. People seek revenge by an accounting that exaggerates their innocence and their adversary's malice; when two sides seek perfect justice, they condemn themselves and their heirs to strife. People can not only overcome their revulsion to hands-on violence but acquire a taste for it; if they indulge it in private, or in cahoots with their peers, they can become sadists. And people can avow a belief they don't hold because they think everyone else avows it; such beliefs can sweep through a closed society and bring it under the spell of a collective delusion.

9

BETTER ANGELS

[It] cannot be disputed that there is some benevolence, however small, infused into our bosom; some spark of friendship for human kind; some particle of the dove, kneaded into our frame, along with the elements of the wolf and serpent. Let these generous sentiments be supposed ever so weak; let them be insufficient to move even a hand or finger of our body; they must still direct the determinations of our mind, and where every thing else is equal, produce a cool preference of what is useful and serviceable to mankind, above what is pernicious and dangerous.

—David Hume, *An Enquiry Concerning the Principles of Morals*

In every era, the way people raise their children is a window into their conception of human nature. When parents believed in children's innate depravity, they beat them when they sneezed; when they believed in innate innocence, they banned the game of dodgeball. The other day when I was riding on my bicycle, I was reminded of the latest fashion in human nature when I passed a mother and her two preschoolers strolling on the side of the road. One was fussing and crying, and the other was being admonished by his mother. As I overtook the trio, I heard a stern Mommy voice enunciating one word: "EMPATHY!"

We live in an age of empathy. So announces a manifesto by the eminent primatologist Frans de Waal, one of a spate of books that have championed this human capability at the end of the first decade of the new millennium.[1] Here is a sample of titles and subtitles that have appeared in just the past two years: *The Age of Empathy, Why Empathy Matters, The Social Neuroscience of Empathy, The Science of Empathy, The Empathy Gap, Why Empathy Is Essential (and Endangered), Empathy in the Global World,* and *How Companies Prosper When They Create Widespread Empathy*. In yet another book, *The Empathic Civilization*, the activist Jeremy Rifkin explains the vision:

> Biologists and cognitive neuroscientists are discovering mirror-neurons—the so-called empathy neurons—that allow human beings and other species to feel and experience another's situation as if it were one's own. We are, it

> appears, the most social of animals and seek intimate participation and companionship with our fellows.
>
> Social scientists, in turn, are beginning to reexamine human history from an empathic lens and, in the process, discovering previously hidden strands of the human narrative which suggest that human evolution is measured not only by the expansion of power over nature, but also by the intensification and extension of empathy to more diverse others across broader temporal and spatial domains. The growing scientific evidence that we are a fundamentally empathic species has profound and far-reaching consequences for society, and may well determine our fate as a species.
>
> What is required now is nothing less than a leap to global empathic consciousness and in less than a generation if we are to resurrect the global economy and revitalize the biosphere. The question becomes this: what is the mechanism that allows empathic sensitivity to mature and consciousness to expand through history?[2]

So it may have been the quest to expand global empathic consciousness, and not just a way to get a brat to stop picking on his sister, that led the roadside mom to drill the concept of empathy into her little boy. Perhaps she was influenced by books like *Teaching Empathy, Teaching Children Empathy,* and *The Roots of Empathy: Changing the World Child by Child,* whose author, according to an endorsement by the pediatrician T. Berry Brazelton, "strives to bring about no less than world peace and protection for our planet's future, starting with schools and classrooms everywhere, one child, one parent, one teacher at a time."[3]

Now, I have nothing against empathy. I think empathy is—in general though not always—a good thing, and I have appealed to it a number of times in this book. An expansion of empathy may help explain why people today abjure cruel punishments and think more about the human costs of war. But empathy today is becoming what love was in the 1960s—a sentimental ideal, extolled in catchphrases (what makes the world go round, what the world needs now, all you need) but overrated as a reducer of violence. When the Americans and Soviets stopped rattling nuclear sabers and stoking proxy wars, I don't think love had much to do with it, or empathy either. And though I like to think I have as much empathy as the next person, I can't say that it's empathy that prevents me from taking out contracts on my critics, getting into fistfights over parking spaces, threatening my wife when she points out I've done something silly, or lobbying for my country to go to war with China to prevent it from overtaking us in economic output. My mind doesn't stop and ponder what it would be like to be the victims of these kinds of violence and then recoil after feeling their pain. My mind never goes in these directions in the first place: they are absurd, ludicrous, unthinkable. Yet options like these clearly were not unthinkable to past generations. The decline of violence may owe something to an expansion of empathy, but it also owes much to harder-boiled faculties like prudence, reason, fairness, self-control, norms and taboos, and conceptions of human rights.

This chapter is about the better angels of our nature: the psychological faculties that steer us away from violence, and whose increased engagement over time can be credited for declines in violence. Empathy is one of these faculties, but it is not the only one. As Hume noted more than 250 years ago, the existence of these faculties cannot be disputed. Though one sometimes still reads that the evolution of beneficence is a paradox for the theory of natural selection, the paradox was resolved decades ago. Controversies remain over the details, but today no biologist doubts that evolutionary dynamics like mutualism, kinship, and various forms of reciprocity can select for psychological faculties that, under the right circumstances, can lead people to coexist peacefully.[4] What Hume wrote in 1751 is certainly true today:

> Nor will those reasoners, who so earnestly maintain the predominant selfishness of human kind, be any wise scandalized at hearing of the weak sentiments of virtue, implanted in our nature. On the contrary, they are found as ready to maintain the one tenet as the other; and their spirit of satire (for such it appears, rather than of corruption) naturally gives rise to both opinions; which have, indeed, a great and almost indissoluble connexion together.[5]

If the spirit of satire leads me to show that empathy has been overhyped, it is not to deny the importance of such sentiments of virtue, nor their indissoluble connection to human nature.

After reading eight chapters on the horrible things that people have done to each other and the darker parts of human nature that spurred them, you have every right to look forward to a bit of uplift in a chapter on our better angels. But I will resist the temptation to please the crowd with *too* happy an ending. The parts of the brain that restrain our darker impulses were also standard equipment in our ancestors who kept slaves, burned witches, and beat children, so they clearly don't make people good by default. And it would hardly be a satisfying explanation of the decline of violence to say that there are bad parts of human nature that make us do bad things and good parts that make us do good things. (War I win; peace you lose.) The exploration of our better angels must show not only how they steer us away from violence, but why they so often fail to do so; not just how they have been increasingly engaged, but why history had to wait so long to engage them fully.

EMPATHY

The word *empathy* is barely a century old. It is often credited to the American psychologist Edward Titchener, who used it in a 1909 lecture, though the

Oxford English Dictionary lists a 1904 usage by the British writer Vernon Lee.[6] Both derived it from the German *Einfühlung* (feeling into) and used it to label a kind of aesthetic appreciation: a "feeling or acting in the mind's muscles," as when we look at a skyscraper and imagine ourselves standing straight and tall. The popularity of the word in English-language books shot up in the mid-1940s, and it soon overtook Victorian virtues such as *willpower* (in 1961) and *self-control* (in the mid-1980s).[7]

The meteoric rise of *empathy* coincided with its taking on a new meaning, one that is closer to "sympathy" or "compassion." The blend of meanings embodies a folk theory of psychology: that beneficence toward other people depends on pretending to be them, feeling what they are feeling, walking a mile in their moccasins, taking their vantage point, or seeing the world through their eyes.[8] This theory is not self-evidently true. In his essay "On a Certain Blindness in Human Beings," William James reflected on the bond between man and man's best friend:

> Take our dogs and ourselves, connected as we are by a tie more intimate than most ties in this world; and yet, outside of that tie of friendly fondness, how insensible, each of us, to all that makes life significant for the other!—we to the rapture of bones under hedges, or smells of trees and lamp-posts, they to the delights of literature and art. As you sit reading the most moving romance you ever fell upon, what sort of a judge is your fox-terrier of your behavior? With all his good will toward you, the nature of your conduct is absolutely excluded from his comprehension. To sit there like a senseless statue, when you might be taking him to walk and throwing sticks for him to catch! What queer disease is this that comes over you every day, of holding things and staring at them like that for hours together, paralyzed of motion and vacant of all conscious life?[9]

So the sense of *empathy* that gets valorized today—an altruistic concern for others—cannot be equated with the ability to think what they are thinking or feel what they are feeling. Let's distinguish several senses of the word that has come to be used for so many mental states.[10]

The original and most mechanical sense of empathy is *projection*—the ability to put oneself into the position of some other person, animal, or object, and imagine the sensation of being in that situation. The example of the skyscraper shows that the object of one's empathy in this sense needn't even *have* feelings, let alone feelings that the empathizer cares about.

Closely related is the skill of *perspective-taking*, namely visualizing what the world looks like from another's vantage point. Jean Piaget famously showed that children younger than about six cannot visualize the arrangement of three toy mountains on a tabletop from the viewpoint of a person seated across from him, a kind of immaturity he called egocentrism. In fairness to children, this ability doesn't come easy to adults either. Reading maps, deciphering "you are here" signs, and mentally rotating three-dimensional objects can tax the best of us, but that should not call our compassion into doubt. More broadly, perspective-taking can embrace guesses about what a person is thinking and feeling as well as what he is seeing, and that brings us to yet another sense of the word *empathy*.

Mind-reading, theory of mind, mentalizing, or *empathic accuracy* is the ability to figure out what someone is thinking or feeling from their expressions, behavior, or circumstances. It allows us to infer, for instance, that a person who has just missed a train is probably upset and is now trying to figure out how to get to his or her destination on time.[11] Mind-reading does not require that we experience the person's experiences ourselves, nor that we care about them, only that we can figure out what they are. Mind-reading may in fact comprise two abilities, one for reading thoughts (which is impaired in autism), the other for reading emotions (which is impaired in psychopathy).[12] Some intelligent psychopaths do learn to read other people's emotional states, the better to manipulate them, though they still fail to appreciate the true emotional texture of those states. An example is a rapist who said of his victims, "They are frightened, right? But, you see, I don't really understand it. I've been frightened myself, and it wasn't unpleasant."[13] And whether or not they truly understand other people's emotional states, they simply don't care. Sadism, schadenfreude, and indifference to the welfare of animals are other cases in which a person may be fully cognizant of the mental states of other creatures but unmoved to sympathize with them.

People do, however, often feel *distress* at witnessing the suffering of another person.[14] This is the reaction that inhibits people from injuring others in a fight, that made the participants in Milgram's experiment anxious about the shocks they thought they were delivering, and that made the Nazi reservists nauseous when they first started shooting Jews at close range. As these examples make all too clear, distress at another's suffering is not the same as a sympathetic concern with their well-being. Instead it can be an unwanted reaction which people may suppress, or an annoyance they may try to escape. Many of us trapped on a plane with a screaming baby feel plenty of distress, but our sympathy is likely to be more with the parent than with the child, and our strongest desire may be to find another seat. For many years a charity called Save the Children ran magazine ads with a heartbreaking photograph of a destitute child and the caption "You can save Juan Ramos for five cents a day. Or you can turn the page." Most people turn the page.

Emotions can be *contagious*. When you're laughing, the whole world laughs with you; that's why situation comedies have laugh tracks and why bad comedians punch up their punch lines with a *bada-bing* rim shot that simulates a staccato burst of laughter.[15] Other examples of emotional contagion are the tears at a wedding or funeral, the urge to dance at a lively party, the panic during a bomb scare, and the spreading nausea on a heaving boat. A weaker version of emotional contagion consists of vicarious responses, as when we wince in sympathy with an injured athlete or flinch when James Bond is tied to a chair and smacked around. Another is motor mimicry, as when we open our mouths when trying to feed applesauce to a baby.

Many empathy fans write as if emotional contagion were the basis of the sense of "empathy" that is most pertinent to human welfare. The sense of empathy we value the most, though, is a distinct reaction that may be called sympathetic concern, or sympathy for short. Sympathy consists in aligning another entity's well-being with one's own, based on a cognizance of their pleasures and pains. Despite the easy equation of sympathy with contagion, it's easy to see why they're not the same.[16] If a child has been frightened by a barking dog and is howling in terror, my sympathetic response is not to howl in terror with her, but to comfort and protect her. Conversely, I may have exquisite sympathy for a person whose suffering I cannot possibly experience vicariously, like a woman in childbirth, a woman who has been raped, or a sufferer of cancer pain. And our emotional reactions, far from automatically duplicating those of other people, can flip 180 degrees depending on whether we feel we are in alliance or competition with them. When a sports fan watches a home game, he is happy when the crowd is happy and dejected when the crowd is dejected. When he watches an away game, he is dejected when the crowd is happy and happy when the crowd is dejected. All too often, sympathy determines contagion, not the other way around.

Today's empathy craze has been set off by scrambling the various senses of the word *empathy*. The confusion is crystallized in the meme that uses *mirror neurons* as a synonym for *sympathy*, in the sense of compassion. Rifkin writes of "so-called empathy neurons that allow human beings and other species to feel and experience another's situation as if it were one's own," and concludes that we are "a fundamentally empathic species" which seeks "intimate participation and companionship with our fellows." The mirror-neuron theory assumes that sympathy (which it blurs with contagion) is hardwired into our brains, a legacy of our primate nature, and has only to be exercised, or at least not repressed, for a new age to dawn. Unfortunately, Rifkin's promise of a "leap to global empathic consciousness and in less than a generation" is based on a dodgy interpretation of the neuroscience.

In 1992 the neuroscientist Giacomo Rizzolatti and his colleagues discovered neurons in the brain of a monkey that fired both when the monkey picked up a raisin and when the monkey watched a person pick up a raisin.[17] Other neurons responded to other actions, whether performed or perceived, such as touching and tearing. Though neuroscientists ordinarily can't impale the brains of human subjects with electrodes, we have reason to believe that people have mirror neurons too: neuroimaging experiments have found areas in the parietal lobe and inferior frontal lobe that light up both when people move and when they see someone else move.[18] The discovery of mirror neurons is important, though not completely unexpected: we could hardly use a verb in both the first person and the third person unless our brains were able to represent an action in the same way regardless of who performs it. But the discovery soon inflated an extraordinary bubble of hype.[19] One neuroscientist claimed that mirror neurons would do for neuroscience what DNA did for biology.[20] Others, aided and abetted by science journalists, have touted mirror neurons as the biological basis of language, intentionality, imitation, cultural learning, fads and fashions, sports fandom, intercessory prayer, and, of course, empathy.

A wee problem for the mirror-neuron theory is that the animals in which the neurons were discovered, rhesus macaques, are a nasty little species with no discernible trace of empathy (or imitation, to say nothing of language).[21] Another problem, as we shall see, is that mirror neurons are mostly found in regions of the brain that, according to neuroimaging studies, have little to do with empathy in the sense of sympathetic concern.[22] Many cognitive neuroscientists suspect that mirror neurons may have a role in mentally representing the concept of an action, though even that is disputed. Most reject the extravagant claims that they can explain uniquely human abilities, and today virtually no one equates their activity with the emotion of sympathy.[23]

There are, to be sure, parts of the brain, particularly the insula, which are metabolically active both when we have an unpleasant experience and when we respond to someone else having an unpleasant experience.[24] The problem is that this overlap is an *effect*, rather than a *cause*, of sympathy with another's well-being. Recall the experiment in which the insula lit up when a participant received a shock and also when he or she watched an innocent person receiving a shock. The same experiment revealed that when the shock victim had cheated the male subjects out of their money, their insulas showed no response, while the striatum and orbital cortex lit up in sweet revenge.[25]

Empathy, in the morally relevant sense of sympathetic concern, is not an automatic reflex of our mirror neurons. It can be turned on and off and even inverted into counterempathy, namely feeling good when someone else feels bad and vice versa. Revenge is one trigger for counterempathy, and the flip-flopping response of the sports fan tells us that competition can trigger it as well. The psychologists John Lanzetta and Basil Englis glued electrodes to the faces and fingers of participants and had them play an investment game with another (bogus) participant.[26] They were told either that the two were working together or that they were in competition (though the actual returns did not depend on what the other participant did). Market gains were signaled by an uptick of a counter; losses were signaled by a mild shock. When the participants thought they were cooperating, the electrodes picked up a visceral calming and a trace of a smile whenever their opposite number gained money, and a burst of sweat and the trace of a frown whenever he was shocked. When they thought they were competing with him, it went the other way: they relaxed and smiled when he suffered, and tensed up and frowned when he did well.

The problem with building a better world through empathy, in the sense of contagion, mimicry, vicarious emotion, or mirror neurons, is that it cannot be counted on to trigger the kind of empathy we want, namely sympathetic concern for others' well-being. Sympathy is endogenous, an effect rather than a cause of how people relate to one another. Depending on how beholders conceive of a relationship, their response to another person's pain may be empathic, neutral, or even counterempathic.

In the previous chapter we explored the circuitry of the brain that underlies our tendencies to violence; now let's see the parts that underlie our better angels. The search for empathy in the human brain has confirmed that vicarious feelings are dimmed or amplified by the rest of the empathizer's beliefs. Claus Lamm, Daniel Batson, and Jean Decety had participants take the perspective of a (fictitious) patient with ringing in his ears while he got "treated" with an experimental cure consisting of blasts of noise over headphones, which made the patient visibly wince.[27] The pattern of activity in the participants' brains as they empathized with the patient overlapped with the pattern that resulted when they themselves heard the noise. One of the active areas was a part of the insula, the island of cortex that, as we have seen, represents literal and metaphorical gut feelings (see figure 8–3). Another was the amygdala, the almond-shaped organ that responds to fearful and distressing stimuli (see figure 8–2). A third was the anterior medial cingulate cortex (see figure 8–4), a strip of cortex on the inward-facing wall of the cerebral hemisphere that is involved in the motivational aspect of pain—not the literal stinging sensation, but the strong desire to turn it off. (Studies of vicarious pain generally don't show activation in the parts of the brain that register the actual bodily sensation; that would be closer to a hallucination than to empathy.) The participants were never put in the kind of situation that evokes counterempathy, like competition or revenge, but their reactions were pushed around by their cognitive construal of the situation. If they had been told that the treatment worked, so the patient's pain had been worthwhile, their brains' vicarious and distressed responses were damped down.

The overall picture that has emerged from the study of the compassionate brain is that there is no empathy center with empathy neurons, but complex patterns of activation and modulation that depend on perceivers' interpretation of the straits of another person and the nature of their relationship with the person. A general atlas of empathy might look more or less as follows.[28] The temporoparietal junction and nearby sulcus (groove) in the superior temporal lobe assess another person's physical and mental state. The dorsolateral prefrontal cortex and the nearby frontal pole (the tip of the frontal lobe) compute the specifics of the situation and one's overall goals in it. The orbital and ventromedial cortex integrate the results of these computations and modulate the responses of the evolutionarily older, more emotional parts of the brain. The amygdala responds to fearful and distressing stimuli, in conjunction with interpretations from the nearby temporal pole (the tip of the temporal lobe). The insula registers disgust, anger, and vicarious pain. The cingulate cortex helps to switch control among brain systems in response to urgent signals, such as those sent by circuits that are calling for incompatible responses, or those that register physical or emotional pain. And unfortunately for the mirror-neuron theory, the areas of the brain richest in mirror neurons, such as parts of the frontal lobe that plan motor movements (the rearmost portions above the Sylvian fissure) and the parts of the parietal lobes that register the body sense, are mostly uninvolved, except for the parts of the parietal lobes that keep track of whose body is where.

In fact, the brain tissue that is closest to empathy in the sense of compassion is neither a patch of cortex nor a subcortical organ but a system of hormonal plumbing. Oxytocin is a small molecule produced by the hypothalamus which acts on the emotional systems of the brain, including the amygdala and striatum, and which is released by the pituitary gland into the bloodstream, where it can affect the rest of the body.[29] Its original evolutionary function was to turn on the components of motherhood, including giving birth, nursing, and nurturing the young. But the ability of the hormone to reduce the fear of closeness to other creatures lent itself over the course of evolutionary history to being co-opted to supporting other forms of affiliation. They include sexual arousal, heterosexual bonding in monogamous species, marital and companionate love, and sympathy and trust among nonrelatives. For these reasons, oxytocin is sometimes called the cuddle hormone. The reuse of the hormone in so many forms of human closeness supports a suggestion by Batson that maternal care is the evolutionary precursor of other forms of human sympathy.[30]

In one of the odder experiments in the field of behavioral economics, Ernst Fehr and his collaborators had people play a Trust game, in which they hand over money to a trustee, who multiplies it and then returns however much he feels like to the participant.[31] Half the participants inhaled a nasal spray containing oxytocin, which can penetrate from the nose to the brain, and the other half inhaled a placebo. The ones who got the oxytocin turned over more of their money to the stranger, and the media had a field day with fantasies of car dealers misting the hormone through their showroom ventilating systems to snooker innocent customers. (So far, no one has proposed spraying it from crop dusters to accelerate global empathic consciousness.) Other experiments have shown that sniffing oxytocin makes people more generous in an Ultimatum game (in which they divide a sum while anticipating the response of a recipient, who can veto the deal for both of them), but not in a Dictator game (where the recipient has to take it or leave it, and the proposer needn't take his reaction into account). It seems likely that the oxytocin network is a vital trigger in the sympathetic response to other people's beliefs and desires.

In chapter 4 I alluded to Peter Singer's hypothesis of an expanding circle of empathy, really a circle of sympathy. Its innermost kernel is the nurturance we feel toward our own children, and the most reliable trigger for this tenderness is the geometry of the juvenile face—the phenomenon of perception we call cuteness. In 1950 the ethologist Konrad Lorenz noted that entities with measurements typical of immature animals evoke feelings of tenderness in the beholder. The lineaments include a large head, cranium, forehead, and eyes, and a small snout, jaw, body, and limbs.[32] The cuteness reflex was originally an adaptation in mothers to care for their own offspring, but the triggering features may have been exaggerated in the offspring themselves (to the extent they are compatible with its own health) to tilt the mother's response toward nurturance and away from infanticide.[33] Species that are lucky enough to possess the geometry of babies may elicit the *awwwww!* response from human beholders and benefit from our sympathetic concern. We find mice and rabbits more adorable than rats and opossums, doves more sympathetic than crows, baby seals more worthy of protection than mink and other weaselly furbearers. Cartoonists exploit the reflex to make their characters more lovable, as do the designers of teddy bears and anime characters. In a famous essay on the evolution of Mickey Mouse, Stephen Jay Gould plotted an increase in the size of the rodent's eyes and cranium during the decades in which his personality changed from an obnoxious brat to a squeaky-clean corporate icon.[34] Gould did not live to see the 2009 makeover in which the Walt Disney Company, concerned that today's children expect "edgier," more "dangerous" characters, unveiled a video game in which Mick's features had de-evolved to a more ratlike anatomy.[35]

As we saw in chapter 8, cuteness is a nuisance to conservation biologists because it attracts disproportionate concern for a few charismatic mammals. One organization figured they might as well put the response to good use and branded itself with the doe-eyed panda. The same trick is used by humanitarian organizations who find photogenic children for their ad campaigns. The psychologist Leslie Zebrowitz has shown that juries treat defendants with more juvenile facial features more sympathetically, a travesty of justice we can attribute to the workings of our sense of sympathy.[36] Physical beauty is yet another sympathy-induced injustice. Unattractive children are punished more harshly by parents and teachers and are more likely to be victims of abuse.[37] Unattractive adults are judged to be less honest, kind, trustworthy, sensitive, and even intelligent.[38]

Of course, we do manage to sympathize with our adult friends and relatives, including the ugly ones. But even then our sympathy is spread not indiscriminately but within a delimited circle within which we apply a suite of moral emotions. Sympathy has to work in concert with these other emotions because social life cannot be a radiation of warm and fuzzy feelings in all directions. Friction is unavoidable in social life: toes get stepped on, noses put out of joint, fur rubbed the wrong way. Together with sympathy we feel guilt and forgiveness, and these emotions tend to apply within the same circle: the people we sympathize with are the people we feel guilty about hurting and the people we find easiest to forgive when they hurt us.[39] Roy Baumeister, Arlene Stillwell, and Todd Heatherton reviewed the social-psychological literature on guilt and found that it went hand in hand with empathy. More empathic people are also more guilt-prone (particularly women, who excel at both emotions), and it is the targets of our empathy who engage our guilt. The effect is enormous: when people are asked to recall incidents that made them feel guilty, 93 percent involved families, friends, and lovers; only 7 percent involved acquaintances or strangers. The proportions were similar when it came to memories of eliciting guilt: we guilt-trip our friends and families, not acquaintances and strangers.

Baumeister and his collaborators explain the pattern with a distinction we will return to in the section on morality. Sympathy and guilt, they note, operate within a circle of *communal* relationships.[40] They are less likely to be felt in *exchange* or equality-matching relationships, the kind we have with acquaintances, neighbors, colleagues, associates, clients, and service providers. Exchange relationships are regulated by norms of fairness and are accompanied by emotions that are cordial rather than genuinely sympathetic. When we harm them or they harm us, we can explicitly negotiate the fines, refunds, and other forms of compensation that rectify the harm. When that is not possible, we reduce our distress by distancing ourselves from them or derogating them. The businesslike quid pro quo negotiations that can repair an exchange relationship are, we shall see, generally taboo in our communal relationships, and the option of severing a communal relationship comes with a high cost.[41] So we repair our communal relationships with the messier but longer-lasting emotional glue of sympathy, guilt, and forgiveness.

So what are the prospects that we can expand the circle of sympathy outward from babies, fuzzy animals, and the people bound to us in communal relationships, to lasso in larger and larger sets of strangers? One set of predictions comes from the theory of reciprocal altruism and its implementation in Tit for Tat and other strategies that are "nice" in the technical sense that they cooperate on the first move and don't defect until defected upon. If people are nice in this sense, they should have some tendency to be sympathetic to strangers, with the ultimate (that is, evolutionary) goal of probing for the possibility of a mutually beneficial relationship.[42] Sympathy should be particularly likely to spring into action when an opportunity presents itself to confer a large benefit to another person at a relatively small cost to oneself, that is, when we come across a person in need. It should also be fired up where there are common interests that grease the skids toward a mutually beneficial relationship, such as having similar values and belonging to a common coalition.

Neediness, like cuteness, is a general elicitor of sympathy. Even toddlers go out of their way to help someone in difficulty or to comfort someone in distress.[43] In his studies of empathy, Batson found that when students are faced with someone in need, such as a patient recovering from leg surgery, they respond with sympathy even when the needy one falls outside their usual social circle. The sympathy is triggered whether the patient is a fellow student, an older stranger, a child, or even a puppy.[44] The other day I came across an overturned horseshoe crab on the beach, its dozen legs writhing uselessly in the air. When I righted it and it slithered beneath the waves, I felt a surge of happiness.

With less easily helped individuals, a perception of shared values and other kinds of similarity makes a big difference.[45] In a seminal experiment, the psychologist Dennis Krebs had student participants watch a second (fake) participant play a perverse game of roulette that paid him whenever the ball landed on an even number and shocked him when it landed on an odd number.[46] The player had been introduced either as a fellow student in the same field who had a similar personality, or as a nonstudent with a dissimilar personality. When the participants thought they were similar to the player, they sweated and their hearts pounded more when they saw him get shocked. They said they felt worse while anticipating his shock and were more willing to get shocked themselves and forgo payments to spare their counterpart additional pain.

Krebs explained the sacrifice of his participants on behalf of their fellows with an idea he called the empathy-altruism hypothesis: empathy encourages altruism.[47] The word *empathy*, as we have seen, is ambiguous, and so we are really dealing with two hypotheses. One, based on the "sympathy" sense, is that our emotional repertoire includes a state in which another person's well-being matters to us—we are pleased when the person is happy, and upset when he or she is not—and that this state motivates us to help them with no ulterior motive. If true, this idea—let's call it the *sympathy*-altruism hypothesis—would refute a pair of old theories called psychological hedonism, according to which people only do things that give them pleasure, and psychological egoism, according to which people only do things that provide them with a benefit. Of course there are circular versions of these theories, in which the very fact that a person helps someone is taken as proof that it *must* feel good or benefit him, if only to scratch an altruistic itch. But any testable version of these cynical theories must identify some *independent* ulterior motive for the help extended, such as assuaging one's own distress, avoiding public censure, or garnering public esteem.

The word *altruism* is ambiguous too. The "altruism" in the empathy-altruism hypothesis is altruism in the psychological sense of a motive to benefit another organism as an end in itself rather than as a means to some other end.[48] This differs from altruism in the evolutionary biologist's sense, which is defined in terms of behavior rather than motives: biological altruism consists of behavior that benefits another organism at a cost to oneself.[49] (Biologists use the term to help distinguish the two ways in which one organism can benefit another. The other way is called mutualism, where an organism benefits another one while also benefiting itself, as with an insect pollinating a plant, a bird eating ticks off the back of a mammal, and roommates with similar tastes enjoying each other's music.)

In practice, the biologist's and psychologist's sense of altruism often coincide, because if we have a motive to do something, we're often willing to incur a cost to do it. And despite a common misunderstanding, evolutionary explanations for biological altruism (such as that organisms benefit their kin or exchange favors, both of which help their genes in the long run) are perfectly compatible with psychological altruism. If natural selection favored costly helping of relatives or of potential reciprocation partners because of the long-term benefits to the genes, it did so by endowing the brain with a direct motive to help those beneficiaries, with no thought of its own welfare. The fact that the altruist's genes may benefit in the long run does not expose the altruist as a hypocrite or undermine her altruistic motives, because the genetic benefit never figures as an explicit goal in her brain.[50]

The first version of the empathy-altruism hypothesis, then, is that psychological altruism exists, and that it is motivated by the emotion we call sympathy. The second version is based on the "projection" and "perspective-taking" senses of empathy.[51] According to this hypothesis, adopting someone's viewpoint, whether by imagining oneself in his or her shoes or imagining what it is like to be that person, induces a state of sympathy for the person (which would then impel the perspective-taker to act altruistically toward the target if the sympathy-altruism hypothesis is true as well). One might call this the perspective-sympathy hypothesis. This is the hypothesis relevant to the question raised in chapters 4 and 5 of whether journalism, memoir, fiction, history, and other technologies of vicarious experience have expanded our collective sense of sympathy and helped drive the Humanitarian Revolution, the Long Peace, the New Peace, and the Rights Revolutions.

Though Batson doesn't always distinguish the two versions of the empathy-altruism hypothesis, his two-decade-long research project has supported both of them.[52]

Let's start with the sympathy-altruism hypothesis and compare it to the cynical alternative in which people help others only to reduce their own distress. Participants in one study watched an ersatz fellow participant, Elaine, get repeatedly shocked in a learning experiment.[53] (The male participants were introduced to Charlie rather than Elaine.) Elaine becomes visibly upset as the session proceeds, and the participant is given an opportunity to take her place. In one condition, the participant has finished her obligation to the experimenter and is free to leave, so taking Elaine's place would be genuinely altruistic. In another, the participant doesn't take Elaine's place and has to watch Elaine get shocked for another eight sessions. Batson reasoned that if the only reason people volunteer to take poor Elaine's place is to reduce their own distress at the sight of her suffering, they won't bother if they are free to leave. Only if they have to endure the sight and sound of her moaning will they prefer to get shocked themselves. As in Krebs's experiment, the participant's sympathy was manipulated by telling her either that she and Elaine had the same values and interests, or that they had incompatible ones (for example, if the participant read *Newsweek*, Elaine would be described as reading *Cosmo* and *Seventeen*). Sure enough, when participants felt themselves to be similar to Elaine, they relieved her of being shocked, whether or not they had to watch her suffer. If they felt themselves to be different, they took her place only when the alternative was to watch the suffering. Together with other studies, the experiment suggests that by default people help others egoistically, to relieve their own distress at having to watch them suffer. But when they sympathize with a victim, they are overcome by a motive to reduce her suffering whether it eases their distress or not.

Another set of experiments tested a second ulterior motive to helping, namely the desire to be seen as doing the socially acceptable thing.[54] This time, rather than manipulating sympathy experimentally, Batson and his collaborators exploited the fact that people spontaneously vary in how sympathetic they feel. After the participants heard Elaine worrying aloud about the impending shocks, they were asked to indicate the degree to which they felt sympathetic, moved, compassionate, tender, warm, and soft-hearted. Some participants wrote high numbers next to these adjectives; others wrote low ones.

Once the procedure began, and long-suffering Elaine started getting zapped and was visibly unhappy about it, the experimenters used sneaky ways of assessing whether any desire on the part of the participants to relieve her distress sprang from pure beneficence or a desire to look good. One study tapped the participants' mood with a questionnaire, and then either gave them the opportunity to relieve Elaine by doing well on a task of their own, or simply dismissed Elaine without the participant being able to claim any credit. The empathizers felt equally relieved in both cases; the nonempathizers only if they were the ones that set her free. In another, the participants had to qualify for an opportunity to take Elaine's place by scoring well in a letter-finding task they had been led to believe was either easy (so there was no way to fake a bad performance and get off the hook) or hard (so they could take a dive and plausibly get out of being asked to make the sacrifice). The nonempathizers took the dive and did worse in the so-called hard task; the empathizers did even *better* on the hard task, where they knew an extra effort would be needed to allow them to suffer in Elaine's stead. The emotion of sympathy, then, can lead to genuine moral concern in Kant's sense of treating a person as an end and not a means to an end—in this case, not even as a means to the end of feeling good about having helped the person.

In these experiments, a person was rescued from a harm caused by someone else, the experimenter. Does sympathy-induced altruism dampen one's *own* tendency to exploit someone, or to retaliate in response to a provocation? It does. In other experiments, Batson had women play a one-shot Prisoner's Dilemma in which they and a (fictitious) fellow participant bid cards that could net them various numbers of raffle tickets, framed as a business transaction.[55] Most of the time they did what game theorists say is the optimal strategy: they defected. They chose to bid a card that protected them against being a sucker and that offered them the chance to exploit their partner, while leaving them with a worse outcome than if the two of them had cooperated by bidding a different card. But when the participant read a personal note from her otherwise anonymous partner and was induced to feel empathy for her, her rate of cooperating jumped from 20 percent to 70 percent. In a second experiment, a new group of women played an *Iterated* Prisoner's Dilemma game, which gave them an opportunity to retaliate against a partner's defection with a defection of their own. They cooperated in response to a defection only 5 percent of the time. But when they were induced beforehand to empathize with their partner, they were far more forgiving, and cooperated 45 percent of the time.[56] Sympathy, then, can mitigate self-defeating exploitation and costly retaliation.

In these experiments, sympathy was manipulated indirectly, by varying the similarity in values between a participant and the target, or it was entirely endogenous: the experimenters counted on some participants spontaneously being more empathic than others, for whatever reason. The key question for understanding the decline of violence is whether sympathy can be pushed around exogenously.

Sympathy, recall, tends to be expressed in communal relationships, the kind that are also accompanied by guilt and forgiveness. Anything that creates a communal relationship, then, should also create sympathy. A prime communality-builder is inducing people to cooperate in a project with a superordinate goal. (The classic example is the warring boys at the Robbers Cave camp, who had to pull together to haul a bus out of the mud.) Many conflict-resolution workshops operate by a similar principle: they bring adversaries together in friendly surroundings where they get to know each other as individuals, and they are tasked with the superordinate goal of figuring out how to resolve the conflict. These circumstances can induce mutual sympathy, and the workshops often try to help it along with exercises in which the participants adopt each other's viewpoints.[57] But in all these cases, cooperation is being forced upon the participants, and it's obviously impractical to get billions of people together in supervised conflict-resolution workshops.

The most powerful exogenous sympathy trigger would be one that is cheap, widely available, and already in place, namely the perspective-taking that people engage in when they consume fiction, memoir, autobiography, and reportage. So the next question in the science of empathy is whether perspective-taking from media consumption actually engages sympathy for the writers and talking heads, and for members of the groups they represent.

In several studies the Batson team convinced participants they were helping with market research for the university radio station.[58] They were asked to evaluate a pilot show called *News from the Personal Side*, a program that aimed to "go beyond the facts of local events to report how these events affect the lives of the individuals involved." One set of participants was asked to "focus on the technical aspects of the broadcast" and "take an objective perspective toward what is described," not getting caught up in the feelings of the interview subject. Another set was asked to "imagine how the person who is interviewed feels about what has happened and how it has affected his or her life"—a manipulation of perspective-taking that ought to instill a state of sympathy. Admittedly, the manipulation is a bit ham-handed: people are not generally told how to think and feel as they read a book or watch the news. But writers know that audiences are most engaged in a story when there is a protagonist whose viewpoint they are seduced into taking, as in the old advice to aspiring scriptwriters, "Find a hero; put him in trouble." So presumably real media also rope their audiences into sympathy with a lead character without the need for explicit orders.

A first experiment showed that the sympathy induced by perspective-taking was as sincere as the kind found in the studies of shocked Elaine.[59] Participants listened to an interview with Katie, who lost her parents in a car crash and was struggling to bring up her younger siblings. They were later presented with an opportunity to help her out in small ways, such as babysitting and giving her lifts. The experimenters manipulated the sign-up sheet so that it looked either as if a lot of students had put their names down, creating peer pressure for them to do the same, or as if only two had, allowing the students to feel comfortable ignoring her plight. The participants who had focused on the technical aspects of the interview signed up to help only if many of their peers had done so; the ones who had listened from Katie's point of view signed up regardless of what their peers had done.

It's one thing to sympathize with a character in need, but it's another to generalize one's sympathy to the group that the character represents. Do readers sympathize just with Uncle Tom or with all African American slaves? With Oliver Twist or with orphaned children in general? With Anne Frank or with all victims of the Holocaust? In an experiment designed to test for such generalizations, students listened to the plight of Julie, a young woman who had contracted AIDS from a blood transfusion after a car accident. (The experiment was run before effective treatments had been discovered for that often-fatal disease.)

> Well, as you can imagine, it's pretty terrifying. I mean, every time I cough or feel a bit run down, I wonder, is this it? Is this the beginning—you know—of the slide? Sometimes I feel pretty good, but in the back of my mind it's always there. Any day I could take a turn for the worse [pause]. And I know that—at least right now—there's no escape. I know they're trying to find a cure—and I know that we all die. But it all seems so unfair. So horrible. Like a nightmare. [pause] I mean, I feel like I was just starting to live, and now, instead, I'm dying. [pause] It can really get you down.[60]

Later, when the students were asked to fill out a questionnaire on attitudes toward people with AIDS, the perspective-takers had become more sympathetic than the technical evaluators, showing that sympathy had indeed spread from the individual to the class she represented. But there was an important twist. The effect of perspective-taking on sympathy was gated by moralization, as we might expect from the fact that sympathy is not an automatic reflex. When Julie confessed to having contracted the disease after a summer of unprotected promiscuous sex, the perspective-takers were more sympathetic to the broad class of victims of AIDS, but they were no more sympathetic to the narrower class of *young women* with AIDS. Similar results came out of a study in which students of both sexes listened to the plight of a man who became homeless either because he had come down with an illness or because he had grown tired of working.

The team of psychologists then pushed the outside of the envelope by seeing how much sympathy they could induce for convicted murderers.[61] It's not that anyone necessarily *wants* people to develop warm feelings toward murderers. But at least some degree of sympathy for the unsympathetic may be necessary to oppose cruel punishments and frivolous executions, and we can imagine that a grain of sympathy of this sort may have led to the reforms of criminal punishment during the Humanitarian Revolution. Batson didn't press his luck by trying to induce sympathy for a psychopathic predator, but he artfully invented a typical crime-blotter homicide in which the perpetrator had been provoked by a victim who was not much more likable than he was. Here is James's story of how he came to kill his neighbor:

> Pretty soon, things went from bad to worse. He'd dump garbage over the fence into my back yard. I sprayed red paint all over the side of his house. Then he set fire to my garage with my car in it. He knew that car was my pride and joy. I really loved it and kept it in great shape. By the time I woke

> up and they got the fire out, the car was ruined—totaled! And he just laughed! I went crazy—not yelling; I didn't say anything, but I was shaking so hard I could hardly stand up, I decided right then that he had to die. That night when he came home, I was waiting on his front porch with my hunting rifle. He laughed at me again and said I was chicken, that I didn't have the guts to do it. But I did. I shot him four times; he died right there on the porch. I was still standing there holding the rifle when the cops came.
>
> [*Interviewer:* Do you regret doing it?]
>
> Now? Sure. I know that murder is wrong and that nobody deserves to die like that, not even him. But at the time all I wanted was to make him pay—big—and to get him out of my life. [Pause] When I shot him, I felt this big sense of relief and release. I felt free. No anger; no fear; no hate. But that feeling lasted only a minute or two. He was the one that was free; I was going to be in prison for the rest of my life. [Pause] And here I am.

The perspective-takers did feel a bit more sympathy for James himself than did the technical evaluators, but it translated into just a sliver of a more positive attitude toward murderers in general.

But then there was a twist on the twist. A week or two later the participants got a phone call out of the blue from a pollster who was doing a survey on prison reform. (The caller was in cahoots with the experimenters, but none of the students figured this out.) Tucked into the opinion poll was an item on attitudes toward murderers, similar to one in the questionnaire that the students had filled out in the lab. At this distance, the effects of perspective-taking made a difference. The students who had tried a couple of weeks before to imagine what James had been feeling showed a noticeable bump in their attitude toward convicted murderers. The delayed influence is what researchers in persuasion call a sleeper effect. When people are exposed to information that changes their attitudes in a way they don't approve of—in this case, warmer feelings toward murderers—they are aware of the unwanted influence and consciously cancel it out. Later, when their guard is down, their change of heart reveals itself. The upshot of the study is that even when a stranger belongs to a group that people are strongly inclined to dislike, listening to his story while taking his perspective can genuinely expand their sympathy for him and for the group he represents, and not just during the few minutes after hearing the story.

People in a connected world are exposed to the stories of strangers through many channels, including face-to-face encounters, interviews in the media, and memoirs and autobiographical accounts. But what about the portion of their information stream that is set in make-believe worlds—the fictional stories, films, and television dramas in which audiences voluntarily lose themselves? The pleasure in a story comes from taking a character's vantage point and in comparing the view to that from other vantage points, such as those of the other characters, of the narrator, and of the reader himself or herself. Could fiction be a stealthy way to expand people's sympathy? In an 1856 essay George Eliot defended this psychological hypothesis:

> Appeals founded on generalizations and statistics require a sympathy ready-made, a moral sentiment already in activity; but a picture of human life such as a great artist can give, surprises even the trivial and the selfish into that attention to what is apart from themselves, which may be called the raw material of moral sentiment. When Scott takes us into Luckie Mucklebackit's cottage, or tells the story of "The Two Drovers,"—when Wordsworth sings to us the reverie of "Poor Susan,"—when Kingsley shows us Alton Locke gazing yearningly over the gate which leads from the highway into the first wood he ever saw,—when Hornung paints a group of chimney-sweepers,—more is done towards linking the higher classes with the lower, towards obliterating the vulgarity of exclusiveness, than by hundreds of sermons and philosophical dissertations. Art is the nearest thing to life; it is a mode of amplifying experience and extending our contact with our fellow-men beyond the grounds of our personal lot.[62]

Today the historian Lynn Hunt, the philosopher Martha Nussbaum, and the psychologists Raymond Mar and Keith Oatley, among others, have championed the reading of fiction as an empathy expander and a force toward humanitarian progress.[63] One might think that literary scholars would line up to join them, eager to show that their subject matter is a force for progress in an era in which students and funding are staying away in droves. But many literary scholars, such as Suzanne Keen in *Empathy and the Novel,* bristle at the suggestion that reading fiction can be morally uplifting. They see the idea as too middlebrow, too therapeutic, too kitsch, too sentimental, too Oprah. Reading fiction can just as easily cultivate schadenfreude, they point out, from gloating over the misfortunes of unsympathetic characters. It can perpetuate condescending stereotypes of "the other." And it can siphon sympathetic concern away from the living beings who could benefit from it and toward appealing victims who don't actually exist. They also note, correctly, that we don't have a trove of good laboratory data showing that fiction expands sympathy. Mar, Oatley, and their collaborators have shown that readers of fiction have higher scores on tests of empathy and social acumen, but that correlation doesn't show whether reading fiction makes people more empathic or empathic people are more likely to read fiction.[64]

It would be surprising if fictional experiences didn't have similar effects to real ones, because people often blur the two in their memories.[65] And a few experiments do suggest that fiction can expand sympathy. One of Batson's radio-show experiments included an interview with a heroin addict who the students had been told was either a real person or an actor.[66] The listeners who were asked to take his point of view became more sympathetic to heroin addicts in general, even when the speaker was fictitious (though the increase was greater when they thought he was real). And in the hands of a skilled narrator, a fictitious victim can elicit even *more* sympathy than a real one. In his book *The Moral Laboratory*, the literary scholar Jèmeljan Hakemulder reports experiments in which participants read similar facts about the plight of Algerian women through the eyes of the protagonist in Malike Mokkeddem's novel *The Displaced* or from Jan Goodwin's nonfiction exposé *Price of Honor*.[67] The participants who read the novel became more sympathetic to Algerian women than those who read the true-life account; they were less likely, for example, to blow off the women's predicament as a part of their cultural and religious heritage. These experiments give us some reason to believe that the chronology of the Humanitarian Revolution, in which popular novels preceded historical reform, may not have been entirely coincidental: exercises in perspective-taking do help to expand people's circle of sympathy.

The science of empathy has shown that sympathy can promote genuine altruism, and that it can be extended to new classes of people when a beholder takes the perspective of a member of that class, even a fictitious one. The research gives teeth to the speculation that humanitarian reforms are driven in part by an enhanced sensitivity to the experiences of living things and a genuine desire to relieve their suffering. And as such, the cognitive process of perspective-taking and the emotion of sympathy must figure in the explanation for many historical reductions in violence. They include institutionalized violence such as cruel punishments, slavery, and frivolous executions; the everyday abuse of vulnerable populations such as women, children, homosexuals, racial minorities, and animals; and the waging of wars, conquests, and ethnic cleansings with a callousness to their human costs.

At the same time, the research reminds us why we should not aim for an "age of empathy" or an "empathic civilization" as the solution to our problems. Empathy has a dark side.[68]

For one thing, empathy can *subvert* human well-being when it runs afoul of a more fundamental principle, fairness. Batson found that when people empathized with Sheri, a ten-year-old girl with a serious illness, they also opted for her to jump a queue for medical treatment ahead of other children who had waited longer or needed it more. Empathy would have consigned these children to death and suffering because they were nameless and faceless. People who learned of Sheri's plight but did not empathize with her acted far more fairly.[69] Other experiments make the point more abstractly. Batson found that in a Public Goods game (where people can contribute to a pool that gets multiplied and redistributed to the contributors), players who were led to empathize with another player (for example, by reading about how she had just broken up with her boyfriend) diverted their contributions to her, starving the public commonwealth to everyone's detriment.[70]

The tradeoff between empathy and fairness is not just a laboratory curiosity; it can have tremendous consequences in the real world. Great harm has befallen societies whose political leaders and government employees act out of empathy by warmly doling out perquisites to kin and cronies rather than heartlessly giving them away to perfect strangers. Not only does this nepotism sap the competence of police, government, and business, but it sets up a zero-sum competition for the necessities of life among clans and ethnic groups, which can quickly turn violent. The institutions of modernity depend on carrying out abstract fiduciary duties that cut across bonds of empathy.

The other problem with empathy is that it is too parochial to serve as a force for a universal consideration of people's interests. Mirror neurons notwithstanding, empathy is not a reflex that makes us sympathetic to everyone we lay eyes upon. It can be switched on and off, or thrown into reverse, by our construal of the relationship we have with a person. Its head is turned by cuteness, good looks, kinship, friendship, similarity, and communal solidarity. Though empathy can be spread outward by taking other people's perspectives, the increments are small, Batson warns, and they may be ephemeral.[71] To hope that the human empathy gradient can be flattened so much that strangers would mean as much to us as family and friends is utopian in the worst 20th-century sense, requiring an unattainable and dubiously desirable quashing of human nature.[72]

Nor is it necessary. The ideal of the expanding circle does not mean that we must feel the pain of everyone else on earth. No one has the time or energy, and trying to spread our empathy that thinly would be an invitation to emotional burnout and compassion fatigue.[73] The Old Testament tells us to love our neighbors, the New Testament to love our enemies. The moral rationale seems to be: Love your neighbors and enemies; that way you won't kill them. But frankly, I don't love my neighbors, to say nothing of my enemies. Better, then, is the following ideal: Don't kill your neighbors or enemies, even if you don't love them.

What really has expanded is not so much a circle of empathy as a circle of *rights*—a commitment that other living things, no matter how distant or dissimilar, be safe from harm and exploitation. Empathy has surely been historically important in setting off epiphanies of concern for members of overlooked groups. But the epiphanies are not enough. For empathy to matter, it must goad changes in policies and norms that determine how the people in those groups are treated. At these critical moments, a newfound sensitivity to the human costs of a practice may tip the decisions of elites and the conventional wisdom of the masses. But as we shall see in the section on reason, abstract moral argumentation is also necessary to overcome the built-in strictures on empathy. The ultimate goal should be policies and norms that become second nature and render empathy unnecessary. Empathy, like love, is in fact not all you need.

SELF-CONTROL

Ever since Adam and Eve ate the apple, Odysseus had himself tied to the mast, the grasshopper sang while the ant stored food, and Saint Augustine prayed "Lord make me chaste—but not yet," individuals have struggled with self-control. In modern societies the virtue is all the more vital, because now that we have tamed the blights of nature most of our scourges are self-inflicted. We eat, drink, smoke, and gamble too much, max out our credit cards, fall into dangerous liaisons, and become addicted to heroin, cocaine, and e-mail.

Violence too is largely a problem of self-control. Researchers have piled up a tall stack of risk factors for violence, including selfishness, insults, jealousy, tribalism, frustration, crowding, hot weather, and maleness. Yet almost half of us are male, and all of us have been insulted, jealous, frustrated, or sweaty without coming to blows. The ubiquity of homicidal fantasies shows that we are not immune to the temptations of violence, but have learned to resist them.

Self-control has been credited with one of the greatest reductions of violence in history, the thirtyfold drop in homicide between medieval and modern Europe. Recall that according to Norbert Elias's theory of the Civilizing Process, the consolidation of states and the growth of commerce did more than just tilt the incentive structure away from plunder. It also inculcated an ethic of self-control that made continence and propriety second nature. People refrained from stabbing each other at the dinner table and amputating each other's noses at the same time as they refrained from urinating in closets, copulating in public, passing gas at the dinner table, and gnawing on bones and returning them to the serving dish. A culture of honor, in which men were respected for lashing out against insults, became a culture of dignity, in which men were respected for controlling their impulses. Reversals in the decline of violence, such as in the developed world in the 1960s and the developing world following decolonization, were accompanied by reversals in the valuation of self-control, from the discipline of elders to the impetuousness of youth.

Lapses of self-control can also cause violence on larger scales. Many stupid wars and riots began when leaders or communities lashed out against some outrage, but come the next morning had reason to regret the outburst. The burning and looting of African American neighborhoods by their own residents following the assassination of Martin Luther King in 1968, and Israel's pulverizing of the infrastructure of Lebanon following a raid by Hezbollah in 2006, are just two examples.[74]

In this section I will examine the science of self-control to see if it supports the theory of the Civilizing Process, in the same way that the preceding section examined the science of empathy to see if it supported the theory of the expanding circle. The theory of the Civilizing Process, like Freud's theory of the id and the ego from which it was derived, makes a number of strong claims about the human nervous system, which we will examine in turn. Does the brain really contain competing systems for impulse and self-control? Is self-control a single faculty in charge of taming every vice, from overeating to promiscuity to procrastination to petty crime to serious aggression? If so, are there ways for individuals to boost their self-control? And could these adjustments proliferate through a society, changing its character toward greater restraint across the board?

Let's begin by trying to make sense of the very idea of self-control and the circumstances in which it is and isn't rational.[75] First we must set aside pure selfishness—doing something that helps oneself but hurts others—and focus on self-indulgence—doing something that helps oneself in the short term but hurts oneself in the long term. Examples abound. Food today, fat tomorrow. Nicotine today, cancer tomorrow. Dance today, pay the piper tomorrow. Sex today, pregnancy, disease, or jealousy tomorrow. Lash out today, live with the damage tomorrow.

There is nothing *inherently* irrational about preferring pleasure now to pleasure later. After all, the You on Tuesday is no less worthy of a chocolate bar than the You on Wednesday. On the contrary, the You on Tuesday is *more* worthy. If the chocolate bar is big enough, it might tide you over, so eating it on Tuesday means that neither You is hungry, whereas saving it for Wednesday consigns you to hunger on Tuesday. Also, if you abstain from chocolate on Tuesday, you might die before you wake, in which case neither the Tuesday You nor the Wednesday You gets to enjoy it. Finally, if you put the chocolate away, it might spoil or be stolen, again depriving both Yous of the pleasure.

All things being equal, it pays to enjoy things now. That is why, when we lend out money, we insist on interest. A dollar tomorrow really is worth less than a dollar today (even if we assume there is no inflation), and interest is the price we put on the difference. Interest is charged at a fixed rate per unit of time, which means that it compounds, or increases exponentially. That compensates you exactly for the decreasing value of the money coming back to you as time elapses, because the decrease in value is also exponential. Why exponential? With every passing day, there is a fixed chance you will die, or that the borrower will abscond or go bankrupt and you'll never see the money again. As the probability that this will not have happened dwindles day by day, the compensation you demand multiplies accordingly. Going back to pleasure, a rational agent, when deciding between indulging today and indulging tomorrow, should indulge tomorrow only if the pleasure would be exponentially greater. In other words, a rational agent *ought to* discount the future and enjoy some pleasure today at the expense of less pleasure tomorrow. It makes no sense to scrimp all your life so that you can have one hell of a ninetieth birthday bash.

Self-indulgence becomes irrational only when we discount the future too *steeply*—when we devalue our future selves way below what they should be worth given the chance that those selves will still be around to enjoy what we've saved for them. There is an optimum rate of discounting the future—mathematically, an optimum interest rate—which depends on how long you expect to live, how likely you will get back what you saved, how long you can stretch out the value of a resource, and how much you would enjoy it at different points in your life (for example, when you're vigorous or frail). "Eat, drink, and be merry, for tomorrow we die" is a completely rational allocation if we are *sure* we are going to die tomorrow. What is not rational is to eat and drink as if there's no tomorrow when there really is a tomorrow. To be overly self-indulgent, to lack self-control, is to devalue our future selves too much, or equivalently, to demand too high an interest rate before we deprive our current selves for the benefit of our future selves. No plausible interest rate would make the pleasure in smoking for a twenty-year-old self outweigh the pain of cancer for her fifty-year-old self.

Much of what looks like a lack of self-control in the modern world may consist of using a discounting rate that was wired into our nervous systems in the iffy world of our pre-state ancestors, when people died much younger and had no institutions that could parlay savings now into returns years later.[76] Economists have noted that when people are left to their own devices, they save far too little for their retirement, as if they expect to die in a few years.[77] That is the basis for the "libertarian paternalism" of Richard Thaler, Cass Sunstein, and other behavioral economists, in which the government would, with people's consent, tilt the playing field between their current and future selves.[78] One example is setting an optimal retirement savings plan as the default, which employees would have to opt out of, rather than as a selection they would have to opt into. Another is to shift the burden of sales taxes onto the least healthy foods.

But weakness of the will is not just a matter of discounting the future too steeply. If we simply devalued our future selves too much, we might make bad choices, but the choices would not change as time passed and the alternatives drew near. If the inner voice shouting "dessert sooner" outvoted the one whispering "fat later," it would do so whether the dessert was available for consumption in five minutes or in five hours. In reality the preference flips with imminence in time, a phenomenon called *myopic* discounting.[79] When we fill in the room service card at night and hang it on the hotel doorknob for the following morning's breakfast, we are apt to tick off the fruit plate with nonfat yogurt. If instead we make our choices at the buffet table, we might go for the bacon and croissants. Many experiments on many species have shown that when two rewards are far away, organisms will sensibly pick a large reward that comes later over a small reward that comes sooner. If, for example, you had a choice between ten dollars in a week and eleven dollars in a week and a day, you'd pick the second. But when the nearer of the two rewards is imminent, self-control fails, the preference flips, and we go for smaller-sooner over larger-later: ten dollars today over eleven dollars tomorrow. Unlike merely discounting the future, which makes sense if the discount rate is properly set, myopic discounting, with its reversal of preferences, is not in any obvious way rational. Yet all organisms are myopic.

Mathematically minded economists and psychologists explain the myopic preference reversal by saying that organisms engage in *hyperbolic* discounting rather than the more rational exponential discounting.[80] When we depreciate our future selves, instead of repeatedly multiplying the subjective value of a reward by a constant fraction with every unit of time we have to wait for it (rendering it half as valuable, then a quarter, then an eighth, then a sixteenth, and so on), we multiply the original subjective value by a smaller and smaller fraction (which renders it half as valuable, then a third, then a quarter, then a fifth, and so on). This insight can also be expressed in a more intuitive, qualitative way. A hyperbola is a mathematical curve with a bit of an elbow, where a steep slope looks like it has been welded onto a shallow one (unlike an exponential curve, which is a smoother ski jump). That jibes with a psychological theory that myopic discounting arises from a handoff between two systems inside the skull, one for rewards that are imminent, another for rewards that are far in the future or entirely hypothetical.[81] As Thomas Schelling put it, "People behave sometimes as if they had two selves, one who wants clean lungs and long life and another who adores tobacco, or one who wants a lean body and another who wants dessert, or one who yearns to improve himself by reading Adam Smith on self-command . . . and another who would rather watch an old movie on television."[82] Freud's theory of the id and the ego, and the older idea that our lapses are the handiwork of inner demons ("The devil made me do it!") are other expressions of the intuition that self-control is a battle of homunculi in the head. The psychologist Walter Mischel, who conducted classic studies of myopic discounting in children (the kids are given the agonizing choice between one marshmallow now and two marshmallows in fifteen minutes), proposed, with the psychologist Janet Metcalfe, that the desire for instant gratification comes from a "hot system" in the brain, whereas the patience to wait comes from a "cool system."[83]

In previous sections we have caught glimpses of what the hot and cool systems might be: the limbic system (whose major parts are exposed in figure 8–2) and the frontal lobes (seen in figure 8–3). The limbic system includes the Rage, Fear, and Dominance circuits that run from the midbrain through the hypothalamus to the amygdala, together with the dopamine-driven Seeking circuit that runs from the midbrain through the hypothalamus to the striatum.

Both have two-way connections to the orbital cortex and other parts of the frontal lobes, which, as we saw, can modulate the activity of these emotional circuits, and which can come between them and the control of behavior. Can we explain self-control as a tug of war between the limbic system and the frontal lobes?

In 2004 the economists David Laibson and George Loewenstein teamed up with the psychologist Samuel McClure and the neuroimager Jonathan Cohen to see if the paradox of myopic discounting could be explained as a give-and-take between two brain systems: as they put it, a limbic grasshopper and a frontal lobe ant.[84] Participants lay in a scanner and chose between a small reward, such as five dollars, which would be available in the nearish future, and a larger reward, such as forty dollars, which would only be available several weeks later. The question was, did the brain treat the choice differently depending on whether it was "$5 now versus $40 in two weeks" or "$5 in two weeks versus $40 in six weeks"? The answer was that it did. Choices that dangled the possibility of immediate gratification in front of a participant lit up the striatum and the medial orbital cortex. All the choices lit up the dorsolateral prefrontal cortex, the part of the frontal lobes involved in cooler, more cognitive calculations. Even better, the neuroimagers could literally read the participants' minds. When their lateral prefrontal cortex was more active than their limbic regions, the participants held off for the larger-later reward; when the limbic regions were as active or more active, they succumbed to the smaller-sooner one.

As the front-heavy brain in figure 8–3 reveals, the frontal lobes are massive structures with many parts, and they carry out self-control of several kinds.[85] The rearmost margin, which abuts the parietal lobe, is called the motor strip, and it controls the muscles. Just in front of it are premotor areas that organize motor commands into more complex programs; these are the regions in which mirror neurons were first discovered. The portion in front of them is called the prefrontal cortex, and it includes the dorsolateral and orbital/ventromedial regions we have already encountered many times, together with the frontal pole at the tip of each hemisphere. The frontal pole is sometimes called "the frontal lobe of the frontal lobe," and together with the dorsolateral prefrontal cortex, it is active when people choose a large late reward over a smaller imminent one.[86]

Traditional neurologists (the doctors who treat patients with brain damage rather than sliding undergraduates into scanners) were not surprised by the discovery that it is the frontal lobes that are most involved in self-control. Many unfortunate patients end up in their clinics because they discounted the future too steeply and drove without a seat belt or bicycled without a helmet. For the small immediate reward of getting on the road a second sooner or feeling the wind in their hair, they gave up a large later reward of walking away from an accident with their frontal lobes intact. It's a bad bargain. Patients with frontal lobe damage are said to be stimulus-driven. Put a comb in front of them, and they will immediately pick it up and comb their hair. Put food in front of them, and they will put it in their mouths. Let them go into the shower, and they won't come out until they are called. Intact frontal lobes are necessary to liberate behavior from stimulus control—to bring people's actions into the service of their goals and plans.

In a collision with a hard surface, the frontal lobe bangs against the front of the skull and is damaged indiscriminately. Phineas Gage's freak accident, which sent a spike cleanly up through his orbital and ventromedial cortex and largely spared the lateral and frontmost parts, tells us that different parts of the frontal lobes implement different kinds of self-control. Gage, recall, was said to have lost the equilibrium "between his intellectual faculties and animal propensities." Neuroscientists today agree that the orbital cortex is a major interface between emotion and behavior. Patients with orbital damage, recall, are impulsive, irresponsible, distractible, socially inappropriate, and sometimes violent. The neuroscientist Antonio Damasio attributes this syndrome to the patients' insensitivity to emotional signals. He has shown that when they gamble on cards with different odds of gaining and losing money, they don't show the cold sweat that normal people experience when they bet on a card with ruinous odds.[87] This emotionally driven self-control—what we might call apprehension—is evolutionarily ancient, as shown by the well-developed orbital cortex in mammals such as rats (see figure 8–1).

But there are also cooler, more rule-driven forms of self-control, and they are implemented in the outer and frontmost parts of the frontal lobe, which are among the parts of the brain that expanded the most in the course of human evolution.[88] We have already seen that the dorsolateral cortex is engaged in rational calculation of costs and benefits, as in the choice between two delayed rewards, or in the choice between diverting a runaway trolley onto a side track with a single worker or letting it proceed to the main track with five.[89] The frontal pole sits even higher in the chain of command, and neuroscientists credit it with our suppleness in negotiating the competing demands of life.[90] It is engaged when we multitask, when we explore a new problem, when we recover from an interruption, and when we switch between daydreaming and focusing on the world around us. It is what allows us to branch to a mental subroutine and then pop back up to the main thing we were trying to accomplish, as when we interrupt ourselves while cooking to run to the store to get a missing ingredient, and then resume the recipe when we return. The neuroscientist Etienne Koechlin summarizes the functioning of the frontal lobe in the following way. The rearmost portions respond to the *stimulus;* the lateral frontal cortex responds to the *context;* and the frontal pole responds to the *episode.* Concretely, when the phone rings and we pick it up, we are responding to the stimulus. When we are at a friend's house and let it ring, we are responding to the context. And when the friend hops into the shower and asks us to pick up the phone if it rings, we are responding to the episode.

Impulsive violence could result from malfunctions in any of these levels of self-control. Take the violent punishment of children. Modern Western parents who have internalized norms against violence might have an automatic, almost visceral aversion to the thought of spanking their children, presumably enforced by the orbital cortex. Parents in earlier times and other subcultures (such as mothers who say, "Wait till your father gets home!") might modulate the spanking depending on how serious the infraction is, whether they are at home or in a public space, and, if they are home, whether there are guests in the house. But if they are weak in self-control, or are inflamed by what they see as egregious naughtiness, they might lose their tempers, which means that the Rage circuit shakes free of frontal lobe control, and they thrash the child in a way they might regret later.

Adrian Raine, who previously showed that psychopaths and impulsive murderers have a small or unresponsive orbital cortex, recently carried out a neuroimaging experiment that supports the idea that violence arises from an imbalance between impulses from the limbic system and self-control from the frontal lobes.[91] He scanned a sample of wife-batterers as they tried to ignore the meanings of printed words for negative emotions such as *anger, hate, terror,* and *fear* and just name the color in which they were printed (a test of attention called the Stroop task). The batterers were slowed down in naming the colors, presumably because their background anger made them hypersensitive to the negative emotions the words spelled out. And compared to the brains of normal people, who can examine the print without getting distracted by the words' meanings, the batterers' limbic structures were more active (including the insula and striatum), while their dorsolateral frontal cortex was less active. We may surmise that in the brains of impulsive assailants, aggressive impulses from the limbic system are stronger, and the self-control exerted by the frontal lobes is weaker.

Most people, of course, are not so lacking in self-control that they ever lash out in violence. But among the nonviolent majority some people have more self-control than others. Aside from intelligence, no other trait augurs as well for a healthy and successful life.[92] Walter Mischel began his studies of delay of gratification (in which he gave children the choice between one marshmallow now and two marshmallows later) in the late 1960s, and he followed the children as they grew up.[93] When they were tested a decade later, the ones who had shown greater willpower in the marshmallow test had now turned into adolescents who were better adjusted, attained higher SAT scores, and stayed in school longer. When they were tested one and two decades after that, the patient children had grown into adults who were less likely to use cocaine, had higher self-esteem, had better relationships, were better at handling stress, had fewer symptoms of borderline personality disorder, obtained higher degrees, and earned more money.

Other studies with large samples of adolescents and adults have documented similar payoffs. Adults can wait indefinitely for two marshmallows, but as we have seen, they can be given equivalent choices such as "Would you rather have five dollars now or forty dollars in two weeks?" Studies by Laibson, Christopher Chabris, Kris Kirby, Angela Duckworth, Martin Seligman, and others have found that people who opt for the later and larger sums get higher grades, weigh less, smoke less, exercise more, and are more likely to pay off their credit card balance every month.[94]

Baumeister and his collaborators measured self-control in a different way.[95] They asked university students to divulge their own powers of self-control by rating sentences such as these:

I am good at resisting temptation.
I blurt out whatever is on my mind.
I never allow myself to lose control.
I get carried away by my feelings.
I lose my temper too easily.
I don't keep secrets very well.
I'd be better off if I stopped to think before acting.
Pleasure and fun sometimes keep me from getting work done.
I am always on time.

After adjusting for any tendency just to tick off socially desirable traits, the researchers combined the responses into a single measure of habitual self-control. They found that the students with higher scores got better grades, had fewer eating disorders, drank less, had fewer psychosomatic aches and pains, were less depressed, anxious, phobic, and paranoid, had higher self-esteem, were more conscientious, had better relationships with their families, had more stable friendships, were less likely to have sex they regretted, were less likely to imagine themselves cheating in a monogamous relationship, felt less of a need to "vent" or "let off steam," and felt more guilt but less shame.[96] Self-controllers are better at perspective-taking and are less distressed when responding to others' troubles, though they are neither more nor less sympathetic in their concern for them. And contrary to the conventional wisdom that says that people with too much self-control are uptight, repressed, neurotic, bottled up, wound up, obsessive-compulsive, or fixated at the anal stage of psychosexual development, the team found that the more self-control people have, the better their lives are. The people at the top of the scale were the mentally healthiest.

Are people with low self-control more likely to perpetrate acts of violence? Circumstantial evidence suggests they are. Recall from chapter 3 the theory of crime (championed by Michael Gottfredson, Travis Hirschi, James Q. Wilson, and Richard Herrnstein) in which the people who commit crimes are those with the least self-control.[97] They opt for small, quick, ill-gotten gains over the longer-term fruits of honest toil, among them the reward of not ending up in jail. Violent adolescents and young adults tend to have a history of misconduct at school, and they tend to get into other kinds of trouble that bespeak a lack of self-control, such as drunk driving, drug and alcohol abuse, accidents, poor school performance, risky sex, unemployment, and nonviolent crimes such as burglary, vandalism, and auto theft. Many violent crimes are strikingly impulsive. A man will walk into a convenience store for some cigarettes and on the spur of the moment pull out a gun and rob the cash register. Or he will react to a curse or insult by pulling out a knife and stabbing the insulter.

To make the case more than circumstantial, one would have to show that the psychologists' conception of self-control (measured by people's choice between smaller-sooner and larger-later rewards or by ratings of their own impulsiveness) match up with the criminologists' conception of self-control (measured by actual outbreaks of violence). Mischel tested children in urban middle schools and in camps for troubled youth and found that the children who waited longer for larger piles of M&Ms were also less likely to get into fights and to pick on their playmates.[98] Many studies of teachers' ratings have confirmed that the children who appear to them to be more impulsive are also the ones who are more aggressive.[99] A particularly informative study by the psychologists Avshalom Caspi and Terri Moffitt followed an entire cohort of children born in the New Zealand city of Dunedin from the year of their birth in 1972–73.[100] Three-year-old children who were rated as undercontrolled—that is, impulsive, restless, negativistic, distractible, and emotionally fluctuating—grew into twenty-one-year-olds who were far more likely to be convicted of a crime. (The study did not distinguish violent from nonviolent crimes, but later studies on the same sample showed that the two kinds of crime tend to go together.)[101] And one of the causes of their greater criminality may have been differences in their anticipation of the consequences of their behavior. In their answers to questionnaires, the less-controlled people gave lower odds that they would get arrested following a string of crimes, and that they would lose the respect of their friends and family if their illegal behavior came to light.

The trajectory of crime in adolescence and young adulthood is related to an increase in self-control, measured in a growing willingness to choose larger late rewards over smaller earlier ones. This change is partly driven by the physical maturation of the brain. The wiring of the prefrontal cortex is not complete until the third decade of life, with the lateral and polar regions developing last.[102] But self-control is not the whole story. If delinquency depended *only* on self-control, young teenagers should become less and less likely to get in trouble as they turn into older teenagers, which is not what happens. The reason is that violence depends not just on self-control but on the urges that self-control has to control.[103] Adolescence is also an age that sees the rise and fall of a motive called sensation-seeking, driven by activity in the Seeking system, which peaks at eighteen.[104] It also sees an increase in male-against-male competitiveness, driven by testosterone.[105] The rise in sensation-seeking and competitiveness can overtake the rise in self-control, making older adolescents and twenty-somethings more violent despite their blossoming frontal lobes. In the long run, self-control gains the upper hand when it is fortified by experience, which teaches adolescents that thrill-seeking and competitiveness have costs and that self-control has rewards. The arc of crime in adolescence is the outcome of these inner forces pushing and pulling in different directions.[106]

Self-control, then, is a stable trait that differentiates one person from another, beginning in early childhood. No one has done the twin and adoption studies that would be needed to show that performance on standard tests of self-control, such as the marshmallow test or the adult equivalent, are heritable. But it's a good bet that they are, because pretty much every psychological trait has turned out to be partly heritable.[107] Self-control is partly correlated with intelligence (with a coefficient of about 0.23 on a scale from –1 to 1), and the two traits depend on the same parts of the brain, though not exactly in the same way.[108] Intelligence itself is highly correlated with crime—duller people commit more violent crimes and are more likely to be the victims of a violent crime—and though we can't rule out the possibility that the effect of self-control is really an effect of intelligence or vice versa, it's likely that both traits contribute independently to nonviolence.[109] Another clue that self-control is heritable is that a syndrome marked by a shortage of self-control, attention deficit hyperactivity disorder (which is also linked with delinquency and crime), is among the most heritable of personality traits.[110]

So far all the evidence that violence is released by a lack of self-control is correlational. It comes from the discovery that some people have less self-control than others, and that those people are likelier to misbehave, get angry, and commit more crimes. But the correlation doesn't prove causation. Perhaps people with low self-control are more crime-prone because they are also less intelligent, or they come from worse environments, or they have some other across-the-board disadvantage. More important, a stable trait that differs from one person to another cannot explain the main thing we're trying to explain: why rates of violence change over the course of history. To show that, we need to show that individual people, when they let up or clamp down on their self-control, get more or less violent as a result. And we must show that people and societies can cultivate the faculty of self-control over time and thereby drive down their rates of violence. Let's see if we can find these missing links.

When a person fights an urge, it feels like a strenuous effort. Many of the idioms for self-control invoke the concept of force, such as *willpower, force of will, strength of will,* and *self-restraint*. The linguist Len Talmy has noticed that the language of self-control borrows from the language of force dynamics, as if self-control were a homunculus that physically impinged on a stubborn antagonist inside the skull.[111] We use the same construction in *Sally forced the door to open* and *Sally forced herself to go to work*; in *Biff controlled his dog* and *Biff controlled his temper.* Like many conceptual metaphors, SELF-CONTROL IS PHYSICAL EFFORT turns out to have a kernel of neurobiological reality.

In a remarkable set of experiments, Baumeister and his collaborators have shown that self-control, like a muscle, can become fatigued. Their laboratory procedure is best introduced by quoting from the Method section of one of their papers:

> *Procedure.* Participants signed up for a study on taste perception. Each participant was contacted to schedule an individual session, and at that time the experimenter requested the participant to skip one meal before the experiment and make sure not to have eaten anything for at least 3 hr.
>
> The laboratory room was carefully set up before participants in the food conditions arrived. Chocolate chip cookies were baked in the room in a small oven, and, as a result, the laboratory was filled with the delicious aroma of fresh chocolate and baking. Two foods were displayed on the table at which the participant was seated. One display consisted of a stack of chocolate chip cookies augmented by some chocolate candies. The other consisted of a bowl of red and white radishes.[112]

The cover story was that the experiment was on sensory memory and that the participants would experience one of two distinctive tastes and have to recall its qualities after a delay. The experimenter told half the participants to eat two or three of the cookies, and the other half to eat two or three of the radishes. She left the room and watched through a one-way mirror to confirm that the participant did not cheat. The article notes: "Several of them did indicate clear interest in the chocolates, to the point of looking longingly at the chocolate display and in a few cases even picking up the cookies to sniff at them." The experimenter then told them they would have to wait fifteen minutes for the test of their memory of the taste. In the interim, they were to solve some puzzles that required tracing a geometric figure with a pencil without either retracing a line or lifting the pencil off the paper. Compounding the sadism, the experimenters had given them puzzles that were unsolvable, and measured how long the participant persisted before giving up. The ones who had eaten the cookies spent 18.9 minutes and made 34.3 attempts to solve the puzzle. The ones who had eaten the radishes spent 8.4 minutes and made 19.4 attempts. Presumably the radish eaters had depleted so much of their mental strength in resisting the cookies that they had little left to persist in solving the puzzles. Baumeister called the effect *ego depletion,* using Freud's sense of *ego* as the mental entity that controls the passions.

The study raises many objections: maybe the radish eaters were just frustrated, or angry, or in a bad mood, or hungry. But the Baumeister team addressed them and over the following decade accumulated a raft of experiments showing that just about any task that requires an exercise of willpower can impede performance in any other task that requires willpower. Here are a few tasks that can deplete the ego:

- Name the color in which a word is displayed (such as the word RED printed in blue ink), ignoring the color it spells out (the Stroop task).
- Track moving boxes on a screen, as if playing a shell game, while ignoring a comedy video on an adjacent screen.
- Write a convincing speech on why tuition fees should be raised.
- Write an essay about the typical day in the life of a fat person without using any stereotypes.
- Watch the scene in *Terms of Endearment* in which a dying Debra Winger says good-bye to her children, without showing emotion.
- For racially prejudiced people, carry on a conversation with an African American.
- Write down all your thoughts, but don't think of a polar bear.[113]

And here are some of the lapses in willpower that result:

- Giving up sooner when squeezing a handgrip, solving anagrams, or watching a movie of a box on a table until something happens.
- Breaking a diet by eating ice cream from a container after rating a spoonful in a taste experiment.
- Drinking more beer in a taste experiment, even when having to take a simulated driving test immediately afterward.
- Failing to stifle sexual thoughts, such as in solving the anagram NISEP as *penis* rather than as *spine*.
- Failing to keep up a running conversation while teaching someone how to putt in golf.
- Being willing to pay more for an attractive watch, car, or boat.
- Blowing your payment for participation in the study on gum, candy, Doritos, or playing cards, which the experimenters had mischievously offered for sale.

Various control conditions allowed the psychologists to rule out alternative explanations such as fatigue, difficulty, mood, and lack of confidence. The only common denominator was the need for self-control.

An important implication of the research is that the exercise of self-control can conceal the differences among individual people.[114] It's no coincidence that 1960s popular culture, which denigrated sobriety and self-control, also denigrated conformity, as in the signature motto "Do your own thing."

Everyone has a different thing, but society insists on just one thing, so we must apply self-control to do it. If self-control flattens individuality, one can predict that when the ego is depleted, individuality will pop back up. And that is what the Baumeister group found. In the ice-cream-tasting experiment, when the participants had not been called on to exercise self-control beforehand, the dieters and the indiscriminate eaters consumed the same amount of ice cream. But when their willpower had been exhausted, the dieters ate more. Other individual differences unmasked by depletion of the ego included the degree of stereotyping by prejudiced and unprejudiced people, the amount of beer drunk by tipplers and moderate drinkers, and the amount of small talk made by shy and outgoing people.

The Baumeister group also vindicated the Victorian idea that some people—particularly men—have to exert their will to control their sexual appetites.[115] In one study, the psychologists assessed how emotionally close a participant had to feel to another person before engaging in casual sex. People of both sexes differ along that dimension, and there is also a robust difference between the sexes, captured in the movie dialogue in which Diane Keaton says, "I believe that sex without love is a meaningless experience" and Woody Allen replies, "Yes, but as meaningless experiences go, it's one of the best." Half the participants in the study went through an ego depletion task (crossing out letters according to shifting rules), and all were then asked to imagine themselves being in a committed romantic relationship and then finding themselves in the hotel room of an attractive acquaintance of the opposite sex. They were then asked whether they imagined themselves succumbing to the temptation. Whether their wills had been tuckered or not, the participants (of both sexes) who had indicated that sex without love was a meaningless experience imagined they would resist the temptation. But a transient weakness of their will affected the ones who were more open to casual sex: if their ego had just been fatigued, their imagined selves were far more likely to say yes.

The pattern for the two sexes was revealing. When the willpower of the participants was fresh, men and women didn't differ: both were resistant to imaginary cheating. When their wills had been weakened, the women were just as resistant, but the men imagined themselves likely to stray. Another sign that gallantry requires self-control came from an analysis that simply compared people who reported having a lot or a little self-control (ignoring momentary ego depletion). Among those with high self-control, neither the men and nor the women imagined cheating on their partners, but among the people with low self-control, the men imagined that they probably would. The pattern suggests that the exercise of self-control hides a deep difference between men and women. Freed from their own willpower, men are more likely to act as evolutionary psychology predicts.

Baumeister and Gailliot pushed their luck in one more experiment, aiming to show that self-control affects real, not just imagined, sexual activity. They invited couples into the lab who were either sexually experienced or just beginning their relationship, separated them, gave them an ego depletion task (concentrating on a boring video while shutting out distractions), reunited them, and invited them to be affectionate with each other for three minutes while the experimenter discreetly left the room. A sense of propriety prevented the experimenters from videotaping the couple or observing them from behind a one-way mirror, so they asked each partner to write a confidential paragraph describing exactly what had gone on between them. Experienced couples, if their wills had been depleted, were a bit *less* physical, as if sex had switched from a passion to a chore. But ego depletion made the inexperienced couples far more physical. According to the write-up, "They kissed open-mouthed for prolonged periods of time, groped and caressed each other (e.g., on the buttocks and woman's chest), and even removed articles of clothing to expose themselves."

According to the theory of the Civilizing Process, a dearth of self-control in medieval Europe underlay many forms of dissoluteness, including slovenliness, petulance, licentiousness, uncouthness, steep discounting of the future, and most important, violence. The science of self-control vindicates the idea that a single capability of mind can counteract many of these forms of dissipation. But it remains to be shown that violence is one of them. We know that people with less self-control are more cantankerous and trouble-prone. But can manipulating self-control in an experiment bring out the beast within?

No one wants a fight to break out in the lab, so Baumeister went to the hot sauce. Hungry participants were asked to take part in a study on the relation between tastes in food and written expression.[116] They indicated their favorite and least favorite flavors, wrote an essay expressing their views on abortion, rated the essay of a bogus fellow participant, rated the taste of a food, and finally read their partner's feedback on their essay. In the taste test, half of them had to rate the taste, texture, and aroma of a donut; half had to rate the taste, texture, and aroma of a radish. But just as they raised the stimulus to their mouth, the experimenter exclaimed, "Wait! I'm sorry; I think I screwed up. This isn't for you. Please don't eat the rest of it. Let me go figure out what's supposed to be next." He then left the participant alone with the donut or the radish for five minutes. Lest there be any doubt that this was a valid test of self-control, one may note the following passage in the write-up:

> *Participants:* Forty undergraduates participated in this study in exchange for course credit. Data from seven participants were discarded from all analyses, four due to expressed suspicion about the feedback and three due to participants having eaten the entire donut.

The participants then read the partner's feedback on their essay, which was scathing. They also were shown his or her taste preferences, which indicated a dislike of spicy foods. And then they were asked to prepare a snack for the partner from a bag of chips and a container of hot sauce prominently labeled SPICY. The amount they applied was measured by weighing the container after they left. The participants were also asked to rate their moods, including anger. The ones whose self-control had been depleted by having to forgo the donut didn't get mad, but they did get even. They applied 62 percent more hot sauce to the chip of their insulting partner, presumably because they could not resist the impulse for revenge. Will-depleted subjects were also more likely to torment a critic by leaning on a button to blast him with an air horn every time he made an error in a computer game.

Another study tapped aggressive fantasies by asking participants to imagine standing at a bar with a beloved girlfriend when a rival shows up and begins to flirt, to her visible enjoyment. (In the scenario given to women, it was a boyfriend who was chatted up by a rival woman.) The participant imagines confronting the rival, who responds by shoving him into the bar. Close at hand is a beer bottle. The participant was asked: "How likely would you be to smash the bottle on the person's head? Indicate your response on a scale from -100 (not at all likely) to 100 (extremely likely)." The participants who were low in self-control, if their wills were ready and rested, indicated they probably would not retaliate. But if their wills had been depleted, they indicated they probably would.

If we combine (1) Baumeister's experiments, which found that reducing self-control in the lab can increase tendencies toward impulsive sex and violence; (2) the correlations across individuals between low self-control on the one hand and childhood misconduct, dissolute behavior, and crime on the other; (3) the neuroimaging studies that showed correlations between frontal lobe activity and self-control; and (4) the neuroimaging studies showing correlations between impulsive violence and impaired frontal lobe function, then we get an empirical picture that supports Elias's conjecture that violence may be caused by weakness in an overarching neural mechanism of self-control.

The picture is still incomplete. The existence of a trait that is stable in an individual over a span of decades, and that can be depleted over a span of minutes, cannot explain how a society can change over a span of centuries. We still need to show that whatever level of self-control a person is born with, he or she has ways to boost it. There is no paradox in the possibility that self-control can both be heritable across individuals and rise over time. That is exactly what happened with stature: genes make some of us taller than others, but over the centuries everyone got taller.[117]

For as long as people have reflected on self-control, they have reflected on ways to enhance it. Odysseus had his sailors tie him to the mast and plug their ears with wax so he could hear the alluring song of the sirens without steering his ship onto the rocks. Techniques in which the present self handicaps the future self are sometimes called Odyssean or Ulyssean in his honor. There are hundreds of examples.[118] We avoid shopping on an empty stomach. We throw out the brownies or the cigarettes or the booze at a time when we aren't craving them, to foil ourselves at a time when we are. We put our alarm clock on the other side of the bedroom so we don't turn it off and fall back asleep. We authorize our employers to invest a part of every paycheck for our retirement. We refrain from buying a magazine or a book or a gadget that will divert our attention from a work project until it is complete. We hand over money to a company like Stickk.com that returns it a fraction at a time if we meet certain goals, or donates it to a political organization we detest if we don't. We make a public resolution to change, so our reputation takes a hit if we don't.

As we saw in chapter 3, one way the early modern Europeans used Odyssean self-control was to keep sharp knives out of reach at the dinner table. The familiar sign in the saloons in old Westerns—"Check your guns at the door"—served the same purpose, as do gun control laws and disarmament agreements today. Another tactic is to keep oneself away from trouble, such as avoiding the place where an aggrieved rival is known to hang out. Brawlers who allow themselves to be pulled apart by bystanders avail themselves of a similar tactic, with the added bonus that they have not conceded weakness or cowardice by disengaging.

Other strategies of self-control are mental rather than physical. Walter Mischel showed that even four-year-olds can wait out a long interval for a double helping of marshmallows if they cover the alluring marshmallow in front of them, look away from it, distract themselves by singing, or even reframe it in their minds as a puffy white cloud rather than a sweet tasty food.[119] An equivalent in the case of violence may be the cognitive reframing of an insult from a devastating blow to one's reputation to an ineffectual gesture or a reflection on the immaturity of the insulter. Such reframing lies behind advice like "Don't take it personally," dismissals like "He's just blowing smoke," "He's only a kid," and "Take it from whom it comes," and proverbs like "Sticks and stones can break my bones but words can never hurt me."

Martin Daly and Margo Wilson, invoking the economists' theory of optimal interest rates and the biologists' theory of optimal foraging, have suggested a third way in which self-control can be manipulated. They propose that organisms are equipped with an internal variable, like an adjustable interest rate, that governs how steeply they discount the future.[120] The setting of the variable is twiddled according to the stability or instability of their environment and an estimate of how long they will live. It doesn't pay to save for tomorrow if tomorrow will never come, or if your world is so chaotic that you have no confidence you would get your savings back. In a quantitative comparison of neighborhoods in a major city, Daly and Wilson found that the shorter the expected life span (from all causes other than violence), the higher the rate of violent crime. The correlation supports the hypothesis that, holding age constant, people are more reckless when they have fewer years of unlived life at risk. A rational adjustment of one's discounting rate in response to the uncertainty of the environment could create a vicious circle, since your own recklessness then figures into the discounting rate of everyone else. The Matthew Effect, in which everything seems to go right in some societies and wrong in others, could be a consequence of environmental uncertainty and psychological recklessness feeding on each other.

A fourth way people in a society might boost their self-control is by improving their nutrition, sobriety, and health. The frontal lobe is a big slab of metabolically demanding tissue with an outsize appetite for glucose and other nutrients. Pushing the metaphor of self-control as physical effort even further, Baumeister found that people's blood glucose level plummets as their ego is depleted by an attention-consuming or a willpower-demanding task.[121] And if they replenish their glucose level by drinking a glass of sugar-sweetened lemonade (but not a glass of aspartame-sweetened lemonade), they avoid the usual slump in the follow-up task. It is not implausible to suppose that real-world conditions that impair the frontal lobes—low blood sugar, drunkenness, druggedness, parasite load, and deficiencies of vitamins and minerals—could sap the self-control of people in an impoverished society and leave them more prone to impulsive violence. Several placebo-controlled studies have suggested that providing prisoners with dietary supplements can reduce their rate of impulsive violence.[122]

Baumeister stretched the metaphor even farther. If willpower is like a muscle that fatigues with use, drains the body of energy, and can be revived by a sugary pick-me-up, can it also be bulked up by exercise? Can people develop their raw strength of will by repeatedly flexing their determination and resolve? The metaphor shouldn't be taken *too* literally—it's unlikely that the frontal lobes literally gain tissue like bulging biceps—but it's possible that the neural connections between the cortex and limbic system may be strengthened with practice. It's also possible that people can learn strategies of self-control, enjoy the feeling of mastery over their impulses, and transfer their newfound tricks of discipline from one part of their behavioral repertoire to another.

Baumeister and other psychologists tested the exercise metaphor by having participants undertake regimens of self-control for several weeks or months before taking part in one of their ego depletion studies.[123] In various studies the regimens required them to keep track of every piece of food they ate; enroll in programs of physical exercise, money management, or study skills; use their nonpreferred hand for everyday tasks like brushing their teeth and using a computer mouse; and one that really gave the students' self-control a workout: avoid curse words, speak in complete sentences, and not begin sentences with *I*. After several weeks of this cross-training, the students indeed turned out to be more resistant to ego depletion tasks in the lab, and they also showed greater self-control in their lives. They smoked fewer cigarettes, drank less alcohol, ate less junk food, spent less money, watched less television, studied more, and washed the dishes more often rather than leaving them in the sink. Score another point for Elias's conjecture that self-control in life's little routines can become second nature and be generalized to the rest of one's comportment.

In addition to being modulated by Ulyssean constraints, cognitive reframing, an adjustable internal discount rate, improvements in nutrition, and the equivalent of muscle gain with exercise, self-control might be at the mercy of whims in fashion.[124] In some eras, self-control defines the paragon of a decent person: a grown-up, a person of dignity, a lady or a gentleman, a mensch. In others it is jeered at as uptight, prudish, stuffy, straitlaced, puritanical. Certainly the crime-prone 1960s were the recent era that most glorified the relaxation of self-control: Do your own thing, Let it all hang out, If it feels good do it, Take a walk on the wild side. The premium on self-indulgence is on full display in concert films from the decade, which show rock musicians working so hard to outdo each other in impulsiveness that it looks as if they put a lot of planning and effort into their spontaneity.

Could these six pathways to self-control proliferate among the members of a society and come to define its global character? That would be the final domino in the chain of explanation that makes up the theory of the Civilizing Process. The exogenous first domino is a change in law enforcement and opportunities for economic cooperation that objectively tilt the payoffs so that a deferral of gratification, in particular, an avoidance of impulsive violence, pays off in the long run. The knock-on effect is a strengthening of people's self-control muscles that allow them (among other things) to inhibit their violent impulses, above and beyond what is strictly necessary to avoid being caught and punished. The process could even feed on itself in a positive feedback loop, "positive" in both the engineering and the human-values sense. In a society in which other people control their aggression, a person has less of a need to cultivate a hair trigger for retaliation, which in turn takes a bit of pressure off everyone else, and so on.

One way to bridge the gap between psychology and history is to look for changes in a society-wide index of self-control. As we have seen, an interest rate is just such an index, because it reveals how much compensation people demand for deferring consumption from the present to the future. To be sure, an interest rate is partly determined by objective factors like inflation, expected income growth, and the risk that the investment will never be returned. But it partly reflects the purely psychological preference for instant over delayed gratification. According to one economist, a six-year-old who prefers to eat one marshmallow now rather than two marshmallows a few minutes from now is in effect demanding an interest rate of 3 percent a day, or 150 percent a month.[125]

Gregory Clark, the economic historian we met in chapter 4, has estimated

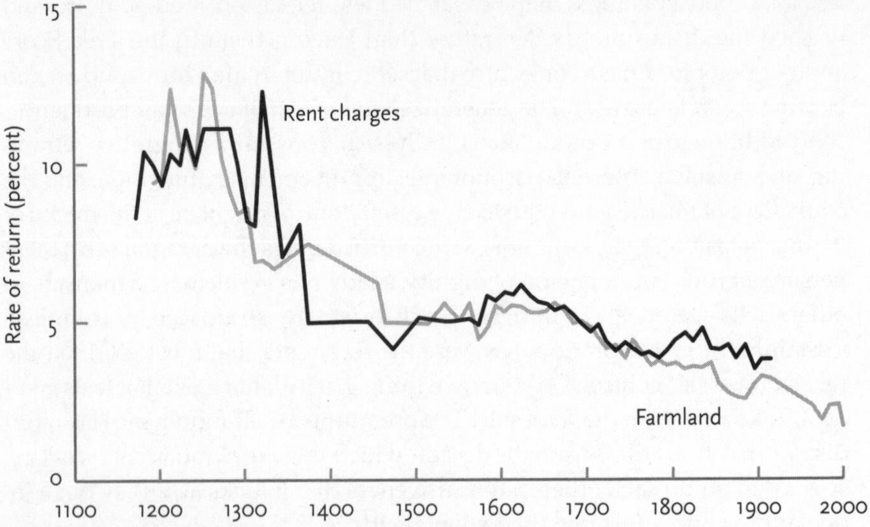

FIGURE 9–1. Implicit interest rates in England, 1170–2000

Source: Graph from Clark, 2007a, p. 33.

the interest rates that Englishmen demanded (in the form of rents on land and houses) from 1170 to 2000, the millennium over which the Civilizing Process took place. Before 1800, he argues, there was no inflation to speak of, incomes were flat, and the risk that an owner would lose his property was low and constant. If so, the effective interest rate was an estimate of the degree to which people favored their present selves over their future selves.

Figure 9–1 shows that during the centuries in which homicide plummeted in England, the effective interest rate also plummeted, from more than 10 percent to around 2 percent. Other European societies showed a similar transition. The correlation does not, of course, prove causation, but it is consistent with Elias's claim that the decline of violence from medieval to modern Europe was part of a broader trend toward self-control and an orientation to the future.

What about more direct measures of a society's aggregate self-control? An annual interest rate is still quite distant from the momentary exercises of forbearance that suppress violent impulses in everyday life. Though there are dangers in essentializing a society by assigning it character traits that really should apply to individuals (remember the so-called fierce people), there may be a grain of truth in the impression that some cultures are marked by more self-control in everyday life than others. Friedrich Nietzsche distinguished between Apollonian and Dionysian cultures, named after the Greek gods of light and wine, and the distinction was used by the anthropologist Ruth Benedict in her classic 1934 ethnography collection *Patterns of Culture*. Apollonian cultures are said to be thinking, self-controlled, rational, logical, and ordered;

Dionysian cultures are said to be feeling, passionate, instinctual, irrational, and chaotic. Few anthropologists invoke the dichotomy today, but a quantitative analysis of the world's cultures by the sociologist Geert Hofstede has rediscovered the distinction in the pattern of survey responses among the middle-class citizens of more than a hundred countries.

According to Hofstede's data, countries differ along six dimensions.[126] One of them is Long-Term versus Short-Term Orientation: "Long-term oriented societies foster pragmatic virtues oriented towards future rewards, in particular saving, persistence, and adapting to changing circumstances. Short-term oriented societies foster virtues related to the past and present such as national pride, respect for tradition, preservation of 'face,' and fulfilling social obligations." Another dimension is Indulgence versus Restraint: "Indulgence stands for a society that allows relatively free gratification of basic and natural human drives related to enjoying life and having fun. Restraint stands for a society that suppresses gratification of needs and regulates it by means of strict social norms." Both, of course, are conceptually related to the faculty of self-control, and not surprisingly, they are correlated with each other (with a coefficient of 0.45 across 110 countries). Elias would have predicted that both of these national traits should correlate with the countries' homicide rates, and the prediction turns out to be true. The citizens of countries with more of a long-term orientation commit fewer homicides, as do the citizens of countries that emphasize restraint over indulgence.[127]

So the theory of the Civilizing Process, like the theory of the expanding circle, has found support in experiments and datasets that are far from its field of origin. Psychology, neuroscience, and economics have confirmed Elias's speculation that humans are equipped with a faculty of self-control that regulates both violent and nonviolent impulses, that can be strengthened and generalized over the lifetime of an individual, and that can vary in strength across societies and historical periods.

So far I have not mentioned yet another explanation for the long-term growth of self-control: that it is a process of evolution in the biologist's sense. Before turning to the last two of our better angels, morality and reason, I need to spend a few pages on this vexed question.

RECENT BIOLOGICAL EVOLUTION?

Many people casually use the word *evolution* to refer both to cultural change (that is, history) and to biological change (that is, a shift in the frequencies of genes across generations). Cultural and biological evolution can certainly interact. For instance, when tribes in Europe and Africa adopted the practice of keeping livestock for milk, they evolved genetic changes that allowed them to digest lactose in adulthood.[128] But the two processes are different. They can always be distinguished, in theory, by experiments in which babies from one society are given up for adoption and brought up in another. If biological evolution has taken place in response to the distinctive culture of either society, then the adopted children should differ, on average, from their native peers.

A frequently asked question about declines in violence is whether they can be attributed to recent biological evolution. In a society that has undergone a Pacification or Civilizing Process, has the genetic makeup of the people changed in response, helping the process along and leaving them permanently less disposed to violence? Any such change would not, of course, be a Lamarckian absorption of the cultural trend into the genome, but a Darwinian response to the altered contingencies of survival and reproduction. The individuals who happen to be genetically suited to the changed culture would outreproduce their neighbors and contribute a larger share of genes to the next generation, gradually shifting the population's genetic makeup.

One can imagine, for example, that in a society that was undergoing a Pacification or Civilizing Process, a tendency toward impulsive violence would begin to pay off less than it did in the days of Hobbesian anarchy, because a hair trigger for retaliation would now be harmful rather than helpful. The psychopaths and hotheads would be weeded out by the Leviathan and sent to the dungeons or gallows, while the empathizers and cooler heads would bring up their children in peace. Genes that fortified empathy and self-control would proliferate, while genes that gave free rein to predation, dominance, and revenge would ebb.

Even a cultural change as simple as a shift from polygyny to monogamy could, in theory, alter the selective landscape. Napoleon Chagnon documented that among the Yanomamö, the men who had killed another man had more wives and more children than the men who had never killed; similar patterns have been found in other tribes, such as the Jivaro (Shuar) of Ecuador.[129] This arithmetic, if it persisted over many generations, would favor a genetic tendency to be willing and able to kill. A society that has shifted to monogamy, in contrast, removes this reproductive jackpot, and conceivably could relax the selection for bellicosity.

Throughout this book I have assumed that human nature, in the sense of the cognitive and emotional inventory of our species, has been constant over the ten-thousand-year window in which declines of violence are visible, and that all differences in behavior among societies have strictly environmental causes. That is a standard assumption in evolutionary psychology, based on the fact that the few centuries and millennia in which societies have separated and changed are a small fraction of the period in which our species has been in existence.[130] Since most adaptive evolutionary change is gradual, the bulk of our biological adaptation must be to the foraging lifestyle that prevailed during those tens of millennia, rather than to the specifics of societies that have departed from that lifestyle and diverged from each other only recently. The assumption is supported by evidence for the psychic unity of humankind—that people in every society have all the basic human faculties such as language, causal reasoning, intuitive psychology, sexual jealousy, fear, anger, love, and disgust, and that the recent mixing of human populations has revealed no qualitative innate differences among them.[131]

But the assumptions of ancient adaptation and psychic unity are only assumptions. The speed of biological evolution depends on many factors, including the strength of the selection pressure (that is, the average difference in the number of surviving offspring between the carriers of two variants of a gene), the demographics of the population, the number of genes required to implement a change, and the patterns of interaction among the genes.[132] Though a complex organ built by a suite of many interacting genes can take aeons to evolve, a quantitative adjustment that can be implemented by one gene, or a small number of independently acting genes, can take place over just a few generations, as long as they have large enough effects on fitness.[133] Nothing rules out the possibility that human populations have undergone some degree of biological evolution in recent millennia or even centuries, long after races, ethnic groups, and nations diverged.

Though people sometimes write that hypotheses about natural selection are just-so stories that can never be verified until someone invents a time machine, in fact natural selection is a distinctive mechanistic process that leaves signs of its handiwork both in the design of organisms' bodies and in the patterning of their genomes. Since the completion of the first phase of the Human Genome Project in 2000, the search for fingerprints of selection has been one of the most exciting research activities in human genetics.[134] One technique juxtaposes the human version of a gene to its counterpart in other species and compares the number of silent changes (which have no effect on the organism and so must have accumulated by random drift) with the number of changes that do have an effect (and thus may have been a target of selection). Another technique looks at the variability of a gene across individuals. A gene that has been subjected to selection should vary less among individual people within the human population than it does between humans as a whole and other mammals. Still other techniques check to see whether a gene lies in the middle of a big chunk of chromosome that is identical across people—the sign of a recent "selective sweep," which carries a stretch of a chromosome along for the ride with a handy gene, before mutations have had a chance to adulterate it or sexual recombination has had a chance to shuffle it into pieces. At least a dozen of these techniques have been devised, and they are continually being refined. Not only can they be aimed at particular genes, but they can be applied to the entire genome to estimate the fraction of our genes that have been targets of recent natural selection.

These analyses have delivered a surprise. As the geneticist Joshua Akey concluded in a stringent 2009 review, "The number of strong selective events thought to exist in the human genome today is considerably more than that imagined less than a decade ago. . . . [Approximately] 8% of the genome has been influenced by positive selection, and an even larger fraction may have been subject to more modest selective pressure."[135] Many of the selected genes involve the functioning of the nervous system, so they could, in theory, affect cognition or emotion. The pattern of selection, moreover, differs among populations.

Some journalists have uncomprehendingly lauded these results as a refutation of evolutionary psychology and what they see as its politically dangerous implication of a human nature shaped by adaptation to a hunter-gatherer lifestyle. In fact the evidence for recent selection, if it applies to genes with effects on cognition and emotion, would license a far more radical form of evolutionary psychology, one in which minds have been biologically shaped by recent environments in addition to ancient ones. And it could have the incendiary implication that aboriginal and immigrant populations are less biologically adapted to the demands of modern life than populations that have lived in literate state societies for millennia. The fact that a hypothesis is politically uncomfortable does not mean that it is false, but it does mean that we should consider the evidence very carefully before concluding that it is true. Is there any reason to believe that declines in violence in particular societies can be attributed to genetic changes in their members?

The neurobiology of violence is a target-rich area for natural selection. Selective breeding of mice for four or five generations can produce a strain that is markedly more or less aggressive than an off-the-shelf lab mouse.[136] Violence in humans, of course, is fantastically more complicated than violence in mice, but if variations among people in their inclinations toward or away from violence are heritable, selection could certainly favor whichever variants result in more surviving offspring, which would change the concentration of the inciting and pacifying genes over time. So first we must establish whether any portion of the variation in aggression among people is caused by variation in their genes, that is, whether aggression is heritable.

Heritability may be measured in at least three ways.[137] One is to look at correlations in traits between identical twins who have been separated at birth and reared apart; they share their genes but not their family environments (within the range of environments in the sample). A second is to see whether there is a higher correlation between identical twins (who share all their genes and most of their family environments) than between fraternal twins (who share only half their variable genes and most of their family environments). The third is to see whether there is a greater correlation between biological siblings (who share half their genes and most of their family environments) than between adopted siblings (who share none of their variable genes and most of their family environments). Each of the methods has strengths and weaknesses (for example, identical twins may become partners in crime more often than fraternal twins), but the strengths and weaknesses are different among the three, so if the methods converge, there are good grounds for believing that the trait in question is heritable.

These methods have shown that having an antisocial personality and getting into trouble with the law have a substantial heritable component, though their effects sometimes depend on features of the environment. In a large 1984 study of Danish adoptees, among the adolescents and young adults who had been brought up in families in which an adoptive parent had been convicted of a crime, about 25 percent of the biological offspring of criminals had been convicted of a crime, whereas only 15 percent of the biological offspring of noncriminals had been convicted.[138] In that study, the effect of biological relatedness was seen only in nonviolent crimes such as auto theft, so in the 1980s many textbooks said that only nonviolent criminality was heritable, not a tendency to violence itself. But the conclusion was premature. There are far fewer criminal convictions of violent than nonviolent crimes, so the sample size for the violent ones is smaller and the ability to detect heritability weaker. Also, conviction rates are buffeted by vicissitudes of the criminal justice system that may overwhelm the effects of an offender's violent tendencies.

Today's studies use more sensitive measures of violence, including confidential self-reports, validated scales of aggression and antisocial behavior, and ratings by teachers, friends, and parents (for example, whether the person being rated "hurts others for his or her own advantage" or "deliberately frightens and causes discomfort in others"). All of the measures correlate with the probability of being convicted of a violent crime, while providing far more plentiful data.[139] When they are analyzed with the tools of behavioral genetics, all three methods reveal substantial heritability of aggressive tendencies.[140]

The analysis of twins separated at birth is the rarest method in behavioral genetics, because nowadays few twins are separated at birth. But the largest of the studies, based at the University of Minnesota, looked at aggression in identical twins reared apart and found a heritability coefficient of 0.38 (which means that about 38 percent of the variation in aggression in the sample can be explained by variation in their genes).[141] Adoption studies are somewhat more common, and one of the better ones estimated the heritability of aggressive behavior in its sample as 0.70.[142] Studies that compare identical and fraternal twins on aggressive tendencies such as arguing, fighting, threatening, destroying property, and disobeying parents and teachers tend to yield estimates of heritability between 0.4 and 0.6, particularly in childhood and adulthood. (In adolescence, the influence of peers often overshadows the influence of genes.)[143]

The behavioral geneticists Soo Hyun Rhee and Irwin Waldman recently reviewed the entire research literature on the genetics of aggression, which includes more than a hundred twin and adoption studies.[144] They selected nineteen that met stringent criteria of research quality and that specifically zeroed in on aggressive actions (such as physical fighting, cruelty to animals, and bullying) rather than the broader category of antisocial tendencies. They also examined all the published twin and adoption studies of criminal arrests and convictions. They estimated the heritability of aggressive behavior at around 0.44, and the heritability of criminality at around 0.75 (of which 0.33 consists of additive heritability, that is, variation that breeds true, and 0.42 consists of nonadditive heritability, variation caused by interactions among genes). Though their dataset on criminality did not distinguish violent from nonviolent crimes, they cited a Danish twin study that separated the two and that yielded a heritability estimate of 0.50 for the violent ones.[145] As in most studies in behavioral genetics, the effects of being brought up in a given family were tiny to nonexistent, though other aspects of the environment that are not easily measurable by these techniques, such as effects of the neighborhood, subculture, or idiosyncratic personal experiences, undoubtedly do have effects. The exact numbers should not be taken too seriously, but the fact that they are all substantially above zero should be. Behavioral genetics confirms that aggressive tendencies can be inherited, and that gives natural selection material to work with in shifting the average violent tendencies of a population.

Heritability is a necessary condition for evolutionary change, but it measures a hodgepodge of diverse contributors to behavior. When we tease them apart, we find many specific pathways by which natural selection could adjust our inclinations toward or away from violence. Let's consider a few.

Self-domestication and pedomorphy. Richard Wrangham has noted that the domestication of animals usually tames them by slowing down components of the developmental timetable to retain juvenile traits into adulthood, a process called pedomorphy or neoteny.[146] Domesticated strains and species tend to have more childlike skulls and faces, to show fewer sex differences, to be more playful, and to be less aggressive. These changes can be seen in farm animals that have been deliberately domesticated, such as horses, cattle, goats, and foxes, and in one species of wolf that was self-domesticated after it started to hang around human campsites thousands of years ago scrounging leftover food and eventually evolved into dogs. And in chapter 2 we saw that bonobos evolved from a chimpanzee-like ancestor via a process of pedomorphy after their foraging ecology had reduced the payoff for aggression in males. Based on pedomorphic changes in the fossils of Paleolithic humans, Wrangham has suggested that a similar process has been taking place in human evolution during the past thirty to fifty thousand years, and may still be taking place.

Brain structure. The neuroscientist Paul Thompson has shown that the distribution of gray matter in the cerebral cortex, including the dorsolateral prefrontal region, is highly heritable: it is almost identical in identical twins and considerably less similar in fraternal twins.[147] So is the distribution of white matter connecting the frontal cortex to other regions of the brain.[148] It is possible, then, that the frontal lobe circuitry that implements self-control varies genetically among individuals, making it eligible for recent natural selection.

Oxytocin, the so-called cuddle hormone which encourages sympathy and trust, acts on receptors in several parts of the brain, and the number and distribution of those receptors can have dramatic effects on behavior. In a famous experiment, biologists inserted a gene for the receptor for vasopressin (a hormone similar to oxytocin that operates in the brains of males) into meadow voles, an aggressive and promiscuous species that lacks it. Higamous, hogamous, the voles were monogamous, just like their evolutionary cousins the prairie voles, which come with the receptors preinstalled.[149] The experiment suggests that simple genetic changes in the oxytocin-vasopressin system can have profound effects on sympathy, bonding, and by extension the inhibition of aggression.

Testosterone. A person's response to a challenge of dominance depends in part on the amount of testosterone released into the bloodstream and on the distribution of receptors for the hormone in his or her brain.[150] The gene for the testosterone receptor varies across individuals, so a given concentration of testosterone can have a stronger effect on the brains of some people than others. Men with genes that code for more sensitive versions of the receptor have a greater surge of testosterone when conversing with an attractive woman (which can lead to reduced fear and greater risk-taking), and in one study were overrepresented in a sample of convicted rapists and murderers.[151] The genetic pathways that regulate testosterone are complicated, but they offer a target by which natural selection could alter people's willingness to take up aggressive challenges.

Neurotransmitters are the molecules that are released from a neuron, seep across a microscopic gap, and lock onto a receptor in the surface of another neuron, changing its activity and thereby allowing patterns of neural firing to propagate through the brain. One major class of neurotransmitters are the amines, which include dopamine, serotonin, and norepinephrine (also called noradrenaline, and related to the adrenaline that triggers the fight-or-flight response). The catecholamines are used in several motivational and emotional systems of the brain, and their concentration is regulated by proteins that break them down or recycle them. One of those enzymes is monoamine oxidase-A, MAO-A for short, which helps to break down these neurotransmitters, preventing them from building up in the brain. When they do build up, the organism can become hyperreactive to threats and more likely to engage in aggression.

The first sign that MAO-A can affect violence in humans was the discovery of a Dutch family that carried a rare mutation that left half the men without a working version of the gene.[152] (The gene is found on the X chromosome, which men have only one of, so if a man's MAO-A gene is defective, he has no backup copy to compensate.) Over at least five generations, the affected men in the family were prone to aggressive outbursts. One, for example, forced his sisters to disrobe at knifepoint; another tried to run over his boss with his car.

A more common kind of variation is found in the part of the gene that determines how much MAO-A is produced. People with a low-activity version of the gene build up higher levels of dopamine, serotonin, and norepinephrine in the brain. They also are more likely to have symptoms of antisocial personality disorder, to report that they have committed acts of violence, to be convicted of a violent crime, to have amygdalas that react more strongly and an orbital cortex that reacts less strongly to angry and fearful faces, and, in the psychology lab, to force a fellow participant to drink hot sauce if they think he has exploited them.[153] Unlike many other genes that affect behavior, the low-activity version of the MAO-A gene seems to be fairly specific to aggression; it does not correlate well with any other personality trait.[154]

The low-activity version of the MAO-A gene makes people more prone to aggression primarily when they have grown up with stressful experiences, such as having been abused or neglected by their parents or having been held back in school.[155] It's hard to pinpoint the exact stressors that have this effect, because stressful lives are often stressful in many ways at once. In fact, the modulating factor may consist of other genes which are shared with an abusive parent, predisposing both parent and child to aggression, and which may also elicit negative reactions from the people around him.[156] But whatever the modifying factor is, it does not turn the effects of the low-activity version of the gene upside down. In all the studies, the gene has an aggregate or main effect in the population that could make it a target of selection. Indeed, Moffitt and Caspi (who first discovered that the effect of the gene depends on stressful experiences) suggest that rather than thinking of the low-activity version of the gene as a contributor to violence, we should think of the high-activity version as an inhibitor of violence: it protects people from overreacting to a stress-filled life. Geneticists have discovered statistical evidence of selection for the MAO-A gene in humans, though the evidence does not single out the low- or high-activity variant; nor does it prove that the gene was selected for its effects on aggression.[157]

Other genes that affect dopamine have been associated with delinquency as well, including a version of a gene that affects the density of dopamine receptors (DRD2) and a version of a gene for a dopamine transporter (DAT1) that mops up excess dopamine from the synapse and transports it back into the neurons that release it.[158] Any of them could be fair game for rapid natural selection.

Genetic tendencies toward or away from violence, then, could have been selected during the historical transitions we have examined. The question is, were they? The mere existence of pathways to evolutionary change does not prove that those pathways were taken. Evolution depends not just on the genetic raw material but on factors such as the demography of a population (including both sheer numbers and the degree to which they have absorbed immigrants from other groups), the roll of the genetic and environmental dice, and the dilution of genetic effects by learned adjustments to the cultural milieu.

So is there evidence that the Pacification or Civilizing Process ever rendered the pacified or civilized peoples constitutionally less susceptible to violence? Casual impressions can be misleading. History offers many examples in which one nation has considered another to be peopled by "savages" or "barbarians," but the impressions were motivated more by racism and observations of differences in societal type than by any attempt to tease apart nature and nurture. Between 1788 and 1868, 168,000 British convicts were sent to penal colonies in Australia, and one might have expected that Australians today would have inherited the obstreperous traits of their founding population. But Australia's homicide rate is lower than that of the mother country; in fact it is one of the lowest in the world. Before 1945 the Germans had a reputation as the most militaristic people on earth; today they may be the most pacifistic.

What about hard evidence from the revolution in evolutionary genomics? In their manifesto *The 10,000 Year Explosion: How Civilization Accelerated Human Evolution*, the physicist Gregory Cochran and the anthropologist Henry Harpending reviewed the evidence for recent selection in humans and speculated that it includes changes in temperament and behavior. But none of the selected genes they describe has been implicated in behavior; all are restricted to digestion, disease resistance, and skin pigmentation.[159]

I know of only two claims of recent evolutionary changes in violence that have at least a soupçon of scientific evidence behind them. One involves the Maori, a Polynesian people who settled in New Zealand about a thousand years ago. Like many nonstate hunter-horticulturalists, the Maori engaged in extensive warfare, including a genocide of the Moriori people in the nearby Chatham Islands. Today their culture preserves many symbols of their warrior past, including the *haka* war dance that fires up New Zealand's All-Blacks rugby team, and an armamentarium of beautiful greenstone weapons. (I have a lovely battle-ax in my office, a gift of the University of Auckland for giving a set of lectures there.) The acclaimed 1994 film *Once Were Warriors* vividly depicts the crime and domestic violence that currently plague some of the Maori communities in New Zealand.

Against this cultural backdrop, the New Zealand press quickly picked up on a 2005 report showing that the low-activity version of the MAO-A gene is more common among the Maori (70 percent) than among descendants of Europeans (40 percent).[160] The lead geneticist, Rod Lea, suggested that the gene had been selected in the Maori because it made them more accepting of risk during the arduous canoe voyages that brought them to New Zealand, and more effective in their internecine tribal battles thereafter. The press dubbed it the

Warrior Gene and speculated that it could help explain the higher levels of social pathology among the Maori in modern New Zealand.

The Warrior Gene theory has not fared well in warfare with skeptical scientists.[161] One problem is that the signatures of selection for the gene could also have been caused by a genetic bottleneck, in which a random assortment of genes that happened to be carried by a few founders of a population was multiplied in their burgeoning descendants. Another is that the low-activity version of the gene is also found in a majority of Chinese men (around 55 percent), and the Chinese are neither descended from warriors in their recent history nor particularly prone to social pathology in modern societies. A third and related problem is that the association between the gene and aggression may differ in European and non-European populations, perhaps because they have evolved somewhat different ways of regulating their catecholamine levels.[162] (Genes often act in networks regulated by feedback loops, so in populations in which a particular gene is less effective, other genes may step up their activity to compensate.) For now, the Warrior Gene theory is staggering around with possibly fatal wounds.

The other claim of a recent evolutionary change appeals to a civilizing rather than a pacification process. In *A Farewell to Alms: A Brief Economic History of the World,* Gregory Clark sought to explain the timing and location of the Industrial Revolution, which for the first time in history increased material well-being faster than the increase could be eaten up by population growth (see, for example, figure 4–7, taken from his book). Why, Clark asked, was it England that hosted this onetime escape from the Malthusian trap?

The answer, he suggested, was that the nature of Englishmen had changed. Starting around 1250, when England began to shift from a knightly society to "a nation of shopkeepers" (as Napoleon would sneeringly call it), the wealthier commoners had more surviving children than the poorer ones, presumably because they married younger and could afford better food and cleaner living conditions. Clark calls it "the survival of the richest": the rich got richer *and* they got children. This upper middle class was also outreproducing the aristocrats, who were splitting heads and chopping off body parts in their tournaments and private wars, as we saw in figure 3–7, also taken from Clark's data. Since the economy as a whole would not begin to expand until the 19th century, the extra surviving children of the wealthier merchants and tradesmen had nowhere to go on the economic ladder but downward. They continuously replaced the poorer commoners, bringing with them their bourgeois traits of thrift, hard work, self-control, patient future discounting, and avoidance of violence. The population of England literally evolved middle-class values. That in turn positioned them to take advantage of the commercial opportunities opened up by the innovations of the Industrial Revolution. Though Clark occasionally dodges the political correctness police by noting that nonviolence and self-control can be passed from parent to child as cultural habits, in a précis of his book entitled "Genetically Capitalist?" he offers the full-strength version of his thesis:

> The highly capitalistic nature of English society by 1800—individualism, low time preference rates, long work hours, high levels of human capital—may thus stem from the nature of the Darwinian struggle in a very stable agrarian society in the long run up to the Industrial Revolution. The triumph of capitalism in the modern world thus may lie as much in our genes as in ideology or rationality.[163]

A Farewell to Alms is filled with illuminating statistics and gripping narratives about the historical precursors to the Industrial Revolution. But the Genetically Capitalist theory has not competed well in the struggle for survival among theories of economic growth.[164] One problem is that until recently, the rich have outreproduced the poor in pretty much *every* society, not just the one that later blasted off in an industrial revolution. Another is that while aristocrats and royals may have had no more legitimate heirs than the bourgeoisie, they more than made up for it in bastards, which could have contributed a disproportionate share of their genes to the next generation. A third is that when institutions change, a nation can vault to spectacular rates of economic growth in the absence of a recent history of selection for middle-class values, such as postwar Japan and post-communist China. And most important, Clark cites no data showing that the English are innately more self-controlled or less violent than the citizenries of countries that did not host an industrial revolution.

So while recent biological evolution may, in theory, have tweaked our inclinations toward violence and nonviolence, we have no good evidence that it actually has. At the same time, we do have good evidence for changes that could not possibly be genetic, because they unfolded on time scales that are too rapid to be explained by natural selection, even with the new understanding of how recently it has acted. The abolition of slavery and cruel punishments during the Humanitarian Revolution; the reduction of violence against minorities, women, children, homosexuals, and animals during the Rights Revolutions; and the plummeting of war and genocide during the Long Peace and the New Peace, all unfolded over a span of decades or even years, sometimes within a single generation. A particularly dramatic decline is the near-halving of the homicide rate during the Great American Crime Decline of the 1990s. The decay rate of that decline, around 7 percent a year, is powerful enough to drag a measure of violence down to 1 percent of its original level over just two generations, all without the slightest change in gene frequencies. Since it is indisputable that cultural and social inputs can adjust the settings of our better angels (such as self-control and empathy) and thereby control our violent inclinations, we have the means to explain all the declines of violence without invoking recent biological evolution. At least for the time being, we have no need for that hypothesis.

MORALITY AND TABOO

The world has far too much morality. If you added up all the homicides committed in pursuit of self-help justice, the casualties of religious and revolutionary wars, the people executed for victimless crimes and misdemeanors, and the targets of ideological genocides, they would surely outnumber the fatalities from amoral predation and conquest. The human moral sense can excuse any atrocity in the minds of those who commit it, and it furnishes them with motives for acts of violence that bring them no tangible benefit. The torture of heretics and conversos, the burning of witches, the imprisonment of homosexuals, and the honor killing of unchaste sisters and daughters are just a few examples. The incalculable suffering that has been visited on the world by people motivated by a moral cause is enough to make one sympathize with the comedian George Carlin when he said, "I think motivation is overrated. You show me some lazy prick who's lying around all day watching game shows and stroking his penis, and I'll show you someone who's *not causing any fucking trouble*!"

Though the net contribution of the human moral sense to human well-being may well be negative, on those occasions when it is suitably deployed it can claim some monumental advances, including the humanitarian reforms of the Enlightenment and the Rights Revolutions of recent decades. When it comes to virulent ideologies, morality may be the disease, but morality is also the cure. The mentality of taboo, like the mentality of morality of which it is part, also can pull in either direction. It can turn religious or sexual nonconformity into an outrage that calls for ghastly punishment, but it can also prevent the mind from sliding into dangerous territory such as wars of conquest, the use of chemical and nuclear weapons, dehumanizing racial stereotypes, casual allusions to rape, and the taking of identifiable human lives.

How can we make sense of this crazy angel—the part of human nature that would seem to have the strongest claim to be the source of our goodness, but that in practice can be more diabolical than our worst inner demon?

To understand the role of the moral sense in the decline of violence, we have to solve a number of psychological enigmas. One is how people in different times and cultures can be driven by goals that they experience as "moral" but that are unrecognizable to our own standards of morality. Another is why the moral sense does not, in general, push toward the reduction of suffering but so often increases it. A third is how the moral sense can be so compartmentalized: why upstanding citizens can beat their wives and children; why liberal democracies can practice slavery and colonial oppression; why Nazi Germany could treat animals with unequaled kindness. A fourth is why, for better and for worse, morality can be extended to thoughts as well as deeds, leading to the paradox of taboo. And the overriding puzzle, of course, is: What changed? What degree of freedom in the human moral sense has been engaged by the processes of history to drive violence downward?

The starting point is to distinguish morality per se, a topic in philosophy (in particular, normative ethics), from the human moral sense, a topic in psychology. Unless one is a radical moral relativist, one believes that people can in some sense be *mistaken* about their moral convictions; that their justifications of genocide, rape, honor killings, and the torture of heretics are erroneous, not just distasteful to our sensibilities.[165] Whether one is a moral realist and believes that moral truths are objectively out there somewhere like mathematical truths, or simply allows that moral statements have some degree of warrant in terms of consistency with universally held convictions or the best understanding that arises out of our collective rational deliberation, one can distinguish questions of morality from questions of moral psychology. The latter ask about the mental processes that people *experience* as moral, and they can be studied in the lab and field, just like any other cognitive or emotional faculty.

The next step in understanding the moral sense is to recognize that it is a distinctive mode of thinking about an action, not just the avoidance of an action. There are important psychological distinctions between avoiding an action because it is deemed immoral ("Killing is wrong") and avoiding it because it is merely disagreeable ("I hate cauliflower"), unfashionable ("Bell-bottoms are out"), or imprudent ("Don't scratch mosquito bites").[166]

One difference is that disapproval of a moralized act is *universalized*. If you think cauliflower is distasteful, then whether other people eat it is of no concern to you. But if you think that murder and torture and rape are immoral, then you cannot simply avoid these activities yourself and be indifferent to whether other people indulge in them. You have to disapprove of *anyone* committing such acts.

Second, moralized beliefs are *actionable*. While people may not unfailingly carry out Socrates' dictum that "To know the good is to do the good," they tacitly aspire to it. People see moral actions as intrinsically worthy goals, which need no ulterior motive. If people believe that murder is immoral, they don't need to be paid or even esteemed to refrain from murdering someone. When people do breach a moral precept, they rationalize the failure by invoking some countervailing precept, by finding an exculpatory excuse, or by acknowledging that the failure is a regrettable personal weakness. Other than devils and storybook villains, no one says, "I believe murder is a heinous atrocity, and I do it whenever it serves my purposes."[167]

Finally, moralized infractions are *punishable*. If one believes that murder is wrong, one is not just entitled to see a murderer punished, but one is *obligated* to make it happen. One may not, as we say, let someone get away with murder. Now just substitute "idolatry" or "homosexuality" or "blasphemy" or "subversion" or "indecency" or "insubordination" for "murder," and you can see how the human moral sense can be a major force for evil.

Another design feature of the moral sense is that many moral convictions operate as norms and taboos rather than as principles the believer can articulate and defend. In the psychologist Lawrence Kohlberg's famous six-stage progression of moral development, from a child's avoidance of punishment to a philosopher's universal principles, the middle two stages (which many people never get out of) consist of conforming to norms to be a good boy or girl, and maintaining conventions to preserve social stability. When reasoning through the moral dilemma that Kohlberg made famous, in which Heinz must break into a drugstore to steal an overpriced drug that will save his dying wife, people in these stages can muster no better justification for their answers than that Heinz shouldn't steal the drug because stealing is bad and illegal and he is not a criminal, or that Heinz should steal the drug because that's what a good husband does.[168] Fewer people can articulate a principled justification, such as that human life is a cardinal value that trumps social norms, social stability, or obedience to the law.

The psychologist Jonathan Haidt has underscored the ineffability of moral norms in a phenomenon he calls moral dumbfounding. Often people have an instant intuition that an action is immoral, and then struggle, often unsuccessfully, to come up with reasons *why* it is immoral.[169] When Haidt asked participants, for example, whether it would be all right for a brother and sister to have voluntary protected sex, for a person to clean a toilet with a discarded American flag, for a family to eat a pet dog that had been killed by a car, for a man to buy a dead chicken and have sex with it, or for a person to break a deathbed vow to visit the grave of his mother, they said no in each case. But when asked for justifications, they floundered ineffectually before giving up and saying, "I don't know, I can't explain it, I just know it's wrong."

Moral norms, even when ineffable, can sometimes be effective brakes on violent behavior. In the modern West, as we have seen, the avoidance of some kinds of violence, such as mercy-killing an abandoned child, retaliating for an insult, and declaring war on another developed state, consist not in weighing the moral issues, empathizing with the targets, or restraining an impulse, but in not having the violent act as a live option in the mind at all. The act is not considered and avoided; it is unthinkable or laughable.

The combination of radical cultural differences in which behaviors are moralized with moral dumbfounding in our own culture may create the impression that norms and taboos are arbitrary—that there may be a culture out there somewhere in which it is immoral to utter a sentence with an even number of words or to deny that the ocean is boiling hot. But the anthropologist Richard Shweder and several of his students and collaborators have found that moral norms across the world cluster around a small number of themes.[170] The intuitions that we in the modern West tend to think of as the core of morality—fairness, justice, the protection of individuals, and the prevention of harm—are just one of several spheres of concern that may attach themselves to the cognitive and emotional paraphernalia of moralization. Even a glance at ancient religions like Judaism, Islam, and Hinduism reminds us that they moralize a slew of other concerns, such as loyalty, respect, obedience, asceticism, and the regulation of bodily functions like eating, sex, and menstruation.

Shweder organized the world's moral concerns in a threefold way.[171] Autonomy, the ethic we recognize in the modern West, assumes that the social world is composed of individuals and that the purpose of morality is to allow them to exercise their choices and to protect them from harm. The ethic of Community, in contrast, sees the social world as a collection of tribes, clans, families, institutions, guilds, and other coalitions, and equates morality with duty, respect, loyalty, and interdependence. The ethic of Divinity posits that the world is composed of a divine essence, portions of which are housed in bodies, and that the purpose of morality is to protect this spirit from degradation and contamination. If a body is merely a container for the soul, which ultimately belongs to or is part of a god, then people do not have the right to do what they want with their bodies. They are obligated to avoid polluting them by refraining from unclean forms of sex, food, and other physical pleasures. The ethic of Divinity lies behind the moralization of disgust and the valorization of purity and asceticism.

Haidt took Shweder's trichotomy and cleaved two of the ethics in two, yielding a total of five concerns that he called moral foundations.[172] Community was bifurcated into In-group Loyalty and Authority/Respect, and Autonomy was sundered into Fairness/Reciprocity (the morality behind reciprocal altruism) and Harm/Care (the cultivation of kindness and compassion, and the inhibition of cruelty and aggression). Haidt also gave Divinity the more secular label Purity/Sanctity. In addition to these adjustments, Haidt beefed up the case that the moral foundations are universal by showing that all five spheres may be found in the moral intuitions of secular Westerners. In his dumbfounding scenarios, for example, Purity/Sanctity underlay the participants' revulsion to incest, bestiality, and the eating of a family pet. Authority/Respect commanded them to visit a mother's grave. And In-group Loyalty prohibited them from desecrating an American flag.

The system I find most useful was developed by the anthropologist Alan Fiske. It proposes that moralization comes out of four relational models, each a distinct way in which people conceive of their relationships.[173] The theory aims to explain how people in a given society apportion resources, where their moral obsessions came from in our evolutionary history, how morality varies across societies, and how people can compartmentalize their morality and protect it with taboos. The relational models line up with the classifications of Shweder and Haidt more or less as shown in the table on page 626.

Shweder's Ethics	Divinity	Community		Autonomy		
Haidt's Moral Foundations	Purity/ Sanctity	In-group Loyalty	Authority/Respect	Harm/Care	Fairness/ Reciprocity	
Fiske's Relational Models	Communal Sharing		Authority Ranking	Equality Matching		Market Pricing/ Rational-Legal

The first model, Communal Sharing (Communality for short), combines In-group Loyalty with Purity/Sanctity. When people adopt the mindset of Communality, they freely share resources within the group, keeping no tabs on who gives or takes how much. They conceptualize the group as "one flesh," unified by a common essence, which must be safeguarded against contamination. They reinforce the intuition of unity with rituals of bonding and merging such as bodily contact, commensal meals, synchronized movement, chanting or praying in unison, shared emotional experiences, common bodily ornamentation or mutilation, and the mingling of bodily fluids in nursing, sex, and blood rituals. They also rationalize it with myths of shared ancestry, descent from a patriarch, rootedness in a territory, or relatedness to a totemic animal. Communality evolved from maternal care, kin selection, and mutualism, and it may be implemented in the brain, at least in part, by the oxytocin system.

Fiske's second relational model, Authority Ranking, is a linear hierarchy defined by dominance, status, age, gender, size, strength, wealth, or precedence. It entitles superiors to take what they want and to receive tribute from inferiors, and to command their obedience and loyalty. It also obligates them to a paternalistic, pastoral, or noblesse oblige responsibility to protect those under them. Presumably it evolved from primate dominance hierarchies, and it may be implemented, in part, by testosterone-sensitive circuits in the brain.

Equality Matching embraces tit-for-tat reciprocity and other schemes to divide resources equitably, such as turn-taking, coin-flipping, matching contributions, division into equal portions, and verbal formulas like eeny-meeny-miney-moe. Few animals engage in clear-cut reciprocity, though chimpanzees have a rudimentary sense of fairness, at least when it comes to themselves being shortchanged. The neural bases of Equality Matching embrace the parts of the brain that register intentions, cheating, conflict, perspective-taking, and calculation, which include the insula, orbital cortex, cingulate cortex, dorsolateral prefrontal cortex, parietal cortex, and temporoparietal junction. Equality Matching is the basis of our sense of fairness and our intuitive economics, and it binds us as neighbors, colleagues, acquaintances, and trading partners rather than as bosom buddies or brothers-in-arms. Many traditional tribes engaged in the ritual exchange of useless gifts, a bit like our Christmas fruitcakes, solely to cement Equality Matching relationships.[174]

(Readers who are comparing and contrasting the taxonomies may wonder why Haidt's category of Harm/Care is adjacent to Fairness and aligned with Fiske's Equality Matching, rather than with more touchy-feely relationships like Community or Sanctity. The reason is that Haidt measures Harm/Care by asking people about the treatment of a generic "someone" rather than the friends and relatives that are the standard beneficiaries of caring. The responses to these questions align perfectly with the responses to his questions about Fairness, and that is no coincidence.[175] Recall that the logic of reciprocal altruism, which implements our sense of fairness, is to be "nice" by cooperating on the first move, by not defecting unless defected on, and by conferring a large benefit to a needy stranger when one can do so at a relatively small cost to oneself. When care and harm are extended outside intimate circles, then they are simply a part of the logic of fairness.)[176]

Fiske's final relational model is Market Pricing: the system of currency, prices, rents, salaries, benefits, interest, credit, and derivatives that powers a modern economy. Market Pricing depends on numbers, mathematical formulas, accounting, digital transfers, and the language of formal contracts. Unlike the other three relational models, Market Pricing is nowhere near universal, since it depends on literacy, numeracy, and other recently invented information technologies. The logic of Market Pricing remains cognitively unnatural as well, as we saw in the widespread resistance to interest and profits until the modern era. One can line up the models, Fiske notes, along a scale that more or less reflects their order of emergence in evolution, child development, and history: Communal Sharing > Authority Ranking > Equality Matching > Market Pricing.

Market Pricing, it seems to me, is specific neither to markets nor to pricing. It really should be lumped with other examples of formal social organization that have been honed over the centuries as a good way for millions of people to manage their affairs in a technologically advanced society, but which may not occur spontaneously to untutored minds.[177] One of these institutions is the political apparatus of democracy, where power is assigned not to a strongman (Authority) but to representatives who are selected by a formal voting procedure and whose prerogatives are delineated by a system of laws. Another is a corporation, university, or nonprofit organization. The people who work in them aren't free to hire their friends and relations (Communality) or to dole out spoils as favors (Equality Matching), but are hemmed in by fiduciary duties and regulations. My emendation of Fiske's theory does not come out of the blue. Fiske notes that one of his intellectual inspirations for Market Pricing was the sociologist Max Weber's concept of a "rational-legal" (as opposed to traditional and charismatic) mode of social legitimation—a system of norms that is worked out by reason and implemented by formal rules.[178] Accordingly, I will sometimes refer to this relational model using the more general term Rational-Legal.

For all their differences in lumping and splitting, the theories of Shweder, Haidt, and Fiske agree on how the moral sense works. No society defines everyday virtue and wrongdoing by the Golden Rule or the Categorical Imperative. Instead, morality consists in respecting or violating one of the relational models (or ethics or foundations): betraying, exploiting, or subverting a coalition; contaminating oneself or one's community; defying or insulting a legitimate authority; harming someone without provocation; taking a benefit without paying the cost; peculating funds or abusing prerogatives.

The point of these taxonomies is not to pigeonhole entire societies but to provide a grammar for social norms.[179] The grammar should reveal common patterns beneath the differences among cultures and periods (including the decline of violence), and should predict people's response to infractions of the reigning norms, including their perverse genius for moral compartmentalization.

Some social norms are merely solutions to coordination games, such as driving on the right-hand side of the road, using paper currency, and speaking the ambient language.[180] But most norms have moral content. Each moralized norm is a compartment containing a relational model, one or more social roles (parent, child, teacher, student, husband, wife, supervisor, employee, customer, neighbor, stranger), a context (home, street, school, workplace), and a resource (food, money, land, housing, time, advice, sex, labor). To be a socially competent member of a culture is to have assimilated a large set of these norms.

Take friendship. Couples who are close friends operate mainly on the model of Communal Sharing. They freely share food at a dinner party, and they do each other favors without keeping score. But they may also recognize special circumstances that call for some other relational model. They may work together on a task in which one is an expert and gives orders to the other (Authority Ranking), split the cost of gas on a trip (Equality Matching), or transact the sale of a car at its blue book value (Market Pricing).

Infractions of a relational model are moralized as straightforwardly wrong. Within the Communal Sharing model that usually governs a friendship, it is wrong for one person to stint on sharing. Within the special case of Equality Matching of gas on a trip, an infraction consists of failing to pay for one's share. Equality Matching, with its assumption of a continuing reciprocal relationship, allows for loose accounting, as when the ranchers of Shasta County compensated each other for damage with roughly equivalent favors and agreed to lump it when a small act of damage went uncompensated.[181] Market Pricing and other Rational-Legal models are less forgiving. A diner who leaves an expensive restaurant without paying cannot count on the owner to let him make it up in the long run, or simply to lump it. The owner is more likely to call the police.

When a person violates the terms of a relational model he or she has tacitly agreed to, the violator is seen as a parasite or cheater and becomes a target of moralistic anger. But when a person applies one relational model to a resource ordinarily governed by another, a different psychology comes into play. That person doesn't violate the rules so much as he or she doesn't "get" them. The reaction can range from puzzlement, embarrassment, and awkwardness to shock, offense, and rage.[182] Imagine, for example, a diner thanking a restaurateur for an enjoyable experience and offering to have him over for dinner at some point in the future (treating a Market Pricing interaction as if it were governed by Communal Sharing). Conversely, imagine the reaction at a dinner party (Communal Sharing) if a guest pulled out his wallet and offered to pay the host for the meal (Market Pricing), or if the host asked a guest to wash the pots while the host relaxed in front of the television (Equality Matching). Likewise, imagine that the guest offered to sell his car to the host, and then drove a hard bargain on the price, or the host suggested that the couples swap partners for a half-hour of sex before everyone went home for the evening.

The emotional response to a relational mismatch depends on whether it is accidental or deliberate, which model is substituted for which, and the nature of the resource. The psychologist Philip Tetlock has suggested that the psychology of taboo—a reaction of outrage to certain thoughts being aired—comes into play with resources that are deemed *sacred*.[183] A sacred value is one that may not be traded off against anything else. Sacred resources are usually governed by the primal models of Communality and Authority, and they trigger the taboo reaction when someone treats them with the more advanced models of Equality Matching or Market Pricing. If someone offered to purchase your child (suddenly thrusting a Communal Sharing relationship under the light of Market Pricing), you would not ask how much he was offering but would be offended at the very idea. The same is true for an offer to buy a personal gift or family heirloom that has been bestowed upon you, or to pay you for betraying a friend, a spouse, or your country. Tetlock found that when students were asked their opinion on the pros and cons of an open market for sacred resources like votes, military service, jury duty, body organs, or babies put up for adoption, most of them did not articulate a good case against the practice (such as that the poor might sell their organs out of desperation) but expressed outrage at being asked. Typical "arguments" were "This is degrading, dehumanizing, and unacceptable" and "What kind of people are we becoming?"

The psychology of taboo is not completely irrational.[184] To maintain precious relationships, it is not enough for us to say and do the right thing. We also have to show that our heart is in the right place, that we don't weigh the costs and benefits of selling out those who trust us. When you are faced with an indecent proposal, anything less than an indignant refusal would betray the awful truth that you don't understand what it means to be a genuine parent or spouse or citizen. And that understanding consists of having absorbed a cultural norm that assigns a sacred value to a primal relational model.

In an old joke, a man asks a woman if she would sleep with him for a million dollars, and she says she would consider it. He then asks if she would sleep with him for a hundred dollars, and she replies, "What kind of a woman do you think I am?" He answers, "We've already established that. We're just haggling over the price." To understand the joke is to appreciate that most sacred values are in fact pseudo-sacred. People can be induced to compromise on them if the tradeoff is obfuscated, spin-doctored, or reframed.[185] (The joke uses the landmark figure "a million dollars" because it reframes a mere exchange of money into a life-transforming opportunity, namely becoming a millionaire.) When life insurance was first introduced, people were outraged at the very idea of assigning a dollar value to a human life, and of allowing wives to bet that their husbands would die, both of which are technically accurate descriptions of what life insurance does.[186] The insurance industry mounted advertising campaigns that reframed the product as an act of responsibility and decency on the part of the husband, who would simply be carrying out his duty to his family during a period in which he happened not to be alive.

Tetlock distinguishes three kinds of tradeoffs. *Routine* tradeoffs are those that fall within a single relational model, such as choosing to be with one friend rather than another, or to purchase one car rather than another. *Taboo* tradeoffs pit a sacred value in one model against a secular value in another, such as selling out a friend, a loved one, an organ, or oneself for barter or cash. *Tragic* tradeoffs pit sacred values against each other, as in deciding which of two needy transplant patients should receive an organ, or the ultimate tragic tradeoff, Sophie's choice between the lives of her two children. The art of politics, Tetlock points out, is in large part the ability to reframe taboo tradeoffs as tragic tradeoffs (or, when one is in the opposition, to do the opposite). A politician who wants to reform Social Security has to reframe it from "breaking our faith with senior citizens" (his opponent's framing) to "lifting the burden on hardworking wage-earners" or "no longer scrimping in the education of our children." Keeping troops in Afghanistan is reframed from "putting the lives of our soldiers in danger" to "guaranteeing our nation's commitment to freedom" or "winning the war on terror." The reframing of sacred values, as we will see, may be an overlooked tactic in the psychology of peacemaking.

The new theories of the moral sense, then, have helped explain moralized emotions, moral compartmentalization, and taboo. Now let's apply them to differences in moralization across cultures and, crucially, over the course of history.

Many assignments of a relational model to a set of social roles feel natural to people in all societies and may be rooted in our biology. They include the Communal Sharing among family members, an Authority Ranking within the family that makes people respect their elders, and the exchange of bulk commodities and routine favors under Equality Matching. But other kinds of assignment of a relational model to a resource and a set of social roles can differ radically across time and culture.[187]

In traditional Western marriages, for example, the husband wielded Authority over the wife. The model was mostly overturned in the 1970s, and some couples influenced by feminism switched to Equality Matching, splitting housework and child-rearing down the middle and strictly auditing the hours devoted to them. Since the businesslike psychology of Equality Matching clashes with the intimacy that most couples crave, most modern marriages have settled on Communal Sharing—with the consequence that many wives feel that the couple's failure to keep tabs on contributions to household duties leaves them overworked and underappreciated. The spouses may also carve out Rational-Legal exceptions, such as a prenuptial agreement, or the stipulation in their wills of separate inheritances for the children of their previous marriages.

Alternative linkages between a relational model and a resource or set of social roles define how cultures differ from one another. The members of one society may allow land to be bartered or sold, and be shocked to learn of another society that does the same with brides—or vice versa. In one culture, a woman's sexuality may fall under the Authority of the males in her family; in another, she is free to share it with her lover in a Communal relationship; in still another, she may barter it for an equivalent favor without stigma, an example of Equality Matching. In some societies, a killing must be avenged by the victim's kinsmen (Equality Matching); in others, it may be compensated with a wergild (Market Pricing); in still others, it is punished by the state (Authority Ranking).

A recognition that someone belongs to a different culture can mitigate, to some extent, the outrage ordinarily triggered by the violation of a relational model. Such violations can even be a source of humor, as in old comedies in which a hapless immigrant or rural bumpkin haggles over the price of a train ticket, grazes his sheep in a public park, or offers to settle a debt by betrothing his daughter in marriage. The formula is reversed in *Borat*, in which the comedian Sacha Baron Cohen pokes fun at the willingness of culturally sensitive Americans to tolerate the outrageous behavior of an obnoxious immigrant in their midst. Tolerance may run out, however, when a violation breaches a sacred value, as when immigrants to Western countries practice female genital cutting, honor killings, or the sale of underage brides, and when Westerners disrespect the prophet Muhammad by depicting him in novels, satirizing him in editorial cartoons, or allowing schoolchildren to name a teddy bear after him.

Differences in the deployment of relational models also define political ideologies.[188] Fascism, feudalism, theocracy, and other atavistic ideologies are based on the primal relational models of Communal Sharing and Authority Ranking. The interests of an individual are submerged within a community (*fascist* comes from an Italian word for "bundle"), and the community is dominated by a military, aristocratic, or ecclesiastical hierarchy. Communism envisioned a Communal Sharing of resources ("From each according to his ability, to each according to his need"), an Equality Matching of the means of production, and an Authority Ranking of political control (in theory, the dictatorship of the proletariat; in practice, a nomenklatura of commissars under a charismatic dictator). A kind of populist socialism seeks Equality Matching for life's necessities, such as land, medicine, education, and child care. At the other pole of the continuum, libertarians would allow people to negotiate virtually any resource under Market Pricing, including organs, babies, medical care, sexuality, and education.

Tucked in between these poles is the familiar liberal-conservative continuum. In several surveys, Haidt has shown that liberals believe that morality is a matter of preventing harm and enforcing fairness (the values that line up with Shweder's Autonomy and Fiske's Equality Matching). Conservatives give equal weight to all five foundations, including In-group Loyalty (values such as stability, tradition, and patriotism), Purity/Sanctity (values such as propriety, decency, and religious observance), and Authority/Respect (values such as respect for authority, deference to God, acknowledgment of gender roles, and military obedience).[189] The American culture war, with its clashes over taxes, medical insurance, welfare, gay marriage, abortion, the size of the military, the teaching of evolution, profanity in the media, and the separation of church and state, is fought in large part over different conceptions of the legitimate moral concerns of the state. Haidt notes that the ideologues at each pole are apt to view their opposite number as amoral, whereas in fact the moral circuitry in all of their brains is burning just as brightly, while filled with different conceptions of what morality comprises.

Before spelling out the connections between moral psychology and violence, let me use the theory of relational models to resolve a psychological puzzle that has been left hanging from earlier chapters. Many moral advances have taken the form of a shift in sensibilities that made an action seem more ridiculous than sinful, such as dueling, bullfighting, and jingoistic war. And many effective social critics, such as Swift, Johnson, Voltaire, Twain, Oscar Wilde, Bertrand Russell, Tom Lehrer, and George Carlin have been smart-ass comedians rather than thundering prophets. What in our psychology allows the joke to be mightier than the sword?

Humor works by confronting an audience with an incongruity, which may be resolved by switching to another frame of reference. And in that alternative frame of reference, the butt of the joke occupies a lowly or undignified status.[190] When Woody Allen says, "I'm very proud of my gold watch. My grandfather sold it to me on his deathbed," listeners are at first surprised that an emotionally precious heirloom would be sold rather than given, particularly by someone who cannot profit from the sale. Then they realize that the Woody Allen character is unloved and comes from a family of venal oddballs. Often the first reference frame, which sets up the incongruity, consists of a prevailing relational model, and to get the joke the audience must step outside it, as in the switch from Communal Sharing to Market Pricing in the Woody Allen joke.

Humor with a political or moral agenda can stealthily challenge a relational model that is second nature to an audience by forcing them to see that it leads to consequences that the rest of their minds recognize as absurd. Rufus T. Firefly's willingness to declare war in response to a wholly imagined insult in *Duck*

Soup deconstructs the Authority Ranking ethos of national grandeur and was appreciated in an era in which the image of war was shifting from thrilling and glorious to wasteful and stupid. Satire also served as an accelerant to recent social changes, such as the 1960s portrayals of racists and sexists as thick-witted Neanderthals and of Vietnam hawks as bloodthirsty psychopaths. The Soviet Union and its satellites also harbored a deep underground current of satire, as in the common definition of the two Cold War ideologies: "Capitalism is the exploitation of man by man; Communism is the exact opposite."

According to the 18th-century writer Mary Wortley Montagu, "Satire should, like a polished razor keen / Wound with a touch that's scarcely felt or seen." But satire is seldom polished that keenly, and the butts of a joke may be all too aware of the subversive power of humor. They may react with a rage that is stoked by the intentional insult to a sacred value, the deflation of their dignity, and a realization that laughter indicates common knowledge of both. The lethal riots in 2005 provoked by the editorial cartoons in the Danish newspaper *Jyllands-Posten* (for example, one showing Muhammad in heaven greeting newly arrived suicide bombers with "Stop, we have run out of virgins!") show that when it comes to the deliberate undermining of a sacred relational model, humor is no laughing matter.

How do the relational models that make up the moral sense license the various kinds of violence that people feel are morally legitimate? And what degree of freedom allows societies to throttle down moralistic violence or, better still, shift it into reverse? All the relational models invite moralistic punishment of the people who violate their rules of engagement. But each model licenses a distinctive kind of violence as well.[191]

Human beings, Fiske notes, need not relate to one another using any of the models at all, a state he calls a null or asocial relationship. People who don't fall under a relational model are *dehumanized:* they are seen as lacking the essential features of human nature and are treated, in effect, like inanimate objects which may be ignored, exploited, or preyed upon at will.[192] An asocial relationship thus sets the stage for the predatory violence of conquest, rape, assassination, infanticide, strategic bombing, colonial expulsions, and other crimes of convenience.

Placing other people under the aegis of a relational model imposes at least some obligation to take their interests into account. Communal Sharing has sympathy and warmth built into it—but only for members of the in-group. Fiske's collaborator Nick Haslam has argued that Communal Sharing can lead to a second kind of dehumanization: not the *mechanistic* dehumanization of an asocial relationship, but an *animalistic* dehumanization that denies to outsiders the traits that are commonly perceived as uniquely human, such as reason, individuality, self-control, morality, and culture.[193] Rather than being treated with callousness or indifference, such outsiders are treated with disgust or contempt. Communal Sharing may encourage this dehumanization because the excluded people are seen as lacking the pure and sacred essence that unites the members of the tribe, and thereby they threaten to pollute it with their animal contaminants. So Communal Sharing, for all its cuddly connotations, supports the mindset behind genocidal ideologies based on tribe, race, ethnicity, and religion.

Authority Ranking also has two sides. It brings a paternalistic responsibility to protect and support one's underlings, and thus may be the psychological basis of the Pacification Process in which overlords protect their subject peoples from internecine violence. In a similar way, it furnishes the moral rationalizations employed by slaveholders, colonial overlords, and benevolent despots. But Authority Ranking also justifies violent punishment for insolence, insubordination, disobedience, treason, blasphemy, heresy, and lèse-majesté. When welded to Communal Sharing, it justifies group-over-group violence, including imperial and jingoistic conquest and the subjugation of subordinate castes, colonies, and slaves.

More benevolent is the obligation of reciprocal exchange in Equality Matching, which gives each party a stake in the continued existence and well-being of the other. Equality Matching also encourages a modicum of perspective-taking, which, as we have seen, can turn into genuine sympathy. The pacifying effect of commerce between individuals and nations may depend on a mindset in which exchange partners, even if they are not genuinely loved, are at least valued. On the other hand, Equality Matching supplies the rationale for tit-for-tat retaliation: an eye for an eye, a tooth for a tooth, a life for a life, blood for blood. As we saw in chapter 8, even people in modern societies are apt to conceive of criminal punishment as just deserts rather than as general or specific deterrence.[194]

Rational-Legal reasoning, the add-on to the moral repertoire in literate and numerate societies, does not come with its own intuitions or emotions, and by itself neither encourages nor discourages violence. Unless all people are explicitly enfranchised and granted ownership of their own bodies and property, the amoral pursuit of profit in a market economy can exploit them in slave markets, human trafficking, and the opening of foreign markets with gunboats. And the deployment of quantitative tools can be used to maximize kill ratios in the waging of high-tech war. Yet Rational-Legal reasoning, as we shall see, can also be put in the service of a utilitarian morality that calculates the greatest good for the greatest number, and that [illegible]tes the amount of legitimate police and military force to the minimum necessary to reduce the aggregate amount of violence.[195]

What, then, are the historical changes in moral psychology that encouraged reductions in violence such as the Humanitarian Revolution, the Long Peace, and the Rights Revolutions?

The direction of the change in prevailing models is clear enough. "Over the last three centuries throughout the world," Fiske and Tetlock observe, "there has been a rapidly accelerating tendency of social systems as a whole to move from Communal Sharing to Authority Ranking to Equality Matching to Market Pricing."[196] And if we use the polling data from chapter 7 as an indication that social liberals are at the leading edge of changes in attitudes that eventually drag along social conservatives as well, then Haidt's data on the moral concerns of liberals and conservatives tell the same story. In judging the importance of moral concerns, recall, social liberals place little weight on In-group Loyalty and Purity/Sanctity (which Fiske lumps under Communal Sharing), and they place little weight on Authority/Respect. Instead they invest all their moral concern in Harm/Care and Fairness/Reciprocity. Social conservatives spread their moral portfolio over all five.[197] The trend toward social liberalism, then, is a trend away from communal and authoritarian values and toward values based on equality, fairness, autonomy, and legally enforced rights. Though both liberals and conservatives may deny that any such a trend has taken place, consider the fact that no mainstream conservative politician today would invoke tradition, authority, cohesion, or religion to justify racial segregation, keeping women out of the workforce, or criminalizing homosexuality, arguments they made just a few decades ago.[198]

Why might a disinvestment of moral resources from community, sanctity, and authority militate against violence? One reason is that communality can legitimize tribalism and jingoism, and authority can legitimize government repression. But a more general reason is that a retrenchment of the moral sense to smaller territories leaves fewer transgressions for which people may legitimately be punished. There is a bedrock of morality based on autonomy and fairness on which everyone, traditional and modern, liberal and conservative, agrees. No one objects to the use of government violence to put assailants, rapists, and murderers behind bars. But defenders of traditional morality wish to heap many nonviolent infractions on top of this consensual layer, such as homosexuality, licentiousness, blasphemy, heresy, indecency, and desecration of sacred symbols. For their moral disapproval to have teeth, traditionalists must get the Leviathan to punish those offenders as well. Expunging these offenses from the law books gives the authorities fewer grounds for clubbing, cuffing, paddling, jailing, or executing people.

The momentum of social norms in the direction of Market Pricing gives many people the willies, but it would, for better or worse, extrapolate the trend toward nonviolence. Radical libertarians, who love the Market Pricing model, would decriminalize prostitution, drug possession, and gambling, and thereby empty the world's prisons of millions of people currently kept there by force (to say nothing of sending pimps and drug lords the way of Prohibition gangsters). The progression toward personal freedom raises the question of whether it is morally *desirable* to trade a measure of socially sanctioned violence for a measure of behavior that many people deem intrinsically wrong, such as blasphemy, homosexuality, drug use, and prostitution. But that's just the point: right or wrong, retracting the moral sense from its traditional spheres of community, authority, and purity entails a reduction of violence. And that retraction is precisely the agenda of classical liberalism: a freedom of individuals from tribal and authoritarian force, and a tolerance of personal choices as long as they do not infringe on the autonomy and well-being of others.

The historical direction of morality in modern societies is not just away from Communality and Authority but toward Rational-Legal organization, and that too is a pacifying development. Fiske notes that utilitarian morality, with its goal of securing the greatest good for the greatest number, is a paradigm case of the Market Pricing model (itself a special case of the Rational-Legal mindset).[199] Recall that it was the utilitarianism of Cesare Beccaria that led to a reengineering of criminal punishment away from a raw hunger for retribution and toward a calibrated policy of deterrence. Jeremy Bentham used utilitarian reasoning to undermine the rationalizations for punishing homosexuals and mistreating animals, and John Stuart Mill used it to make an early case for feminism. The national reconciliation movements of the 1990s, in which Nelson Mandela, Desmond Tutu, and other peacemakers abjured in-kind retributive justice for a cocktail of truth-telling, amnesty, and measured punishment of the most atrocious perpetrators, was another accomplishment of violence reduction via calculated proportionality. So is the policy of responding to international provocations with economic sanctions and tactics of containment rather than retaliatory strikes.

If the recent theories of moral psychology are on the right track, then intuitions of community, authority, sacredness, and taboo are part of human nature and will probably always be with us, even if we try to sequester their influence. That is not necessarily a cause for alarm. Relational models can be combined and embedded, and Rational-Legal reasoning that seeks to minimize overall violence can strategically deploy the other mental models in benign ways.[200]

If a version of Communal Sharing is assigned to the resource of human life, and applied to a community consisting of the entire species rather than a family, tribe, or nation, it can serve as an emotional undergirding of the abstract principle of human rights. We are all one big family, and no one within it may usurp the life or freedom of anyone else. Authority Ranking may authorize the state's monopoly on the use of violence in order to prevent greater violence. And the authority of the state over its citizens can be embedded in other authority rankings in the form of democratic checks and balances, as when the president can veto the bills of Congress while at the same time Congress can impeach and remove the president. Sacred values, and the taboos that protect them, can be attached to resources that we decide are genuinely precious, such as identifiable lives, national borders, and the nonuse of chemical and nuclear weapons.

An ingenious rerouting of the psychology of taboo in the service of peace has recently been explored by Scott Atran, working with the psychologists Jeremy Ginges and Douglas Medin and the political scientist Khalil Shikaki.[201] In theory, peace negotiations should take place within a framework of Market Pricing. A surplus is generated when adversaries lay down their arms—the so-called peace dividend—and the two sides get to yes by agreeing to divide it. Each side compromises on its maximalist demand in order to enjoy a portion of that surplus, which is greater than what they would end up with if they walked away from the table and had to pay the price of continuing conflict.

Unfortunately, the mindset of sacredness and taboo can confound the best-laid plans of rational deal-makers. If a value is sacred in the minds of one of the antagonists, then it has infinite value, and may not be traded away for any other good, just as one may not sell one's child for any other good. People inflamed by nationalist and religious fervor hold certain values sacred, such as sovereignty over hallowed ground or an acknowledgment of ancient atrocities. To compromise them for the sake of peace or prosperity is taboo. The very thought unmasks the thinker as a traitor, a quisling, a mercenary, a whore.

In a daring experiment, the researchers did not simply avail themselves of the usual convenience sample of a few dozen undergraduates who fill out questionnaires for beer money. They surveyed real players in the Israel-Palestine dispute: more than six hundred Jewish settlers in the West Bank, more than five hundred Palestinian refugees, and more than seven hundred Palestinian students, half of whom identified with Hamas or Palestinian Islamic Jihad. The team had no trouble finding fanatics within each group who treated their demands as sacred values. Almost half the Israeli settlers indicated that it would never be permissible for the Jewish people to give up part of the Land of Israel, including Judea and Samaria (which make up the West Bank), no matter how great the benefit. Among the Palestinians, more than half the students indicated that it was impermissible to compromise on sovereignty over Jerusalem, no matter how great the benefit, and 80 percent of the refugees held that no compromise was possible on the "right of return" of Palestinians to Israel.

The researchers divided each group into thirds and presented them with a hypothetical peace deal that required all sides to compromise on a sacred value. The deal was a two-state solution in which the Israelis would withdraw from 99 percent of the West Bank and Gaza but would not have to absorb Palestinian refugees. Not surprisingly, the proposal did not go over well. The absolutists on both sides reacted with anger and disgust and said that they would, if necessary, resort to violence to oppose the deal.

With a third of the participants, the deals were sweetened with cash compensation from the United States and the European Union, such as a billion dollars a year for a hundred years, or a guarantee that the people would live in peace and prosperity. With these sweeteners on the table, the nonabsolutists, as expected, softened their opposition a bit. But the absolutists, forced to contemplate a taboo tradeoff, were even *more* disgusted, angry, and prepared to resort to violence. So much for the rational-actor conception of human behavior when it comes to politico-religious conflict.

All this would be pretty depressing were it not for Tetlock's observation that many ostensibly sacred values are really pseudo-sacred and may be compromised if a taboo tradeoff is cleverly reframed. In a third variation of the hypothetical peace deal, the two-state solution was augmented with a purely symbolic declaration by the enemy in which it compromised one of *its* sacred values. In the deal presented to the Israeli settlers, the Palestinians "would give up any claims to their right of return, which is sacred to them," or "would be required to recognize the historic and legitimate right of the Jewish people to Eretz Israel." In the deal presented to the Palestinians, Israel would "recognize the historic and legitimate right of the Palestinians to their own state and would apologize for all of the wrongs done to the Palestinian people," or would "give up what they believe is their sacred right to the West Bank," or would "symbolically recognize the historic legitimacy of the right of return" (while not actually granting it). The verbiage made a difference. Unlike the bribes of money or peace, the symbolic concession of a sacred value by the enemy, especially when it acknowledges a sacred value on one's own side, reduced the absolutists' anger, disgust, and willingness to endorse violence. The reductions did not shrink the absolutists' numbers to a minority of their respective sides, but the proportions were large enough to have potentially reversed the outcomes of their recent national elections.

The implications of this manipulation of people's moral psychology are profound. To find *anything* that softens the opposition of Israeli and Palestinian fanatics to what the rest of the world recognizes as the only viable solution to their conflict is something close to a miracle. The standard tools of diplomacy wonks, who treat the disputants as rational actors and try to manipulate the costs and benefits of a peace agreement, can backfire. Instead they must treat the disputants as *moralistic* actors, and manipulate the symbolic framing of the peace agreement, if they want a bit of daylight to open up. The human moral sense is not always an obstacle to peace, but it can be when the mindset of sacredness and taboo is allowed free rein. Only when that mindset is redeployed under the direction of rational goals will it yield an outcome that can truly be called moral.

What exogenous causes are shifting the allocation of moral intuitions away from community, authority, and purity and toward fairness, autonomy, and rationality?

One obvious force is geographic and social mobility. People are no longer confined to the small worlds of family, village, and tribe, in which conformity and solidarity are essential to daily life, and ostracism and exile are a form of social death. They can seek their fortunes in other circles, which expose them to alternative worldviews and lead them into a more ecumenical morality, which gravitates to the rights of individuals rather than chauvinistic veneration of the group.

By the same token, open societies, where talent, ambition, or luck can dislodge people from the station in which they were born, are less likely to see an Authority Ranking as an inviolable law of nature, and more likely to see it as a historical artifact or a legacy of injustice.

When diverse individuals mingle, engage in commerce, and find themselves on professional or social teams that cooperate to attain a superordinate goal, their intuitions of purity can be diluted. One example, mentioned in chapter 7, is the greater tolerance of homosexuality among people who personally know homosexuals. Haidt observes that when one zooms in on an electoral map of the United States, from the coarse division into red and blue states to a finer-grained division into red and blue *counties*, one finds that the blue counties, representing the regions that voted for the more liberal presidential candidate, cluster along the coasts and major waterways. Before the advent of jet airplanes and interstate highways, these were the places where people and their ideas most easily mixed. That early advantage installed them as hubs of transportation, commerce, media, research, and education, and they continue to be pluralistic—and liberal—zones today. Though American political liberalism is by no means the same as classical liberalism, the two overlap in their weighting of the moral spheres. The micro-geography of liberalism suggests that the moral trend away from community, authority, and purity is indeed an effect of mobility and cosmopolitanism.[202]

Another subverter of community, authority, and purity is the objective study of history. The mindset of Communality, Fiske notes, conceives of the group as eternal: the group is held together by an immutable essence, and its traditions stretch back to the dawn of time.[203] Authority Rankings too are naturally portrayed as everlasting. They were ordained by the gods, or are inherent in a great chain of being that organizes the universe. And both models boast an abiding nobility and purity as part of their essential nature.

In this tissue of rationalizations, a real historian is about as welcome as a skunk at a garden party. Donald Brown, before he embarked on his survey of human universals, wanted to explain why the Hindus of India had produced so little in the way of serious historical scholarship, unlike the neighboring civilizations of China.[204] The elites of a hereditary caste society, he suspected, figured that no good could come from scholars nosing around in archives where they might stumble upon evidence that undermined their claims to have descended from heroes and gods. Brown looked at twenty-five civilizations in Asia and Europe and found that the ones that were stratified into hereditary classes favored myth, legend, and hagiography and discouraged history, social science, natural science, biography, realistic portraiture, and uniform education. More recently, the nationalist movements of the 19th and 20th centuries recruited cadres of hacks to write potted histories of their nations' timeless values and glorious pasts.[205] Beginning in the 1960s, many democracies were traumatized by revisionist histories that unearthed their nations' shallow roots and exposed their sordid misdeeds. The declines of patriotism, tribalism, and trust in hierarchies are in part a legacy of the new historiography. Many liberal-conservative battles continue to be fought over school curricula and museum exhibits.

Though historical fact is the best antidote to self-serving legend, the imaginative projections of fiction can also reorient an audience's moral sense. The protagonists of many plots have struggled with conflicts between a morality defined by loyalty, obedience, patriotism, duty, law, or convention and the course of action that is morally defensible. In the 1967 film *Cool Hand Luke,* a prison guard is about to punish Paul Newman by locking him in a sweltering box and explains, "Sorry, Luke, I'm just doing my job. You gotta appreciate that." Luke replies, "Nah—calling it your job don't make it right, Boss."

Less often, an author can shake readers into realizing that conscience itself can be an untrustworthy guide to what is right. Huckleberry Finn, while drifting down the Mississippi, is suddenly racked with guilt over having helped Jim escape from his lawful owner and reach a free state.

> Jim talked out loud all the time while I was talking to myself. He was saying how the first thing he would do when he got to a free State he would go to saving up money and never spend a single cent, and when he got enough he would buy his wife, which was owned on a farm close to where Miss Watson lived; and then they would both work to buy the two children, and if their master wouldn't sell them, they'd get an Ab'litionist to go and steal them.
>
> It most froze me to hear such talk. . . . Thinks I, this is what comes of my not thinking. Here was this nigger, which I had as good as helped to run away, coming right out flat-footed and saying he would steal his children—children that belonged to a man I didn't even know; a man that hadn't ever done me no harm. . . .
>
> My conscience got to stirring me up hotter than ever, until at last I says to it, "Let up on me—it ain't too late yet—I'll paddle ashore at the first light and tell." . . . All my troubles was gone. I went to looking out sharp for a light, and sort of singing to myself. By and by one showed. . . .
>
> [Jim] jumped and got the canoe ready, and put his old coat in the bottom for me to set on, and give me the paddle; and as I shoved off, he says, "Pooty soon I'll be a-shout'n' for joy, en I'll say, it's all on accounts o' Huck; I's a free man, en I couldn't ever ben free ef it hadn' ben for Huck; Huck done it. Jim

won't ever forgit you, Huck; you's de bes' fren' Jim's ever had; en you's de ONLY fren' ole Jim's got now."

I was paddling off, all in a sweat to tell on him; but when he says this, it seemed to kind of take the tuck all out of me.

In this heart-stopping sequence, a conscience guided by principle, obedience, reciprocity, and sympathy for a stranger pulls Huck in the wrong direction, and an immediate tug on his sympathy from a friend (bolstered in the reader's mind by a conception of human rights) pulls him in the right one. It is perhaps the finest portrayal of the vulnerability of the human moral sense to competing convictions, most of which are morally wrong.

REASON

Reason appears to have fallen on hard times. Popular culture is plumbing new depths of dumbth, and American political discourse has become a race to the bottom.[206] We are living in an era of scientific creationism, New Age flimflam, 9/11 conspiracy theories, psychic hotlines, and resurgent religious fundamentalism.

As if the proliferation of unreason weren't bad enough, many commentators have been mustering their powers of reason to argue that reason is overrated. During the honeymoon following George W. Bush's inauguration in 2001, editorialists opined that a great president need not be intelligent, because a good heart and steadfast moral clarity are superior to the triangulations and equivocations of overeducated mandarins. After all, they said, it was the Harvard-educated best and the brightest who dragged America into the quagmire of Vietnam. "Critical theorists" and postmodernists on the left, and defenders of religion on the right, agree on one thing: that the two world wars and the Holocaust were the poisoned fruit of the West's cultivation of science and reason since the Enlightenment.[207]

Even the scientists are piling on. Human beings are led by their passions, say many psychologists, and deploy their puny powers of reason only to rationalize their gut feelings after the fact. Behavioral economists exult in showing how human behavior departs from the rational-actor theory, and the journalists who publicize their work waste no opportunity to smack the theory around. The implication is that since irrationality is inevitable, we may as well lie back and enjoy it.

In this section, the last before the concluding chapter, I will try to convince you that both the pessimistic assessment of the state of reason in the world, and any sentiment that this would not be such a bad thing, are mistaken. For all their foolishness, modern societies have been getting smarter, and all things being equal, a smarter world is a less violent world.

Before we turn to this evidence, let me sweep away some of the prejudices against reason. Now that the presidency of George W. Bush is over, the theory that we are better off with unintellectual leaders is just embarrassing, and the reasons for the embarrassment may be quantified. Measuring the psychological traits of public figures, to be sure, has a sketchy history, but the psychologist Dean Simonton has developed several historiometric measures that are reliable and valid (in the psychometrician's technical sense) and politically nonpartisan.[208] He analyzed a dataset of 42 presidents from GW to GWB and found that both raw intelligence and openness to new ideas and values are significantly correlated with presidential performance as it has been assessed by nonpartisan historians.[209] Though Bush himself is well above the average of the population in intelligence, he is third-lowest among the presidents, and comes in dead last in openness to experience, with a rock-bottom score of 0.0 on the 0–100 scale. Simonton published his work in 2006, while Bush was still in office, but the three historians' surveys conducted since then bear out the correlation: Bush was ranked 37th, 36th, and 39th among the 42 presidents.[210]

As for Vietnam, the implication that the United States would have avoided the war if only the advisors of Kennedy and Johnson had been less intelligent seems unlikely in light of the fact that after they left the scene, the war was ferociously prosecuted by Richard Nixon, who was neither the best nor the brightest.[211] The relationship between presidential intelligence and war may also be quantified. Between 1946 (when the PRIO dataset begins) and 2008, a president's IQ is negatively correlated with the number of battle deaths in wars involving the United States during his presidency, with a coefficient of -0.45.[212] One could say that for every presidential IQ point, 13,440 fewer people die in battle, though it's more accurate to say that the three smartest postwar presidents, Kennedy, Carter, and Clinton, kept the country out of destructive wars.

The idea that the Holocaust was a product of the Enlightenment is ludicrous, if not obscene. As we saw in chapter 6, what changed in the 20th century was not so much the occurrence of genocide as the recognition that genocide was something bad. The technological and bureaucratic trappings of the Holocaust are a sideshow in the reckoning of its human costs and are unnecessary to the perpetration of mass murder, as the bloody machetes of the Rwandan genocide remind us. Nazi ideology, like the nationalist, romantic militarist, and communist movements of the same era, was a fruit of the 19th-century counter-Enlightenment, not the line of thinking that connects Erasmus, Bacon, Hobbes, Spinoza, Locke, Hume, Kant, Bentham, Jefferson, Madison, and Mill. The scientific pretensions of Nazism were risible pseudoscience, as real science easily showed. In a brilliant recent essay, the philosopher Yaki Menschenfreund reviews the theory that Enlightenment rationality is responsible for the Holocaust:

> It is impossible to understand so destructive a policy without recognizing that Nazi ideology was, for the most part, not only irrational—but

> antirational. It cherished the pagan, pre-Christian past of the German nation, adopted romantic ideas of a return to nature and a more "organic" existence, and nurtured an apocalyptic expectation of an end of days, whence the eternal struggle between the races would be resolved. . . . The contempt for rationalism and its association with the despised Enlightenment stood at the core of Nazi thought; the movement's ideologues emphasized the contradiction between *weltanschauung* ("worldview"), the natural and direct experience of the world, and *welt-an-denken* ("thinking about the world"), the "destructive" intellectual activity that breaks reality down through conceptualization, calculation, and theorization. Against the "degenerate" liberal bourgeois' worship of reason, the Nazis championed the idea of a vital, spontaneous life, unhindered and undimmed by compromises or dilemmas.[213]

Finally, let's consider the suggestion that reason is incapable of contending against the brawn of the emotions, a tail that tries to wag the dog. The psychologists David Pizarro and Paul Bloom have argued that this is an overinterpretation of the laboratory phenomena of moral dumbfounding and other visceral reactions to moral dilemmas.[214] Even if a decision is guided by intuition, the intuition itself may be a legacy of moral reasoning that had taken place beforehand, whether in private reflection, in dinner table debates, or through the assimilation of norms that were the output of past debates. Case studies reveal that at critical moments in an individual's life (such as a woman's decision to have an abortion), or at critical moments in a society's history (such as the struggles over civil, women's, and gay rights and the nation's participation in war), people can be consumed in agonizing reflection and deliberation. We have seen many historic moral changes that originated in painstaking intellectual briefs, which were in turn met with furious rebuttals. Once the debate had been settled, the winning side entrenched itself in people's sensibilities and erased its own tracks. Today, for example, people might be dumbfounded when asked whether we should burn heretics, keep slaves, whip children, or break criminals on the wheel, yet those very debates took place several centuries ago. We even saw a neuroanatomical basis for the give-and-take between intuition and reasoning in Joshua Greene's studies of trolley problems in the brain scanner: each of these moral faculties has distinct neurobiological hubs.[215]

When Hume famously wrote that "reason is, and ought to be, only the slave of the passions," he was not advising people to shoot from the hip, blow their stack, or fall head over heels for Mr. Wrong.[216] He was basically making the logical point that reason, by itself, is just a means of getting from one true proposition to the next and does not care about the value of those propositions. Nonetheless there are many reasons why reason, working in conjunction with "some particle of the dove, kneaded into our frame," must "direct the determinations of our mind, and where every thing else is equal, produce a cool preference of what is useful and serviceable to mankind, above what is pernicious and dangerous." Let's consider some of the ways the application of reason might be expected to reduce the rate of violence.

The chronological sequence in which the Scientific Revolution and the Age of Reason preceded the Humanitarian Revolution reminds us of one big reason, the one captured in Voltaire's quip that absurdities lead to atrocities. A debunking of hogwash—such as the ideas that gods demand sacrifices, witches cast spells, heretics go to hell, Jews poison wells, animals are insensate, children are possessed, Africans are brutish, and kings rule by divine right—is bound to undermine many rationales for violence.

A second pacifying effect of exercising the faculty of reason is that it goes hand in hand with self-control. Recall that the two traits are statistically correlated in individuals and that their physiological substrates overlap in the brain.[217] It is reason—a deduction of the long-term consequences of an action—that gives the self reasons to control the self.

Self-control, moreover, involves more than just avoiding rash choices that will damage one's future self. It can also mean suppressing some of our base instincts in the service of motives that we are better able to justify. Sneaky laboratory techniques, such as measurements of how quickly people associate white and black faces with words like *good* and *bad*, and neuroimaging experiments that monitor activity in the amygdala, have shown that many white people have small, visceral negative reactions to African Americans.[218] Yet the sea change in explicit attitudes toward African Americans that we saw in figures 7–6, 7–7, and 7–8, and the obvious comity with which whites and blacks live and work with each other today, show that people can allow their better judgments to overcome these biases.

Reasoning can also interact with the moral sense. Each of the four relational models from which moral impulses spring comes with a particular style of reasoning. Each of these modes of reasoning may be matched with a mathematical scale, and each is implemented by a distinctive family of cognitive intuitions.[219] Communal Sharing thinks in all-or-none categories (also called a nominal scale): a person is either in the hallowed group or out of it. The cognitive mindset is that of intuitive biology, with its pure essences and potential contaminants. Authority Ranking uses an ordinal scale: the linear ranking of a dominance hierarchy. Its cognitive gadget is an intuitive physics of space, force, and time: higher-ranking people are deemed bigger, stronger, higher, and first in the series. Equality Matching is measured on a scale of intervals, which allows two quantities to be compared to see which is larger but not entered into proportions. It reckons by concrete procedures such as lining things up, counting them off, or comparing them with a balance scale. Only Market Pricing (and the Rational-Legal mindset of which it is part) allows one to reason in terms of *proportionality.* The Rational-Legal model requires the nonintuitive tools of symbolic mathematics, such as fractions, percentages, and exponentiation. And as I have mentioned, it is far from universal, and depends on the cognition-enhancing skills of literacy and numeracy.

It's no coincidence that the word *proportionality* has a moral as well as a mathematical sense. Only preachers and pop singers profess that violence will someday vanish off the face of the earth. A measured degree of violence, even if only held in reserve, will always be necessary in the form of police forces and armies to deter predation or to incapacitate those who cannot be deterred. Yet there is a vast difference between the minimal violence necessary to prevent greater violence and the bolts of fury that an uncalibrated mind is likely to deliver in acts of rough justice. A coarse sense of tit-for-tat payback, especially with the thumb of self-serving biases on the scale, produces many kinds of excess violence, including cruel and unusual punishments, savage beatings of naughty children, destructive retaliatory strikes in war, lethal reprisals for trivial insults, and brutal repression of rebellions by crappy governments in the developing world. By the same token, many moral advances have consisted not of eschewing force across the board but of applying it in carefully measured doses. Some examples include the reform of criminal punishment following Beccaria's utilitarian arguments, the measured punishments of children by enlightened parents, civil disobedience and passive resistance that stop just short of violence, the calibrated responses to provocations by modern democracies (military exercises, warning shots, surgical strikes on military installations), and the partial amnesties in postconflict conciliation. These reductions in violence required a sense of proportionality, a habit of mind that does not come naturally and must be cultivated by reason.

Reason can also be a force against violence when it abstracts violence itself as a mental category and construes it as a problem to be solved rather than a contest to be won. The Greeks of Homer conceived of their devastating wars as the handiwork of sadistic puppeteers on high.[220] That, to be sure, required a feat of abstraction: they lifted themselves out of a vantage point from which war is the fault of one's eternally treacherous enemies. Yet blaming the gods for war does not open up many practical opportunities for mere mortals to reduce it. Moralistic denunciations of war also single it out as an entity, but they provide few guidelines on what to do when an invading army is at one's doorstep. A real change came in the writings of Grotius, Hobbes, Kant, and other modern thinkers: war was intellectualized as a game-theoretic problem, to be solved by proactive institutional arrangements. Centuries later some of these arrangements, such as Kant's triad of democratization, trade, and an international community, helped to drive down the rate of war in the Long Peace and the New Peace. And the Cuban Missile Crisis was defused when Kennedy and Khrushchev consciously reframed it as a trap for the two of them to escape without either side losing face.

None of these rationales for rationality speaks to Hume's point that rationality is merely a means to an end, and that the end depends on the thinker's passions. Reason can lay out a road map to peace and harmony, if the reasoner wants peace and harmony. But it can also lay out a road map to war and strife, if the reasoner delights in war and strife. Do we have any reason to expect that rationality should orient a reasoner to *wanting* less violence?

On the grounds of austere logic, the answer is no. But it doesn't take much to switch it to yes. All you need are two conditions. The first is that the reasoners care about their own well-being: that they prefer to live rather than die, keep their body parts intact rather than have them maimed, and spend their days in comfort rather than in pain. Mere logic does not force them to have those prejudices. Yet any product of natural selection—indeed, any agent that manages to endure the ravages of entropy long enough to be reasoning in the first place—will in all likelihood have them.

The second condition is that the reasoners be part of a community of other agents who can impinge on their well-being and who can exchange messages and comprehend each other's reasoning. That assumption too is not logically necessary. One could imagine a Robinson Crusoe who reasons in solitude, or a Galactic Overlord who is untouchable by his subjects. But natural selection could not have manufactured a solitary reasoner, because evolution works on populations, and *Homo sapiens* in particular is not just a rational animal but a social and language-using one. As for the Overlord, uneasy lies the head that wears a crown. Even he, in principle, must worry about the possibility of a fall from power that would require him to deal with his erstwhile underlings.

As we saw at the end of chapter 4, the assumptions of self-interest and sociality combine with reason to lay out a morality in which nonviolence is a goal. Violence is a Prisoner's Dilemma in which either side can profit by preying on the other, but both are better off if neither one tries, since mutual predation leaves each side bruised and bloodied if not dead. In the game theorist's definition of the dilemma, the two sides are not allowed to talk, and even if they were, they would have no grounds for trusting each other. But in real life people can confer, and they can bind their promises with emotional, social, or legal guarantors. And as soon as one side tries to prevail on the other not to injure him, he has no choice but to commit himself not to injure the other side either. As soon as he says, "It's bad for you to hurt me," he's committed to "It's bad for me to hurt you," since logic cannot tell the difference between "me" and "you." (After all, their meaning switches with every turn in the conversation.) As the philosopher William Godwin put it, "What magic is there in the pronoun 'my' that should justify us in overturning the decisions of impartial truth?"[221] Nor can reason distinguish between Mike and Dave, or Lisa and Amy, or any other set of individuals, because as far as logic is concerned, they're just a bunch of x's and y's. So as soon as you try to persuade someone to avoid harming you by appealing to *reasons* why he shouldn't, you're sucked into a commitment to the avoidance of harm as a general goal. And to the extent that you pride yourself on the quality of your reason, work to apply it far and wide, and use it to persuade others, you will be forced to deploy that reason in pursuit of universal interests, including an avoidance of violence.[222]

Humans, of course, were not created in a state of Original Reason. We descended from apes, spent hundreds of millennia in small bands, and evolved our cognitive processes in the service of hunting, gathering, and socializing. Only gradually, with the appearance of literacy, cities, and long-distance travel and communication, could our ancestors cultivate the faculty of reason and apply it to a broader range of concerns, a process that is still ongoing. One would expect that as collective rationality is honed over the ages, it will progressively whittle away at the shortsighted and hot-blooded impulses toward violence, and force us to treat a greater number of rational agents as we would have them treat us.

Our cognitive faculties need not have evolved to go in this direction. But once you have an open-ended reasoning system, even if it evolved for mundane problems like preparing food and securing alliances, you can't keep it from entertaining propositions that are consequences of other propositions. When you acquired your mother tongue and came to understand *This is the cat that killed the rat*, nothing could prevent you from understanding *This is the rat that ate the malt.* When you learned how to add 37 + 24, nothing could prevent you from deriving the sum of 32 + 47. Cognitive scientists call this feat systematicity and attribute it to the combinatorial power of the neural systems that underlie language and reasoning.[223] So if the members of species have the power to reason with one another, and enough opportunities to exercise that power, sooner or later they will stumble upon the mutual benefits of nonviolence and other forms of reciprocal consideration, and apply them more and more broadly.

This is the theory of the expanding circle as Peter Singer originally formulated it.[224] Though I have co-opted his metaphor as a name for the historical process in which increased opportunities for perspective-taking led to sympathy for more diverse groups of people, Singer himself did not have the emotions in mind so much as the intellect. He is a philosopher's philosopher, and argued that over the aeons people had the power to literally *think their way* into greater respect for the interests of others. And that respect cannot be confined to the interests of the people with whom we rub shoulders in a small social circle. Just as you can't favor yourself over someone else when holding up ideals on how to behave, you can't favor members of your group over the members of another group. For Singer, it is hardheaded reason more than softhearted empathy that expands the ethical circle ever outward:

> Beginning to reason is like stepping onto an escalator that leads upward and out of sight. Once we take the first step, the distance to be traveled is

> independent of our will and we cannot know in advance where we shall end. . . .
>
> If we do not understand what an escalator is, we might get on it intending to go a few meters, only to find that once we are on, it is difficult to avoid going all the way to the end. Similarly, once reasoning has got started it is hard to tell where it will stop. The idea of a disinterested defense of one's conduct emerges because of the social nature of human beings and the requirements of group living, but in the thought of reasoning beings, it takes on a logic of its own which leads to its extension beyond the bounds of the group.[225]

In the historical sequence that Singer adduces, the moral circle of the early Greeks was confined to the city-state, as in this unintentionally comical epitaph from the mid-5th century CE:

> This memorial is set over the body of a very good man. Pythion, from Megara, slew seven men and broke off seven spear points in their bodies. . . . This man, who saved three Athenian regiments . . . having brought sorrow to no one among all men who dwell on earth, went down to the underworld felicitated in the eyes of all.[226]

Plato widened the circle a bit by arguing that Greeks should spare other Greeks from devastation and enslavement, visiting these fates only on non-Greeks. In modern times Europeans expanded the no-taking-slaves rule to other Europeans, but Africans were fair game. Today, of course, slavery is illegal for everyone.

The only problem with Singer's metaphor is that the history of moral concern looks less like an escalator than an elevator that gets stuck on a floor for a seeming eternity, then lurches up to the next floor, gets stuck there for a while, and so on. Singer's history finds just four circle sizes in almost two and a half millennia, which works out to one ascent every 625 years. That feels a bit jerky for an escalator. Singer acknowledges the bumpiness of moral progress and attributes it to the rarity of great thinkers:

> Insofar as the timing and success of the emergence of a questioning spirit is concerned, history is a chronicle of accidents. Nevertheless, if reasoning flourishes within the confines of customary morality, progress in the long run is not accidental. From time to time, outstanding thinkers will emerge who are troubled by the boundaries that custom places on their reasoning, for it is in the nature of reasoning that it dislikes notices saying "off limits." Reasoning is inherently expansionist. It seeks universal application. Unless crushed by countervailing forces, each new application will become part of the territory of reasoning bequeathed to future generations.[227]

But it remains puzzling that these outstanding thinkers have appeared so rarely on the world's stage, and that the expansion of reason should have dawdled so. Why did human rationality need thousands of years to arrive at the conclusion that something might be a wee bit wrong with slavery? Or with beating children, raping unattached women, exterminating native peoples, imprisoning homosexuals, or waging wars to assuage the injured vanity of kings? It shouldn't take an Einstein to figure it out.

One possibility is that the theory of an escalator of reason is historically incorrect, and that humanity was led up the incline of moral progress by the heart rather than the head. A different possibility is that Singer is right, at least in part, but the escalator is powered not just by the sporadic appearance of outstanding thinkers but by a rise in the quality of *everyone's* thinking. Perhaps we're getting better because we're getting smarter.

Believe it or not, we *are* getting smarter. In the early 1980s the philosopher James Flynn had a Eureka! moment when he noticed that the companies that sell IQ tests periodically renorm the scores.[228] The average IQ has to be 100 by definition, but the percentage of questions answered correctly is an arbitrary number that depends on how hard the questions are. The test-mongers have to map the percentage-correct scale onto the IQ scale by formula, but the formula kept getting out of whack. The average scores on the tests had been creeping up for decades, so to keep the average at 100, every once in a while they jiggered the formula so the test-takers would need a larger number of correct answers to earn a given IQ. Otherwise there would be IQ inflation.

This inflation, Flynn realized, is not a kind that one should try to whip, but is telling us something important about recent history and the human mind. Later generations, given the same set of questions as earlier ones, got more of them correct. Later generations must be getting better at whatever skills IQ tests measure. Since IQ tests have been administered in massive numbers all over the world for much of the 20th century, in some countries, down to the last schoolchild and draftee, one can plot a country's change in measured intelligence over time. Flynn scoured the world for datasets in which the same IQ test was given over many years, or the scoring norms were available to keep the numbers commensurate. The result was the same in every sample: IQ scores increased over time.[229] In 1994 Richard Herrnstein and the political scientist Charles Murray christened the phenomenon the Flynn Effect, and the name has stuck.[230]

The Flynn Effect has been found in thirty countries, including some in the developing world, and it has been going on ever since IQ tests were first given en masse around the time of World War I.[231] An even older dataset from Britain suggests that the Flynn Effect may even have begun with the cohort of

Britons who were born in 1877 (though of course they were tested as adults).[232] The gains are not small: an average of three IQ points (a fifth of a standard deviation) per decade.

The implications are stunning. An average teenager today, if he or she could time-travel back to 1950, would have had an IQ of 118. If the teenager went back to 1910, he or she would have had an IQ of 130, besting 98 percent of his or her contemporaries. Yes, you read that right: if we take the Flynn Effect at face value, a typical person today is smarter than 98 percent of the people in the good old days of 1910. To state it in an even more jarring way, a typical person of 1910, if time-transported forward to the present, would have a mean IQ of 70, which is at the border of mental retardation. With the Raven's Progressive Matrices, a test that is sometimes considered the purest measure of general intelligence, the rise is even steeper. An ordinary person of 1910 would have an IQ of 50 today, which is smack in the middle of mentally retarded territory, between "moderate" and "mild" retardation.[233]

Obviously we can't take the Flynn Effect at face value. The world of 1910 was not populated by people who today we would consider mentally retarded. Commentators have looked for ways to make the Flynn Effect go away, but none of the obvious ones work. Writers on the egalitarian left and on the lift-yourself-up-by-the-bootstraps right have long tried to undermine the very idea of intelligence and the instruments that claim to measure it. But the scientists who study human individual differences are virtually unanimous that intelligence can be measured, that it is fairly stable over a lifetime of an individual, and that it predicts academic and professional success at every level of the scale.[234] Perhaps, you might think, children got more quiz-savvy over the decades as schools began to test the living daylights out of them. But as Flynn points out, the gains have been steady over time, while the popularity of testing has waxed and waned.[235] Could it be, then, that the content of the test questions, like "Who wrote *Romeo and Juliet*?" have become common knowledge, or that the words in the vocabulary section have spread into everyday parlance, or that the arithmetic problems have been taught earlier in school? Unfortunately the biggest gains in the IQ tests are found in exactly those items that do *not* tap knowledge, vocabulary, or arithmetic.[236] They are found in the items that tap abstract reasoning, such as similarities ("What do a pound and an inch have in common?"), analogies ("BIRD is to EGG as TREE is to what?"), and visual matrices (where geometric patterns fill the rows and columns of a grid, and the test-taker has to figure out how to fill a gap at the bottom right: for example, going from left to right in each row, a shape may acquire a border, lose a vertical line, and then have a hollow area blacked in). The subtests on vocabulary and math have shown the *smallest* rise over time, and others tests that are filled with them, like the SAT, have even shown a bit of a decline in some age groups in some years.[237] Figure 9–2 shows the increases on IQ and its various subtests in the United States since the late 1940s.

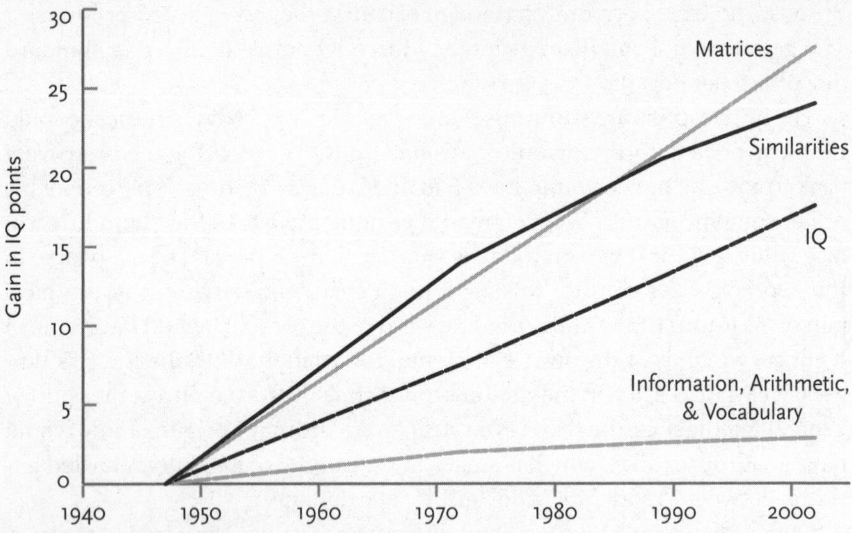

FIGURE 9–2. The Flynn Effect: Rising IQ scores, 1947–2002
Source: Graph from Flynn, 2007, p. 8.

The Flynn Effect was a scientific bombshell, because if one were to focus only on the improvement in the Matrices and Similarities, one would think that what rose over the decades was *general intelligence.* These subtests are considered the purest measures of general intelligence because they correlate well with the tendency of people to score well or poorly on a large battery of dissimilar tests. That tendency is called *g*, and the existence of *g* is often considered the most important discovery in the science of mental testing.[238] If you give people just about any test you can think of that taps the commonsense notion of intelligence—math, vocabulary, geometry, logic, text comprehension, factual knowledge—the people who do well on one also tend to do well on the others. That was not a foregone conclusion. We all know the inarticulate math whiz and the eloquent poet who can't balance a checkbook, and one might have thought that the different kinds of intelligence trade off resources in the brain, so that the more neural tissue you have for math the less you have for language and vice versa. Not so. Some people are indeed relatively better at math than others, and some relatively better at language, but compared to the population as a whole, the two talents—and every other talent that we associate with the concept of intelligence—tend to go together.

General intelligence, moreover, is highly heritable, and mostly unaffected by the family environment (though it may be affected by the cultural environment).[239] We know this because measures of *g* in adults are strongly correlated in identical twins separated at birth and are not at all correlated in adopted siblings who have been raised in the same family. General intelligence is also correlated with several measures of neural structure and functioning, including the speed of information processing, the overall size of the brain, the thickness of the gray matter in the cerebral cortex, and the integrity of the white matter connecting one cortical region to another.[240] Most likely *g* represents the summed effects of many genes, each of which affects brain functioning in a small way.

The bombshell is that the Flynn Effect is almost certainly environmental. Natural selection has a speed limit measured in generations, but the Flynn Effect is measurable on the scale of decades and years. Flynn was also able to rule out increases in nutrition, overall health, and outbreeding (marrying outside one's local community) as explanations for his eponymous effect.[241] Whatever propels the Flynn Effect, then, is likely to be in people's *cognitive* environments, not in their genes, diets, vaccines, or dating pools.

A breakthrough in the mystery of the Flynn Effect was the realization that the increases are *not* gains in general intelligence.[242] If they were, they would have lifted the scores on all the subtests, including vocabulary, math, and raw memory power, with a rate related to the degree each test correlates with *g*. In fact the boost was concentrated in subtests like Similarities and Matrices. Whatever the mystery factor in the environment may be, it is highly selective in the components of intelligence it is enhancing: not raw brainpower, but the abilities needed to score well on the subtests of abstract reasoning.

The best guess is that the Flynn Effect has several causes, which may have acted with different strengths at different times in the century. The improvements on visual matrices may have been fueled by an increasingly high-tech and symbol-rich environment that forced people to analyze visual patterns and connect them to arbitrary rules.[243] But proficiency with visuals is a sideshow to understanding the gains in intelligence that might be relevant to moral reasoning. Flynn identifies the newly rising ability as *postscientific* (as opposed to prescientific) thinking.[244] Consider a typical question from the Similarities section of an IQ test: "What do dogs and rabbits have in common?" The answer, obvious to us, is that they are both mammals. But an American in 1900 would have been just as likely to say, "You use dogs to hunt rabbits." The difference, Flynn notes, is that today we spontaneously classify the world with the categories of science, but not so long ago the "correct" answer would seem abstruse and irrelevant. " 'Who cares that they are both mammals?' " Flynn imagines the test-taker asking in 1900. "That is the least important thing about them from his point of view. What is important is orientation in space and time, what things are useful, and what things are under one's control."[245]

Flynn was putting words in the mouths of the dead, but that style of reasoning has been documented in studies of premodern peoples by psychologists such as Michael Cole and Alexander Luria. Luria transcribed interviews with Russian peasants in remote parts of the Soviet Union who were given similarities questions like the ones on IQ tests:

Q: What do a fish and a crow have in common?
A: A fish—it lives in water. A crow flies. If the fish just lies on top of the water, the crow would peck at it. A crow can eat a fish but a fish can't eat a crow.
Q: Could you use one word for them both [such as "animals"]?
A: If you call them "animals," that wouldn't be right. A fish isn't an animal and a crow isn't either. . . . A person can eat a fish but not a crow.

Luria's informants also rejected a purely hypothetical mode of thinking—the stage of cognition that Jean Piaget called formal (as opposed to concrete) operations.

Q: All bears are white where there is always snow. In Novaya Zemlya there is always snow. What color are the bears there?
A: I have seen only black bears and I do not talk of what I have not seen.
Q: But what do my words imply?
A: If a person has not been there he cannot say anything on the basis of words. If a man was 60 or 80 and had seen a white bear there and told me about it, he could be believed.[246]

Flynn remarks, "The peasants are entirely correct. They understand the difference between analytic and synthetic propositions: pure logic cannot tell us anything about facts; only experience can. But this will do them no good on current IQ tests." That is because current IQ tests tap abstract, formal reasoning: the ability to detach oneself from parochial knowledge of one's own little world and explore the implications of postulates in purely hypothetical worlds.

If Flynn is right that much of the Flynn Effect is caused by an increasing tendency to see the world through "scientific spectacles," as he puts it, what are the exogenous causes of the availability of those spectacles? An obvious one is schooling. We know that schooling coaxes adolescents from Piaget's stage of concrete operation to his stage of formal operations, and that even with schooling, not everyone makes the transition.[247] Over the course of the 20th century, and all over the world, children came to spend more time in school. In 1900 an average American adult had seven years of schooling, and a quarter of them had less than four years.[248] Only in the 1930s did high school become compulsory.

And during this transition, the nature of the schooling changed. Early in the century reading consisted of standing and reciting aloud from books. As the education researcher Richard Rothstein observed, "Many World War I recruits failed a basic written intelligence test partly because even if they had attended a few years of school and learned how to read aloud, they were being asked by the Army to understand and interpret what they had read, a skill that many of them had never learned."[249] Another researcher, Jeremy Genovese, documented the changing goals of education in the 20th century by analyzing the content of the high school entrance exams in 1902–13 and comparing them to high-school proficiency tests given to students of the same age in the 1990s.[250] As far as factual knowledge is concerned, less is expected of adolescents today. For example, in the geography section of today's high-stakes test, the students were asked to pick out the United States on a map of the world! Their great-grandparents were required to "name the states you would pass through in traveling on a meridian from Columbus [Ohio] to the Gulf of Mexico and name and locate the capital of each." On the other hand, a typical test item today requires students to grapple with rates, amounts, multiple contingencies, and basic economics:

> A community is located in a region where very little drinking water is available. In order to manage their water resources, which of the following should the community NOT do?
>
> A. Increase their water usage.
> B. Buy water from another community.
> C. Install water-saving devices in homes.
> D. Charge higher rates for water.

Anyone who understands the phrase *law of supply and demand* realizes that option D cannot be the correct answer. But if you simply have an image of a pool of water and people drinking from it, the connection between how much it costs and how quickly it shrinks is not immediately apparent.

Flynn suggests that over the course of the 20th century, scientific reasoning infiltrated from the schoolhouse and other institutions into everyday thinking. More people worked in offices and the professions, where they manipulated symbols rather than crops, animals, and machines. People had more time for leisure, and they spent it in reading, playing combinatorial games, and keeping up with the world. And, Flynn suggests, the mindset of science trickled down to everyday discourse in the form of shorthand abstractions. A shorthand abstraction is a hard-won tool of technical analysis that, once grasped, allows people to effortlessly manipulate abstract relationships. Anyone capable of reading this book, even without training in science or philosophy, has probably assimilated hundreds of these abstractions from casual reading, conversation, and exposure to the media, including *proportional, percentage, correlation, causation, control group, placebo, representative sample, false positive, empirical, post hoc, statistical, median, variability, circular argument, trade-off,* and *cost-benefit analysis.* Yet each of them—even a concept as second-nature to us as *percentage*—at one time trickled down from the academy and other highbrow sources and increased in popularity in printed usage over the course of the 20th century.[251]

It isn't just the chattering classes that have absorbed the shorthand abstractions of the technocracy. The linguist Geoffrey Nunberg has commented on Bruce Springsteen's lyric in "The River": "I got a job working construction for the Johnstown Company / But lately there ain't been much work on account of the economy." Only in the last forty years, Nunberg notes, would ordinary people have talked about "the economy" as a natural force with causal powers like the weather.[252] Earlier they might have said "on account of times are hard." Or, he might have added, on account of the Jews, the Negroes, or the rich peasants.

We can now put together the two big ideas of this section: the pacifying effects of reason, and the Flynn Effect. We have several grounds for supposing that enhanced powers of reason—specifically, the ability to set aside immediate experience, detach oneself from a parochial vantage point, and frame one's ideas in abstract, universal terms—would lead to better moral commitments, including an avoidance of violence. And we have just seen that over the course of the 20th century, people's reasoning abilities—particularly their ability to set aside immediate experience, detach themselves from a parochial vantage point, and think in abstract terms—were steadily enhanced. Can we put these two ideas together to help explain the documented declines of violence in the second half of the 20th century: the Long Peace, New Peace, and Rights Revolutions? Could there be a *moral* Flynn Effect, in which an accelerating escalator of reason carried us away from impulses that lead to violence?

The idea is not crazy. The cognitive skill that is most enhanced in the Flynn Effect, abstraction from the concrete particulars of immediate experience, is precisely the skill that must be exercised to take the perspectives of others and expand the circle of moral consideration. Flynn himself drew the connection in recounting a conversation he had with his Irish father, who was born in 1884 and was highly intelligent but relatively unschooled.

> My father had so much hatred for the English that there was little room left over for prejudice against any other group. But he harbored a bit of racism against blacks, and my brother and I tried to talk him out of it. "What if you woke up one morning and discovered your skin had turned black? Would that make you any less of a human being?" He shot back, "Now that's the stupidest thing you've ever said. Who ever heard of a man's skin turning black overnight?"[253]

Like the Russian peasant considering the color of bears, Flynn's father was stuck in a concrete, prescientific mode of thinking. He refused to enter a hypothetical world and explore its consequences, which is one of the ways people can rethink their moral commitments, including their tribalism and racism.

Or consider the high school test question about the water usage in a particular town, which requires, among other things, thinking about proportions. Flynn notes that proportionality questions are surprisingly difficult for many adolescents, and are among the skills that rose as part of the Flynn Effect.[254] As we have seen, the mindset of proportionality is essential to calibrating the just use of violence, as in criminal punishment and military action. One has only to replace "manage their water resources" with "manage their crime rate" in the test question to see how an increase in intelligence could translate into more humane policies. A recent study by the psychologist Michael Sargent showed that people with a high "need for cognition"—the trait of enjoying mental challenges—have less punitive attitudes toward criminal justice, even after taking into account their age, sex, race, education, income, and political orientation.[255]

Before we test the idea that the Flynn Effect accelerated an escalator of reason and led to greater moral breadth and less violence, we need a sanity check on the Flynn Effect itself. Could the people of today really be that much smarter than the people of yesterday? Flynn himself, in an early paper, noted incredulously that in some countries, if earlier scoring norms were applied today a quarter of the students should now be classified as "gifted," and the number of certifiable "geniuses" should have increased sixtyfold. "The result," he said skeptically, "should be a cultural renaissance too great to be overlooked."[256] But of course there *has* been an intellectual renaissance in recent decades, perhaps not in culture but certainly in science and technology. Cosmology, particle physics, geology, genetics, molecular biology, evolutionary biology, and neuroscience have made vertiginous leaps in understanding, while technology has given us secular miracles such as replaceable body parts, routine genome scans, stunning photographs of outer planets and distant galaxies, and tiny gadgets that allow one to chat with billions of people, take photographs, pinpoint one's location on the globe, listen to vast music collections, read from vast libraries, and access the wonders of the World Wide Web. These miracles have come at such a rapid pace that they have left us blasé about the ideas that made them possible. But no historian who takes in the sweep of human history on the scale of centuries could miss the fact that we are now living in a period of extraordinary brainpower.

We tend to be blasé about moral progress as well, but historians who take the long view have also marveled at the moral advances of the past six decades. As we saw, the Long Peace has had the world's most distinguished military historians shaking their heads in disbelief. The Rights Revolutions too have given us ideals that educated people today take for granted but that are virtually unprecedented in human history, such as that people of all races and creeds have equal rights, that women should be free from all forms of coercion, that children should never, ever be spanked, that students should be protected from bullying, and that there's nothing wrong with being gay. I don't find it at all implausible that these are gifts, in part, of a refined and widening application of reason.

The other half of the sanity check is to ask whether our recent ancestors can really be considered morally retarded. The answer, I am prepared to argue, is yes. Though they were surely decent people with perfectly functioning brains, the collective moral sophistication of the culture in which they lived was as primitive by modern standards as their mineral spas and patent medicines are by the medical standards of today. Many of their beliefs can be considered not just monstrous but, in a very real sense, stupid. They would not stand up to intellectual scrutiny as being consistent with other values they claimed to hold, and they persisted only because the narrower intellectual spotlight of the day was not routinely shone on them.

Lest you think this judgment a slander on our forebears, consider some of the convictions that were common in the decades before the effects of rising abstract intelligence began to accumulate. A century ago dozens of great writers and artists extolled the beauty and nobility of war, and eagerly looked forward to World War I. One "progressive" president, Theodore Roosevelt, wrote that the decimation of Native Americans was necessary to prevent the continent from becoming a "game preserve for squalid savages," and that in nine out of ten cases, "the only good Indians are the dead Indians."[257] Another, Woodrow Wilson, was a white supremacist who kept black students out of Princeton when he was president of the university, praised the Ku Klux Klan, cleansed the federal government of black employees, and said of ethnic immigrants, "Any man who carries a hyphen about with him carries a dagger that he is ready to plunge into the vitals of this Republic whenever he gets ready."[258] A third, Franklin Roosevelt, drove a hundred thousand American citizens into concentration camps because they were of the same race as the Japanese enemy.

On the other side of the Atlantic, the young Winston Churchill wrote of taking part in "a lot of jolly little wars against barbarous peoples" in the British Empire. In one of those jolly little wars, he wrote, "we proceeded systematically, village by village, and we destroyed the houses, filled up the wells, blew down the towers, cut down the shady trees, burned the crops and broke the reservoirs in punitive devastation." Churchill defended these atrocities on the grounds that "the Aryan stock is bound to triumph," and he said he was "strongly in favor of using poisoned gas against uncivilized tribes." He blamed the people of India for a famine caused by British mismanagement because they kept "breeding like rabbits," adding, "I hate Indians. They are a beastly people with a beastly religion."[259]

Today we are stunned by the compartmentalized morality of these men, who in many ways were enlightened and humane when it came to their own race. Yet they never took the mental leap that would have encouraged them to treat the people of other races with the same consideration. I still remember gentle lessons from my mother when my sister and I were children in the early

1960s, lessons that millions of children have received in the decades since: "There are bad Negroes and there are good Negroes, just like there are bad white people and good white people. You can't tell whether a person is good or bad by looking at the color of his skin." "Yes, the things those people do look funny to us. But the things we do look funny to them." Such lessons are not indoctrination but guided reasoning, leading children to conclusions they can accept by their own lights. Surely this reasoning was within the ken of the neural hardware of the great statesmen of a century ago. The difference is that today's children have been encouraged to take these cognitive leaps, and the resulting understanding has become second nature. Shorthand abstractions like *freedom of speech, tolerance, human rights, civil rights, democracy, peaceful coexistence,* and *nonviolence* (and their antitheses such as *racism, genocide, totalitarianism,* and *war crimes*) spread outward from their origins in abstract political discourse and became a part of everyone's mental tool kit. The advances can fairly be called a gain in intelligence, not completely different from the ones that drove scores in abstract reasoning upward.

Moral stupidity was not confined to the policies of leaders; it was written into the law of the land. Within the lifetimes of many readers of this book, the races in much of the country were forcibly segregated, women could not serve on juries in rape trials because they would be embarrassed by the testimony, homosexuality was a felony crime, and men were allowed to rape their wives, confine them to the house, and sometimes kill them and their adulterous lovers. And if you think that today's congressional proceedings are dumb, consider this 1876 testimony from a lawyer representing the city of San Francisco in hearings on the rights of Chinese immigrants:

> In relation to [the Chinese] religion, it is not our religion. That is enough to say about it; because if ours is right theirs must necessarily be wrong. [Question: What is our religion?] Ours is a belief in the existence of a Divine Providence that holds in its hands the destinies of nations. The Divine Wisdom has said that He would divide the country and the world as the heritage of five great families; that to the blacks He would give Africa; to the whites He would give Europe; to the red man He would give America, and Asia He would give to the yellow races. He inspires us with the determination not only to have preserved our own inheritance, but to have stolen from the red man America; and it is settled now that the Saxon, American or European groups of families, the white race, is to have the inheritance of Europe and of America and that the yellow races of China are to be confined to what God Almighty originally gave them; and as they are not a favorite people they are not to be permitted to steal from us what we robbed the American savage of.[260]

Nor was it just lawmakers who were intellectually challenged when it came to moral reasoning. I mentioned in chapter 6 that in the decades surrounding the turn of the 20th century, many literary intellectuals (including Yeats, Shaw, Flaubert, Wells, Lawrence, Woolf, Bell, and Eliot) expressed a contempt for the masses that bordered on the genocidal.[261] Many others would come to support fascism, Nazism, and Stalinism.[262] John Carey quotes from an essay by Eliot in which the poet comments on the spiritual superiority of the great artist: "It is better, in a paradoxical way, to do evil than to do nothing: at least we exist." Carey comments from a later era: "This appalling sentence leaves out of account, we notice, the effect of evil on its victims."[263]

The idea that the changes behind the Flynn Effect have also expanded the moral circle passes a sanity check, but that does not mean it is true. To show that rising intelligence has led to less violence, at the very least one needs to establish this intermediate link: that on average, and all else being equal, people with more sophisticated reasoning abilities (as assessed by IQ or other measures) are more cooperative, have larger moral circles, and are less sympathetic to violence. Better still, one would like to show that entire societies of better-reasoning individuals adopt policies that are less conducive to violence. If smarter people and smarter societies are less likely to be violent, then perhaps the recent rise in intelligence can help explain the recent decline of violence.

Before we examine the evidence for this hypothesis, let me clarify what it is not. The kind of reasoning relevant to moral progress is not general intelligence in the sense of raw brainpower, but the cultivation of abstract reasoning, the aspect of intelligence that has been pulled upward by the Flynn Effect. The two are highly correlated, so measures of IQ will, in general, track abstract reasoning, but it's the latter that is relevant to the escalator hypothesis. For the same reason, the specific differences in reasoning that I will focus on are not necessarily heritable (even though general intelligence is highly heritable), and I will stick with the assumption that all differences among groups are environmental in origin.

It's also important to note that the escalator hypothesis is about the influence of *rationality*—the level of abstract reasoning in a society—and not about the influence of *intellectuals*. Intellectuals, in the words of the writer Eric Hoffer, "cannot operate at room temperature."[264] They are excited by daring opinions, clever theories, sweeping ideologies, and utopian visions of the kind that caused so much trouble during 20th century. The kind of reason that expands moral sensibilities comes not from grand intellectual "systems" but from the exercise of logic, clarity, objectivity, and proportionality. These habits of mind are distributed unevenly across the population at any time, but the Flynn Effect lifts all the boats, and so we might expect to see a tide of mini- and micro-enlightenments across elites and ordinary citizens alike.

Let me present seven links, varying in directness, between reasoning ability and peaceable values.

Intelligence and Violent Crime. The first link is the most direct: smarter people commit fewer violent crimes, and they are the victims of fewer violent crimes, holding socioeconomic status and other variables constant.[265] We have no way to pinpoint the causal arrow—whether smarter people realize that violence is wrong or pointless, whether they exercise more self-control, or whether they keep themselves out of situations in which violence takes place. But all else being equal (setting aside, for example, the oscillations in crime from the 1960s through the 1980s), as people get smarter, there should be less violence.

Intelligence and Cooperation. At the other end of the abstractness scale, we can consider the purest model of how abstract reasoning might undermine the temptations of violence, the Prisoner's Dilemma. In his popular *Scientific American* column, the computer scientist Douglas Hofstadter once agonized over the fact that the seemingly rational response in a one-shot Prisoner's Dilemma was to defect.[266] You cannot trust the other player to cooperate, because he has no grounds for trusting you, and cooperating while he defects will bring you the worst outcome. Hofstadter's agony came from the observation that if both sides looked down on their dilemma from a single Olympian vantage point, stepping out of their parochial stations, they should both deduce that the best outcome is for both to cooperate. If each has confidence that the other realizes that, and that the other realizes that he or she realizes it, ad infinitum, both should cooperate and reap the benefits. Hofstadter envisioned a *superrationality* in which both sides were certain of the other's rationality, and certain that the other was certain of theirs, and so on, though he wistfully acknowledged that it was not easy to see how to get people to be superrational.

Can higher intelligence at least nudge people in the direction of superrationality? That is, are better reasoners likely to reflect on the fact that mutual cooperation leads to the best joint outcome, assume that the other guy is reflecting on it as well, and profit from the resulting simultaneous leap of trust? No one has given people of different levels of intelligence a true one-shot Prisoner's Dilemma, but a recent study came close by using a *sequential* one-shot Prisoner's Dilemma, in which the second player acts only after seeing the first player's move. The economist Stephen Burks and his collaborators gave a thousand trainee truck drivers a Matrices IQ test and a Prisoner's Dilemma, using money for the offers and payoffs.[267] The smarter truckers were more likely to cooperate on the first move, even after controlling for age, race, gender, schooling, and income. The investigators also looked at the response of the second player to the first player's move. This response has nothing to do with superrationality, but it does reflect a willingness to cooperate in response to the other player's cooperation in such a way that both players would benefit if the game were iterated. Smarter truckers, it turned out, were more likely to respond to cooperation with cooperation, and to defection with defection.

The economist Garrett Jones connected intelligence to the Prisoner's

Dilemma by a different route. He scoured the literature for all the Iterated Prisoner's Dilemma experiments that had been conducted in colleges and universities from 1959 to 2003.[268] Across thirty-six experiments involving thousands of participants, he found that the higher a school's mean SAT score (which is strongly correlated with mean IQ), the more its students cooperated. Two very different studies, then, agree that intelligence enhances mutual cooperation in the quintessential situation in which its benefits can be foreseen. A society that gets smarter, then, may be a society that becomes more cooperative.

Intelligence and Liberalism. Now we get to a finding that sounds more tendentious than it is: smarter people are more liberal. The statement will make conservatives see red, not just because it seems to impugn their intelligence but because they can legitimately complain that many social scientists (who are overwhelmingly liberal or leftist) use their research to take cheap shots at the right, studying conservatism as if it were a mental defect. (Tetlock and Haidt have both called attention to this politicization.)[269] So before turning to the evidence that links intelligence to liberalism, let me qualify the connection.

For one thing, since intelligence is correlated with social class, any correlation with liberalism, if not statistically controlled, could simply reflect the political prejudices of the upper middle classes. But the key qualification is that the escalator of reason predicts only that intelligence should be correlated with *classical* liberalism, which values the autonomy and well-being of individuals over the constraints of tribe, authority, and tradition. Intelligence is expected to correlate with classical liberalism because classical liberalism is itself a consequence of the interchangeability of perspectives that is inherent to reason itself. Intelligence need not correlate with other ideologies that get lumped into contemporary left-of-center political coalitions, such as populism, socialism, political correctness, identity politics, and the Green movement. Indeed, classical liberalism is sometimes congenial to the libertarian and anti-political-correctness factions in today's right-of-center coalitions. But on the whole, Haidt's surveys show that it is the people who identify their politics with the word *liberal* who are more likely to emphasize fairness and autonomy, the cardinal virtues of classical liberalism, over community, authority, and purity.[270] And as we saw in chapter 7, the self-described liberals are ahead of the curve on issues of personal autonomy, and the positions they pioneered decades ago have been increasingly accepted by conservatives today.

The psychologist Satoshi Kanazawa has analyzed two large American datasets and found that in both, intelligence correlates with the respondents' political liberalism, holding age, sex, race, education, earnings, and religion statistically constant.[271] Among more than twenty thousand young adults who had participated in the National Longitudinal Study of Adolescent Health, average IQ increased steadily from those who identified themselves as "very conservative" (94.8) to those who identified themselves as "very liberal" (106.4). The General Social Survey shows a similar correlation, while also containing a hint that intelligence tracks classical liberalism more closely than left-liberalism. The smarter respondents in the survey were *less* likely to agree with the statement that the government has a responsibility to redistribute income from the rich to the poor (leftist but not classically liberal), while being more likely to agree that the government should help black Americans to compensate for the historical discrimination against them (a formulation of a liberal position which is specifically motivated by the value of fairness).

A better case that intelligence causes, rather than merely correlates with, classical liberal attitudes comes from analyses by the psychologist Ian Deary and his colleagues on a dataset that includes every child born in Britain in a particular week in 1970. The title of their paper says it all: "Bright children become enlightened adults."[272] By "enlightened" they mean the mindset of the Enlightenment, which they define, following the *Concise Oxford Dictionary*, as "a philosophy emphasizing reason and individualism rather than tradition." They found that children's IQ at the age of ten (including tests of abstract reasoning) predicted their endorsement of antiracist, socially liberal, and pro-working-women attitudes at the age of thirty, holding constant their education, their social class, and their parents' social class. The socioeconomic controls, together with the twenty-year lag between the measurement of intelligence and the measurement of attitudes, make a prima facie case that the causal arrow goes from intelligence to classical liberalism. A second analysis discovered that brighter ten-year-olds were more likely to vote when they grew up, and more likely to vote for the Liberal Democrats (a center-left/libertarian coalition) or the Greens, and less likely to vote for nationalist and anti-immigrant parties. Again, there is a suggestion that intelligence leads to classical rather than left-liberalism: when social class was controlled, the IQ-Green correlation vanished, but the IQ–Lib Dem correlation survived.

Intelligence and Economic Literacy. And now for a correlation that will annoy the left as much as the correlation with liberalism annoyed the right. The economist Bryan Caplan also looked at data from the General Social Survey and found that smarter people tend to think more like economists (even after statistically controlling for education, income, sex, political party, and political orientation).[273] They are more sympathetic to immigration, free markets, and free trade, and less sympathetic to protectionism, make-work policies, and government intervention in business. Of course none of these positions is directly related to violence. But if one zooms out to the full continuum on which these positions lie, one could argue that the direction that is aligned with intelligence is also the direction that has historically pointed peaceward. To think like an economist is to accept the theory of gentle commerce from classical liberalism, which touts the positive-sum payoffs of exchange and its knock-on benefit of expansive networks of cooperation.[274] That sets it in opposition to populist, nationalist, and communist mindsets that see the world's wealth as zero-sum and infer that the enrichment of one group must come at the expense of another. The historical result of economic illiteracy has often been ethnic and class violence, as people conclude that the have-nots can improve their lot only by forcibly confiscating wealth from the haves and punishing them for their avarice.[275] As we saw in chapter 7, ethnic riots and genocides have declined since World War II, especially in the West, and a greater intuitive appreciation of economics may have played a part (lately there ain't been much work on account of the economy). At the level of international relations, trade has been superseding beggar-thy-neighbor protectionism over the past half-century and, together with democracy and an international community, has contributed to a Kantian Peace.[276]

Education, Intellectual Proficiency, and Democracy. Speaking of the Kantian Peace, the democracy leg of the tripod may also be fortified by reasoning. One of the great puzzles of political science is why democracy takes root in some countries but not others—why, for example, the former satellites and republics of the Soviet Union in Europe made the transition, but the -stans in Central Asia did not. The wobbliness of the democracies imposed on Iraq and Afghanistan makes the problem all the more acute.

Theorists have long speculated that a literate, knowledgeable populace is a prerequisite to a functioning democracy. Down the road from where I sit, the Boston Public Library displays on its entablature the stirring words "The Commonwealth requires the education of the people as the safeguard of order and liberty." Presumably by "education" what the carvers had in mind was not an ability to name the capitals of all the states one would pass through on a trip from Columbus, Ohio, to the Gulf of Mexico, but literacy and numeracy, an understanding of the principles behind democratic government and civil society, an ability to evaluate leaders and their policies, an awareness of other peoples and their diverse cultures, and an expectation that one is part of a commonwealth of educated citizens who share these understandings.[277] These competencies require a modicum of abstract reasoning, and they overlap with the abilities that have risen with the Flynn Effect, presumably because the Flynn Effect itself has been driven by education.

But the Boston Public Library theory of democracy-readiness has not, until recently, been tested. It's long been known that mature democracies have better educated and smarter populations, but mature democracies have more of everything that is good in life, and we can't tell what causes what. Perhaps more democratic countries are also richer and can afford more schools and libraries, which make their citizens better educated and smarter, rather than the other way around.

The psychologist Heiner Rindermann tried to cut the correlational knot with a social science technique called cross-lagged correlation (we saw an example of the technique in the British study showing that bright children become enlightened adults).[278] Several datasets assign countries numerical scores on their levels of democracy and rule of law. Also available for many countries are the number of years of education attained by their children. In a subsample of countries, Rindermann also obtained data on their average scores on widely used intelligence tests, together with performance on internationally administered tests of academic achievement; he combined the two into a measure of intellectual ability. Rindermann tested whether a country's level of education and intellectual ability in one era (1960–72) predicted its level of prosperity, democracy, and rule of law in a later one (1991–2003). If the Boston Public Library theory is true, these correlations should be strong even when other variables, like the nation's wealth in the earlier period, are held constant. And crucially, they should be far stronger than the correlation between democracy and rule of law in an earlier period and education and intellectual ability in the later one, because the past affects the present, not the other way around.

Let's tip our hats to the stonecarvers of the Boston Public Library. Education and intellectual abilities in the past indeed predicted democracy and rule of law (together with prosperity) in the recent present, holding all else constant. Wealth in the past, in contrast, did not predict democracy in the present (though it did mildly predict rule of law). Intellectual ability was a more powerful predictor of democracy than the number of years of schooling, and Rindermann showed that schooling was predictive only because of its correlation with intellectual ability. It is not a big leap to conclude that an education-fueled rise in reasoning ability made at least some parts of the world safe for democracy. Democracy by definition is associated with less government violence, and we know that it is statistically associated with an aversion to interstate war, deadly ethnic riots, and genocide, and with a reduction in the severity of civil wars.[279]

Education and Civil War. What about the developing world? Average scores on intelligence tests, though they started from lower levels, have been steeply rising in the countries in which the trends have been measured, such as Kenya and Dominica.[280] Can we attribute any part of the New Peace to rising levels of reasoning in those countries? Here the evidence is circumstantial but suggestive. Earlier we saw that the New Peace has been led, in part, by a greater acceptance of democracy and open economies, which, as we have just seen, smarter people tend to favor. Put the two together, and we can entertain the possibility that more education can lead to smarter citizens (in the sense of "smart" we care about here), which can prepare the way for democracy and open economies, which can favor peace.

It's difficult to verify every link in that chain, but the first and last links have been correlated in a recent paper whose title is self-explanatory: "ABC's,

123's, and the Golden Rule: The Pacifying Effect of Education on Civil War, 1980–1999."[281] The political scientist Clayton Thyne analyzed 160 countries and 49 civil wars taken from the dataset of James Fearon and David Laitin, which we visited in chapter 6. Thyne discovered that four indicators of a country's level of education—the proportion of its gross domestic product invested in primary education, the proportion of its school-age population enrolled in primary schools, the proportion of its adolescent population that was enrolled in secondary schools (especially the males), and (marginally) the level of adult literacy—all reduced the chance the country would be embroiled in a civil war a year later. The effects were sizable: compared to a country that is a standard deviation below the average in primary-school enrollment, a country that is a standard deviation above the average was 73 percent less likely to fight a civil war the following year, holding constant prior wars, per capita income, population, mountainous terrain, oil exports, the degree of democracy and anocracy, and ethnic and religious fractionation.

Now, we cannot conclude from these correlations that schooling makes people smarter, which makes them more averse to civil war. Schooling has other pacifying effects. It increases people's confidence in their government by showing that it can do at least one thing right. It gives them skills that they can parlay into jobs rather than brigandage and warlording. And it keeps teenage boys off the streets and out of the militias. But the correlations are tantalizing, and Thyne argues that at least a part of the pacifying effect of education consists of "giving people tools with which they can resolve disputes peacefully."[282]

Sophistication of Political Discourse. Finally, let's have a look at political discourse, which most people believe has been getting dumb and dumber. There's no such thing as the IQ of a speech, but Tetlock and other political psychologists have identified a variable called integrative complexity that captures a sense of intellectual balance, nuance, and sophistication.[283] A passage that is low in integrative complexity stakes out an opinion and relentlessly hammers it home, without nuance or qualification. Its minimal complexity can be quantified by counting words like *absolutely, always, certainly, definitively, entirely, forever, indisputable, irrefutable, undoubtedly,* and *unquestionably.* A passage gets credit for some degree of integrative complexity if it shows a touch of subtlety with words like *usually, almost, but, however,* and *maybe.* It is rated higher if it acknowledges two points of view, higher still if it discusses connections, tradeoffs, or compromises between them, and highest of all if it explains these relationships by reference to a higher principle or system. The integrative complexity of a passage is not the same as the intelligence of the person who wrote it, but the two are correlated, especially, according to Simonton, among American presidents.[284]

Integrative complexity is related to violence. People whose language is less integratively complex, on average, are more likely to react to frustration with violence and are more likely to go to war in war games.[285] Working with the psychologist Peter Suedfeld, Tetlock tracked the integrative complexity of the speeches of national leaders in a number of political crises of the 20th century that ended peacefully (such as the Berlin blockade in 1948 and the Cuban Missile Crisis) or in war (such as World War I and the Korean War), and found that when the complexity of the leaders' speeches declined, war followed.[286] In particular, they found a linkage between rhetorical simple-mindedness and military confrontations in speeches by Arabs and Israelis, and by the Americans and Soviets during the Cold War.[287] We don't know exactly what the correlations mean: whether mule-headed antagonists cannot think their way to an agreement, or bellicose antagonists simplify their rhetoric to stake out an implacable bargaining position. Reviewing both laboratory and real-world studies, Tetlock suggests that both dynamics are in play.[288]

Has the integrative complexity of political discourse been pulled upward by the Flynn Effect? A study by the political scientists James Rosenau and Michael Fagen suggests it may have.[289] The investigators coded the integrative complexity of American congressional testimony and press coverage in the early (1916–32) and late (1970–93) decades of the 20th century. They looked at the verbiage surrounding controversies in the two eras with roughly similar content, such as the Smoot-Hawley Act, which clamped down on free trade, and then the NAFTA agreement, which opened it up, and the granting of women's suffrage and then the passage of the Equal Rights Amendment. In almost every case, contrary to the worst fears of today's political buffs, the integrative complexity of the political discourse *increased* from early to late in the century. The sole exception lay in congressmen's statements on women's rights. Here is an example of the quality of argument used to support voting rights for women in 1917:

> In the great Lone Star State, 58 counties of which it is my honor to represent, a State which is the largest in this Union, every person over 21 years of age may vote except a convict, a lunatic, and a woman. I am not willing that woman shall be placed in the same class and category in the Lone Star State with a convict and a lunatic.[290]

And here is an example of an argument used in 1972 to oppose the Equal Rights Amendment, from Senator Sam Ervin, born in 1896:

> [The ERA] says that men and women are identical and equal legal human beings. It gives consideration for many foolish things like that. It is absolutely ridiculous to talk about taking a mother away from her children so that she may go out to fight the enemy and leave the father at home to nurse the children. The Senator from Indiana may think that is wise, but I do not. I think it is foolish.[291]

But the unchanging inanity of the senators' arguments on women's rights was overshadowed by twenty-eight other comparisons that found an increase in sophistication over the course of the century. Ervin, by the way, was no troglodyte, but a respected senator who would soon be lionized for chairing the committee on Watergate that brought down Richard Nixon. The fact that his words sound so fatuous today, even by the low standards of senatorial speechifying, reminds us not to get too nostalgic for the political discourse of decades past.

In one arena, however, politicians really do seem to be swimming against the Flynn Effect: American presidential debates. To those who followed these debates in 2008, three words are enough to make the point: Joe the Plumber. The psychologists William Gorton and Janie Diels quantified the trend by scoring the sophistication of candidates' language in the debates from 1960 through 2008.[292] They found that the overall sophistication declined from 1992 to 2008, and the quality of remarks on economics began a free fall even earlier, in 1984. Ironically, the decrease in sophistication in presidential debates may be the product of an *increase* in the sophistication of political strategists. Televised debates in the waning weeks of a campaign are aimed at a sliver of undecided voters who are among the least informed and least engaged sectors of the electorate. They are apt to make their choice based on sound bites and one-liners, so the strategists advise the candidates to aim low. The level of sophistication cratered in 2000 and 2004, when Bush's Democratic opponents matched him platitude for platitude. This exploitable vulnerability of the American political system might help explain how the country found itself in two protracted wars during an era of increasing peace.

There is a reason that I made reason the last of the better angels of our nature. Once a society has a degree of civilization in place, it is reason that offers the greatest hope for further reducing violence. The other angels have been with us for as long as we have been human, but during most of our long existence they have been unable to prevent war, slavery, despotism, institutionalized sadism, and the oppression of women. As important as they are, empathy, self-control, and the moral sense have too few degrees of freedom, and too restricted a range of application, to explain the advances of recent decades and centuries.

Empathy is a circle that may be stretched, but its elasticity is limited by kinship, friendship, similarity, and cuteness. It reaches a breaking point long before it encircles the full set of people that reason tells us should fall within our moral concern. Also, empathy is vulnerable to being dismissed as mere sentimentality. It is reason that teaches us the tricks for expanding our empathy, and it is reason that tells us how and when we should parlay our compassion for a pathetic stranger into an actionable policy.

Self-control is a muscle that may be strengthened, but it can prevent only the harms for which we ourselves harbor inner temptations. Also, the 1960s slogans were right about one thing: there are moments in life when one really should cut loose and do one's thing. Reason tells us what those moments are: the times when doing your thing does not impinge on other people's freedom to do their thing.

The moral sense offers three ethics that can be assigned to social roles and resources. But most applications of the moral sense are not particularly moral but rather tribal, authoritarian, or puritanical, and it is reason that tells us which of the other applications we should entrench as norms. And the one ethic that we can design to bring about the greatest good for the greatest number, the Rational-Legal mindset, is not part of the natural moral sense at all.

Reason is up to these demands because it is an open-ended combinatorial system, an engine for generating an unlimited number of new ideas. Once it is programmed with a basic self-interest and an ability to communicate with others, its own logic will impel it, in the fullness of time, to respect the interests of ever-increasing numbers of others. It is reason too that can always take note of the shortcomings of previous exercises of reasoning, and update and improve itself in response. And if you detect a flaw in this argument, it is reason that allows you to point it out and defend an alternative.

Adam Smith, friend of Hume and fellow luminary of the Scottish Enlightenment, first made this argument in *The Theory of Moral Sentiments*, using a poignant example that resonates today. Smith asked us to imagine our reaction to reading about a dreadful calamity befalling a large number of strangers, such as a hundred million Chinese perishing in an earthquake. If we're honest, we will admit that our reaction would run more or less as follows. We would feel bad for a while, pitying the victims and perhaps reflecting on the fragility of life. Perhaps today we would write a check or click on a Web site to aid the survivors. And then we would get back to work, have dinner, and go to bed as if nothing had happened. But if an accident befell us personally, even if it were trivial in comparison, such as losing a little finger, we would be immensely more upset, and would not be able to put the misfortune out of our minds.

This all sounds terribly cynical, but Smith continues. Consider a different scenario. This time you are presented with a choice: you can lose your little finger, or a hundred million Chinese will be killed. Would you sacrifice a hundred million people to save your little finger? Smith predicts, and I agree, that almost no one would select this monstrous option. But why not, Smith asks, given that our empathy for strangers is so much less compelling than our distress at a personal misfortune? He resolves the paradox by comparing our better angels:

> It is not the soft power of humanity, it is not that feeble spark of benevolence which Nature has lighted up in the human heart, that is thus capable of

counteracting the strongest impulses of self-love. It is a stronger power, a more forcible motive, which exerts itself upon such occasions. It is reason, principle, conscience, the inhabitant of the breast, the man within, the great judge and arbiter of our conduct. It is he who, whenever we are about to act so as to affect the happiness of others, calls to us, with a voice capable of astonishing the most presumptuous of our passions, that we are but one of the multitude, in no respect better than any other in it; and that when we prefer ourselves so shamefully and so blindly to others, we become the proper objects of resentment, abhorrence, and execration. It is from him only that we learn the real littleness of ourselves, and of whatever relates to ourselves, and the natural misrepresentations of self-love can be corrected only by the eye of this impartial spectator. It is he who shows us the propriety of generosity and the deformity of injustice; the propriety of resigning the greatest interests of our own, for the yet greater interests of others, and the deformity of doing the smallest injury to another, in order to obtain the greatest benefit to ourselves.[293]

10

ON ANGELS' WINGS

As man advances in civilization, and small tribes are united into larger communities, the simplest reason would tell each individual that he ought to extend his social instincts and sympathies to all the members of the same nation, though personally unknown to him. This point being once reached, there is only an artificial barrier to prevent his sympathies extending to the men of all nations and races.

—Charles Darwin, *The Descent of Man*

This book grew out of an answer to the question "What are you optimistic about?" and I hope that the numbers I have marshaled have lifted your assessment of the state of the world from the lugubrious conventional wisdom. But having documented dozens of declines and abolitions and zeroes, my mood is one not so much of optimism as of gratitude. Optimism requires a touch of arrogance, because it extrapolates the past to an uncertain future. Though I am confident that human sacrifice, chattel slavery, breaking on the wheel, and wars between democracies will not make a comeback anytime soon, to predict that the current levels of crime, civil war, or terrorism will endure is to sally into territory where angels fear to tread. What we *can* feel sure about is that many kinds of violence have declined up to the present, and we can try to understand why that has happened. As a scientist, I must be skeptical of any mystical force or cosmic destiny that carries us ever upward. Declines of violence are a product of social, cultural, and material conditions. If the conditions persist, violence will remain low or decline even further; if they don't, it won't.

In this final chapter I will not try to make predictions; nor will I offer advice to politicians, police chiefs, or peacemakers, which given my qualifications would be a form of malpractice. What I will try to do is identify the broad forces that have pushed violence downward. My quarry will be developments that repeatedly turned up in the historical chapters (2 through 7) and that engage the faculties of mind that were explored in the psychological chapters (8 and 9). That is, I will look for common threads in the Pacification Process, the Civilizing Process, the Humanitarian Revolution, the Long Peace, the New Peace, and the Rights Revolutions. Each should represent a way in which predation, dominance, revenge, sadism, or ideology has been overpowered by self-control, empathy, morality, or reason.

We should not expect these forces to fall out of a grand unified theory. The declines we seek to explain unfolded over vastly different scales of time and damage: the taming of chronic raiding and feuding, the reduction of vicious interpersonal violence such as cutting off noses, the elimination of cruel practices like human sacrifice, torture-executions, and flogging, the abolition of institutions such as slavery and debt bondage, the falling out of fashion of blood sports and dueling, the eroding of political murder and despotism, the recent decline of wars, pogroms, and genocides, the reduction of violence against women, the decriminalization of homosexuality, the protection of children and animals. The only thing these superseded practices have in common is that they physically hurt a victim, and so it is only from a generic victim's perspective—which, as we saw, is also the perspective of the moralist—that we could even dream of a final theory. From the scientist's perspective, the motives of the perpetrators may be motley, and so will the explanations for the forces that pushed against those motives.

At the same time, all these developments undeniably point in the same direction. It's a good time in history to be a potential victim. One can imagine a historical narrative in which different practices went in different directions: slavery stayed abolished, for example, but parents decided to bring back savage beatings of their children; or states became increasingly humane to their citizens but more likely to wage war on one another. That hasn't happened. Most practices have moved in the less violent direction, too many to be a coincidence.

To be sure, some developments went the other way: the destructiveness of European wars through World War II (overshadowing the decrease in their frequency until both fell in tandem), the heyday of genocidal dictators in the middle decades of the 20th century, the rise of crime in the 1960s, and the bulge of civil wars in the developing world following decolonization. Yet every one of these developments has been systematically reversed, and from where we sit on the time line, most trends point peaceward. We may not be entitled to a theory of everything, but we do need a theory that explains why so many somethings point the same way.

IMPORTANT BUT INCONSISTENT

Let me begin by noting a few forces that one might have thought would be important to the processes, peaces, and revolutions of chapters 2–7, but as best I can tell turned out not to be. It's not that these forces are by any means minor; it's just that they have not consistently worked to reduce violence.

Weaponry and Disarmament. Writers who are engrossed by violence and those who are repelled by it have one thing in common: they are fixated on weaponry. Military histories, written by and for guys, obsess over longbows, stirrups, artillery, and tanks. Many movements for nonviolence have been disarmament movements: the demonization of "merchants of war," the antinuclear demonstrations, the campaigns for gun control. And then there is the contrary though equally weaponcentric prescription according to which the invention of unthinkably destructive weapons (dynamite, poison gas, nuclear bombs) would make war unthinkable.

The technology of weaponry has obviously changed the course of history many times by determining winners and losers, making deterrence credible, and multiplying the destructive power of certain antagonists. No one would argue, for example, that the proliferation of automatic weapons in the developing world has been good for peace. Yet it's hard to find any correlation over history between the destructive power of weaponry and the human toll of deadly quarrels. Over the millennia weapons, just like every technology, got better and better, yet rates of violence have not gone steadily up but rather have lurched up and down the slope of an inclined sawtooth. The spears and arrows of pre-state peoples notched up higher proportional body counts than has anything since (chapter 2), and the pikemen and cavalry of the Thirty Years' War did more human damage than the artillery and gas of World War I (chapter 5). Though the 16th and 17th centuries saw a military revolution, it was less an arms race than an *armies* race, in which governments beefed up the size and efficiency of their armed forces. The history of genocide shows that people can be slaughtered as efficiently with primitive weapons as they can with industrial technology (chapters 5 and 6).

Nor did precipitous drops in violence, such as those of the Long Peace, the New Peace, and the Great American Crime Decline, originate with the antagonists melting down their weapons. The historical sequence has usually gone the other way, as in the dismantling of armamentaria that was part of the peace dividend after the end of the Cold War. As for the nuclear peace, we have seen that nuclear weapons may have made little difference to the course of world events, given their uselessness in battle and the massive destructive power of conventional forces (chapter 5). And the popular (if bizarre) argument that nuclear weapons would inevitably be used by the great powers to justify the cost of developing them turned out to be flat wrong.

The failure of technological determinism as a theory of the history of violence should not be that surprising. Human behavior is goal-directed, not stimulus-driven, and what matters most to the incidence of violence is whether one person wants another one dead. The cliché of gun control opponents is literally true: guns don't kill people; people kill people (which is not to endorse the arguments for or against gun control). Anyone who is equipped to hunt, harvest crops, chop firewood, or prepare salad has the means to damage a lot of human flesh. With necessity being the mother of invention, people can upgrade their technology to the extent their enemies force them to. Weaponry, in other words, appears to be largely endogenous to the historical dynamics that result in large declines in violence. When people are rapacious or terrified, they develop the weapons they need; when cooler heads prevail, the weapons rust in peace.

Resources and Power. When I was a student in the 1970s, I had a professor who shared with anyone who would listen the truth about the Vietnam War: it was really about tungsten. The South China Sea, he discovered, had the world's largest deposits of the metal used in lightbulb filaments and superhard steel. The debates on communism and nationalism and containment were all a smokescreen for the superpowers' battle to control the source of this vital resource.

The tungsten theory of the Vietnam War is an example of resource determinism, the idea that people inevitably fight over finite resources like land, water, minerals, and strategic terrain. One version holds that conflict arises from an unequal allocation of resources, and that peace will come when they are distributed more equitably. Another feeds into "realist" theories that see conflict over land and resources as a permanent feature of international relations, and peace as the outcome of a balance of power in which each side is deterred from encroaching on the other's sphere of influence.

While contests over resources are a vital dynamic in history, they offer little insight into grand trends in violence. The most destructive eruptions of the past half millennium were fueled not by resources but by ideologies, such as religion, revolution, nationalism, fascism, and communism (chapter 5). Though no one can prove that each of these cataclysms wasn't really about tungsten or some other ulterior resource, any effort to show that they are is bound to look like a nutball conspiracy theory. As for the balance of power, the upending of the pans after the Soviet Union collapsed and the Germanys were unified did not send the world into a mad scramble. Rather, it had no discernible effect on the Long Peace among developed countries, and it presaged a New Peace among developing ones. Nor did either of these pleasant surprises originate in the discovery or redistribution of resources. In fact, resources in the developing world often turn out to be a curse rather than a blessing. Countries rich in oil and minerals, despite having a larger pie to divide among their citizens, are among those with the most violence (chapter 6).

The looseness of the connection between resource control and violence should also come as no surprise. Evolutionary psychologists tell us that no matter how rich or poor men are, they can always fight over women, status, and dominance. Economists tell us that wealth originates not from land with stuff in it but from the mobilization of ingenuity, effort, and cooperation to turn that stuff into usable products. When people divide the labor and exchange its fruits, wealth can grow and everyone wins. That means that resource competition is not a constant of nature but is endogenous to the web of societal forces that includes violence. Depending on their infrastructure and mindset, people in different times and places can choose to engage in positive-sum exchanges of finished products or in zero-sum contests over raw materials—indeed, negative-sum contests, because the costs of war have to be subtracted from the value of the plundered materials. The United States could invade Canada to seize its shipping lane to the Great Lakes or its precious deposits of nickel, but what would be the point, when it already enjoys their benefits through trade?

Affluence. Over the millennia, the world has become more prosperous, and it has also become less violent. Do societies become more peaceful as they get richer? Perhaps the daily pains and frustrations of poverty make people more ornery and give them more to fight over, and the bounty of an affluent society gives them more reasons to value their lives, and by extension, the lives of others.

Nonetheless tight correlations between affluence and nonviolence are hard to find, and some correlations go the other way. Among pre-state peoples, it is often the sedentary tribes living in temperate regions flush with fish and game, such as the Pacific Northwest, that had slaves, castes, and a warrior culture, while the materially modest San and Semai are at the peaceable end of the distribution (chapter 2). And it was the glorious ancient empires that had slaves, crucifixions, gladiators, ruthless conquest, and human sacrifice (chapter 1).

The ideas behind democracy and other humanitarian reforms blossomed in the 18th century, but upsurges in material well-being came considerably later (chapter 4). Wealth in the West began to surge only with the Industrial Revolution of the 19th century, and health and longevity took off with the public health revolution at the end of the 19th. Smaller-scale fluctuations in prosperity also appear to be out of sync with a concern for human rights. Though it has been suggested that lynchings in the American South went up when cotton prices went down, the overwhelming historical trend was an exponential decay of lynchings in the first half of the 20th century, without a deflection in either the Roaring Twenties or the Great Depression (chapter 7). As far as we can tell, the Rights Revolutions that started in the late 1950s did not pick up steam or run out of it in tandem with the ups and downs of the business cycle. And they are not automatic outcomes of modern affluence, as we see in the relatively high tolerance of domestic violence and the spanking of children in some well-to-do Asian states (chapter 7).

Nor does violent crime closely track the economic indicators. The careenings of the American homicide rate in the 20th century were largely uncorrelated with measures of prosperity: the murder rate plunged in the midst of the Great Depression, soared during the boom years of the 1960s, and hugged new lows during the Great Recession that began in 2007 (chapter 3). The poor correlation could have been predicted by the police blotters, which show that homicides are driven by moralistic motives like payback for insults and infidelity rather than by material motives such as cash or food.

Wealth and violence do show a powerful connection in one comparison: differences among countries at the bottom of the economic scale (chapter 6). The likelihood that a country will be torn by violent civil unrest, as we saw, starts to soar as its annual per capita domestic product falls below $1,000. It's hard, though, to pinpoint the causes behind the correlation. Money can buy many things, and it's not obvious which of the things that a country cannot afford is responsible for its violence. It may be deprivations of individual people, such as nutrition and health care, but it also may be deprivations of the entire country, such as decent schools, police, and governments (chapter 6). And since war is development in reverse, we cannot even know the degree to which poverty causes war or war causes poverty.

And though extreme poverty is related to civil war, it does not seem to be related to genocide. Recall that poor countries have more political crises, and political crises can lead to genocides, but once a country has a crisis, poverty makes it no more likely to host a genocide (chapter 6). At the other end of the affluence scale, late 1930s Germany had the worst of the Great Depression behind it and was becoming an industrial powerhouse, yet that was when it brewed the atrocities that led to the coining of the word *genocide*.

The tangled relationship between wealth and violence reminds us that humans do not live by bread alone. We are believing, moralizing animals, and a lot of our violence comes from destructive ideologies rather than not enough wealth. For better or worse—usually worse—people are often willing to trade off material comfort for what they see as spiritual purity, communal glory, or perfect justice.

Religion. Speaking of ideologies, we have seen that little good has come from ancient tribal dogmas. All over the world, belief in the supernatural has authorized the sacrifice of people to propitiate bloodthirsty gods, and the murder of witches for their malevolent powers (chapter 4). The scriptures present a God who delights in genocide, rape, slavery, and the execution of nonconformists, and for millennia those writings were used to rationalize the massacre of infidels, the ownership of women, the beating of children, dominion over animals, and the persecution of heretics and homosexuals (chapters 1, 4, and 7). Humanitarian reforms such as the elimination of cruel punishment, the dissemination of empathy-inducing novels, and the abolition of slavery were met with fierce opposition in their time by ecclesiastical authorities and their apologists (chapter 4). The elevation of parochial values to the realm of the sacred is a license to dismiss other people's interests, and an imperative to reject the possibility of compromise (chapter 9). It inflamed the combatants in the European Wars of Religion, the second-bloodiest period in modern Western history, and it continues to inflame partisans in the Middle East and parts of the Islamic world today. The theory that religion is a force for peace, often heard among the religious right and its allies today, does not fit the facts of history.

Defenders of religion claim that the two genocidal ideologies of the 20th century, fascism and communism, were atheistic. But the first claim is mistaken and the second irrelevant (chapter 4). Fascism happily coexisted with Catholicism in Spain, Italy, Portugal, and Croatia, and though Hitler had little use for Christianity, he was by no means an atheist, and professed that he was carrying out a divine plan.[1] Historians have documented that many of the Nazi elite melded Nazism with German Christianity in a syncretic faith, drawing on its millennial visions and its long history of anti-Semitism.[2] Many Christian clerics and their flocks were all too happy to sign up, finding common cause with the Nazis in their opposition to the tolerant, secular, cosmopolitan culture of the Weimar era.[3]

As for godless communism, godless it certainly was. But the repudiation of one illiberal ideology does not automatically grant immunity from others. Marxism, as Daniel Chirot observed (see page 330), helped itself to the worst idea in the Christian Bible, a millennial cataclysm that will bring about a utopia and restore prelapsarian innocence. And it violently rejected the humanism and liberalism of the Enlightenment, which placed the autonomy and flourishing of individuals as the ultimate goal of political systems.[4]

At the same time, *particular* religious movements at particular times in history *have* worked against violence. In zones of anarchy, religious institutions have sometimes served as a civilizing force, and since many of them claim to hold the morality franchise in their communities, they can be staging grounds for reflection and moral action. The Quakers parlayed Enlightenment arguments against slavery and war into effective movements for abolition and pacifism, and in the 19th century other liberal Protestant denominations joined them (chapter 4). Protestant churches also helped to tame the wild frontier of the American South and West (chapter 3). African American churches supplied organizational infrastructure and rhetorical power to the civil rights movement (though as we saw, Martin Luther King rejected mainstream Christian theology and drew his inspiration from Gandhi, secular Western philosophy, and renegade humanistic theologians). These churches also worked with the police and community organizations to lower crime in African American inner cities in the 1990s (chapter 3). In the developing world, Desmond Tutu and other church leaders worked with politicians and nongovernmental organizations in the reconciliation movements that healed countries following apartheid and civil unrest (chapter 8).

So the subtitle of Christopher Hitchens's atheist bestseller, *How Religion Poisons Everything,* is an overstatement. Religion plays no single role in the history of violence because religion has not been a single force in the history of anything. The vast set of movements we call religions have little in common but their distinctness from the secular institutions that are recent appearances on the human stage. And the beliefs and practices of religions, despite their claims to divine provenance, are endogenous to human affairs, responding to their intellectual and social currents. When the currents move in enlightened directions, religions often adapt to them, most obviously in the discreet neglect of the bloodthirsty passages of the Old Testament. Not all of the accommodations are as naked as those of the Mormon church, whose leaders had a revelation from Jesus Christ in 1890 that the church should cease polygamy (around the time that polygamy was standing in the way of Utah's joining the Union), and another one in 1978 telling them that the priesthood should accept black men, who were previously deemed to bear the mark of Cain. But subtler accommodations instigated by breakaway denominations, reform movements, ecumenical councils, and other liberalizing forces have allowed other religions to be swept along by the humanistic tide. It is when fundamentalist forces stand athwart those currents and impose tribal, authoritarian, and puritanical constraints that religion becomes a force for violence.

THE PACIFIST'S DILEMMA

Let me turn from the historical forces that don't seem to be consistent reducers of violence to those that do. And let me try to place these forces into a semblance of an explanatory framework so that, rather than ticking off items on a list, we can gain insight into what they might have in common. What we seek is an understanding of why violence has always been so tempting, why people have always yearned to reduce it, why it has been so hard to reduce, and why certain kinds of changes eventually did reduce it. To be genuine explanations, these changes should be exogenous: they should not be a part of the very decline we are trying to explain, but independent developments that preceded and caused it.

A good way to make sense of the changing dynamics of violence is to think back to the paradigmatic model of the benefits of cooperation (in this case, refraining from aggression), namely the Prisoner's Dilemma (chapter 8). Let's change the labels and call it the Pacifist's Dilemma. A person or coalition may be tempted by the gains of a victory in predatory aggression (the equivalent of defecting against a cooperator), and certainly wants to avoid the sucker's payoff of being defeated by an adversary who acts on the same temptation. But if they both opt for aggression, they will fall into a punishing war (mutual defection), which will leave them both worse off than if they had opted for the rewards of peace (mutual cooperation). Figure 10–1 is a depiction of the

Own choices	Other's choices: Pacifist	Other's choices: Aggressor
Pacifist	Peace (5) Peace (5)	Defeat (–100) Victory (10)
Aggressor	Victory (10) Defeat (–100)	War (–50) War (–50)

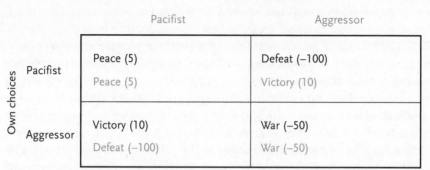

FIGURE 10–1. The Pacifist's Dilemma

Pacifist's Dilemma; the numbers for the gains and losses are arbitrary, but they capture the dilemma's tragic structure.

The Pacifist's Dilemma is by no stretch of the imagination a mathematical model, but I will keep pointing to it to offer a second way of conveying the ideas I will try to explain in words. The numbers capture the twofold tragedy of violence. The first part of the tragedy is that when the world has these payoffs, it is irrational to be a pacifist. If your adversary is a pacifist, you are tempted to exploit his vulnerability (the 10 points of victory are better than the 5 points of peace), whereas if he is an aggressor, you are better off enduring the punishment of a war (a loss of 50 points) than being a sucker and letting him exploit you (a devastating loss of 100). Either way, aggression is the rational choice.

The second part of the tragedy is that the costs to a victim (–100, in this case) are vastly disproportionate to the benefits to the aggressor (10). Unless two adversaries are locked in a fight to the death, aggression is not zero-sum but negative-sum; they are collectively better off not doing it, despite the advantage to the victor. The advantage to a conqueror in gaining a bit more land is swamped by the disadvantage to the family he kills in stealing it, and the few moments of drive reduction experienced by a rapist are obscenely out of proportion to the suffering he causes his victim. The asymmetry is ultimately a consequence of the law of entropy: an infinitesimal fraction of the states of the universe are orderly enough to support life and happiness, so it's easier to destroy and cause misery than to cultivate and cause happiness. All of this means that even the most steely-eyed utilitarian calculus, in which a disinterested observer tots up the total happiness and unhappiness, will deem violence undesirable, because it creates more unhappiness in its victims than happiness in its perpetrators, and lowers the aggregate amount of happiness in the world.

But when we descend from the lofty vantage point of the disinterested observer to the earthly one of the players, we can see why violence is so hard to eliminate. Each side would be crazy to be the only one to opt for pacifism, because if his adversary was tempted by aggression, he would pay a terrible cost. The other-guy problem explains why pacifism, turning the other cheek, beating swords into plowshares, and other moralistic sentiments have not been a consistent reducer of violence: they only work if one's adversary is overcome by the same sentiments at the same time. It also, I think, helps us to understand why violence can spiral upward or downward so unpredictably at various times in history. Each side has to be aggressive enough not to be a sitting duck for its adversary, and often the best defense is a good offense. The resulting mutual fear of attack—the Hobbesian trap or security dilemma—can escalate everyone's belligerence (chapter 2). Even when the game is played repeatedly and the threat of reprisals can (in theory) deter both sides, the strategic advantage of overconfidence and other self-serving biases can lead instead to cycles of feuding. By the same logic, a credible goodwill gesture can occasionally be reciprocated, unwinding the cycle and sending violence downward when everyone least expects it.

And here is the key to identifying a common thread that might tie together the historical reducers of violence. Each should change the payoff structure of the Pacifist's Dilemma—the numbers in the checkerboard—in a way that attracts the two sides into the upper left cell, the one that gives them the mutual benefits of peace.

In light of the history and psychology we have reviewed, I believe we can identify five developments that have pushed the world in a peaceful direction. Each shows up, to varying degrees, in a number of historical sequences, quantitative datasets, and experimental studies. And each can be shown to move around the numbers in the Pacifist's Dilemma in a way that entices people into the precious cell of peace. Let's go through them in the order in which they were introduced in the preceding chapters.

THE LEVIATHAN

A state that uses a monopoly on force to protect its citizens from one another may be the most consistent violence-reducer that we have encountered in this book. Its simple logic was depicted in the aggressor-victim-bystander triangle in figure 2–1 and may be restated in terms of the Pacifist's Dilemma. If a government imposes a cost on an aggressor that is large enough to cancel out his gains—say, a penalty that is three times the advantage of aggressing over being peaceful—it flips the appeal of the two choices of the potential aggressor, making peace more attractive than war (figure 10–2).

In addition to changing the rational-actor arithmetic, a Leviathan—or his female counterpart Justitia, the goddess of justice—is a disinterested third party whose penalties are not inflated by the self-serving biases of the

Own choices	Other's choices: Pacifist	Other's choices: Aggressor
Pacifist	Peace (5) Peace (5)	Defeat (–100) Victory – Penalty (10 – 15 = –5)
Aggressor	Victory – Penalty (10 – 15 = –5) Defeat (–100)	War – Penalty (–50 – 150 = –200) War – Penalty (–50 – 150 = –200)

FIGURE 10–2. How a Leviathan resolves the Pacifist's Dilemma

participants, and who is not a deserving target of revenge. A referee hovering over the game gives one's opponent less of an incentive to strike preemptively or self-defensively, reducing one's own desire to maintain an aggressive stance, putting the adversary at ease, and so on, and thus can ramp down the cycle of belligerence. And thanks to the generalized effects of self-control that have been demonstrated in the psychology lab, refraining from aggression can become a habit, so the civilized parties will inhibit their temptation to aggress even when Leviathan's back is turned.

Leviathan effects lay behind the Pacification and Civilizing Processes that gave chapters 2 and 3 their names. When bands, tribes, and chiefdoms came under the control of the first states, the suppression of raiding and feuding reduced their rates of violent death fivefold (chapter 2). And when the fiefs of Europe coalesced into kingdoms and sovereign states, the consolidation of law enforcement eventually brought down the homicide rate another thirtyfold (chapter 3). Pockets of anarchy that lay beyond the reach of government retained their violent cultures of honor, such as the peripheral and mountainous backwaters of Europe, and the frontiers of the American South and West (chapter 3). The same is true of the pockets of anarchy in the socioeconomic landscape, such as the lower classes who are deprived of consistent law enforcement and the purveyors of contraband who cannot avail themselves of it (chapter 3). When law enforcement retreats, such as in instant decolonization, failed states, anocracies, police strikes, and the 1960s, violence can come roaring back (chapters 3 and 6). Inept governance turns out to be among the biggest risk factors for civil war, and is perhaps the principal asset that distinguishes the violence-torn developing world from the more peaceful developed world (chapter 6). And when the citizens of a country with a weak rule of law are invited into the lab, they indulge in gratuitous spiteful punishment that leaves everyone worse off (chapter 8).

Leviathan, in the depiction that Hobbes commissioned, and Justitia, as represented in courthouse statuary, are both armed with swords. But sometimes the blindfold and the scales are enough. People avoid hits to their reputations as well as to their bodies and bank accounts, and occasionally the soft power of influential third parties or the threat of shaming and ostracism can have the same effect as police or armies that threaten them with force. This soft power is crucial in the international arena, where world government has always been a fantasy, but in which judgments by third parties, intermittently backed by sanctions or symbolic displays of force, can go a long way. The lowered risk of war when countries belong to international organizations or host international peacekeepers are two quantifiable examples of the pacifying effects of unarmed or lightly armed third parties (chapters 5 and 6).

When Leviathan does brandish a sword, the benefit depends on its applying the force judiciously, adding penalties only to the "aggression" cells in its subjects' decision matrix. When the Leviathan adds penalties indiscriminately to all four cells, brutalizing its subjects to stay in power, it can cause as much harm as it prevents (chapters 2 and 4). The benefits of democracies over autocracies and anocracies come when a government carefully eyedrops just enough force into the right cells of the decision matrix to switch the pacifist option from an agonizingly unattainable ideal to the irresistible choice.

GENTLE COMMERCE

The idea that an exchange of benefits can turn zero-sum warfare into positive-sum mutual profit was one of the key ideas of the Enlightenment, and it was revived in modern biology as an explanation of how cooperation among nonrelatives evolved. It changes the Pacifist's Dilemma by sweetening the outcome of mutual pacifism with the mutual gains of exchange (figure 10–3).

Though gentle commerce does not eliminate the disaster of being defeated in an attack, it eliminates the adversary's incentive to attack (since he benefits

Own choices	Other's choices: Pacifist	Other's choices: Aggressor
Pacifist	Peace + Profit (5 + 100 = 105) Peace + Profit (5 + 100 = 105)	Defeat (–100) Victory (10)
Aggressor	Victory (10) Defeat (–100)	War (–50) War (–50)

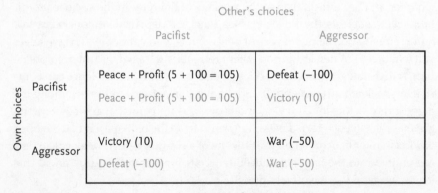

FIGURE 10–3. How commerce resolves the Pacifist's Dilemma

from peaceful exchange too) and so takes that worry off the table. The profitability of mutual cooperation is at least partly exogenous because it depends on more than the agents' willingness to trade: it depends as well on whether each one specializes in producing something the other one wants, and on the presence of an infrastructure that lubricates their exchange, such as transportation, finance, record-keeping, and the enforcement of contracts. And once people are enticed into voluntary exchange, they are encouraged to take each other's perspectives to clinch the best deal ("the customer is always right"), which in turn may lead them to respectful consideration of each other's interests, if not necessarily to warmth.

In the theory of Norbert Elias, the Leviathan and gentle commerce were the two drivers of the European Civilizing Process (chapter 3). Beginning in the late Middle Ages, expanding kingdoms not only penalized plunder and nationalized justice, but supported an infrastructure of exchange, including money and the enforcement of contracts. This infrastructure, together with technological advances such as in roads and clocks, and the removal of taboos on interest, innovation, and competition, made commerce more attractive, and as a result merchants, craftsmen, and bureaucrats displaced knightly warriors. The theory has been supported by historical data showing that commerce did start to expand in the late Middle Ages, and by criminological data showing that rates of violent death really did plunge (chapters 9 and 3).

Among larger entities such as cities and states, commerce was enhanced by oceangoing ships, new financial institutions, and a decline in mercantilist policies. These developments have been credited in part with the 18th-century domestication of warring imperial powers such as Sweden, Denmark, the Netherlands, and Spain into commercial states that made less trouble (chapter 5). Two centuries later the transformation of China and Vietnam from authoritarian communism to authoritarian capitalism was accompanied by a decreased willingness to indulge in the all-out ideological wars that in the preceding decades had made both countries the deadliest places on earth (chapter 6). In other parts of the world as well, the tilting of values away from national glory and toward making money may have taken the wind out of the sails of cantankerous revanchist movements (chapters 5 and 6). Part of the tilt may have come from a relaxation of the grip of ideologies that came to be seen as morally bankrupt, but another part may have come from a seduction by the lucrative rewards of the globalized economy.

These narratives have been supported by quantitative studies. During the postwar decades that saw the Long Peace and the New Peace, international trade skyrocketed, and we saw that countries that trade with each other are less likely to cross swords, holding all else constant (chapter 5). Recall as well that countries that are more open to the world economy are less likely to host genocides and civil wars (chapter 6). Pulling in the other direction, governments that base their nation's wealth on digging oil, minerals, and diamonds out of the ground rather than adding value to it via commerce and trade are more likely to fall into civil wars (chapter 6).

The theory of gentle commerce is not only supported by numbers from international datasets but is consistent with a phenomenon long known to anthropologists: that many cultures maintain active networks of exchange, even when the goods exchanged are useless gifts, because they know it helps keep the peace among them.[5] This is one of the phenomena in the ethnographic record that led Alan Fiske and his collaborators to suggest that people in a relationship of Equality Matching or Market Pricing feel that they are bound by mutual obligations and are less likely to dehumanize each other than when they are in a null or asocial relationship (chapter 9).

The mindset behind gentle commerce, unlike that of the other pacifying forces I review in this chapter, has not been directly tested in the psychology lab. We do know that when people (and for that matter, monkeys) are joined in a positive-sum game requiring them to collaborate in order to achieve a goal that benefits them both, hostile tensions can dissolve (chapter 8). We also know that exchange in the real world can be a lucrative positive-sum game. But we don't know whether exchange itself reduces hostile tensions. As far as I know, in the vast literature on empathy and cooperation and aggression, no one has tested whether people who have consummated a mutually profitable exchange are less likely to shock each other or to spike each other's food with three-alarm hot sauce. I suspect that among researchers, gentle commerce is just not a sexy idea. Cultural and intellectual elites have always felt superior to businesspeople, and it doesn't occur to them to credit mere merchants with something as noble as peace.[6]

FEMINIZATION

Depending on how you look at it, the late Tsutomu Yamaguchi is either the world's luckiest man or the world's unluckiest man. Yamaguchi survived the atomic blast at Hiroshima, but then made an unfortunate choice as to where to go to flee the devastation: Nagasaki. He survived that blast as well and lived another sixty-five years, passing away in 2010 at the age of ninety-three. A man who survived the only two nuclear strikes in history deserves our respectful attention, and before he died he offered a prescription for peace in the nuclear age: "The only people who should be allowed to govern countries with nuclear weapons are mothers, those who are still breast-feeding their babies."[7]

Yamaguchi was invoking the most fundamental empirical generalization about violence, that it is mainly committed by men. From the time they are boys, males play more violently than females, fantasize more about violence, consume more violent entertainment, commit the lion's share of violent crimes, take more delight in punishment and revenge, take more foolish risks in aggressive attacks, vote for more warlike policies and leaders, and plan and carry out almost all the wars and genocides (chapters 2, 3, 7, and 8). Even when the sexes overlap and the difference between their averages is small, the difference can decide a close election, or set off a spiral of belligerence in which each side has to be a bit more bellicose than the other. Historically, women have taken the leadership in pacifist and humanitarian movements out of proportion to their influence in other political institutions of the time, and recent decades, in which women and their interests have had an unprecedented influence in all walks of life, are also the decades in which wars between developed states became increasingly unthinkable (chapters 5 and 7). James Sheehan's characterization of the postwar transformation of the mission of the European state, from military prowess to cradle-to-grave nurturance, is almost a caricature of traditional gender roles.

Yamaguchi's exact prescription, of course, can be debated. George Shultz recalls that when he told Margaret Thatcher in 1986 that he had stood by as Ronald Reagan suggested to Mikhail Gorbachev that they abolish nuclear weapons, she clobbered him with her handbag.[8] But, Yamaguchi might reply, Thatcher's own children were already grown up, and in any case her views were tuned to a world that was run by men. Since the world's nuclear states will not all be governed by women anytime soon, let alone by nursing mothers, we will never know whether Yamaguchi's prescription is right. But he had a point when he speculated that a more feminized world is a more peaceful world.

Female-friendly values may be expected to reduce violence because of the psychological legacy of the basic biological difference between the sexes, namely that males have more of an incentive to compete for sexual access to females, while females have more of an incentive to stay away from risks that would make their children orphans. Zero-sum competition, whether it takes the form of the contests for women in tribal and knightly societies or the contests for honor, status, dominance, and glory in modern ones, is more a man's obsession than a woman's. Suppose that in the Pacifist's Dilemma, some portion of the rewards of victory and the costs of defeat—say, 80 percent—consists of the swelling and bruising of the male ego. And suppose that the choices are now made by female actors, so these psychic payoffs are reduced accordingly (figure 10–4; I have omitted the symmetrical Other's Choices for clarity). Now peace is more tempting than victory, and war more costly than defeat. The pacifist option wins hands-down. The reversal would be even more dramatic if we adjusted the war cell to reflect a greater cost of violent conflict to women than to men.

To be sure, a shift from male to female influence in decision-making may not be completely exogenous. In a society in which rapacious invaders may swoop in at any moment, the costs of defeat to both sexes can be catastrophic, and anything short of the most truculent martial values may be suicidal. A female-tilted value system may be a luxury enjoyed by a society that is already

		Other's choices	
		Pacifist	Aggressor
Own choices	Pacifist	Peace (5)	Defeat without humiliation (−100 + 80 = −20)
	Aggressor	Victory without glory (10 − 8 = 2)	War (−50)

FIGURE 10–4. How feminization can resolve the Pacifist's Dilemma

safe from predatory invasion. But a relative tilt in power toward women's interests can also be caused by exogenous forces that have nothing to do with violence. In traditional societies, one of these forces is living arrangements: women are better off in societies in which they stay with their birth family under the wing of their fathers and brothers, and their husbands are visitors, than in societies in which they move in with their husband's clan and are dominated by their husbands and his kin (chapter 7). In modern societies, the exogenous forces include technological and economic advances that freed women from chronic child-rearing and domestic duties, such as store-bought food, labor-saving devices, contraception, longer life spans, and the shift to an information economy.

Societies in which women get a better deal, both traditional and modern, tend to be societies that have less organized violence (chapter 8). This is obvious enough in the tribes and chiefdoms that literally go to war to abduct women or avenge past abductions, such as the Yanomamö and the Homeric Greeks (chapters 1 and 2). But it may also be seen among contemporary countries in the contrast between the low levels of political and judicial violence in the über-feminist democracies of Western Europe and the high levels in the genital-cutting, adulteress-stoning, burqa-cladding Sharia states of Islamic Africa and Asia (chapter 6).

Feminization need not consist of women literally wielding more power in decisions on whether to go to war. It can also consist in a society moving away from a culture of manly honor, with its approval of violent retaliation for insults, toughening of boys through physical punishment, and veneration of martial glory (chapter 8). This has been the trend in the democracies of Europe and the developed world and in the bluer states of America (chapters 3 and 7). Several conservative scholars have ruefully suggested to me that the modern West has been diminished by the loss of virtues like bravery and valor and the ascendancy of materialism, frivolity, decadence, and effeminacy. Now, I have been assuming that violence is always a bad thing except when it prevents greater violence, but these men are correct that this is a value judgment, and that no logical argument inherently favors peace over honor and glory. But I would think that the potential victims of all this manliness deserve a say in this discussion, and they may not agree that their lives and limbs are a price worth paying for the glorification of masculine virtues.

Feminization is a pacifying development for yet another reason. Social and sexual arrangements that favor the interests of women tend to drain the swamps where violent male-male competition proliferates. One of these arrangements is marriage, in which men commit themselves to investing in the children they sire rather than competing with each other for sexual opportunities. Getting married reduces men's testosterone and their likelihood of living a life of crime, and we saw that American homicide rates plunged in the marriage-happy 1940s and 1950s, rose in the marriage-delaying 1960s and 1970s, and remain high in African American communities that have particularly low rates of marriage (chapter 3).

Another swamp-drainer is equality in numbers. Unpoliced all-male social milieus, such as the cowboy and mining camps of the American frontier, are almost always violent (chapter 3). The West was wild because it was young men who went there while the young women stayed behind in the East. But societies can become stacked with males for a more sinister reason, namely that their female counterparts were aborted or killed at birth. In an article called "A Surplus of Men, a Deficit of Peace," the political scientists Valerie Hudson and Andrea den Boer show that the traditional killing of baby girls in China has long resulted in large numbers of unattached men.[9] They are always poor men, because the richer ones attract the scarce women. These "bare branches," as they are called in China, congregate in gangs of drifters who brawl and duel among themselves and rob and terrorize settled populations. They can even grow into armies that menace local or national governments. A leader can clamp down on the gangs by violent repression, or he can try to co-opt them, which usually requires adopting a macho ruling philosophy that is congenial to their mores. Best of all, he can export their destructive energy by sending them to other territories as migrant workers, colonists, or soldiers. When the leaders of rival countries all try to dispose of their excess men, the result can be a grinding war of attrition. As Hudson and den Boer put it, "Each society has plenty of bare branches to spare in such a conflict—and the respective governments might be happy to spare them."[10]

Traditional gynecide, joined in the 1980s by the female-abortion industry, injected a bolus of excess males into the population structures of Afghanistan, Bangladesh, China, Pakistan, and parts of India (chapter 7).[11] These surpluses of men bode poorly for the immediate prospects of peace and democracy in those regions. Over the longer term, the sex ratio may eventually be rebalanced by the feminist and humanitarian concern with the right of female fetuses to take their first breath, together with political leaders' finally grasping the demographic arithmetic and enhancing the incentives to raise daughters. The resulting boon for baby girls would translate into less violent societies. But until the first fifty-fifty cohorts are born and grow up, those societies may be in for a bumpy ride.

A society's respect for the interests of women has one more connection to its rate of violence. Violence is a problem not just of too many males but of too many *young* males. At least two large studies have suggested that countries with a larger proportion of young men are more likely to fight interstate and civil wars (chapter 6).[12] A population pyramid with a thick base of young people is dangerous not just because young men like to raise hell, and in bottom-heavy societies will outnumber their more prudent elders. It's also dangerous because these young men are likely to be deprived of status and mates. The sclerotic economies of countries in the developing world cannot nimbly put a youth bulge to work, leaving many of the men unemployed or underemployed. And if the society has a degree of official or de facto polygyny, with many young women being usurped by older or richer men, the surfeit of marginalized young people will turn into a surfeit of marginalized young men. These men have nothing to lose, and may find work and meaning in militias, warlord gangs, and terrorist cells (chapter 6).

The title *Sex and War* sounds like the ultimate guy bait, but this recent book is a manifesto for the empowerment of women.[13] The reproductive biologist Malcolm Potts, writing with the political scientist Martha Campbell and the journalist Thomas Hayden, has amassed evidence that when women are given access to contraception and the freedom to marry on their own terms, they have fewer offspring than when the men of their societies force them to be baby factories. And that, in turn, means that their countries' populations will be less distended by a thick slab of young people at the bottom. (Contrary to an earlier understanding, a country does not have to become affluent before its rate of population growth comes down.) Potts and his coauthors argue that giving women more control over their reproductive capacity (always the contested territory in the biological battle of the sexes) may be the single most effective way of reducing violence in the dangerous parts of the world today. But this empowerment often must proceed in the teeth of opposition from traditional men who want to preserve their control over female reproduction, and from religious institutions that oppose contraception and abortion.

Several varieties of feminization, then—direct political empowerment, the deflation of manly honor, the promotion of marriage on women's terms, the right of girls to be born, and women's control over their own reproduction—have been forces in the decline of violence. The parts of the world that lag in this historical march are the parts that lag in the decline of violence. But worldwide polling data show that even in the most benighted countries there is considerable pent-up demand for female empowerment, and many international organizations are committed to hurrying it along (chapters 6 and 7). These are hopeful signs in the long term, if not the immediate term, for further reductions in violent conflict in the world.

THE EXPANDING CIRCLE

The last two pacifying forces scramble the psychological payoffs of violence. The first is the expansion of the circle of sympathy. Suppose that living in a more cosmopolitan society, one that puts us in contact with a diverse sample of other people and invites us to take their points of view, changes our emotional response to their well-being. Imagine taking this change to its logical conclusion: our own well-being and theirs have become so intermingled that we literally love our enemies and feel their pain. Our potential adversary's payoffs would simply be summed with our own (and vice versa), and pacifism would become overwhelmingly preferable to aggression (figure 10–5).

Of course, a perfect fusion of the interests of every living human is an unattainable nirvana. But smaller increments in the valuation of other people's interests—say, a susceptibility to pangs of guilt when thinking about enslaving, torturing, or annihilating others—could shift the likelihood of aggressing against them.

We have seen evidence for both links in this causal chain: exogenous events that expanded opportunities for perspective-taking, and a psychological response that turns perspective-taking into sympathy (chapters 4 and 9). Beginning in the 17th century, technological advances in publishing and transportation created a Republic of Letters and a Reading Revolution in which the seeds of the Humanitarian Revolution took root (chapter 4). More people read books, including fiction that led them to inhabit the minds of other people, and satire that led them to question their society's norms. Vivid depictions of the suffering wrought by slavery, sadistic punishments, war, and cruelty to children

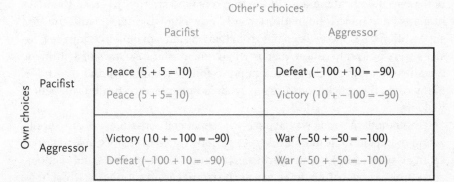

Own choices	Other's choices: Pacifist	Other's choices: Aggressor
Pacifist	Peace (5 + 5 = 10) Peace (5 + 5= 10)	Defeat (–100 + 10 = –90) Victory (10 + –100 = –90)
Aggressor	Victory (10 + –100 = –90) Defeat (–100 + 10 = –90)	War (–50 + –50 = –100) War (–50 + –50 = –100)

FIGURE 10–5. How empathy and reason resolve the Pacifist's Dilemma

and animals preceded the reforms that outlawed or reduced those practices. Though chronology does not prove causation, the laboratory studies showing that hearing or reading a first-person narrative can enhance people's sympathy for the narrator at least make it plausible (chapter 9).

Literacy, urbanization, mobility, and access to mass media continued their rise in the 19th and 20th centuries, and in the second half of the 20th a Global Village began to emerge that made people even more aware of others unlike themselves (chapters 5 and 7). Just as the Republic of Letters and the Reading Revolution helped to kindle the Humanitarian Revolution of the 18th century, the Global Village and the electronics revolution may have helped along the Long Peace, New Peace, and Rights Revolutions of the 20th. Though we cannot prove the common observation that media coverage accelerated the civil rights movement, antiwar sentiment, and the fall of communism, the perspective-sympathy studies are suggestive, and we saw several statistical links between the cosmopolitan mixing of peoples and the endorsement of humanistic values (chapters 7 and 9).[14]

THE ESCALATOR OF REASON

The expanding circle and the escalator of reason are powered by some of the same exogenous causes, particularly literacy, cosmopolitanism, and education.[15] And their pacifying effect may be depicted by the same fusion of interests in the Pacifist's Dilemma. But the expanding circle (as I have been using the term) and the escalator of reason are conceptually distinct (chapter 9). The first involves occupying another person's vantage point and imagining his or her emotions as if they were one's own. The second involves ascending to an Olympian, superrational vantage point—the perspective of eternity, the view from nowhere—and considering one's own interests and another person's as equivalent.

The escalator of reason has an additional exogenous source: the nature of reality, with its logical relationships and empirical facts that are independent of the psychological makeup of the thinkers who attempt to grasp them. As humans have honed the institutions of knowledge and reason, and purged superstitions and inconsistencies from their systems of belief, certain conclusions were bound to follow, just as when one masters the laws of arithmetic certain sums and products are bound to follow (chapters 4 and 9). And in many cases the conclusions are ones that led people to commit fewer acts of violence.

Throughout the book we have seen the beneficial consequences of an application of reason to human affairs. At various times in history superstitious killings, such as in human sacrifice, witch hunts, blood libels, inquisitions, and ethnic scapegoating, fell away as the factual assumptions on which they rested crumbled under the scrutiny of a more intellectually sophisticated populace (chapter 4). Carefully reasoned briefs against slavery, despotism, torture, religious persecution, cruelty to animals, harshness to children, violence against women, frivolous wars, and the persecution of homosexuals were not just hot air but entered into the decisions of the people and institutions who attended to the arguments and implemented reforms (chapters 4 and 7).

Of course it's not always easy to distinguish empathy from reason, the heart from the head. But the limited reach of empathy, with its affinity for people like us and people close to us, suggests that empathy needs the universalizing boost of reason to bring about changes in policies and norms that actually reduce violence in the world (chapter 9). These changes include not just legal prohibitions against acts of violence but institutions that are engineered to reduce the temptations of violence. Among these wonkish contraptions are democratic government, the Kantian safeguards against war, reconciliation movements in the developing world, nonviolent resistance movements, international peacekeeping operations, the crime prevention reforms and civilizing offensives of the 1990s, and tactics of containment, sanctions, and wary engagement designed to give national leaders more options than just the game of chicken that led to the First World War or the appeasement that led to the Second (chapters 3 to 8).

A broader effect of the escalator of reason, albeit one with many stalls, reversals, and holdouts, is the movement away from tribalism, authority, and purity in moral systems and toward humanism, classical liberalism, autonomy, and human rights (chapter 9). A humanistic value system, which privileges human flourishing as the ultimate good, is a product of reason because it can be *justified:* it can be mutually agreed upon by any community of thinkers who value their own interests and are engaged in reasoned negotiation, whereas communal and authoritarian values are parochial to a tribe or hierarchy (chapters 4 and 9).

When cosmopolitan currents bring diverse people into discussion, when freedom of speech allows the discussion to go where it pleases, and when history's failed experiments are held up to the light, the evidence suggests that value systems evolve in the direction of liberal humanism (chapters 4 to 9). We saw this in the recent decline of totalitarian ideologies and the genocides and wars they ignited, and we saw it in the contagion of the Rights Revolutions, when the indefensibility of oppressing racial minorities was generalized to the oppression of women, children, homosexuals, and animals (chapter 7). We saw it as well in the way that these revolutions eventually swept up the conservatives who first opposed them. The exception that proves the rule is the insular societies that are starved of ideas from the rest of the world and muzzled by governmental and clerical repression of the press: these are also the societies that most stubbornly resist humanism and cling to their tribal, authoritarian, and religious ideologies (chapter 6). But even these societies may not be able to withstand the liberalizing currents of the new electronic Republic of Letters forever.

The metaphor of an escalator, with its implication of directionality superimposed on the random walk of ideological fashion, may seem Whiggish and presentist and historically naïve. Yet it is a kind of Whig history that is supported by the facts. We saw that many liberalizing reforms that originated in Western Europe or on the American coasts have been emulated, after a time lag, by the more conservative parts of the world (chapters 4, 6, and 7). And we saw correlations, and even a causal relation or two, between a well-developed ability to reason and a receptiveness to cooperation, democracy, classical liberalism, and nonviolence (chapter 9).

REFLECTIONS

The decline of violence may be the most significant and least appreciated development in the history of our species. Its implications touch the core of our beliefs and values—for what could be more fundamental than an understanding of whether the human condition, over the course of its history, has gotten steadily better, steadily worse, or has not changed? Hanging in the balance are conceptions of a fall from innocence, of the moral authority of religious scripture and hierarchy, of the innate wickedness or benevolence of human nature, of the forces that drive history, and of the moral valuation of nature, community, tradition, emotion, reason, and science. My attempt to document and explain declines of violence has filled many pages, and this is not the place to fill many more in exploring their implications. But I will end with two reflections on what one might take away from the historical decline of violence.

The first concerns the way we should view modernity—the transformation of human life by science, technology, and reason, with the attendant diminishment of custom, faith, community, traditional authority, and embeddedness in nature.

A loathing of modernity is one of the great constants of contemporary social criticism. Whether the nostalgia is for small-town intimacy, ecological sustainability, communitarian solidarity, family values, religious faith, primitive communism, or harmony with the rhythms of nature, everyone longs to turn back the clock. What has technology given us, they say, but alienation, despoliation, social pathology, the loss of meaning, and a consumer culture that is destroying the planet to give us McMansions, SUVs, and reality television?

Lamentations of a fall from Eden have a long history in intellectual life, as the historian Arthur Herman has shown in *The Idea of Decline in Western History*.[16] And ever since the 1970s, when romantic nostalgia became the conventional wisdom, statisticians and historians have marshaled facts against it. The titles of their books tell the story: *The Good News Is the Bad News Is Wrong*,

It's Getting Better All the Time, The Good Old Days—They Were Terrible!, The Case for Rational Optimism, The Improving State of the World, The Progress Paradox, and most recently, Matt Ridley's *The Rational Optimist* and Charles Kenny's *Getting Better*.[17]

These defenses of modernity recount the trials of daily living before the advent of affluence and technology. Our ancestors, they remind us, were infested with lice and parasites and lived above cellars heaped with their own feces. Food was bland, monotonous, and intermittent. Health care consisted of the doctor's saw and the dentist's pliers. Both sexes labored from sunrise to sundown, whereupon they were plunged into darkness. Winter meant months of hunger, boredom, and gnawing loneliness in snowbound farmhouses.

But it was not just mundane physical comforts that our recent ancestors did without. It was also the higher and nobler things in life, such as knowledge, beauty, and human connection. Until recently most people never traveled more than a few miles from their place of birth. Everyone was ignorant of the vastness of the cosmos, the prehistory of civilization, the genealogy of living things, the genetic code, the microscopic world, and the constituents of matter and life. Musical recordings, affordable books, instant news of the world, reproductions of great art, and filmed dramas were inconceivable, let alone available in a tool that can fit in a shirt pocket. When children emigrated, their parents might never see them again, or hear their voices, or meet their grandchildren. And then there are modernity's gifts of life itself: the additional decades of existence, the mothers who live to see their newborns, the children who survive their first years on earth. When I stroll through old New England graveyards, I am always struck by the abundance of tiny plots and poignant epitaphs. "Elvina Maria, died July 12, 1845; aged 4 years, and 9 months. *Forgive this tear, a parent weeps. 'Tis here, the faded floweret sleeps.*"

Even with all these reasons why no romantic would really step into a time machine, the nostalgic have always been able to pull out one moral card: the profusion of modern violence. At least, they say, our ancestors did not have to worry about muggings, school shootings, terrorist attacks, holocausts, world wars, killing fields, napalm, gulags, and nuclear annihilation. Surely no Boeing 747, no antibiotic, no iPod is worth the suffering that modern societies and their technologies can wreak.

And here is where unsentimental history and statistical literacy can change our view of modernity. For they show that nostalgia for a peaceable past is the biggest delusion of all. We now know that native peoples, whose lives are so romanticized in today's children's books, had rates of death from warfare that were greater than those of our world wars. The romantic visions of medieval Europe omit the exquisitely crafted instruments of torture and are innocent of the thirtyfold greater risk of murder in those times. The centuries for which people are nostalgic were times in which the wife of an adulterer could have her nose cut off, children as young as eight could be hanged for property crimes, a prisoner's family could be charged for easement of irons, a witch could be sawn in half, and a sailor could be flogged to a pulp. The moral commonplaces of our age, such as that slavery, war, and torture are wrong, would have been seen as saccharine sentimentality, and our notion of universal human rights almost incoherent. Genocide and war crimes were absent from the historical record only because no one at the time thought they were a big deal. From the vantage point of almost seven decades after the world wars and genocides of the first half of the 20th century, we see that they were not harbingers of worse to come, nor a new normal to which the world would grow inured, but a local high from which it would bumpily descend. And the ideologies behind them were not woven into modernity but atavisms that ended up in the dustbin of history.

The forces of modernity—reason, science, humanism, individual rights—have not, of course, pushed steadily in one direction; nor will they ever bring about a utopia or end the frictions and hurts that come with being human. But on top of all the benefits that modernity has brought us in health, experience, and knowledge, we can add its role in the reduction of violence.

To writers who *have* noticed declines of violence, the sheer abundance of them, operating on so many scales of time and magnitude, has an aura of mystery. James Payne wrote of a temptation to allude to "a higher power at work," of a process that seems "almost magical."[18] Robert Wright nearly succumbs to the temptation, wondering whether the decline of zero-sum competition is "evidence of divinity," signs of a "divinely imparted meaning," or a story with a "cosmic author."[19]

I can easily resist the temptation, but agree that the multiplicity of datasets in which violence meanders downward is a puzzle worth pondering. What do we make of the impression that human history contains an arrow? Where is this arrow, we are entitled to wonder, and who posted it? And if the alignment of so many historical forces in a beneficial direction does not imply a divine sign painter, might it vindicate some notion of moral realism—that moral truths are out there somewhere for us to discover, just as we discover the truths of science and mathematics?[20]

My own view is that the Pacifist's Dilemma at least clarifies the mystery, and shows how the nonrandom direction of history is rooted in an aspect of reality that informs our conceptions of morality and purpose. Our species was born into the dilemma because our ultimate interests are distinct, because our vulnerable bodies make us sitting ducks for exploitation, and because the enticements to being the exploiter rather than the exploited will sentence all sides to punishing conflict. Unilateral pacifism is a losing strategy, and joint peace is out of everyone's reach. These maddening contingencies are inherent in the mathematical structure of the payoffs, and in that sense they are in the nature of reality. It is no wonder that the ancient Greeks blamed their wars on the caprice of the gods, or that the Hebrews and Christians appealed to a moralistic deity who might jigger the payoffs in the next world and thereby change the perceived incentive structure in this one.

Human nature, as evolution left it, is not up to the challenge of getting us into the blessedly peaceful cell in the upper left corner of the matrix. Motives like greed, fear, dominance, and lust keep drawing us toward aggression. And though a major work-around, the threat of tit-for-tat vengeance, has the potential to bring about cooperation if the game is repeated, in practice it is miscalibrated by self-serving biases and often results in cycles of feuding rather than stable deterrence.

But human nature also contains motives to climb into the peaceful cell, such as sympathy and self-control. It includes channels of communication such as language. And it is equipped with an open-ended system of combinatorial reasoning. When the system is refined in the crucible of debate, and its products are accumulated through literacy and other forms of cultural memory, it can think up ways of changing the payoff structure and make the peaceful cell increasingly attractive. Not least among those tactics is the super-rational appeal to another abstract feature of reality: the interchangeability of perspectives, the nonspecialness of our parochial vantage points, which corrodes the dilemma by blending the payoffs of the two antagonists.

Only an inflated sense of our own importance could turn our desire to escape the Pacifist's Dilemma into a grand purpose of the cosmos. But the desire does seem to tap into contingencies of the world that are not exactly physical, and so it is different from the desires that were the mothers of other inventions such as refined sugar or central heating. The maddening structure of a Pacifist's Dilemma is an abstract feature of reality. So is its most comprehensive solution, the interchangeability of perspectives, which is the principle behind the Golden Rule and its equivalents that have been rediscovered in so many moral traditions. Our cognitive processes have been struggling with these aspects of reality over the course of our history, just as they have struggled with the laws of logic and geometry.

Though our escape from destructive contests is not a cosmic purpose, it *is* a human purpose. Defenders of religion have long claimed that in the absence of divine edicts, morality can never be grounded outside ourselves. People can pursue only selfish interests, perhaps tweaked by taste or fashion, and are sentenced to lives of relativism and nihilism. We can now appreciate why this line of argument is mistaken. Discovering earthly ways in which human beings can flourish, including stratagems to overcome the tragedy of the inherent appeal of aggression, should be purpose enough for anyone. It is a goal that is nobler than joining a celestial choir, melting into a cosmic spirit, or being reincarnated into a higher life-form, because the goal can be justified to any fellow thinker rather than being inculcated to arbitrary factions by charisma, tradition, or force. And the data we have seen in this book show that it is a goal on which progress can be made—progress that is halting and incomplete, but unmistakable nonetheless.

A final reflection. In writing this book I have adopted a voice that is analytic, and at times irreverent, because I believe the topic has inspired too much piety and not enough understanding. But at no point have I been unaware of the reality behind the numbers. To review the history of violence is to be repeatedly astounded by the cruelty and waste of it all, and at times to be overcome with anger, disgust, and immeasurable sadness. I know that behind the graphs there is a young man who feels a stab of pain and watches the life drain slowly out of him, knowing he has been robbed of decades of existence. There is a victim of torture whose contents of consciousness have been replaced by unbearable agony, leaving room only for the desire that consciousness itself should cease. There is a woman who has learned that her husband, her father, and her brothers lie dead in a ditch, and who will soon "fall into the hand of hot and forcing violation."[21] It would be terrible enough if these ordeals befell one person, or ten, or a hundred. But the numbers are not in the hundreds, or the thousands, or even the millions, but in the hundreds of millions—an order of magnitude that the mind staggers to comprehend, with deepening horror as it comes to realize just how much suffering has been inflicted by the naked ape upon its own kind.[22]

Yet while this planet has gone cycling on according to the fixed law of gravity, that species has also found ways to bring the numbers down, and allow a greater and greater proportion of humanity to live in peace and die of natural causes.[23] For all the tribulations in our lives, for all the troubles that remain in the world, the decline of violence is an accomplishment we can savor, and an impetus to cherish the forces of civilization and enlightenment that made it possible.

NOTES

Preface

1. Estimating probability from availability in memory: Slovic, 1987; Tversky & Kahneman, 1973.
2. Long Peace: Coined by Gaddis, 1986.
3. Discussions of the decline of violence in my previous books: Pinker, 1997, pp. 518–19; Pinker, 2002, pp. 166–69, 320, 330–36.
4. Other books on the decline of violence: Elias, 1939/2000; Human Security Report Project, 2011; Keeley, 1996; Muchembled, 2009; Mueller, 1989; Nazaretyan, 2010; Payne, 2004; Singer, 1981/2011; Wright, 2000; Wood, 2004.

Chapter 1: A Foreign Country

1. Survey data: Bennett Haselton and I presented 265 Internet users with five pairs of historical periods and asked them which they thought had higher rates of violent death: prehistoric hunter-gatherer bands or the first states; contemporary hunter-gatherer bands or modern Western societies; homicide in 14th-century England or 20th-century England; warfare in the 1950s or the 2000s; homicide in the United States in the 1970s or the 2000s. In each case respondents thought the later culture was more violent, by a factor of 1.1 to 4.6. In each case, as we shall see, the earlier culture was more violent, by a factor of 1.6 to more than 30.
2. Ötzi: B. Cullen, "Testimony from the Iceman," *Smithsonian*, Feb. 2003; C. Holden, "Iceman's final hours," *Science, 316*, Jun. 1, 2007, p. 1261.
3. Kennewick Man: McManamon, 2004; C. Holden, "Random samples," *Science, 279*, Feb. 20, 1998, p. 1137.
4. Lindow Man: Joy, 2009.
5. Severed skull with preserved brain: "2000-year-old brain found in Britain," *Boston Globe*, Dec. 13, 2008.
6. Murdered family: C. Holden, "A family affair," *Science, 322*, Nov. 21, 2008, p. 1169.
7. Homeric warfare: Gottschall, 2008.
8. Agamemnon encourages genocide: Homer, 2003, p. 101.
9. "Fast ships with shallow drafts": Gottschall, 2008, p. 1.
10. "Breached with surprising ease": Gottschall, 2008, pp. 143–44.
11. "Many sleepless nights": *Iliad* 9.325–27, quoted in Gottschall, p. 58.
12. The Hebrew Bible: Kugel, 2007.
13. "Should our sister": Genesis 34:25–31.
14. "kill every male": Numbers 31.
15. "Thou shalt save alive nothing": Deuteronomy 20:16–17.
16. "utterly destroyed" Jericho: Joshua 6.
17. "destroyed all that breathed": Joshua 10:40–41.
18. "Now go and smite Amalek": 1 Samuel 15:3.
19. Saul plots to have him killed: 1 Samuel 18:7.
20. "wasted the country": 1 Chronicles 20:1–3.
21. Solomon's "divine wisdom": 1 Kings 3:23–28.
22. Quantifying biblical homicide: Schwager, 2000, pp. 47, 60.
23. Victims of the Noachian flood: Biblical literalists date the flood to around 2300 BCE. McEvedy & Jones, 1978, estimate that the world contained around 14 million people in 3000 BCE and 27 million in 2000 BCE.

24. Real and fictitious biblical history: Kugel, 2007.
25. Authorship of the Christian Bible: Ehrman, 2005.
26. Pagan Jesuses: B. G. Walker, "The other Easters," *Freethought Today*, Apr. 2008, pp. 6–7; Smith, 1952.
27. Roman entertainment: Kyle, 1998.
28. Forensics of crucifixion: Edwards, Gabel, & Hosmer, 1986.
29. Martyrologies: Gallonio, 1903/2004; Kay, 2000.
30. "The mother was present": Quoted in Gallonio, 1903/2004, p. 133.
31. Punishment for seven deadly sins: Lehner & Lehner, 1971.
32. Inquisition: Grayling, 2007; Rummel, 1994.
33. "As the levers bent forward": Quoted in Bronowski, 1973, p. 216.
34. Witch-burning statistics: Rummel, 1994.
35. Augustine on snakes and rotten branches: Grayling, 2007, p. 25.
36. "Burning men": John 15:6.
37. "Limiting ourselves to quantifiable instances": Kaeuper, 2000, p. 24.
38. "never killing a knight who begged for mercy": Quoted in Kaeuper, 2000, p. 31.
39. "take a lady or maiden in any way he desires": Quoted in Kaeuper, 2000, p. 30.
40. Henry V's ultimatum: *Henry V*, Act 3, Scene III.
41. Witch burned to death: Tatar, 2003, p. 207.
42. Grimm's fairy tales: Tatar, 2003.
43. Punch and Judy: Schechter, 2005, pp. 83–84.
44. Violence in nursery rhymes: Davies, Lee, Fox, & Fox, 2004.
45. Hamilton: Chernow, 2004.
46. Hamilton's duel: Krystal, 2007.
47. History of dueling: Krystal, 2007; Schwartz, Baxter, & Ryan, 1984.
48. Humor as a weapon against honor: Pinker, 1997, chap. 8.
49. Ridicule ended dueling: Stevens, 1940, pp. 280–83, quoted in Mueller, 1989, p. 10.
50. Martial culture and its decline: Sheehan, 2008; van Creveld, 2008.
51. Nonviolent German games: A. Curry, "Monopoly killer," *Wired*, Apr. 2009.
52. Decline of elite violence: Cooney, 1997.
53. Misogynistic ads: Ad Nauseam, 2000. The Chase & Sanborn ad ran in *Life* magazine on Aug. 11, 1952.
54. Changes in reactions to the word *rape:* Tom Jones, e-mail to the author, Nov. 19, 2010, reprinted with his permission.
55. Beating schoolchildren: Personal communication from British and Catholic friends; also S. Lyall, "Blaming church, Ireland details scourge of abuse: Report spans 60 years," *New York Times*, May 21, 2009.

Chapter 2: The Pacification Process

1. From a cartoon by Bob Mankoff.
2. Darwin, genetics, and game theory: Maynard Smith, 1998; Maynard Smith & Szathmáry, 1997.
3. "To a survival machine": Dawkins, 1976/1989, p. 66.
4. Animal violence: Williams, 1988; Wrangham, 1999a.
5. "three principal causes of quarrel": Hobbes, 1651/1957, p. 185.
6. Competition over females: Darwin, 1874; Trivers, 1972.
7. Misunderstandings of sexual selection: Pinker, 1997, 2002.
8. Security dilemma: Schelling, 1960.
9. "nothing can be more gentle than [man] in his primitive state": Rousseau, 1755/1994, pp. 61–62.
10. Peace and Harmony Mafia: Van der Dennen, 1995, 2005.
11. Chimpanzee violence: Goodall, 1986; Wilson & Wrangham, 2003; Wrangham, 1999a; Mitani, Watts, & Amsler, 2010.
12. Aggressive displays in animals: Maynard Smith, 1988; Wrangham, 1999a.
13. Goodall's shocking discovery: Goodall, 1986.
14. Lethal violence in chimpanzees: Wilson & Wrangham, 2003; Wrangham, 1999a; Wrangham, Wilson, & Muller, 2006.
15. Adaptive value of chimpicide: Wilson & Wrangham, 2003; Wrangham, 1999a; Wrangham & Peterson, 1996; Mitani et al., 2010.
16. Bonobos: de Waal & Lanting, 1997; Furuichi & Thompson, 2008; Wrangham & Peterson, 1996. Bonobos and popular culture: I. Parker, "Swingers," *New Yorker*, Jul. 30, 2007; M. Dowd, "The Baby Bust," *New York Times*, Apr. 10, 2002.
17. Bonobos as model of human ancestor: de Waal, 1996; de Waal & Lanting, 1997.
18. Bonobos in the wild: Furuichi & Thompson, 2008; Wrangham & Peterson, 1996; I. Parker, "Swingers," *New Yorker*, Jul. 30, 2007.
19. Bonobos as outliers: Wrangham & Pilbeam, 2001.
20. Sexual dimorphism and male-male competition: Plavcan, 2000.

21. *Ardipithecus ramidus:* White et al., 2009.
22. Sexual dimorphism and male-male competition in *Homo:* Plavcan, 2000; Wrangham & Peterson, 1996, pp. 178–82.
23. Neolithic Revolution: Diamond, 1997; Gat, 2006; Otterbein, 2004.
24. Wave of farming: Cavalli-Sforza, 2000; Gat, 2006.
25. Kinds of societies: Gat, 2006.
26. First states: Diamond, 1997; Gat, 2006; Kurtz, 2001; Otterbein, 2004.
27. Modern chiefdoms: Goldstein, 2011.
28. Early states as protection rackets: Gat, 2006; Kurtz, 2001; North, Wallis, & Weingast, 2009; Otterbein, 2004; Steckel & Wallis, 2009; Tilly, 1985.
29. Effeminate Chambri: Daly & Wilson, 1988, p. 152.
30. Dirty tricks against anthropologists: Freeman, 1999; Pinker, 2002, chap. 6; Dreger, 2011; C. C. Mann, "Chagnon critics overstepped bounds, historian says," *Science*, Dec. 11, 2009.
31. Myths of harmless primitive warfare: Keeley, 1996.
32. "little to fight about": Eckhardt, 1992, p. 1.
33. Pre-state violence: Keeley, 1996; LeBlanc, 2003; Gat, 2006; Van der Dennen, 1995; Thayer, 2004; Wrangham & Peterson, 1996.
34. Raids in primitive warfare: Chagnon, 1996; Gat, 2006; Keeley, 1996; LeBlanc, 2003; Thayer, 2004; Wrangham & Peterson, 1996.
35. Primitive weaponry: Keeley, 1996.
36. "[they] delight to torment men": Quoted in Schechter, 2005, p. 2.
37. Yanomamö raid: Valero & Biocca, 1970.
38. Wathaurung raid: Morgan, 1852/1979, pp. 43–44.
39. Iñupiaq raid: Burch, 2005, p. 110.
40. Cannibalism: Fernández-Jalvo et al., 1996; Gibbons, 1997.
41. Prion diseases: E. Pennisi, "Cannibalism and prion disease may have been rampant in ancient humans," *Science*, Apr. 11, 2003, pp. 227–28.
42. Maori warrior taunt: A. Vayda's *Maori warfare* (1960), quoted in Keeley, 1996, p. 100.
43. Motives for primitive warfare: Chagnon, 1988; Daly & Wilson, 1988; Gat, 2006; Keeley, 1996; Wiessner, 2006.
44. Yanomamö man "tired of fighting": Quoted in Wilson, 1978, pp. 119–20.
45. Universality of revenge: Daly & Wilson, 1988; McCullough, 2008.
46. Body counts in nonstate societies: Bowles, 2009; Gat, 2006; Keeley, 1996.
47. Forensic archaeology: Keeley, 1996; McCall & Shields, 2007; Steckel & Wallis, 2009; Thorpe, 2003; Walker, 2001.
48. Death by violence in prehistoric societies: Bowles, 2009; Keeley, 1996.
49. Death by violence in hunter-gatherers: Bowles, 2009.
50. Death by violence in hunter-horticulturalists and tribal farmers: Gat, 2006; Keeley, 1996.
51. Death by violence in state societies: Keeley, 1996.
52. Rates of death in war: The 3 percent estimate comes from Wright's sixteen-hundred-page *A study of war:* Wright, 1942, p. 245. The first edition was completed in November 1941, before World War II's most destructive years. The figure was unchanged, however, in the 1965 revision (Wright, 1942/1965, p. 245), and in the 1964 abridgment (Wright, 1942/1964, p. 60), though the latter mentions Dresden, Hiroshima, and Nagasaki in the same paragraph. I assume the unchanged estimate was intentional, and that the additional world war deaths were offset by the billion people added to the world during the more fecund and less lethal postwar decades.
53. 20th-century U.S. and Europe estimate: Keeley, 1996, from Harris, 1975.
54. Battle deaths are summed for the years 1900 through 1945 inclusive from the three Correlates of War datasets (Inter-State, Extrastate, and Intrastate), using the larger figure of the two columns "State Deaths" and "Total Deaths" (Sarkees, 2000; http://www.correlatesofwar.org), together with the geometric mean of the "Battle Dead Low" and "Battle Dead High" estimates for the years 1946 to 2000 inclusive from the PRIO Battle Deaths Dataset (Gleditsch, Wallensteen, Eriksson, Sollenberg, & Strand, 2002; Lacina & Gleditsch, 2005; http://www.prio.no/Data/).
55. 20th-century battle deaths: The denominator of 6 billion deaths comes from an estimate that 12 billion people lived in the 20th century (Mueller, 2004b, p. 193) and that about 5.75 billion were alive at the century's end.
56. 180 million violent deaths: White, in press; the 3 percent figure comes from using 6.25 billion as the estimate of the total number of deaths; see note 55.
57. Iraq and Afghanistan casualties: Iraq Coalition Casualty Count, www.icasualties.org.
58. War battle death rate: Human Security Report Project, 2008, p. 29. The worldwide death estimate of 56.5 million deaths is from the World Health Organization. The twentyfold multiplier is based on the WHO estimate of 310,000 "war-related deaths" in 2000, the most recent year available in the *World report on violence and health.* See Krug, Dahlberg, Mercy, Zwi, & Lozano, 2002, p. 10.
59. Death by violence in pre-Columbian nonstate and state societies: Steckel & Wallis, 2009.

60. Homicide rates in modern Europe: Eisner, 2001.
61. United States homicide rates in the 1970s and 1980s: Daly & Wilson, 1988.
62. Aztecs: Keeley, 1996, table 6.1, p. 195.
63. Death rates for France, Russia, Germany, and Japan: Keeley, 1996, table 6.1, p. 195; 20th-century figures are pro-rated for missing years.
64. Deaths in American wars: Leland & Oboroceanu, 2010, "Total Deaths" column. Population figures are from U.S. Census, http://www.census.gov/compendia/statab/hist_stats.html.
65. All violent 20th-century deaths: Based on 180 million deaths estimated by White, in press, and an average annual world population for the 20th century of 3 billion.
66. 2005 battle deaths: United States: www.icasualties.org. World: UCDP/PRIO Armed Conflict Dataset, Human Security Report Project, 2007; see Human Security Centre, 2005, based in part on data from Gleditsch et al., 2002, and Lacina & Gleditsch, 2005.
67. Prevalence of war among hunter-gatherers: Divale, 1972; Ember, 1978; Keeley, 1996. See also Chagnon, 1988; Gat, 2006; Knauft, 1987; Otterbein, 2004. Van der Dennen, 2005, cites eight estimates of the proportion of nonstate societies that have rarely or never engaged in war; the median is 15 percent.
68. Andamanese: "Noble or savage? The era of the hunter-gatherer was not the social and environmental Eden that some suggest," *Economist*, Dec. 19, 2007.
69. Defeated refugees: Gat, 2006; Keeley, 1996; Van der Dennen, 2005.
70. Violent !Kung San past: Goldstein, 2001, p. 28.
71. Semai violence: Knauft, 1987.
72. !Kung and Central Arctic Inuit: Gat, 2006; Lee, 1982.
73. United States homicide rates: Fox & Zawitz, 2007; Zahn & McCall, 1999; Pax Botswaniana: Gat, 2006.
74. Pax Canadiana: Chirot & McCauley, 2006, p. 114.
75. "Life was better since the government came": Quoted in Thayer, 2004, p. 140.
76. Retribution, feuds, and adjudication: Ericksen & Horton, 1992.
77. Increase in violence with decolonization: Wiessner, 2006.
78. Health hazards of civilization: Steckel & Wallis, 2009; Diamond, 1997.
79. Fall from Eden and early civilization: Kugel, 2007.
80. Hunting-farming tradeoff: Gat, 2006; North et al., 2009; Steckel & Wallis, 2009.
81. First states: Steckel & Wallis, 2009.
82. Cruelty and despotism in early states: Betzig, 1986; Otterbein, 2004; Spitzer, 1975.

Chapter 3: The Civilizing Process

1. Norbert Elias: Fletcher, 1997.
2. Gurr graph of homicide in England: Gurr, 1981.
3. Survey of violence perception: See note 1 to chapter 1.
4. History of homicide: Cockburn, 1991; Eisner, 2001, 2003; Johnson & Monkkonen, 1996; Monkkonen, 1997; Spierenburg, 2008.
5. Long-term homicide trends: Eisner, 2003.
6. Homicides in Kent: Cockburn, 1991.
7. Correlation of homicide with other violence: Eisner, 2003, pp. 93–94; Zimring, 2007; Marvell, 1999; Daly & Wilson, 1988.
8. Quacks: Keeley, 1996, pp. 94–97; Eisner, 2003, pp. 94–95.
9. Constancies in homicide: Eisner, 2003, 2009; Daly & Wilson, 1988.
10. Decline in elite violence: Eisner, 2003; Clark, 2007a, p. 122; Cooney, 1997.
11. Verkko's Law: Daly & Wilson, 1988; Eisner, 2003; Eisner, 2008.
12. *The medieval housebook*: Elias, 1939/2000, pp. 513–16; discussion on pp. 172–82; Graf zu Waldburg Wolfegg, 1988.
13. Furious gusto of knightly destruction: Tuchman, 1978, p. 8.
14. Everyday violence in the Middle Ages: Elias, 1939/2000, p. 168.
15. "his brains flowed forth": Hanawalt, 1976, pp. 311–12, quoted in Monkkonen, 2001, p. 154.
16. Violent medieval entertainment: Tuchman, 1978, p. 135.
17. Cutting off noses: Groebner, 1995.
18. "whether a nose once cut off can grow back": Groebner, 1995, p. 4.
19. Impetuousness in the Middle Ages: Elias, 1939/2000, pp. 168–69.
20. "childishness noticeable in medieval behavior": Tuchman, 1978, p. 52.
21. "slight compression of the lips": D. L. Sayers, introduction, *The song of Roland* (New York: Viking, 1957), p. 15, quoted in Kaeuper, 2000, p. 33.
22. "pearls and rubies" in handkerchief: Elias, 1939/2000, p. 123.
23. "anything purulent": Elias, 1939/2000, p. 130.
24. Disgust as an adaptation: Curtis & Biran, 2001; Pinker, 1997, chap. 6; Rozin & Fallon, 1987.
25. Changes in swearing: Pinker, 2007b, chap. 7.

26. Pissabeds and windfuckers: Hughes, 1991/1998, p. 3.
27. Self-control: Daly & Wilson, 2000; Pinker, 1997, chap. 6; Schelling, 1984.
28. Universal propriety: Brown, 1991; Duerr, 1988–97, but see Mennell & Goudsblom, 1997.
29. "no zero point": Elias, 1939/2000, pp. 135, 181, 403, 421.
30. Number of political units in Europe: Wright, 1942, p. 215; Richardson, 1960, pp. 168–69.
31. Military revolution: Levy, Walker, & Edwards, 2001.
32. "States make war and vice-versa": Tilly, 1985.
33. King's peace: Daly & Wilson, 1988, p. 242.
34. Blood money and coroners: Daly & Wilson, 1988, pp. 241–45.
35. "The Christian attitude . . . held that money was evil": Tuchman, 1978, p. 37.
36. "commercial law prohibited innovation": Tuchman, 1978, p. 37.
37. Evolution of cooperation: Cosmides & Tooby, 1992; Ridley, 1997; Trivers, 1971.
38. Free markets and empathy: Mueller, 1999, 2010b.
39. Doux commerce: Quoted in Fukuyama, 1999, p. 254.
40. Major transitions in evolution: Maynard Smith & Szathmáry, 1997. For a review, see Pinker, 2000.
41. Positive-sum games and progress: Wright, 2000.
42. Decivilizing process in Nazi Germany: de Swaan, 2001; Fletcher, 1997; Krieken, 1998; Mennell, 1990; Steenhuis, 1984.
43. Continuation of homicide decline in Nazi Germany: Eisner, 2008.
44. Problems for the Civilizing Process: Eisner, 2003.
45. State legitimacy and nonviolence: Eisner, 2003; Roth, 2009.
46. Informal norms of cooperation: Ellickson, 1991; Fukuyama, 1999; Ridley, 1997.
47. Equality matching: Fiske, 1992; see also "Morality and Taboo" in chap. 9 of this book.
48. Homicide rate in ranchers: Roth, 2009, p. 355. The rate per 100,000 adults for ranchers, taken from figure 7.2, is multiplied by 0.65 to convert it to a rate per 100,000 people, as suggested on p. 495 of Roth's book.
49. Changing socioeconomic profile of violence: Cooney, 1997; Eisner, 2003.
50. "I have beat many a fellow": Quoted in Wouters, 2007, p. 37.
51. "There are men to whom nothing": Quoted in Wouters, 2007, p. 37.
52. "With the battle-ax": S. Sailer, 2004, "More diversity = Less welfare?" http://www.vdare.com/sailer/diverse.htm.
53. Civilization of the middle and working classes: Spierenburg, 2008; Wiener, 2004; Wood, 2004.
54. Crime as self-help justice: Black, 1983; Wood, 2003.
55. Motives for homicide: Black, 1983; Daly & Wilson, 1988; Eisner, 2009.
56. Murderers as martyrs: Black, 1983, p. 39.
57. Violence as a public health problem: See Pinker, 2002, chap. 17.
58. Definition of a mental disorder: Wakefield, 1992.
59. Police and African Americans: Black, 1980, 134–41, quoted in Cooney, 1997, p. 394.
60. Judicial system uninterested in low-status people: Cooney, 1997, p. 394.
61. neighborhood knucklehead: MacDonald, 2006.
62. Self-help in the inner city: Wilkinson, Beaty, & Lurry, 2009.
63. Persistence of clan violence in Europe: Eisner, 2003; Gat, 2006.
64. Fine line between civil war and organized crime: Mueller, 2004a.
65. Reliability of cross-national crime statistics: LaFree, 1999; LaFree & Tseloni, 2006.
66. The homicide rates for individual countries come from UN Office on Drugs and Crime, 2009. If a WHO estimate is listed, I used that; if not, I report the geometric mean between the high and low estimates.
67. World homicide rate: Krug et al., 2002, p. 10.
68. Europeans eat with swords: Elias, 1939/2000, p. 107.
69. Low crime in autocracies and established democracies: LaFree & Tseloni, 2006; Patterson, 2008; O. Patterson, "Jamaica's bloody democracy," *New York Times*, May 26, 2010. Civil war in anocracies: Gleditsch, Hegre, & Strand, 2009; Hegre, Ellingsen, Gates, & Gleditsch, 2001; Marshall & Cole, 2008. Civil war shades into crime: Mueller, 2004a.
70. Post-decolonization violence in New Guinea: Wiessner, 2006.
71. Enga proverbs: Wiessner, 2006, p. 179.
72. Civilizing offensives: Spierenburg, 2008; Wiener, 2004; Wood, 2003, 2004.
73. Civilizing offensive in New Guinea: Wiessner, 2010.
74. Vacuous explanations of American violence: see Pinker, 2002, pp. 308–9.
75. Americans more violent even without guns: Monkkonen, 1989, 2001. Approximately 65 percent of American homicides are committed with firearms, Cook & Moore, 1999, p. 279; U.S. Department of Justice, 2007, Expanded Homicide Data, Table 7, http://www2.fbi.gov/ucr/cius2007/offenses/expanded_information/data/shrtable_07.html. This means that the American homicide rate without firearms is higher than the total homicide rates of most European countries.
76. Homicide statistics for countries and regions: See note 66.
77. Black and white homicide rates: Fox & Zawitz, *Homicide trends in the US*, 2007.

78. Race gap in crime surveys: Skogan, 1989, pp. 240–41.
79. North-South difference not just a black-white difference: Courtwright, 1996, p. 61; Nisbett & Cohen, 1996.
80. 19th-century American violence: Gurr, 1981; Gurr, 1989a; Monkkonen, 1989, 2001; Roth, 2009.
81. Irish American homicide: Gurr, 1981, 1989a; Monkkonen, 1989, 2001.
82. Decline of northeastern urban homicide: Gurr, 1981, 1989a.
83. History of the race gap in violence: Monkkonen, 2001; Roth, 2009. Increasing black-white homicide gap in New York: Gurr, 1989b, p. 39.
84. Code of the streets: Anderson, 1999.
85. Democracy came too early: Spierenburg, 2006.
86. "the South had a deliberately weak state": Monkkonen, 2001, p. 157.
87. Most killings were reasonable: Monkkonen, 1989, p. 94.
88. Jacksonian justice: quoted in Courtwright, 1996, p. 29.
89. More violence in the South: Monkkonen, 2001, pp. 156–57; Nisbett & Cohen, 1996; Gurr, 1989a, pp. 53–54, note 74.
90. Southern culture of honor: Nisbett & Cohen, 1996.
91. Honor killing versus auto theft: Cohen & Nisbett, 1997.
92. Insulted southerners: Cohen, Nisbett, Bowdle, & Schwarz, 1996.
93. Lumping it: Ellickson, 1991. Herding and violence: Chu, Rivera, & Loftin, 2000.
94. Ox-stunning fisticuffs: Nabokov, 1955/1997, pp. 171–72.
95. Drunken cowboys: Courtwright, 1996, p. 89.
96. Homicide rates in the Wild West: Courtwright, 1996, pp. 96–97. Ellsworth: Roth, 2009, p. 381.
97. Ineffective justice in the Wild West: Courtwright, 1996, p. 100.
98. He Called Bill Smith a Liar: Courtwright, 1996, p. 29.
99. Dirty deck, dirty neck: Courtwright, 1996, p. 92.
100. Gold rush property rights: Umbeck, 1981, p. 50.
101. Gomorrah: Courtwright, 1996, pp. 74–75.
102. Violence is a young man's game: Daly & Wilson, 1988; Eisner, 2009; Wrangham & Peterson, 1996.
103. Evolutionary psychology of male violence: Buss, 2005; Daly & Wilson, 1988; Geary, 2010; Gottschall, 2008.
104. On track to reproductive failure: Daly & Wilson, 1988, p. 163.
105. Alcohol and violence: Bushman, 1997; Bushman & Cooper, 1990.
106. Women pacified the West: Courtwright, 1996.
107. Pacifying effects of marriage: Sampson, Laub, & Wimer, 2006.
108. Uptick in violence in the 1960s: Eisner, 2003; Eisner, 2008; Fukuyama, 1999; Wilson & Herrnstein, 1985.
109. Homicide boom: U.S. Bureau of Justice Statistics, Fox & Zawitz, 2007.
110. Black homicide rate: Zahn & McCall, 1999.
111. Marriage boom: Courtwright, 1996.
112. Baby boom can't explain crime boom: Zimring, 2007, pp. 59–60; Skogan, 1989.
113. Perennial invasion of barbarians: Wilson, 1974, pp. 58–59, quoted in Zimring, 2007, pp. 58–59.
114. Relative cohort size not enough: Zimring, 2007, pp. 58–59.
115. Common knowledge and solidarity: Chwe, 2001; Pinker, 2007b, chap. 8.
116. Informalization in dress and manners: Lieberson, 2000. Informalization in forms of address: Pinker, 2007b, chap. 8.
117. Decline of trust in institutions: Fukuyama, 1999.
118. *Proletarianization* from Arnold Toynbee; *defining deviancy down* from Daniel Patrick Moynihan; quoted in Charles Murray, "Prole Models," *Wall Street Journal*, Feb. 6, 2001.
119. Timekeeping and self-control: Elias, 1939/2000, p. 380.
120. Conning the intellectuals: See, e.g., Pinker, 2002, pp. 261–62.
121. Rape as radical chic: See Brownmiller, 1975, pp. 248–55, and chap. 7, for numerous examples.
122. Rape as insurrection: Cleaver, 1968/1999, p. 33. See also Brownmiller, 1975, pp. 248–53.
123. Intelligent and eloquent rapist: Jacket and interior blurbs in Cleaver, 1968/1999.
124. Retreat of the justice system: Wilson & Herrnstein, 1985, pp. 424–25. See also Zimring, 2007, figure 3.2, p. 47.
125. Decriminalization of public disorder: Fukuyama, 1999.
126. Failure to protect African Americans: Kennedy, 1997.
127. Paranoia about the police: Wilkinson et al., 2009.
128. Moynihan report: Massey & Sampson, 2009.
129. Perverse incentives: Fukuyama, 1999; Murray, 1984.
130. Skepticism of parenting effects: Harris, 1998/2008; Pinker, 2002, chap. 19; Wright & Beaver, 2005.
131. American homicide rates: FBI *Uniform crime reports*, 1950–2005, U.S. Federal Bureau of Investigation, 2010b.
132. Canadian homicide: Gartner, 2009.

133. European homicide: Eisner, 2008.
134. Decline in other crimes: U.S. Bureau of Justice Statistics, *National crime victimization survey*, 1990 and 2000, reported in Zimring, 2007, p. 8.
135. Cloud beyond the horizon: Quoted in Zimring, 2007, p. 21.
136. Blood bath: Quoted in Levitt, 2004, p. 169.
137. Super-predators: Quoted in Levitt, 2004, p. 169.
138. Gotham City without Batman: Quoted in Gardner, 2010, p. 225.
139. Smaller crime-prone cohort: Zimring, 2007, pp. 22, 61–62.
140. Different unemployment trends in Canada and the United States: Zimring, 2007.
141. Unemployment and violence going in different directions: Eisner, 2008.
142. Unemployment doesn't predict violent crime: Zimring, 2007, p. 63; Levitt, 2004; Raphael & Winter-Ebmer, 2001.
143. "never right to begin with": Quoted in A. Baker, "In this recession, bad times do not bring more crime (if they ever did)," *New York Times*, Nov. 30, 2009.
144. Inequality and violence: Daly, Wilson, & Vasdev, 2001; LaFree, 1999.
145. Gini index for the United States: U.S. Census Bureau, 2010b.
146. Inequality may not cause crime: Neumayer, 2003, 2010.
147. Claim that abortion lowers crime: Donohue & Levitt, 2001.
148. Abortion one of four causes of crime decline: Levitt, 2004.
149. Problems for the abortion-crime connection: Joyce, 2004; Lott & Whitley, 2007; Zimring, 2007; Foote & Goetz, 2008; S. Sailer & S. Levitt, "Does abortion prevent crime?" *Slate*, Aug. 23, 1999, http://www.slate.com/id/33569/entry/33571/. Levitt's reply: Levitt, 2004; see also his responses to Sailer in *Slate*.
150. More at-risk children following *Roe v. Wade*: Lott & Whitley, 2007; Zimring, 2007.
151. Women who have abortions more responsible: Joyce, 2004.
152. Peers trump parents: Harris, 1998/2008, chaps. 9, 12, 13; Wright & Beaver, 2005.
153. Wrong age-cohort predictions: Foote & Goetz, 2008; Lott & Whitley, 2007; S. Sailer & S. Levitt, "Does abortion prevent crime?" *Slate*, Aug. 23, 1999, http://www.slate.com/id/33569/entry/33571/.
154. Explaining the 1990s crime decline: Blumstein & Wallman, 2006; Eisner, 2008; Levitt, 2004; Zimring, 2007.
155. American incarceration mania: J. Webb, "Why we must fix our prisons," *Parade*, Mar. 29, 2009.
156. Imprisoned Americans: Zimring, 2007, figure 3.2, p. 47; J. Webb, "Why we must fix our prisons," *Parade*, Mar. 29, 2009.
157. Small number commit many crimes: Wolfgang, Figlio, & Sellin, 1972.
158. Criminals have low self-control: Gottfredson & Hirschi, 1990; Wilson & Herrnstein, 1985.
159. Deterrence works: Levitt & Miles, 2007; Lott, 2007; Raphael & Stoll, 2007.
160. Montreal police strike: "City without cops," *Time*, Oct. 17, 1969, p. 47; reproduced in Kaplan, 1973, p. 20.
161. Problems with the imprisonment explanation: Eisner, 2008; Zimring, 2007.
162. Diminishing returns in imprisonment: Johnson & Raphael, 2006.
163. Effectiveness of additional police: Levitt, 2004.
164. Boston policing: F. Butterfield, "In Boston, nothing is something," *New York Times*, Nov. 21, 1996; Winship, 2004.
165. New York policing: MacDonald, 2006.
166. Broken Windows theory: Wilson & Kelling, 1982.
167. New York success story: Zimring, 2007; MacDonald, 2006.
168. Biggest crime prevention achievement in history: Zimring, 2007, p. 201.
169. Problems with Broken Windows: Levitt, 2004; B. E. Harcourt, "Bratton's 'broken windows': No matter what you've heard, the chief's policing method wastes precious funds," *Los Angeles Times*, Apr. 20, 2006.
170. Dutch Broken Windows: Keizer, Lindenberg, & Steg, 2008.
171. Hard-headed statisticians: Eisner, 2008; Rosenfeld, 2006. See also Fukuyama, 1999.
172. Black women and clergy as civilizing forces: Anderson, 1999; Winship, 2004.
173. Boston Miracle: Winship, 2004; P. Shea, "Take us out of the old brawl game," *Boston Globe*, Jun. 30, 2008; F. Butterfield, "In Boston, nothing is something," *New York Times*, Nov. 21, 1996.
174. M. Cramer, "Homicide rate falls to lowest level since '03," *Boston Globe*, Jan. 1, 2010.
175. Small predictable versus large capricious punishments: J. Rosen, "Prisoners of parole," *New York Times Magazine*, Jan. 10, 2010.
176. Discounting the future: Daly & Wilson, 2000; Hirschi & Gottfredson, 2000; Wilson & Herrnstein, 1985. See also "Self-Control" in chap. 9 of this book.
177. Dominance versus fairness: Fiske, 1991, 1992, 2004a. See also "Morality and Taboo" in chap. 9 of this book.
178. "My probation officer doesn't like me": J. Rosen, "Prisoners of parole," *New York Times Magazine*, Jan. 10, 2010.

179. Kant and crime prevention: J. Seabrook, "Don't shoot: A radical approach to the problem of gang violence," *New Yorker*, Jun. 22, 2009.
180. House on fire: J. Seabrook, "Don't shoot: A radical approach to the problem of gang violence," *New Yorker*, Jun. 22, 2009, pp. 37–38.
181. Reconstitution of social order: Fukuyama, 1999, p. 271; "Positive trends recorded in U.S. data on teenagers," *New York Times*, Jul. 13, 2007.
182. Informalization and third nature: Wouters, 2007.

Chapter 4: The Humanitarian Revolution

1. Torture museum: http://www.torturamuseum.com/this.html.
2. Coffee table books on torture: Held, 1986; Puppi, 1990.
3. Medieval torture: Held, 1986; Levinson, 2004b; Mannix, 1964; Payne, 2004; Puppi, 1990.
4. Pope Paul IV a sainted torturer: Held, 1986, p. 12.
5. Illogical torture: Mannix, 1964, pp. 123–24.
6. Universality of torture: Davies, 1981; Mannix, 1964; Payne, 2004; Spitzer, 1975.
7. Critical theorists: Menschenfreund, 2010. Theoconservatism: Linker, 2007.
8. Humanism in Asia: Bourgon, 2003; Kurlansky, 2006; Sen, 2000.
9. Human sacrifice: Davies, 1981; Mannix, 1964; Otterbein, 2004; Payne, 2004.
10. "that no one might burn": 2 Kings 23:10.
11. Number of Aztec sacrifices: White, in press.
12. Suttee deaths: White, in press.
13. Superstitious rationale for human sacrifice: Payne, 2004, pp. 40–41.
14. Quoted in M. Gerson, "Europe's burqa rage," *Washington Post*, May 26, 2010.
15. Shortchanging the gods: Payne, 2004, p. 39.
16. Witchcraft and hunter-gatherer warfare: Chagnon, 1997; Daly & Wilson, 1988; Gat, 2006; Keeley, 1996; Wiessner, 2006.
17. Overactive cause detection: Atran, 2002.
18. Witchcraft accusations: Daly & Wilson, 1988, pp. 237, 260–61.
19. Enforcing unpopular norms: Willer, Kuwabara, & Macy, 2009. See also McKay, 1841/1995.
20. *Malleus Maleficarum*: Mannix, 1964; A. Grafton, "Say anything," *New Republic*, Nov. 5, 2008.
21. 60,000 witch-hunt victims: White, in press. 100,000 witch-hunt victims: Rummel, 1994, p. 70.
22. Witch hunts: Rummel, 1994, p. 62; A. Grafton, "Say anything," *New Republic*, Nov. 5, 2008.
23. Blood libels: Rummel, 1994, p. 56.
24. Witchcraft skeptics: Mannix, 1964, pp. 133–34.
25. Witchcraft experiments: Mannix, 1964, pp. 134–35, also recounted in McKay, 1841/1995.
26. Decline of witchcraft: Thurston, 2007; Mannix, 1964, p. 137.
27. Atrocitology: Rummel, 1994; Rummel, 1997; White, in press; White, 2010b.
28. Death tolls of Christian wars and massacres: White, in press, provides the following estimates: Crusades, 3 million; Albigensian suppression, 1 million; Huguenot Wars, 2–4 million; Thirty Years' War, 7.5 million. He does not provide his own estimate of the Inquisition's death toll using multiple sources but cites an estimate from the general secretary of the Inquisition in 1808 of 32,000.
29. 400 million people: Estimate of world population in 1200 CE from *Historical estimates of world population*, U.S. Census Bureau, 2010a.
30. Albigensian Crusade: Rummel, 1994, p. 46.
31. Albigensian Crusade as genocide: Chalk & Jonassohn, 1990; Kiernan, 2007; Rummel, 1994.
32. Tortured for clean underwear: Mannix, 1964, pp. 50–51.
33. Death toll from the Spanish Inquisition: Rummel, 1994, p. 70.
34. Religious persecution: Grayling, 2007.
35. Luther and the Jews: Lull, 2005.
36. John Calvin, "Sermon on Deuteronomy": Quoted in Grayling, 2007, p. 41.
37. Murderous Calvin: Grayling, 2007.
38. Henry VIII burned 3.25 heretics: Payne, 2004, p. 17.
39. European Wars of Religion: Wright, 1942, p. 198.
40. Death toll of Wars of Religion: See the table on p. 195 for similar estimates and comparisons from Matthew White.
41. English Civil War death toll: Schama, 2001, p. 13. Schama cites "at least a quarter of a million" dead in England, Wales, and Scotland and guesstimates another 200,000 in Ireland, out of a population of 5 million in the British Isles at the time.
42. Papal pique: Holsti, 1991, p. 25.
43. Decline of the Inquisition: Perez, 2006.
44. Erasmus and other skeptics: Popkin, 1979.
45. Scrutiny of religious persecution: Grayling, 2007.
46. "Calvin says that he is certain": *Concerning heretics, whether they are to be persecuted*, quoted in Grayling, 2007, pp. 53–54.

47. Skeptical Francis Bacon: Quoted in Grayling, 2007, p. 102.
48. Cat-burning: Quoted in Payne, 2004, p. 126.
49. Pepys: Quoted in Clark, 2007a, p. 182.
50. Lethal pillorying: Mannix, 1964, pp. 132–33.
51. Lethal flogging: Mannix, 1964, pp. 146–47. See also Payne, 2004, chap. 9.
52. Cruel prisons: Payne, 2004, p. 122.
53. Prison reform: Payne, 2004, p. 122.
54. Infamous burning at the stake: Mannix, 1964, p. 117.
55. Breaking on the wheel: *Trewlicher Bericht eynes scrocklichen Kindermords beym Hexensabath*. Hamburg, Jun. 12, 1607. http://www.borndigital.com/wheeling.htm.
56. Infamous breaking on the wheel: Hunt, 2007, pp. 70–76.
57. Compassion for wheel victim: Hunt, 2007, p. 99.
58. Voltaire on torture: Quoted in Hunt, 2007, p. 75.
59. Christian hypocrisy: Montesquieu, 1748/2002.
60. "the principle of sympathy": Quoted in Hunt, 2007, pp. 112, 76.
61. Rush on reforming criminals: Quoted in Hunt, 2007, p. 98.
62. "blunt the sentiments": Quoted in Hunt, 2007, p. 98.
63. Beccaria: Hunt, 2007.
64. Religious defense of torture: Hunt, 2007, chap. 2.
65. Early animal rights movements: Gross, 2009; Shevelow, 2008.
66. Frivolous capital infractions: Rummel, 1994, p. 66; Payne, 2004.
67. Speedy trials: Payne, 2004, p. 120.
68. Frivolous execution count: Rummel, 1994, p. 66.
69. Decline of capital punishment: Payne, 2004, p. 119.
70. UN's nonbinding moratorium: E. M. Lederer, "UN General Assembly calls for death penalty moratorium," *Boston Globe*, Dec. 18, 2007.
71. United States are outliers: Capital punishment has been abolished in Alaska, Hawaii, Illinois, Iowa, Maine, Massachusetts, Michigan, Minnesota, New Hampshire, New Jersey, New Mexico, New York, North Dakota, Rhode Island, Vermont, West Virginia, Wisconsin, and the District of Columbia. Kansas's last nonmilitary execution was in 1965.
72. American execution rate infinitesimal: Every year during the 2000s around 16,500 people were murdered and around 55 were executed.
73. Decline in American executions in the 2000s: Death Penalty Information Center, 2010b.
74. Capital punishment for crimes other than murder: Death Penalty Information Center, 2010a.
75. "In reform after reform": Payne, 2004, p. 132.
76. History of slavery: Davis, 1984; Patterson, 1985; Payne, 2004; Sowell, 1998.
77. Recent abolitions: Rodriguez, 1999.
78. Quote from "Report on the coast of Africa made by Captain George Collier, 1918–19," reproduced in Eltis & Richardson, 2010.
79. Slave trade statistics: Rummel, 1994, pp. 48, 70. White, in press, estimates that 16 million people died in the Atlantic slave trade, and another 18.5 million in the Mideast slave trade.
80. Slavery as a negative-sum game: Smith, 1776/2009, p. 281.
81. "slaveholders were bad businessmen": Mueller, 1989, p. 12.
82. Economics of antebellum slavery: Fogel & Engerman, 1974.
83. British abolition of slave trade: Nadelmann, 1990, p. 492.
84. Humanitarian motives for British slave trade ban: Nadelmann, 1990, p. 493; Ray, 1989, p. 415.
85. Humanitarian motives for abolition of slavery: Davis, 1984; Grayling, 2007; Hunt, 2007; Mueller, 1989; Payne, 2004; Sowell, 1998.
86. "mankind has not been born with saddles": Thomas Jefferson, "To Roger C. Weightman," Jun. 24, 1826, in *Portable Thomas Jefferson*, p. 585.
87. Debt bondage: Payne, 2004, pp. 193–99.
88. Shuddering at debt bondage: Quoted in Payne, 2004, p. 196.
89. Debt collection as force: Payne, 2004, p. 197.
90. Trafficking: Feingold, 2010, p. 49.
91. Dubious statistic: Free the Slaves (http://www.freetheslaves.net/, accessed Oct. 19, 2010) claims that "there are 27 million slaves in the world today," a figure that is orders of magnitude higher than those from the UNESCO Trafficking Statistics Project; Feingold, 2010. Bales on progress in fighting slavery: S. L. Leach, "Slavery is not dead, just less recognizable," *Christian Science Monitor*, Sept. 1, 2004.
92. Despotism: Betzig, 1986.
93. Summary executions by despots: Davies, 1981, p. 94.
94. Political murder: Payne, 2004, chap. 7; Woolf, 2007.
95. Regicide: Eisner, 2011.
96. Death by government: Rummel, 1994, 1997.
97. Decline of political murder: Payne, 2004, pp. 88–94; Eisner, 2011.

98. Lay down this right to all things if others do too: Hobbes, 1651/1957, p. 190.
99. "exempt themselves from the obedience": Locke, *Two treatises on government*, quoted in Grayling, 2007, p. 127.
100. Framers and human nature: Pinker, 2002, chap. 16; McGinnis, 1996, 1997.
101. "If men were angels": Federalist Papers No. 51, in Rossiter, 1961, p. 322.
102. "ambition must be made to counteract ambition": Federalist Papers No. 51, in Rossiter, 1961, pp. 331–32.
103. Framers and positive-sum cooperation: McGinnis, 1996, 1997.
104. Mozi on war and evil: Quoted in the epigraph of Kurlansky, 2006.
105. Swords into plowshares: Isaiah 2:4.
106. Turning the other cheek: Luke 6:27–29.
107. War as the medieval default: G. Schwarzenberger, "International law," *New Encyclopaedia Britannica*, 15th ed., quoted in Nadelmann, 1990.
108. Rates of war: See figure 5–17, which is based on Peter Brecke's Conflict Catalog, discussed in chapter 5; Brecke, 1999, 2002; Long & Brecke, 2003.
109. Persecution of pacifists: Kurlansky, 2006.
110. Falstaff on honor: *Henry IV, Part I*, Act 5, scene 1.
111. Johnson on European war: *Idler*, No. 81 [82], Nov. 3, 1759, in Greene, 2000, pp. 296–97.
112. "pernicious Race of little odious Vermin": *Gulliver's travels*, part II, chap. 6.
113. Antiwar *pensée*: "Justice and the reason of effects," *Pensées*, 293.
114. Trade as an antiwar tactic: Bell, 2007a; Mueller, 1989, 1999; Russett & Oneal, 2001; Schneider & Gleditsch, 2010.
115. "The spirit of commerce": Kant, 1795/1983.
116. Entrepreneurial Quakers: Mueller, 1989, p. 25.
117. "Perpetual Peace": Kant, 1795/1983.
118. Kingly declarations of love of peace: Luard, 1986, 346–47.
119. "No longer was it possible": Mueller, 1989, p. 18, based on research in Luard, 1986.
120. Dropping out of the conquest game: Mueller, 1989, pp. 18–21.
121. Great power wars less frequent but more damaging: Levy, 1983.
122. Mysterious decline of force: Payne, 2004, p. 29.
123. Nonviolence a prerequisite to democracy: Payne, 2004, 2005.
124. Moral agitation and the slave trade: Nadelmann, 1990.
125. Homicidal and sexual fantasies: Buss, 2005; Symons, 1979.
126. Equation of visceral disgust with moral disgust: Haidt, Björklund, & Murphy, 2000; Rozin, 1997.
127. Degradation and mistreatment: Glover, 1999.
128. Medieval reintroduction of Roman judicial torture: Langbein, 2005.
129. Life was cheap: Payne, 2004, p. 28.
130. Productivity in book publishing: Clark, 2007a, pp. 251–52.
131. Libraries with novels: Keen, 2007, p. 45.
132. Increasing literacy: Clark, 2007a, pp. 178–80; Vincent, 2000; Hunt, 2007, pp. 40–41.
133. Increase in French literacy: Blum & Houdailles, 1985. Other European countries: Vincent, 2000, pp. 4, 9.
134. Reading Revolution: Darnton, 1990; Outram, 1995.
135. Turning point in reading: Darnton, 1990, p. 166.
136. Expanding circle: Singer, 1981.
137. History of the novel: Hunt, 2007; Price, 2003. Number of novels published: Hunt, 2007, p. 40.
138. Weeping officer: Quoted in Hunt, 2007, pp. 47–48.
139. Diderot's eulogy for Richardson: Quoted in Hunt, 2007, p. 55.
140. Novels denounced: Quoted in Hunt, 2007, p. 51.
141. Morally influential novels: Keen, 2007.
142. Global campus: Lodge, 1988, pp. 43–44.
143. Republic of Letters: P. Cohen, "Digital keys for unlocking the humanities' riches," *New York Times*, Nov. 16, 2010.
144. Combinatorial mind: Pinker, 1999, chap. 10; Pinker, 1997, chap. 2; Pinker, 2007b, chap. 9.
145. Spinoza: Goldstein, 2006.
146. Subversive cities: E. L. Glaeser, "Revolution of urban rebels," *Boston Globe*, Jul. 4, 2008.
147. Skepticism as the origin of modern thought: Popkin, 1979.
148. Interchangeability of perspectives as the basis for morality: Nagel, 1970; Singer, 1981.
149. Humanitarian revolutions in Asia: Bourgon, 2003; Sen, 2000. See also Kurlansky, 2006.
150. Tragic vision of human condition: Burke, 1790/1967; Sowell, 1987.
151. Readiness for democracy: Payne, 2005; Rindermann, 2008.
152. Madison, government, and human nature: Federalist Papers No. 51, in Rossiter, 1961, p. 322. See also McGinnis, 1996, 1997; Pinker, 2002, chap. 16.
153. French people as a different species: Quoted in Bell, 2007a, p. 77.

154. Tragic and Utopian visions: Originally from Sowell, 1987, who called them the "constrained" and "unconstrained" visions; see Pinker, 2002, chap 16.
155. Counter-Enlightenments: Berlin, 1979; Garrard, 2006; Howard, 2001, 2007; Chirot, 1995; Menschenfreund, 2010.
156. *Schwerpunkt*: Berlin, 1979, p. 11.
157. Cosmopolitanism as a bad thing: Berlin, 1979, p. 12.
158. Warmth! Blood! Life!: Quoted in Berlin, 1979, p. 14.
159. Super-personal organism: Berlin, 1979, p. 18.
160. "Men desire harmony": Quoted in Bell, 2007a, p. 81.
161. The myth of social Darwinism: Claeys, 2000; Johnson, 2010; Leonard, 2009. The myth originated in a politicized history by Richard Hofstadter titled *Social Darwinism in American thought.*
162. War is noble: Mueller, 1989, p. 39.

Chapter 5: The Long Peace

1. World War II not the climax: *War and civilization* (1950), p. 4, quoted in Mueller, 1995, p. 191.
2. Predictions of doomsday: Mueller, 1989, 1995.
3. Lewis F. Richardson: Hayes, 2002; Richardson, 1960; Wilkinson, 1980.
4. A long future without a world war: Richardson, 1960, p. 133.
5. Hemoclysm: White, 2004.
6. Grim diagnoses of modernity: Menschenfreund, 2010. Prominent examples include Sigmund Freud, Michel Foucault, Zygmunt Bauman, Edmund Husserl, Theodor Adorno, Max Horkheimer, and Jean-François Lyotard.
7. Long Peace: Gaddis, 1986, 1989. Gaddis was referring to the absence of war between the United States and the Soviet Union, but I have broadened it to include peace among great powers and developed states.
8. Overoptimistic turkey: Usually credited to the management scholar Nassim Nicholas Taleb.
9. The population of the world: Historical population estimates from McEvedy & Jones, 1978.
10. Subjective probability through mnemonic availability: Tversky & Kahneman, 1973, 1974.
11. Misperceptions of risk: Gardner, 2008; Ropeik & Gray, 2002; Slovic, Fischof, & Lichtenstein, 1982.
12. Worst Things People Have Done: White, 2010a. See also White, in press, for narratives of the events and more recent estimates. The Web site lists the numbers and sources that went into his estimates.
13. An Lushan Revolt: White notes that the figure is controversial. Some historians attribute it to migration or the breakdown of the census; others treat it as credible, because subsistence farmers would have been highly vulnerable to a disruption of irrigation infrastructure.
14. Assyrian chariots: Keegan, 1993, p. 166.
15. Revolting savagery: Saunders, 1979, p. 65.
16. "The greatest joy a man can know": Quoted in numerous sources, including Gat, 2006, p. 427.
17. Genghis's Y chromosome: Zerjal et al., 2003.
18. Towers of skulls: Rummel, 1994, p. 51.
19. Forgotten massacres: White, in press.
20. List of wars since 3000 BCE: Eckhardt, 1992.
21. Hockey-stick graph of wars: Eckhardt, 1992, p. 177.
22. Associated Press versus 16th-century monks: Payne, 2004, p. 69.
23. Historical myopia as a distorter of war trends: Payne, 2004, pp. 67–70.
24. Measuring historical myopia with a ruler: Taagepera & Colby, 1979.
25. Military horizon: Keegan, 1993, pp. 121–22.
26. Evil in war outweighs good: Richardson, 1960, p. xxxvii.
27. "to condemn much is to understand little": Richardson, 1960, p. xxxv.
28. "Thinginess fails": Richardson, 1960, p. 35.
29. Slave trade as war: Richardson, 1960, p. 113.
30. Excluded small wars: Richardson, 1960, pp. 112, 135–36.
31. Portia versus Richardson: Richardson, 1960, p. 130.
32. Bird's-eye view and tests of hypotheses: Richardson, 1960; Wilkinson, 1980.
33. Cluster illusion: Feller, 1968.
34. Gambler's fallacy: Kahneman & Tversky, 1972; Tversky & Kahneman, 1974.
35. Glowworms and constellations: Gould, 1991.
36. Random onsets of war have also been found in Singer and Small's Correlates of War Project, Singer & Small, 1972, pp. 205–6 (see also Helmbold, 1998); Quincy Wright's *A study of war* database, Richardson, 1960, p. 129; Pitirim Sorokin's 2,500-year list of wars, Sorokin, 1957, p. 561; and Levy's Great Power War database, Levy, 1983, pp. 136–37.
37. Exponential distribution of war durations was also found in the Conflict Catalog, Brecke, 1999, 2002.
38. Wars most likely to end in first year: See Wilkinson, 1980, for refinements.

39. No good cycles: Richardson, 1960, pp. 140–41; Wilkinson, 1980, pp. 30–31; Levy, 1983, pp. 136–38; Sorokin, 1957, pp. 559–63; Luard, 1986, p. 79.
40. History is not an engine: Sorokin, 1957, p. 563.
41. Most important person of the 20th century: White, 1999.
42. World War I need not have happened: Lebow, 2007.
43. Historians on Hitler: quoted in Mueller, 2004a, p. 54.
44. No Hitler, no World War II: Mueller, 2004a, p. 54.
45. No Hitler, no Holocaust: Goldhagen, 2009; Himmelfarb, 1984, p. 81; Fischer, 1998, p. 288; Valentino, 2004.
46. Probability of heads: Keller, 1986. Persi Diaconis, a statistician and magician, can throw heads ten times in a row; see E. Landuis, "Lifelong debunker takes on arbiter of neutral choices," *Stanford Report*, Jun. 7, 2004.
47. Henri Poincaré: *Science and method*, quoted in Richardson, 1960, p. 131.
48. "mankind has become less warlike": Richardson, 1960, p. 167.
49. Other datasets point to same conclusion: Sorokin, 1957, p. 564: "As in the data presented here is nothing to support the claim of disappearance of war in the past, so is there nothing to support the claim, in spite of the exceptionally high figures for the twentieth century, that there has been (or will be) any steady trend toward increase of war. No, the curve fluctuates, and that is all." Singer & Small, 1972, p. 201: "Is war on the increase, as many scholars and laymen of our generation have been inclined to believe? The answer would seem to be a very unambiguous negative." Luard, 1986, p. 67: "The overall frequency of war [from 1917 to 1986] was not very different from that in the previous age [1789 to 1917]. . . . The average amount of war per state, the more significant measure, is now lower if the comparison is made with the entire period 1789–1914. But if the comparison is made with 1815–1914 alone there is little decline."
50. Fewer but more lethal wars: Richardson, 1960, p. 142.
51. Technically they are not "proportional," since there is usually a nonzero intercept, but are "linearly related."
52. Power-law distribution in the Correlates of War Dataset: Cederman, 2003.
53. Plotting power-law distributions: Newman, 2005.
54. Power-law distributions, theory and data: Mitzenmacher, 2004, 2006; Newman, 2005.
55. Zipf's laws: Zipf, 1935.
56. Word type and token frequencies: Francis & Kucera, 1982.
57. Things with power-law distributions: Hayes, 2002; Newman, 2005.
58. Examples of normal and power-law distributions: Newman, 2005.
59. Newman presented the percentages of cities with an exact population size, rather than in a range of population sizes, to keep the units commensurable in the linear and logarithmic graphs (personal communication, February 1, 2011).
60. Causes of war: Levy & Thompson, 2010; Vasquez, 2009.
61. Mechanisms that generate power-law distributions: Mitzenmacher, 2004; Newman, 2005.
62. Power laws for deadly quarrels versus cities: Richardson, 1960, pp. 154–56.
63. Self-organized criticality and war sizes: Cederman, 2003; Roberts & Turcotte, 1998.
64. War of Attrition game: Maynard Smith, 1982, 1988; see also Dawkins, 1976/1989.
65. Loss aversion: Kahneman & Renshon, 2007; Kahneman & Tversky, 1979, 1984; Tversky & Kahneman, 1981. Sunk costs in nature: Dawkins & Brockmann, 1980.
66. More lethal wars last longer: Richardson, 1960, p. 130; Wilkinson, 1980, pp. 20–30.
67. Nonlinear increase in the costliness of longer wars: The death tolls of the 79 wars in the Correlates of War Inter-State War Dataset (Sarkees, 2000) are better predicted by an exponential function of duration (which accounts for 48 percent of the variance) than by the durations themselves (which account for 18 percent of the variance).
68. Weber's Law for perceived deaths: Richardson, 1960, p. 11.
69. Rediscovery of Weber's Law for perceived deaths: Slovic, 2007.
70. Attrition with sunk costs generates a power-law distribution: Wilkinson, 1980, pp. 23–26; Weiss, 1963; Jean-Baptiste Michel, personal communication.
71. 80:20 rules: Newman, 2005.
72. Interpolation of small quarrels: Richardson, 1960, pp. 148–50.
73. Homicides in the United States: Fox & Zawitz, 2007; with data for the years 2006–2009 extrapolated as 17,000 per year, the estimated total is 955,603.
74. Homicides outnumber war-related deaths: Krug et al., 2002, p. 10; see also note 76.
75. Wars less deadly than disease: Richardson, 1960, p. 153.
76. Wars still less deadly than disease: In 2000, according to the WHO *World report on violence and health*, there were 520,000 homicides and 310,000 "war-related deaths." With approximately 56 million deaths from all causes, this works out to an overall violent death rate of around 1.5 percent. The figure cannot be compared directly to Richardson's since his estimates for the 1820–1952 period were radically less complete.

77. 80:2 rule: Based on the 94 wars of magnitude 4–7 for which Richardson had plottable death data.
78. Great powers: Levy, 1983; Levy et al., 2001.
79. A few great powers fight a lot of wars: Levy, 1983, p. 3.
80. Great powers still fight a lot of wars: Gleditsch et al., 2002; Lacina & Gleditsch, 2005; http://www.prio.no/Data/.
81. Great power wars kill the most people: Levy, 1983, p. 107.
82. Data from last quarter of the 20th century: Correlates of War Inter-State War Dataset, 1816–1997 (v3.0), http://www.correlatesofwar.org/, Sarkees, 2000.
83. Recall that Levy excluded colonial wars, unless a great power was fighting with an insurgent movement against the colonial government.
84. Short wars in the late 20th century: Correlates of War Inter-State War Dataset, 1816–1997 (v3.0), http://www.correlatesofwar.org/, Sarkees, 2000; and for the Kosovo war, PRIO Battle Deaths Dataset, 1946–2008, Version 3.0, http://www.prio.no/CSCW/Datasets/Armed-Conflict/Battle-Deaths/, Gleditsch et al., 2002; Lacina & Gleditsch, 2005.
85. Conflict Catalog: Brecke, 1999, 2002; Long & Brecke, 2003.
86. Definition of "Europe": Though I have stuck with the coding scheme of the Conflict Catalog, other datasets code these countries as Asian (see, for example, Human Security Report Project, 2008).
87. The same data plotted on a logarithmic scale look something like figure 5–15, reflecting the power-law distribution in which the largest wars (those involving the great powers, most of which are European) account for the majority of the deaths. But because of the massive decline in deaths in European wars after 1950, the logarithmic scale magnifies the small bounce in the last quarter of the century.
88. Conflict Catalog and pre-1400 European Conflict Catalog: Long & Brecke, 2003; Brecke, 1999, 2002.
89. Obscure wars: None of these were mentioned in the thousand or so responses to a survey in which I asked one hundred Internet users to name as many wars as they could think of.
90. War as part of the natural order of things: Howard, 2001, pp. 12, 13.
91. Casualties didn't weigh heavily: Luard, 1986, p. 240.
92. Sex life of European kings: Betzig, 1986, 1996a, 2002.
93. Overlordship: Luard, 1986, p. 85.
94. Royal pissing contests: Luard, 1986, pp. 85–86, 97–98, 105–6.
95. Habsburgs: *Black lamb and grey falcon* (1941), quoted in Mueller, 1995, p. 177.
96. Biology and dynasties: Betzig, 1996a, 1996b, 2002.
97. Killing people who worshipped the wrong God: Luard, 1986, pp. 42–43.
98. Religion preempts diplomacy: Mattingly, 1958, p. 154, quoted in Luard, 1986, p. 287.
99. Declining number of political units in Europe: Wright, 1942, p. 215; Gat, 2006, p. 456.
100. Military revolution: Gat, 2006; Levy & Thompson, 2010; Levy et al., 2001; Mueller, 2004a.
101. Soldiers and outlaws: Tilly, 1985, p. 173.
102. Take the money and run: Mueller, 2004a, p. 17.
103. Military evolution: Gat, 2006, pp. 456–80.
104. Napoleon and total war: Bell, 2007a.
105. Relatively pacific 18th century: Brecke, 1999; Luard, 1986, p. 52; Bell, 2007a, p. 5. Bell's "performing poodles" quote is from Michael Howard.
106. Ideology, nationalism, and Enlightenment: Howard, 2001.
107. Fervid Napoleonic France: Bell, 2007a.
108. Disfigurement of Enlightenment: Bell, 2007a, p. 77.
109. Dialectic between Enlightenment and Counter-Enlightenment: Howard, 2001, p. 38.
110. Concert of Europe as product of Enlightenment: Howard, 2001, p. 41; see also Schroeder, 1994.
111. Nations must fight their way into existence: Howard, 2001, p. 45.
112. Conservatives and nationalists merged: Howard, 2001, p. 54.
113. Hegelian nationalism: Luard, 1986, p. 355.
114. Marxism and nationalism: Glover, 1999.
115. Self-determination as dynamite: Quoted in Moynihan, 1993, p. 83.
116. War is good: Quotes are from Mueller, 1989, pp. 38–51.
117. Wallowing in materialism: Quoted in Mueller, 1995, p. 187.
118. Peace is bad: Quotes are from Mueller, 1989, pp. 38–51.
119. Moral equivalent of war: James, 1906/1971.
120. Belloc wanted war: Mueller, 1989, p. 43.
121. Valéry wanted war: Bell, 2007a, p. 311.
122. Sherlock Holmes wanted war: Gopnik, 2004.
123. Why the Great War happened: Ferguson, 1998; Gopnik, 2004; Lebow, 2007; Stevenson, 2004.
124. 8.5 million: Correlates of War Inter-State War Dataset, Sarkees, 2000; 15 million: White, in press.

125. Anti-Enlightenment ideologies in Germany, Italy, and Japan: Chirot, 1995; Chirot & McCauley, 2006.
126. Cold War as containment of communist expansionism: Mueller, 1989, 2004a.
127. 19th-century peace movements: Howard, 2001; Kurlansky, 2006; Mueller, 1989, 2004a; Payne, 2004.
128. Ridiculing pacifists: Mueller, 1989, p. 30.
129. Shavian accompaniment: Quoted in Wearing, 2010, p. viii.
130. What Angell really wrote: Ferguson, 1998; Gardner, 2010; Mueller, 1989.
131. War no longer justified: Luard, 1986, p. 365.
132. *All Quiet on the Western Front*: Remarque, 1929/1987, pp. 222–25.
133. A mountain . . . cannot offend a mountain: Remarque, 1929/1987, p. 204.
134. War aversion among most Germans in the 1930s: Mueller, 1989, 2004a.
135. Alternatives to Hitler would not have started World War II: Turner, 1996.
136. Hitler's demonic genius: Mueller, 1989, p. 65. Hitler manipulating the world: Mueller, 1989, p 64.
137. We're doomed: See Mueller, 1989, p. 271, notes 2 and 4, and p. 98.
138. Morgenthau on World War III: Quoted in Mueller, 1995, p. 192.
139. Cold War superpowers stayed out of each other's way: This included Korea, where the Soviet Union provided only limited air support to its North Korean ally, and never closer than sixty miles from the battlefront.
140. Longest great power peace since the Roman Empire: Mueller, 1989, pp. 3–4; Gaddis, 1989.
141. No army crossing the Rhine: B. DeLong, "Let us give thanks (Wacht am Rhein Department)," Nov. 12, 2004, http://www.j-bradford-delong.net/movable_type/2004-2_archives/000536.html.
142. No interstate wars in Western Europe: Correlates of War Inter-State War Dataset (v3.0), Sarkees, 2000.
143. No interstate wars in Eastern Europe: Correlates of War Inter-State War Dataset (v3.0), Sarkees, 2000. This follows the CoW definition of *interstate war* as a conflict with one thousand casualties in a year and members of the interstate system on each side. The NATO bombing of Yugoslavia in 1999 is not included in the current CoW databases, which end in 1997; the PRIO Dataset counts it as an internationalized intrastate (civil) war, because NATO entered in support of the Kosovo Liberation Army. Note that Levy's criteria would include it as a war involving a great power.
144. No interstate wars between developed countries: Mueller, 1989, pp. 4 and 271, note 5.
145. The decline of conquest since 1948: An exhaustive review by the political scientist Mark Zacher (Zacher, 2001) lists seven: India-Goa (1961), Indonesia–West Irian (1961–62), China–Northeast Frontier (1962), Israel-Jerusalem/West Bank/Gaza/Golan (1967), North Vietnam–South Vietnam (1975), Iran–Strait of Hormuz Islands (1971), and China–Paracel Islands (1974). A few other successful aggressions resulted in minor changes or in the establishment of new political entities.
146. Greatest transfer of power in history: Sheehan, 2008, pp. 167–71.
147. No more colonial or imperial wars: Human Security Centre, 2005; Human Security Report Project, 2008.
148. No states eliminated: Zacher, 2001.
149. Twenty-two states occupied in first half of 20th century, none in second half: Russett & Oneal, 2001, p. 180.
150. Danes roaring for a fight: Mueller, 1989, p. 21.
151. Longest period of great power peace: Levy et al., 2001, p. 18.
152. Chance of one great power war in sixty-five years: In 1991 Levy had to exclude the Korean War to calculate that the probability of the number of observed great power wars since the end of World War II was just 0.005 (see Levy et al., 2001, note 11). Two decades later we don't have to make that judgment call to get massive statistical significance.
153. Improbability of the Long Peace: The rates of onsets per year of wars between great powers, and of onsets per year for wars with a great power on at least one side, between 1495 and 1945, were taken from Levy, 1983, table 4.1, pp. 88–91. The rates of onsets per year for wars between European states from 1815 through 1945 were taken from the Correlates of War Dataset, Sarkees, 2000. These were multiplied by 65 to generate the lambda parameter for a Poisson distribution, and the probability of drawing the number of observed wars or fewer was calculated from that distribution.
154. Precocious assessments of war decline: Levi, 1981; Gaddis, 1986; Holsti, 1986; Luard, 1988; Mueller, 1989; Fukuyama, 1989; Ray, 1989; Kaysen, 1990.
155. "Postwar" world: Jervis, 1988, p. 380.
156. Anxious prediction: Kaysen, 1990, p. 64.
157. War no longer desirable: Keegan, 1993, p. 59.
158. War may not recur: Howard, 1991, p. 176.
159. Striking discontinuity: Luard, 1986, p. 77.

160. Nothing like it in history: Gat, 2006, p. 609.
161. Press gangs: Payne, 2004, p. 73.
162. Shrinking conscription: Sheehan, 2008, p. 217.
163. Conscription in forty-eight nations: Payne, 2004; International Institute for Strategic Studies, 2010; Central Intelligence Agency, 2010.
164. Proportion of population in uniform as best indicator of militarism: Payne, 1989.
165. Average proportion of soldiers: Unweighted average of 63 countries in existence during the entire timespan, from Correlates of War National Material Capabilities Dataset (1816–2001), Sarkees, 2000, http://www.correlatesofwar.org.
166. Skittishness about Universal Declaration of Human Rights: Hunt, 2007, pp. 202–3.
167. Downgrading the nation-state: V. Havel, "How Europe could fail," *New York Review of Books*, Nov. 18, 1993, p. 3.
168. Territory most important issue in war: Vasquez, 2009, pp. 165–66.
169. Territorial integrity norm: Zacher, 2001.
170. Cognitive landmarks in positive-sum negotiation: Schelling, 1960.
171. Higher value on life: Luard, 1986, p. 268.
172. Khrushchev: Quoted in Mueller, 2004a, p. 74.
173. Carter's restraint: "Carter defends handling of hostage crisis," *Boston Globe*, Nov. 17, 2009.
174. Saving face during the Cuban Missile Crisis: Glover, 1999, p. 202.
175. RFK on Cuban Missile Crisis: Kennedy, 1969/1999, p. 49.
176. Khrushchev and Kennedy pulling on a knot: Quoted in Glover, 1999, p. 202.
177. Cold War as ladder versus escalator: Mueller, 1989.
178. Military aversion to gratuitous killing: Hoban, 2007; Jack Hoban, personal communication, Nov. 14, 2009.
179. The Ethical Marine Warrior: Hoban, 2007, 2010.
180. "The Hunting Story": Humphrey, 1992.
181. Relatively few battle deaths: According to the PRIO dataset (Lacina & Gleditsch, 2005), there were 14,200 battle deaths in Afghanistan from 2001 to 2008, and 13,500 battle deaths in Iraq. Deaths from intercommunal conflict are higher; they will be discussed in the next chapter.
182. New anti-insurgency campaign: N. Shachtman, "The end of the air war," *Wired*, Jan. 2010, pp. 102–9.
183. Smart targeting kills fewer civilians: Goldstein, 2011.
184. Afghanistan civilian deaths: Bohannon, 2011. My estimate of 5,300 for 2004–10 is the sum of the 4,024 civilian deaths reported for 2004–9 (p. 1260) and 1,152 deaths estimated for 2010 (55 percent of the total of 2,537 for 2009 and 2010, p. 1257). Civilian battle deaths in Vietnam estimated at 843,000 by Rummel (1997), table 6.1.A. This jibes with the total battle-death estimate (civilians plus soldiers) of 1.6 million from the PRIO New War Dataset (Gleditsch et al., 2002; Lacina, 2009) combined with the assumption, discussed in chap. 6, that approximately half of all battle deaths represent civilians.
185. Europeans are from Venus: Kagan, 2002.
186. "Down with This Sort of Thing": Sheehan, 2008.
187. This ain't the Wehrmacht: E-mail from Kabul, Dec. 11, 2003.
188. Scared straight by nuclear terror: See Mueller, 1989, p. 271, note 3. For a recent example, see van Creveld, 2008.
189. Sturdy child of terror: W. Churchill, "Never Despair," speech to House of Commons, Mar. 1, 1955.
190. Nobel for the nukes: Quoted in Mueller, 2004a, p. 164.
191. Long Peace not a nuclear peace: Mueller, 1989, chap. 5; Ray, 1989, pp. 429–31.
192. Dynamite peace: Quoted in Ray, 1989, p. 429.
193. Failure of new weapons to implement peace: Ray, 1989, pp. 429–30.
194. Poison gas from planes: Mueller, 1989.
195. Nondeterrence by weapons of mass destruction: Luard, 1986, p. 396.
196. Calling nuclear powers' bluff: Ray, 1989, p. 430; Huth & Russett, 1984; Kugler, 1984; Gochman & Maoz, 1984, pp. 613–15.
197. Nuclear taboo: Schelling, 2000, 2005; Tannenwald, 2005b.
198. Neutron bomb compatible with just war: Tannenwald, 2005b, p. 31.
199. Not quite a taboo: Paul, 2009; Tannenwald, 2005b.
200. Daisy ad: "Daisy: The Complete History of an Infamous and Iconic Ad," http://www.conelrad.com/daisy/index.php.
201. Mark of Cain: Quoted in Tannenwald, 2005b, p. 30.
202. Sanctification of Hiroshima: Quoted in Schelling, 2005, p. 373.
203. Gradual emergence of nuclear taboo: Schelling, 2005; Tannenwald, 2005b. Dulles quote from Schelling, 2000, p. 1.
204. Eisenhower on nukes: Schelling, 2000, p. 2.
205. Johnson ratifies taboo: Schelling, 2000, p. 3.
206. Kennedy's prediction of nuclear proliferation: Mueller, 2010a, p. 90.

207. Predictions that Germany and Japan would go nuclear: See Mueller, 2010a, p. 92.
208. Condemned by the civilized world: Quoted in Price, 1997, p. 91.
209. Churchill on chemical weapons: Quoted in Mueller, 1989, p. 85.
210. Accidental release of poison gas: Mueller, 1989, p. 85; Price, 1997, p. 112.
211. Saddam hanged for chemical attack: "Charges facing Saddam Hussein," BBC News, Jul. 1, 2004, http://news.bbc.co.uk/2/hi/middle_east/3320293.stm.
212. Small death toll from gas: Mueller, 1989, p. 84; Mueller, 2004a, p. 43.
213. *Venomous woman:* Hallissy, 1987, pp. 5–6, quoted in Price, 1997, p. 23.
214. "A World Free of Nuclear Weapons": Perry, Shultz, Kissinger, & Nunn, 2008; Shultz, Perry, Kissinger, & Nunn, 2007.
215. Three-quarters of former secretaries of defense and state and NSAs: Shultz, 2009, p. 81. Global Zero: www.globalzero.org.
216. Road map for Global Zero: Global Zero Commission, 2010.
217. Global Zero skeptics: Schelling, 2009; H. Brown & J. Deutch, "The nuclear disarmament fantasy," *Wall Street Journal*, Nov. 19, 2007, p. A19. Global Zero planning: B. Blechman, "Stop at Start," *New York Times*, Feb. 19, 2010.
218. Democracy is obsolete: D. Moynihan, "The American experiment," *Public Interest*, Fall 1975, quoted in Mueller, 1995, p. 192. See Gardner, 2010, for other examples.
219. Democracies and autocracies: Marshall & Cole, 2009.
220. Democratic Peace debate: Mueller, 1989, 2004a; Ray, 1989; Rosato, 2003; White, 2005b.
221. Democratic Peace is a Pax Americana: Rosato, 2003.
222. Democratic Peace strikes back: Russett & Oneal, 2001; White, 2005b.
223. A statistical Democratic Peace: Russett & Oneal, 2001.
224. Militarized interstate disputes: Gochman & Maoz, 1984; Jones, Bremer, & Singer, 1996.
225. Copious militarized interstate disputes: The analysis presented in Russett & Oneal, 2001, was based on the Correlates of War Project's Militarized Interstate Dispute 2.1 Dataset (Jones et al., 1996), which runs from 1885 to 1992 (see also Gochman & Maoz, 1984). Russett, 2008, has since extended it with data from the 3.0 database (Ghosn, Palmer, & Bremer, 2004), which runs through 2001.
226. Support for the Democratic Peace: Russett & Oneal, 2001, pp. 108–11; Russett, 2008, 2010.
227. Democracies more peaceful across the board: Russett & Oneal, 2001, p. 116.
228. No Autocratic Peace: Russett & Oneal, 2001, p. 115.
229. Democratic Peace not a Pax Americana: Russett & Oneal, 2001, p. 112.
230. No Pax Americana or Pax Britannica: Russett & Oneal, 2001, pp. 188–89.
231. Peace among new democracies: Russett & Oneal, 2001, p. 121.
232. No Democratic Peace in 19th century: Russett & Oneal, 2001, p. 114.
233. Liberal Peace: Gleditsch, 2008; Goldstein & Pevehouse, 2009; Schneider & Gleditsch, 2010.
234. Golden Arches Peace: The idea is usually attributed to the journalist Thomas Friedman. An earlier marginal exception was the U.S. attack on Panama in 1989, but its death count falls short of the minimum required for a war according to the standard definition. The 1999 Kargil War between Pakistan and India may be another exception, depending on whether Pakistani forces are counted as independent guerrillas or government soldiers; see White, 2005b.
235. Skeptics of trade-peace connection: Gaddis, 1986, p. 111; Ray, 1989.
236. Raiding and trading: Keegan, 1993, e.g., p. 126.
237. Trade between England and Germany before World War I: Ferguson, 2006.
238. Trade explains World War I after all: Gat, 2006, pp. 554–57; Weede, 2010.
239. Trade reduces conflict: Russett & Oneal, 2001, pp. 145–48. Even controlling for economic growth: p. 153.
240. Trade and development: Hegre, 2000.
241. Openness to the global economy: Russett & Oneal, 2001, p. 148.
242. Democracy is only dyadic; commerce is monadic too: McDonald, 2010; Russett, 2010.
243. Capitalist Peace: Gartzke, 2007; Gartzke & Hewitt, 2010; McDonald, 2010; Mousseau, 2010; Mueller, 1999, 2010b; Rosecrance, 2010; Schneider & Gleditsch, 2010; Weede, 2010.
244. Pitting democracy and capitalism against each other: Gartzke & Hewitt, 2010; McDonald, 2010; Mousseau, 2010; but see also Russett, 2010.
245. "Make money, not war": Gleditsch, 2008.
246. Scientists and world government: Mueller, 1989, p. 98.
247. Bertrand Russell and preemptive nuclear war: Mueller, 1989, pp. 109–10; Sowell, 2010, chap. 8.
248. European economic communities: Sheehan, 2008, pp. 158–59.
249. Peace and European IGOs: Sheehan, 2008.
250. All three variables: Russett, 2008.
251. International relations and moral norms: Cederman, 2001; Mueller, 1989, 2004a, 2007; Nadelmann, 1990; Payne, 2004; Ray, 1989.
252. "Realism" in international relations: See Goldstein & Pevehouse, 2009; Ray, 1989; Thayer, 2004.

253. Ideas and the end of the Cold War: Bennett, 2005; English, 2005; Tannenwald, 2005a; Tannenwald & Wohlforth, 2005; Thomas, 2005.
254. *Glasnost:* A. Brown, "When Gorbachev took charge," *New York Times*, Mar. 11, 2010.
255. Kant on nations' learning from experience: Kant, 1784/1970, p. 47, quoted in Cederman, 2001.
256. Back to Kant: Cederman, 2001.
257. Learning from mistakes: See also Dershowitz, 2004a.

Chapter 6: The New Peace

1. Expert predictions: Gardner, 2010; Mueller, 1995, 2010a.
2. "great power rivalries": Quoted in S. McLemee, "What price Utopia?" (review of J. Gray's *Black mass*), *New York Times Book Review*, Nov. 25, 2007, p. 20.
3. "blood-soaked course": S. Tanenhaus, "The end of the journey: From Whittaker Chambers to George W. Bush," *New Republic*, Jul. 2, 2007, p. 42.
4. "nothing but terrorism... and genocides": Quoted in C. Lambert, "Le Professeur," *Harvard Magazine*, Jul.–Aug., 2007, p. 36.
5. "more dangerous place than ever": Quoted in C. Lambert, "Reviewing 'reality,'" *Harvard Magazine*, Mar.–Apr. 2007, p. 45.
6. "the same damned thing": M. Kinsley, "The least we can do," *Atlantic*, Oct. 2010.
7. False sense of insecurity: Leif Wenar, quoted in Mueller, 2006, p. 3.
8. "new wars": Kaldor, 1999.
9. Existential threats: For quotations, see Mueller, 2006, pp. 6, 45.
10. Decline of deaths and displacements in "new wars": Melander, Oberg, & Hall, 2009; Goldstein, 2011; Human Security Report Project, 2011.
11. Conflict Catalog: Brecke, 1999, 2002; Long & Brecke, 2003.
12. PRIO Battle Deaths Dataset: Lacina & Gleditsch, 2005; http://www.prio.no/CSCW/Datasets/Armed-Conflict/Battle-Deaths/.
13. Related conflict datasets: UCDP: http://www.prio.no/CSCW/Datasets/Armed-Conflict/UCDP-PRIO/. SIPRI: www.sipri.org, Stockholm International Peace Research Institute, 2009. Human Security Report Project: http://www.hsrgroup.org/; Human Security Centre, 2005, 2006; Human Security Report Project, 2007, 2008, 2009.
14. Categories of armed conflict: Human Security Report Project, 2008, p. 10; Hewitt, Wilkenfeld, & Gurr, 2008; Lacina, 2009.
15. Genocide more destructive: The calculation that genocide killed more people in the 20th century than war was first performed by Rummel, 1994, replicated by White, 2005a, and captured in the title of Goldhagen's 2009 book on genocide, *Worse than war.* Matthew White (in press) points out that the comparison depends on how one classifies wartime genocides, which make up half the genocide deaths. Most deaths in the Holocaust, for example, depended on Germany's conquest of Europe. If wartime genocides are added to battle deaths, then wars are worse, 105 million to 40 million. If they are lumped with peacetime genocides, then genocides are worse, 81 million to 64 million. (Neither figure includes deaths from famines.)
16. Genocide databases: Eck & Hultman, 2007; Harff, 2003, 2005; Rummel, 1994, 1997; "One-Sided Violence Dataset" in http://www.pcr.uu.se/research/ucdp/datasets/.
17. Death categories: PRIO Documentation of Coding Decisions, Lacina, 2009, pp. 5–6; Human Security Report Project, 2008.
18. Concept of causation: Pinker, 2007b, pp. 65–73, 208–25.
19. World War I and the 1918 flu pandemic: Oxford et al., 2002.
20. Low rate of battle deaths: Average for 2000–2005, from the state-based battle-death figures reported in Human Security Report Project, 2007, based on the UCDP/PRIO dataset, Gleditsch et al., 2002. Population figures are from *International Data Base*, U.S. Census Bureau, 2010c.
21. Mean homicide rate: Krug et al., 2002, p. 10.
22. All death figures in these paragraphs are taken from the PRIO dataset: Gleditsch et al., 2002; Lacina, 2009; Lacina & Gleditsch, 2005. The dataset differs slightly from the UCDP/PRIO dataset that went into the three graphs: Human Security Centre, 2006; Human Security Report Project, 2007.
23. Interstate wars: PRIO New war dataset, "Best Estimates" for battle fatalities. Gleditsch et al., 2002; Lacina, 2009.
24. China's peaceful rise: Bijian, 2005; Weede, 2010; Human Security Report Project, 2011. Turkey's "zero problems with neighbors" policy: "Ahmet Davutoglu," *Foreign Policy*, Dec. 2010, p. 45. Brazilian boast: S. Glasser, "The FP Interview: The Soft-Power Power" (interview with Celso Amorim), *Foreign Policy*, Dec. 2010, p. 43.
25. East Asian peace: Human Security Report Project, 2011, chaps. 1, 3.
26. Decreasing lethality of civil wars: Marshall & Cole, 2009, p. 114.
27. Declining lethality of all wars: Human Security Report Project, 2009, p. 2.

28. Poverty and war: Human Security Centre, 2005, p. 152, using data from Macartan Humphreys and Ashutosh Varshney.
29. Poverty may not cause violent resource competition: Fearon & Laitin, 2003; Theisen, 2008.
30. War as development in reverse: Human Security Report Project, 2008, p. 5; Collier, 2007.
31. Richer governments can keep the peace: Human Security Report Project, 2011, chaps. 1, 3.
32. Structural variables change slowly: Human Security Report Project, 2007, p. 27.
33. Police-induced love: Quoted in Goldhagen, 2009, p. 212.
34. Bad policing: Fearon & Laitin, 2003; Mueller, 2004a.
35. He's our S.O.B.: Roosevelt may not have originated the trope; see http://message.snopes.com/showthread.php?t=8204.
36. Proxy wars only part of the decline: Human Security Centre, 2005, p. 153.
37. Mao indifferent to deaths: Quoted in Glover, 1999, p. 297.
38. Mao OK with half of humanity dying: Quoted in Mueller, 2010a.
39. Vietnamese willingness to absorb casualties: Mueller, 2004a, pp. 76–77. American misestimation: Blight & Lang, 2007.
40. Grow beards or remain in Europe: C. J. Chivers & M. Schwirtz, "Georgian president vows to rebuild army," *New York Times*, Aug. 24, 2008.
41. Anocracies: Human Security Report Project, 2007, 2008; Marshall & Cole, 2009.
42. The trouble with anocracies: Marshall & Cole, 2008. See also Pate, 2008, p. 31.
43. Distribution of anocracies: Human Security Report Project, 2008, pp. 48–49.
44. Resource curse: Collier, 2007; Faris, 2007; Ross, 2008.
45. Bottom billion in 14th century: Collier, 2007, p. 1.
46. Remnants of war: Mueller, 2004a, p. 1.
47. General Butt Naked: Mueller, 2004a, p. 103.
48. Statistics of civil wars: Fearon & Laitin, 2003, p. 76.
49. Improvement in African governments: Human Security Report Project, 2007, pp. 26–27.
50. African democratic leaders: R. Rotberg, "New breed of African leader," *Christian Science Monitor*, Jan. 9, 2002.
51. International pressure: Human Security Report Project, 2007, pp. 28–29; Human Security Centre, 2005, pp. 153–55.
52. Democracies don't have large civil wars: Gleditsch, 2008; Lacina, 2006.
53. Globalization reduces civil conflict: Blanton, 2007; Bussman & Schneider, 2007; Gleditsch, 2008, pp. 699–700.
54. Peacekeeping: Fortna, 2008; Goldstein, 2011.
55. Civil wars accumulated: Hewitt et al., 2008, p. 24; Human Security Report Project, 2008, p. 45. Rates of onset and termination: Fearon & Laitin, 2003.
56. Rise of peacekeeping and international activism: Human Security Centre, 2005, pp. 153–55; Fortna, 2008; Gleditsch, 2008; Goldstein, 2011.
57. Blessed are the peacekeepers: Fortna, 2008, p. 173.
58. Peace and cocktail parties: Fortna, 2008, p. 129.
59. "If Kabbah go": Fortna, 2008, p. 140.
60. Peacekeepers save face: Fortna, 2008, p. 153.
61. Lack of record-keeping by the UN: Human Security Centre, 2005, p. 19.
62. Warlord death count: Rummel, 1994, p. 94.
63. Decline of deaths in nonstate conflicts: Human Security Report Project, 2007, pp. 36–37; Human Security Report Project, 2011.
64. Iraq body counts: Fischer, 2008.
65. Democratic Republic of the Congo: The DRC total of 147,618 is the sum of the "best estimate" battle deaths from 1998 through 2008. The all-wars total of 9.4 million is the geometric mean of the sums of the low and high battle-death estimates. Both from the PRIO Battle Deaths Dataset, 1946–2008, Version 3.0, http://www.prio.no/CSCW/Datasets/Armed-Conflict/Battle-Deaths/, Lacina & Gleditsch, 2005.
66. Myth of reversal in civilian war deaths: Human Security Centre, 2005, p. 75; Goldstein, 2011; Roberts, 2010; White, in press.
67. Civilian deaths in the Civil War: Faust, 2008.
68. *Lancet* study: Burnham et al., 2006.
69. Bias in epidemiological studies: Human Security Report Project, 2009; Johnson et al., 2008; Spagat, Mack, Cooper, & Kreutz, 2009.
70. Fudge factor: Bohannon, 2008.
71. Retrospective surveys of war deaths: Obermeyer, Murray, & Gakidou, 2008.
72. The trouble with surveys: Spagat et al., 2009.
73. Claim of 5.4 million deaths in DRC: Coghlan et al., 2008.
74. Problems with DRC estimate: Human Security Report Project, 2009.
75. Famine and disease decline during war: Human Security Report Project, 2009.
76. Lives saved by vaccination: Human Security Report Project, 2009, p. 3.

77. Indirect deaths have probably decreased: Human Security Report Project, 2009, p. 27.
78. Definitions of *genocide, politicide, democide:* Rummel, 1994, p. 31. Reviews of genocide: Chalk & Jonassohn, 1990; Chirot & McCauley, 2006; Glover, 1999; Goldhagen, 2009; Harff, 2005; Kiernan, 2007; Payne, 2004; Power, 2002; Rummel, 1994; Valentino, 2004.
79. 170 million victims of democide: Rummel, 1994, 1997.
80. Rummel's highball estimates: White, 2010c, note 4; Dulić, 2004a, 2004b; Rummel, 2004.
81. More conservative estimate of democide victims: White, 2005a, in press.
82. Democides in wartime: White, in press; see also note 15, this chapter.
83. Mass drownings: Bell, 2007a, pp. 182–83; Payne, 2004, p. 54.
84. *Einsatzgruppen* more lethal than gas chambers: Goldhagen, 2009, p. 124. Earlier mobile killing squads: Keegan, 1993, p. 166.
85. Tutsi techniques of mass killing: Goldhagen, 2009, p. 120.
86. What "razing" meant: Chalk & Jonassohn, 1990, p. 7.
87. Biblical starvation: Deuteronomy 28:52–57; translation from Kugel, 2007, pp. 346–47.
88. Torture and mutilation during democides: Goldhagen, 2009; Power, 2002; Rummel, 1994.
89. Artistically cruel: Dostoevsky, 1880/2002, p. 238.
90. Cultural Revolution death toll: Rummel, 1994, p. 100. Though see also Harff, 2003, who gives a more conservative estimate.
91. Red Guards ransacking: Glover, 1999, p. 290.
92. Christian Wirth: Glover, 1999, p. 342.
93. Psychology of categories: Pinker, 1997, pp. 306–13; Pinker, 1999/2011, chap. 10.
94. Accuracy of stereotypes: Jussim, McCauley, & Lee, 1995; Lee, Jussim, & McCauley, 1995; McCauley, 1995.
95. Applying stereotypes categorically when pressured: Jussim et al., 1995.
96. Categories reflect attitudes: Jussim et al., 1995; Lee et al., 1995; McCauley, 1995.
97. Essentializing social groups: Gelman, 2005; Gil-White, 1999; Haslam, Rothschild, & Ernst, 2000; Hirschfeld, 1996; Prentice & Miller, 2007.
98. Taxonomies of motives for democide: Chalk & Jonassohn, 1990; Chirot & McCauley, 2006; Goldhagen, 2009; Harff, 2003; Valentino, 2004.
99. Genocide for convenience: Goldhagen, 2009.
100. Roman massacre of Alexandrian Jews: Kiernan, 2007, p. 14.
101. Hobbesian trap in former Yugoslavia: Glover, 1999; Goldhagen, 2009.
102. Aristotle on hatred: Quoted in Chirot & McCauley, 2006, pp. 72–73.
103. Jivaro genocide: Quoted in Daly & Wilson, 1988, p. 232.
104. Fandi genocide: Quoted in Daly & Wilson, 1988, pp. 231–32.
105. Metaphor and analogy in cognition: Pinker, 2007b, chap. 5.
106. Vermin metaphors: Chirot & McCauley, 2006; Goldhagen, 2009; Kane, 1999; Kiernan, 2007.
107. "Kill the nits": Kane, 1999. Quote from Kiernan, 2007, p. 606.
108. Yuki: Quoted in Chalk & Jonassohn, 1990, p. 198.
109. Cheyenne: Quoted in Kiernan, 2007, p. 606; Kane, 1999.
110. Biological metaphors for Jews: Chirot & McCauley, 2006, pp. 16, 42; Goldhagen, 2009.
111. Psychology of disgust: Curtis & Biran, 2001; Rozin & Fallon, 1987; Rozin, Markwith, & Stoess, 1997.
112. Moralization of disgust: Haidt, 2002; Haidt et al., 2000; Haidt, Koller, & Dias, 1993; Rozin et al., 1997; Shweder, Much, Mahapatra, & Park, 1997. See also "Morality and Taboo" in chap. 9 of this book.
113. Primo Levi, *The drowned and the saved*, quoted in Glover, 1999, pp. 88–89.
114. Dehumanization and demonization in genocide: Goldhagen, 2009. See also Haslam, 2006.
115. Murderous ideology: Epigraph for this chapter, from Solzhenitsyn, 1973/1991, pp. 173–74.
116. Ancestry myths: Geary, 2002.
117. Intuitive economics: Caplan, 2007; Caplan & Miller, 2010; Fiske, 1991, 1992, 2004a; Sowell, 1980, 2005.
118. Mobility of middlemen minorities: Sowell, 1996.
119. Violence against middlemen minorities: Chirot, 1994; Courtois et al., 1999; Glover, 1999; Horowitz, 2001; Sowell, 1980, 2005.
120. Marxism and Christianity: Chirot & McCauley, 2006, pp. 142–43.
121. Nazism and the book of Revelation: Chirot & McCauley, 2006, p. 144. See also Ericksen & Heschel, 1999; Goldhagen, 1996; Heschel, 2008; Steigmann-Gall, 2003.
122. Psychological traits of utopian tyrants: Chirot, 1994; Glover, 1999; Oakley, 2007.
123. Mao's callousness: Glover, 1999, p. 291.
124. Mao's harebrained schemes: Chirot & McCauley, 2006, p. 144; Glover, 1999, pp. 284–86.
125. Neighboring ethnic groups usually don't commit genocide: Brown, 1997; Fearon & Laitin, 1996; Harff, 2003; Valentino, 2004.
126. Germans anti-Semitic but not genocidal: Valentino, 2004, p. 24.
127. Democides committed by armed minority: Mueller, 2004a; Payne, 2005; Valentino, 2004.
128. Division of labor in democides: Valentino, 2004. See also Goldhagen, 2009.

129. Genocides end when leaders die or are toppled: Valentino, 2004.
130. Ancient historians: Chalk & Jonassohn, 1990, p. 58.
131. Genocide as old as history: Chalk & Jonassohn, 1990, p. xvii.
132. Table of contents of genocides: Kiernan, 2007, p. 12.
133. Democide counts: Rummel, 1994, pp. 45, 70; see also Rummel, 1997, for the raw data. The numerical precision in his estimates is not intended to be an accurate count but to allow others to verify his sources and calculations.
134. Increase Mather praises genocide: Chalk & Jonassohn, 1990, p. 180.
135. Moabites return the favor: Payne, 2004, p. 47.
136. Immortal soul justifies murder: Bhagavad-Gita, 1983, pp. 74, 87, 106, 115, quoted in Payne, 2004, p. 51.
137. Cromwell: Quoted in Payne, 2004, p. 53.
138. Parliamentary reaction: Quoted in Payne, 2004, p. 53.
139. Voice in the wilderness: Quoted in Kiernan, 2007, pp. 82–85.
140. Military codes of honor: Chirot & McCauley, 2006, pp. 101–2. Nothing wrong with genocide: Payne, 2004, pp. 54–55; Chalk & Jonassohn, 1990, pp. 199, 213–14; Goldhagen, 2009, p. 241.
141. Roosevelt on Indians: Courtwright, 1996, p. 109.
142. Lawrence and lethal chamber: Carey, 1993, p. 12.
143. Extermination of the Japanese: Mueller, 1989, p. 88.
144. Origins of the word *genocide:* Chalk & Jonassohn, 1990.
145. Holocaust denial: Payne, 2004, p. 57.
146. Novelty of genocide memoirs: Chalk & Jonassohn, 1990, p. 8.
147. Conception of history: Chalk & Jonassohn, 1990, p. 8.
148. Rummel's estimation methods: Rummel, 1994, pp. xvi–xx; Rummel, 1997. See also White, 2010c, note 4, for caveats.
149. Definition of "democide": Rummel, 1994, chap 2.
150. Great Leap Forward: Rummel has since changed his mind because of revelations that Mao knew about the devastation as it was taking place (Rummel, 2002), but I will stick with his original numbers.
151. Pseudo-governments versus recognized governments: White, 2010c, note 4.
152. Governments prevent more deaths than they cause: White, 2007.
153. Death tolls from major democides: Rummel, 1994, p. 4.
154. Democratic, authoritarian, and totalitarian governments and their death tolls: Rummel, 1997, p. 367.
155. Death rates from democratic, authoritarian, and totalitarian governments: Rummel, 1994, p. 15.
156. Totalitarian versus democratic governments: Rummel, 1994, p. 2.
157. Form of government matters: Rummel, 1997, pp. 6–10; see also Rummel, 1994.
158. Problem and solution of democide: Rummel, 1994, p. xxi.
159. Hemoclysm: Note that figure 6–7 double-counts some of the deaths in figure 6–1, because Rummel classifies many civilian battle deaths as democides. Also double-counted are some of the deaths in Matthew White's table on p. 195, which folds wartime genocides into the total count for wars.
160. Trends in democide and democracy: Rummel, 1997, p. 471. Rummel's regression analyses supporting this point, however, are problematic.
161. Perpetrators of Rwanda genocide: Mueller, 2004a, p. 100.
162. Rwanda genocide was preventable: Goldhagen, 2009; Mueller, 2004a; Power, 2002.
163. New genocide dataset: Harff, 2003, 2005; Marshall et al., 2009.
164. UCDP One-sided Violence Dataset: Kreutz, 2008; Kristine & Hultman, 2007; http://www.pcr.uu.se/research/ucdp/datasets/.
165. Mass killings in mid-1960s to late 1970s: Death figures are geometric means of the ranges in table 8.1 in Harff, 2005, except Darfur, which is taken by converting the magnitude entries from the PITF database to the geometric means of the ranges spelled out in Marshall et al., 2009, and summing them for the years 2003 through 2008.
166. Overestimated genocide death tolls: The Bosnian massacres, for example, probably killed closer to 100,000 than 200,000 people; Nettelfield, 2010. On conflict numbers, see Andreas & Greenhill, 2010.
167. Risk factors for democide: Harff, 2003, 2005.
168. Effects of instability: Harff, 2003, p. 62.
169. Pathways to democide: Harff, 2003, p. 61.
170. Hitler read Marx: Watson, 1985. Fraternal twins: P. Chaunu, cited in Besançon, 1998. See also Bullock, 1991; Courtois et al., 1999; Glover, 1999.
171. Decline of democide and fade of communism: Valentino, 2004, p. 150.
172. Utopian omelets: Sometimes attributed to the journalist Walter Duranty, a *New York Times* correspondent in the Soviet Union in the 1930s, but identified as "Anonymous: French" by *Bartlett's Familiar Quotations*, 17th ed.

173. Humans are not eggs: Pipes, 2003, p. 158.
174. Fewer democides in the coming century: Valentino, 2004, p. 151.
175. No Hitler, no Holocaust: Himmelfarb, 1984.
176. No Stalin, no Purge: Quoted in Valentino, 2004, p. 61.
177. No Mao, no Cultural Revolution: Quoted in Valentino, 2004, p. 62.
178. Worldwide terrorism toll: Memorial Institute for the Prevention of Terrorism (dataset no longer publicly available), reported in Human Security Centre, 2006, p. 16.
179. Existential threats: See Mueller, 2006, for quotations.
180. Prophecy of terrorist doom: R. A. Clarke, "Ten years later," *Atlantic*, Jan.–Feb. 2005.
181. Terrorist attacks fall into a power-law distribution: Clauset, Young, & Gleditsch, 2007.
182. Rarity of destructive terrorist attacks: Global Terrorism Database, START (National Consortium for the Study of Terrorism and Responses to Terrorism, 2010; http://www.start.umd.edu/gtd/), accessed on Apr. 21, 2010. The figure excludes terrorist attacks associated with the Rwanda genocide.
183. Overblown terrorist fears: Mueller, 2006.
184. Rates of death: National Vital Statistics for the year 2007, Xu, Kochanek, & Tejada-Vera, 2009, table 2.
185. More deaths from deer than terrorists: Mueller, 2006, note 1, pp. 199–200; National Safety Council statistics conveniently summarized in http://danger.mongabay.com/injury death.htm.
186. Excess deaths from avoiding plane travel: Gigerenzer, 2006.
187. Psychology of risk: Slovic, 1987; Slovic et al., 1982. See also Gigerenzer, 2006; Gigerenzer & Murray, 1987; Kahneman, Slovic, & Tversky, 1982; Ropeik & Gray, 2002; Tversky & Kahneman, 1973, 1974, 1983.
188. Other benefits of exaggerated fears: Daly & Wilson, 1988, pp. 231–32, 237, 260–61.
189. Policy distortions from cognitive illusions: Mueller, 2006; Slovic, 1987; Slovic et al., 1982; Tetlock, 1999.
190. Bin Laden bluster: "Timeline: The al-Qaida tapes," *Guardian*, http://www.guardian.co.uk/alqaida/page/0,12643,839823,00.html. See also Mueller, 2006, p. 3.
191. Kerry gaffe: Quoted in M. Bai, "Kerry's undeclared war," *New York Times*, Oct. 10, 2004.
192. History of terrorism: Payne, 2004, pp. 137–40; Cronin, 2009, p. 89.
193. Terrorists of the 1970s: Abrahms, 2006; Cronin, 2009; Payne, 2004.
194. Most terrorist groups fail; all die: Abrahms, 2006; Cronin, 2009; Payne, 2004.
195. Terrorism fails: See also Cronin, 2009, p. 91.
196. States immortal, terrorist groups not: Cronin, 2009, p. 110.
197. Terrorist groups never take over states: Cronin, 2009, p. 93. 6 percent success rate: Cronin, 2009, p. 215.
198. Decency as an international language: Cronin, 2009, p. 114. The Oklahoma City death toll of 165 is taken from the Global Terrorism Database (see n. 182).
199. Global Terrorism Database, START (National Consortium for the Study of Terrorism and Responses to Terrorism), 2010; accessed on Apr. 6, 2010.
200. Inclusion criteria: National Consortium for the Study of Terrorism and Responses to Terrorism, 2009.
201. Increase in Islamist suicide terrorism: Atran, 2006.
202. Proportion of deaths from Sunni terrorism: National Counterterrorism Center, 2009.
203. Lethality of suicide terrorism: Cronin, 2009, p. 67; note 145, p. 242. Minority of attacks, majority of casualties: Atran, 2006.
204. Ingredients of a suicide terrorist attack: Quoted in Atran, 2003.
205. Playing the battlefield odds: Tooby & Cosmides, 1988.
206. Cowardly warriors: Chagnon, 1997.
207. "between the devil and the deep blue sea": Valentino, 2004, p. 59.
208. Support for kin selection: Gaulin & McBurney, 2001; Lieberman, Tooby, & Cosmides, 2002.
209. Yanomamö village mates were related: Chagnon, 1988, 1997.
210. Chimpanzee troops: Wilson & Wrangham, 2003.
211. Perceived versus actual kinship: Daly, Salmon, & Wilson, 1997; Lieberman et al., 2002; Pinker, 1997.
212. Cues to kinship: Johnson, Ratwik, & Sawyer, 1987; Lieberman et al., 2002; Salmon, 1998.
213. Soldiers fight for bands of brothers: Mueller, 2004a; Thayer, 2004.
214. Manchester on brothers in arms: Quoted in Thayer, 2004, p. 183.
215. Why men love war: Broyles, 1984.
216. Better odds for suicide missions: Rapoport, 1964, pp. 88–89; Tooby & Cosmides, 1988.
217. Psychological profiles of suicide terrorists: Atran, 2003, 2006, 2010.
218. Tamil Tiger file closers: Atran, 2006.
219. Hamas carrots: Atran, 2003; Blackwell & Sugiyama, in press.
220. Men do foolish things in groups: Willer et al., 2009.
221. Sacred values and suicide terrorism: Atran, 2006, 2010; Ginges et al., 2007; McGraw & Tetlock, 2005.

222. Testimony on suicide terrorism: Atran, 2010.
223. Fate of suicide terrorism in Israel: Cronin, 2009, pp. 48–57, 66–67.
224. Physical barriers to terrorism: Cronin, 2009, p. 67. Other examples include the United States in Baghdad, and governments in Northern Ireland, Cyprus, and Lebanon.
225. Palestinian nonviolence: E. Bronner, "Palestinians try a less violent path to resistance," *New York Times*, Apr. 6, 2010.
226. Recent decline of suicide terrorism: S. Shane, "Rethinking which terror groups to fear," *New York Times*, Sept. 26, 2009.
227. Sinking favorables for Al Qaeda: Human Security Report Project, 2007.
228. al-Odah, "fostering a culture": Quoted in F. Zakaria, "The jihad against the jihadis," *Newsweek*, Feb. 12, 2010.
229. al-Odah, "My brother Osama": Quoted in P. Bergen & P. Cruickshank, "The unraveling: Al Qaeda's revolt against bin Laden," *New Republic*, Jun. 11, 2008.
230. Favorable response to criticism of Al Qaeda: P. Bergen & P. Cruickshank, "The unraveling: Al Qaeda's revolt against bin Laden," *New Republic*, Jun. 11, 2008.
231. Mufti's fatwa: F. Zakaria, "The jihad against the jihadis," *Newsweek*, Feb. 12, 2010.
232. Jihad as Sharia violation: Quoted in P. Bergen & P. Cruickshank, "The unraveling: Al Qaeda's revolt against bin Laden," *New Republic*, Jun. 11, 2008.
233. Killing innocents: Quoted in P. Bergen & P. Cruickshank, "The unraveling: Al Qaeda's revolt against bin Laden," *New Republic*, Jun. 11, 2008.
234. Endorsing violence against civilians: Quoted in F. Zakaria, "The jihad against the jihadis," *Newsweek*, Feb. 12, 2010.
235. Public opinion in war zones: Human Security Report Project, 2007, p. 19.
236. Cratering: F. Zakaria, "The only thing we have to fear . . . ," *Newsweek*, Jun. 2, 2008.
237. Opposition to attacks on civilians: Human Security Report Project, 2007, p. 15.
238. Iraq death toll: Iraq Body Count, http://www.iraqbodycount.org/database/, accessed Nov. 24, 2010. See also Human Security Report Project, 2007, p. 14.
239. Sunni Awakening: Human Security Report Project, 2007, p. 15.
240. The experts speak: Gardner, 2010; Mueller, 1995, 2010a.
241. Muslim armed conflicts: Nineteen out of 36 armed conflicts for 2008 in the PRIO database involved a Muslim country: Israel-Hamas, Iraq–Al-Mahdi, Philippines-MILF, Sudan-JEM, Pakistan-BLA, Afghanistan-Taliban, Somalia–Al-Shabaab, Iran-Jandulla, Turkey-PKK, India–Kashmir Insurgents, Mali-ATNMC, Algeria-AQIM, Pakistan-TTP, United States–Al Qaeda, Thailand–Pattani insurgents, Niger-MNJ, Russia–Caucasus Emirate, India-PULF, Djibouti-Eritrea. Thirty of 44 foreign terrorist organizations in the U.S. State Department's Country Reports on Terrorism 2008: http://www.state.gov/s/ct/rls/crt/2008/122449.htm, accessed Apr. 21, 2010.
242. Larger Muslim armies: Payne, 1989.
243. Muslim terrorist organizations: Thirty of 44 foreign terrorist organizations in the U.S. State Department's Country Reports on Terrorism 2008, http://www.state.gov/s/ct/rls/crt/2008/122449.htm, accessed Apr. 21, 2010.
244. Rarity of Muslim democracies: Esposito & Mogahed, 2007, p. 30.
245. Dubious democracies: Esposito & Mogahed, 2007, p. 30.
246. Fewer rights in Muslim countries: Pryor, 2007, pp. 155–56.
247. Cruel punishments: Payne, 2004, p. 156.
248. Oppression of women: Esposito & Mogahed, 2007, p. 117.
249. Slavery in the Muslim world: Payne, 2004, p. 156.
250. Witchcraft prosecution: A. Sandels, "Saudi Arabia: Kingdom steps up hunt for 'witches' and 'black magicians,'" *Los Angeles Times*, Nov. 26, 2009.
251. Exaggerated culture of honor: Fattah & Fierke, 2009; Ginges & Atran, 2008.
252. Genocidal quotes: see Goldhagen, 2009, pp. 494–504; Mueller, 1989, pp. 255–56.
253. *Arab Human Development Report:* United Nations Development Programme, 2003; see also R. Fisk, "UN highlights uncomfortable truths for Arab world," *Independent*, Jul. 3, 2002.
254. Insularity: "A special report on the Arab world," *Economist*, Jul. 23, 2009.
255. Muslim tolerance: Lewis, 2002, p. 114.
256. Printing as desecration: Lewis, 2002, p. 142.
257. Subversive humanities: "Iran launches new crackdown on universities," *Radio Free Europe/Radio Liberty*, August 26, 2010; http://www.rferl.org/content/Iran_Launches_New_Crackdown_On_Universities/2138387.html.
258. Clash of civilizations: Huntington, 1993.
259. What a billion Muslims really think: Esposito & Mogahed, 2007.
260. Muslim political organizations that endorse violence: Asal, Johnson, & Wilkenfeld, 2008.
261. Power law for terrorist attacks: Clauset & Young, 2005; Clauset et al., 2007.
262. Time to plan terrorist attacks: Mueller, 2006, p. 179.
263. Conjunction fallacy: Tversky & Kahneman, 1983.
264. Counting scenarios versus summing probabilities: Slovic et al., 1982.

265. Pay more for terrorism insurance: Johnson et al., 1993.
266. Taylor prediction: Mueller, 2010a, p. 162.
267. Allison prediction: Mueller, 2010a, p. 181.
268. Falkenrath prediction: Quoted in Parachini, 2003.
269. Negroponte and Garwin predictions: Mueller, 2010a, p. 181.
270. Expert reputations immune to falsified predictions: Gardner, 2010.
271. Assessing the risks of nuclear terrorism: Levi, 2007; Mueller, 2010a; Parachini, 2003; Schelling, 2005.
272. Not so mass destruction: Mueller, 2006; Mueller, 2010a.
273. Only three nonconventional terrorist attacks: Parachini, 2003.
274. Nations take their nukes seriously: Quoted in Mueller, 2010a, p. 166.
275. Pakistan won't fall to extremists: Human Security Report Project, 2007, p. 19.
276. Radioactive scrap metal: Mueller, 2010a, p. 166.
277. Herculean challenges to nuclear terrorism: Quoted in Mueller, 2010a, p. 185.
278. Murphy's Law of Nuclear Terrorism: Levi, 2007, p. 8.
279. Inevitable war with Iran: J. T. Kuhner, "The coming war with Iran: Real question is not if, but when," *Washington Times*, Oct. 4, 2009.
280. Not the end of the world: Mueller, 2010a, pp. 153–55; Lindsay & Takeyh, 2010; Procida, 2009; Riedel, 2010; P. Scoblic, "What are nukes good for?" *New Republic*, Apr. 7, 2010.
281. Fatwa against Islamic bomb: "Iran breaks seals at nuclear plant," CNN, Aug. 10, 2005, http://edition.cnn.com/2005/WORLD/europe/08/10/iran.iaea.1350/index.html.
282. Deadlines have passed: C. Krauthammer, "In Iran, arming for Armageddon," *Washington Post*, Dec. 16, 2005. By 2009: Y. K. Halevi & M. Oren, "Contra Iran," *New Republic*, Feb. 5, 2007.
283. Ahmadinejad interview: M. Ahmadinejad, interview by A. Curry, NBC News, Sept. 18, 2009, http://www.msnbc.msn.com/id/32913296/ns/world_news-mideastn_africa/print/1/displaymode/1098/.
284. "Wiping Israel off the map": E. Bronner, "Just how far did they go, those words against Israel?" *New York Times*, Jun. 11, 2006.
285. Predictions of consequences of North Korean nukes: Mueller, 2010a, p. 150.
286. Unlikelihood of nuclear proxies: Mueller, 2010a; Procida, 2009.
287. "Too precious to waste killing people": Schelling, 2005.
288. Climate stress as bad as nukes: T. F. Homer-Dixon, "Terror in the weather forecast," *New York Times*, Apr. 24, 2007.
289. Climate change justifies war on terror: Quoted in S. Giry, "Climate conflicts," *New York Times*, Apr. 9, 2007; see also Salehyan, 2008.
290. Climate change may not lead to war: Buhaug, 2010; Gleditsch, 1998; Salehyan, 2008; Theisen, 2008.
291. Terrorists are not subsistence farmers: Atran, 2003.

Chapter 7: The Rights Revolutions

1. Rough-and-tumble play: Boulton & Smith, 1992; Geary, 2010; Maccoby & Jacklin, 1987.
2. Recreational fighting: Geary, 2010; Ingle, 2004; Nisbett & Cohen, 1996.
3. Deadly ethnic riots: Horowitz, 2001.
4. Anatomy of a pogrom: Horowitz, 2001, chap. 1.
5. Reading the Riot Act: Payne, 2004, pp. 173–75.
6. American pogroms: Payne, 2004, pp. 180–81.
7. History of American lynching: Waldrep, 2002.
8. Antiblack communal violence: Payne, 2004, p. 180.
9. Decline of rioting: Payne, 2004, pp. 174, 180–82; Horowitz, 2001, p. 300.
10. Strange Fruit: "The best of the century," *Time*, Dec. 31, 1999.
11. FBI hate crime statistics: http://www.fbi.gov/hq/cid/civilrights/hate.htm.
12. No more deadly ethnic riots in the West: Horowitz, 2001, p. 561.
13. 1960s riots weren't race riots: Horowitz, 2001, pp. 300–301.
14. End of deadly ethnic riots in the West: Horowitz, 2001, p. 561.
15. Ruby Nell Bridges: Steinbeck, 1962/1997, p. 194.
16. No epidemic of hate crimes or church burnings: La Griffe du Lion, 2000; M. Fumento, "A church arson epidemic? It's smoke and mirrors," *Wall Street Journal*, Jul. 8, 1996.
17. No fatal violence against Muslims: Human Rights First, 2008. The closest examples were (1) the fatal assault of a Danish teenager of Turkish descent, but the police ruled out racism as a motive, and (2) video footage of the execution-style slayings of two men, possibly Dagestani and Tajik, by a Russian neo-Nazi group.
18. Policing riots: Horowitz, 2001, pp. 518–21.
19. Discriminatory policies predict ethnic violence: Gurr & Monty, 2000; Asal & Pate, 2005, pp. 32–33.

20. Decline of ethnic discrimination: Asal & Pate, 2005.
21. Discrimination around the world: Asal & Pate, 2005, pp. 35–36.
22. Discrimination in decline: Asal & Pate, 2005, p. 38.
23. Prediction of African American rampages: A. Hacker, *The end of the American era*, quoted in Gardner, 2010, p. 96.
24. "huge racial chasm": Quote from p. 219.
25. Bell on racism: Quoted in Bobo, 2001.
26. Attitudes of whites toward blacks: Bobo, 2001; see also Patterson, 1997.
27. Dropped from questionnaires: Bobo, 2001.
28. Religious tolerance: Caplow, Hicks, & Wattenberg, 2001, p. 116.
29. Racist Bugs Bunny: Search for "racist Bugs Bunny" on Youtube.com.
30. Funny Face: http://theimaginaryworld.com/ffpac.html.
31. Speech codes on campus: Kors & Silverglate, 1998. See also Foundation for Individual Rights in Education, www.thefire.org.
32. Self-parody: "Political correctness versus freedom of thought—The Keith John Sampson story," http://www.thefire.org/article/10067.html; "Brandeis University: Professor found guilty of harassment for protected speech," http://www.thefire.org/case/755.html.
33. Racial "insensitivity": Kors & Silverglate, 1998.
34. Rape in riots and genocides: Goldhagen, 2009; Horowitz, 2001; Rummel, 1994. Rape in war: Brownmiller, 1975; Rummel, 1994.
35. Traditional conceptions of rape: Brownmiller, 1975; Wilson & Daly, 1992.
36. Rape wisecrack by legal scholars: Brownmiller, 1975, p. 312.
37. Rape wisecracks by the police: Brownmiller, 1975, pp. 364–66.
38. Women barred from juries: Brownmiller, 1975, p. 296.
39. Interested parties: Thornhill & Palmer, 2000; Wilson & Daly, 1992; Jones, 1999.
40. Sexual economics: A. Dworkin, 1993, p. 119.
41. Sexual selection: Archer, 2009; Clutton-Brock, 2007; Symons, 1979; Trivers, 1972.
42. Harassment and rape in other species: Jones, 1999.
43. Risk factors for rape: Jones, 1999, 2000; Thornhill & Palmer, 2000.
44. Pregnancies from rape: Gottschall & Gottschall, 2003; Jones, 1999.
45. Sexual jealousy: Buss, 2000; Symons, 1979; Wilson & Daly, 1992.
46. Sex differences in jealousy: Buss, 2000.
47. Women as the property of men: Brownmiller, 1975; Wilson & Daly, 1992.
48. Changes in rape laws in the Middle Ages: Brownmiller, 1975. Current vestiges: Wilson & Daly, 1992.
49. Illicit jury deliberation: Brownmiller, 1975, p. 374.
50. Husband of rape victim: quote from Wilson & Daly, 1992. Divorce not uncommon: Brownmiller, 1975.
51. Evolution of abhorrence to rape: Symons, 1979; Thornhill & Palmer, 2000.
52. Agony of violation: Buss, 1989; Thornhill & Palmer, 2000.
53. Principle of autonomy: Hunt, 2007; Macklin, 2003.
54. Changes in rape law: Brownmiller, 1975.
55. Mary Astell: Quoted in Jaggar, 1983, p. 27.
56. *Newsweek* review of *A clockwork orange:* Quoted in Brownmiller, 1975, p. 302.
57. Kubrick: Quoted in Brownmiller, 1975, p. 302.
58. Marital rape: United Nations Development Fund for Women, 2003.
59. Rape taboo in video games: Personal communication from F. X. Shen, Jan. 13, 2007. Outrage at violations: http://www.gamegrene.com/node/447; http://www.cnn.com/2010/WORLD/asiapcf/03/30/japan.video.game.rape/index.html.
60. Rape may be overreported: Taylor & Johnson, 2008.
61. Junk rape statistics: Sommers, 1994, chap. 10; MacDonald, 2008.
62. Victimization survey: U.S. Bureau of Justice Statistics, Maston, 2010.
63. Changes in attitudes toward women: Spence, Helmreich, & Stapp, 1973; Twenge, 1997.
64. Abyss between the sexes: Salmon & Symons, 2001, p. 4.
65. Men underestimate, women overestimate: Buss, 1989.
66. Function of rape: Brownmiller, 1975, p. 15.
67. Myrmidons: Brownmiller, 1975, p. 209.
68. Problems with myrmidon theory: Check & Malamuth, 1985; Gottschall & Gottschall, 2001; Jones, 1999, 2000; MacDonald, 2008; Sommers, 1994; Thornhill & Palmer, 2000; Pinker, 2002, chap. 18.
69. Marxist inspiration for myrmidon theory: Brownmiller, 1975, pp. 210–11.
70. What guys want: MacDonald, 2008.
71. Motives for spousal abuse: Kimmel, 2002; Wilson & Daly, 1992.
72. Mate-guarding and domestic violence: Buss, 2000; Symons, 1979; Wilson & Daly, 1992.
73. Women as property across the world: Wilson & Daly, 1992.

74. Laws allowing husbands to chastise wives: Suk, 2009, p. 13.
75. Changes in laws allowing mate-guarding: Wilson & Daly, 1992.
76. Mandatory protection orders and prosecution: Suk, 2009, p. 10.
77. Wife-beating no big deal, selling LSD worse than rape: Rossi, Waite, Bose, & Berk, 1974.
78. When a man attacks a woman: Shotland & Straw, 1976.
79. Wife-beating now important: Johnson & Sigler, 2000.
80. Beating a wife with a belt: Johnson & Sigler, 2000.
81. Sexual symmetry in domestic violence: Archer, 2009; Straus, 1977/1978; Straus & Gelles, 1988.
82. Stereotype of rolling pin: Straus, 1977/1978, pp. 447–48.
83. Kinds of marital violence: Johnson, 2006; Johnson & Leone, 2005.
84. Men commit more serious domestic violence: Dobash et al., 1992; Graham-Kevan & Archer, 2003; Johnson, 2006; Johnson & Leone, 2005; Kimmel, 2002; Saunders, 2002.
85. Little change in small stuff: Straus, 1995; Straus & Kantor, 1994.
86. Decline began in 1985: Straus & Kantor, 1994, figure 2.
87. Women's shelters save abusive men's lives: Browne & Williams, 1989.
88. No reported trends for domestic violence in British Crime Survey Report: see Jansson, 2007.
89. Cross-national differences in domestic violence: Archer, 2006a.
90. WHO on domestic violence worldwide: Heise & Garcia-Moreno, 2002.
91. A fifth to a half: United Nations Population Fund, 2000.
92. Laws on violence against women: United Nations Development Fund for Women, 2003, appendix 1.
93. Atrocities against women: Kristof & WuDunn, 2009; United Nations Development Fund for Women, 2003.
94. Gender empowerment and individualism correlate with less violence against women: Archer, 2006a. The correlations reported in this paper did not control for affluence, but in a personal communication on May 18, 2010, Archer confirmed that both remain statistically significant when GDP per capita is entered into the regression.
95. Pressure to end violence against women: Kristof & WuDunn, 2009; United Nations Development Fund for Women, 2003.
96. International shaming campaigns: Nadelmann, 1990.
97. Momentum against violence against women: Statement by I. Alberdi, "10th Anniversary Statement on the UN International Day for the Elimination of Violence Against Women," UNIFEM, http://www.unifem.org/news_events/story_detail.php?StoryID=976.
98. Universal embrace of gender equality: Pew Research Center, 2010.
99. Muslim attitudes toward women's empowerment: Esposito & Mogahed, 2007; Mogahed, 2006.
100. Legendary exposures: Milner, 2000, pp. 206–8.
101. Infanticide: Breiner, 1990; Daly & Wilson, 1988; deMause, 1998; Hrdy, 1999; Milner, 2000; Piers, 1978; Resnick, 1970; Williamson, 1978.
102. Infanticide across cultures: Williamson, 1978. See also Daly & Wilson, 1988; Divale & Harris, 1976; Hrdy, 1999; Milner, 2000.
103. Ten to 50 percent: Milner, 2000, p. 3; see also Williamson, 1978.
104. "All families": deMause, 1974, quoted in Milner, 2000, p. 2.
105. Histories of childhood: Breiner, 1990; deMause, 1974, 1998, 2008; Heywood, 2001; Hrdy, 1999; Milner, 2000.
106. Infanticide as "natural": Milner, 2000, p. 537.
107. Life history theory: Daly & Wilson, 1988; Hagen, 1999; Hawkes, 2006; Hrdy, 1999; Maynard Smith, 1988, 1998.
108. More investment in suckling than bearing: Hagen, 1999.
109. Sunk-cost avoidance: Maynard Smith, 1998. Exception to sunk-cost avoidance: Dawkins & Brockmann, 1980.
110. Infanticide as triage: Daly & Wilson, 1988; Hagen, 1999; Hrdy, 1999.
111. Empirical test of triage theory: Daly & Wilson, 1988, pp. 37–60.
112. Yanomamö infanticide: Quoted in Daly & Wilson, 1988, p. 51.
113. Lack of correlation between infanticide and warring: Williamson, 1978, p. 64.
114. Tylor: Quoted in Milner, 2000, p. 12.
115. Filthy newborns: Plutarch, "On affection for children," quoted in Milner, 2000, p. 508.
116. Postpartum depression as adaptation: Hagen, 1999; Daly & Wilson, 1988, pp. 61–77.
117. Personhood ceremonies: Daly & Wilson, 1988; Milner, 2000.
118. Hundred million missing girls: Sen, 1990; Milner, 2000, chap. 8; N. D. Kristof, "Stark data on women: 100 million are missing," *New York Times*, Nov. 5, 1991.
119. Female infanticide is ancient: Milner, 2000, pp. 236–45; see also Hudson & den Boer, 2002.
120. "if a boy gets sick": Quoted in N. D. Kristof, "Stark data on women: 100 million are missing," *New York Times*, Nov. 5, 1991.
121. Female infanticide: Milner, 2000, chap. 8; Hrdy, 1999; Hawkes, 1981; Daly & Wilson, 1988, pp. 53–56.
122. Infants in dunghills: Breiner, 1990, pp. 6–7.

123. Infanticide in medieval and early modern Europe: Milner, 2000; Hanlon, 2007; Hynes, in press.
124. Evolution of sex ratios: Maynard Smith, 1988, 1998.
125. ZPG theory: Divale & Harris, 1976. Problems for the ZPG theory: Chagnon, 1997; Daly & Wilson, 1988; Hawkes, 1981.
126. Trivers-Willard theory: Trivers & Willard, 1973. Problems with the theory: Hawkes, 1981; Hrdy, 1999.
127. Moderate support from wills: Hrdy, 1999.
128. Rarity of son-killing: Exceptions are the Shensi of China: Milner, 2000, p. 238; the Rendille of Kenya: Williamson, 1978, note 33; poor urban workers in 17th-century Parma: Hynes, in press.
129. Female infanticide as a free rider problem: Gottschall, 2008.
130. Female infanticide as a vicious circle: Chagnon, 1997; Gottschall, 2008.
131. Female infanticide and inheritance: Hawkes, 1981; Sen, 1990.
132. Daughter = water: Quoted in Milner, 2000, p. 130.
133. India and China today; Milner, 2000, pp. 236–45.
134. Gynecide today, trouble tomorrow: Hudson & den Boer, 2002.
135. Infanticides in the United States today: FBI Uniform Crime Reports, "2007: Crime in the United States," U.S. Department of Justice, 2007, http://www2.fbi.gov/ucr/cius2007/offenses/expanded_information/data/shrtable_02.html.
136. Infanticidal mothers in America: Milner, 2000, p. 124; Daly & Wilson, 1988; Resnick, 1970.
137. Prohibition of infanticide in Judaism and Christianity: Milner, 2000, chap. 2; Breiner, 1990.
138. Mercier on newborns: Quoted in Milner, 2000, p. 512.
139. Morality of infanticide and the human life taboo: Brock, 1993; Glover, 1977; Green, 2001; Kohl, 1978; Singer, 1994; Tooley, 1972.
140. Infanticide distinguished from other child killings: Milner, 2000, p. 16.
141. Leniency in infanticide: Resnick, 1970.
142. Human mercy killing: From a memoir by Benjamin Franklin Bonney, quoted in Courtwright, 1996, pp. 118–19.
143. Slippery slope in the Holocaust: Glover, 1999.
144. Fuzziness of human life: Brock, 1993; Gazzaniga, 2005; Green, 2001; Singer, 1994.
145. Frigid indifference: W. Langer, quoted by Milner, 2000, p. 68. See also Hanlon, 2007; Hynes, in press.
146. Infanticide in the Middle Ages: Milner, 2000, p. 70.
147. Babies in latrines: Quoted in deMause, 1982, p. 31.
148. Inquisitions of servant girls: Milner, 2000, p. 71.
149. De facto infanticide in Europe: Milner, 2000, pp. 99–107; chaps. 3–5.
150. British coroner: Quoted in Milner, 2000, p. 100.
151. Statistics on abortion: Henshaw, 1990; Sedgh et al., 2007.
152. Onset of neural activity: Gazzaniga, 2005.
153. Conceptions of the minds of fetuses and other things: Gray, Gray, & Wegner, 2007.
154. Less corporal punishment among hunter-gatherers: Levinson, 1989; Milner, 2000, p. 267.
155. Spare the rod: Milner, 2000, p. 257.
156. "Spare the rod" variations: Heywood, 2001, p. 100.
157. Corporal punishment in medieval Europe: deMause, 1998.
158. Proportion of beaten children: Heywood, 2001, p. 100.
159. Executing children: Capital Punishment U.K., "The execution of children and juveniles," http://www.capitalpunishmentuk.org/child.html.
160. German parenting: deMause, 2008, p. 10.
161. Ancient punishments: Milner, 2000, p. 267.
162. Japanese parenting: deMause, 2008.
163. English lullaby: Piers, 1978, quoted in Milner, 2000, p. 266.
164. Yiddish verse: Milner, 2000, pp. 386–89; see also Heywood, 2001, pp. 94–97; Daly & Wilson, 1999; Tatar, 2003.
165. Parent-offspring conflict: Dawkins, 1976/1989; Hrdy, 1999; Trivers, 1974, 1985.
166. German preacher: Quoted in Heywood, 2001, p. 33.
167. Locke, Rousseau, and the revolution in childhood: Heywood, 2001, pp. 23–24.
168. Locke: Quotes from Heywood, 2001, p. 23.
169. Rousseau: Quotes from Heywood, 2001, p. 24.
170. Reforms around the turn of the 20th century: Heywood, 2001; Zelizer, 1985.
171. Economically worthless, emotionally priceless: Zelizer, 1985.
172. Child Study movement and child welfare policies: White, 1996.
173. Child abuse analogized to animal abuse: H. Markel, "Case shined first light on abuse of children," *New York Times*, Dec. 15, 2009.
174. Complexities in history of childhood: Heywood, 2001.
175. Advisories against spanking: Harris, 1998/2008; Straus, 1999.
176. Children should never, ever be spanked: Straus, 2005.

177. Skepticism about the harmfulness of spanking: Harris, 1998/2008.
178. Culture of honor: Nisbett & Cohen, 1996.
179. Spanking survey: www.surveyusa.com/50StateDisciplineChild0805SortedbyTeacher.htm. "Red states" and "blue states" defined by votes in the 2004 presidential election.
180. State-by-state differences in spanking: www.surveyusa.com/50StateDisciplineChild0805SortedbyTeacher.htm.
181. Trends in spanking approval: Data from the General Social Survey, http://www.norc.org/GSS+Website/.
182. Corporal punishment decline: Straus, 2001, pp. 27–29; Straus, 2009; Straus & Kantor, 1995.
183. Declines in Europe: Straus, 2009.
184. Corporal punishment around the world: Straus, 2009.
185. Ethnic differences in spanking: Harris, 1998/2008.
186. Decline in approval of spanking in all ethnic groups: Data from the General Social Survey, http://www.norc.org/GSS+Website/.
187. Outlawing spanking: Straus, 2009.
188. Disapproval of paddling in American schools: www.surveyusa.com/50StateDisciplineChild0805SortedbyTeacher.htm.
189. International condemnation of corporal punishment: Human Rights Watch, 2008.
190. Child abuse checklist: Straus & Kantor, 1995.
191. Child abuse as a serious problem: 1976 and 1985 polls: Straus & Gelles, 1986. 1999 poll: PR Newswire, http://www.nospank.net/n-e62.htm.
192. Rate of violent deaths of children: A. Gentleman, "'The fear is not in step with reality,'" *Guardian*, Mar. 4, 2010.
193. Columbine: Cullen, 2009.
194. Effects of bullying: P. Klass, "At last, facing down bullies (and their enablers)," *New York Times*, Jun. 9, 2009.
195. Anti-bullying movement: J. Saltzman, "Antibully law may face free speech challenges," *Boston Globe*, May 4, 2010; W. Hu, "Schools' gossip girls and boys get some lessons in empathy," *New York Times*, Apr. 5, 2010; P. Klass, "At last, facing down bullies (and their enablers)," *New York Times*, Jun. 9, 2009.
196. Oprah against bullying: "The truth about bullying," *Oprah Winfrey Show*, May 6, 2009; http://www.oprah.com/relationships/School-Bullying.
197. School crime and safety: DeVoe et al., 2004.
198. Girls going wild: M. Males & M. Lind, "The myth of mean girls," *New York Times*, Apr. 2, 2010; W. Koch, "Girls have not gone wild, juvenile violence study says," *USA Today*, Nov. 20, 2008. See also Girls Study Group, 2008, for data through 2004.
199. Nurture Assumption: Harris, 1998/2008; see also Pinker, 2002, chap. 19; Harris, 2006; Wright & Beaver, 2005.
200. Food fight: S. Saulny, "25 Chicago students arrested for a middle-school food fight," *New York Times*, Nov. 11, 2009.
201. Cub Scout and 12-year-old: I. Urbina, "It's a fork, it's a spoon, it's a . . . weapon?" *New York Times*, Oct. 12, 2009; I. Urbina, "After uproar on suspension, district will rewrite rules," *New York Times*, Oct. 14, 2009. Eagle Scout: "Brickbats," *Reason*, Apr. 2010.
202. Recess coaches: W. Hu, "Forget goofing around: Recess has a new boss," *New York Times*, Mar. 14, 2010.
203. Digital disarming: Schechter, 2005.
204. Zombies no, carrots yes: J. Steinhauer, "Drop the mask! It's Halloween, kids, you might scare somebody," *New York Times*, Oct. 30, 2009.
205. Hate crime: Skenazy, 2009, p. 161.
206. Nightmare on *Sesame Street:* Skenazy, 2009, p. 69.
207. Child abduction panic: Skenazy, 2009; Finkelhor, Hammer, & Sedlak, 2002; "Phony numbers on child abduction," *STATS at George Mason University*; http://stats.org/stories/2002/phony_aug01_02.htm.
208. *Playdate* origin: Google Books, analyzed with Bookworm, Michel et al., 2011; see the caption to figure 7–1.
209. Changes in walking and playing: Skenazy, 2009.
210. Free-range children: Skenazy, 2009.
211. Cairns calculation: Cited in Skenazy, 2009, p. 16.
212. Crime-control theater: D. Bennett, "Abducted: The Amber Alert system is more effective as theater than as a way to protect children," *Boston Globe*, Jul. 20, 2008.
213. Kids hit by parents driving kids: Skenazy, 2009, p. 176.
214. Counterproductive kidnapping alerts: D. Bennett, "Abducted: The Amber Alert system is more effective as theater than as a way to protect children," *Boston Globe*, Jul. 20, 2008.
215. Alan Turing: Hodges, 1983.
216. Turing machines: Turing, 1936.

217. Can machines think?: Turing, 1950.
218. State-sponsored homophobia, past: Fone, 2000. Present: Ottosson, 2009.
219. More homophobia against gay men: Fone, 2000. More laws against male homosexuality: Ottosson, 2006.
220. More hate crimes against men: U.S. Department of Justice, FBI, *2008 Hate crime statistics*, table 4, http://www2.fbi.gov/ucr/hc2008/data/table_04.html.
221. Biology of homosexuality: Bailey, 2003; Hamer & Copeland, 1994; LeVay, 2010; Peters, 2006.
222. Flexible females: Baumeister, 2000.
223. Advantage in female fertility: Hamer & Copeland, 1994.
224. Cross-cultural uncommonness of homosexuality: Broude & Greene, 1976.
225. Cross-cultural disapproval of homosexuality: Broude & Greene, 1976.
226. Confusing morality with disgust: Haidt, 2002; Rozin, 1997.
227. History of homophobia: Fone, 2000.
228. Enlightenment rethinking of homosexuality: Fone, 2000.
229. State-sponsored homophobia: Ottosson, 2006, 2009.
230. Pillay: Quoted in Ottosson, 2009.
231. Kennedy opinion: *Lawrence v. Texas* (02–102), 2003, http://www.law.cornell.edu/supct/html/02–102.ZO.html.
232. Roger Brown: Pinker, 1998.
233. Attitudes on homosexuality: Gallup, 2008.
234. Effects of knowing a gay person: Gallup, 2009.
235. "Gay? Whatever, dude": Gallup, 2002.
236. FBI hate-crime statistics: http://www.fbi.gov/hq/cid/civilrights/hate.htm.
237. Problems with FBI hate-crime statistics: Harlow, 2005.
238. Hate-crime statistics for 2008: FBI, *2008 Hate crime statistics,* http://www2.fbi.gov/ucr/hc2008/index.html. Crime statistics for 2008: FBI, *2008 Crime in the United States,* http://www2.fbi.gov/ucr/cius2008/index.html.
239. Hate-crime homicides: FBI, *2008 Hate crime statistics,* http://www2.fbi.gov/ucr/hc2008/index.html. There were 115 hate-crime homicides between 1996 and 2005, and about a fifth of them targeted homosexuals.
240. Scientists believe animals feel pain: Herzog, 2010, p. 209.
241. History of treatment of animals: Gross, 2009; Harris, 1985; Herzog, 2010; Spencer, 2000; Stuart, 2006.
242. Human carnivory: Boyd & Silk, 2006; Harris, 1985; Herzog, 2010; Wrangham, 2009a.
243. Carnivory and human evolution: Boyd & Silk, 2006; Cosmides & Tooby, 1992; Tooby & DeVore, 1987.
244. Meat hunger, feasting, and carnality: Boyd & Silk, 2006; Harris, 1985; Symons, 1979.
245. Cuteness and conservation: Herzog, 2010.
246. Hopi cruelty to animals: Brandt, 1974.
247. Roast turtle recipe: www.nativetech.org/recipes/recipe.php?recipeid=211.
248. Larders on the hoof: Wrangham, 2009a.
249. Kicking and eating dogs: Gray & Young, 2011; quote from C. Turnbull.
250. Aristotle: Quoted in Stuart, 2006, p. xviii.
251. Celsus: Quoted in Spencer, 2000, p. 210.
252. Galen: Quoted in Gross, 2009.
253. Aquinas: Quoted in Gross, 2009.
254. Mind has no parts: Descartes, 1641/1967.
255. Early modern vivisection: Spencer, 2000, p. 210.
256. Seventeenth-century tenderization: P.C.D. Brears, *The gentlewoman's kitchen,* 1984, quoted in Spencer, 2000, p. 205.
257. Seventeenth-century factory farming: P. Pullar, *Consuming passions,* 1970, quoted in Spencer, 2000, p. 206.
258. Motives for vegetarianism: Herzog, 2010; Rozin et al., 1997; Spencer, 2000; Stuart, 2006.
259. Morality equated with cleanliness and asceticism: Haidt, 2002; Rozin et al., 1997; Shweder et al., 1997.
260. You are what you eat: Rozin, 1996.
261. Vegetarianism and romanticism: Spencer, 2000; Stuart, 2006.
262. Cockfighting and class warfare: Herzog, 2010.
263. Jewish dietary laws: Schechter, Greenstone, Hirsch, & Kohler, 1906.
264. Early vegetarianism: Spencer, 2000.
265. Sacred cows: Harris, 1985.
266. Nazism and animal rights: Herzog, 2010; Stuart, 2006.
267. Voltaire: Quoted in Spencer, 2000, p. 210.
268. Animal rights as laughing matter: N. Kristof, "Humanity toward animals," *New York Times,* Apr. 8, 2009.

269. Nineteenth-century animal rights: Gross, 2009; Herzog, 2010; Stuart, 2006.
270. Orwell on food cranks: *The road to Wigan Pier*, quoted in Spencer, 2000, pp. 278–79.
271. Takeoff in the 1970s: Singer, 1975/2009; Spencer, 2000.
272. Brophy: Quoted in Spencer, 2000, p. 303.
273. *Animal Liberation:* Singer, 1975/2009.
274. *Expanding Circle:* Singer, 1981/2011.
275. V-Frog: K. W. Burton, "Virtual dissection," *Science*, Feb. 22, 2008.
276. Cockfighting: Herzog, 2010, pp. 155–62.
277. Bullfight blackout: D. Woolls, "Tuning out tradition: Spain pulls live bullfights off state TV," *Boston Globe*, Aug. 23, 2007.
278. Increasing age of hunters: K. Johnson, "For many youths, hunting loses the battle for attention," *New York Times*, Sept. 25, 2010.
279. Hunting versus watching: U.S. Fish and Wildlife Service, 2006.
280. Locavore hunters: S. Rinella, "Locavore, get your gun," *New York Times*, Dec. 14, 2007.
281. Humane fishing: P. Bodo, "Hookless fly-fishing is a humane advance," *New York Times*, Nov. 7, 1999.
282. No animals were harmed: American Humane Association Film and Television Unit, 2010; http://www.americanhumane.org/protecting-animals/programs/no-animals-were-harmed/.
283. Not harming animals: American Humane Association Film and Television Unit, 2009.
284. Exterminator in chief: M. Leibovich, "What's white, has 132 rooms, and flies?" *New York Times*, Jun. 18, 2009.
285. Broiler Chicken Revolution: Herzog, 2010.
286. More chicken eaten than beef: U.S. Department of Agriculture, Economic Research Service, graphed at http://www.humanesociety.org/assets/pdfs/farm/Per-Cap-Cons-Meat-1.pdf.
287. Two hundred chickens = 1 cow: Herzog, 2010, p. 193.
288. Meat without feet: J. Temple, "The no-kill carnivore," *Wired*, Feb. 2009.
289. Three times as many ex-vegetarians: Herzog, 2010, p. 200.
290. Loose definitions of vegetarianism: Herzog, 2010; C. Stahler, "How many vegetarians are there?" *Vegetarian Journal*, Jul.–Aug. 1994.
291. Vegetarianism and eating disorders: Herzog, 2010, pp. 198–99.
292. Declining consumption of mammals: U.S. Department of Agriculture, Economic Research Service, graphed at http://www.humanesociety.org/assets/pdfs/farm/Per-Cap-Cons-Meat-1.pdf.
293. Gassing chickens: W. Neuman, "New way to help chickens cross to other side," *New York Times*, Oct. 21, 2010.
294. Eighty percent of Britons want better conditions: 2000 Taylor Nelson Poll for the RSPCA, cited in Vegetarian Society, 2010.
295. Gallup poll on animal protection: Gallup, 2003.
296. Arizona, Colorado, Florida, Oregon: N. D. Kristof, "A farm boy reflects," *New York Times*, Jul. 31, 2008.
297. European Union regulations: http://ec.europa.eu/food/animal/index_en.htm.
298. L. Hickman, "The lawyer who defends animals," *Guardian*, Mar. 5, 2010.
299. Poll on animal welfare: Gallup, 2003.
300. Vegetarianism among Dean activists: "The Dean activists: Their profile and prospects," Pew Research Center for the People & the Press, 2005, http://people-press.org/report/?pageid=936.
301. Rethinking human life and death: Singer, 1994.
302. Hard problem of consciousness: Pinker, 1997, chaps. 2, 8.
303. Extinguishing carnivores: J. McMahan, "The meat eaters," *New York Times*, Sept. 19, 2010.
304. Leftward shift of conservatism: Nash, 2009, p. 329; Courtwright, 2010.
305. Fivefold increase in books: Caplow et al., 2001, p. 267.
306. Catchment area for innovations: Diamond, 1997; Sowell, 1994, 1996, 1998.
307. "Pilgrimage to Nonviolence": King, 1963/1995.
308. Arc of justice: Parker, 1852/2005, "Of Justice and Conscience," in *Ten Sermons of Religion*.

Chapter 8: Inner Demons

1. Resistance to acknowledging the dark side: See Pinker, 2002.
2. Terrible twos: Côté et al., 2006.
3. Tremblay on babies with guns: Quoted in C. Holden, "The violence of the lambs," *Science*, *289*, 2000, pp. 580–81.
4. Homicidal fantasies: Kenrick & Sheets, 1994; Buss, 2005, pp. 5–8.
5. Revenge fantasy: Quoted in Buss, 2005, pp. 6–7.
6. Special effects: Schechter, 2005, p. 81.
7. History of violent entertainment: Schechter, 2005.
8. Fiction as an instruction manual for life: Pinker, 1997, chap. 8. Morbid curiosity about violence: Baumeister, 1997; Tiger, 2006.

9. Sex and consciousness: Symons, 1979.
10. Attack cat: Panksepp, 1998, p. 194.
11. Urge to bite surgeon: Quoted in Hitchcock & Cairns, 1973, pp. 897, 898.
12. Swearing: Pinker, 2007b, chap. 7.
13. Soldiers who don't fire weapons: Collins, 2008; Grossman, 1995; Marshall, 1947/1978.
14. Problems with claims of nonshooting soldiers: Bourke, 1999; Spiller, 1988.
15. Boring fights: Collins, 2008.
16. Ecstasy in battle: Bourke, 1999; Collins, 2008; Thayer, 2004.
17. Trigger of chimpanzee violence: Wrangham, 1999a.
18. Violence against violence researchers: Pinker, 2002, chap. 6; Dreger, 2011.
19. Capacity for evil: Baumeister, 1997; Baumeister & Campbell, 1999.
20. Narratives of harm: Baumeister, Stillwell, & Wotman, 1990.
21. Rate of anger: Baumeister et al., 1990.
22. Harm narratives with harm controlled: Stillwell & Baumeister, 1997.
23. Self-serving biases: Goffman, 1959; Tavris & Aronson, 2007; Trivers, in press; von Hippel & Trivers, 2011; Kurzban, 2011.
24. Cognitive dissonance: Festinger, 1957. Lake Wobegon Effect and other positive illusions: Taylor, 1989.
25. Moral emotions as the basis for cooperation: Haidt, 2002; Pinker, 2008; Trivers, 1971.
26. Advantages of the Moralization Gap: Baumeister, 1997; Baumeister et al., 1990; Stillwell & Baumeister, 1997.
27. Self-deception as an adaptation: Trivers, 1976, 1985, in press; von Hippel & Trivers, 2011.
28. Orwell: Quoted in Trivers, 1985.
29. Problems with self-deception: Pinker, 2011.
30. True self-deception: Valdesolo & DeSteno, 2008.
31. Competing historical narratives: Baumeister, 1997.
32. Aggrieved Serbs: Baumeister, 1997, pp. 50–51; van Evera, 1994.
33. Truthiness of Roosevelt's infamy speech: Mueller, 2006.
34. Evildoers' point of view: Baumeister, 1997, chap. 2.
35. Hitler as an idealist: Baumeister, 1997, chap. 2; Bullock, 1991; Rosenbaum, 1998.
36. Spree killer: Quoted in J. McCormick & P. Annin, "Alienated, marginal, and deadly," *Newsweek*, Sept. 19, 1994.
37. Serial rapist: Quoted in Baumeister, 1997, p. 41.
38. Gacy: Quoted in Baumeister, 1997, p. 49.
39. Crime as social control: Black, 1983. Provocations to domestic violence: Buss, 2005; Collins, 2008; Straus, 1977/1978.
40. Perspectives of the victim, perpetrator, scientist, and moralist: Baumeister, 1997.
41. Immorality of explaining the Holocaust: Shermer, 2004, pp. 76–79; Rosenbaum, 1998.
42. Banality of evil: Arendt, 1963.
43. Eichmann: Goldhagen, 2009.
44. Research inspired by Arendt: Milgram, 1974.
45. Predation versus aggression in mammals: Adams, 2006; Panksepp, 1998.
46. Sham versus real rage: Panksepp, 1998.
47. Modularity of motor program versus emotional state: Adams, 2006.
48. Rage circuit: Panksepp, 1998.
49. Aggression in the rat: Adams, 2006; Panksepp, 1998.
50. Pain, frustration, and aggression: Renfrew, 1997, chap. 6.
51. Orbitofrontal and ventromedial cortex: Damasio, 1994; Fuster, 2008; Jensen et al., 2007; Kringelbach, 2005; Raine, 2008; Scarpa & Raine, 2007; Seymour, Singer, & Dolan, 2007.
52. Seeking system: Panksepp, 1998.
53. Self-stimulation: Olds & Milner, 1954.
54. Offensive versus defensive attacks: Adams, 2006; Panksepp, 1998.
55. Fear versus rage: Adams, 2006; Panksepp, 1998.
56. Dominance system: Panksepp, 1998.
57. Male sexuality and aggressiveness: Panksepp, 1998, p. 199.
58. Testosterone: Archer, 2006b; Dabbs & Dabbs, 2000; Panksepp, 1998.
59. Phineas Gage: Damasio, 1994; Macmillan, 2000.
60. Gage was no longer Gage: Quoted in Macmillan, 2000.
61. Orbital and ventromedial cortex connected to amygdala: Damasio, 1994; Fuster, 2008; Jensen et al., 2007; Kringelbach, 2005; Raine, 2008; Scarpa & Raine, 2007; Seymour et al., 2007.
62. Anger at unfairness lights up the insula: Sanfey et al., 2003.
63. Orbitofrontal versus ventromedial cortex: Jensen et al., 2007; Kringelbach, 2005; Raine, 2008; Seymour et al., 2007.
64. Modern Phineas Gages: Séguin, Sylvers, & Lilienfeld, 2007, p. 193.

65. Traits of patients with frontal-lobe damage: Scarpa & Raine, 2007, p. 153.
66. Brains of psychopaths, murderers, and antisocials: Blair & Cipolotti, 2000; Blair, 2004; Raine, 2008; Scarpa & Raine, 2007.
67. Orbital lesions affect dominance hierarchies: Séguin et al., 2007, p. 193.
68. Orbital damage, faux pas, and empathy: Stone, Baron-Cohen, & Knight, 1998.
69. Orbital damage and shame: Raine et al., 2000.
70. *Mens rea* and fMRI: Young & Saxe, 2009.
71. Theory of mind and temporoparietal junction: Saxe & Kanwisher, 2003.
72. Crying babies and runaway trolleys: Greene, in press; Greene & Haidt, 2002; Pinker, 2008.
73. Your brain on morality: Greene, in press; Greene & Haidt, 2002; Greene et al., 2001.
74. Dorsolateral prefrontal cortex: Fuster, 2008.
75. Taxonomy of violence: Baumeister, 1997.
76. Adaptations for exploitation: Buss & Duntley, 2008.
77. Castrating a horse: From F. Zimring, quoted in Kaplan, 1973, p. 23.
78. Empathy and tracking: Liebenberg, 1990.
79. Ugandan atrocity: Baumeister, 1997, p. 125.
80. Psychopaths: Hare, 1993; Lykken, 1995; Mealey, 1995; Raine, 2008; Scarpa & Raine, 2007.
81. Proportion of psychopaths among violent criminals: G. Miller, "Investigating the psychopathic mind," *Science*, Sept. 5, 2008, pp. 1284–86; Hare, 1993; Baumeister, 1997, p. 138.
82. Psychopaths' brains: Raine, 2008.
83. Heritability of psychopathic symptoms: Hare, 1993; Lykken, 1995; Mealey, 1995; Raine, 2008. Psychopathy as a cheater strategy: Kinner, 2003; Lalumière, Harris, & Rice, 2001; Mealey, 1995; Rice, 1997.
84. Psychopaths in genocides and civil wars: Mueller, 2004a; Valentino, 2004.
85. Backsliding into emotional categories: Cosmides & Tooby, 1992; Pinker, 2007b, chaps. 5 and 9.
86. Disgust, hatred, anger: Tooby & Cosmides, 2010.
87. Positive illusions: Johnson, 2004; Tavris & Aronson, 2007; Taylor, 1989.
88. Self-deception and lie detection: von Hippel & Trivers, 2011.
89. Advantages of self-deception: Trivers, 1976, in press; von Hippel & Trivers, 2011.
90. Churchill: Quoted in Johnson, 2004, p. 1.
91. Deluded war leaders: Luard, 1986, pp. 204, 212, 268–69.
92. Initiators often lose their wars: Johnson, 2004, p. 4; Lindley & Schildkraut, 2005; Luard, 1986, p. 268.
93. Military incompetence as self-deception: Wrangham, 1999b.
94. War game: Johnson et al., 2006.
95. Groupthink in the Bush administration: K. Alter, "Is Groupthink driving us to war?" *Boston Globe*, Sept. 21, 2002.
96. Groupthink: Janis, 1982.
97. Logic of trivial altercations: Daly & Wilson, 1988, p. 127.
98. Logic of dominance: Daly & Wilson, 1988; Dawkins, 1976/1989; Maynard Smith, 1988.
99. Alliances in dominance: Boehm, 1999; de Waal, 1998.
100. Displays in dominance: Dawkins, 1976/1989; Maynard Smith, 1988.
101. Common knowledge: Chwe, 2001; Lee & Pinker, 2010; Lewis, 1969; Pinker, 2007b.
102. Culture of honor predicts violence: Brezina, Agnew, Cullen, & Wright, 2004.
103. Audience effect: Felson, 1982; Baumeister, 1997, pp. 155–56. See also McCullough, 2008; McCullough, Kurzban, & Tabak, 2010.
104. Dominance and group size: Baumeister, 1997, p. 167.
105. Forgiveness in primates: de Waal, 1996; McCullough, 2008.
106. Forgiveness only among kin or cooperators: McCullough, 2008.
107. Chimps don't reconcile across community lines: Van der Dennen, 2005; Wrangham & Peterson, 1996; Wrangham et al., 2006.
108. Men more obsessed by status: Browne, 2002; Susan M. Pinker, 2008; Rhoads, 2004.
109. Men take more risks: Byrnes, Miller, & Schafer, 1999; Daly & Wilson, 1988; Johnson, 2004; Johnson et al., 2006; Rhoads, 2004.
110. Labor economists and gender gaps: Browne, 2002; Susan M. Pinker, 2008; Rhoads, 2004.
111. Gender gap in violence: Archer, 2006b, 2009; Buss, 2005; Daly & Wilson, 1988; Geary, 2010; Goldstein, 2001.
112. Biological basis of sex difference: Geary, 2010; Pinker, 2002, chap. 18; Archer, 2009; Blum, 1997; Browne, 2002; Halpern, 2000.
113. Relational aggression: Geary, 2010; Crick, Ostrov, & Kawabata, 2007.
114. Dominance and sex appeal: Buss, 1994; Daly & Wilson, 1988; Ellis, 1992; Symons, 1979.
115. Ancient perquisites of dominance: Betzig, 1986; Betzig, Borgerhoff Mulder, & Turke, 1988.
116. Modern sex appeal of dominance: Buss, 1994; Ellis, 1992.
117. Sex differences in the brain: Blum, 1997; Geary, 2010; Panksepp, 1998.

118. Guyness: N. Angier, "Does testosterone equal aggression? Maybe not," *New York Times*, Jun. 20, 1995.
119. Testosterone and challenge: Archer, 2006b; Dabbs & Dabbs, 2000; Johnson et al., 2006; McDermott, Johnson, Cowden, & Rosen, 2007.
120. Parenting versus mating effort: Buss, 1994; Buss & Schmitt, 1993.
121. Paradox of greater risk-taking in youth: Daly & Wilson, 2005.
122. Violence over the life cycle: Daly & Wilson, 1988, 2000; Rogers, 1994.
123. The self-esteem myth: Baumeister, 1997; Baumeister, Smart, & Boden, 1996.
124. "multitalented superachievers": Quoted in Baumeister, 1997, p. 144.
125. Explaining Hitler: Rosenbaum, 1998, p. xii.
126. Narcissistic personality disorder in *DSM-IV:* American Psychiatric Association, 2000.
127. Narcissism, borderline, and psychopathic disorders in tyrants: Bullock, 1991; Oakley, 2007; Shermer, 2004. See also Chirot, 1994; Glover, 1999.
128. Social identity: Brown, 1985; Pratto, Sidanius, & Levin, 2006; Sidanius & Pratto, 1999; Tajfel, 1981; Tooby, Cosmides, & Price, 2006.
129. Mood and sports teams: Brown, 1985.
130. Testosterone after a sports match: Archer, 2006b; Dabbs & Dabbs, 2000; McDermott et al., 2007.
131. Testosterone after an election: Stanton et al., 2009.
132. In-group favoritism: Brown, 1985; Hewstone, Rubin, & Willis, 2002; Pratto et al., 2006; Sidanius & Pratto, 1999; Tajfel, 1981.
133. Little racists: Aboud, 1989. Babies, race, and accent: Kinzler, Shutts, DeJesus, & Spelke, 2009.
134. Social dominance: Pratto et al., 2006; Sidanius & Pratto, 1999.
135. Races versus coalitions: Kurzban, Tooby, & Cosmides, 2001; Sidanius & Pratto, 1999.
136. Accent and prejudice: Tucker & Lambert, 1969; Kinzler et al., 2009.
137. *Ressentiment*: Chirot, 1994, chap. 12; Goldstein, 2001, p. 409; Baumeister, 1997, p. 152.
138. German ressentiment: Chirot, 1994, chap. 12; Goldstein, 2001, p. 409; Baumeister, 1997.
139. Islamic ressentiment: Fattah & Fierke, 2009.
140. Hollandization: Mueller, 1989.
141. Skepticism about ancient hatreds: Brown, 1997; Fearon & Laitin, 1996; Fearon & Laitin, 2003; Lacina, 2006; Mueller, 2004a; van Evera, 1994.
142. Number of languages: Pinker, 1994, chap. 8.
143. Ethnic comity in the developed world: Brown, 1997.
144. Policing loose cannons: Fearon & Laitin, 1996.
145. Rubegoldbergian governments: Asal & Pate, 2005; Bell, 2007b; Brown, 1997; Mnookin, 2007; Sowell, 2004; Tyrrell, 2007. Rugby team as national unifier: Carlin, 2008.
146. Identity and violence: Appiah, 2006; Sen, 2006.
147. Men at both ends of racism: Pratto et al., 2006; Sidanius & Pratto, 1999; Sidanius & Veniegas, 2000.
148. *War and Gender:* Goldstein, 2001.
149. Queens who waged war: Luard, 1986, p. 194.
150. War as a man's game: Gottschall, 2008.
151. Feminism and pacifism: Goldstein, 2001; Mueller, 1989.
152. Gender gap in opinion polls: Goldstein, 2001, pp. 329–30.
153. Gender gap in presidential elections: "Exit polls, 1980–2008," *New York Times*, http://elections.nytimes.com/2008/results/president/exit-polls.html.
154. Gender gap smaller than society gap: Goldstein, 2001, pp. 329–30.
155. Feminist gap in the Middle East: Goldstein, 2001, pp. 329–30.
156. Treatment of women and war across cultures: Goldstein, 2001, pp. 396–99.
157. Women and war in modern countries: Goldstein, 2001, p. 399.
158. Women's empowerment and rootless men: Hudson & den Boer, 2002; Potts & Hayden, 2008.
159. Dominance jargon: Google Books, analyzed by Bookworm (see the caption to figure 7–1), Michel et al., 2011.
160. *Glorious* and *honorable:* Google Books, analyzed by Bookworm (see the caption to figure 7–1), Michel et al., 2011.
161. "glow in our eyes": Quoted in Daly & Wilson, 1988, p. 228.
162. "developing wings": Quoted in J. Diamond, "Vengeance is ours," *New Yorker*, Apr. 21, 2008.
163. "hot with . . . joy": Quoted in Daly & Wilson, 1988, p. 230.
164. Universality of revenge: McCullough, 2008, pp. 74–76; Daly & Wilson, 1988, pp. 221–27. Revenge in tribal warfare: Chagnon, 1997; Daly & Wilson, 1988; Keeley, 1996; Wiessner, 2006.
165. Revenge in homicides, shootings, bombings: McCullough et al., 2010.
166. Revenge in terrorism, riots, and wars: Atran, 2003; Horowitz, 2001; Mueller, 2006.
167. Declaring war in anger: Luard, 1986, p. 269.
168. Cataclysmic fury: G. Prange, quoted in Mueller, 2006, p. 59.
169. Alternatives to reprisal not considered after Pearl Harbor or 9/11: Mueller, 2006.

170. Bin Laden: "Full text: Bin Laden's 'Letter to America,'" *Observer*, Nov. 24, 2002; http://www.guardian.co.uk/world/2002/nov/24/theobserver.
171. Revenge fantasies: Buss, 2005; Kenrick & Sheets, 1994.
172. Revenge in the lab: McCullough, 2008.
173. Drinking to salve frustrated revenge: Giancola, 2000.
174. Rage circuit: Panksepp, 1998.
175. Anger in the insula: Sanfey et al., 2003.
176. Neuroscience of revenge: de Quervain et al., 2004.
177. Neuroscience of revenge, empathy, and gender: Singer et al., 2006.
178. Men are from justice, women are from mercy: Gilligan, 1982.
179. Relational aggression in women: Crick et al., 2007; Geary, 2010.
180. Revenge as disease, forgiveness as cure: McCullough, 2008; McCullough et al., 2010.
181. Logic of deterrence: Daly & Wilson, 1988, p. 128.
182. Models of the evolution of cooperation: Axelrod, 1984/2006; Axelrod & Hamilton, 1981; McCullough, 2008; Nowak, 2006; Ridley, 1997; Sigmund, 1997.
183. Prisoner's Dilemma as great idea: Poundstone, 1992.
184. First Iterated PD tournament: Axelrod, 1984/2006; Axelrod & Hamilton, 1981.
185. Reciprocal altruism: Trivers, 1971.
186. More recent tournaments: McCullough, 2008; Nowak, May, & Sigmund, 1995; Ridley, 1997; Sigmund, 1997.
187. Components of Tit for Tat: Axelrod, 1984/2006.
188. Indirect reciprocity: Nowak, 2006; Nowak & Sigmund, 1998.
189. Public Goods game: Fehr & Gächter, 2000; Herrmann, Thöni, & Gächter, 2008a; Ridley, 1997.
190. Tragedy of the Commons: Hardin, 1968.
191. Effectiveness of deterrence in economic games: Fehr & Gächter, 2000; Herrmann, Thöni, & Gächter, 2008b; McCullough, 2008; McCullough et al., 2010; Ridley, 1997.
192. Fear of revenge mitigates revenge: Diamond, 1977; see also Ford & Blegen, 1992.
193. Implacability of revenge: Frank, 1988; Schelling, 1960. Self-help justice: Black, 1983; Daly & Wilson, 1988.
194. Vengeful anger as a recalibration mechanism: Sell, Tooby, & Cosmides, 2009.
195. Target must know he has been singled out: Gollwitzer & Denzler, 2009.
196. Audience effects on revenge: Bolton & Zwick, 1995; Brown, 1968; Kim, Smirth, & Brigham, 1998.
197. Audience effects on fights: Felson, 1982.
198. Ultimatum game: Bolton & Zwick, 1995; Fehr & Gächter, 2000; Ridley, 1997; Sanfey et al., 2003.
199. Ultimatum game in the scanner: Sanfey et al., 2003.
200. Moralization Gap and escalation of revenge: Baumeister, 1997.
201. Boys in backseat: D. Gilbert, "He who cast the first stone probably didn't," *New York Times*, Jul. 24, 2006.
202. Two eyes for an eye: Shergill, Bays, Frith, & Wolpert, 2003.
203. Just deserts as justification of criminal punishment: Kaplan, 1973.
204. Bafflegab about justice: Daly & Wilson, 1988, p. 256.
205. Deterrence versus just deserts: Carlsmith, Darley, & Robinson, 2002.
206. Pure justice as an antigaming strategy: Pinker, 2002, chap. 10.
207. Informal cooperation in Shasta County: Ellickson, 1991.
208. Spiteful punishment across societies: Herrmann et al., 2008a, 2008b.
209. Forgiveness as the dimmer of revenge: McCullough, 2008; McCullough et al., 2010.
210. Forgiveness in primates: de Waal, 1996.
211. Boys at war at Robbers Cave: Sherif, 1966.
212. Guilt, shame, embarrassment: Baumeister, Stillwell, & Heatherton, 1994; Haidt, 2002; Trivers, 1971.
213. Common knowledge: Chwe, 2001; Lee & Pinker, 2010; Lewis, 1969; Pinker, 2007b; Pinker, Nowak, & Lee, 2008.
214. Political apologies: Dodds, 2003b, accessed Jun. 28, 2010. See also Dodds, 2003a.
215. Tolerating injustice: Long & Brecke, 2003, pp. 70–71.
216. "If you want peace, work for peace": Goldstein, 2011.
217. Reconciliation gestures: Long & Brecke, 2003, p. 72.
218. Shakespearean and Chekhovian tragedies: Oz, 1993, p. 260.
219. Occasional effectiveness of torture: Levinson, 2004a, p. 34; P. Finn, J. Warrick, & J. Tate, "Detainee became an asset," *Washington Post*, Aug. 29, 2009.
220. Occasional justifiability of torture: Levinson, 2004a; Posner, 2004; Walzer, 2004.
221. Ineffectiveness of most torture: A. Grafton, "Say anything," *New Republic*, Nov. 5, 2008.
222. Execution as entertainment: Tuchman, 1978.
223. Serial killers versus mass murderers: Schechter, 2003.
224. More serial murder scholars than serial murderers: Fox & Levin, 1999, p. 166.

225. Decline in the number of serial killers: C. Beam, "Blood loss: The decline of the serial killer," *Slate*, Jan. 5, 2011.
226. Number of serial killers and victims: Fox & Levin, 1999, p. 167; J. A. Fox, cited in Schechter, 2003, p. 286.
227. No identifiable cause of serial killers: Schechter, 2003.
228. Morbid fascination: Nell, 2006; Tiger, 2006; Baumeister, 1997.
229. Sadism as dominance: Potegal, 2006.
230. Schadenfreude in the scanner: Takahashi et al., 2009.
231. Revenge turns off empathy: Singer et al., 2006. Revenge requires knowledge of victim's awareness: Gollwitzer & Denzler, 2009.
232. More M than S: Baumeister, 1997; Baumeister & Campbell, 1999.
233. Intertwined circuits for sex and aggression: Panksepp, 1998.
234. Gun as hard-on: Quoted in Thayer, 2004, p. 191.
235. Killing as orgasm: Quoted in Baumeister, 1997, p. 224.
236. Martyrologies: Gallonio, 1903/2004; Puppi, 1990.
237. Women in peril in macabre entertainment: Schechter, 2005.
238. Aroused during flogging: Theweleit, 1977/1987, quoted in deMause, 2002, p. 217.
239. Male serial killers: Schechter, 2003, p. 31.
240. Female serial killers: Schechter, 2003, p. 31.
241. Guilt is anticipatory: Baumeister, 1997, chap. 10; Baumeister et al., 1994.
242. Prohibitions of torture: Levinson, 2004b.
243. Torture warrants: Dershowitz, 2004b.
244. Response to torture warrants: Dershowitz, 2004b; Levinson, 2004a.
245. Taboo against torture is useful: Levinson, 2004a; Posner, 2004.
246. Aversiveness of conspecifics in pain: de Waal, 1996; Preston & de Waal, 2002.
247. Reasons for aversiveness of pain displays: Hauser, 2000, pp. 219–23.
248. Anxiety while hurting others: Milgram, 1974.
249. Trolley Problem: Greene & Haidt, 2002; Greene et al., 2001.
250. Aversion to direct violence: Collins, 2008.
251. Ordinary Germans: Browning, 1992.
252. Nausea not soul-searching: Baumeister, 1997, p. 211.
253. Distinguishing fiction from reality: Sperber, 2000.
254. Blunted emotions in psychopathy: Blair, 2004; Hare, 1993; Raine et al., 2000.
255. Variability among guards: Baumeister, 1997, chap. 7.
256. Sadism as an acquired taste: Baumeister, 1997, chap. 7; Baumeister & Campbell, 1999.
257. Escalating sadism and serial killers: Baumeister, 1997; Schechter, 2003.
258. Opponent-process theory of motivation: Solomon, 1980.
259. Sadism and opponent-process theory: Baumeister, 1997, chap. 7; Baumeister & Campbell, 1999.
260. Benign masochism: Rozin, 1996. Benign masochism as an adaptation: Pinker, 1997, pp. 389, 540.
261. Ideological violence: Baumeister, 1997, chap. 6; Chirot & McCauley, 2006; Glover, 1999; Goldhagen, 2009; Kiernan, 2007; Valentino, 2004.
262. I thank Jennifer Sheehy-Skeffington for this insight.
263. Group polarization: Myers & Lamm, 1976.
264. Groupthink: Janis, 1982.
265. Group animosity: Hoyle, Pinkley, & Insko, 1989; see also Baumeister, 1997, 193–94.
266. Obedience experiments: Milgram, 1974.
267. Fact and fiction about Kitty Genovese: Manning, Levine, & Collins, 2007. Bystander apathy: Latané & Darley, 1970.
268. Stanford Prison Experiment: Zimbardo, 2007; Zimbardo, Maslach, & Haney, 2000.
269. No Germans punished for disobedience: Goldhagen, 2009.
270. Milgram replication: Burger, 2009. See Reicher & Haslam, 2006, for a partial replication of the Stanford Prison Experiment, but with too many differences to allow a test of trends over time.
271. Obedience might be even lower: Twenge, 2009.
272. Advantages of conformity: Deutsch & Gerard, 1955.
273. Positive feedback loops in popularity: Salganik, Dodds, & Watts, 2006.
274. Pluralistic ignorance: Centola, Willer, & Macy, 2005; Willer et al., 2009.
275. Spiral of silence and Basque terrorism: Spencer & Croucher, 2008.
276. Asch conformity experiment: Asch, 1956.
277. Enforcement and pluralistic ignorance: Centola et al., 2005; Willer et al., 2009.
278. Too terrified to stop clapping: Glover, 1999, p. 242.
279. Thought control in Maoist China: Glover, 1999, pp. 292–93.
280. Simulated pluralistic ignorance: Centola et al., 2005.
281. Growth of fascism: Payne, 2005.
282. Six degrees of separation: Travers & Milgram, 1969.

283. Pluralistic ignorance in the lab: Willer et al., 2009.
284. Sokal Hoax: Sokal, 2000.
285. Cognitive dissonance: Festinger, 1957.
286. Moral disengagement: Bandura, 1999; Bandura, Underwood, & Fromson, 1975; Kelman, 1973; Milgram, 1974; Zimbardo, 2007; Baumeister, 1997, part 3.
287. Politics and the English Language: Orwell, 1946/1970.
288. Burke: Quoted in Nunberg, 2006, p. 20.
289. Euphemism, framing, and plausible deniability: Pinker, 2007b; Pinker et al., 2008.
290. Gradual slide into barbarism: Glover, 1999; Baumeister, 1997, chaps. 8 and 9.
291. Milgram experiment as Escalation game: Katz, 1987.
292. Diffusion of responsibility: Bandura et al., 1975; Milgram, 1974.
293. Diffusion of responsibility in military units and bureaucracies: Arendt, 1963; Baumeister, 1997; Browning, 1992; Glover, 1999.
294. Resistance to up-close harm: Greene, in press.
295. Psychic numbing through large numbers: Slovic, 2007.
296. Derogating the victim: Bandura et al., 1975.
297. Advantageous comparison: Bandura, 1999; Gabor, 1994.
298. A little psychology goes a long way: See also Kahneman & Renshon, 2007.

Chapter 9: Better Angels

1. Age of empathy: de Waal, 2009.
2. Empathic civilization: Rifkin, 2009. Excerpt from http://www.huffingtonpost.com/jeremy-rifkin/the-empathic-civilization_b_416589.html.
3. Building peace one child at a time: Gordon, 2009.
4. Faculties of peaceful coexistence: Dawkins, 1976/1989; McCullough, 2008; Nowak, 2006; Ridley, 1997.
5. Sentiments of virtue: Hume, 1751/2004.
6. Titchener on empathy: Titchener, 1909/1973.
7. Popularity of *empathy, willpower, self-control:* Based on an analysis of Google Books by the Bookworm program, Michel et al., 2011; see the caption to figure 7–1.
8. Meanings of *empathy:* Batson, Ahmad, Lishmer, & Tsang, 2002; Hoffman, 2000; Keen, 2007; Preston & de Waal, 2002.
9. James on sympathy and fox terriers: James, 1977.
10. Senses of empathy: Batson et al., 2002; Hoffman, 2000; Keen, 2007; Preston & de Waal, 2002.
11. Empathy as mind-reading: Baron-Cohen, 1995.
12. Dissociability of reading thoughts and emotions: Blair & Perschardt, 2002.
13. Psychopaths read but don't feel emotions: Hare, 1993; Mealey & Kinner, 2002.
14. Empathy versus distress at others' suffering: Batson et al., 2002.
15. Emotional contagion: Preston & de Waal, 2002.
16. Sympathy not the same as contagion: Bandura, 2002.
17. Discovery of mirror neurons: di Pellegrino, Fadiga, Fogassi, Gallese, & Rizzolatti, 1992.
18. Possible mirror neurons in humans: Iacoboni et al., 1999.
19. Mirror mania: Iacoboni, 2008; J. Lehrer, "Built to be fans," *Seed*, Feb. 10, 2006, pp. 119–20; C. Buckley, "Why our hero leapt onto the tracks and we might not," *New York Times*, Jan. 7, 2007; S. Vedantam, "How brain's 'mirrors' aid our social understanding," *Washington Post*, Sept. 25, 2006.
20. Mirror neurons as DNA: Ramachandran, 2000.
21. Nasty macaques: McCullough, 2008, p. 125.
22. Empathy in the brain: Lamm, Batson, & Decety, 2007; Moll, de Oliveira-Souza, & Eslinger, 2003; Moll, Zahn, de Oliveira-Souza, Krueger, & Grafman, 2005.
23. Skepticism about mirror neurons: Csibra, 2008; Alison Gopnik, 2007; Hickok, 2009; Hurford, 2004; Jacob & Jeannerod, 2005.
24. Overlap in insula: Singer et al., 2006; Wicker et al., 2003.
25. No overlap in insula when feeling revenge: Singer et al., 2006.
26. Counterempathy in competition: Lanzetta & Englis, 1989.
27. Empathy in the brain: Lamm et al., 2007.
28. Atlas of empathy: Damasio, 1994; Lamm et al., 2007; Moll et al., 2003; Moll et al., 2005; Raine, 2008.
29. Oxytocin: Pfaff, 2007.
30. Maternal care as precursor to sympathy: Batson et al., 2002; Batson, Lishner, Cook, & Sawyer, 2005.
31. Oxytocin induces trust: Kosfeld et al., 2005; Zak, Stanton, Ahmadi, & Brosnan, 2007.
32. Cuteness: Lorenz, 1950/1971.
33. Babies exploit cuteness response: Hrdy, 1999.

34. Evolution of Mickey Mouse: Gould, 1980.
35. Dangerous Mick: B. Barnes, "After Mickey's makeover, less Mr. Nice Guy," *New York Times*, Nov. 4, 2009.
36. Baby-faced litigants: Zebrowitz & McDonald, 1991.
37. Ugly children punished more: Berkowitz & Frodi, 1979.
38. Unattractive adults judged more harshly: Etcoff, 1999.
39. Forgiveness, sympathy, guilt: Baumeister et al., 1994; Hoffman, 2000; McCullough, 2008; McCullough et al., 2010.
40. Communal versus exchange relationships: Baumeister et al., 1994; Clark, Mills, & Powell, 1986; Fiske, 1991; Fiske, 1992, 2004a.
41. Taboos in communal relationships: Fiske & Tetlock, 1997; McGraw & Tetlock, 2005.
42. Modicum of sympathy to strangers as default: Axelrod, 1984/2006; Baumeister et al., 1994; Trivers, 1971.
43. Toddlers aid and comfort people in distress: Warneken & Tomasello, 2007; Zahn-Waxler, Radke-Yarrow, Wagner, & Chapman, 1992.
44. Sympathy for those in need: Batson et al., 2005b.
45. Similarity matters: Preston & de Waal, 2002, p. 16; Batson, Turk, Shaw, & Klein, 1995c.
46. Shared traits and relief from shock: Krebs, 1975.
47. Empathy-altruism hypothesis: Batson & Ahmad, 2001; Batson et al., 2002; Batson, Ahmad, & Stocks, 2005a; Batson, Duncan, Ackerman, Buckley, & Birch, 1981; Batson et al., 1988; Krebs, 1975.
48. Psychological definition of altruism: Batson et al., 2002; Batson et al., 1981; Batson et al., 1988.
49. Evolutionary definition of altruism: Dawkins, 1976/1989; Hamilton, 1963; Maynard Smith, 1982.
50. Confusions about altruism: Pinker, 1997, chaps. 1, 6; Pinker, 2006.
51. Empathy-altruism hypotheses: Batson & Ahmad, 2001; Batson et al., 2002; Batson et al., 2005a; Batson et al., 1981; Batson et al., 1988.
52. Batson's empathy-altruism research: Batson et al., 2002; Batson et al., 2005a.
53. Similarity, empathy, and ease of escape: Batson et al., 1981.
54. Empathy and social acceptability: Batson et al., 1988.
55. Empathy and a one-shot Prisoner's Dilemma: Batson & Moran, 1999.
56. Empathy and an iterated Prisoner's Dilemma: Batson & Ahmad, 2001.
57. Empathy from superordinate goals, and in conflict resolution workshops: Batson et al., 2005a, pp. 367–68; Stephan & Finlay, 1999.
58. Sympathy toward groups via perspective-taking: Batson et al., 1997.
59. Taking a victim's perspective induces altruism: Batson et al., 1988.
60. Taking a victim's perspective induces altruism toward group: Batson et al., 1997.
61. Sympathy for convicted murderers: Batson et al., 1997.
62. George Eliot on empathy through fiction: From "The natural history of German life," quoted in Keen, 2007, p. 54.
63. Fiction as an empathy expander: Hunt, 2007; Mar & Oatley, 2008; Mar et al., 2006; Nussbaum, 1997, 2006.
64. Empathizers read fiction: Mar et al., 2006.
65. Confusing fact with fiction: Strange, 2002.
66. Empathy for a fictitious character and his group: Batson, Chang, Orr, & Rowland, 2008.
67. Fiction as a moral laboratory: Hakemulder, 2000.
68. The dark side of empathy: Batson et al., 2005a; Batson et al., 1995a; Batson, Klein, Highberger, & Shaw, 1995b; Prinz, in press.
69. Empathy subverts fairness: Batson et al., 1995b.
70. Empathy and public goods: Batson et al., 1995a.
71. Ephemeral benefits from empathy: Batson et al., 2005a, p. 373.
72. Utopia versus human nature: Pinker, 2002.
73. Burnout and fatigue from excess empathy: Batson et al., 2005a.
74. Lebanon war as a lapse of self-control: Mueller & Lustick, 2008.
75. The logic of self-control: Ainslie, 2001; Daly & Wilson, 2000; Kirby & Herrnstein, 1995; Schelling, 1978, 1984, 2006.
76. Ancestral versus modern discounting rates: Daly & Wilson, 1983, 2000, 2005; Wilson & Daly, 1997.
77. Myopic retirement planning: Akerlof, 1984; Frank, 1988.
78. Libertarian paternalism: Thaler & Sunstein, 2008.
79. Myopic discounting: Ainslie, 2001; Kirby & Herrnstein, 1995.
80. Hyperbolic discounting: Ainslie, 2001; Kirby & Herrnstein, 1995.
81. Hyperbolic discounting as composite of two mechanisms: Pinker, 1997, p. 396; Laibson, 1997.
82. Two selves: Schelling, 1984, p. 58.
83. Hot and cool brain systems: Metcalfe & Mischel, 1999.
84. Limbic grasshopper and frontal lobe ant: McClure, Laibson, Loewenstein, & Cohen, 2004.
85. Frontal lobes: Fuster, 2008.

86. Frontal lobes in temporal discounting: Shamosh et al., 2008.
87. Gage and his modern counterparts: Anderson et al., 1999; Damasio, 1994; Macmillan, 2000; Raine, 2008; Raine et al., 2000; Scarpa & Raine, 2007.
88. Cortical expansion during evolution: Hill et al., 2010.
89. Dorsolateral prefrontal cortex in cost-benefit analyses: Greene et al., 2001; McClure et al., 2004.
90. Frontal pole: Gilbert et al., 2006; Koechlin & Hyafil, 2007; L. Helmuth, "Brain model puts most sophisticated regions front and center," *Science, 302*, p. 1133.
91. Limbic and prefrontal responses in batterers: Lee, Chan, & Raine, 2008.
92. Importance of intelligence: Gottfredson, 1997a, 1997b; Neisser et al., 1996.
93. Marshmallow test: Metcalfe & Mischel, 1999; Mischel et al., in press.
94. Shallow discounting and life outcomes: Chabris et al., 2008; Duckworth & Seligman, 2005; Kirby, Winston, & Santiesteban, 2005.
95. Self-reports on self-control: Tangney, Baumeister, & Boone, 2004.
96. Benefits of self-control: Tangney et al., 2004.
97. Crime and self-control: Gottfredson, 2007; Gottfredson & Hirschi, 1990; Wilson & Herrnstein, 1985.
98. Delay of gratification and aggression: Rodriguez, Mischel, & Shoda, 1989.
99. Teacher ratings of impulsiveness and aggressiveness: Dewall et al., 2007; Tangney et al., 2004.
100. Longitudinal study of temperament: Caspi, 2000. See also Beaver, DeLisi, Vaughn, & Wright, 2008.
101. Violent and nonviolent crimes correlated in New Zealand sample: Caspi et al., 2002.
102. Maturation of frontal lobes: Fuster, 2008, pp. 17–19.
103. Delay discounting doesn't correlate with juvenile delinquency: Wilson & Daly, 2006.
104. Sensation-seeking peaks at eighteen: Romer, Duckworth, Sznitman, & Park, 2010.
105. Testosterone: Archer, 2006b.
106. Pushes and pulls in adolescent brains: Romer et al., 2010.
107. All psychological traits are heritable: Bouchard & McGue, 2003; Harris, 1998/2008; McCrae et al., 2000; Pinker, 2002; Plomin, DeFries, McClearn, & McGuffin, 2008; Turkheimer, 2000.
108. Self-control correlated with intelligence: Burks, Carpenter, Goette, & Rustichini, 2009; Shamosh & Gray, 2008. Self-control and intelligence in the frontal lobes: Shamosh et al., 2008.
109. Intelligence and committing crimes: Herrnstein & Murray, 1994; Neisser et al., 1996. Intelligence and getting murdered: Batty, Deary, Tengstrom, & Rasmussen, 2008.
110. Heritability of ADHD and links to crime: Beaver et al., 2008; Wright & Beaver, 2005.
111. Force dynamics metaphor and self-control: Talmy, 2000; Pinker, 2007b, chap. 4.
112. Fatiguing self-control: Baumeister et al., 1998; quote from p. 1254.
113. Ego depletion studies: Baumeister et al., 1998; Baumeister, Gailliot, Dewall, & Oaten, 2006; Dewall et al., 2007; Gailliot & Baumeister, 2007; Gailliot et al., 2007; Hagger, Wood, Stiff, & Chatzisarantis, 2010.
114. Self-control masks individual differences: Baumeister et al., 2006.
115. Self-control and male sexuality: Gailliot & Baumeister, 2007.
116. Ego depletion and violence: Dewall et al., 2007.
117. Heritability of height: Weedon & Frayling, 2008.
118. Odyssean self-control: Schelling, 1984, 2006.
119. Self-control strategies in children: Metcalfe & Mischel, 1999.
120. Discounting rate as an internal variable: Daly & Wilson, 2000, 2005; Wilson & Daly, 1997, 2006.
121. Self-control and glucose: Gailliot et al., 2007.
122. Alcohol and violence: Baumeister, 1997; Bushman, 1997. Nutritional supplements in prisons: J. Bohannon, "The theory? Diet causes violence. The lab? Prison," *Science, 325*, Sept. 25, 2009.
123. Exercising the will: Baumeister et al., 2006.
124. Fashions in self-control and dignity: Eisner, 2008; Wiener, 2004; Wouters, 2007.
125. Children's interest rates: Clark, 2007a, p. 171.
126. Variation across cultures: Hofstede & Hofstede, 2010.
127. Long-term Orientation and homicide: The correlation between Long-Term Orientation and homicide rates across the 95 countries for which data are available is −0.325. The correlation between Indulgence and homicide is 0.25. Both are statistically significant. Long-Term Orientation and Indulgence scores are taken from http://www.geerthofstede.nl/research-vsm/dimension-data-matrix.aspx. Homicide data are the high estimates taken from *International homicide statistics* figures, United Nations Office on Drugs and Crime, 2009.
128. Lactose tolerance in adulthood: Tishkoff et al., 2006.
129. Yanomamö killers: Chagnon, 1988; Chagnon, 1997. Jivaro killers: Redmond, 1994.
130. Assumption of little recent evolutionary change: Pinker, 1997; Tooby & Cosmides, 1990a, 1990b.
131. Psychic unity of humankind: Brown, 1991, 2000; Tooby & Cosmides, 1990a, 1992.
132. Mechanics of natural selection: Maynard Smith, 1998.
133. Quantitative and single-gene evolution and evolutionary psychology: Tooby & Cosmides, 1990a.

134. Genomic tests of selection: Akey, 2009; Kreitman, 2000; Przeworski, Hudson, & Di Rienzo, 2000.
135. Recent selection in humans: Akey, 2009, p. 717.
136. Selective breeding of aggression in mice: Cairns, Gariépy, & Hood, 1990.
137. Measuring heritability: Plomin et al., 2008; Pinker, 2002, chap. 19.
138. Danish adoption study: Mednick, Gabrielli, & Hutchings, 1984.
139. Measures of aggression correlate with violent crime: Caspi et al., 2002; Guo, Roettger, & Cai, 2008b.
140. Heritability of aggression: Plomin et al., 2008, chap. 13; Bouchard & McGue, 2003; Eley, Lichtenstein, & Stevenson, 1999; Ligthart et al., 2005; Lykken, 1995; Raine, 2002; Rhee & Waldman, 2007; Rowe, 2002; Slutske et al., 1997; van Beijsterveldt, Bartels, Hudziak, & Boomsma, 2003; van den Oord, Boomsma, & Verhulst, 1994.
141. Aggression in separated twins: Bouchard & McGue, 2003, table 6.
142. Aggression in adoptees: van den Oord et al., 1994; see also Rhee & Waldman, 2007.
143. Aggression in twins: Cloninger & Gottesman, 1987; Eley et al., 1999; Ligthart et al., 2005; Rhee & Waldman, 2007; Slutske et al., 1997; van Beijsterveldt et al., 2003.
144. Meta-analysis of behavioral genetics of aggression: Rhee & Waldman, 2007.
145. Violent crime in twins: Cloninger & Gottesman, 1987.
146. Pedomorphy and self-domestication: Wrangham, 2009b; Wrangham & Pilbeam, 2001.
147. Heritability of gray matter distribution: Thompson et al., 2001.
148. Heritability of white matter connectivity: Chiang et al., 2009.
149. Making voles monogamous: McGraw & Young, 2010.
150. Testosterone and aggressive challenges: Archer, 2006b; Dabbs & Dabbs, 2000.
151. Testosterone receptor variation: Rajender et al., 2008; Roney, Simmons, & Lukaszewski, 2009.
152. MAO-A knockout and violence in humans: Brunner et al., 1993.
153. MAO-A variants and aggression: Alia-Klein et al., 2008; Caspi et al., 2002; Guo, Ou, Roettger, & Shih, 2008a; Guo et al., 2008b; McDermott et al., 2009; Meyer-Lindenberg, 2006.
154. MAO-A specific to violence: N. Alia-Klein, quoted in Holden, 2008, p. 894; Alia-Klein et al., 2008.
155. Effects of MAO-A depend on experience: Caspi et al., 2002; Guo et al., 2008b.
156. Modulating factor for MAO-A may be other genes: Harris, 2006; Guo et al., 2008b, p. 548.
157. Selection of MAO-A gene: Gilad, 2002.
158. Dopamine receptor & transporter genes: Guo et al., 2008b; Guo, Roettger, & Shih, 2007.
159. No evidence cited of recent selection of genes for behavior: Cochran & Harpending, 2009. See also Wade, 2006.
160. Warrior Gene furor: Holden, 2008; Lea & Chambers, 2007; Merriman & Cameron, 2007.
161. Problems for the Warrior Gene hypothesis: Merriman & Cameron, 2007.
162. Failure to replicate MAO-A–violence link in nonwhites: Widom & Brzustowicz, 2006.
163. "Genetically Capitalist?": Clark, 2007b, p. 1. See also Clark, 2007a, p. 187.
164. Problems for Genetically Capitalist theory: Betzig, 2007; Bowles, 2007; Pomeranz, 2008.
165. Morality as a matter of fact: Harris, 2010; Nagel, 1970; Railton, 1986; Sayre-McCord, 1988.
166. Moralized versus nonmoralized preferences: Haidt, 2002; Rozin, 1997; Rozin et al., 1997.
167. Moral rationalization: Bandura, 1999; Baumeister, 1997.
168. Unjustified norms in moral development: Kohlberg, 1981.
169. Moral dumbfounding: Haidt, 2001.
170. Cross-culturally recurring moral themes: Fiske, 1991; Haidt, 2007; Rai & Fiske, 2011; Shweder et al., 1997.
171. Three ethics: Shweder et al., 1997.
172. Five foundations: Haidt, 2007.
173. Four relational models: Fiske, 1991, 1992, 2004a, 2004b; Haslam, 2004; Rai & Fiske, 2011.
174. Ritual gift exchange: Mauss, 1924/1990.
175. Harm/Care judgments track Fairness/Reciprocity judgments: Haidt, 2007.
176. Why Harm/Care to strangers equals Fairness/Reciprocity: Axelrod, 1984/2006; Trivers, 1971.
177. Extension of Market Pricing to formal institutions: Pinker, 2007b, chaps. 8 & 9; Lee & Pinker, 2010; Pinker et al., 2008; Pinker, 2010.
178. Rational-Legal reasoning and Market Pricing: Fiske, 1991, pp. 435, 47; Fiske, 2004b, p. 17.
179. Grammar of social and moral norms: Fiske, 2004b.
180. Nonmoralized social norms: Fiske, 2004b.
181. Norms in Shasta County: Ellickson, 1991.
182. Not "getting" social norms: Fiske & Tetlock, 1999; Tetlock, 1999.
183. Sacred values and the psychology of taboo: Fiske & Tetlock, 1999; Tetlock, 1999; Tetlock et al., 2000.
184. Rationale for taboos: Fiske & Tetlock, 1999; Tetlock, 2003.
185. Reframing taboo tradeoffs: Fiske & Tetlock, 1997; McGraw & Tetlock, 2005; Tetlock, 1999, 2003.
186. Life insurance: Zelizer, 2005.

187. Cultural differences in relational models: Fiske, 1991, 1992, 2004a; Rai & Fiske, 2011.
188. Political ideologies and relational models: Fiske & Tetlock, 1999; McGraw & Tetlock, 2005; Tetlock, 2003.
189. Moral foundations and the liberal-conservative culture war: Haidt, 2007; Haidt & Graham, 2007; Haidt & Hersh, 2001.
190. Logic of humor: Koestler, 1964; Pinker, 1997, chap. 8.
191. Relational models and violence: Fiske, 1991, pp. 46–47, 130–33.
192. Null/asocial relational model: Fiske, 2004b.
193. Two kinds of dehumanization: Haslam, 2006.
194. Criminal justice as in-kind retribution: Carlsmith et al., 2002; see also Sargent, 2004.
195. Rational-Legal reasoning, Market Pricing, and utilitarianism: Rai & Fiske, 2011; Fiske, 1991, p. 47; McGraw & Tetlock, 2005.
196. Historical shift from Community Sharing to Market Pricing: Fiske & Tetlock, 1997, p. 278, note 3.
197. Liberals and conservatives: Haidt, 2007; Haidt & Graham, 2007; Haidt & Hersh, 2001.
198. We are all liberals now: Courtwright, 2010; Nash, 2009.
199. Market Pricing and utilitarianism: Rai & Fiske, 2011; Fiske, 1991, p. 47; McGraw & Tetlock, 2005.
200. Grammar of relational models: Fiske, 2004b; Fiske & Tetlock, 1999; Rai & Fiske, 2011.
201. Taboo, sacred values, and the Israel-Palestine conflict: Ginges et al., 2007.
202. Red counties and blue counties: Haidt & Graham, 2007; see also http://elections.nytimes.com/2008/results/president/map.html.
203. Communal Sharing implies the group is eternal: Fiske, 1991, p. 44.
204. Hierarchy and history writing: Brown, 1988.
205. Nationalism and potted histories: Bell, 2007b; Scheff, 1994; Tyrrell, 2007; van Evera, 1994.
206. The word *dumbth* was coined by Steve Allen.
207. Blaming the Enlightenment for the Holocaust: See Menschenfreund, 2010. Examples from the left include Zygmunt Bauman, Michel Foucault, and Theodor Adorno; examples from defenders of religion include Dinesh D'Souza in *What's so great about Christianity?* and theoconservatives such as Richard John Neuhaus; see Linker, 2007.
208. Historiometry: Simonton, 1990.
209. Intelligence, openness, and performance of U.S. presidents: Simonton, 2006.
210. Bush third-lowest among presidents: C-SPAN 2009 Historians presidential leadership survey, C-SPAN, 2010; J. Griffin & N. Hines, "Who's the greatest? The Times U.S. presidential rankings," *New York Times*, Mar. 24, 2010; Siena Research Institute, 2010.
211. Neither the best nor the brightest: Nixon is ranked 38th, 27th, and 30th among the 42 presidents in the historians' polls cited in note 210, and 25th in intelligence; see Simonton, 2006, table 1, p. 516, column I-C (chosen because the IQ numbers are the most plausible).
212. The correlation and slope estimates are from a statistical analysis where battle deaths for all war-years in which the United States was a primary or secondary participant were regressed against the president's IQ. Battle deaths are the "Best Estimate" figures from the PRIO Battle Deaths Dataset (Lacina, 2009); IQ estimates from Simonton, 2006, table 1, p. 516, column I-C.
213. Rationality and the Holocaust: Menschenfreund, 2010.
214. Emotional dog and rational tail: Haidt, 2001. Moral reasoning and moral intuition: Pizarro & Bloom, 2003.
215. Moral intuition and moral reasoning in the brain: Greene, in press; Greene et al., 2001.
216. Reason as slave of the passions: Hume, 1739/2000, p. 266.
217. Correlation of intelligence with self-control: Burks et al., 2009; Shamosh & Gray, 2008. Self-control and intelligence in the brain: Shamosh et al., 2008.
218. Emotional reaction to other races: Phelps et al., 2000.
219. Relational models, mathematical scales, and cognition: Fiske, 2004a.
220. Sadistic puppeteers: Gottschall, 2008.
221. Godwin: Quoted in Singer, 1981/2011, pp. 151–52.
222. Logic of morality: Nagel, 1970; Singer, 1981/2011.
223. Systematicity of reasoning: Fodor & Pylyshyn, 1988; Pinker, 1994, 1997, 1999, 2007b.
224. Expanding circle: Singer, 1981/2011.
225. Escalator of reason: Singer, 1981/2011, pp. 88, 113–14.
226. Greek epitaph: Quoted in Singer, 1981/2011, p. 112.
227. Outstanding thinkers: Singer, 1981/2011, pp. 99–100.
228. Flynn's Eureka: Flynn, 1984; Flynn, 2007.
229. Rising IQ around the world: Flynn, 2007, p. 2; Flynn, 1987.
230. Naming the Flynn Effect: Herrnstein & Murray, 1994.
231. Thirty countries: Flynn, 2007, p. 2.
232. Flynn Effect began in 1877: Flynn, 2007, p. 23.
233. Adult of 1910 would be retarded today: Flynn, 2007, p. 23.
234. Scientists' consensus on intelligence: Deary, 2001; Gottfredson, 1997a; Neisser et al., 1996. Intelligence as a predictor of life success: Gottfredson, 1997b; Herrnstein & Murray, 1994.

235. Flynn Effect uncorrelated with testing fads: Flynn, 2007, p. 14.
236. Flynn Effect not in math, vocabulary, knowledge: Flynn, 2007; Greenfield, 2009. See also Wicherts et al., 2004.
237. Slight declines in SAT: Flynn, 2007, p. 20; Greenfield, 2009.
238. General intelligence: Deary, 2001; Flynn, 2007; Neisser et al., 1996.
239. Heritability of intelligence, lack of family influence: Bouchard & McGue, 2003; Harris, 1998/2008; Pinker, 2002; Plomin et al., 2008; Turkheimer, 2000.
240. General intelligence and the brain: Chiang et al., 2009; Deary, 2001; Thompson et al., 2001.
241. Flynn Effect not from hybrid vigor: Flynn, 2007, pp. 101–2. Flynn Effect not from gains in health and nutrition: Flynn, 2007, pp. 102–6.
242. Flynn Effect not in *g*: Flynn, 2007; Wicherts et al., 2004.
243. Visual complexity and IQ: Greenfield, 2009.
244. Prescientific versus postscientific reasoning: Flynn, 2007. See also Neisser, 1976; Tooby & Cosmides, in press; Pinker, 1997, pp. 302–6.
245. Dogs and rabbits: Flynn, 2007, p. 24.
246. Dialogues on similarity and hypotheticals: Cole, Gay, Glick, & Sharp, 1971; Luria, 1976; Neisser, 1976.
247. Schooling and formal operations: Flynn, 2007, p. 32.
248. Increases in schooling: Flynn, 2007, p. 32.
249. Reading comprehension: Rothstein, 1998, p. 19.
250. Changes in school tests: Genovese, 2002.
251. Shorthand abstractions: All of these terms increased in frequency during the 20th century, according to analyses of Google Books by the Bookworm program: Michel et al., 2011; see the caption to figure 7–1.
252. "on account of the economy": G. Nunberg, Language commentary segment on *Fresh Air*, National Public Radio, 2001.
253. Concrete operations in Flynn's father: J. Flynn, "What is intelligence: Beyond the Flynn effect," Harvard Psychology Department Colloquium, Dec. 5, 2007; see also "The world is getting smarter," *Economist/ Intelligent Life*, Dec. 2007; http://moreintelligentlife.com/node/654.
254. Difficulty with proportions: Flynn, 2007, p. 30.
255. Thinking people are less punitive: Sargent, 2004.
256. "cultural renaissance": Flynn, 1987, p. 187.
257. "squalid savages": Roosevelt, *The winning of the West* (Whitefish, Mont.: Kessinger), vol. 1, p. 65. "Dead Indians": Quoted in Courtwright, 1996, p. 109.
258. Woodrow Wilson's racism: Loewen, 1995, pp. 22–31.
259. Churchill's racism: Toye, 2010; quotes excerpted in J. Hari, "The two Churchills," *New York Times*, Aug. 12, 2010.
260. Dumb congressional testimony: Quoted in Courtwright, 1996, pp. 155–56.
261. Literary intellectuals' contempt for the masses: Carey, 1993.
262. Intellectuals' support of totalitarianism: Carey, 1993; Glover, 1999; Lilla, 2001; Sowell, 2010; Wolin, 2004.
263. Appalling Eliot: Carey, 1993, p. 85.
264. Intellectuals' troublemaking: Carey, 1993; Glover, 1999; Lilla, 2001; Sowell, 2010; Wolin, 2004.
265. Smarter people are less violent: Herrnstein & Murray, 1994; Wilson & Herrnstein, 1985; Farrington, 2007, pp. 22–23, 26–27.
266. Superrationality in the Prisoner's Dilemma: Hofstadter, 1985.
267. Truckers' IQ and the Prisoner's Dilemma: Burks et al., 2009.
268. University SATs and the Prisoner's Dilemma: Jones, 2008.
269. Politicization of social science: Haidt & Graham, 2007; Tetlock, 1994.
270. Political liberals and fairness: Haidt, 2007; Haidt & Graham, 2007.
271. Liberalism and intelligence in the United States: Kanazawa, 2010.
272. Bright children become enlightened adults: Deary, Batty, & Gale, 2008.
273. Smarter people think like economists: Caplan & Miller, 2010.
274. Market economies as a pacifying force: Kant, 1795/1983; Mueller, 1999; Russett & Oneal, 2001; Schneider & Gleditsch, 2010; Wright, 2000 Mueller, 2010b.
275. Zero-sum mindset & ethnic violence: Sowell, 1980, 1996.
276. Kantian Peace: Gleditsch, 2008; Russett, 2008; Russett & Oneal, 2001.
277. Cognitive prerequisites for democracy: Rindermann, 2008.
278. Intellectual abilities and democracy: Rindermann, 2008.
279. Democracy and violence: Gleditsch, 2008; Harff, 2003, 2005; Lacina, 2006; Pate, 2008; Rummel, 1994; Russett, 2008; Russett & Oneal, 2001.
280. Flynn Effect in Kenya and Dominica: Flynn, 2007, p. 144.
281. Education and civil war: Thyne, 2006.
282. Pacifying effect of education: Thyne, 2006, p. 733.
283. Integrative complexity: Suedfeld & Coren, 1992; Tetlock, 1985; Tetlock, Peterson, & Lerner, 1996.

284. Low but significant correlation between IQ and integrative complexity: Suedfeld & Coren, 1992. Correlation of 0.58 among presidents: Simonton, 2006.
285. Integrative complexity and violence: Tetlock, 1985, pp. 1567–68.
286. Integrative complexity and major wars: Suedfeld & Tetlock, 1977.
287. Integrative complexity and Arab-Israeli wars: Suedfeld, Tetlock, & Ramirez, 1977. Integrative complexity and U.S.-Soviet actions: Tetlock, 1985.
288. Teasing apart effects of integrative complexity: Tetlock, 1985; Tetlock et al., 1996.
289. Integrative complexity of American political discourse: Rosenau & Fagen, 1997.
290. Inane 1917 congressman: Quoted in Rosenau & Fagen, 1997, p. 676.
291. Inane 1972 senator: Quoted in Rosenau & Fagen, 1997, p. 677.
292. Sophistication of U.S. presidential debates: Gorton & Diels, 2010.
293. Inhabitant of the breast: Smith, 1759/1976, p. 136.

Chapter 10: On Angels' Wings

1. Hitler was not an atheist: Murphy, 1999.
2. Nazism and Christianity: Ericksen & Heschel, 1999; Goldhagen, 1996; Heschel, 2008; Steigmann-Gall, 2003; Chirot & McCauley, 2006, p. 144.
3. Common cause between Nazism and the church: Ericksen & Heschel, 1999, p. 11.
4. Marxism and Christianity: Chirot & McCauley, 2006, pp. 142–43; Chirot, 1995.
5. Gifts and amity: Mauss, 1924/1990.
6. Unsexy merchants: Mueller, 1999, 2010b.
7. Only nursing mothers should command nuclear forces: D. Garner, "After atom bomb's shock, the real horrors began unfolding," *New York Times*, Jan. 20, 2010.
8. Handbagged: Shultz, 2009.
9. Surplus of men, deficit of peace: Hudson & den Boer, 2002.
10. Sparing bare branches: Hudson & den Boer, 2002, p. 26.
11. Bolus of men: Hudson & den Boer, 2002.
12. Number of young men and odds of war: Fearon & Laitin, 2003; Mesquida & Wiener, 1996.
13. *Sex and War*: Potts, Campbell, & Hayden, 2008.
14. Mixing of peoples fosters humanism: Examples include the findings that countries in which women are more influential have less domestic violence (Archer, 2006a); that people who know gay people are less homophobic (see note 232 to chap. 7); and that American counties along coasts and waterways are more liberal (Haidt & Graham, 2007).
15. Escalator of reason, expanding circle: Both are taken from Singer, 1981/2011.
16. The idea of decline: Herman, 1997.
17. Rational optimists: Bettmann, 1974; Easterbrook, 2003; Goklany, 2007; Kenny, 2011; Ridley, 2010; Robinson, 2009; Wattenberg, 1984.
18. A higher power?: Payne, 2004, p. 29.
19. Evidence of divinity?: Wright, 2000, p. 319. Divinely imparted meaning?: Wright, 2000, p. 320. Cosmic author?: Wright, 2000, p. 334.
20. Moral realism: Nagel, 1970; Railton, 1986; Sayre-McCord, 1988; Shafer-Landau, 2003; Harris, 2010.
21. "hot and forcing violation": *Henry V*, act 3, scene 3.
22. Hundreds of millions murdered: Rummel, 1994, 1997. "Naked ape": from Desmond Morris.
23. While this planet has gone cycling on: Charles Darwin, *The Origin of Species*, final sentence.

REFERENCES

ABC News. 2002. Most say spanking's OK by parents but not by grade-school teachers. *ABC News Poll.* New York. http://abcnews.go.com/sections/US/DailyNews/spanking_poll021108.html.
Aboud, F. E. 1989. *Children and prejudice.* Cambridge, Mass.: Blackwell.
Abrahms, M. 2006. Why terrorism does not work. *International Security, 31,* 42–78.
Ad Nauseam. 2000. *"You mean a woman can open it . . . ?": The woman's place in the classic age of advertising.* Holbrook, Mass.: Adams Media.
Adams, D. 2006. Brain mechanisms of aggressive behavior: An updated review. *Neuroscience & Biobehavioral Reviews, 30,* 304–18.
Ainslie, G. 2001. *Breakdown of will.* New York: Cambridge University Press.
Akerlof, G. A. 1984. *An economic theorist's book of tales.* New York: Cambridge University Press.
Akey, J. M. 2009. Constructing genomic maps of positive selection in humans: Where do we go from here? *Genome Research, 19,* 711–22.
Alia-Klein, N., Goldstein, R. Z., Kriplani, A., Logan, J., Tomasi, D., Williams, B., Telang, F., Shumay, E., Biegon, A., Craig, I. W., Henn, F., Wang, G. J., Volkow, N. D., & Fowler, J. S. 2008. Brain monoamine oxidase-A activity predicts trait aggression. *Journal of Neuroscience, 28,* 5099–5104.
American Humane Association Film and Television Unit. 2009. *No animals were harmed: Guidelines for the safe use of animals in filmed media.* Englewood, Colo.: American Humane Association.
American Humane Association Film and Television Unit. 2010. No animals were harmed: A legacy of protection. http://www.americanhumane.org/protecting-animals/programs/no-animals-were-harmed/legacy-of-protection.html.
American Psychiatric Association. 2000. *Diagnostic and statistical manual of mental disorders: DSM-IV-TR,* 4th ed. Washington, D.C.: American Psychiatric Association.
Amnesty International. 2010. Abolitionist and retentionist countries. http://www.amnesty.org/en/death-penalty/abolitionist-and-retentionist-countries.
Anderson, E. 1999. *The code of the street: Violence, decency, and the moral life of the inner city.* New York: Norton.
Anderson, S. W., Bechara, A., Damasio, H., Tranel, D., & Damasio, A. R. 1999. Impairment of social and moral behavior related to early damage in human prefrontal cortex. *Nature Neuroscience, 2,* 1032–37.
Andreas, P., & Greenhill, K. M., eds. 2010. *Sex, drugs, and body counts: The politics of numbers in global crime and conflict.* Ithaca, N.Y.: Cornell University Press.
Appiah, K. A. 2006. *Cosmopolitanism: Ethics in a world of strangers.* New York: Norton.
Archer, J. 2006a. Cross-cultural differences in physical aggression between partners: A social role analysis. *Personality and Social Psychology Bulletin, 10,* 133–53.
Archer, J. 2006b. Testosterone and human aggression: An evaluation of the challenge hypothesis. *Neuroscience & Biobehavioral Reviews, 30,* 319–45.
Archer, J. 2009. Does sexual selection explain human sex differences in aggression? *Behavioral & Brain Sciences, 32,* 249–311.
Arendt, H. 1963. *Eichmann in Jerusalem: A report on the banality of evil.* New York: Viking Press.
Asal, V., Johnson, C., & Wilkenfeld, J. 2008. Ethnopolitical violence and terrorism in the Middle East. In J. J. Hewitt, J. Wilkenfeld & T. R. Gurr, eds., *Peace and conflict 2008.* Boulder, Colo.: Paradigm.
Asal, V., & Pate, A. 2005. The decline of ethnic political discrimination 1950–2003. In M. G. Marshall & T. R. Gurr, eds., *Peace and conflict 2005: A global survey of armed conflicts, self-determination movements, and democracy.* College Park: Center for International Development and Conflict Management, University of Maryland.

Asch, S. E. 1956. Studies of independence and conformity I: A minority of one against a unanimous majority. *Psychological Monographs: General and Applied, 70,* Whole No. 416.

Atran, S. 2002. *In gods we trust: The evolutionary landscape of supernatural agency.* New York: Oxford University Press.

Atran, S. 2003. Genesis of suicide terrorism. *Science, 299,* 1534–39.

Atran, S. 2006. The moral logic and growth of suicide terrorism. *Washington Quarterly, 29,* 127–47.

Atran, S. 2010. Pathways to and from violent extremism: The case for science-based field research (Statement Before the Senate Armed Services Subcommittee on Emerging Threats & Capabilities, March 10, 2010). *Edge.* http://www.edge.org/3rd_culture/atran10/atran10_index.html.

Axelrod, R. 1984/2006. *The evolution of cooperation.* New York: Basic Books.

Axelrod, R., & Hamilton, W. D. 1981. The evolution of cooperation. *Science, 211,* 1390–96.

Bailey, M. 2003. *The man who would be queen: The science of gender-bending and transsexualism.* Washington, D.C.: National Academies Press.

Bandura, A. 1999. Moral disengagement in the perpetration of inhumanities. *Personality & Social Psychology Review, 3,* 193–209.

Bandura, A. 2002. Reflexive empathy: On predicting more than has ever been observed. *Behavioral & Brain Sciences, 25,* 24–25.

Bandura, A., Underwood, B., & Fromson, M. E. 1975. Disinhibition of aggression through diffusion of responsibility and dehumanization of victims. *Journal of Research in Personality, 9,* 253–69.

Baron-Cohen, S. 1995. *Mindblindness: An essay on autism and theory of mind.* Cambridge, Mass.: MIT Press.

Batson, C. D., & Ahmad, N. 2001. Empathy-induced altruism in a prisoner's dilemma II: What if the partner has defected? *European Journal of Social Psychology, 31,* 25–36.

Batson, C. D., Ahmad, N., Lishmer, D. A., & Tsang, J.-A. 2002. Empathy and altruism. In C. R. Snyder & S. J. Lopez, eds., *Handbook of positive psychology.* Oxford, U.K.: Oxford University Press.

Batson, C. D., Ahmad, N., & Stocks, E. L. 2005. Benefits and liabilities of empathy-induced altruism. In A. G. Miller ed., *The social psychology of good and evil.* New York: Guilford Press.

Batson, C. D., Batson, J. G., Todd, M., Brummett, B. H., Shaw, L. L., & Aldeguer, C. M. R. 1995. Empathy and the collective good: Caring for one of the others in a social dilemma. *Journal of Personality & Social Psychology, 68,* 619–31.

Batson, C. D., Chang, J., Orr, R., & Rowland, J. 2008. Empathy, attitudes, and action: Can feeling for a member of a stigmatized group motivate one to help the group? *Personality & Social Psychology Bulletin, 28,* 1656–66.

Batson, C. D., Duncan, B. D., Ackerman, P., Buckley, T., & Birch, K. 1981. Is empathic emotion a source of altruistic motivation? *Journal of Personality & Social Psychology, 40,* 292–302.

Batson, C. D., Dyck, J. L., Brandt, R. B., Batson, J. G., Powell, A. L., McMaster, M. R., & Griffitt, C. 1988. Five studies testing two new egoistic alternatives to the empathy-altruism hypothesis. *Journal of Personality & Social Psychology, 55,* 52–77.

Batson, C. D., Klein, T. R., Highberger, L., & Shaw, L. L. 1995b. Immorality from empathy-induced altruism: When compassion and justice conflict. *Journal of Personality & Social Psychology, 68,* 1042–54.

Batson, C. D., Lishner, D. A., Cook, J., & Sawyer, S. 2005b. Similarity and nurturance: Two possible sources of empathy for strangers. *Basic & Applied Social Psychology, 27,* 15–25.

Batson, C. D., & Moran, T. 1999. Empathy-induced altruism in a prisoner's dilemma. *European Journal of Social Psychology, 29,* 909–24.

Batson, C. D., Polycarpou, M. R., Harmon-Jones, E., Imhoff, H. J., Mitchener, E. C., Bednar, L. L., Klein, T. R., & Highberger, L. 1997. Empathy and attitudes: Can feeling for a member of a stigmatized group improve feelings toward the group? *Journal of Personality & Social Psychology, 72,* 105–18.

Batson, C. D., Turk, C. L., Shaw, L. L., & Klein, T. R. 1995. Information function of empathic emotion: Learning that we value the other's welfare. *Journal of Personality and Social Psychology, 68,* 300–313.

Batty, G. D., Deary, I. J., Tengstrom, A., & Rasmussen, F. 2008. IQ in early adulthood and later risk of death by homicide: Cohort study of 1 million men. *British Journal of Psychiatry, 193,* 461–65.

Baumeister, R. F. 1997. *Evil: Inside human violence and cruelty.* New York: Holt.

Baumeister, R. F. 2000. Gender differences in erotic plasticity: The female sex drive as socially flexible and responsive. *Psychological Bulletin, 126,* 347–74.

Baumeister, R. F., Bratslavsky, E., Muraven, M., & Tice, D. M. 1998. Ego depletion: Is the active self a limited resource? *Journal of Personality & Social Psychology, 24,* 1252–65.

Baumeister, R. F., & Campbell, W. K. 1999. The intrinsic appeal of evil: Sadism, sensational thrills, and threatened egotism. *Personality & Social Psychology Review, 3,* 210–21.

Baumeister, R. F., Gailliot, M., Dewall, C. N., & Oaten, M. 2006. Self-regulation and personality: How interventions increase regulatory success, and how depletion moderates the effects of traits on behavior. *Journal of Personality, 74,* 1773–1801.

Baumeister, R. F., Smart, L., & Boden, J. M. 1996. Relation of threatened egotism to violence and aggression: The dark side of high self-esteem. *Psychological Review, 103,* 5–33.

Baumeister, R. F., Stillwell, A. M., & Wotman, S. R. 1990. Victim and perpetrator accounts of interpersonal conflict: Autobiographical narratives about anger. *Journal of Personality & Social Psychology, 59,* 994–1005.

Baumeister, R. F., Stillwell, A. M., & Heatherton, T. F. 1994. Guilt: An interpersonal approach. *Psychological Bulletin, 115,* 243–67.

Beaver, K. M., DeLisi, M., Vaughn, M. G., & Wright, J. P. 2008. The intersection of genes and neuropsychological deficits in the prediction of adolescent delinquency and low self-control. *International Journal of Offender Therapy & Comparative Criminology, 54,* 22–42.

Bell, D. A. 2007a. *The first total war: Napoleon's Europe and the birth of warfare as we know it.* Boston: Houghton Mifflin.

Bell, D. A. 2007b. Pacific nationalism. *Critical Review, 19,* 501–10.

Bennett, A. 2005. The guns that didn't smoke: Ideas and the Soviet non-use of force in 1989. *Journal of Cold War Studies, 7,* 81–109.

Berkowitz, L., & Frodi, A. 1979. Reactions to a child's mistakes as affected by her/his looks and speech. *Social Psychology Quarterly, 42,* 420–25.

Berlin, I. 1979. The counter-enlightenment. *Against the current: Essays in the history of ideas.* Princeton, N.J.: Princeton University Press.

Besançon, A. 1998. Forgotten communism. *Commentary,* Jan., 24–27.

Bettmann, O. L. 1974. *The good old days—they were terrible!* New York: Random House.

Betzig, L. L. 1986. *Despotism and differential reproduction.* Hawthorne, N.Y.: Aldine de Gruyter.

Betzig, L. L. 1996a. Monarchy. In D. Levinson & M. Ember, eds., *Encyclopedia of Cultural Anthropology.* New York: Holt.

Betzig, L. L. 1996b. Political succession. In D. Levinson & M. Ember, eds., *Encyclopedia of Cultural Anthropology.* New York: Holt.

Betzig, L. L. 2002. British polygyny. In M. Smith, ed., *Human biology and history.* London: Taylor & Francis.

Betzig, L. L. 2007. The son also rises. *Evolutionary Psychology, 5,* 733–39.

Betzig, L. L., Borgerhoff Mulder, M., & Turke, P. 1988. *Human reproductive behavior: A Darwinian perspective.* New York: Cambridge University Press.

Bhagavad-Gita. 1983. *Bhagavad-Gita as it is.* Los Angeles: Bhaktivedanta Book Trust.

Bijian, Z. 2005. China's "peaceful rise" to great-power status. *Foreign Affairs,* Sept./Oct.

Black, D. 1983. Crime as social control. *American Sociological Review, 48,* 34–45.

Blackwell, A. D., & Sugiyama, L. S. In press. When is self-sacrifice adaptive? In L. S. Sugiyama, M. Scalise Sugiyama, F. White, D. Kenneth, & H. Arrow, eds., *War in Evolutionary Perspective.*

Blair, R.J.R. 2004. The roles of orbital frontal cortex in the modulation of antisocial behavior. *Brain & Cognition, 55,* 198–208.

Blair, R.J.R., & Cipolotti, L. 2000. Impaired social response reversal: A case of "acquired sociopathy." *Brain, 123,* 1122–41.

Blair, R.J.R., & Perschardt, K. S. 2002. Empathy: A unitary circuit or a set of dissociable neurocognitive systems? *Behavioral & Brain Sciences, 25,* 27–28.

Blanton, R. 2007. Economic globalization and violent civil conflict: Is openness a pathway to peace? *Social Science Journal, 44,* 599–619.

Blight, J. G., & Lang, J. M. 2007. Robert McNamara: Then and now. *Daedalus, 136,* 120–31.

Blum, A., & Houdailles, J. 1985. L'alphabétisation au XVIIIe et XIXeme siècle. L'illusion parisienne. *Population, 6.*

Blum, D. 1997. *Sex on the brain: The biological differences between men and women.* New York: Viking.

Blumstein, A., & Wallman, J., eds. 2006. *The crime drop in America,* rev. ed. New York: Cambridge University Press.

Bobo, L. D. 2001. Racial attitudes and relations at the close of the twentieth century. In N. J. Smelser, W. J. Wilson, & F. Mitchell, eds., *America becoming: Racial trends and their consequences.* Washington, D.C.: National Academies Press.

Bobo, L. D., & Dawson, M. C. 2009. A change has come: Race, politics, and the path to the Obama presidency. *Du Bois Review, 6,* 1–14.

Boehm, C. 1999. *Hierarchy in the forest: The evolution of egalitarian behavior.* Cambridge, Mass.: Harvard University Press.

Bohannon, J. 2008. Calculating Iraq's death toll: WHO study backs lower estimate. *Science, 319,* 273.

———. 2011. Counting the dead in Afghanistan. *Science, 331,* 1256–60.

Bolton, G. E., & Zwick, R. 1995. Anonymity versus punishment in ultimatum bargaining. *Games & Economic Behavior, 10,* 95–121.

Bouchard, T. J., Jr., & McGue, M. 2003. Genetic and environmental influences on human psychological differences. *Journal of Neurobiology, 54,* 4–45.

Boulton, M. J., & Smith, P. K. 1992. The social nature of play fighting and play chasing: Mechanisms and strategies underlying cooperation and compromise. In J. Barkow, L. Cosmides, & J. Tooby, eds., *The adapted mind: Evolutionary psychology and the generation of culture.* New York: Oxford University Press.

Bourgon, J. 2003. Abolishing "cruel punishments": A reappraisal of the Chinese roots and long-term efficiency of the Xinzheng legal reforms. *Modern Asian Studies, 37,* 851–62.

Bourke, J. 1999. *An intimate history of killing: Face-to-face killing in 20th-century warfare.* New York: Basic Books.

Bowles, S. 2007. Genetically capitalist? *Science, 318,* 394–95.

Bowles, S. 2009. Did warfare among ancestral hunter-gatherers affect the evolution of human social behaviors? *Science, 324,* 1293–98.

Boyd, R., & Silk, J. B. 2006. *How humans evolved,* 4th ed. New York: Norton.

Brandt, R. B. 1974. *Hopi ethics: A theoretical analysis.* Chicago: University of Chicago Press.

Brecke, P. 1999. *Violent conflicts 1400 A.D. to the present in different regions of the world.* Paper presented at the 1999 Meeting of the Peace Science Society (International).

Brecke, P. 2002. Taxonomy of violent conflicts. http://www.inta.gatech.edu/peter/power.html.

Breiner, S. J. 1990. *Slaughter of the innocents: Child abuse through the ages and today.* New York: Plenum.

Brezina, T., Agnew, R., Cullen, F. T., & Wright, J. P. 2004. The code of the street: A quantitative assessment of Elijah Anderson's subculture of violence thesis and its contribution to youth violence research. *Youth Violence & Juvenile Justice, 2,* 303–28.

Brock, D. W. 1993. *Life and death: Philosophical essays in biomedical ethics.* New York: Cambridge University Press.

Bronowski, J. 1973. *The ascent of man.* Boston: Little, Brown.

Broude, G. J., & Greene, S. J. 1976. Cross-cultural codes on twenty sexual attitudes and practices. *Ethnology, 15,* 409–29.

Brown, B. R. 1968. The effects of need to maintain face on interpersonal bargaining. *Journal of Experimental Social Psychology, 4,* 107–22.

Brown, D. E. 1988. *Hierarchy, history, and human nature: The social origins of historical consciousness.* Tucson: University of Arizona Press.

Brown, D. E. 1991. *Human universals.* New York: McGraw-Hill.

Brown, D. E. 2000. Human universals and their implications. In N. Roughley, ed., *Being humans: Anthropological universality and particularity in transdisciplinary perspectives.* New York: Walter de Gruyter.

Brown, M. E. 1997. The impact of government policies on ethnic relations. In M. Brown & S. Ganguly, eds., *Government policies and ethnic relations in Asia and the Pacific.* Cambridge, Mass.: MIT Press.

Brown, R. 1985. *Social psychology: The second edition.* New York: Free Press.

Browne, A., & Williams, K. R. 1989. Exploring the effect of resource availability and the likelihood of female-perpetrated homicides. *Law & Society Review, 23,* 75–94.

Browne, K. 2002. *Biology at work.* New Brunswick, N.J.: Rutgers University Press.

Browning, C. R. 1992. *Ordinary men: Reserve police battalion 101 and the final solution in Poland.* New York: HarperCollins.

Brownmiller, S. 1975. *Against our will: Men, women, and rape.* New York: Fawcett Columbine.

Broyles, W. J. 1984. Why men love war. *Esquire* (November).

Brunner, H. G., Nelen, M., Breakfield, X. O., Ropers, H. H., & van Oost, B. A. 1993. Abnormal behavior associated with a point mutation in the structural gene for monoamine oxidase A. *Science, 262,* 578–80.

Buhaug, H. 2010. Climate not to blame for African civil wars. *Proceedings of the National Academy of Sciences, 107,* 16477–82.

Bullock, A. 1991. *Hitler and Stalin: Parallel lives.* London: HarperCollins.

Burch, E. S. 2005. *Alliance and conflict.* Lincoln: University of Nebraska Press.

Burger, J. M. 2009. Replicating Milgram: Would people still obey today? *American Psychologist, 64,* 1–11.

Burke, E. 1790/1967. *Reflections on the revolution in France.* London: J. M. Dent & Sons.

Burks, S. V., Carpenter, J. P., Goette, L., & Rustichini, A. 2009. Cognitive skills affect economic preferences, strategic behavior, and job attachment. *Proceedings of the National Academy of Sciences, 106,* 7745–50.

Burnham, G., Lafta, R., Doocy, S., & Roberts, L. 2006. Mortality after the 2003 invasion of Iraq: A cross-sectional cluster sample survey. *Lancet, 368,* 1421–28.

Bushman, B. J. 1997. Effects of alcohol on human aggression: Validity of proposed explanations. *Recent Developments in Alcoholism, 13,* 227–43.

Bushman, B. J., & Cooper, H. M. 1990. Effects of alcohol on human aggression: An integrative research review. *Psychological Bulletin, 107,* 341–54.

Buss, D. M. 1989. Conflict between the sexes. *Journal of Personality & Social Psychology, 56,* 735–47.

Buss, D. M. 1994. *The evolution of desire.* New York: Basic Books.

Buss, D. M. 2000. *The dangerous passion: Why jealousy is as necessary as love and sex.* New York: Free Press.

Buss, D. M. 2005. *The murderer next door: Why the mind is designed to kill.* New York: Penguin.

Buss, D. M., & Duntley, J. D. 2008. Adaptations for exploitation. *Group Dynamics: Theory, Research, & Practice, 12,* 53–62.

Buss, D. M., & Schmitt, D. P. 1993. Sexual strategies theory: An evolutionary perspective on human mating. *Psychological Review, 100,* 204–32.

Bussman, M., & Schneider, G. 2007. When globalization discontent turns violent: Foreign economic liberalization and internal war. *International Studies Quarterly, 51*, 79–97.
Byrnes, J. P., Miller, D. C., & Schafer, W. D. 1999. Gender differences in risk-taking: A meta-analysis. *Psychological Bulletin, 125*, 367–83.
C-SPAN. 2010. C-SPAN 2009 Historians presidential leadership survey. http://legacy.c-span.org/PresidentialSurvey/presidential-leadership-survey.aspx.
Cairns, R. B., Gariépy, J.-L., & Hood, K. E. 1990. Development, microevolution, and social behavior. *Psychological Review, 97*, 49–65.
Capital Punishment U.K. 2004. *The end of capital punishment in Europe.* http://www.capitalpunishmentuk.org/europe.html.
Caplan, B. 2007. *The myth of the rational voter: Why democracies choose bad policies.* Princeton, N.J.: Princeton University Press.
Caplan, B., & Miller, S. C. 2010. Intelligence makes people think like economists: Evidence from the General Social Survey. *Intelligence, 38*, 636–47.
Caplow, T., Hicks, L., & Wattenberg, B. 2001. *The first measured century: An illustrated guide to trends in America, 1900–2000.* Washington, D.C.: AEI Press.
Carey, J. 1993. *The intellectuals and the masses: Pride and prejudice among the literary intelligentsia, 1880–1939.* New York: St. Martin's Press.
Carlin, J. 2008. *Playing the enemy: Nelson Mandela and the game that made a nation.* New York: Penguin.
Carlsmith, K. M., Darley, J. M., & Robinson, P. H. 2002. Why do we punish? Deterrence and just deserts as motives for punishment. *Journal of Personality & Social Psychology, 83*, 284–99.
Carswell, S. 2001. *Survey on public attitudes towards the physical punishment of children.* Wellington: New Zealand Ministry of Justice.
Caspi, A. 2000. The child is father of the man: Personality continuities from childhood to adulthood. *Journal of Personality & Social Psychology, 78*, 158–72.
Caspi, A., McClay, J., Moffitt, T. E., Mill, J., Martin, J., Craig, I. W., Taylor, A., & Poulton, R. 2002. Evidence that the cycle of violence in maltreated children depends on genotype. *Science, 297*, 727–42.
Cavalli-Sforza, L. L. 2000. *Genes, peoples, and languages.* New York: North Point Press.
Cederman, L.-E. 2001. Back to Kant: Reinterpreting the democratic peace as a macrohistorical learning process. *American Political Science Review, 95*, 15–31.
Cederman, L.-E. 2003. Modeling the size of wars: From billiard balls to sandpiles. *American Political Science Review, 97*, 135–50.
Center for Systemic Peace. 2010. Integrated network for societal conflict research data page. http://www.systemicpeace.org/inscr/inscr.htm.
Centola, D., Willer, R., & Macy, M. 2005. The emperor's dilemma: A computational model of self-enforcing norms. *American Journal of Sociology, 110*, 1009–40.
Central Intelligence Agency. 2010. *The world factbook.* https://www.cia.gov/library/publications/the-world-factbook/.
Chabris, C. F., Laibson, D., Morris, C. L., Schuldt, J. P., & Taubinsky, D. 2008. Individual laboratory-measured discount rates predict field behavior. *Journal of Risk & Uncertainty, 37*, 237–69.
Chagnon, N. A. 1988. Life histories, blood revenge, and warfare in a tribal population. *Science, 239*, 985–92.
Chagnon, N. A. 1996. Chronic problems in understanding tribal violence and warfare. In G. Bock & J. Goode, eds., *The genetics of criminal and antisocial behavior.* New York: Wiley.
Chagnon, N. A. 1997. *Yanomamö*, 5th ed. New York: Harcourt Brace.
Chalk, F., & Jonassohn, K. 1990. *The history and sociology of genocide: Analyses and case studies.* New Haven, Conn.: Yale University Press.
Check, J.V.P., & Malamuth, N. 1985. An empirical assessment of some feminist hypotheses about rape. *International Journal of Women's Studies, 8*, 414–23.
Chernow, R. 2004. *Alexander Hamilton.* New York: Penguin.
Chiang, M. C., Barysheva, M., Shattuck, D. W., Lee, A. D., Madsen, S. K., Avedissian, C., Klunder, A. D., Toga, A. W., McMahon, K. L., de Zubicaray, G. I., Wright, M. J., Srivastava, A., Balov, N., & Thompson, P. M. 2009. Genetics of brain fiber architecture and intellectual performance. *Journal of Neuroscience, 29*, 2212–24.
China (Taiwan), Republic of, Department of Statistics, Ministry of the Interior. 2000. The analysis and comparison on statistics of criminal cases in various countries. http://www.moi.gov.tw/stat/english/topic.asp.
Chirot, D. 1994. *Modern tyrants.* Princeton, N.J.: Princeton University Press.
Chirot, D. 1995. Modernism without liberalism: The ideological roots of modern tyranny. *Contention, 5*, 141–82.
Chirot, D., & McCauley, C. 2006. *Why not kill them all? The logic and prevention of mass political murder.* Princeton, N.J.: Princeton University Press.
Chu, R., Rivera, C., & Loftin, C. 2000. Herding and homicide: An examination of the Nisbett-Reeves hypothesis. *Social Forces, 78*, 971–87.

Chwe, M. S.-Y. 2001. *Rational ritual: Culture, coordination, and common knowledge.* Princeton, N.J.: Princeton University Press.

Claeys, G. 2000. "The survival of the fittest" and the origins of Social Darwinism. *Journal of the History of Ideas, 61,* 223–40.

Clark, G. 2007a. *A farewell to alms: A brief economic history of the world.* Princeton, N.J.: Princeton University Press.

Clark, G. 2007b. Genetically capitalist? The Malthusian era and the formation of modern preferences. http://www.econ.ucdavis.edu/faculty/gclark/papers/capitalism%20genes.pdf.

Clark, M. S., Mills, J., & Powell, M. C. 1986. Keeping track of needs in communal and exchange relationships. *Journal of Personality & Social Psychology, 51,* 333–38.

Clauset, A., & Young, M. 2005. Scale invariance in global terrorism. arXiv:physics/0502014v2 [physics.soc-ph].

Clauset, A., Young, M., & Gleditsch, K. S. 2007. On the frequency of severe terrorist events. *Journal of Conflict Resolution, 51,* 58–87.

Cleaver, E. 1968/1999. *Soul on ice.* New York: Random House.

Cloninger, C. R., & Gottesman, I. I. 1987. Genetic and environmental factors in antisocial behavior disorders. In S. A. Mednick, T. E. Moffitt, & S. A. Stack, eds., *The causes of crime: New biological approaches.* New York: Cambridge University Press.

Clutton-Brock, T. 2007. Sexual selection in males and females. *Science, 318,* 1882–85.

Cochran, G., & Harpending, H. 2009. *The 10,000 year explosion: How civilization accelerated human evolution.* New York: Basic Books.

Cockburn, J. S. 1991. Patterns of violence in English society: Homicide in Kent, 1560–1985. *Past & Present, 130,* 70–106.

Coghlan, B., Ngoy, P., Mulumba, F., Hardy, C., Bemo, V. N., Stewart, T., Lewis, J., & Brennan, R. 2008. *Mortality in the Democratic Republic of Congo: An ongoing crisis.* New York: International Rescue Committee. http://www.theirc.org/sites/default/files/migrated/resources/2007/2006-7_congomortalitysurvey.pdf.

Cohen, D., & Nisbett, R. E. 1997. Field experiments examining the culture of honor: The role of institutions in perpetuating norms about violence. *Personality & Social Psychology Bulletin, 23,* 1188–99.

Cohen, D., Nisbett, R. E., Bowdle, B., & Schwarz, N. 1996. Insult, aggression, and the southern culture of honor: An "experimental ethnography." *Journal of Personality & Social Psychology, 20,* 945–60.

Cole, M., Gay, J., Glick, J., & Sharp, D. W. 1971. *The cultural context of learning and thinking.* New York: Basic Books.

Collier, P. 2007. *The bottom billion: Why the poorest countries are failing and what can be done about it.* New York: Oxford University Press.

Collins, R. 2008. *Violence: A micro-sociological theory.* Princeton, N.J.: Princeton University Press.

Cook, P. J., & Moore, M. H. 1999. Guns, gun control, and homicide. In M. D. Smith & M. A. Zahn, eds., *Homicide: A sourcebook of social research.* Thousand Oaks, Calif.: Sage.

Cooney, M. 1997. The decline of elite homicide. *Criminology, 35,* 381–407.

Cosmides, L., & Tooby, J. 1992. Cognitive adaptations for social exchange. In J. H. Barkow, L. Cosmides, & J. Tooby, eds., *The adapted mind: Evolutionary psychology and the generation of culture.* New York: Oxford University Press.

Côté, S. M., Vaillancourt, T., LeBlanc, J. C., Nagin, D. S., & Tremblay, R. E. 2006. The development of physical aggression from toddlerhood to pre-adolescence: A nationwide longitudinal study of Canadian children. *Journal of Abnormal Child Psychology, 34,* 71–85.

Courtois, S., Werth, N., Panné, J.-L., Paczkowski, A., Bartosek, K., & Margolin, J.-L. 1999. *The black book of communism: Crimes, terror, repression.* Cambridge, Mass.: Harvard University Press.

Courtwright, D. T. 1996. *Violent land: Single men and social disorder from the frontier to the inner city.* Cambridge, Mass.: Harvard University Press.

Courtwright, D. T. 2010. *No right turn: Conservative politics in a liberal America.* Cambridge, Mass.: Harvard University Press.

Crick, N. R., Ostrov, J. M., & Kawabata, Y. 2007. Relational aggression and gender: An overview. In D. J. Flannery, A. T. Vazsonyi, & I. D. Waldman, eds., *The Cambridge handbook of violent behavior and aggression.* New York: Cambridge University Press.

Cronin, A. K. 2009. *How terrorism ends: Understanding the decline and demise of terrorist campaigns.* Princeton, N.J.: Princeton University Press.

Csibra, G. 2008. Action mirroring and action understanding: An alternative account. In P. Haggard, Y. Rosetti, & M. Kawamoto, eds., *Attention and performance XXII: Sensorimotor foundations of higher cognition.* New York: Oxford Univeristy Press.

Cullen, D. 2009. *Columbine.* New York: Twelve.

Curtis, V., & Biran, A. 2001. Dirt, disgust, and disease: Is hygiene in our genes? *Perspectives in Biology & Medicine, 44,* 17–31.

Dabbs, J. M., & Dabbs, M. G. 2000. *Heroes, rogues, and lovers: Testosterone and behavior.* New York: McGraw Hill.

Daly, M., Salmon, C., & Wilson, M. 1997. Kinship: The conceptual hole in psychological studies of social cognition and close relationships. In J. Simpson & D. Kenrick, eds., *Evolutionary social psychology*. Mahwah, N.J.: Erlbaum.
Daly, M., & Wilson, M. 1983. *Sex, evolution, and behavior*, 2nd ed. Belmont, Calif.: Wadsworth.
Daly, M., & Wilson, M. 1988. *Homicide*. New York: Aldine De Gruyter.
Daly, M., & Wilson, M. 1999. *The truth about Cinderella: A Darwinian view of parental love*. New Haven, Conn.: Yale University Press.
Daly, M., & Wilson, M. 2000. Risk-taking, intrasexual competition, and homicide. *Nebraska Symposium on Motivation*. Lincoln: University of Nebraska Press.
Daly, M., & Wilson, M. 2005. Carpe diem: Adaptation and devaluing of the future. *Quarterly Review of Biology, 80*, 55–60.
Daly, M., Wilson, M., & Vasdev, S. 2001. Income inequality and homicide rates in Canada and the United States. *Canadian Journal of Criminology, 43*, 219–36.
Damasio, A. R. 1994. *Descartes' error: Emotion, reason, and the human brain*. New York: Putnam.
Darnton, R. 1990. *The kiss of Lamourette: Reflections in cultural history*. New York: Norton.
Darwin, C. R. 1874. *The descent of man, and selection in relation to sex*, 2nd ed. New York: Hurst & Co.
Davies, N. 1981. *Human sacrifice in history and today*. New York: Morrow.
Davies, P., Lee, L., Fox, A., & Fox, E. 2004. Could nursery rhymes cause violent behavior? A comparison with television viewing. *Archives of Diseases of Childhood, 89*, 1103–5.
Davis, D. B. 1984. *Slavery and human progress*. New York: Oxford University Press.
Dawkins, R. 1976/1989. *The selfish gene*, new ed. New York: Oxford University Press.
Dawkins, R., & Brockmann, H. J. 1980. Do digger wasps commit the Concorde fallacy? *Animal Behavior, 28*, 892–96.
de Quervain, D. J.-F., Fischbacher, U., Treyer, V., Schellhammer, M., Schnyder, U., Buck, A., & Fehr, E. 2004. The neural basis of altruistic punishment. *Science, 305*, 1254–58.
de Swaan, A. 2001. Dyscivilization, mass extermination, and the state. *Theory, Culture, & Society, 18*, 265–76.
de Waal, F.B.M. 1996. *Good natured: The origins of right and wrong in humans and other animals*. Cambridge, Mass.: Harvard University Press.
de Waal, F.B.M. 1998. *Chimpanzee politics: Power and sex among the apes*. Baltimore: Johns Hopkins University Press.
de Waal, F.B.M. 2009. *The age of empathy: Nature's lessons for a kinder society*. New York: Harmony Books.
de Waal, F.B.M., & Lanting, F. 1997. *Bonobo: The forgotten ape*. Berkeley: University of California Press.
Deary, I. J. 2001. *Intelligence: A very short introduction*. New York: Oxford University Press.
Deary, I. J., Batty, G. D., & Gale, C. R. 2008. Bright children become enlightened adults. *Psychological Science, 19*, 1–6.
Death Penalty Information Center. 2010a. *Death penalty for offenses other than murder*. http://www.deathpenaltyinfo.org/death-penalty-offenses-other-murder.
Death Penalty Information Center. 2010b. *Facts about the death penalty*. http://www.deathpenaltyinfo.org/documents/FactSheet.pdf.
deMause, L. 1974. *The history of childhood*. New York: Harper Torchbooks.
deMause, L. 1982. *Foundations of psychohistory*. New York: Creative Roots.
deMause, L. 1998. The history of child abuse. *Journal of Psychohistory, 25*, 216–36.
deMause, L. 2002. *The emotional life of nations*. New York: Karnac.
deMause, L. 2008. The childhood origins of World War II and the Holocaust. *Journal of Psychohistory, 36*, 2–35.
Dershowitz, A. M. 2004a. *Rights from wrongs: A secular theory of the origins of rights*. New York: Basic Books.
Dershowitz, A. M. 2004b. Tortured reasoning. In S. Levinson, ed., *Torture: A collection*. New York: Oxford University Press.
Descartes, R. 1641/1967. Meditations on first philosophy. In R. Popkin, ed., *The philosophy of the 16th and 17th centuries*. New York: Free Press.
Deutsch, M., & Gerard, G. B. 1955. A study of normative and informational social influence upon individual judgment. *Journal of Abnormal & Social Psychology, 51*, 629–36.
DeVoe, J. F., Peter, K., Kaufman, P., Miller, A., Noonan, M., Snyder, T. D., & Baum, K. 2004. *Indicators of school crime and safety: 2004*. Washington, D.C.: National Center for Education Statistics and Bureau of Justice Statistics.
Dewall, C., Baumeister, R. F., Stillman, T., & Gailliot, M. 2007. Violence restrained: Effects of self-regulation and its depletion on aggression. *Journal of Experimental Social Psychology, 43*, 62–76.
Dewar Research. 2009. *Government statistics on domestic violence: Estimated prevalence of domestic violence*. Ascot, U.K.: Dewar Research. http://www.dewar4research.org/DOCS/DVGovtStatsAug09.pdf.
di Pellegrino, G., Fadiga, L., Fogassi, L., Gallese, V., & Rizzolatti, G. 1992. Understanding motor events: a neurophysiological study. *Experimental Brain Research, 91*, 176–80.
Diamond, J. M. 1997. *Guns, germs, and steel: The fates of human societies*. New York: Norton.

Diamond, S. R. 1977. The effect of fear on the aggressive responses of anger-aroused and revenge-motivated subjects. *Journal of Psychology, 95,* 185–88.

Divale, W. T. 1972. System population control in the Middle and Upper Paleolithic: Inferences based on contemporary hunter-gatherers. *World Archaeology, 4,* 222–43.

Divale, W. T., & Harris, M. 1976. Population, warfare, and the male supremacist complex. *American Anthropologist, 78,* 521–38.

Dobash, R. P., Dobash, R. E., Wilson, M., & Daly, M. 1992. The myth of sexual symmetry in marital violence. *Social Problems, 39,* 71–91.

Dodds, G. G. 2003a. Political apologies and public discourse. In J. Rodin & S. P. Steinberg, eds., *Public discourse in America.* Philadelphia: University of Pennsylvania Press.

Dodds, G. G. 2003b. Political apologies: Chronological list. http://reserve.mg2.org/apologies.htm.

Dodds, G. G. 2005. Political apologies since update of 01/23/03. Concordia University.

Donohue, J. J., & Levitt, S. D. 2001. The impact of legalized abortion on crime. *Quarterly Journal of Economics, 116,* 379–420.

Dostoevsky, F. 1880/2002. *The Brothers Karamazov.* New York: Farrar, Straus & Giroux.

Dreger, A. 2011. Darkness's descent on the American Anthropological Association: A cautionary tale. *Human Nature.* DOI 10.1007/s12110-011-9103-y.

Duckworth, A. L., & Seligman, M.E.P. 2005. Self-discipline outdoes IQ in predicting academic performance of adolescents. *Psychological Science, 16,* 939–44.

Duerr, H.-P. 1988–97. *Der Mythos vom Zivilisationsprozess,* vols. 1–4. Frankfurt: Suhrkamp.

Dulić, T. 2004a. A reply to Rummel. *Journal of Peace Research, 41,* 105–6.

Dulić, T. 2004b. Tito's slaughterhouse: A critical analysis of Rummel's work on democide. *Journal of Peace Research, 41,* 85–102.

Dworkin, A. 1993. *Letters from a war zone.* Chicago: Chicago Review Press.

Easterbrook, G. 2003. *The progress paradox: How life gets better while people feel worse.* New York: Random House.

Eck, K., & Hultman, L. 2007. Violence against civilians in war. *Journal of Peace Research, 44,* 233–46.

Eckhardt, W. 1992. *Civilizations, empires, and wars.* Jefferson, N.C.: McFarland & Co.

Edwards, W. D., Gabel, W. J., & Hosmer, F. 1986. On the physical death of Jesus Christ. *Journal of the American Medical Association, 255,* 1455–63.

Ehrman, B. D. 2005. *Misquoting Jesus: The story behind who changed the Bible and why.* New York: HarperCollins.

Eisner, M. 2001. Modernization, self-control, and lethal violence: The long-term dynamics of European homicide rates in theoretical perspective. *British Journal of Criminology, 41,* 618–38.

Eisner, M. 2003. Long-term historical trends in violent crime. *Crime & Justice, 30,* 83–142.

Eisner, M. 2008. Modernity strikes back? A historical perspective on the latest increase in interpersonal violence 1960–1990. *International Journal of Conflict & Violence, 2,* 288–316.

Eisner, M. 2009. The uses of violence: An examination of some cross-cutting issues. *International Journal of Conflict & Violence, 3,* 40–59.

Eisner, M. 2011. Killing kings: Patterns of regicide in Europe, 600–1800. *British Journal of Criminology, 51,* 556–77.

Eley, T. C., Lichtenstein, P., & Stevenson, J. 1999. Sex differences in the etiology of aggressive and nonaggressive antisocial behavior: Results from two twin studies. *Child Development, 70,* 155–68.

Elias, N. 1939/2000. *The civilizing process: Sociogenetic and psychogenetic investigations,* rev. ed. Cambridge, Mass.: Blackwell.

Ellickson, R. C. 1991. *Order without law: How neighbors settle disputes.* Cambridge, Mass.: Harvard University Press.

Ellis, B. J. 1992. The evolution of sexual attraction: Evaluative mechanisms in women. In J. Barkow, L. Cosmides, & J. Tooby, eds., *The adapted mind: Evolutionary psychology and the generation of culture.* New York: Oxford University Press.

Eltis, D., & Richardson, D. 2010. *Atlas of the transatlantic slave trade.* New Haven, Conn.: Yale University Press.

Ember, C. 1978. Myths about hunter-gatherers. *Ethnology, 27,* 239–48.

English, R. 2005. The sociology of new thinking: Elites, identity change, and the end of the Cold War. *Journal of Cold War Studies, 7,* 43–80.

Ericksen, K. P., & Horton, H. 1992. "Blood feuds": Cross-cultural variation in kin group vengeance. *Behavioral Science Research, 26,* 57–85.

Ericksen, R. P., & Heschel, S. 1999. *Betrayal: German churches and the Holocaust.* Minneapolis: Fortress Press.

Esposito, J. L., & Mogahed, D. 2007. *Who speaks for Islam? What a billion Muslims really think.* New York: Gallup Press.

Espy, M. W., & Smykla, J. O. 2002. Executions in the United States, 1608–2002. Death Penalty Information Center. http://www.deathpenaltyinfo.org/executions-us-1608-2002-espy-file.

Etcoff, N. L. 1999. *Survival of the prettiest: The science of beauty.* New York: Doubleday.

Faris, S. 2007. Fool's gold. *Foreign Policy,* July.

Farrington, D. P. 2007. Origins of violent behavior over the life span. In D. J. Flannery, A. T. Vazsonyi, & I. D. Waldman, eds., *The Cambridge handbook of violent behavior and aggression*. New York: Cambridge University Press.

Fattah, K., & Fierke, K. M. 2009. A clash of emotions: The politics of humiliation and political violence in the Middle East. *European Journal of International Relations, 15*, 67–93.

Faust, D. 2008. *The republic of suffering: Death and the American Civil War*. New York: Vintage.

Fearon, J. D., & Laitin, D. D. 1996. Explaining interethnic cooperation. *American Political Science Review, 90*, 715–35.

Fearon, J. D., & Laitin, D. D. 2003. Ethnicity, insurgency, and civil war. *American Political Science Review, 97*, 75–90.

Fehr, E., & Gächter, S. 2000. Fairness and retaliation: The economics of reciprocity. *Journal of Economic Perspectives, 14*, 159–81.

Feingold, D. A. 2010. Trafficking in numbers: The social construction of human trafficking data. In P. Andreas & K. M. Greenhill, eds., *Sex, drugs, and body counts: The politics of numbers in global crime and conflict*. Ithaca, N.Y.: Cornell University Press.

Feller, W. 1968. *An introduction to probability theory and its applications*. New York: Wiley.

Felson, R. B. 1982. Impression management and the escalation of aggression and violence. *Social Psychology Quarterly, 45*, 245–54.

Ferguson, N. 1998. *The pity of war: Explaining World War I*. New York: Basic Books.

Ferguson, N. 2006. *The war of the world: Twentieth-century conflict and the descent of the West*. New York: Penguin.

Fernández-Jalvo, Y., Diez, J. C., Bermúdez de Castro, J. M., Carbonell, E., & Arsuaga, J. L. 1996. Evidence of early cannibalism. *Science, 271*, 277–78.

Festinger, L. 1957. *A theory of cognitive dissonance*. Stanford, Calif.: Stanford University Press.

Finkelhor, D., Hammer, H., & Sedlak, A. J. 2002. *Nonfamily abducted children: National estimates and characteristics*. Washington, D.C.: U.S. Department of Justice, Office of Justice Programs: Office of Juvenile Justice and Delinquency Prevention.

Finkelhor, D., & Jones, L. 2006. Why have child maltreatment and child victimization declined? *Journal of Social Issues, 62*, 685–716.

Fischer, H. 2008. *Iraqi civilian deaths estimates*. Washington, D.C.: Library of Congress.

Fischer, K. P. 1998. *The history of an obsession: German Judeophobia and the Holocaust*. New York: Continuum.

Fiske, A. P. 1991. *Structures of social life: The four elementary forms of human relations*. New York: Free Press.

Fiske, A. P. 1992. The four elementary forms of sociality: Framework for a unified theory of social relations. *Psychological Review, 99*, 689–723.

Fiske, A. P. 2004a. Four modes of constituting relationships: Consubstantial assimilation; space, magnitude, time, and force; concrete procedures; abstract symbolism. In N. Haslam, ed., *Relational models theory: A contemporary overview*. Mahwah, N.J.: Erlbaum.

Fiske, A. P. 2004b. Relational models theory 2.0. In N. Haslam, ed., *Relational models theory: A contemporary overview*. Mahwah, N.J.: Erlbaum.

Fiske, A. P., & Tetlock, P. E. 1997. Taboo trade-offs: Reactions to transactions that transgress the spheres of justice. *Political Psychology, 18*, 255–97.

Fiske, A. P., & Tetlock, P. E. 1999. Taboo tradeoffs: Constitutive prerequisites for social life. In S. A. Renshon & J. Duckitt, eds., *Political psychology: Cultural and cross-cultural perspectives*. London: Macmillan.

Fletcher, J. 1997. *Violence and civilization: An introduction to the work of Norbert Elias*. Cambridge, U.K.: Polity Press.

Flynn, J. R. 1984. The mean IQ of Americans: Massive gains 1932 to 1978. *Psychological Bulletin, 95*, 29–51.

Flynn, J. R. 1987. Massive IQ gains in 14 nations: What IQ tests really measure. *Psychological Bulletin, 101*, 171–91.

Flynn, J. R. 2007. *What is intelligence?* Cambridge, U.K.: Cambridge University Press.

Fodor, J. A., & Pylyshyn, Z. 1988. Connectionism and cognitive architecture: A critical analysis. *Cognition, 28*, 3–71.

Fogel, R. W., & Engerman, S. L. 1974. *Time on the cross: The economics of American Negro slavery*. Boston: Little, Brown & Co.

Fone, B. 2000. *Homophobia: A history*. New York: Picador.

Foote, C. L., & Goetz, C. F. 2008. The impact of legalized abortion on crime: Comment. *Quarterly Journal of Economics, 123*, 407–23.

Ford, R., & Blegen, M. A. 1992. Offensive and defensive use of punitive tactics in explicit bargaining. *Social Psychology Quarterly, 55*, 351–62.

Fortna, V. P. 2008. *Does peacekeeping work? Shaping belligerents' choices after civil war*. Princeton, N.J.: Princeton University Press.

Fox, J. A., & Levin, J. 1999. Serial murder: Popular myths and empirical realities. In M. D. Smith & M. A. Zahn, eds., *Homicide: A sourcebook of social research*. Thousand Oaks, Calif.: Sage.

Fox, J. A., & Zawitz, M. W. 2007. Homicide trends in the United States. http://bjs.ojp.usdoj.gov/content/homicide/homtrnd.cfm.

Francis, N., & Kučera, H. 1982. *Frequency analysis of English usage: Lexicon and grammar.* Boston: Houghton Mifflin.

Frank, R. H. 1988. *Passions within reason: The strategic role of the emotions.* New York: Norton.

Freeman, D. 1999. *The fateful hoaxing of Margaret Mead: A historical analysis of her Samoan research.* Boulder, Colo.: Westview Press.

French Ministry of Foreign Affairs. 2007. *The death penalty in France.* http://ambafrance-us.org/IMG/pdf/Death_penalty.pdf.

Fukuyama, F. 1989. The end of history? *National Interest*, Summer.

Fukuyama, F. 1999. *The great disruption: Human nature and the reconstitution of social order.* New York: Free Press.

Furuichi, T., & Thompson, J. M. 2008. *The bonobos: Behavior, ecology, and conservation.* New York: Springer.

Fuster, J. M. 2008. *The prefrontal cortex,* 4th ed. New York: Elsevier.

Gabor, T. 1994. *Everybody does it! Crime by the public.* Toronto: University of Toronto Press.

Gaddis, J. L. 1986. The long peace: Elements of stability in the postwar international system. *International Security, 10,* 99–142.

Gaddis, J. L. 1989. *The long peace: Inquiries into the history of the Cold War.* New York: Oxford University Press.

Gailliot, M. T., & Baumeister, R. F. 2007. Self-regulation and sexual restraint: Dispositionally and temporarily poor self-regulatory abilities contribute to failures at restraining sexual behavior. *Personality & Social Psychology Bulletin, 33,* 173–86.

Gailliot, M. T., Baumeister, R. F., Dewall, C. N., Maner, J. K., Plant, E. A., Tice, D. M., Brewer, L. E., & Schmeichel, B. J. 2007. Self-control relies on glucose as a limited energy source: Willpower is more than a metaphor. *Journal of Personality & Social Psychology, 92,* 325–36.

Gallonio, A. 1903/2004. *Tortures and torments of the Christian martyrs.* Los Angeles: Feral House.

Gallup. 2001. American attitudes toward homosexuality continue to become more tolerant. http://www.gallup.com/poll/4432/American-Attitudes-Toward-Homosexuality-Continue-Become-More-Tolerant.aspx.

Gallup. 2002. Acceptance of homosexuality: A youth movement. http://www.gallup.com/poll/5341/Acceptance-Homosexuality-Youth-Movement.aspx.

Gallup. 2003. Public lukewarm on animal rights. http://www.gallup.com/poll/8461/Public-Lukewarm-Animal-Rights.aspx.

Gallup. 2008. Americans evenly divided on morality of homosexuality. http://www.gallup.com/poll/108115/Americans-Evenly-Divided-Morality-Homosexuality.aspx.

Gallup. 2009. Knowing someone gay/lesbian affects views of gay issues. http://www.gallup.com/poll/118931/Knowing-Someone-Gay-Lesbian-Affects-Views-Gay-Issues.aspx.

Gallup. 2010. Americans' acceptance of gay relations crosses 50% threshold. http://www.gallup.com/poll/135764/Americans-Acceptance-Gay-Relations-Crosses-Threshold.aspx.

Gallup, A. 1999. *The Gallup poll cumulative index: Public opinion, 1935–1997.* Wilmington, Del.: Scholarly Resources.

Gardner, D. 2008. *Risk: The science and politics of fear.* London: Virgin Books.

Gardner, D. 2010. *Future babble: Why expert predictions fail—and why we believe them anyway.* New York: Dutton.

Garrard, G. 2006. *Counter-Enlightenments: From the eighteenth century to the present.* New York: Routledge.

Gartner, R. 2009. Homicide in Canada. In J. I. Ross, ed., *Violence in Canada: Sociopolitical perspectives.* Piscataway, N.J.: Transaction.

Gartzke, E. 2007. The capitalist peace. *American Journal of Political Science, 51,* 166–91.

Gartzke, E., & Hewitt, J. J. 2010. International crises and the capitalist peace. *International Interactions, 36,* 115–45.

Gat, A. 2006. *War in human civilization.* New York: Oxford University Press.

Gaulin, S.J.C., & McBurney, D. H. 2001. *Psychology: An evolutionary approach.* Upper Saddle River, N. J.: Prentice Hall.

Gazzaniga, M. S. 2005. *The ethical brain.* New York: Dana Press.

Geary, D. C. 2010. *Male, female: The evolution of human sex differences,* 2nd ed. Washington, D.C.: American Psychological Association.

Geary, P. J. 2002. *The myth of nations: The medieval origins of Europe.* Princeton, N.J.: Princeton University Press.

Gelman, S. A. 2005. *The essential child: Origins of essentialism in everyday thought.* New York: Oxford University Press.

Genovese, J.E.C. 2002. Cognitive skills valued by educators: Historical content analysis of testing in Ohio. *Journal of Educational Research, 96,* 101–14.

Ghosn, F., Palmer, G., & Bremer, S. 2004. The MID 3 Data Set, 1993–2001: Procedures, coding rules, and description. *Conflict Management & Peace Science, 21*, 133–54.

Giancola, P. R. 2000. Executive functioning: A conceptual framework for alcohol-related aggression. *Experimental & Clinical Psychopharmacology, 8*, 576–97.

Gibbons, A. 1997. Archeologists rediscover cannibals. *Science, 277*, 635–37.

Gigerenzer, G. 2006. Out of the frying pan into the fire: Behavioral reactions to terrorist attacks. *Risk Analysis, 26*, 347–51.

Gigerenzer, G., & Murray, D. J. 1987. *Cognition as intuitive statistics*. Hillsdale, N.J.: Erlbaum.

Gil-White, F. 1999. How thick is blood? *Ethnic & Racial Studies, 22*, 789–820.

Gilad, Y. 2002. Evidence for positive selection and population structure at the human MAO-A gene. *Proceedings of the National Academy of Sciences, 99*, 862–67.

Gilbert, S. J., Spengler, S., Simons, J. S., Steele, J. D., Lawrie, S. M., Frith, C. D., & Burgess, P. W. 2006. Functional specialization within rostral prefrontal cortex (Area 10): A meta-analysis. *Journal of Cognitive Neuroscience, 18*, 932–48.

Gilligan, C. 1982. *In a different voice: Psychological theory and women's development*. Cambridge, Mass.: Harvard University Press.

Ginges, J., & Atran, S. 2008. Humiliation and the inertia effect: Implications for understanding violence and compromise in intractable intergroup conflicts. *Journal of Cognition & Culture, 8*, 281–94.

Ginges, J., Atran, S., Medin, D., & Shikaki, K. 2007. Sacred bounds on rational resolution of violent political conflict. *Proceedings of the National Academy of Sciences, 104*, 7357–60.

Girls Study Group. 2008. *Violence by teenage girls: Trends and context*. Washington, D.C.: U.S. Department of Justice: Office of Justice Programs. http://girlsstudygroup.rti.org/docs/OJJDP_GSG_Violence_Bulletin.pdf.

Gleditsch, K. S. 2002. Expanded trade and GDP data. *Journal of Conflict Resolution, 46*, 712–24.

Gleditsch, N. P. 1998. Armed conflict and the environment: A critique of the literature. *Journal of Peace Research, 35*, 381–400.

Gleditsch, N. P. 2008. The liberal moment fifteen years on. *International Studies Quarterly, 52*, 691–712.

Gleditsch, N. P., Hegre, H., & Strand, H. 2009. Democracy and civil war. In M. Midlarsky, ed., *Handbook of war studies III*. Ann Arbor: University of Michigan Press.

Gleditsch, N. P., Wallensteen, P., Eriksson, M., Sollenberg, M., & Strand, H. 2002. Armed conflict 1946–2001: A new dataset. *Journal of Peace Research, 39*, 615–37.

Global Zero Commission. 2010. *Global Zero action plan*. http://static.globalzero.org/files/docs/GZAP_6.0.pdf.

Glover, J. 1977. *Causing death and saving lives*. London: Penguin.

Glover, J. 1999. *Humanity: A moral history of the twentieth century*. London: Jonathan Cape.

Gochman, C. S., & Maoz, Z. 1984. Militarized interstate disputes, 1816–1976: Procedures, patterns, and insights. *Journal of Conflict Resolution, 28*, 585–616.

Goffman, E. 1959. *The presentation of self in everyday life*. New York: Doubleday.

Goklany, I. M. 2007. *The improving state of the world: Why we're living longer, healthier, more comfortable lives on a cleaner planet*. Washington, D.C.: Cato Institute.

Goldhagen, D. J. 1996. *Hitler's willing executioners: Ordinary Germans and the Holocaust*. New York: Knopf.

Goldhagen, D. J. 2009. *Worse than war: Genocide, eliminationism, and the ongoing assault on humanity*. New York: PublicAffairs.

Goldstein, J. S. 2001. *War and gender*. Cambridge, U.K.: Cambridge University Press.

Goldstein, J. S. 2011. *Winning the war on war: The surprising decline in armed conflict worldwide*. New York: Dutton.

Goldstein, J. S., & Pevehouse, J. C. 2009. *International relations*, 8th ed., 2008–2009 update. New York: Pearson Longman.

Goldstein, R. N. 2006. *Betraying Spinoza: The renegade Jew who gave us modernity*. New York: Nextbook/Schocken.

Gollwitzer, M., & Denzler, M. 2009. What makes revenge sweet: Seeing the offender suffer or delivering a message? *Journal of Experimental Social Psychology, 45*, 840–44.

Goodall, J. 1986. *The chimpanzees of Gombe: Patterns of behavior*. Cambridge, Mass.: Harvard University Press.

Gopnik, Adam. 2004. The big one: Historians rethink the war to end all wars. *New Yorker* (August 23).

Gopnik, Alison. 2007. Cells that read minds? What the myth of mirror neurons gets wrong about the human brain. *Slate* (April 26).

Gordon, M. 2009. *Roots of empathy: Changing the world child by child*. New York: Experiment.

Gorton, W., & Diels, J. 2010. Is political talk getting smarter? An analysis of presidential debates and the Flynn effect. *Public Understanding of Science* (March 18).

Gottfredson, L. S. 1997a. Mainstream science on intelligence: An editorial with 52 signatories, history, and bibliography. *Intelligence, 24*, 13–23.

Gottfredson, L. S. 1997b. Why *g* matters: The complexity of everyday life. *Intelligence, 24*, 79–132.

Gottfredson, M. R. 2007. Self-control and criminal violence. In D. J. Flannery, A. T. Vazsonyi, & I. D. Waldman, eds., *The Cambridge handbook of violent behavior and aggression*. New York: Cambridge University Press.

Gottfredson, M. R., & Hirschi, T. 1990. *A general theory of crime*. Stanford, Calif.: Stanford University Press.

Gottschall, J. 2008. *The rape of Troy: Evolution, violence, and the world of Homer*. New York: Cambridge University Press.

Gottschall, J., & Gottschall, R. 2003. Are per-incident rape-pregnancy rates higher than per-incident consensual pregnancy rates? *Human Nature, 14*, 1–20.

Gould, S. J. 1980. A biological homage to Mickey Mouse. *The panda's thumb: More reflections in natural history*. New York: Norton.

Gould, S. J. 1991. Glow, big glowworm. *Bully for brontosaurus*. New York: Norton.

Graf zu Waldburg Wolfegg, C. 1988. *Venus and Mars: The world of the medieval housebook*. New York: Prestel.

Graham-Kevan, N., & Archer, J. 2003. Intimate terrorism and common couple violence: A test of Johnson's predictions in four British samples. *Journal of Interpersonal Violence, 18*, 1247–70.

Gray, H. M., Gray, K., & Wegner, D. M. 2007. Dimensions of mind perception. *Science, 315*, 619.

Gray, P. B., & Young, S. M. 2010. Human-pet dynamics in cross-cultural perspective. *Anthrozoos 1, 24*, 17–30.

Grayling, A. C. 2007. *Toward the light of liberty: The struggles for freedom and rights that made the modern Western world*. New York: Walker.

Green, R. M. 2001. *The human embryo research debates: Bioethics in the vortex of controversy*. New York: Oxford University Press.

Greene, D., ed. 2000. *Samuel Johnson: The major works*. New York: Oxford University Press.

Greene, J. D. In press. *The moral brain and what to do with it*.

Greene, J. D., & Haidt, J. 2002. How (and where) does moral judgment work? *Trends in Cognitive Science, 6*, 517–23.

Greene, J. D., Sommerville, R. B., Nystrom, L. E., Darley, J. M., & Cohen, J. D. 2001. An fMRI investigation of emotional engagement in moral judgment. *Science, 293*, 2105–8.

Greenfield, P. M. 2009. Technology and informal education: What is taught, and what is learned. *Science, 323*, 69–71.

Groebner, V. 1995. Losing face, saving face: Noses and honour in the late medieval town. *History Workshop Journal, 40*, 1–15.

Gross, C. G. 2009. Early steps toward animal rights. *Science, 324*, 466–67.

Grossman, L.C.D. 1995. *On killing: The psychological cost of learning to kill in war and society*. New York: Back Bay Books.

Guo, G., Ou, X.-M., Roettger, M., & Shih, J. C. 2008a. The VNTR 2 repeat in MAOA and delinquent behavior in adolescence and young adulthood: Associations and MAOA promoter activity. *European Journal of Human Genetics, 16*, 626–34.

Guo, G., Roettger, M. E., & Cai, T. 2008b. The integration of genetic propensities into social-control models of delinquency and violence among male youths. *American Sociological Review, 73*, 543–68.

Guo, G., Roettger, M. E., & Shih, J. C. 2007. Contributions of the DAT1 and DRD2 genes to serious and violent delinquency among adolescents and young adults. *Human Genetics, 121*, 125–36.

Gurr, T. R. 1981. Historical trends in violent crime: A critical review of the evidence. In N. Morris & M. Tonry, eds., *Crime and justice*, vol. 3. Chicago: University of Chicago Press.

Gurr, T. R. 1989a. Historical trends in violent crime: Europe and the United States. In T. R. Gurr, ed., *Violence in America*, vol. 1: *The history of crime*. Newbury Park, Calif.: Sage.

Gurr, T. R., ed. 1989b. *Violence in America*, vol. 1. London: Sage.

Gurr, T. R., & Monty, M. G. 2000. Assessing the risks of future ethnic wars. In T. R. Gurr, ed., *Peoples versus states*. Washington, D.C.: United States Institute of Peace Press.

Hagen, E. H. 1999. The functions of postpartum depression. *Evolution & Human Behavior, 20*, 325–59.

Hagger, M., Wood, C., Stiff, C., & Chatzisarantis, N.L.D. 2010. Ego depletion and the strength model of self-control: A meta-analysis. *Psychological Bulletin, 136*, 495–525.

Haidt, J. 2001. The emotional dog and its rational tail: A social intuitionist approach to moral judgment. *Psychological Review, 108*, 813–34.

Haidt, J. 2002. The moral emotions. In R. J. Davidson, K. R. Scherer, & H. H. Goldsmith, eds., *Handbook of affective sciences*. New York: Oxford University Press.

Haidt, J. 2007. The new synthesis in moral psychology. *Science, 316*, 998–1002.

Haidt, J., Bjorklund, F., & Murphy, S. 2000. Moral dumbfounding: When intuition finds no reason. University of Virginia.

Haidt, J., & Graham, J. 2007. When morality opposes justice: Conservatives have moral intuitions that liberals may not recognize. *Social Justice Research, 20*, 98–116.

Haidt, J., & Hersh, M. A. 2001. Sexual morality: The cultures and emotions of conservatives and liberals. *Journal of Applied Social Psychology, 31*, 191–221.

Haidt, J., Koller, H., & Dias, M. G. 1993. Affect, culture, and morality, or is it wrong to eat your dog? *Journal of Personality & Social Psychology, 65*, 613–28.

Hakemulder, J. F. 2000. *The moral laboratory: Experiments examining the effects of reading literature on social perception and moral self-concept.* Philadelphia: J. Benjamins.

Hallissy, M. 1987. *Venomous woman: Fear of the female in literature.* Westport, Conn.: Greenwood Press.

Halpern, D. F. 2000. *Sex differences in cognitive abilities,* 3rd ed. Mahwah, N.J.: Erlbaum.

Hamer, D. H., & Copeland, P. 1994. *The science of desire: The search for the gay gene and the biology of behavior.* New York: Simon & Schuster.

Hamilton, W. D. 1963. The evolution of altruistic behavior. *American Naturalist, 97,* 354–56.

Hanawalt, B. A. 1976. Violent death in 14th and early 15th-century England. *Contemporary Studies in Society & History, 18,* 297–320.

Hanlon, G. 2007. *Human nature in rural Tuscany: An early modern history.* New York: Palgrave Macmillan.

Hardin, G. 1968. The tragedy of the commons. *Science, 162,* 1243–48.

Hare, R. D. 1993. *Without conscience: The disturbing world of the psychopaths around us.* New York: Guilford Press.

Harff, B. 2003. No lessons learned from the Holocaust? Assessing the risks of genocide and political mass murder since 1955. *American Political Science Review, 97,* 57–73.

Harff, B. 2005. Assessing risks of genocide and politicide. In M. G. Marshall & T. R. Gurr, eds., *Peace and conflict 2005: A global survey of armed conflicts, self-determination movements, and democracy.* College Park, Md.: Center for International Development & Conflict Management, University of Maryland.

Harlow, C. W. 2005. *Hate crime reported by victims and police.* Washington, D.C.: U.S. Department of Justice, Bureau of Justice Statistics. http://bjs.ojp.usdoj.gov/content/pub/pdf/hcrvp.pdf.

Harris, J. R. 1998/2008. *The nurture assumption: Why children turn out the way they do,* 2nd ed. New York: Free Press.

Harris, J. R. 2006. *No two alike: Human nature and human individuality.* New York: Norton.

Harris, M. 1975. *Culture, people, nature,* 2nd ed. New York: Crowell.

Harris, M. 1985. *Good to eat: Riddles of food and culture.* New York: Simon & Schuster.

Harris, S. 2010. *The moral landscape: How science can determine human values.* New York: Free Press.

Haslam, N. 2006. Dehumanization: An integrative review. *Personality & Social Psychology Review, 10,* 252–64.

Haslam, N., ed. 2004. *Relational models theory: A contemporary overview.* Mahwah, N.J.: Erlbaum.

Haslam, N., Rothschild, L., & Ernst, D. 2000. Essentialist beliefs about social categories. *British Journal of Social Psychology, 39,* 113–27.

Hauser, M. D. 2000. *Wild minds: What animals really think.* New York: Henry Holt.

Hawkes, K. 1981. A third explanation for female infanticide. *Human Ecology, 9,* 79–96.

Hawkes, K. 2006. Life history theory and human evolution. In K. Hawkes and R. Paine, eds., *The evolution of human life history.* Oxford: SAR Press.

Hayes, B. 2002. Statistics of deadly quarrels. *American Scientist, 90,* 10–15.

Hegre, H. 2000. Development and the liberal peace: What does it take to be a trading state? *Journal of Peace Research, 37,* 5–30.

Hegre, H., Ellingsen, T., Gates, S., & Gleditsch, N. P. 2001. Toward a democratic civil peace? Democracy, political change, and civil war 1816–1992. *American Political Science Review, 95,* 33–48.

Heise, L., & Garcia-Moreno, C. 2002. Violence by intimate partners. In E. G. Krug, L. L. Dahlberg, J. A. Mercy, A. B. Zwi, & R. Lozano, eds., *World report on violence and health.* Geneva: World Health Organization.

Held, R. 1986. *Inquisition: A selected survey of the collection of torture instruments from the Middle Ages to our times.* Aslockton, Notts, U.K.: Avon & Arno.

Helmbold, R. 1998. How many interstate wars will there be in the decade 2000–2009? *Phalanx, 31,* 21–23.

Henshaw, S. K. 1990. Induced abortion: A world review, 1990. *Family Planning Perspectives, 22,* 76–89.

Herman, A. 1997. *The idea of decline in Western history.* New York: Free Press.

Herrmann, B., Thöni, C., & Gächter, S. 2008a. Antisocial punishment across societies. *Science, 319,* 1362–67.

Herrmann, B., Thöni, C., & Gächter, S. 2008b. Antisocial punishment across societies: Supporting online material. http://www.sciencemag.org/content/319/5868/1362/supp/DC1.

Herrnstein, R. J., & Murray, C. 1994. *The bell curve: Intelligence and class structure in American life.* New York: Free Press.

Herzog, H. 2010. *Some we love, some we hate, some we eat: Why it's so hard to think straight about animals.* New York: HarperCollins.

Heschel, S. 2008. *The Aryan Jesus: Christian theologians and the Bible in Nazi Germany.* Princeton, N.J.: Princeton University Press.

Hewitt, J. J., Wilkenfeld, J., & Gurr, T. R., eds. 2008. *Peace and conflict 2008.* Boulder, Colo.: Paradigm.

Hewstone, M., Rubin, M., & Willis, H. 2002. Intergroup bias. *Annual Review of Psychology, 53,* 575–604.

Heywood, C. 2001. *A history of childhood.* Malden, Mass.: Polity.

Hickok, G. 2009. Eight problems for the mirror neuron theory of action understanding in monkeys and humans. *Journal of Cognitive Neuroscience, 21,* 1229–43.

Hill, J., Inder, T., Neil, J., Dierker, D., Harwell, J., & Van Essen, D. 2010. Similar patterns of cortical expansion during human development and evolution. *Proceedings of the National Academy of Sciences, 107*, 13135–40.

Himmelfarb, M. 1984. No Hitler, no Holocaust. *Commentary*, March, 37–43.

Hirschfeld, A. O. 1996. *Race in the making: Cognition, culture, and the child's construction of human kinds*. Cambridge, Mass.: MIT Press.

Hirschi, T., & Gottfredson, M. R. 2000. In defense of self-control. *Theoretical Criminology, 4*, 55–69.

Hitchcock, E., & Cairns, V. 1973. Amygdalotomy. *Postgraduate Medical Journal, 49*, 894–904.

Hoban, J. E. 2007. The ethical marine warrior: Achieving a higher standard. *Marine Corps Gazette* (September), 36–40.

Hoban, J. E. 2010. Developing the ethical marine warrior. *Marine Corps Gazette* (June), 20–25.

Hobbes, T. 1651/1957. *Leviathan*. New York: Oxford University Press.

Hodges, A. 1983. *Alan Turing: The enigma*. New York: Simon & Schuster.

Hoffman, M. L. 2000. *Empathy and moral development: Implications for caring and justice*. Cambridge, U.K.: Cambridge University Press.

Hofstadter, D. R. 1985. Dilemmas for superrational thinkers, leading up to a Luring lottery. In *Metamagical themas: Questing for the essence of mind and pattern*. New York: Basic Books.

Hofstede, G., & Hofstede, G. J. 2010. Dimensions of national cultures. http://www.geerthofstede.nl/culture/dimensions-of-national-cultures.aspx, retrieved July 19, 2010.

Holden, C. 2008. Parsing the genetics of behavior. *Science, 322*, 892–95.

Holsti, K. J. 1986. The horsemen of the apocalypse: At the gate, detoured, or retreating? *International Studies Quarterly, 30*, 355–72.

Holsti, K. J. 1991. *Peace and war: Armed conflicts and international order 1648–1989*. Cambridge, U.K.: Cambridge University Press.

Homer. 2003. *The Iliad*, trans. E. V. Rieu & P. Jones. New York: Penguin.

Horowitz, D. L. 2001. *The deadly ethnic riot*. Berkeley: University of California Press.

Howard, M. 1991. *The lessons of history*. New Haven, Conn.: Yale University Press.

Howard, M. 2001. *The invention of peace and the reinvention of war*. London: Profile.

Howard, M. 2007. *Liberation or catastrophe? Reflections on the history of the twentieth century*. London: Continuum.

Hoyle, R. H., Pinkley, R. L., & Insko, C. A. 1989. Perceptions of social behavior: Evidence of differing expectations for interpersonal and intergroup interaction. *Personality & Social Psychology Bulletin, 15*, 365–76.

Hrdy, S. B. 1999. *Mother nature: A history of mothers, infants, and natural selection*. New York: Pantheon.

Hudson, V. M., & den Boer, A. D. 2002. A surplus of men, a deficit of peace: Security and sex ratios in Asia's largest states. *International Security, 26*, 5–38.

Hughes, G. 1991. *Swearing: A social history of foul language, oaths, and profanity in English*. New York: Penguin.

Human Rights First. 2008. *Violence against Muslims: 2008 hate crime survey*. New York: Human Rights First.

Human Rights Watch. 2008. *A violent education: Corporal punishment of children in U.S. public schools*. New York: Human Rights Watch.

Human Security Centre. 2005. *Human Security Report 2005: War and peace in the 21st century*. New York: Oxford University Press.

Human Security Centre. 2006. *Human Security Brief 2006*. Vancouver, B.C.: Human Security Centre.

Human Security Report Project. 2007. *Human Security Brief 2007*. Vancouver, B.C.: Human Security Report Project.

Human Security Report Project. 2008. *Miniatlas of human security*. Washington, D.C.: World Bank.

Human Security Report Project. 2009. *Human Security Report 2009: The shrinking costs of war*. New York: Oxford University Press.

Human Security Report Project. 2011. *Human Security Report 2009/2010: The causes of peace and the shrinking costs of war*. New York: Oxford University Press.

Hume, D. 1739/2000. *A treatise of human nature*. New York: Oxford University Press.

Hume, D. 1751/2004. *An enquiry concerning the principles of morals*. Amherst, N.Y.: Prometheus Books.

Humphrey, R. L. 1992. *Values for a new millennium*. Maynardville, Tenn.: Life Values Press.

Hunt, L. 2007. *Inventing human rights: A history*. New York: Norton.

Huntington, S. P. 1993. The clash of civilizations? *Foreign Affairs*, Summer.

Hurford, J. R. 2004. Language beyond our grasp: What mirror neurons can, and cannot, do for language evolution. In D. K. Oller & U. Griebel, eds., *Evolution of communication systems: A comparative approach*. Cambridge, Mass.: MIT Press.

Huth, P., & Russett, B. 1984. What makes deterrence work? Cases from 1900 to 1980. *World Politics, 36*, 496–526.

Hynes, L. In press. Routine infanticide by married couples in Parma, 16th–18th century. *Journal of Early Modern History*.

Iacoboni, M. 2008. *Mirroring people: The new science of how we connect with others*. New York: Farrar, Straus & Giroux.

Iacoboni, M., Woods, R. P., Brass, M., Bekkering, H., Mazziotta, J. C., & Rizzolatti, G. 1999. Cortical mechanisms of human imitation. *Science, 286*, 2526–28.

Ingle, D. 2004. Recreational fighting. In G. S. Cross, ed., *Encyclopedia of recreation and leisure in America*. New York: Scribner.

International Institute for Strategic Studies. 2010. *The military balance 2010*. London: Routledge.

Jacob, P., & Jeannerod, M. 2005. The motor theory of social cognition: A critique. *Trends in Cognitive Sciences, 9*, 21–25.

Jaggar, A. M. 1983. *Feminist politics and human nature*. Lanham, Md.: Rowman & Littlefield.

James, W. 1906/1971. The moral equivalent of war. *The moral equivalent of war and other essays*. New York: Harper & Row.

James, W. 1977. On a certain blindness in human beings. In J. J. McDermott, ed., *The writings of William James*. Chicago: University of Chicago Press.

Janis, I. L. 1982. *Groupthink: Psychological studies of policy decisions and fiascoes*, 2nd ed. Boston: Houghton Mifflin.

Jansson, K. 2007. *British crime survey: Measuring crime for 25 years*. London: U.K. Home Office.

Jensen, J., Smith, A. J., Willeit, M., Crawley, A. P., Mikulis, D. J., Vitcu, I., & Kapur, S. 2007. Separate brain regions code for salience versus valence during reward prediction in humans. *Human Brain Mapping, 28*, 294–302.

Jervis, R. 1988. The political effects of nuclear weapons. *International Security, 13*, 80–90.

Johnson, D.D.P. 2004. *Overconfidence and war: The havoc and glory of positive illusions*. Cambridge, Mass.: Harvard University Press.

Johnson, D.D.P., McDermott, R., Barrett, E. S., Cowden, J., Wrangham, R., McIntyre, M. H., & Rosen, S. P. 2006. Overconfidence in wargames: Experimental evidence on expectations, aggression, gender and testosterone. *Proceedings of the Royal Society B, 273*, 2513–20.

Johnson, E. A., & Monkkonen, E. H. 1996. *The civilization of crime*. Urbana: University of Illinois Press.

Johnson, E. J., Hershey, J., Meszaros, J., & Kunreuther, H. 1993. Framing, probability distortions, and insurance decisions. *Journal of Risk & Uncertainty, 7*, 35–41.

Johnson, E. M. 2010. Deconstructing social Darwinism. *Primate Diaries*. scienceblogs.com/primatediaries/2010/01/deconstructing_social_darwinis.php

Johnson, G. R., Ratwik, S. H., & Sawyer, T. J. 1987. The evocative significance of kin terms in patriotic speech. In V. Reynolds, V. Falger & I. Vine, eds., *The sociobiology of ethnocentrism*. London: Croon Helm.

Johnson, I. M., & Sigler, R. T. 2000. The stability of the public's endorsements of the definition and criminalization of the abuse of women. *Journal of Criminal Justice, 28*, 165–79.

Johnson, M. P. 2006. Conflict and control: Gender symmetry and asymmetry in domestic violence. *Violence Against Women, 12*, 1003–18.

Johnson, M. P., & Leone, J. M. 2005. The differential effects of intimate terrorism and situational couple violence: Findings from the National Violence Against Women Survey. *Journal of Family Issues, 26*, 322–49.

Johnson, N. F., Spagat, M., Gourley, S., Onnela, J.-P., & Reinert, G. 2008. Bias in epidemiological studies of conflict mortality. *Journal of Peace Research, 45*, 653–63.

Johnson, R., & Raphael, S. 2006. How much crime reduction does the marginal prisoner buy? University of California, Berkeley.

Jones, D. M., Bremer, S., & Singer, J. D. 1996. Militarized interstate disputes, 1816–1992: Rationale, coding rules, and empirical patterns. *Conflict Management & Peace Science, 15*, 163–213.

Jones, G. 2008. Are smarter groups more cooperative? Evidence from prisoner's dilemma experiments, 1959–2003. *Journal of Economic Behavior & Organization, 68*, 489–97.

Jones, L., & Finkelhor, D. 2007. *Updated trends in child maltreatment, 2007*. Durham, N.H.: Crimes Against Children Research Center, University of New Hampshire.

Jones, O. D. 1999. Sex, culture, and the biology of rape: Toward explanation and prevention. *California Law Review, 87*, 827–942.

Jones, O. D. 2000. Reconsidering rape. *National Law Journal*, A21.

Joy, J. 2009. *Lindow Man*. London: British Museum Press.

Joyce, T. 2004. Did legalized abortion lower crime? *Journal of Human Resources, 39*, 1–28.

Jussim, L. J., McCauley, C. R., & Lee, Y.-T. 1995. Why study stereotype accuracy and inaccuracy? In Y.-T. Lee, L. J. Jussim, & C. R. McCauley, eds., *Stereotype accuracy: Toward appreciating group differences*. Washington, D.C.: American Psychological Association.

Kaeuper, R. W. 2000. Chivalry and the "civilizing process." In R. W. Kaeuper, ed., *Violence in medieval society*. Rochester, N.Y.: Boydell & Brewer.

Kagan, R. 2002. Power and weakness. *Policy Review, 113*, 1–21.

Kahneman, D., & Renshon, J. 2007. Why hawks win. *Foreign Policy*, December.

Kahneman, D., Slovic, P., & Tversky, A. 1982. *Judgment under uncertainty: Heuristics and biases*. New York: Cambridge University Press.

Kahneman, D., & Tversky, A. 1972. Subjective probability: A judgment of representativeness. *Cognitive Psychology, 3*, 430–54.

Kahneman, D., & Tversky, A. 1979. Prospect theory: An analysis of decisions under risk. *Econometrica, 47,* 313–27.

Kahneman, D., & Tversky, A. 1984. Choices, values, and frames. *American Psychologist, 39,* 341–50.

Kaldor, M. 1999. *New and old wars: Organized violence in a global era.* Stanford, Calif.: Stanford University Press.

Kanazawa, S. 2010. Why liberals and atheists are more intelligent. *Social Psychology Quarterly, 73,* 33–57.

Kane, K. 1999. Nits make lice: Drogheda, Sand Creek, and the poetics of colonial extermination. *Cultural Critique, 42,* 81–103.

Kant, I. 1784/1970. Idea for a universal history with a cosmopolitan purpose. In H. Reiss, ed., *Kant's political writings.* New York: Cambridge University Press.

Kant, I. 1795/1983. Perpetual peace: A philosophical sketch. In *Perpetual peace and other essays.* Indianapolis: Hackett.

Kaplan, J. 1973. *Criminal justice: Introductory cases and materials.* Mineola, N.Y.: Foundation Press.

Katz, L. 1987. *Bad acts and guilty minds: Conundrums of criminal law.* Chicago: University of Chicago Press.

Kay, S. 2000. The sublime body of the martyr. In R. W. Kaeuper, ed., *Violence in medieval society.* Woodbridge, U.K.: Boydell.

Kaysen, C. 1990. Is war obsolete? *International Security, 14,* 42–64.

Keegan, J. 1993. *A history of warfare.* New York: Vintage.

Keeley, L. H. 1996. *War before civilization: The myth of the peaceful savage.* New York: Oxford University Press.

Keen, S. 2007. *Empathy and the novel.* Oxford, U.K.: Oxford University Press.

Keizer, K., Lindenberg, S., & Steg, L. 2008. The spreading of disorder. *Science, 322,* 1681–85.

Keller, J. B. 1986. The probability of heads. *American Mathematical Monthly, 93,* 191–97.

Kelman, H. C. 1973. Violence without moral restraint: Reflections on the dehumanization of victims and victimizers. *Journal of Social Issues, 29,* 25–61.

Kennedy, R. 1997. *Race, crime, and the law.* New York: Vintage.

Kennedy, R. F. 1969/1999. *Thirteen days: A memoir of the Cuban missile crisis.* New York: Norton.

Kenny, C. 2011. *Getting better: Why global development is succeeding—and how we can improve the world even more.* New York: Basic Books.

Kenrick, D. T., & Sheets, V. 1994. Homicidal fantasies. *Ethology & Sociobiology, 14,* 231–46.

Kiernan, B. 2007. *Blood and soil: A world history of genocide and extermination from Sparta to Darfur.* New Haven, Conn.: Yale University Press.

Kim, S. H., Smirth, R. H., & Brigham, N. L. 1998. Effects of power imbalance and the presence of third parties on reactions to harm: Upward and downward revenge. *Personality & Social Psychology Bulletin, 24,* 353–61.

Kimmel, M. S. 2002. "Gender symmetry" in domestic violence. *Violence Against Women, 8,* 1332–63.

King, M. L., Jr. 1963/1995. Pilgrimage to nonviolence. In S. Lynd & A. Lynd, eds., *Nonviolence in America: A documentary history.* Maryknoll, N.Y.: Orbis Books.

Kinner, S. 2003. Psychopathy as an adaptation: Implications for society and social policy. In R. W. Bloom & N. Dess, eds., *Evolutionary psychology and violence: A primer for policymakers and public policy advocates.* Westport, Conn.: Praeger.

Kinzler, K. D., Shutts, K., DeJesus, J., & Spelke, E. S. 2009. Accent trumps race in guiding children's social preferences. *Social Cognition, 27,* 623–34.

Kirby, K. N., & Herrnstein, R. J. 1995. Preference reversals due to myopic discounting of delayed reward. *Psychological Science, 6,* 83–89.

Kirby, K. N., Winston, G., & Santiesteban, M. 2005. Impatience and grades: Delay-discount rates correlate negatively with college GPA. *Learning and Individual Differences, 15,* 213–22.

Knauft, B. 1987. Reconsidering violence in simple human societies. *Current Anthropology, 28,* 457–500.

Koechlin, E., & Hyafil, A. 2007. Anterior prefrontal function and the limits of human decision-making. *Science, 318,* 594–98.

Koestler, A. 1964. *The act of creation.* New York: Dell.

Kohl, M., ed. 1978. *Infanticide and the value of life.* Buffalo, N.Y.: Prometheus Books.

Kohlberg, L. 1981. *The philosophy of moral development: Moral stages and the idea of justice.* San Francisco: Harper & Row.

Kors, A. C., & Silverglate, H. A. 1998. *The shadow university: The betrayal of liberty on America's campuses.* New York: Free Press.

Kosfeld, M., Heinrichs, M., Zak, P. J., Fischbacher, U., & Fehr, E. 2005. Oxytocin increases trust in humans. *Nature, 435,* 673–76.

Krebs, D. L. 1975. Empathy and altruism. *Journal of Personality & Social Psychology, 32,* 1134–46.

Kreitman, M. 2000. Methods to detect selection in populations with applications to the human. *Annual Review of Genomics & Human Genetics, 1,* 539–59.

Kreutz, J. 2008. *UCDP one-sided violence codebook version 1.3.* http://www.pcr.uu.se/digitalAssets/19/19256_UCDP_One-sided_violence_Dataset_Codebook_v1.3.pdf.

Krieken, R. V. 1998. Review article: What does it mean to be civilised? Norbert Elias on the Germans and modern barbarism. *Communal/Plural: Journal of Transnational & Cross-Cultural Studies, 6,* 225–33.

Kringelbach, M. L. 2005. The human orbitofrontal cortex: Linking reward to hedonic experience. *Nature Reviews Neuroscience, 6,* 691–702.

Kristine, E., & Hultman, L. 2007. One-sided violence against civilians in war: Insights from new fatality data. *Journal of Peace Research, 44,* 233–46.

Kristof, N. D., & WuDunn, S. 2009. *Half the sky: Turning oppression into opportunity for women worldwide.* New York: Random House.

Krug, E. G., Dahlberg, L. L., Mercy, J. A., Zwi, A. B., & Lozano, R., eds. 2002. *World report on violence and health.* Geneva: World Health Organization.

Krystal, A. 2007. En garde! The history of dueling. *New Yorker* (March 12).

Kugel, J. L. 2007. *How to read the Bible: A guide to scripture, then and now.* New York: Free Press.

Kugler, J. 1984. Terror without deterrence. *Journal of Conflict Resolution, 28,* 470–506.

Kurlansky, M. 2006. *Nonviolence: Twenty-five lessons from the history of a dangerous idea.* New York: Modern Library.

Kurtz, D. V. 2001. Anthropology and the study of the state. *Political anthropology: Power and paradigms.* Boulder, Colo.: Westview Press.

Kurzban, R. 2011. *Why everyone (else) is a hypocrite: Evolution and the modular mind.* Princeton, N.J.: Princeton University Press.

Kurzban, R., Tooby, J., & Cosmides, L. 2001. Can race be erased? Coalitional computation and social categorization. *Proceedings of the National Academy of Sciences, 98,* 15387–92.

Kyle, D. G. 1998. *Spectacles of death in ancient Rome.* New York: Routledge.

La Griffe du Lion. 2000. Analysis of hate crime. *La Griffe du Lion,* 2(5). http://lagriffedulion.f2s.com/hatecrime.htm.

Lacina, B. 2006. Explaining the severity of civil wars. *Journal of Conflict Resolution, 50,* 276–89.

Lacina, B. 2009. *Battle Deaths Dataset 1946–2009: Codebook for version 3.0.* Center for the Study of Civil War, and International Peace Research Institute Oslo (PRIO).

Lacina, B., & Gleditsch, N. P. 2005. Monitoring trends in global combat: A new dataset in battle deaths. *European Journal of Population, 21,* 145–66.

Lacina, B., Gleditsch, N. P., & Russett, B. 2006. The declining risk of death in battle. *International Studies Quarterly, 50,* 673–80.

LaFree, G. 1999. A summary and review of cross-national comparative studies of homicide. In M. D. Smith & M. A. Zahn, eds., *Homicide: A sourcebook of social research.* Thousand Oaks, Calif.: Sage.

LaFree, G., & Tseloni, A. 2006. Democracy and crime: A multilevel analysis of homicide trends in forty-four countries, 1950–2000. *Annals of the American Academy of Political and Social Science, 605,* 25–49.

Laibson, D. 1997. Golden eggs and hyperbolic discounting. *Quarterly Journal of Economics, 112,* 443–77.

Lalumière, M. L., Harris, G. T., & Rice, M. E. 2001. Psychopathy and developmental instability. *Evolution and Human Behavior, 22,* 75–92.

Lamm, C., Batson, C. D., & Decety, J. 2007. The neural substrate of human empathy: Effects of perspective-taking and cognitive appraisal. *Journal of Cognitive Neuroscience, 19,* 42–58.

Lane, R. 1989. On the social meaning of homicide trends in America. In T. R. Gurr, ed., *Violence in America,* vol. 1: *The history of crime.* Newbury Park, Calif.: Sage.

Langbein, J. H. 2005. The legal history of torture. In S. Levinson, ed., *Torture: A collection.* New York: Oxford University Press.

Lanzetta, J. T., & Englis, B. G. 1989. Expectations of cooperation and competition and their effects on observers' vicarious emotional responses. *Journal of Personality & Social Psychology, 56,* 543–54.

Latané, B., & Darley, J. M. 1970. *The unresponsive bystander: Why doesn't he help?* New York: Appleton-Century Crofts.

Lea, R., & Chambers, G. 2007. Monoamine oxidase, addiction, and the "warrior" gene hypothesis. *Journal of the New Zealand Medical Association, 120.*

LeBlanc, S. A. 2003. *Constant battles: The myth of the noble savage and a peaceful past.* New York: St. Martin's Press.

Lebow, R. N. 2007. Contingency, catalysts, and nonlinear change: The origins of World War I. In J. S. Levy & G. Goertz, eds., *Explaining war and peace: Case studies and necessary condition counterfactuals.* New York: Routledge.

Lee, J. J., & Pinker, S. 2010. Rationales for indirect speech: The theory of the strategic speaker. *Psychological Review, 117,* 785–807.

Lee, R. 1982. Politics, sexual and non-sexual, in egalitarian society. In R. Lee & R. E. Leacock, eds., *Politics and history in band societies.* New York: Cambridge University Press.

Lee, T.M.C., Chan, S. C., & Raine, A. 2008. Strong limbic and weak frontal activation to aggressive stimuli in spouse abusers. *Molecular Psychiatry, 13,* 655–60.

Lee, Y.-T., Jussim, L. J., & McCauley, C. R., eds. 1995. *Stereotype accuracy: Toward appreciating group differences.* Washington, D.C.: American Psychological Association.

Lehner, E., & Lehner, J. 1971. *Devils, demons, and witchcraft.* Mineola, N.Y.: Dover.

Leiter, R. A., ed. 2007. *National survey of state laws,* 6th ed. Detroit: Thomson/Gale.

Leland, A., & Oboroceanu, M.-J. 2010. *American war and military operations casualties: Lists and statistics.* http://fpc.state.gov/documents/organization/139347.pdf.

Leonard, T. 2009. Origins of the myth of Social Darwinism: The ambiguous legacy of Richard Hofstadter's "Social Darwinism in American thought." *Journal of Economic Behavior & Organization, 71*, 37–59.

LeVay, S. 2010. *Gay, straight, and the reason why: The science of sexual orientation*. New York: Oxford University Press.

Levi, M. A. 2007. *On nuclear terrorism*. Cambridge, Mass.: Harvard University Press.

Levi, W. 1981. *The coming end of war*. Beverly Hills, Calif.: Sage Publications.

Levinson, D. 1989. *Family violence in cross-cultural perspective*. Thousand Oaks, Calif.: Sage.

Levinson, S., ed. 2004a. Contemplating torture: An introduction. In S. Levinson, ed., *Torture: A collection*. New York: Oxford University Press.

Levinson, S. 2004b. *Torture: A collection*. New York: Oxford University Press.

Levitt, S. D. 2004. Understanding why crime fell in the 1990s: Four factors that explain the decline and six that do not. *Journal of Economic Perspectives, 18*, 163–90.

Levitt, S. D., & Miles, T. J. 2007. Empirical study of criminal punishment. In A. M. Polinsky & S. Shavell, eds., *Handbook of law and economics*, vol. 1. Amsterdam: Elsevier.

Levy, J. S. 1983. *War in the modern great power system 1495–1975*. Lexington: University Press of Kentucky.

Levy, J. S., & Thompson, W. R. 2010. *Causes of war*. Malden, Mass.: Wiley-Blackwell.

Levy, J. S., & Thompson, W. R. 2011. *The arc of war: Origins, escalation, and transformation*. Chicago: University of Chicago Press.

Levy, J. S., Walker, T. C., & Edwards, M. S. 2001. Continuity and change in the evolution of warfare. In Z. Maoz & A. Gat, eds., *War in a changing world*. Ann Arbor: University of Michigan Press.

Lewis, B. 2002. *What went wrong? The clash between Islam and modernity in the Middle East*. New York: HarperPerennial.

Lewis, D. K. 1969. *Convention: A philosophical study*. Cambridge, Mass.: Harvard University Press.

Liebenberg, L. 1990. *The art of tracking: The origin of science*. Cape Town, South Africa: David Philip.

Lieberman, D., Tooby, J., & Cosmides, L. 2002. Does morality have a biological basis? An empirical test of the factors governing moral sentiments relating to incest. *Proceedings of the Royal Society of London B, 270*, 819–26.

Lieberson, S. 2000. *A matter of taste: How names, fashions, and culture change*. New Haven, Conn.: Yale University Press.

Ligthart, L., Bartels, M., Hoekstra, R. A., Hudziak, J. J., & Boomsma, D. I. 2005. Genetic contributions to subtypes of aggression. *Twin Research & Human Genetics, 8*, 483–91.

Lilla, M. 2001. *The reckless mind: Intellectuals in politics*. New York: New York Review of Books.

Lindley, D., & Schildkraut, R. 2005. *Is war rational? The extent of miscalculation and misperception as causes of war*. Paper presented at the International Studies Association. http://www.allacademic.com/meta/p71904_index.html.

Lindsay, J. M., & Takeyh, R. 2010. After Iran gets the bomb. *Foreign Affairs*.

Linker, D. 2007. *The theocons: Secular America under siege*. New York: Anchor.

Lodge, D. 1988. *Small world*. New York: Penguin.

Loewen, J. W. 1995. *Lies my teacher told me: Everything your American history textbook got wrong*. New York: New Press.

Long, W. J., & Brecke, P. 2003. *War and reconciliation: Reason and emotion in conflict resolution*. Cambridge, Mass.: MIT Press.

Lorenz, K. 1950/1971. Part and parcel in animal and human societies. *Studies in animal and human behavior*, vol. 2. Cambridge, Mass.: Harvard University Press.

Lott, J. 2007. Crime and punishment. *Freedomnomics: Why the free market works and other half-baked theories don't*. Washington, D.C.: Regnery.

Lott, J. R., Jr., & Whitley, J. E. 2007. Abortion and crime: Unwanted children and out-of-wedlock births. *Economic Inquiry, 45*, 304–24.

Luard, E. 1986. *War in international society*. New Haven, Conn.: Yale University Press.

Luard, E. 1988. *The blunted sword: The erosion of military power in modern world politics*. New York: New Amsterdam Books.

Lull, T. F., ed. 2005. *Martin Luther's basic theological writings*. Minneapolis: Augsburg Fortress.

Luria, A. R. 1976. *Cognitive development: Its cultural and social foundations*. Cambridge, Mass.: Harvard University Press.

Lykken, D. T. 1995. *The antisocial personalities*. Mahwah, N.J.: Erlbaum.

Maccoby, E. E., & Jacklin, C. N. 1987. *The psychology of sex differences*. Stanford, Calif.: Stanford University Press.

MacDonald, H. 2006. New York cops: Still the finest. *City Journal, 16*.

MacDonald, H. 2008. The campus rape myth. *City Journal, 18*.

Macklin, R. 2003. Human dignity is a useless concept. *British Medical Journal, 327*, 1419–20.

Macmillan, M. 2000. *An odd kind of fame: Stories of Phineas Gage*. Cambridge, Mass.: MIT Press.

Manning, R., Levine, M., & Collins, A. 2007. The Kitty Genovese murder and the social psychology of helping: The parable of the 38 witnesses. *American Psychologist, 62*, 555–62.

Mannix, D. P. 1964. *The history of torture*. Sparkford, U.K.: Sutton.

Mar, R. A., & Oatley, K. 2008. The function of fiction is the abstraction and simulation of social experience. *Perspectives on Psychological Science, 3*, 173–92.

Mar, R. A., Oatley, K., Hirsh, J., dela Paz, J., & Peterson, J. B. 2006. Bookworms versus nerds: Exposure to fiction versus non-fiction, divergent associations with social ability, and the simulation of fictional social worlds. *Journal of Research in Personality, 40*, 694–717.

Marshall, M. G., & Cole, B. R. 2008. Global report on conflict, governance, and state fragility, 2008. *Foreign Policy Bulletin, 18*, 3–21.

Marshall, M. G., & Cole, B. R. 2009. *Global report 2009: Conflict, governance, and state fragility*. Arlington, Va.: George Mason University Center for Global Policy.

Marshall, M. G., Gurr, T. R., & Harff, B. 2009. *PITF State Failure Problem Set: Internal wars and failures of governance 1955–2008. Dataset and coding guidelines*. Political Instability Task Force. http://globalpolicy.gmu.edu/pitf/pitfdata.htm.

Marshall, S.L.A. 1947/1978. *Men against fire: The problem of battle command in future war*. Gloucester, Mass.: Peter Smith.

Marvell, T. B. 1999. Homicide trends 1947–1996: Short-term versus long-term factors. *Proceedings of the Homicide Research Working Group Meetings, 1997 and 1998*. Washington, D.C.: U.S. Department of Justice.

Massey, D. S., & Sampson, R., eds. 2009. Special issue: The Moynihan report revisited: Lessons and reflections after four decades. *Annals of the American Academy of Political & Social Science, 621*, 6–326.

Maston, C. T. 2010. Survey methodology for criminal victimization in the United States, 2007. http://bjs.ojp.usdoj.gov/content/pub/pdf/cvus/cvus07mt.pdf.

Mattingly, G. 1958. International diplomacy and international law. In R. B. Wernham, ed., *The New Cambridge Modern History*, vol. 3. New York: Cambridge University Press.

Mauss, M. 1924/1990. *The gift: The form and reason for exchange in archaic societies*. New York: Norton.

Maynard Smith, J. 1982. *Evolution and the theory of games*. New York: Cambridge University Press.

Maynard Smith, J. 1988. *Games, sex, and evolution*. New York: Harvester Wheatsheaf.

Maynard Smith, J. 1998. *Evolutionary genetics*, 2nd ed. New York: Oxford University Press.

Maynard Smith, J., & Szathmáry, E. 1997. *The major transitions in evolution*. New York: Oxford University Press.

McCall, G. S., & Shields, N. 2007. Examining the evidence from small-scale societies and early prehistory and implications for modern theories of aggression and violence. *Aggression and Violent Behavior, 13*, 1–9.

McCauley, C. R. 1995. Are stereotypes exaggerated? A sampling of racial, gender, academic, occupational, and political stereotypes. In Y.-T. Lee, L. J. Jussim, & C. R. McCauley, eds., *Stereotype accuracy: Toward appreciating group differences*. Washington, D.C.: American Psychological Association.

McClure, S. M., Laibson, D., Loewenstein, G., & Cohen, J. D. 2004. Separate neural systems value immediate and delayed monetary rewards. *Science, 306*, 503–7.

McCrae, R. R., Costa, P. T., Ostendorf, F., Angleitner, A., Hrebickova, M., Avia, M. D., Sanz, J., Sanchez-Bernardos, M. L., Kusdil, M. E., Woodfield, R., Saunders, P. R., & Smith, P. B. 2000. Nature over nurture: Temperament, personality, and life span development. *Journal of Personality & Social Psychology, 78*, 173–86.

McCullough, M. E. 2008. *Beyond revenge: The evolution of the forgiveness instinct*. San Francisco: Jossey-Bass.

McCullough, M. E., Kurzban, R., & Tabak, B. A. 2010. Revenge, forgiveness, and evolution. In M. Mikulincer & P. R. Shaver, eds., *Understanding and reducing aggression, violence, and their consequences*. Washington, D.C.: American Psychological Association.

McDermott, R., Johnson, D., Cowden, J., & Rosen, S. 2007. Testosterone and aggression in a simulated crisis game. *Annals of the American Association for Political & Social Science, 614*, 15–33.

McDermott, R., Tingley, D., Cowden, J., Frazzetto, G., & Johnson, D.D.P. 2009. Monoamine oxidase A gene (MAOA) predicts behavioral aggression following provocation. *Proceedings of the National Academy of Sciences, 106*, 2118–23.

McDonald, P. J. 2010. Capitalism, commitment, and peace. *International Interactions, 36*, 146–68.

McEvedy, C., & Jones, R. 1978. *Atlas of world population history*. London: A. Lane.

McGinnis, J. O. 1996. The original constitution and our origins. *Harvard Journal of Law & Public Policy, 19*, 251–61.

McGinnis, J. O. 1997. The human constitution and constitutive law: A prolegomenon. *Journal of Contemporary Legal Issues, 8*, 211–39.

McGraw, A. P., & Tetlock, P. E. 2005. Taboo trade-offs, relational framing, and the acceptability of exchanges. *Journal of Consumer Psychology, 15*, 2–15.

McGraw, L. A., & Young, L. J. 2010. The prairie vole: An emerging model organism for understanding the social brain. *Trends in Neurosciences, 33*, 103–9.

McKay, C. 1841/1995. *Extraordinary popular delusions and the madness of crowds*. New York: Wiley.

McManamon, F. P. 2004. Kennewick Man. *Archeology Program*. http://www.nps.gov/archeology/kennewick/index.htm.

Mealey, L. 1995. The sociobiology of sociopathy: An integrated evolutionary model. *Behavioral & Brain Sciences, 18*, 523–41.

Mealey, L., & Kinner, S. 2002. The perception-action model of empathy and psychopathic "cold-heartedness." *Behavioral & Brain Sciences, 42*, 42–43.

Mednick, S. A., Gabrielli, W. F., & Hutchings, B. 1984. Genetic factors in criminal behavior: Evidence from an adoption cohort. *Science, 224*, 891–93.

Melander, E., Oberg, M., & Hall, J. 2009. *Are "new wars" more atrocious?* Paper presented at the annual meeting of the International Studies Association: Exploring the Past, Anticipating the Future.

Mennell, S. 1990. Decivilising processes: Theoretical significance and some lines of research. *International Sociology, 5*, 205–23.

Mennell, S., & Goudsblom, J. 1997. Civilizing processes—myth or reality? A comment on Duerr's critique of Elias. *Comparative Studies in Society & History, 39*, 729–33.

Menschenfreund, Y. 2010. The holocaust and the trial of modernity. *Azure, 39*, 58–83.

Merriman, T., & Cameron, V. 2007. Risk-taking: Behind the warrior gene story. *Journal of the New Zealand Medical Association, 120*.

Mesquida, C. G., & Wiener, N. I. 1996. Human collective aggression: A behavioral ecology perspective. *Ethology & Sociobiology, 17*, 247–62.

Metcalfe, J., & Mischel, W. 1999. A hot/cool system analysis of delay of gratification: Dynamics of willpower. *Psychological Review, 106*, 3–19.

Meyer-Lindenberg, A. 2006. Neural mechanisms of genetic risk for impulsivity and violence in humans. *Proceedings of the National Academy of Sciences, 103*, 6269–74.

Michel, J.-B., Shen, Y. K., Aiden, A. P., Veres, A., Gray, M. K., The Google Books Team, Pickett, J. P., Hoiberg, D., Clancy, D., Norvig, P., Orwant, J., Pinker, S., Nowak, M., & Lieberman-Aiden, E. 2011. Quantitative analysis of culture using millions of digitized books. *Science, 331*, 176–82.

Milgram, S. 1974. *Obedience to authority: An experimental view*. New York: Harper & Row.

Milner, L. S. 2000. *Hardness of heart / Hardness of life: The stain of human infanticide*. New York: University Press of America.

Mischel, W., Ayduk, O., Berman, M. G., Casey, B. J., Gotlib, I., Jonides, J., Kross, E., Teslovich, T., Wilson, N., Zayas, V., & Shoda, Y. I. In press. "Willpower" over the life span: Decomposing impulse control. *Social Cognitive & Affective Neuroscience*.

Mitani, J. C., Watts, D. P., & Amsler, S. J. 2010. Lethal intergroup aggression leads to territorial expansion in wild chimpanzees. *Current Biology, 20*, R507–8.

Mitzenmacher, M. 2004. A brief history of generative models for power laws and lognormal distributions. *Internet Mathematics, 1*, 226–51.

Mitzenmacher, M. 2006. Editorial: The future of power law research. *Internet Mathematics, 2*, 525–34.

Mnookin, R. H. 2007. Ethnic conflicts: Flemings and Walloons, Palestinians and Israelis. *Daedalus, 136*, 103–19.

Mogahed, D. 2006. *Perspectives of women in the Muslim world*. Washington, D.C.: Gallup.

Moll, J., de Oliveira-Souza, R., & Eslinger, P. J. 2003. Morals and the human brain: A working model. *NeuroReport, 14*, 299–305.

Moll, J., Zahn, R., de Oliveira-Souza, R., Krueger, F., & Grafman, J. 2005. The neural basis of human moral cognition. *Nature Reviews Neuroscience, 6*, 799–809.

Monkkonen, E. 1989. Diverging homicide rates: England and the United States, 1850–1875. In T. R. Gurr, ed., *Violence in America*, vol. 1: *The history of crime*. Newbury Park, Calif.: Sage.

Monkkonen, E. 1997. Homicide over the centuries. In L. M. Friedman & G. Fisher, eds., *The crime conundrum: Essays on criminal justice*. Boulder, Colo.: Westview Press.

Monkkonen, E. 2001. *Murder in New York City*. Berkeley: University of California Press.

Montesquieu. 1748/2002. *The spirit of the laws*. Amherst, N.Y.: Prometheus Books.

Moore, S., & Simon, J. L. 2000. *It's getting better all the time: Greatest trends of the last 100 years*. Washington, D.C.: Cato Institute.

Morgan, J. 1852/1979. *The life and adventures of William Buckley: Thirty-two years as a wanderer amongst the aborigines*. Canberra: Australia National University Press.

Mousseau, M. 2010. Coming to terms with the capitalist peace. *International Interactions, 36*, 185–92.

Moynihan, D. P. 1993. *Pandaemonium: Ethnicity in international politics*. New York: Oxford University Press.

Muchembled, R. 2009. *Une histoire de la violence*. Paris: Seuil.

Mueller, J. 1989. *Retreat from doomsday: The obsolescence of major war*. New York: Basic Books.

Mueller, J. 1995. *Quiet cataclysm: Reflections on the recent transformation of world politics*. New York: HarperCollins.

Mueller, J. 1999. *Capitalism, democracy, and Ralph's Pretty Good Grocery*. Princeton, N.J.: Princeton University Press.

Mueller, J. 2004a. *The remnants of war*. Ithaca, N.Y.: Cornell University Press.

Mueller, J. 2004b. Why isn't there more violence? *Security Studies, 13*, 191–203.

Mueller, J. 2006. *Overblown: How politicians and the terrorism industry inflate national security threats, and why we believe them*. New York: Free Press.

Mueller, J. 2007. The demise of war and of speculations about the causes thereof. Paper presented at the national convention of the International Studies Association.

Mueller, J. 2010a. *Atomic obsession: Nuclear alarmism from Hiroshima to Al-Qaeda*. New York: Oxford University Press.

Mueller, J. 2010b. Capitalism, peace, and the historical movement of ideas. *International Interactions, 36*, 169–84.

Mueller, J., & Lustick, I. 2008. Israel's fight-or-flight response. *National Interest* (November 1).

Murphy, J.P.M. 1999. Hitler was *not* an atheist. *Free Inquiry, 9*.

Murray, C. A. 1984. *Losing ground: American social policy, 1950–1980*. New York: Basic Books.

Myers, D. G., & Lamm, H. 1976. The group polarization phenomenon. *Psychological Bulletin, 83*, 602–27.

Nabokov, V. V. 1955/1997. *Lolita*. New York: Vintage.

Nadelmann, E. A. 1990. Global prohibition regimes: The evolution of norms in international society. *International Organization, 44*, 479–526.

Nagel, T. 1970. *The possibility of altruism*. Princeton, N.J.: Princeton University Press.

Nash, G. H. 2009. *Reappraising the right: The past and future of American conservatism*. Wilmington, Del.: Intercollegiate Studies Institute.

National Consortium for the Study of Terrorism and Responses to Terrorism. 2009. *Global terrorism database: GTD variables and inclusion criteria*. College Park: University of Maryland.

National Consortium for the Study of Terrorism and Responses to Terrorism. 2010. Global Terrorism Database. http://www.start.umd.edu/gtd/.

National Counterterrorism Center. 2009. *2008 Report on terrorism*. Washington, D.C.: National Counterterrorism Center. http://wits-classic.nctc.gov/ReportPDF.do?f=crt2008nctcannexfinal.pdf.

Nazaretyan, A. P. 2010. *Evolution of non-violence: Studies in big history, self-organization, and historical psychology*. Saarbrucken: Lambert Academic Publishing.

Neisser, U. 1976. General, academic, and artificial intelligence: Comments on the papers by Simon and by Klahr. In L. Resnick, ed., *The nature of intelligence*. Mahwah, N.J.: Erlbaum.

Neisser, U., Boodoo, G., Bouchard, T. J. Jr., Boykin, A. W., Brody, N., Ceci, S. J., Halpern, D. F., Loehlin, J. C., Perloff, R., Sternberg, R. J., & Urbina, S. 1996. Intelligence: Knowns and unknowns. *American Psychologist, 51*, 77–101.

Nell, V. 2006. Cruelty's rewards: The gratifications of perpetrators and spectators. *Behavioral & Brain Sciences, 29*, 211–57.

Nettelfield, L. J. 2010. Research and repercussions of death tolls: The case of the Bosnian book of the dead. In P. Andreas & K. M. Greenhill, eds., *Sex, drugs, and body counts*. Ithaca, N.Y.: Cornell University Press.

Neumayer, E. 2003. Good policy can lower violent crime: Evidence from a cross-national panel of homicide rates, 1980–97. *Journal of Peace Research, 40*, 619–40.

Neumayer, E. 2010. Is inequality really a major cause of violent crime? Evidence from a cross-national panel of robbery and violent theft rates. London School of Economics.

Newman, M.E.J. 2005. Power laws, Pareto distributions and Zipf's law. *Contemporary Physics, 46*, 323–51.

Nisbett, R. E., & Cohen, D. 1996. *Culture of honor: The psychology of violence in the South*. New York: HarperCollins.

North, D. C., Wallis, J. J., & Weingast, B. R. 2009. *Violence and social orders: A conceptual framework for interpreting recorded human history*. New York: Cambridge University Press.

Nowak, M. A. 2006. Five rules for the evolution of cooperation. *Science, 314*, 1560–63.

Nowak, M. A., May, R. M., & Sigmund, K. 1995. The arithmetic of mutual help. *Scientific American, 272*, 50–55.

Nowak, M. A., & Sigmund, K. 1998. Evolution of indirect reciprocity by image scoring. *Nature, 393*, 573–77.

Nunberg, G. 2006. *Talking right: How conservatives turned liberalism into a tax-raising, latte-drinking, sushi-eating, Volvo-driving,* New York Times*–reading, body-piercing, Hollywood-loving, left-wing freak show*. New York: PublicAffairs.

Nussbaum, M. 1997. *Cultivating humanity: A classical defense of reform in liberal education*. Cambridge, Mass.: Harvard University Press.

Nussbaum, M. 2006. Arts education: Teaching humanity. *Newsweek* (August 21–28).

Oakley, B. 2007. *Evil genes: Why Rome fell, Hitler rose, Enron failed, and my sister stole my mother's boyfriend*. Amherst, N.Y.: Prometheus Books.

Obermeyer, Z., Murray, C.J.L., & Gakidou, E. 2008. Fifty years of violent war deaths from Vietnam to Bosnia: Analysis of data from the World Health Survey Programme. *BMJ, 336*, 1482–86.

Olds, J., & Milner, P. 1954. Positive reinforcement produced by electrical stimulation of septal area and other regions of rat brain. *Journal of Comparative & Physiological Psychology, 47*, 419–27.

Orwell, G. 1946/1970. Politics and the English language. In *A collection of essays*. Boston: Mariner Books.

Otterbein, K. F. 2004. *How war began*. College Station, Tex.: Texas A&M University Press.

Ottosson, D. 2006. *LGBT world legal wrap up survey*. Brussels: International Lesbian and Gay Association.

Ottosson, D. 2009. *State-sponsored homophobia*. Brussels: International Lesbian, Gay, Bisexual, Trans, and Intersex Association.

Outram, D. 1995. *The enlightenment*. New York: Cambridge University Press.

Oxford, J. S., Sefton, A., Jackson, R., Innes, W., Daniels, R. S., & Johnson, N. P. 2002. World War I may have allowed the emergence of "Spanish" influenza. *Lancet Infectious Diseases, 2*, 111–14.

Oz, A. 1993. A postscript ten years later. In A. Oz, *In the land of Israel*. New York: Harcourt.

Panksepp, J. 1998. *Affective neuroscience: The foundations of human and animal emotions*. New York: Oxford University Press.

Parachini, J. 2003. Putting WMD terrorism into perspective. *Washington Quarterly, 26*, 37–50.

Parker, T. 1852/2005. *Ten sermons of religion*. Ann Arbor: University of Michigan Library.

Pate, A. 2008. Trends in democratization: A focus on instability in anocracies. In J. J. Hewitt, J. Wilkenfeld, & T. R. Gurr, eds., *Peace and conflict 2008*. Boulder, Colo.: Paradigm.

Patterson, O. 1985. *Slavery and social death*. Cambridge, Mass.: Harvard University Press.

Patterson, O. 1997. *The ordeal of integration*. Washington, D.C.: Civitas.

Patterson, O. 2008. Democracy, violence, and development in Jamaica: A comparative analysis. Harvard University.

Paul, T. V. 2009. *The tradition of non-use of nuclear weapons*. Stanford, Calif.: Stanford University Press.

Payne, J. L. 1989. *Why nations arm*. New York: Blackwell.

Payne, J. L. 2004. *A history of force: Exploring the worldwide movement against habits of coercion, bloodshed, and mayhem*. Sandpoint, Idaho: Lytton.

Payne, J. L. 2005. The prospects for democracy in high-violence societies. *Independent Review, 9*, 563–72.

Perez, J. 2006. *The Spanish Inquisition: A history*. New Haven, Conn.: Yale University Press.

Perry, W. J., Shultz, G. P., Kissinger, H. A., & Nunn, S. 2008. Toward a nuclear-free world. *Wall Street Journal*, A13 (January 15).

Peters, N. J. 2006. *Conundrum: The evolution of homosexuality*. Bloomington, Ind.: AuthorHouse.

Pew Research Center. 2010. *Gender equality universally embraced, but inequalities acknowledged*. Washington, D.C.: Pew Research Center. http://pewglobal.org/files/pdf/Pew-Global-Attitudes-2010-Gender-Report.pdf.

Pfaff, D. W. 2007. *The neuroscience of fair play: Why we (usually) follow the golden rule*. New York: Dana Press.

Phelps, E. A., O'Connor, K. J., Cunningham, W. A., Funayama, E. S., Gatenby, J. C., Gore, J. C., & Banaji, M. R. 2000. Performance on indirect measures of race evaluation predicts amygdala activation. *Journal of Cognitive Neuroscience, 12*, 729–38.

Piers, M. W. 1978. *Infanticide: Past and present*. New York: Norton.

Pinker, S. 1994. *The language instinct*. New York: HarperCollins.

Pinker, S. 1997. *How the mind works*. New York: Norton.

Pinker, S. 1998. Obituary: Roger Brown. *Cognition, 66*, 199–213.

Pinker, S. 1999. *Words and rules: The ingredients of language*. New York: HarperCollins.

Pinker, S. 2000. Review of John Maynard Smith and Eörs Szathmáry's "The origins of life: From the birth of life to the origin of language." *Trends in Evolution & Ecology, 15*, 127–28.

Pinker, S. 2002. *The blank slate: The modern denial of human nature*. New York: Viking.

Pinker, S. 2006. Deep commonalities between life and mind. In A. Grafen & M. Ridley, eds., *Richard Dawkins: How a scientist changed the way we think*. New York: Oxford University Press.

Pinker, S. 2007a. A history of violence. *New Republic* (March 19).

Pinker, S. 2007b. *The stuff of thought: Language as a window into human nature*. New York: Viking.

Pinker, S. 2008. The moral instinct. *New York Times Sunday Magazine* (January 13).

Pinker, S. 2010. The cognitive niche: Coevolution of intelligence, sociality, and language. *Proceedings of the National Academy of Sciences, 107*, 8993–99.

Pinker, S. 2011. Two problems with invoking self-deception too easily: Self-serving biases versus genuine self-deception, and distorted representations versus adjusted decision criteria. *Behavioral & Brain Sciences, 34*, 35–37.

Pinker, S., Nowak, M. A., & Lee, J. J. 2008. The logic of indirect speech. *Proceedings of the National Academy of Sciences USA, 105*, 833–38.

Pinker, Susan M. 2008. *The sexual paradox: Men, women, and the real gender gap*. New York: Scribner.

Pipes, R. 2003. *Communism: A history*. New York: Modern Library.

Pizarro, D. A., & Bloom, P. 2003. The intelligence of the moral intuitions: A comment on Haidt (2001). *Psychological Review, 110*, 193–96.

Plavcan, J. M. 2000. Inferring social behavior from sexual dimorphism in the fossil record. *Journal of Human Evolution, 39*, 327–44.

Plomin, R., DeFries, J. C., McClearn, G. E., & McGuffin, P. 2008. *Behavior genetics*, 5th ed. New York: Worth.

Pomeranz, K. 2008. A review of "A farewell to alms" by Gregory Clark. *American Historical Review, 113*, 775–79.

Popkin, R. 1979. *The history of skepticism from Erasmus to Spinoza*. Berkeley: University of California Press.

Posner, R. A. 2004. Torture, terrorism, and interrogation. In S. Levinson, ed., *Torture: A collection*. New York: Oxford University Press.

Potegal, M. 2006. Human cruelty is rooted in the reinforcing effects of intraspecific aggression that subserves dominance motivation. *Behavioral & Brain Sciences, 29*, 236–37.

Potts, M., & Hayden, T. 2008. *Sex and war: How biology explains warfare and terrorism and offers a path to a safer world*. Dallas, Tex.: Benbella.

Poundstone, W. 1992. *Prisoner's dilemma: Paradox, puzzles, and the frailty of knowledge*. New York: Anchor.

Power, S. 2002. *A problem from hell: America and the age of genocide*. New York: HarperPerennial.

Pratto, F., Sidanius, J., & Levin, S. 2006. Social dominance theory and the dynamics of intergroup relations: Taking stock and looking forward. *European Review of Social Psychology, 17*, 271–320.

Prentice, D. A., & Miller, D. T. 2007. Psychological essentialism of human categories. *Current Directions in Psychological Science, 16*, 202–6.

Preston, S. D., & de Waal, F.B.M. 2002. Empathy: Its ultimate and proximate bases. *Behavioral & Brain Sciences, 25*, 1–72.

Price, L. 2003. *The anthology and the rise of the novel: From Richardson to George Eliot*. New York: Cambridge University Press.

Price, R. M. 1997. *The chemical weapons taboo*. Ithaca, N.Y.: Cornell University Press.

Prinz, J. J. In press. Is empathy necessary for morality? In P. Goldie & A. Coplan, eds., *Empathy: Philosophical and psychological perspectives*. Oxford: Oxford University Press.

Procida, F. 2009. Overblown: Why an Iranian nuclear bomb is not the end of the world. *Foreign Affairs*.

Pryor, F. L. 2007. Are Muslim countries less democratic? *Middle East Quarterly, 14*, 53–58.

Przeworski, M., Hudson, R. R., & Di Rienzo, A. 2000. Adjusting the focus on human variation. *Trends in Genetics, 16*, 296–302.

Puppi, L. 1990. *Torment in art: Pain, violence, and martyrdom*. New York: Rizzoli.

Rai, T., & Fiske, A. P. 2011. Moral psychology is relationship regulation: Moral motives for unity, hierarchy, equality, and proportionality. *Psychological Review, 118*, 57–75.

Railton, P. 1986. Moral realism. *Philosophical Review, 95*, 163–207.

Raine, A. 2002. The biological basis of crime. In J. Q. Wilson & J. Petersilia, eds., *Crime: Public policies for crime control*. Oakland, Calif.: ICS Press.

Raine, A. 2008. From genes to brain to antisocial behavior. *Current Directions in Psychological Science, 17*, 323–28.

Raine, A., Lencz, T., Bihrle, S., LaCasse, L., & Colletti, P. 2000. Reduced prefrontal gray matter volume and reduced autonomic activity in antisocial personality disorder. *Archives of General Psychiatry, 57*, 119–29.

Rajender, S., Pandu, G., Sharma, J. D., Gandhi, K.P.C., Singh, L., & Thangaraj, K. 2008. Reduced CAG repeats length in androgen receptor gene is associated with violent criminal behavior. *International Journal of Legal Medicine, 122*, 367–72.

Ramachandran, V. S. 2000. Mirror neurons and imitation learning as the driving force behind "the great leap forward" in human evolution. *Edge*. http://www.edge.org/3rd_culture/ramachandran/ramachandran_index.html.

Raphael, S., & Stoll, M. A. 2007. *Why are so many Americans in prison?* Berkeley: University of California Press.

Raphael, S., & Winter-Ebmer, R. 2001. Identifying the effect of unemployment on crime. *Journal of Law & Economics, 44*, 259–83.

Rapoport, A. 1964. *Strategy and conscience*. New York: Harper & Row.

Ray, J. L. 1989. The abolition of slavery and the end of international war. *International Organization, 43*, 405–39.

Redmond, E. M. 1994. *Tribal and chiefly warfare in South America*. Ann Arbor: University of Michigan Museum.

Reicher, S., & Haslam, S. A. 2006. Rethinking the psychology of tyranny: The BBC prison study. *British Journal of Social Psychology, 45*, 1–40.

Remarque, E. M. 1929/1987. *All quiet on the western front*. New York: Ballantine.

Renfrew, J. W. 1997. *Aggression and its causes: A biopsychosocial approach*. New York: Oxford University Press.

Resnick, P. J. 1970. Murder of the newborn: A psychiatric review of neonaticide. *American Journal of Psychiatry, 126*, 58–64.

Rhee, S. H., & Waldman, I. D. 2007. Behavior-genetics of criminality and aggression. In D. J. Flannery, A. T. Vazsonyi, & I. D. Waldman, eds., *The Cambridge handbook of violent behavior and aggression*. New York: Cambridge University Press.

Rhoads, S. E. 2004. *Taking sex differences seriously*. San Francisco: Encounter Books.

Rice, M. 1997. Violent offender research and implications for the criminal justice system. *American Psychologist, 52*, 414–23.

Richardson, L. F. 1960. *Statistics of deadly quarrels*. Pittsburgh: Boxwood Press.

Ridley, M. 1997. *The origins of virtue: Human instincts and the evolution of cooperation*. New York: Viking.

Ridley, M. 2010. *The rational optimist: How prosperity evolves*. New York: HarperCollins.

Riedel, B. 2010. If Israel attacks. *National Interest* (August 24).
Rifkin, J. 2009. *The empathic civilization: The race to global consciousness in a world in crisis.* New York: J. P. Tarcher/Penguin.
Rindermann, H. 2008. Relevance of education and intelligence for the political development of nations: Democracy, rule of law and political liberty. *Intelligence, 36,* 306–22.
Roberts, A. 2010. Lives and statistics: Are 90% of war victims civilians? *Survival, 52,* 115–36.
Roberts, D. C., & Turcotte, D. L. 1998. Fractality and self-organized criticality of wars. *Fractals, 6,* 351–57.
Robinson, F. S. 2009. *The case for rational optimism.* New Brunswick, N.J.: Transaction.
Rodriguez, J. P. 1999. *Chronology of world slavery.* Santa Barbara, Calif.: ABC-CLIO.
Rodriguez, M. L., Mischel, W., & Shoda, Y. 1989. Cognitive person variables in the delay of gratification of older children at risk. *Journal of Personality & Social Psychology, 57,* 358–67.
Rogers, A. R. 1994. Evolution of time preference by natural selection. *American Economic Review, 84,* 460–81.
Romer, D., Duckworth, A. L., Sznitman, S., & Park, S. 2010. Can adolescents learn self-control? Delay of gratification in the development of control over risk taking. *Prevention Science, 11,* 319–30.
Roney, J. R., Simmons, Z. L., & Lukaszewski, A. W. 2009. Androgen receptor gene sequence and basal cortisol concentrations predict men's hormonal responses to potential mates. *Proceedings of the Royal Society B: Biological Sciences, 277,* 57–63.
Ropeik, D., & Gray, G. 2002. *Risk: A practical guide for deciding what's really safe and what's really dangerous in the world around you.* Boston: Houghton Mifflin.
Rosato, S. 2003. The flawed logic of democratic peace theory. *American Political Science Review, 97,* 585–602.
Rosecrance, R. 2010. Capitalist influences and peace. *International Interactions, 36,* 192–98.
Rosenau, J. N., & Fagen, W. M. 1997. A new dynamism in world politics: Increasingly skillful individuals? *International Studies Quarterly, 41,* 655–86.
Rosenbaum, R. 1998. *Explaining Hitler: The search for the origins of his evil.* New York: Random House.
Rosenfeld, R. 2006. Patterns in adult homicide: 1980–1995. In A. Blumstein & J. Wallman, eds., *The crime drop in America,* rev. ed. New York: Cambridge University Press.
Ross, M. L. 2008. Blood barrels: Why oil wealth fuels conflict. *Foreign Affairs.*
Rossi, P. H., Waite, E., Bose, C., & Berk, R. A. 1974. The structuring of normative judgements concerning the seriousness of crimes. *American Sociological Review, 39,* 224–37.
Rossiter, C., ed. 1961. *The Federalist Papers.* New York: New American Library.
Roth, R. 2001. Homicide in early modern England, 1549–1800: The need for a quantitative synthesis. *Crime, History & Societies, 5,* 33–67.
Roth, R. 2009. *American homicide.* Cambridge, Mass.: Harvard University Press.
Rothstein, R. 1998. *The way we were? The myths and realities of America's student achievement.* New York: Century Foundation Press.
Rousseau, J.-J. 1755/1994. *Discourse upon the origin and foundation of inequality among mankind.* New York: Oxford University Press.
Rowe, D. C. 2002. *Biology and crime.* Los Angeles: Roxbury.
Rozin, P. 1996. Towards a psychology of food and eating: From motivation to module to model to marker, morality, meaning, and metaphor. *Current Directions in Psychological Science, 5,* 18–24.
Rozin, P. 1997. Moralization. In A. Brandt & P. Rozin, eds., *Morality and health.* New York: Routledge.
Rozin, P., & Fallon, A. 1987. A perspective on disgust. *Psychological Review, 94,* 23–41.
Rozin, P., Markwith, M., & Stoess, C. 1997. Moralization and becoming a vegetarian: The transformation of preferences into values and the recruitment of disgust. *Psychological Science, 8,* 67–73.
Rummel, R. J. 1994. *Death by government.* Piscataway, N.J.: Transaction.
Rummel, R. J. 1997. *Statistics of democide.* Piscataway, N.J.: Transaction.
Rummel, R. J. 2002. 20th century democide. http://www.hawaii.edu/powerkills/20th.htm.
Rummel, R. J. 2004. One-thirteenth of a data point does not a generalization make: A reply to Dulić. *Journal of Peace Research, 41,* 103–4.
Russett, B. 2008. Peace in the twenty-first century? The limited but important rise of influences on peace. Yale University.
Russett, B. 2010. Capitalism *or* democracy? Not so fast. *International Interactions, 36,* 198–205.
Russett, B., & Oneal, J. 2001. *Triangulating peace: Democracy, interdependence, and international organizations.* New York: Norton.
Sagan, S. D. 2009. The global nuclear future. *Bulletin of the American Academy of Arts & Sciences, 62,* 21–23.
Sagan, S. D. 2010. Nuclear programs with sources. Stanford University.
Salehyan, I. 2008. From climate change to conflict? No consensus yet. *Journal of Peace Research, 45,* 315–26.
Salganik, M. J., Dodds, P. S., & Watts, D. J. 2006. Experimental study of inequality and unpredictability in an artificial cultural market. *Science, 311,* 854–56.
Salmon, C. A. 1998. The evocative nature of kin terminology in political rhetoric. *Politics & the Life Sciences, 17,* 51–57.

Salmon, C. A., & Symons, D. 2001. *Warrior lovers: Erotic fiction, evolution, and female sexuality*. New Haven, Conn.: Yale University Press.

Sampson, R. J., Laub, J. H., & Wimer, C. 2006. Does marriage reduce crime? A counterfactual approach to within-individual causal effects. *Criminology, 44*, 465–508.

Sanfey, A. G., Rilling, J. K., Aronson, J. A., Nystrom, L. E., & Cohen, J. D. 2003. The neural basis of economic decision-making in the ultimatum game. *Science, 300*, 1755–58.

Sargent, M. J. 2004. Less thought, more punishment: Need for cognition predicts support for punitive responses to crime. *Personality & Social Psychology Bulletin, 30*, 1485–93.

Sarkees, M. R. 2000. The Correlates of War data on war: An update to 1997. *Conflict Management & Peace Science, 18*, 123–44.

Saunders, D. G. 2002. Are physical assaults by wives and girlfriends a major social problem? A review of the literature. *Violence Against Women, 8*, 1424–48.

Saunders, J. J. 1979. *The history of the Mongol conquests*. London: Routledge & Kegan Paul.

Saxe, R., & Kanwisher, N. 2003. People thinking about thinking people: The role of the temporo-parietal junction in "theory of mind." *NeuroImage, 19*, 1835–42.

Sayre-McCord, G. 1988. *Essays on moral realism*. Ithaca, N.Y.: Cornell University Press.

Scarpa, A., & Raine, A. 2007. Biosocial bases of violence. In D. J. Flannery, A. T. Vazsonyi, & I. D. Waldman, eds., *The Cambridge handbook of violent behavior and aggression*. New York: Cambridge University Press.

Schama, S. 2001. *A history of Britain*, vol. 2: *The wars of the British 1603–1776*. New York: Hyperion.

Schechter, H. 2003. *The serial killer files: The who, what, where, how, and why of the world's most terrifying murderers*. New York: Ballantine.

Schechter, H. 2005. *Savage pastimes: A cultural history of violent entertainment*. New York: St. Martin's Press.

Schechter, S., Greenstone, J. H., Hirsch, E. G., & Kohler, K. 1906. Dietary laws. *Jewish encyclopedia*.

Scheff, T. J. 1994. *Bloody revenge: Emotions, nationalism, and war*. Lincoln, Neb.: iUniverse.com.

Schelling, T. C. 1960. *The strategy of conflict*. Cambridge, Mass.: Harvard University Press.

Schelling, T. C. 1978. *Micromotives and macrobehavior*. New York: Norton.

Schelling, T. C. 1984. The intimate contest for self-command. *Choice and consequence: Perspectives of an errant economist*. Cambridge, Mass.: Harvard University Press.

Schelling, T. C. 2000. The legacy of Hiroshima: A half-century without nuclear war. *Philosophy & Public Policy Quarterly, 20*, 1–7.

Schelling, T. C. 2005. An astonishing sixty years: The legacy of Hiroshima. In K. Grandin, ed., *Les Prix Nobel*. Stockholm: Nobel Foundation.

Schelling, T. C. 2006. *Strategies of commitment, and other essays*. Cambridge, Mass.: Harvard University Press.

Schelling, T. C. 2009. A world without nuclear weapons? *Daedalus, 138*, 124–29.

Schneider, G., & Gleditsch, N. P. 2010. The capitalist peace: The origins and prospects of a liberal idea. *International Interactions, 36*, 107–14.

Schroeder, P. W. 1994. *The transformation of European politics, 1763–1848*. New York: Oxford University Press.

Schuman, H., Steeh, C., & Bobo, L. D. 1997. *Racial attitudes in America: Trends and interpretations*. Cambridge, Mass.: Harvard University Press.

Schwager, R. 2000. *Must there be scapegoats? Violence and redemption in the Bible*. New York: Crossroad.

Schwartz, W. F., Baxter, K., & Ryan, D. 1984. The duel: Can these men be acting efficiently? *Journal of Legal Studies, 13*, 321–55.

Sedgh, G., Henshaw, S. K., Singh, S., Bankole, A., & Drescher, J. 2007. Legal abortion worldwide: Incidence and recent trends. *International Family Planning Perspectives, 33*, 106–16.

Séguin, J. R., Sylvers, P., & Lilienfeld, S. O. 2007. The neuropsychology of violence. In D. J. Flannery, A. T. Vazsonyi, & I. D. Waldman, eds., *The Cambridge handbook of violent behavior and aggression*. New York: Cambridge University Press.

Sell, A., Tooby, J., & Cosmides, L. 2009. Formidability and the logic of human anger. *Proceedings of the National Academy of Sciences, 106*, 15073–78.

Sen, A. 1990. More than 100 million women are missing. *New York Review of Books* (December 20).

Sen, A. 2000. East and West: The reach of reason. *New York Review of Books* (July 20).

Sen, A. 2006. *Identity and violence: The illusion of destiny*. New York: Norton.

Seymour, B., Singer, T., & Dolan, R. 2007. The neurobiology of punishment. *Nature Reviews Neuroscience, 8*, 300–311.

Shafer-Landau, R. 2003. *Moral realism: A defence*. Oxford: Clarendon Press.

Shamosh, N. A., De Young, C. G., Green, A. E., Reis, D. L., Johnson, M. R., Conway, A.R.A., Engle, R. W., Braver, T. S., & Gray, J. R. 2008. Individual differences in delay discounting: Relation to intelligence, working memory, and anterior prefrontal cortex. *Psychological Science, 19*, 904–11.

Shamosh, N. A., & Gray, J. R. 2008. Delay discounting and intelligence: A meta-analysis. *Intelligence, 38*, 289–305.

Sheehan, J. J. 2008. *Where have all the soldiers gone? The transformation of modern Europe*. Boston: Houghton Mifflin.

Shergill, S. S., Bays, P. M., Frith, C. D., & Wolpert, D. M. 2003. Two eyes for an eye: the neuroscience of force escalation. *Science, 301*, 187.

Sherif, M. 1966. *Group conflict and cooperation: Their social psychology*. London: Routledge & Kegan Paul.

Shermer, M. 2004. *The science of good and evil: Why people cheat, gossip, care, share, and follow the golden rule*. New York: Holt.

Shevelow, K. 2008. *For the love of animals*. New York: Holt.

Shotland, R. L., & Straw, M. K. 1976. Bystander response to an assault: When a man attacks a woman. *Journal of Personality & Social Psychology, 34*, 990–99.

Shultz, G. P. 2009. A world free of nuclear weapons. *Bulletin of the American Academy of Arts & Sciences, 62*, 81–82.

Shultz, G. P., Perry, W. J., Kissinger, H. A., & Nunn, S. 2007. A world free of nuclear weapons. *Wall Street Journal* (January 4).

Shweder, R. A., Much, N. C., Mahapatra, M., & Park, L. 1997. The "big three" of morality (autonomy, community, and divinity) and the "big three" explanations of suffering. In A. Brandt & P. Rozin, eds., *Morality and health*. New York: Routledge.

Sidanius, J., & Pratto, F. 1999. *Social dominance*. Cambridge, U.K.: Cambridge University Press.

Sidanius, J., & Veniegas, R. C. 2000. Gender and race discrimination: The interactive nature of disadvantage. In S. Oskamp, ed., *Reducing prejudice and discrimination: The Claremont symposium on applied social psychology*. Mahwah, N.J.: Erlbaum.

Siena Research Institute. 2010. *American presidents: Greatest and worst. Siena's 5th presidential expert poll*. Loudonville, N.Y.: Siena College. http://www.siena.edu/uploadedfiles/home/parents_and_community/community_page/sri/independent_research/Presidents%20Release_2010_final.pdf.

Sigmund, K. 1997. Games evolution plays. In A. Schmitt, K. Atzwanger, K. Grammer, & K. Schäfer, eds., *Aspects of human ethology*. New York: Plenum.

Simons, O. 2001. *Marteaus Europa oder Der Roman, bevor er Literatur wurde*. Amsterdam: Rodopi.

Simonton, D. K. 1990. *Psychology, science, and history: An introduction to historiometry*. New Haven, Conn.: Yale University Press.

Simonton, D. K. 2006. Presidential IQ, openness, intellectual brilliance, and leadership: Estimates and correlations for 42 U.S. chief executives. *Political Psychology, 27*, 511–26.

Singer, D. J., & Small, M. 1972. *The wages of war 1816–1965: A statistical handbook*. New York: Wiley.

Singer, P. 1975/2009. *Animal liberation: The definitive classic of the animal movement*, updated ed. New York: HarperCollins.

Singer, P. 1981/2011. *The expanding circle: Ethics and sociobiology*. Princeton, N.J.: Princeton University Press.

Singer, P. 1994. *Rethinking life and death: The collapse of our traditional ethics*. New York: St. Martin's Press.

Singer, T., Seymour, B., O'Doherty, J. P., Stephan, K. E., Dolan, R. J., & Frith, C. D. 2006. Empathic neural responses are modulated by the perceived fairness of others. *Nature, 439*, 466–69.

Skenazy, L. 2009. *Free-range kids: Giving our children the freedom we had without going nuts with worry*. San Francisco: Jossey-Bass.

Skogan, W. 1989. Social change and the future of violent crime. In T. R. Gurr, ed., *Violence in America*, vol. 1: *The history of crime*. Newbury Park, Calif.: Sage.

Slovic, P. 1987. Perception of risk. *Science, 236*, 280–85.

Slovic, P. 2007. "If I look at the mass I will never act": Psychic numbing and genocide. *Judgment & Decision Making, 2*, 79–95.

Slovic, P., Fischof, B., & Lichtenstein, S. 1982. Facts versus fears: Understanding perceived risk. In D. Kahneman, P. Slovic, & A. Tversky, eds., *Judgment under uncertainty: Heuristics and biases*. New York: Cambridge University Press.

Slutske, W. S., Heath, A. C., Dinwiddie, S. H., Madden, P.A.F., Bucholz, K. K., Dunne, M. P., Statham, D. J., & Martin, N. G. 1997. Modeling genetic and environmental influences in the etiology of conduct disorder: A study of 2,682 adult twin pairs. *Journal of Abnormal Psychology, 106*, 266–79.

Smith, A. 1759/1976. *The theory of moral sentiments*. Indianapolis: Liberty Classics.

Smith, A. 1776/2009. *The wealth of nations*. New York: Classic House Books.

Smith, H. 1952. *Man and his gods*. Boston: Little, Brown.

Sokal, A. D. 2000. *The Sokal hoax: The sham that shook the academy*. Lincoln: University of Nebraska Press.

Solomon, R. L. 1980. The opponent-process theory of acquired motivation. *American Psychologist, 35*, 691–712.

Solzhenitsyn, A. 1973/1991. *The Gulag archipelago*. New York: HarperPerennial.

Sommers, C. H. 1994. *Who stole feminism?* New York: Simon & Schuster.

Sorokin, P. 1957. *Social and cultural dynamics: A study of change in major systems of art, truth, ethics, law, and social relationships*. Boston: Extending Horizons.

Sowell, T. 1980. *Knowledge and decisions*. New York: Basic Books.

Sowell, T. 1987. *A conflict of visions: Ideological origins of political struggles*. New York: Quill.

Sowell, T. 1994. *Race and culture: A world view*. New York: Basic Books.

Sowell, T. 1996. *Migrations and cultures: A world view*. New York: Basic Books.

Sowell, T. 1998. *Conquests and cultures: An international history.* New York: Basic Books.

Sowell, T. 2004. *Affirmative action around the world: An empirical study.* New Haven, Conn.: Yale University Press.

Sowell, T. 2005. Are Jews generic? In T. Sowell, *Black rednecks and white liberals.* New York: Encounter.

Sowell, T. 2010. *Intellectuals and society.* New York: Basic Books.

Spagat, M., Mack, A., Cooper, T., & Kreutz, J. 2009. Estimating war deaths: An arena of contestation. *Journal of Conflict Resolution, 53,* 934–50.

Spence, J. T., Helmreich, R., & Stapp, J. 1973. A short version of the Attitudes toward Women Scale (AWS). *Bulletin of the Psychonomic Society,* 2, 219–20.

Spencer, A. T., & Croucher, S. M. 2008. Basque nationalism and the spiral of silence: An analysis of public perceptions of ETA in Spain and France. *International Communication Gazette, 70,* 137–53.

Spencer, C. 2000. *Vegetarianism: A history.* New York: Four Walls Eight Windows.

Sperber, D., ed. 2000. *Metarepresentations: A multidisciplinary perspective.* New York: Oxford University Press.

Spierenburg, P. 2006. Democracy came too early: A tentative explanation for the problem of American homicide. *American Historical Review, 111,* 104–14.

Spierenburg, P. 2008. *A history of murder: Personal violence in Europe from the Middle Ages to the present.* Cambridge, U.K.: Polity.

Spiller, R. J. 1988. S.L.A. Marshall and the ratio of fire. *RUSI Journal, 133,* 63–71.

Spitzer, S. 1975. Punishment and social organization: A study of Durkheim's theory of penal evolution. *Law & Society Review, 9,* 613–38.

Stanton, S. J., Beehner, J. C., Saini, E. K., Kuhn, C. M., & LaBar, K. S. 2009. Dominance, politics, and physiology: Voters' testosterone changes on the night of the 2008 United States presidential election. *PLoS ONE, 4,* e7543.

Statistics Canada. 2008. Table 1: Homicide rates by province/territory, 1961 to 2007. http://www.statcan.gc.ca/pub/85-002-x/2008009/article/t/5800411-eng.htm.

Statistics Canada. 2010. Homicide offences, number and rate, by province and territory. http://www40.statcan.ca/l01/cst01/legal12a-eng.htm.

Steckel, R. H., & Wallis, J. 2009. Stones, bones, and states: A new approach to the Neolithic Revolution. http://www.nber.org/~confer/2007/daes07/steckel.pdf.

Steenhuis, A. 1984. We have not learnt to control nature and ourselves enough: An interview with Norbert Elias. *De Groene Amsterdammer* (May 16), 10–11.

Steigmann-Gall, R. 2003. *The Holy Reich: Nazi conceptions of Christianity, 1919–1945.* New York: Cambridge University Press.

Steinbeck, J. 1962/1997. *Travels with Charley and later novels, 1947–1962.* New York: Penguin.

Stephan, W. G., & Finlay, K. 1999. The role of empathy in improving intergroup relations. *Journal of Social Issues, 55,* 729–43.

Stevens, W. O. 1940. *Pistols at ten paces: The story of the code of honor in America.* Boston: Houghton Mifflin.

Stevenson, D. 2004. *Cataclysm: The first world war as political tragedy.* New York: Basic Books.

Stillwell, A. M., & Baumeister, R. F. 1997. The construction of victim and perpetrator memories: Accuracy and distortion in role-based accounts. *Personality & Social Psychology Bulletin, 23,* 1157–72.

Stockholm International Peace Research Institute. 2009. *SIPRI yearbook 2009: Armaments, disarmaments, and international security.* New York: Oxford University Press.

Stone, V. E., Baron-Cohen, S., & Knight, R. T. 1998. Frontal lobe contributions to theory of mind. *Journal of Cognitive Neuroscience, 10,* 640–56.

Strange, J. J. 2002. How fictional tales wag real-world beliefs: Models and mechanisms of narrative influence. In M. C. Green, J. J. Strange, & T. C. Brock, eds., *Narrative impact: Social and cognitive foundations.* New York: Routledge.

Straus, M. A. 1977/1978. Wife-beating: How common, and why? *Victimology, 2,* 443–58.

Straus, M. A. 1995. Trends in cultural norms and rates of partner violence: An update to 1992. In S. M. Stith & M. A. Straus, eds., *Understanding partner violence: Prevalence, causes, consequences, and solutions.* Minneapolis: National Council on Family Relations.

Straus, M. A. 1999. Corporal punishment by American parents: National data on prevalence, chronicity, severity, and duration, in relation to child, and family characteristics. *Clinical Child & Family Psychology Review, 2,* 55–70.

Straus, M. A. 2001. *Beating the devil out of them: Corporal punishment in American families and its effects on children,* rev. ed. New Brunswick, N.J.: Transaction.

Straus, M. A. 2005. Children should never, ever be spanked no matter what the circumstances. In D. R. Loseke, R. J. Gelles & M. M. Cavanaugh eds., *Current controversies about family violence.* Thousand Oaks, Calif.: Sage.

Straus, M. A. 2009. Differences in corporal punishment by parents in 32 nations and its relation to national differences in IQ. Paper presented at the 14th International Conference on Violence, Abuse, and Trauma. http://pubpages.unh.edu/~mas2/Cp98D%20CP%20%20IQ%20world-wide.pdf.

Straus, M. A., & Gelles, R. J. 1986. Societal change and change in family violence from 1975 to 1985 as revealed by two national surveys. *Journal of Marriage & the Family, 48,* 465–80.

Straus, M. A., & Gelles, R. J. 1988. How violent are American families? Estimates from the National Family Violence Resurvey and other studies. In G. T. Hotaling, D. Finkelhor, J. T. Kirkpatrick, & M. A. Straus, eds., *Family abuse and its consequences: New directions in research.* Thousand Oaks, Calif.: Sage.

Straus, M. A., & Kantor, G. K. 1994. Change in spouse assault rates from 1975 to 1992: A comparison of three national surveys in the United States. Paper presented at the 13th World Congress of Sociology. http://pubpages.unh.edu/~mas2/V55.pdf.

Straus, M. A., & Kantor, G. K. 1995. Trends in physical abuse by parents from 1975 to 1992: A comparison of three national surveys. Paper presented at the American Society of Criminology.

Straus, M. A., Kantor, G. K., & Moore, D. W. 1997. Changes in cultural norms approving marital violence from 1968 to 1994. In G. K. Kantor & J. L. Jasinski, eds., *Out of the darkness: Contemporary perspectives on family violence.* Thousand Oaks, Calif.: Sage.

Stuart, T. 2006. *The bloodless revolution: A cultural history of vegetarianism from 1600 to modern times.* New York: Norton.

Suedfeld, P., & Coren, S. 1992. Cognitive correlates of conceptual complexity. *Personality & Individual Differences, 13,* 1193–99.

Suedfeld, P., & Tetlock, P. E. 1977. Integrative complexity of communications in international crises. *Journal of Conflict Resolution, 21,* 169–84.

Suedfeld, P., Tetlock, P. E., & Ramirez, C. 1977. War, peace, and integrative complexity: UN speeches on the Middle East problem 1947–1976. *Journal of Conflict Resolution, 21,* 427–42.

Suk, J. 2009. *At home in the law: How the domestic violence revolution is transforming privacy.* New Haven, Conn.: Yale University Press.

Symons, D. 1979. *The evolution of human sexuality.* New York: Oxford University Press.

Taagepera, R., & Colby, B. N. 1979. Growth of western civilization: Epicyclical or exponential? *American Anthropologist, 81,* 907–12.

Tajfel, H. 1981. *Human groups and social categories.* New York: Cambridge University Press.

Takahashi, H., Kato, M., Matsuura, M., Mobbs, D., Suhara, T., & Okubo, Y. 2009. When your gain is my pain and your pain is my gain: Neural correlates of envy and schadenfreude. *Science, 323,* 937–39.

Talmy, L. 2000. Force dynamics in language and cognition. *Toward a cognitive semantics 1: Concept structuring systems.* Cambridge, Mass.: MIT Press.

Tangney, J. P., Baumeister, R. F., & Boone, A. L. 2004. High self-control predicts good adjustment, less pathology, better grades, and interpersonal success. *Journal of Personality, 72,* 272–324.

Tannenwald, N. 2005a. Ideas and explanation: Advancing the theoretical agenda. *Journal of Cold War Studies, 7,* 13–42.

Tannenwald, N. 2005b. Stigmatizing the bomb: Origins of the nuclear taboo. *International Security, 29,* 5–49.

Tannenwald, N., & Wohlforth, W. C. 2005. Introduction: The role of ideas and the end of the Cold War. *Journal of Cold War Studies, 7,* 3–12.

Tatar, M. 2003. *The hard facts of the Grimm's fairy tales,* 2nd rev. ed. Princeton, N.J.: Princeton University Press.

Tavris, C., & Aronson, E. 2007. *Mistakes were made (but not by me): Why we justify foolish beliefs, bad decisions, and hurtful acts.* Orlando, Fla.: Harcourt.

Taylor, S., & Johnson, K. C. 2008. *Until proven innocent: Political correctness and the shameful injustices of the Duke lacrosse rape case.* New York: St. Martin's Press.

Taylor, S. E. 1989. *Positive illusions: Creative self-deception and the healthy mind.* New York: Basic Books.

Tetlock, P. E. 1985. Integrative complexity of American and Soviet foreign policy rhetoric: A time-series analysis. *Journal of Personality & Social Psychology, 49,* 1565–85.

Tetlock, P. E. 1994. Political psychology or politicized psychology: Is the road to scientific hell paved with good moral intentions? *Political Psychology, 15,* 509–29.

Tetlock, P. E. 1999. Coping with tradeoffs: Psychological constraints and political implications. In A. Lupia, M. McCubbins, & S. Popkin, eds., *Political reasoning and choice.* Berkeley: University of California Press.

Tetlock, P. E. 2003. Thinking the unthinkable: Sacred values and taboo cognitions. *Trends in Cognitive Sciences, 7,* 320–24.

Tetlock, P. E., Kristel, O. V., Elson, B., Green, M. C., & Lerner, J. 2000. The psychology of the unthinkable: Taboo tradeoffs, forbidden base rates, and heretical counterfactuals. *Journal of Personality & Social Psychology, 78,* 853–70.

Tetlock, P. E., Peterson, R. S., & Lerner, J. S. 1996. Revising the value pluralism model: Incorporating social content and context postulates. In C. Seligman, J. M. Olson, & M. P. Zanna, eds., *The psychology of values: The Ontario symposium,* vol. 8. Mahwah, N.J.: Erlbaum.

Thaler, R. H., & Sunstein, C. R. 2008. *Nudge: Improving decisions about health, wealth, and happiness.* New Haven, Conn.: Yale University Press.

Thayer, B. A. 2004. *Darwin and international relations: On the evolutionary origins of war and ethnic conflict*. Lexington: University Press of Kentucky.

Theisen, O. M. 2008. Blood and soil? Resource scarcity and internal armed conflict revisited. *Journal of Peace Research, 45*, 801–18.

Theweleit, M. 1977/1987. *Male fantasies*. Minneapolis: University of Minnesota Press.

Thomas, D. C. 2005. Human rights ideas, the demise of communism, and the end of the Cold War. *Journal of Cold War Studies, 7*, 110–41.

Thompson, P. M., Cannon, T. D., Narr, K. L., van Erp, T.G.M., Poutanen, V.-P., Huttunen, M., Lönnqvist, J., Standertskjöld-Nordenstam, C.-G., Kaprio, J., Khaledy, M., Dail, R., Zoumalan, C. I., & Toga, A. W. 2001. Genetic influences on brain structure. *Nature Neuroscience, 4*, 1–6.

Thornhill, R., & Palmer, C. T. 2000. *A natural history of rape: Biological bases of sexual coercion*. Cambridge, Mass.: MIT Press.

Thorpe, I.J.N. 2003. Anthropology, archaeology, and the origin of war. *World Archaeology, 35*, 145–65.

Thurston, R. 2007. *Witch hunts: A history of the witch persecutions in Europe and North America*. New York: Longman.

Thyne, C. L. 2006. ABC's, 123's, and the golden rule: The pacifying effect of education on civil war, 1980–1999. *International Studies Quarterly, 50*, 733–54.

Tiger, L. 2006. Torturers, horror films, and the aesthetic legacy of predation. *Behavioral & Brain Sciences, 29*, 244–45.

Tilly, C. 1985. War making and state making as organized crime. In P. Evans, D. Rueschemeyer, & T. Skocpol, eds., *Bringing the state back in*. New York: Cambridge University Press.

Tishkoff, S. A., Reed, F. A., Ranciaro, A., Voight, B. F., Babbitt, C. C., Silverman, J. S., Powell, K., Mortensen, H. M., Hirbo, J. B., Osman, M., Ibrahim, M., Omar, S. A., Lema, G., Nyambo, T. B., Ghori, J., Bumptstead, S., Pritchard, J. K., Wray, G. A., & Deloukas, P. 2006. Convergent adaptation of human lactase persistence in Africa and Europe. *Nature Genetics, 39*, 31–40.

Titchener, E. B. 1909/1973. *Lectures on the experimental psychology of the thought-processes*. New York: Arno Press.

Tooby, J., & Cosmides, L. 1988. The evolution of war and its cognitive foundations. *Institute for Evolutionary Studies Technical Report*.

Tooby, J., & Cosmides, L. 1990a. On the universality of human nature and the uniqueness of the individual: The role of genetics and adaptation. *Journal of Personality, 58*, 17–67.

Tooby, J., & Cosmides, L. 1990b. The past explains the present: Emotional adaptations and the structure of ancestral environments. *Ethology & sociobiology, 11*, 375–424.

Tooby, J., & Cosmides, L. 1992. Psychological foundations of culture. In J. Barkow, L. Cosmides, & J. Tooby, eds., *The adapted mind: Evolutionary psychology and the generation of culture*. New York: Oxford University Press.

Tooby, J., & Cosmides, L. 2010. Groups in mind: The coalitional roots of war and morality. In H. Høgh-Oleson, ed., *Human morality and sociality: Evolutionary and comparative perspectives*. New York: Palgrave Macmillan.

Tooby, J., & Cosmides, L. In press. Ecological rationality and the multimodular mind: Grounding normative theories in adaptive problems. In K. I. Manktelow & D. E. Over, eds., *Reasoning and rationality*. London: Routledge.

Tooby, J., Cosmides, L., & Price, M. E. 2006. Cognitive adaptations for n-person exchange: The evolutionary roots of organizational behavior. *Managerial & Decision Economics, 27*, 103–29.

Tooby, J., & DeVore, I. 1987. The reconstruction of hominid evolution through strategic modeling. In W. G. Kinzey, ed., *The evolution of human behavior: Primate models*. Albany, N.Y.: SUNY Press.

Tooley, M. 1972. Abortion and infanticide. *Philosophy & Public Affairs, 2*, 37–65.

Toye, R. 2010. *Churchill's empire: The world that made him and the world he made*. New York: Henry Holt.

Travers, J., & Milgram, S. 1969. An experimental study of the small-world problem. *Sociometry, 32*, 425–43.

Trivers, R. L. 1971. The evolution of reciprocal altruism. *Quarterly Review of Biology, 46*, 35–57.

Trivers, R. L. 1972. Parental investment and sexual selection. In B. Campbell, ed., *Sexual selection and the descent of man*. Chicago: Aldine.

Trivers, R. L. 1974. Parent-offspring conflict. *American Zoologist, 14*, 249–64.

Trivers, R. L. 1976. Foreword. In R. Dawkins, ed., *The selfish gene*. New York: Oxford University Press.

Trivers, R. L. 1985. *Social evolution*. Reading, Mass.: Benjamin/Cummings.

Trivers, R. L. In press. *Deceit and self-deception*.

Trivers, R. L., & Willard, D. E. 1973. Natural selection of parental ability to vary the sex ratio of offspring. *Science, 179*, 90–91.

Tuchman, B. W. 1978. *A distant mirror: The calamitous 14th century*. New York: Knopf.

Tucker, G. R. & Lambert, W. E. 1969. White and Negro listeners' reactions to various American-English dialects. *Social Forces, 47*, 465–68.

Turing, A. M. 1936. On computable numbers, with an application to the Entscheidungsproblem. *Proceedings of the London Mathematical Society, 42*, 230–65.

Turing, A. M. 1950. Computing machinery and intelligence. *Mind, 59*, 433–60.

Turkheimer, E. 2000. Three laws of behavior genetics and what they mean. *Current Directions in Psychological Science, 5,* 160–64.

Turner, H. A. 1996. *Hitler's thirty days to power: January 1933.* New York: Basic Books.

Tversky, A., & Kahneman, D. 1973. Availability: A heuristic for judging frequency and probability. *Cognitive Psychology, 4,* 207–32.

Tversky, A., & Kahneman, D. 1974. Judgment under uncertainty: Heuristics and biases. *Science, 185,* 1124–31.

Tversky, A., & Kahneman, D. 1981. The framing of decisions and the psychology of choice. *Science, 211,* 453–58.

Tversky, A., & Kahneman, D. 1983. Extensions versus intuitive reasoning: The conjunction fallacy in probability judgment. *Psychological Review, 90,* 293–315.

Twenge, J. M. 1997. Attitudes toward women, 1970–1995: A meta-analysis. *Psychology of Women Quarterly, 21,* 35–51.

Twenge, J. M. 2009. Change over time in obedience: The jury's still out, but it might be decreasing. *American Psychologist, 64,* 28–31.

Tyrrell, M. 2007. Homage to Ruritania: Nationalism, identity, and diversity. *Critical Review, 19,* 511–12.

Umbeck, J. 1981. Might makes rights: A theory of the formation and initial distribution of property rights. *Economic Inquiry, 19,* 38–59.

United Kingdom. Home Office. 2010. Research development statistics: Crime. http://rds.homeoffice.gov.uk/rds/bcs1.html.

United Kingdom. Office for National Statistics. 2009. Population estimates for U.K., England and Wales, Scotland, and Northern Ireland—Current datasets. http://www.statistics.gov.uk/statbase/Product.asp?vlnk=15106.

United Nations. 2008. *World population prospects*: Population database, 2008 rev. http://esa.un.org/unpp/.

United Nations Development Fund for Women. 2003. *Not a minute more: Ending violence against women.* New York: United Nations.

United Nations Development Programme. 2003. *Arab Human Development Report 2002: Creating opportunities for future generations.* New York: Oxford University Press.

United Nations Office on Drugs and Crime. 2009. *International homicide statistics.* http://www.unodc.org/documents/data-and-analysis/IHS-rates-05012009.pdf.

United Nations Population Fund. 2000. *The state of world population: Lives together, worlds apart—Men and women in a time of change.* New York: United Nations.

U.S. Bureau of Justice Statistics. 2009. *National crime victimization survey spreadsheet.* http://bjs.ojp.usdoj.gov/content/glance/sheets/viortrd.csv.

U.S. Bureau of Justice Statistics. 2010. *Intimate partner violence in the U.S.* http://bjs.ojp.usdoj.gov/content/intimate/victims.cfm.

U.S. Bureau of Justice Statistics. 2011. *Homicide trends in the U.S.: Intimate homicide.* http://bjs.ojp.usdoj.gov/content/homicide/intimates.cfm.

U.S. Census Bureau. 2010a. *Historical estimates of world population.* http://www.census.gov/ipc/www/worldhis.html.

U.S. Census Bureau. 2010b. *Income. Families.* Table F-4: Gini ratios of families by race and Hispanic origin of householder. http://www.census.gov/hhes/www/income/data/historical/families/index.html.

U.S. Census Bureau. 2010c. *International data base (IDB): Total midyear population for the world: 1950–2020.* http://www.census.gov/ipc/www/idb/worldpop.php.

U.S. Federal Bureau of Investigation. 2007. *Crime in the United States.* http://www.fbi.gov/ucr/cius2007/index.html.

U.S. Federal Bureau of Investigation. 2010a. *Hate crimes.* http://www.fbi.gov/about-us/investigate/civilrights/hate_crimes/hate_crimes.

U.S. Federal Bureau of Investigation. 2010b. *Uniform crime reports.* http://www.fbi.gov/about-us/cjis/ucr/ucr.

U.S. Federal Bureau of Investigation. 2011. *Preliminary annual uniform crime report, January–December 2010.* http://www.fbi.gov/about-us/cjis/ucr/crime-in-the-u.s/2010/preliminary-annual-ucr-jan-dec-2010.

U.S. Fish and Wildlife Service. 2006. *National survey of fishing, hunting, and wildlife-associated recreation.* https://docs.google.com/viewer?url=http://library.fws.gov/Pubs/nat_survey2006.pdf.

Valdesolo, P., & DeSteno, D. 2008. The duality of virtue: Deconstructing the moral hypocrite. *Journal of Experimental Social Psychology, 44,* 1334–38.

Valentino, B. 2004. *Final solutions: Mass killing and genocide in the 20th century.* Ithaca, N.Y.: Cornell University Press.

Valero, H., & Biocca, E. 1970. *Yanoama: The narrative of a white girl kidnapped by Amazonian Indians.* New York: Dutton.

van Beijsterveldt, C.E.M., Bartels, M., Hudziak, J. J., & Boomsma, D. I. 2003. Causes of stability of aggression from early childhood to adolescence: A longitudinal genetic analysis in Dutch twins. *Behavior Genetics, 33,* 591–605.

van Creveld, M. 2008. *The culture of war.* New York: Ballantine.
van den Oord, E.J.C.G., Boomsma, D. I., & Verhulst, F. C. 1994. A study of problem behaviors in 10- to 15-year-old biologically related and unrelated international adoptees. *Biological Genetics, 24,* 193–205.
Van der Dennen, J.M.G. 1995. *The origin of war: The evolution of a male-coalitional reproductive strategy.* Groningen, Netherlands: Origin Press.
Van der Dennen, J.M.G. 2005. *Querela pacis:* Confession of an irreparably benighted researcher on war and peace. An open letter to Frans de Waal and the "Peace and Harmony Mafia." University of Groningen.
van Evera, S. 1994. Hypotheses on nationalism and war. *International Security, 18,* 5–39.
Vasquez, J. A. 2009. *The war puzzle revisited.* New York: Cambridge University Press.
Vegetarian Society. 2010. Information sheet. http://www.vegsoc.org/info/statveg.html.
Vincent, D. 2000. *The rise of mass literacy: Reading and writing in modern Europe.* Malden, Mass.: Blackwell.
von Hippel, W., & Trivers, R. L. 2011. The evolution and psychology of self-deception. *Behavioral & Brain Sciences, 34,* 1–56.
Wade, N. 2006. *Before the dawn: Recovering the lost history of our ancestors.* New York: Penguin.
Wakefield, J. C. 1992. The concept of mental disorder: On the boundary between biological facts and social values. *American Psychologist, 47,* 373–88.
Waldrep, C. 2002. *The many faces of Judge Lynch: Extralegal violence and punishment in America.* New York: Palgrave Macmillan.
Walker, A., Flatley, J., Kershaw, C., & Moon, D. 2009. *Crime in England and Wales 2008/09.* London: U. K. Home Office.
Walker, P. L. 2001. A bioarchaeological perspective on the history of violence. *Annual Review of Anthropology, 30,* 573–96.
Walzer, M. 2004. Political action: The problem of dirty hands. In S. Levinson, ed., *Torture: A collection.* New York: Oxford University Press.
Warneken, F., & Tomasello, M. 2007. Helping and cooperation at 14 months of age. *Infancy, 11,* 271–94.
Watson, G. 1985. *The idea of liberalism.* London: Macmillan.
Wattenberg, B. J. 1984. *The good news is the bad news is wrong.* New York: Simon & Schuster.
Wearing, J. P., ed. 2010. *Bernard Shaw on war.* London: Hesperus Press.
Weede, E. 2010. The capitalist peace and the rise of China: Establishing global harmony by economic independence. *International Interactions, 36,* 206–13.
Weedon, M., & Frayling, T. 2008. Reaching new heights: Insights into the genetics of human stature. *Trends in Genetics, 24,* 595–603.
Weiss, H. K. 1963. Stochastic models for the duration and magnitude of a "deadly quarrel." *Operations Research, 11,* 101–21.
White, M. 1999. Who's the most important person of the twentieth century? http://users.erols.com/mwhite28/20c-vip.htm.
White, M. 2004. 30 worst atrocities of the 20th century. http://users.erols.com/mwhite28/atrox.htm.
White, M. 2005a. Deaths by mass unpleasantness: Estimated totals for the entire 20th century. http://users.erols.com/mwhite28/warstat8.htm.
White, M. 2005b. Democracies do not make war on each other . . . or do they? http://users.erols.com/mwhite28/demowar.htm.
White, M. 2007. Death tolls for the man-made megadeaths of the 20th century: FAQ. http://users.erols.com/mwhite28/war-faq.htm.
White, M. 2010a. Selected death tolls for wars, massacres and atrocities before the 20th century. http://users.erols.com/mwhite28/warstato.htm.
White, M. 2010b. Selected death tolls for wars, massacres and atrocities before the 20th century, page 2. http://users.erols.com/mwhite28/warstatv.htm#Primitive.
White, M. 2010c. Death tolls for the man-made megadeaths of the twentieth century. http://users.erols.com/mwhite28/battles.htm.
White, M. 2011. *The great big book of horrible things. The definitive chronicle of history's 100 worst atrocities.* New York: Norton.
White, S. H. 1996. The relationships of developmental psychology to social policy. In E. Zigler, S. L. Kagan, & N. Hall, eds., *Children, family, and government: Preparing for the 21st century.* New York: Cambridge University Press.
White, T. D., Asfaw, B., Beyene, Y., Haile-Selassie, Y., Lovejoy, C. O., Suwa, G., & WoldeGabriel, G. 2009. *Ardipithecus ramidus* and the paleobiology of early hominids. *Science, 326,* 64–86.
Wicherts, J. M., Dolan, C. V., Hessen, D. J., Oosterveld, P., Van Baal, G. C. M., Boomsma, D. I., & Span, M. M. 2004. Are intelligence tests measurement invariant over time? Investigating the nature of the Flynn effect. *Intelligence, 32,* 509–37.
Wicker, B., Keysers, C., Plailly, J., Royet, J.-P., Gallese, V., & Rizzolatti, G. 2003. Both of us are disgusted in my insula: The common neural basis of seeing and feeling disgust. *Neuron, 40,* 655–64.
Widom, C., & Brzustowicz, L. 2006. MAOA and the "cycle of violence": Childhood abuse and neglect, MAOA genotype, and risk for violent and antisocial behavior. *Biological Psychiatry, 60,* 684–89.

Wiener, M. J. 2004. *Men of blood: Violence, manliness, and criminal justice in Victorian England*. New York: Cambridge University Press.

Wiessner, P. 2006. From spears to M-16s: Testing the imbalance of power hypothesis among the Enga. *Journal of Anthropological Research, 62*, 165–91.

Wiessner, P. 2010. Youth, elders, and the wages of war in Enga Province, PNG. *Working Papers in State, Society, and Governance in Melanesia*. Canberra: Australian National University.

Wilkinson, D. 1980. *Deadly quarrels: Lewis F. Richardson and the statistical study of war*. Berkeley: University of California Press.

Wilkinson, D. L., Beaty, C. C., & Lurry, R. M. 2009. Youth violence—crime or self help? Marginalized urban males' perspectives on the limited efficacy of the criminal justice system to stop youth violence. *Annals of the American Association for Political Science, 623*, 25–38.

Willer, R., Kuwabara, K., & Macy, M. 2009. The false enforcement of unpopular norms. *American Journal of Sociology, 115*, 451–90.

Williams, G. C. 1988. Huxley's evolution and ethics in sociobiological perspective. *Zygon: Journal of Religion and Science, 23*, 383–407.

Williamson, L. 1978. Infanticide: An anthropological analysis. In M. Kohl, ed., *Infanticide and the value of life*. Buffalo, N.Y.: Prometheus Books.

Wilson, E. O. 1978. *On human nature*. Cambridge, Mass.: Harvard University Press.

Wilson, J. Q. 1974. *Thinking about crime*. New York: Basic Books.

Wilson, J. Q., & Herrnstein, R. J. 1985. *Crime and human nature*. New York: Simon & Schuster.

Wilson, J. Q., & Kelling, G. 1982. Broken windows: The police and neighborhood safety. *Atlantic Monthly, 249*, 29–38.

Wilson, M., & Daly, M. 1992. The man who mistook his wife for a chattel. In J. H. Barkow, L. Cosmides, & J. Tooby, eds., *The adapted mind: Evolutionary psychology and the generation of culture*. New York: Oxford University Press.

Wilson, M., & Daly, M. 1997. Life expectancy, economic inequality, homicide, and reproductive timing in Chicago neighborhoods. *British Medical Journal, 314*, 1271–74.

Wilson, M., & Daly, M. 2006. Are juvenile offenders extreme future discounters? *Psychological Science, 17*, 989–94.

Wilson, M. L., & Wrangham, R. W. 2003. Intergroup relations in chimpanzees. *Annual Review of Anthropology, 32*, 363–92.

Winship, C. 2004. The end of a miracle? Crime, faith, and partnership in Boston in the 1990's. In R. D. Smith, ed., *Long march ahead: The public influences of African American churches*. Raleigh, N.C.: Duke University Press.

Wolfgang, M., Figlio, R., & Sellin, T. 1972. *Delinquency in a birth cohort*. Chicago: University of Chicago Press.

Wolin, R. 2004. *The seduction of unreason: The intellectual romance with fascism from Nietzsche to postmodernism*. Princeton, N.J.: Princeton University Press.

Wood, J. C. 2003. Self-policing and the policing of the self: Violence, protection, and the civilizing bargain in Britain. *Crime, History, & Societies, 7*, 109–28.

Wood, J. C. 2004. *Violence and crime in nineteenth-century England: The shadow of our refinement*. London: Routledge.

Woolf, G. 2007. *Et tu, Brute? A short history of political murder*. Cambridge, Mass.: Harvard University Press.

Wouters, C. 2007. *Informalization: Manners and emotions since 1890*. Los Angeles: Sage.

Wrangham, R. W. 1999a. Evolution of coalitionary killing. *Yearbook of Physical Anthropology, 42*, 1–30.

Wrangham, R. W. 1999b. Is military incompetence adaptive? *Evolution & Human Behavior, 20*, 3–17.

Wrangham, R. W. 2009a. *Catching fire: How cooking made us human*. New York: Basic Books.

Wrangham, R. W. 2009b. The evolution of cooking: A talk with Richard Wrangham. *Edge*. http://www.edge.org/3rd_culture/wrangham/wrangham_index.html.

Wrangham, R. W., & Peterson, D. 1996. *Demonic males: Apes and the origins of human violence*. Boston: Houghton Mifflin.

Wrangham, R. W., & Pilbeam, D. 2001. African apes as time machines. In B.M.F. Galdikas, N. E. Briggs, L. K. Sheeran, G. L. Shapiro, & J. Goodall, eds., *All apes great and small*. New York: Kluwer.

Wrangham, R. W., Wilson, M. L., & Muller, M. N. 2006. Comparative rates of violence in chimpanzees and humans. *Primates, 47*, 14–26.

Wright, J. P., & Beaver, K. M. 2005. Do parents matter in creating self-control in their children? A genetically informed test of Gottfredson and Hirschi's theory of low self-control. *Criminology, 43*, 1169–202.

Wright, Q. 1942. *A study of war*, vol. 1. Chicago: University of Chicago Press.

Wright, Q. 1942/1964. *A study of war*, 2nd ed. Abridged by Louise Leonard Wright. Chicago: University of Chicago Press.

Wright, Q. 1942/1965. *A study of war*, 2nd ed., with a commentary on war since 1942. Chicago: University of Chicago Press.

Wright, R. 2000. *Nonzero: The logic of human destiny*. New York: Pantheon.

Xu, J., Kochanek, M. A., & Tejada-Vera, B. 2009. *Deaths: Preliminary data for 2007*. Hyattsville, Md.: National Center for Health Statistics.

Young, L., & Saxe, R. 2009. Innocent intentions: A correlation between forgiveness for accidental harm and neural activity. *Neuropsychologia, 47*, 2065–72.

Zacher, M. W. 2001. The territorial integrity norm: International boundaries and the use of force. *International Organization, 55*, 215–50.

Zahn, M. A., & McCall, P. L. 1999. Trends and patterns of homicide in the 20th century United States. In M. D. Smith & M. A. Zahn, eds., *Homicide: A sourcebook of social research*. Thousand Oaks, Calif.: Sage.

Zahn-Waxler, C., Radke-Yarrow, M., Wagner, E., & Chapman, M. 1992. Development of concern for others. *Developmental Psychology, 28*, 126–36.

Zak, P. J., Stanton, A. A., Ahmadi, S., & Brosnan, S. 2007. Oxytocin increases generosity in humans. *PLoS ONE, 2*, e1128.

Zebrowitz, L. A., & McDonald, S. M. 1991. The impact of litigants' babyfacedness and attractiveness on adjudications in small claims courts. *Law & Human Behavior, 15*, 603–23.

Zelizer, V. A. 1985. *Pricing the priceless child: The changing social value of children*. New York: Basic Books.

Zelizer, V. A. 2005. *The purchase of intimacy*. Princeton, N.J.: Princeton University Press.

Zerjal, T., Xue, Y., Bertorelle, G., Wells, R. S., Bao, W., Zhu, S., Qamar, R., Ayub, Q., Mohyuddin, A., Fu, S., Li, P., Yuldasheva, N., Ruzibakiev, R., Xu, J., Shu, Q., Du, R., Yang, H., Hurles, M. E., Robinson, E., Gerelsaikhan, T., Dashnyam, B., Mehdi, S. Q., & Tyler-Smith, C. 2003. The genetic legacy of the Mongols. *American Journal of Human Genetics, 72*, 717–21.

Zimbardo, P. G. 2007. *The Lucifer effect: Understanding how good people turn evil*. New York: Random House.

Zimbardo, P. G., Maslach, C., & Haney, C. 2000. Reflections on the Stanford prison experiment: Genesis, transformations, consequences. In T. Blass, ed., *Current perspectives on the Milgram paradigm*. Mahwah, N.J.: Erlbaum.

Zimring, F. E. 2007. *The great American crime decline*. New York: Oxford University Press.

Zipf, G. K. 1935. *The psycho-biology of language: An introduction to dynamic philology*. Boston: Houghton Mifflin.

INDEX

Note: Page numbers in *italics* refer to figures.

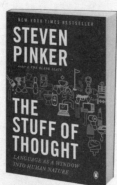

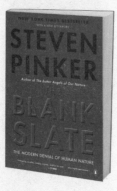